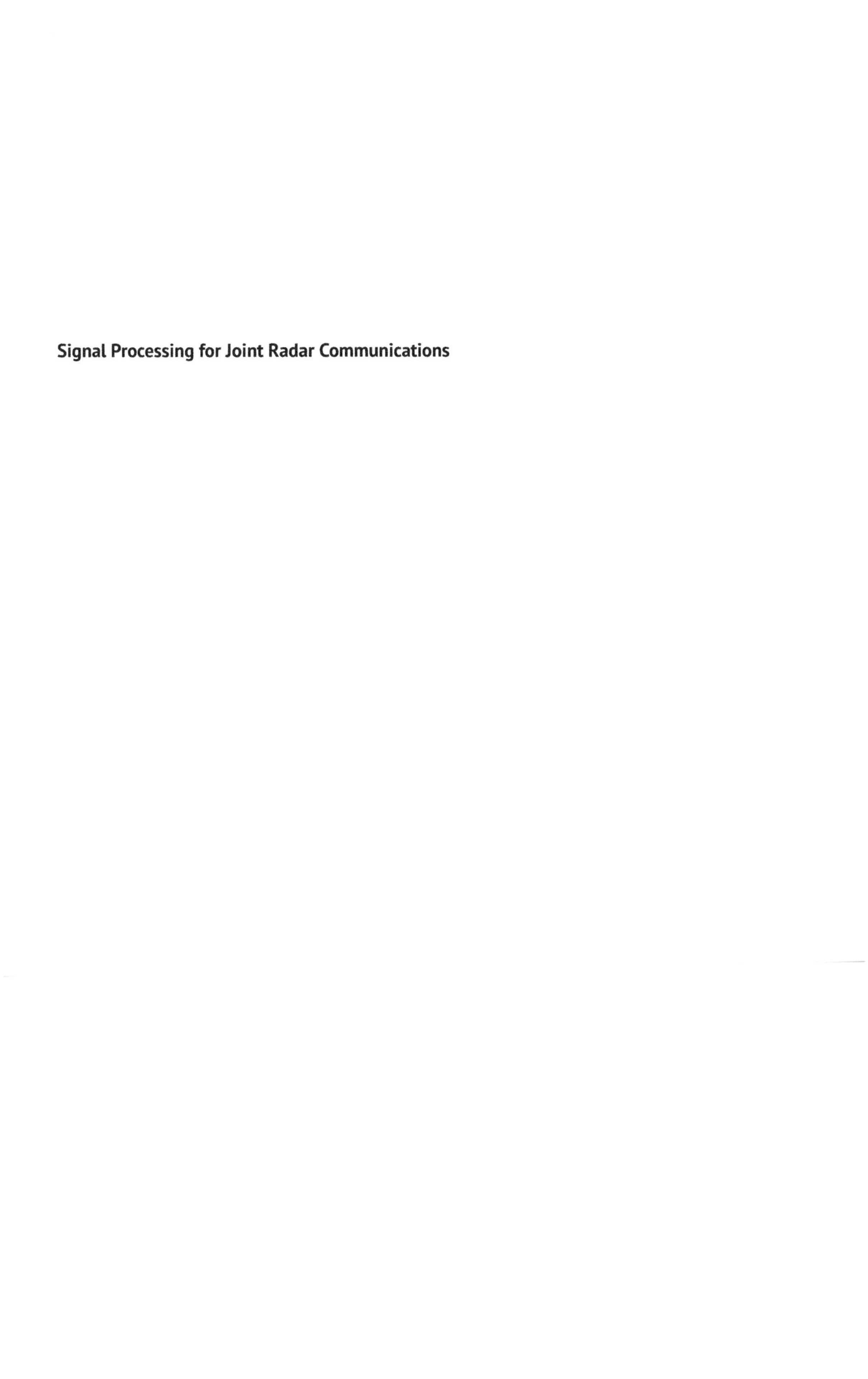

Signal Processing for Joint Radar Communications

Signal Processing for Joint Radar Communications

Edited by

Kumar Vijay Mishra, M. R. Bhavani Shankar, Björn Ottersten, and A. Lee Swindlehurst

Published by John Wiley & Sons, Inc., Hoboken, New Jersey.

Published simultaneously in Canada.

Library of Congress Cataloging-in-Publication Data:

Names: Mishra, Kumar Vijay, editor. | Shankar, M. R. Bhavani, editor. | Ottersten, Björn, editor. | Swindlehurst, A. Lee, editor.
Title: Signal processing for joint radar communications / edited by Kumar Vijay Mishra, M. R. Bhavani Shankar, Björn Ottersten, and A. Lee Swindlehurst.
Description: Hoboken, NJ : Wiley, 2024. | Includes index.
Identifiers: LCCN 2024008702 (print) | LCCN 2024008703 (ebook) | ISBN 9781119795537 (hardback) | ISBN 9781119795544 (adobe pdf) | ISBN 9781119795551 (epub)
Subjects: LCSH: Signal processing. | Radar.
Classification: LCC TK5102.9 .S5426 2024 (print) | LCC TK5102.9 (ebook) | DDC 621.382/2–dc23/eng/20240321
LC record available at https://lccn.loc.gov/2024008702
LC ebook record available at https://lccn.loc.gov/2024008703

Cover Design: Wiley
Cover Image: © Jackie Niam/Shutterstock

Set in 9.5/12.5pt STIXTwoText by Straive, Chennai, India

"Dedicated, with supreme reverence and humility, to Saraswati, goddess of wisdom and knowledge. To my Mom Shraddha Mishra and the memory of my Dad Shyam Bihari Mishra." – K. V. M.

"To my late father Mr. M. N. Rama Rao." – M. R. B. S.

"To the students I have had the privilege to work with." – B. O.

"To the students, friends, and colleagues who have helped me on my lucky journey." – A. L. S.

Contents

List of Editors

Kumar Vijay Mishra, United States CCDC Army Research Laboratory, Adelphi, MD, USA

M. R. Bhavani Shankar, Interdisciplinary Centre for Security, Reliability and Trust (SnT), University of Luxembourg, Luxembourg

Björn Ottersten, KTH Royal Institute of Technology, Stockholm, Sweden

A. Lee Swindlehurst, University of California, Irvine, CA, USA

List of Contributors

Ammar Ahmed
Aptiv
Advanced Safety and User Experience
Agoura Hills, CA
USA

Tuomas Aittomäki
Aalto University
Department of Signal Processing and Acoustics
Espoo
Finland

Mohammad Alaee-Kerahroodi
Interdisciplinary Centre for Security,
Reliability and Trust (SnT)
University of Luxembourg
Luxembourg

Anum Ali
Standards and Mobility Innovation Laboratory
Samsung Research America
Plano, TX
USA

Moeness G. Amin
Villanova University
Center for Advanced Communications
Villanova, PA
USA

Augusto Aubry
University of Naples, Federico II
Department of Electrical and Information
Technology Engineering
Napoli
Italy

Rick S. Blum
Electrical and Computer Engineering
Department
Lehigh University
Bethlehem, PA
USA

Giuseppe Caire
Technical University of Berlin
Chair of Communications and Information
Theory
Berlin
Germany

Jonathon A. Chambers
University of Leicester
School of Engineering
Leicestershire, Leicester
UK

Gaojie Chen
University of Surrey
5GIC & 6GIC, Institute for Communication
Systems
Department of Electrical and Electronic
Engineering (ICS)
Guildford
UK

Yun Chen
Department of Electrical and Computer
Engineering
North Carolina State University
Raleigh, NC
USA

Matthew A. Clark
The Aerospace Corporation
El Segundo, CA
USA

Guolong Cui
University of Electronic Science and Technology of China
School of Information and Communication Engineering
Chengdu
China

Yuanhao Cui
Aalto University
Department of Signal Processing and Acoustics
Espoo
Finland

and

Beijing University of Posts and Telecommunications
School of Information and Communication Engineering
Beijing
China

Antonio DeMaio
University of Naples, Federico II
Department of Electrical and Information Technology Engineering
Napoli
Italy

Sisai Fang
University of Surrey, 5GIC & 6GIC
Institute for Communication Systems
Department of Electrical and Electronic Engineering (ICS)
Guildford
UK

Nuria González-Prelcic
Department of Electrical and Computer Engineering
North Carolina State University
Raleigh, NC
USA

Yingjie J. Guo
University of Technology Sydney
Global Big Data Technologies Centre, and
School of Electrical and Data Engineering
Faculty of Engineering and IT
Sydney, New South Wales
Australia

Aboulnasr Hassanien
Aptiv PLC
Radar Systems Group
Agoura Hills, CA
USA

Qian He
Yangtze Delta Region Institute Quzhou
University of Electronic Science and Technology of China
Zhejiang, Quzhou
China

and

Electronic Engineering Department
University of Electronic Science and Technology of China
Sichuan, Chengdu
China

Xiaojing Huang
University of Technology Sydney
Global Big Data Technologies Centre, and
School of Electrical and Data Engineering
Faculty of Engineering and IT
Sydney, New South Wales
Australia

Jeremy Johnston
Columbia University
Electrical Engineering Department
New York, NY
USA

Mari Kobayashi
Technical University of Munich
Chair of Communications Engineering
Munich
Germany

Visa Koivunen
Aalto University
Department of Signal Processing and Acoustics
Espoo
Finland

Sangarapillai Lambotharan
Loughborough University
School of Mechanical, Manufacturing and Electrical Engineering
Leicestershire, Loughborough
UK

Bo Li
Aurora Innovation, Inc.
Research and Development Department
Pittsburgh, PA
USA

Kumar Vijay Mishra
United States CCDC Army Research Laboratory
Adelphi, MD
USA

Björn Ottersten
KTH Royal Institute of Technology
Stockholm
Sweden

Cunhua Pan
Queen Mary University of London
School of Electronic Engineering and Computer Science
London
UK

Athina P. Petropulu
Rutgers, the State University of New Jersey
Department of Electrical and Computer Engineering
Piscataway, NJ
USA

Konstantinos Psounis
University of Southern California
Los Angeles, CA
USA

Junhui Qian
Chongqing University
School of Microelectronic and Communication Engineering
Chongqing
China

M. R. Bhavani Shankar
Interdisciplinary Centre for Security, Reliability and Trust (SnT)
University of Luxembourg
Luxembourg

A. Lee Swindlehurst
University of California
Irvine, CA
USA

Xiaodong Wang
Columbia University
Electrical Engineering Department
New York, NY
USA

Zhen Wang
Yangtze Delta Region Institute Quzhou
University of Electronic Science and Technology of China
Zhejiang, Quzhou
China

and

Electronic Engineering Department
University of Electronic Science and Technology of China
Sichuan, Chengdu
China

Kai Wu
University of Technology Sydney
Global Big Data Technologies Centre, and School of Electrical and Data Engineering
Faculty of Engineering and IT
Sydney, New South Wales
Australia

Jing Yang
University of Electronic Science and
Technology of China
School of Information and Communication
Engineering
Chengdu
China

Xianxiang Yu
University of Electronic Science and
Technology of China
School of Information and Communication
Engineering
Chengdu
China

Jian A. Zhang
University of Technology Sydney
Global Big Data Technologies Centre, and
School of Electrical and Data Engineering
Faculty of Engineering and IT
Sydney, New South Wales
Australia

Yimin D. Zhang
Temple University
Department of Electrical and Computer
Engineering
Philadelphia, PA
USA

Junze Zhu
Yangtze Delta Region Institute Quzhou
University of Electronic Science and
Technology of China
Zhejiang, Quzhou
China

and

Electronic Engineering Department
University of Electronic Science and
Technology of China
Sichuan, Chengdu
China

Foreword

We congratulate the editors and authors on the timely publication of this book discussing the important problem of coexistence of radar and communication systems, as well as many related topics. The radiofrequency spectrum is a precious resource, which has become more and more congested in recent years. Let us take automotive radar as an example, which is capable of functioning in extreme adverse weather and light conditions: autonomous driving vehicles are equipped with ten or more radar sensors. Therefore, mutual interference among automotive radars is becoming increasingly severe, and the current practice is to disregard the parts of data that are corrupted. Interestingly, through joint radar and communication coordination, it may, in fact, be possible to translate mutual interference into collaboration by sharing information.

However, the coordination of radar and communications can be rather challenging in practical applications. For autonomous driving, for example, latency allowed for this coordination is extremely small. Moreover, for mass markets, radar sensors must be of low cost, which makes it difficult to use high-precision converters at the transmitter and receiver over a wide frequency band. Furthermore, multi-input multi-out radars have become a standard in the automotive radar industry. However, to realize a large virtual aperture for high angular resolution, through exploiting diversity, waveform orthogonality requirements must be considered.

Dual-function systems, which simultaneously support radar and communications, are the topic of this volume. Dual-function systems have many advantages over traditional systems: the number of antennas and the radar cross section of the platform are reduced; the weight of the platform is decreased and maneuverability is enhanced; and last but definitely not least, mutual interference among nearby systems is minimized and spectral compatibility is improved. Due to their great potential, dual-function systems have recently received considerable attention from both academia and industry, and researchers dealing with these systems in research and in practice will find this book to be a worthwhile addition to their library.

As innovative ideas and discoveries are generated in this new field of joint radar and communications, they can be adopted to tackle the challenges encountered in practical applications, including the aforementioned ones. This book is an excellent collection of recent advances on joint radar and communications systems, and it will, without doubt, inspire further research on and developments for addressing radiofrequency spectrum congestion.

Jian Li
University of Florida, Gainesville, Florida, USA
Petre Stoica
Uppsala University, Uppsala, Sweden

Preface

We are delighted to edit this new book, *Signal Processing for Joint Radar and Communications*, published by the prestigious Wiley-IEEE Press. The book comprises 14 chapters that are written by exceptionally well-qualified experts from academia and research laboratories across the globe. The focus of this book is on signal processing aspects of joint radar and communications (JRC)–one of the most actively researched problems across as many as 15 IEEE societies/communities. The chapters were selected by the editors, who have a combined experience of 120 years in prestigious research laboratories.

The future of spectrum access will be increasingly shared, dynamic, and secure. The International Telecommunication Union (ITU) began allocating radio bands at least as early as 1937. The IEEE Radar Band Standard 521, which has been maintained since 1976, follows the spectrum designation practices that originated during World War II. Over the interceding multiple decades of bandwidth expansion for sensing, navigation, wireless communications, timing, and positioning, policy planners and technologists have faced pressures of inefficient and excessively cautious use of the spectrum. Conventional spectrum sharing rules have been framed more for worst-case scenarios than for an optimal utilization of available frequencies. This approach inevitably leads to conflicts because the electromagnetic spectrum is a scarce resource. In 2021, the US Federal Communications Commission (FCC) was sued by AT&T over FCC's allocation of 6–7 GHz for dynamic allocation of allowed channels to Wi-Fi access points for indoor Wi-Fi 7 protocol. The US Federal Aviation Administration (FAA) was recently embroiled in a highly public battle with the airline industry over the use of the C-band for fifth-generation (5G) wireless services near airports. At lower bands, the FCC has been challenged to update their propagation models for their TV-band rulings so as to allow the reuse of TV bands for other services.

Consequently, sensing systems (radar, lidar, or sonar) that share the spectrum with wireless communications (radio-frequency/RF, optical, or acoustical) and still operate without any significant performance losses have captured significant research interest. Although a large fraction of these bands remains underutilized, radars need to maintain constant access to these bands for target sensing and detection as well as to increase the spectrum to accomplish missions such as secondary surveillance, multi-function integrated RF operations, communications-enabled autonomous driving, and cognitive capabilities. On the other hand, the wireless industry's demand for spectrum for providing new services and accommodating a massive number of users with high data rate requirement continues to increase. The present spectrum is used very inefficiently due to its highly fragmented allocation. Emerging wireless systems such as commercial Long-Term Evolution (LTE) communications technology, fifth-generation (5G), WiFi, Internet-of-Things (IoT), and Citizens Broadband Radio Services (CBRS) are already causing spectral interference to legacy military, weather, astronomy, and aircraft surveillance radars. Similarly, radar signals in adjacent bands leak to spectrum allocated for communications and deteriorate the service quality. Therefore, it is essential and beneficial for radar and communications to develop strategies to simultaneously and opportunistically operate in the same spectral bands in a mutually beneficial manner.

The interference from other emitters and its mitigation has been of interest within IEEE for decades. Until the early 1960s, IEEE used to publish *IEEE Transactions on Radio Frequency Interference* led by *IRE*

Professional Technical Group on Radio Frequency Interference. The journal ceased publication in 1963. The periodical *Frontiers of Technology: Trends in Electronics Research* by Mattraw and Moyer focused on radar–communications interference in its Volume 63 published in 1958. Scientific literature on JRC systems remained largely scattered until the 2000s. However, the spectral overlap of centimeter-wave radars with a number of wireless systems at the 3.5 GHz frequency band led to the 2012 U.S. President's Council of Advisors on Science and Technology (PCAST) report on spectrum sharing. Thereafter, changes in regulation for this band became a driver for spectrum-sharing research programs of multiple agencies including the Defense Advanced Research Projects-Agency (DARPA) and National Science Foundation (NSF). Today, it is the higher end of the RF spectrum–millimeter-wave and terahertz band–that requires concerted efforts for spectrum management. Some recent studies also mention joint visible light communications (VLC) and visible light positioning (VLP). Conceptual articles have also been reported on joint quantum communications and quantum sensing.

At the time of the publication of this book, journals/conferences sponsored by the following 15 IEEE societies have published JRC studies: IEEE Aerospace and Electronic Systems Society, IEEE Antennas & Propagation Society, IEEE Circuits and Systems Society, IEEE Communications Society, IEEE Computer Society, IEEE Control Systems Society, IEEE Engineering in Medicine and Biology Society, IEEE Geoscience and Remote Sensing Society, IEEE Information Theory Society, IEEE Instrumentation & Measurement Society, IEEE Microwave Theory and Techniques Society, IEEE Photonics Society, IEEE Signal Processing Society, IEEE Solid-State Circuits Society, and IEEE Vehicular Technology Society. In addition, spectrum sharing and/or JRC working groups and task forces have been reported from various IEEE societies, ITU, International Union of Radio Science (URSI), and American Meteorological Society (AMS).

In the context of overwhelmingly fast-paced developments in JRC, the goal of this book is twofold:

- to provide a list of references to JRC researchers and engineers, helping them to find information required for their current research, and
- to serve as a reference text in an advanced graduate level course on JRC.

The book consists of three parts: fundamental limits and background; physical layer signal processing; networking and hardware implementation.

Part I: Fundamental Limits and Background

We begin with a historical context, nomenclature, classification, and the principles of JRC systems through the first five chapters.

Chapter 1: A Signal Processing Outlook Toward Joint Radar-Communications

This chapter provides an overview of the JRC origins and subsequent developments. It details various signal processing (SP) techniques employed to achieve JRC systems by exploiting different degrees of freedom at transmitter and receiver as well as information on the channel state and target scenarios. The chapter summarizes various current JRC topologies and state-of-the-art solutions. It describes in detail the generic technical issues related to both transmit and receive signal processing, with examples of each prevailing design topology.

Chapter 2: Principles of Radar-Centric Dual-Function Radar-Communication Systems

This chapter provides an overview of existing radar-centric DFRC techniques and discusses their potential and future directions. It focuses on radar as an incumbent service defining the system resources while the communications functionality is seen as an added service carried out using the same system resources under the conditions of minimal overhead and changes to radar parameters. This dual functionality is enabled through multiple signal and system design strategies, each of which is clearly elucidated. The techniques presented leverage the developments in concepts, algorithms, design, and implementation of radar-centric waveforms, and the chapter offers an ideal platform for the reader on the technical details of DFRC.

Chapter 3: Interference, Clutter, and Jamming Suppression in Joint Radar–Communications Systems – Coordinated and Uncoordinated Designs
Expanding the scope of the DFRC design, this chapter introduces interference due to the coexistence of radar and communications systems and explores different interference suppression methods. It starts with the joint design of transmission and reception strategies in MIMO radar and MIMO communications systems wherein information exchange is enabled for coordinated interference suppression. Subsequently, an uncoordinated approach is pursued whereby interference suppression is integrated into the processing method used for the system's primary function. The chapter leverages constrained maximization techniques including iterative optimization, sparse modeling, and model-based deep learning architecture.

Chapter 4: Beamforming and Interference Management in Joint Radar–Communications Systems
This chapter continues the discussion of interference management and focuses on beamforming, multicarrier waveform design, and interference management using precoder and decoder designs in joint radar and communications systems. Cooperative settings where information about the state of the radio spectrum and awareness of the environment can be exchanged among the users and subsystems for mutual benefit are considered. A variety of precoding and decoding methods that allow for managing interference to avoid significant deterioration of sensing or communications performance are studied. Orthogonal multicarrier designs facilitating simultaneous communications and sensing and efficient Doppler estimation are also designed.

Chapter 5: Information Theoretic Aspects of Joint Sensing and Communications
An understanding of the limits of joint sensing and communications needs to be understood to set a path for research and offer a benchmark for design. This chapter precisely addresses this aspect in a simplified yet representative point-to-point communications scenario. The joint system is abstracted as a transmitter, equipped with onboard sensors, wishing to convey a message to its intended receiver and simultaneously estimate the corresponding parameters of interest from the backscattered signals modeled as generalized feedback. The chapter brings out the relation between the information theoretic quantities of capacity and distortion measures for the two systems and illustrates the findings with examples.

Part II: Physical-Layer Signal Processing
The earlier chapters dealt with the presentation of the background to JRC systems and brought out some interesting fundamental trade-offs. The subsequent four chapters deal with the various types of physical layer signal processing for JRC, largely focusing on enabling spectrum sharing.

Chapter 6: Radar-Aided Millimeter Wave Communication
This chapter takes a different look at the integration of the two systems, radar and communications. It presents the idea of radar-aided MIMO communication, motivated by the need to reduce the high overhead associated with the set-up and maintenance of millimeter wave links involving large arrays. The chapter exploits radar sensors mounted at the base station/access points or the user terminals as a source of out-of-band information for overhead reduction. It presents some of the initial works on strategies that leverage diverse information extracted from radar sensors to significantly reduce MIMO communications overhead, both in line-of-sight and non-line-of-sight scenarios. Classical optimization and learning-based approaches are presented in this context.

Chapter 7: Design of Constant-Envelope Radar Signals Under Multiple Spectral Constraints
Spectrum sharing, in addition to hardware reuse, has been a key motivating factor for joint radar and communications systems. This aspect is elucidated in this chapter, which focuses on the synthesis of radar waveforms optimizing surveillance system performance while guaranteeing coexistence with the surrounding radio frequency emitters, via multiple spectral compatibility constraints. The coexistence with the communications system is controlled by guaranteeing a certain quality of service explicitly

and through interference energy management implicitly. The spectrum-sharing problem is then cast in an optimization framework and solved using an iterative optimization procedure with established convergence and linear complexity in certain design parameters.

Chapter 8: Spectrum Sharing Between MIMO Radar and MIMO Communication Systems
Continuing on the topic of spectrum sharing, this chapter starts with a detailed canvas of the current spectrum-sharing scenario and then considers a sparse-sensing-based MIMO radar architecture and a MIMO communications system, operating in the vicinity of each other. In this cooperative setting, the joint design in the spectrum-sharing paradigm is formulated as a constraint optimization problem where the sparse sensing notion is embedded in the radar system design. In particular, suitable precoded random unitary waveforms are transmitted and a sub-sampling strategy is employed at the receiver. These novel elements have been jointly designed by the control center to maximize the radar performance while meeting certain constraints for the communications system.

Chapter 9: Performance and Design for Cooperative MIMO Radar and MIMO Communications
This chapter brings out the necessity and advantages of cooperation in coexisting radar and communications systems. Such a coexisting system offers fewer constraints and more design freedom to enhance performance, unlike a jointly designed system. In this context, a hybrid active–passive radar employs target returns contributed from both the radar and communications transmitters for target localization and localization. In a similar vein, signals reflected from the radar target are exploited along with those received directly from communications transmitters. The chapter derives the performance metrics and mechanisms to optimize the cooperative system fully using the available observations in the system to surpass the performance of the existing systems.

Part III: Networking and Hardware Implementations
Having considered the physical layer contributions, the subsequent chapters deal with higher layer aspects including classical resource allocation in the new paradigm as well as the emerging topics of secrecy and privacy. To establish the credibility of the research, hardware demonstration is essential, and Chapter 12 brings out the efforts in that direction.

Chapter 10: Frequency-Hopping MIMO Radar-based Data Communications
This chapter considers a different degree of freedom for a JRC system that exploits a frequency-hopping MIMO radar to achieve a reasonably high-speed and long-distance data communications link. Several strategies to embed the information to be transmitted in the frequency-hopping system are detailed along with the appropriate information demodulation techniques. This has been encapsulated in a comprehensive presentation of the system and signal model. Starting with ideal assumptions, the chapter then progresses to detail some of the recent breakthroughs in timing offset and channel estimations as well as the challenges. The chapter also highlights signaling strategies for higher data rates, the impact of the multipath channels, and security issues.

Chapter 11: Optimized Resource Allocation for Joint Radar-Communications
An aspect of joint design involves the allocation of available resources to the two systems so as to optimize the performance of each of them. In this context, this chapter considers resource allocation, a key concept in enabling three different JRC architectures. A single transmit antenna-based system using OFDM jointly optimizes radar and communications by subcarrier and power allocation. Optimal antenna selection and power is considered for multi-antenna systems, and the problem of optimal power allocation in distributed systems is also addressed based on the target localization performance and the communications capacity. The chapter opens up a resource allocation over a broader range of constraints, scenarios, and emerging joint sensing and communications paradigms.

Chapter 12: Emerging Prototyping Activities in Joint Radar-Communications
The previous chapters have discussed the canvas of JRC, highlighting the key approaches of radar-centric, communications-centric, and dual-function radar–communications systems. A hardware

validation of these techniques would lend credence to the results while enabling their embrace by industry. To this end, this chapter presents some of the prototyping initiatives that address some salient aspects of JRC systems. The chapter describes some existing prototypes to highlight the challenges in the design and performance of JRC. In particular, a coexistence prototype is detailed and two other developments are summarized.

Chapter 13: Secrecy Rate Maximization for Intelligent Reflective Surface-Assisted MIMO Communication Radar
The intelligent reflective surface is an emerging technology for enhancing coverage and performance of largely communications systems, but of late, also radar systems. This chapter investigates the joint transmitter beampattern and phase shifts of the intelligent surfaces in the context of a MIMO joint radar and communications system with an eavesdropping target. It brings in the concept of secrecy rate to JRC and shows that the use of reflective surfaces enhances the secrecy rate performance compared to ordinary MIMO radar. In particular, secrecy rate maximization at the legitimate user and the transmit power minimization lead to non-convex problems, which are then handled in the chapter using iterative optimization to derive two beamforming vectors, one to detect the target and the other to serve legitimate users.

Chapter 14: Privacy in Spectrum Sharing Systems with Applications to Communications and Radar
The concluding chapter of the book looks at another emerging topic of significant ramifications given the awareness about privacy. Since the two systems share the spectrum, which invariably involves information exchange, concerns of user privacy arise. This chapter explores techniques for the design and operation of radar and communications systems in shared spectrum environments, including analytical methods, models, and optimization to achieve performance and user privacy objectives. In particular, the chapter explores the characteristics of the privacy–performance trade-off with privacy evaluated in terms of inference attacks. The general evaluation framework enables the formulation of optimal privacy preservation strategies to offer a benchmark for future system design.

The book concludes with an epilogue that captures the future directions in this field. In summary, this book highlights some pioneering leaps in a very wide spectrum of JRC technologies. The plethora of JRC applications, their possible variants, and their relative benefits make it difficult to conclude which JRC design and algorithm will always be the best. The only foregone conclusion is that the JRC community will continue to complement the strengths of each technique and customize them as per application-specific requirements. We hope that this book helps demonstrate this rapidly evolving signal processing technology.

We thank all contributing authors for submitting their high-quality contributions. We sincerely acknowledge the support and help from all the reviewers for their timely and comprehensive evaluations of the manuscripts that have improved the quality of this book.

Finally, we are grateful to the IEEE Press Editorial Board and the Wiley staff members Ms. Aileen Storry and Kimberly Monroe-Hill for their support, feedback, and guidance.

Kumar Vijay Mishra
Adelphi, Maryland, USA

M. R. Bhavani Shankar
Luxembourg

Björn Ottersten
Stockholm, Sweden

A. Lee Swindlehurst
Irvine, California, USA

Acknowledgments

K. V. M. acknowledges support from the National Academies of Sciences, Engineering, and Medicine via Army Research Laboratory Harry Diamond Distinguished Fellowship.

B. O. and M. R. B. S. acknowledge the funding under the European Research Council Advanced Grant, Actively Enhanced Cognition based Framework for Design of Complex Systems (AGNOSTIC), bearing the reference

EC/H2020/ERC2016ADG/742648/AGNOSTIC.

Part I

Fundamental Limits and Background

1

A Signal Processing Outlook Toward Joint Radar-Communications

Kumar Vijay Mishra[1], M. R. Bhavani Shankar[2], Björn Ottersten[3], and A. Lee Swindlehurst[4]

[1] *United States CCDC Army Research Laboratory, Adelphi, MD, USA*
[2] *Interdisciplinary Centre for Security, Reliability and Trust (SnT), University of Luxembourg, Luxembourg*
[3] *KTH Royal Institute of Technology, Stockholm, Sweden*
[4] *University of California, Irvine, CA, USA*

1.1 Introduction

In recent years, sensing systems (radar, lidar, or sonar) that share the spectrum with wireless communications (radio-frequency/RF, optical, or acoustical) and still operate without any significant performance losses have captured significant research interest [Paul et al., 2017; Hassanien et al., 2017]. The interest in such spectrum-sharing systems is largely because the spectrum required by the wireless media is a scarce resource, while performance of both communications and remote sensing systems improves by exploiting a wider spectrum.

Several portions of frequency bands – from Very High Frequency (VHF) to Terahertz (THF) – are allocated exclusively for different radar applications [Cohen et al., 2018]. Although a large fraction of these bands remains underutilized, radars need to maintain constant access to these bands for target sensing and detection as well as obtain more spectrum to accomplish missions such as secondary surveillance, multi-function integrated RF operations, communications-enabled autonomous driving and cognitive capabilities. On the other hand, the wireless industry's demand for spectrum continues to increase for providing new services and accommodating a massive number of users with a high data rate requirement. The present spectrum is used very inefficiently due to its highly fragmented allocation. Emerging wireless systems such as commercial Long Term-Evolution (LTE) communications technology, fifth-generation (5G), Wi-Fi, Internet-of-Things (IoT), and Citizens Broadband Radio Services (CBRS) already cause spectral interference to legacy military, weather, astronomy, and aircraft surveillance radars (ASR) [Paul et al., 2017; Cohen et al., 2018]. Similarly, radar signals in adjacent bands leak to spectrum allocated for communications and deteriorate the service quality. Therefore, it is essential and beneficial for radar and communications to develop strategies to simultaneously and opportunistically operate in the same spectral bands in a mutually beneficial manner.

At the lower end of the spectrum in the VHF (30–300 MHz), Ultra High Frequency (UHF) (300–1000 MHz), and L-bands (1–2 GHz), radar systems such as FOliage PENetration (FOPEN) radar, astronomy/ionosphere radars, and Air Route Surveillance Radar (ARSR) have been encountering and managing interference from the broadcast and TV stations for decades now. The spectral congestion in centimeter-wave (cmWave) bands (S-, C-, X-, Ku-, and K-) arose later, primarily due to LTE waveforms, e.g. 802.11b/g/n (2.4 GHz) Wide-band Code Division Multiplexing Access (WCDMA), WiMAX LTE, LTE Global System for Mobile (GSM) communication, Enhanced

Signal Processing for Joint Radar Communications, First Edition.
Edited by Kumar Vijay Mishra, M. R. Bhavani Shankar, Björn Ottersten, and A. Lee Swindlehurst.

Data rates for GSM Evolution (EDGE), 802.11a/ac Very High Throughput (VHT) wireless LAN (WLAN), and commercial flight communications that now share spectrum with radars such as ASR, Terminal Weather Doppler Radar (TDWR) network, and other weather radars.

The spectral overlap of cmWave radars with a number of wireless systems at the 3.5 GHz frequency band led to the 2012 U. S. President's Council of Advisors on Science and Technology (PCAST) report on spectrum-sharing [2012], and changes in regulation for this band became a driver for spectrum-sharing research programs of multiple agencies [Cohen et al., 2018].

Today, it is the higher end of the RF spectrum, i.e., the millimeter-wave (mmWave), formally defined with the frequency range 30–300 GHz, that requires concerted efforts for spectrum management because its technologies are in an early development stage. Increasingly, the mmWave systems [Rappaport et al., 2015] are the preferred technology for near-field communications since they provide transmission bandwidth that is several GHz wide and currently unlicensed. This enables applications that require huge data rates such as 5G wireless backhaul, uncompressed high-definition (HD) video, in-room gaming, intra-large-vehicle communications, inter-vehicular communications, indoor positioning systems, and IoT-enabled wearable technologies [Daniels and Heath Jr, 2007]. There has also been a spurt of novel sensing systems in the mmWave band. Although these devices typically have short ranges because of heavy attenuation by physical barriers, weather, and atmospheric absorption, they provide high-range resolution resulting from the wide bandwidth. Typical mmWave radar applications include autonomous vehicles [Dokhanchi et al., 2019a], gesture recognition [Lien et al., 2016], cloud observation [Mishra et al., 2018], RF identification [Decarli et al., 2014], indoor localization [Mishra and Eldar, 2017b], and health monitoring [Fortino et al., 2012].

A recent rise of both radar and communications applications at terahertz (THz) band has also led to the development of integration of radar sensing and communications functionalities at these frequencies [Elbir et al., 2021]. The precise definition of THz band varies among different community members. Recent works in wireless communications generally define this band in the range 0.03–10 THz with an obvious overlap with the conventional mmWave frequencies. For the radar, microwave, and remote sensing engineers, THz band starts at the upper-mmWave limit of 100 GHz, and, in particular, low-THz term is used for the range 0.1–1 THz. In optics, on the other hand, THz spectrum is defined to end at 10 THz, beyond which frequencies are considered far-infrared. The Terahertz Technology and Applications Committee of the IEEE Microwave Theory and Techniques Society (MTT-S) focuses on the 0.3–3 THz range, while the IEEE Transactions on Terahertz Science and Technology Journal targets 0.3–10 THz.

Table 1.1 summarizes some of the co-existing communications services across various IEEE radar bands. In this chapter, we provide an overview of signal processing techniques and aspects of spectrum-sharing across different bands.

Table 1.1 Co-existing radar systems and communications services at different IEEE radar bands

IEEE radar band	VHF/UHF [30 MHz–1 GHz]	L [1–2 GHz]	S [2–4 GHz]	C [4–8 GHz]	X [8–12 GHz]	Ku, K, Ka, V, W, THz [12–300 GHz]
Example radar systems	FOPEN	ARSR	ASR, Next-Generation Weather Radar (NEXRAD)	TDWR	Mobile weather radars	Automotive radars, cloud radars
Co-existing communications	TV/broadcast/ 802.11ah/f	WiMAX, Joint Tactical Information Distribution System (JTIDS)	LTE	802.11a/ac	LTE	802.11ad, mmWave and THz communications

1.2 Policy and Licensing Issues

In the United States, a 75 MHz bandwidth within the 5.9 GHz band (specifically ranging from 5.850 to 5.925 GHz) has been exclusively assigned to intelligent transportation systems (ITSs) and car safety purposes for the past two decades, employing the dedicated short-range communications (DSRCs) technology. However, due to the lack of advancements in the DSRC band and the exponential growth of Wi-Fi technology, the Federal Communications Commission (FCC) recently designated the lower 45 MHz (5.850–5.895 GHz) for unlicensed applications, such as Wi-Fi. Consequently, only the upper 30 MHz spectrum (5.895–5.925 GHz band) is presently allocated for dedicated usage by ITS technologies. While the FCC has reserved 30 MHz of spectrum for critical safety services, advanced connected vehicle services will necessitate additional spectrum resources, particularly for data-intensive applications like augmented reality. Fortunately, the FCC has made available the 5 GHz unlicensed frequency band (~500 MHz) for 3GPP technologies, which can be utilized by cellular vehicle-to-everything (C-V2X) communications.

DSRC is an ad hoc communication system that operates independently of network infrastructure. Due to its widespread availability, many automobile manufacturers seeking to adopt V2X communications have favored the IEEE 802.11p standard, which serves as the foundation for DSRC. However, the implementation of 802.11p (and its successor, 802.11bd) necessitates the installation of numerous new access points (APs) and gateways, resulting in extended deployment time and increased costs. Given the absence of a clear business model, it is challenging to find an operator willing to bear the expenses associated with deploying numerous new APs based on freely available 802.11p/bd technologies. Consequently, the FCC has endorsed and prioritized C-V2X technologies based on the 3GPP standards, utilizing the 5.9 GHz frequency band, to pave the way for future connected vehicular systems. This includes requirements for over-the-air (OTA) software updates and the establishment of a car-to-cloud ecosystem.

The CBRS band serves as an intriguing case study due to the unequal access priority and the necessity of accurately detecting incumbents with 99% accuracy, even in the presence of lower-priority transmissions. Presently, tier 1 naval radar is safeguarded by a threshold limit on the combined power of noise and transmissions within the designated surrounding whisper zones. This necessitates a reduction in transmission power levels for LTE/5G priority access license (PAL) grantees, which consequently impacts consumer devices. The requirement for low power usage by PAL users becomes further complicated in the context of mobile radar, as the whisper zones may dynamically shift around the radar's moving location.

At higher frequencies, mmWave is generally considered to be unlicensed, depending on the jurisdiction. International Telecommunications Union (ITU) has completed guidelines for allocation of spectrum up to 275 GHz, and active discussions are currently underway to allot spectrum beyond 275 GHz.

1.3 Legal Challenges

Recent developments in spectrum-sharing allocation policies have highlighted the challenges that arise from overly cautious and inefficient utilization. For instance, the adoption of an outdated terrain-agnostic propagation model in the FCC TV Band ruling has artificially restricted the reuse of TV bands in areas where it would not interfere with broadcast TV receivers. In a highly publicized dispute, the Federal Aviation Administration (FAA) and the airline industry opposed the allocation of the C-band for 5G service deployment. The FAA relied on theoretical propagation models to suggest potential interference with airplane radar altimeters, while the FCC conducted extensive measurements over a five-year rule-making process to demonstrate that the interference between C-band 5G devices and altimeters would be negligible in practical scenarios. More recently, the FCC granted permission for indoor usage of the 6–7 GHz band for indoor Wi-Fi 7 through a centralized automatic frequency coordination (AFC) system. This system employs the location of Wi-Fi APs to predict interference to nearby microwave relay receivers, which are the primary users of the band. It dynamically assigns permissible channels to the APs. However, AT&T and other microwave relay tower operators filed a lawsuit against the FCC. According to AT&T, "The FCC has no plan to mitigate the interference when it inevitably occurs. Once millions of these new unlicensed devices are released and in use, it will be impracticable, if not impossible, for the FCC to identify and remove specific devices causing interference."

1.4 Agency-Driven Projects

Significant synergistic efforts are currently underway for efficient radio spectrum utilization by multiple entities. The National Science Foundation (NSF) has sponsored the Enhancing Access to the Radio Spectrum (EARS) project that brings together many different users for a flexible access to the spectrum [Bernhard et al., 2010]. A significant advancement is due to the Defense Advanced Research Projects Agency's (DARPA) Shared Spectrum Access for Radar and Communications (SSPARC) program focused on spectrum sharing for S-band military radars [Jacyna et al., 2016]. However, with new and emerging communications systems and novel radar applications such as next-generation automobiles and medical and health monitoring devices, the spectral congestion issues now extend far beyond classical communications and military radar. Realization of such emerging versatile systems requires a holistic approach from multiple perspectives such as physical layer transmission/reception, protocols, and inter-system coordination, among others.

The significance of wireless spectrum in the nation's pursuit of economic prosperity, infrastructure development, and national security cannot be overstated. Recent strategic plans, such as the "Electromagnetic Spectrum Superiority Strategy" released by the Department of Defense in late 2020 and the "Presidential Memorandum on Developing a Sustainable Spectrum Strategy for America's Future" from 2018, emphasize the need to utilize RF spectrum efficiently and effectively to fulfill current and future economic, national security, scientific, safety, and federal mission objectives. An important advancement in spectrum sharing within the sub-6 GHz band is the opening of the CBRS band, which presents significant opportunities for deploying 4G LTE and 5G networks in the desirable mid-band frequencies ranging from 3.55 to 3.7 GHz. However, utilizing these frequency bands requires adherence to certain conditions, including the implementation of an FCC-certified sensing architecture to protect federal incumbents such as ship-borne radar. Consequently, environmental sensing capability (ESC) sensors are employed to detect these signals and subsequently update the spectrum access server (SAS), which manages spectrum allocations.

Apart from EARS, NSF has recently expanded on spectrum-sharing efforts through a volley of programs: Spectrum and Wireless Innovation enabled by Future Technologies (SWIFT), Platforms for Advanced Wireless Research (PAWR), RF Data Factory, and Spectrum Innovation Center (SpectrumX), to name a few. In 2023, the US government released a presidential memorandum to establish a National Spectrum Strategy Implementation Plan (NSSIP) through the National Telecommunications and Information Administration (NTIA).

1.5 Channel Considerations

The characteristics of cmWave, mmWave, and THz channels differ significantly, necessitating careful considerations in system design and deployment. In this section, we will explore the channel considerations of various bands, highlighting their key features and implications for communication systems. Understanding these channel aspects is crucial for developing effective strategies to overcome challenges and harness the full potential of these frequency bands in wireless communication applications.

1.5.1 cmWave vs mmWave

Strong Attenuation Compared to sub-6 GHz transmissions envisaged in 5G, mmWave signals encounter a more complex propagation environment characterized by higher scattering, severe penetration losses, and lower diffraction. These losses result in mmWave communications links being near line-of-sight (LoS) with fewer non-line-of-sight (NLoS) clusters and smaller coverage areas. Similarly, lower diffraction results in poorer coverage around corners. High attenuation also implies that mmWave radars are useful only at short ranges and, as a result, multipath is a less severe problem.

High Path-Loss and Large Arrays Quite naturally, the mmWave signals suffer from higher path-loss for fixed transmitter (TX) and receiver (RX) gains. By the Friis transmission formula, compensating for these losses while keeping the same effective antenna aperture (or increasing the gain) imposes constraints on the transceiver hardware. Since the received power is contingent on the beams of the transmitter and receiver being oriented toward each other, the same aperture is accomplished by using steerable antenna arrays whose elements are spaced by at most half the wavelength ($\lambda/2$) of the transmitted signal to prevent undesirable grating lobes. This inter-element spacing varies between 0.5 and 5 mm for mmWave carriers. Such narrow spacings impact the choice of RF and intermediate frequency (IF) elements because they should fit in the limited space available, and precise mounting may be difficult in, for instance, vehicular platforms.

Wide Bandwidths The unlicensed, wide mmWave bandwidth enables higher data rates for communications as well as the range resolution in radar. In automotive radar, this ensures detection of distinct, informative micro-motions of targets such as pedestrians and cyclists [Duggal et al., 2020]. The mmWave receivers sampling at the Nyquist rate require expensive, high-rate analog-to-digital converters (ADCs). Large bandwidths also imply that the use of low-complexity algorithms in transmitter and receiver processing is critical [Dokhanchi et al., 2019a]. Further, mmWave channels are sparse in both time and angular dimensions – a property exploited for low-complexity, low-rate reconstruction using techniques such as compressed sensing [Mishra and Eldar, 2017b, 2019]. It is crucial to consider if relevant narrowband assumptions hold in a mmWave application; otherwise, the signal bandwidth is very broad with respect to the center frequency and the steering vectors become frequency-dependent.

Power Consumption The power consumption of an ADC increases linearly with the sampling frequency. At baseband, each full-resolution ADC consumes 15–795 mW at 36 MHz–1.8 GHz bandwidths. In addition, power consumed by other RF elements such as power amplifiers and data interface circuits in conjunction with the narrow spacing between antenna elements renders it infeasible to utilize a separate RF-IF chain for each element. Thus, a feasible multi-antenna TX/RX structure and beamformers should be analog or hybrid (wherein the potential array gain is exploited without using a dedicated RF chain per antenna and phase shifter (PS)) [Méndez-Rial et al., 2016] because fully digital beamforming is infeasible.

Short Coherence Times The mmWave environments such as indoor and vehicular communications are highly variable with typical channel coherence times of nanoseconds [Rappaport et al., 2015]. The reliability and coverage of dynamic mmWave vehicular links are severely affected by the use of narrow beams. The intermittent blockage necessitates frequent beam re-alignment to maintain high data rates. Also, mmWave radar requires a wide Doppler range to detect both fast vehicles and slow pedestrians [Duggal et al., 2020]. Short coherence times impact the use of feedback and waveform adaptation in many joint radar-communication (JRC) designs, where the channel knowledge may be invalid or outdated when transmit waveform optimization takes place.

1.5.2 Toward THz Band

Compared to the mmWave channel, the THz channel exhibits certain unique characteristics (Figure 1.1). The JRC-specific challenges are listed below [Elbir et al., 2021].

Path Loss The THz channel experiences substantial path loss, approximately 120 dB/100 m at 0.6 THz. This path loss is primarily influenced by spreading loss and molecular absorption, which are more significant compared to mmWave frequencies. To compensate for the high path loss, a large number of antennas are deployed in a user-centric manner, allowing for beamforming gain that generates multiple beams toward communication users and radar targets.

Multipath In mmWave frequencies, both LoS and NLoS multipath components play significant roles. However, at THz frequencies, NLoS paths, especially in outdoor scenarios, are negligible. For instance, the first- and second-order reflected paths are attenuated by an average of 5 and 15 dB, respectively. Consequently, the THz channel is primarily characterized as LoS-dominant and NLoS-assisted. The THz JRC system can benefit from the presence of NLoS paths to enhance diversity, particularly in communication scenarios with low-resolution beamforming. Simultaneously, the insignificance of NLoS paths is advantageous for THz sensing/radar applications, where an explicit LoS path between the transmitter and the target is required.

Transmission Range THz systems have shorter transmission ranges compared to mmWave frequencies due to significant attenuation. In the THz JRC system, there is a trade-off between the requirement for long transmission distances, such as up to 200 m, for sensing applications and the need for shorter ranges, like 20 m, to achieve a data rate of 100 Gbps in communication tasks. Recent automotive radar applications have reported a transmission distance of up to 200 m with specific attenuation of approximately X dB/km over the 0.1–0.3 THz range, with a path loss of 4 dB. However, shorter distances are necessary for communication scenarios, such as around 20 m, to achieve a data rate of 100 Gbps.

Channel Sparsity The utilization of a large number of antennas and the predominance of LoS characteristics make the THz channels highly sparse in the angular domain. Compared to their millimeter-wave counterparts, THz channels exhibit a smaller angular spread, approximately 10–15° at 140 GHz, as opposed to 20–100° at 60 GHz. This sparsity can be leveraged to employ subarray models such as array-of-subarrays (AoSA) and group-of-subarrays (GoSA), which help

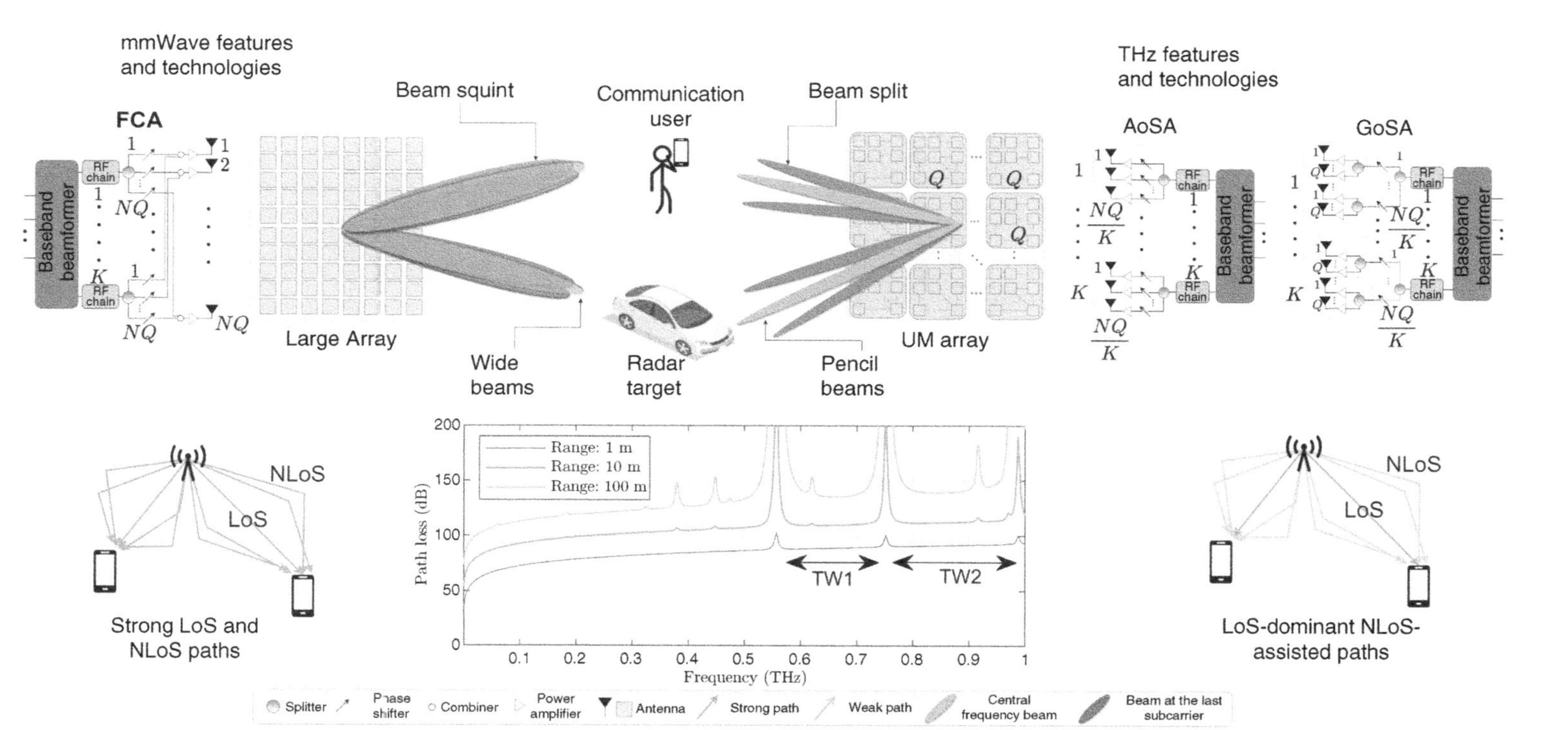

Figure 1.1 Comparison of millimeter-wave and THz band characteristics for JRC design including distance-dependent path loss, multipath components, beam alignment, and antenna array structures [Elbir et al., 2022].

reduce the hardware and computational complexities associated with high-frequency systems. Additionally, sparsity-based receivers can be designed to benefit both sensing and communication tasks in THz JRC systems.

Wideband Beam Split THz channels exhibit unique characteristics, including misalignment and phase uncertainties in PSs. The wideband mmWave systems commonly employ subcarrier-independent analog beamformers, but in THz channels, these beamformers can cause a phenomenon known as beam split. Due to the ultra-wide bandwidth, the generated beams split into different physical directions at each subcarrier. This effect, also referred to as beam squint in mmWave research, has more severe implications on achievable communication rates. In sensing applications, the beam split is approximately 4° (1.4°) for 0.3 THz with a bandwidth of 30 GHz (60 GHz with 2 GHz), respectively, for a broadside target direction-of-arrival (DoA).

Near-Field Effect Due to the shorter transmission distance in THz systems, the wave emitted from the transmitter and impinging on the receive array may no longer be a plane wave. In the near-field, specifically when the transmission range is shorter than the Fraunhofer distance, the wavefront becomes spherical. As a result, channel acquisition algorithms need to consider both direction and range information. While near-field signal processing is relatively new in the context of communication, it is a well-established concept in sensing/radar applications such as through-the-wall sensing, near-field imaging, and strip-map synthetic aperture radar (SAR). The range-dependent beampattern is also observed in certain far-field applications like frequency diverse array (FDA) radars, which use linear frequency offsets across antennas to achieve range-dependent beampatterns. However, the FDA wavefront is not spherical. Our focus is on the spherical wavefront near-field in a THz communication system that employs an extremely large array and transmits multiple subcarriers using orthogonal frequency-division multiplexing (OFDM) signaling.

Distance-Dependent Bandwidth In the low-THz band (below 1 THz), although there are regions of local attenuation minima, the frequency range exhibits distance-dependent spectral windows. While the entire band can be considered a single transmission window with a bandwidth on the order of a THz at distances below 1 m, at longer ranges, there are multiple transmission windows that are only tens or hundreds of GHz wide due to increased molecular absorption. The bandwidth of each transmission window decreases with range due to more severe absorption. For example, when the distance increases from 1 to 10 m, the bandwidth reduces by an order of magnitude. Therefore, in the THz band, there is a critical trade-off between operating radar or communication at high bandwidth and maintaining an adequate maximum detectable range or high data rates.

Doppler Shift In wideband THz systems, the Doppler spread can cause significant inter-carrier interference (ICI), especially in high mobility scenarios. At 0.3 THz, the Doppler shift becomes 10 times larger than that at 30 GHz. This severe Doppler effect undermines the orthogonality among subcarriers, leading to ICI and posing challenges for OFDM.

Table 1.2 summarizes the differences between the mmWave and THz channels.

Table 1.2 Comparison of mmWave and THz transmission characteristics [Elbir et al., 2022]

Phenomenon	mmWave	THz
Path loss	The path loss exponent ≈2. Massive array structures are used to mitigate path loss	The path loss exponent doubles. Ultra-massive arrays are employed to mitigate path loss
Channel model	Superposition of LoS and NLoS paths. Channel estimation may solely exploit sparse reconstruction techniques, e.g. compressed sensing (CS) and orthogonal matching pursuit (OMP)	A dominant LoS path with multiple NLoS paths. Channel estimation may require joint OMP and true-time-delay (TTD) techniques
Beam alignment	Beams are *squinted* but still cover the user across the entire bandwidth. This effect is largely dependent on the large number of antennas. Corrected via TTD processing at each antenna	Beams become totally *split* and cannot cover the user with their main lobes. This split is a function of both bandwidth and array size. Corrected by delay-phase precoding (DPP) or beam split phase correction
Array structure	Large arrays with fully-connected or subarray structures, where the latter has lower hardware complexity	Much larger array structures with additional subarray levels to reduce the hardware complexity, e.g. AoSA, widely spaced multi-subarray (WSMS) or GoSA structures
Bandwidth	Wideband throughout the entire range	Distance-dependent bandwidth arising from peculiarities in molecular absorption
Wavefronts	Planar wave in far-field	Spherical wavefront in near-field below Fraunhofer distance

We now present details of channel models commonly used in communications and radars.

1.5.3 Communications Channel

Consider a transmitter that employs an antenna array or a single directional antenna with carrier frequency f and TX (RX) antenna gain G_{TX} (G_{RX}). The LoS communications channel with a delay spread comprising $L_c - 1$ delay taps is $h_c(t,f) = G_c \sum_{\ell=0}^{L_c-1} \alpha_\ell e^{-j2\pi\tau_\ell f} e^{j2\pi\nu_\ell t}$, where G_c is the large-scale communications channel gain at the reception, and α_ℓ is the path loss coefficient of the lth path with time delay τ_ℓ and Doppler shift ν_ℓ. The free space attenuation model yields $G_c = \frac{G_{\mathrm{TX}} G_{\mathrm{RX}} \lambda^2}{(4\pi)^2 \rho_c^\gamma}$, where γ is the path loss (PL) exponent. Further, $\gamma \approx 2$ for mmWave LoS outdoor urban [Rappaport et al., 2015] and rural scenarios [MacCartney et al., 2016].

1.5.4 Radar Channel

The doubly selective (time- and frequency-selective) mmWave radar channel is modeled after TX/RX beamforming using virtual representation obtained by uniformly sampling in the range dimension [Kumari et al., 2018]. Assume L uniformly sampled range bins and that the ℓth range bin consists of a few, (say) K_ℓ, virtual scattering centers. Each (ℓ, k)th virtual scattering center is characterized by its distance ρ_ℓ, delay τ_ℓ, velocity $v_{\ell,k}$, Doppler shift $\nu_{\ell,k} = 2v_{\ell,k}/\lambda$, large-scale channel gain $G_{\ell,k}$, and small-scale fading gain $\beta_{\ell,k}$. Then, the multi-target radar channel model is $h_r(t,f) = \sum_{\ell=0}^{L-1}\sum_{k=0}^{K_\ell-1} G_{\ell,k}\beta_{\ell,k}e^{-j2\pi\tau_\ell f}\cdot e^{-j2\pi\nu_{\ell,k}t}$. The large-scale channel gain corresponding to the (ℓ, k)th virtual target scattering center is $G_{\ell,k} = \frac{\lambda^2\sigma_{\ell,k}}{64\pi^3\rho_\ell^4}$, where $\sigma_{\ell,k}$ is the corresponding scatterer's radar cross section (RCS). The small-scale gain is assumed to be a superposition of a complex Gaussian component and a fixed LoS component leading to Rician fading. Similarly, the corresponding frequency-selective models can also include Rician fading. They capture, as a special case, the spiky model used in prior works on mmWave communications/radar. In this case, the corresponding radar target models are approximated by the Swerling III/IV scatterers [Skolnik, 2008].

Further, clustered channel models can be considered to incorporate correlations and extended target scenarios although they remain unexamined in detail. For instance, the conventional mmWave automotive target model assumes a single non-fluctuating (i.e., constant RCS) scatterer based on the Swerling 0 model. This greatly simplifies the development and analysis of receive processing algorithms and tracking filters [Dokhanchi et al., 2019a]. However, when the target is located within the close range of high-resolution radar, the received signal is composed of multiple reflections from different parts of the same object. This *extended* target model is more appropriate for mmWave applications and may also include correlated RCS [Duggal et al., 2020].

It is typical to assume a frequency-selective Rayleigh fading model for both communications and radar channels during the dwell time comprising N_{CPI} coherent processing intervals (CPI). In radar terminology, this corresponds to Swerling I/II target models. In each CPI with M frames, the channel amplitude of each tap is considered to be constant, i.e., a block fading model is assumed. Moreover, constant velocity and quasi-stationarity conditions are imposed on the target model.

1.5.5 Channel-Sharing Topologies

The existing mmWave JRC systems could be classified by the joint use of the channel [Paul et al., 2017; Geng et al., 2018] (Figure 1.2). In the *spectral coexistence* approach, radar and communications operate as separate entities and focus on devising strategies to adjust transmit parameters and mitigate the interference adaptively for the other [Cohen et al., 2018]. To this end, some information exchange between the two systems, i.e. spectral cooperation, may be allowed but with minimal changes in the standardization, system hardware and processing. In *spectral co-design* [Dokhanchi et al., 2019a; Paul et al., 2017], new *joint* radio-frequency sensing and communications techniques are developed where a single unit is employed for both purposes while also accessing the spectrum in an opportunistic manner. New fully-adaptive, software-defined systems are attempting to integrate these systems into the same platform to minimize circuitry and maximize flexibility. Here, each transmitter and receiver may have multiple antennas in a phased array or Multiple-Input Multiple-Output (MIMO) configuration.

Some spectrum-sharing solutions also involve cooperation between radar and communications. The exchange of information, such as the channel state information (CSI), may also be facilitated via a fusion center [Li and Petropulu, 2017]. The spectral cooperation enables both systems to benefit from increased degrees of freedom (DoFs) and allows joint optimization of system parameters through one [He et al., 2019] or more [Dokhanchi et al., 2020] objective functions. Sometimes, sensing (communications) may be opportunistically present because of flexibility in the communications (radar) transmitter. Table 1.3 shows various other categories of JRC solutions. Apart from the abovementioned hardware-sharing criterion, it is possible to classify the JRC systems as follows:

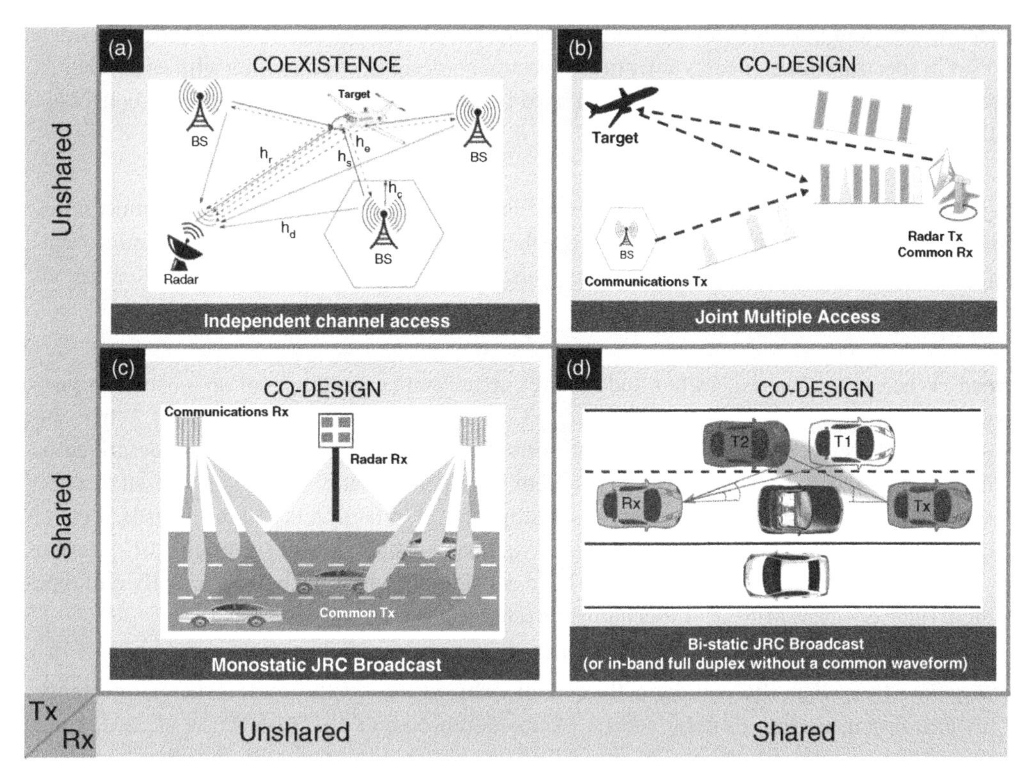

Figure 1.2 (a) Spectral coexistence system where radar and communications subsystems are independently located and access the associated radio channels such as radar target channel h_r, communications channel h_c, radar-to-communications interference h_s, and communications-to-radar interference h_d [Bică et al., 2016]. (b) Co-design system where only Rx are shared. In this *joint multiple access* channel, the radar operates in monostatic mode and both systems transmit different waveforms that are orthogonal in spectrum, code or time [Mishra et al., 2017]. (c) In TX-shared co-design, the monostatic radar functions as a communications transmitter emitting a common JRC waveform [Muns et al., 2019]. (d) A bi-static broadcast co-design with common TX, RX, and a joint waveform [Dokhanchi et al., 2019a]. The joint waveform transmitted by the TX vehicle bounces off from targets such as T1 and T2 and received by the Rx vehicle. A variant is an in-band full duplex system with different waveforms but common TX and Rx [Donnet and Longstaff, 2006]. The term "BS" stands for "base station."

Table 1.3 A comprehensive classification of JRC systems

Channel access	**Hardware**	**Waveform**
• Independent • Coordinated • Joint • Shared	• Separate Tx and Rx • Same Tx, Common Rx • Common Tx, Same Rx • Common Tx and Rx	• Separate • Common • Resource-shared
Location	**Performance/functionality**	**Specialized**
• Colocated • Bi-static • Distributed • Networked • Heterogeneous	• Radar-centric • Comms-centric • Joint radar-comms • Dual-Function Radar-Comms	• MRMC • IBFD ISAC • IRS-Aided ISAC • mmWave, THz, VLC, quantum

Channel Access Radar and communications systems may access channels independent of each other (as in spectral coexistence), jointly (as in spectral co-design), coordinating with each other as per requirements [Cohen et al., 2018], or on a fixed resource-shared basis [Alaee-Kerahroodi et al., 2019].

Waveforms Irrespective of whether the hardware is shared between radar and communications, both systems may employ separate waveforms [Cohen et al., 2018; Wu et al., 2022], a common waveform [Dokhanchi et al., 2019a], or the same waveform on a shared resource basis [Alaee-Kerahroodi et al., 2019].

Location When the radar transmitter and receiver are close to each other (or on a common platform), it is termed a monostatic or colocated JRC system [Dokhanchi et al., 2019b]. When the transmitter and receiver are on different platforms, e.g. in a vehicular environment described in Dokhanchi et al. [2019a], then the resulting system is termed bi-static. This is generalized to a distributed or multi-static JRC, which comprises multiple radar transmitters and receivers [Liu et al., 2020]. The multiple units may also be networked via fusion center [Li and Petropulu 2017]. In Wu et al. [2022], a heterogamous JRC system has been considered, where each radar is a different type of system (phased array, MIMO, or mechanically scanning).

Functionality In a selfish JRC paradigm, the overall architecture usually promotes the performance of only one system leading to radar-centric [Alaee-Kerahroodi et al., 2019; Wu et al., 2022] and communications-centric [Ayyar and Mishra, 2019] co-existence solutions. On the other hand, the holistic solution relies on extensive cooperation between the two systems in transmitting strategies and receiving processing [Wu et al., 2022]. This has also been termed as integrated sensing and communications (ISAC) system. A specific system is dual-function radar-communications (DFRC), which refers to any JRC system that employs the same hardware for radar or communications functions.

Specialized When both radar and communications are deployed in MIMO configuration with several transmit and receive antennas to benefit from high spectrum efficiency and spatial diversity, the resulting joint MIMO-radar-MIMO-communications (MRMC) systems are more challenging to optimize [Alaee-Kerahroodi et al., 2019; Liu et al., 2020]. Further, recent developments in massive MIMO utilize uplink/downlink (UL/DL) channel reciprocity via a vast number of service antennas to serve a lower number of mobile users with time-division duplexing. More recently, the MRMC systems have considered in-band full-duplexing (IBFD) communications that enable UL and DL to function in a single time/frequency channel through advanced self-interference cancellation techniques [Liu et al., 2020]. More recently, reconfigurable intelligent surface (RIS) or intelligent reflecting surfaces (IRS) [Hodge et al., 2023] have introduced unprecedented flexibility and adaptability toward smart wireless channels. Recent research [Wei et al., 2023a,b; Esmaeilbeig et al., 2024; Elbir et al., 2023] on JRC systems has demonstrated that RIS platforms enable enhanced signal quality, coverage, and link capacity. Finally, apart from mmWave and THz spectrum, sensing and communications have also been explored in the visible light [Fragner et al., 2022] and quantum regimes [Wang et al., 2022].

In the next section, we provide some examples of transmit and receive strategies for two major JRC topologies: coexistence and co-design.

1.6 JRC Coexistence

Interference management is central to spectral coexistence of different radio systems. This typically requires sensing the state of the shared spectrum and adjusting TX and RX parameters so that the impact of interference is sufficiently reduced and individual system performance is enhanced. We now present the figures of merit qualifying system performance and then discuss methodologies for coexistence.

1.6.1 Communications Performance Criteria

Since the goal of communications systems is to transfer data at a high rate error-free for a given bandwidth, the commonly used performance criteria include quality of service (QoS) indicators such as spectral efficiency, mutual information, channel capacity, pairwise error probability, bit/symbol error rates (BER/SER), and signal-to-interference-and-noise ratio (SINR). Given a communications signal model, the achievable spectral efficiency can be used as a universal communications performance criterion. In practice, the achievable spectral efficiency r is an upper bound, while the effective spectral efficiency r_{eff} depends on the implemented receiver (e.g. minimum mean square error or MMSE [Shi et al., 2004], decision feedback [Takizawa et al., 2012] or time-domain equalizer [Liu et al., 2013]), and is a fraction of the achievable spectral efficiency. The effective communications rate is then the product of the signal bandwidth W and r_{eff}.

1.6.2 Radar Performance Criteria

Radar systems, by virtue of their use in both detection and estimation, lend themselves to a plethora of performance criteria depending on the specific task. Target detection performance is characterized by probabilities of correct detection, mis-detection, and false alarm. In a parameter estimation task, mean square error (MSE) or variance in comparison to the Cramér-Rao Lower Bound (CRLB) is commonly considered. The CRLB defines the lower bound for estimation error variance for unbiased estimators. There are also several radar design parameters such as range/Doppler/angular resolution/coverage and the number of targets a radar can simultaneously resolve. In particular, the radar's ability to discriminate in both range and velocity is completely characterized by the *ambiguity function* (AF) of its transmit waveform; it is obtained by correlating the waveform with its Doppler-shifted and delayed replicas.

1.6.3 Interference Mitigation

The mmWave radar and communications TX and RX can use all of their DoFs such as different antennas, frequency, coding, transmission slots, power, or polarization to mitigate or avoid mutual interference. Interference may also be caused by leakage of signals from adjacent channels because of reusing identical frequencies in different locations. In general, the higher the frequency in mmWave bands, the weaker the multipath effects. The transmitters can adjust their parameters so that the level of interference is reduced at the receiver. To this end, awareness about the dynamic state of the radio spectrum and interference experienced in different locations, subbands, and time instances is desired. This may be in the form of feedback provided by the receivers to the transmitter about the channel response and SINR. Both the TX and RX can be optimized such that the SINR is maximized at the receivers for both subsystems.

1.6.3.1 Receiver Techniques

Interference mitigation may be performed only at the RX rendering CSI exchange optional. Typically, this requires multiple antenna at RX, a common feature at mmWave, and processing of the received signals in the spatial and/or temporal domain. These techniques employ receive array covariance matrix $\mathbf{m}\Sigma$ (or its estimate $\mathbf{m}\hat{\Sigma}$) in certain interference canceling RX structures. Here, the received signal space spanned by eigenvectors of $\mathbf{m}\Sigma$ is divided into two orthogonal subspaces of signal and interference-plus-noise. The received signal is then projected to a subspace orthogonal to the interference-and-noise subspace to enable processing of practically interference-free signals. If the interference impinges the receiver from angles different from the desired signal, RX beamforming is commonly used [Geng et al., 2018]. The beampattern design ensures high gains toward the desired signals and steers nulls toward the interference. Common solutions include Minimum Variance Distortionless Response (MVDR), Linearly Constrained Minimum Variance (LCMV), and diagonal loading [Vorobyov, 2014].

Advanced interference cancellation receivers estimate CSI, use feedback about channel response, or sense other properties of the state of the radio spectrum. These estimates are later used to cancel the interference contribution from the overall received signal. The coherence time of the channels should be sufficiently long that the feedback or channel estimates are not outdated during the interference cancellation process. These techniques either require knowledge of modulation schemes employed by coexisting radio systems or are applied to digital modulation methods only. A prime example is the Successive Interference Cancellation (SIC) method that decodes and subtracts the strongest signal first from the overall received signals and then repeats the same procedure by extracting the next weaker signal from the residual signal and so on [Paul et al., 2017]. In the absence of CSI, non-traditional radar interference models are used for robust communications signal decoders [Ayyar and Mishra, 2019].

1.6.3.2 Transmitter Techniques

Adapting transmitters and optimizing transmit waveforms may be used to minimize the impact of interferences in coexistence systems. In a considered coexistence scenario, for example, the optimization objective could be maximizing the SINR at each receiver while providing the desired data rate for each communications user and target Neyman-Pearson detector performance for radar users. Designing a precoder for each transmitter or/and decoders for each receiver achieves this goal by steering the interferences to a different space than the desired a signals.

One such example design in the context of MIMO communications and MIMO radar is the Switched Small Singular Value Space Projection (SSSVSP) method [Mahal et al., 2017] in which the interference is steered to space spanned by singular vectors corresponding to zero or negligible singular values. This method requires information exchange between the radar subsystem and communications base-stations. Another example of a precoder-decoder design for interference management in radar-communications coexistence is via Interference Alignment (IA) [Cui et al., 2018] where IA coordinates coexisting multiple transmitters such that their mutual interference aligns at the receivers and occupies only a portion of the signal space. The interference-free signal space is then used for radar and communications purposes.

1.7 JRC Co-Design

Central toward facilitating the co-design of radar and communications systems are waveform design and their optimization exploiting available DoFs (spatial, temporal, spectral, polarization). The optimization is based on the system performance criteria and availability of channel state information (CSI), awareness about target scene, and the levels of unintentional or intentional interference at the receivers.

1.7.1 JRC Performance Criteria

In co-design, JRC waveforms are modeled to simultaneously improve the functionalities of both subsystems with some quantifiable trade-off. In Bliss [2014], a radar round-trip delay estimation rate is developed and coupled with the communications information rate. This radar estimation, however, is not drawn from the same class of distributions as that of communications data symbols and, therefore, provides only an approximate representation of the radar performance. However, potential invalidity of some assumptions limits the extension of this to estimation of other target parameters.

The mmWave designs in Kumari et al. [2017, 2020] for single- and multiple-target scenarios suggest an interesting JRC performance criterion that attempts to parallel the radar CRLB performance with a new effective communications symbol MMSE criteria as a function of effective maximum achievable communications spectral efficiency, r_{eff}. The MMSE communications criteria here is analogous to the mean-squared error distortion in the rate distortion theory. Let MMSE_c be the MMSE of a communications system with spectral efficiency r. Then MMSE_c and r are related to each other through the equation $\frac{1}{N}\text{Tr}\left[\log_2 \text{MMSE}_c\right] = -r$, where N is the code length. Therefore, the effective communications distortion MMSE (DMSE) that satisfies $\frac{1}{N}\text{Tr}\left[\log_2 \text{DMSE}_{\text{eff}}\right] = -r_{\text{eff}} = -\delta \cdot r$ can be defined as $\text{DMSE}_{\text{eff}} \triangleq \text{MMSE}_c^{\delta}$, where δ is a constant fraction of communications symbols transmitted in a CPI with the channel capacity C. The performance trade-off between communications and radar is quantified in terms of a weighted combination of the scalar quantities $\frac{1}{N}\text{Tr}\left[\log_2 \text{DMSE}_{\text{eff}}\right]$ and $\frac{1}{Q}\text{Tr}\left[\log_2 \text{CRLB}\right]$, respectively, where the log-scale is used to achieve proportional fairness between the communications distortion and radar CRLB values and Q is the number of detected targets. Pareto-optimal solutions that assign weights to different design goals have also been explored in this context [Ciuonzo et al., 2015].

Mutual information (MI) is also a popular waveform optimization criterion. At the radar receiver, depending on whether the communications signal reflected off the target is treated as useful energy or interference or ignored altogether, a different MI-based criterion results [Bică et al., 2016]. Although MI maximization enhances the characterizing capacity of a radar system, it does not maximize the probability of detection. The optimal radar signals for target characterization and detection tasks are generally different [Bică et al., 2016; Cohen et al., 2018].

1.7.2 Radar-Centric Waveform Design

We first consider the appropriate radar-centric waveforms here. These range from conventional signals to emerging multi-carrier waveforms.

Conventional Continuous Wave and Modulated Waveforms A simple continuous-wave (CW) radar provides information about only Doppler velocity. To extract range information, either the frequency/phase of the CW signal is modulated or very short duration pulses are transmitted. In practice, the well-known Frequency Modulated Continuous Wave (FMCW) and Phase Modulated Continuous Wave (PMCW) radars are used. A typical FMCW radar transmits one or multiple chirp signals wherein the frequency increases or decreases linearly in time and then the chirps reflected off the targets are captured at the receiver. Chirp bandwidth of a few GHz may be used to provide a range resolution of a few centimeters, e.g., a 4 GHz chirp achieves a range resolution of 3.75 cm. For

PMCW, binary pseudorandom sequences with desirable auto-correlation/cross-correlation properties are typically used. The AF of PMCW has lower sidelobes than FMCW, and PMCW is also easier to implement in hardware [Dokhanchi et al., 2019a].

A general bi-static, PMCW-JRC system using a uniform linear array (ULA) [Dokhanchi et al., 2019a] follows the topology shown in Figure 1.2d. The transmitter sends M repetitions of the PMCW code of length L from each of its N_t transmit antennas. The Doppler shift and flight time for the paths are assumed to be fixed over the CPI. The reflections from Q targets impinge on N_r receive antennas. Let t_c be chip time (time for transmitting one element of one PMCW code sequence, i.e., fast time). The Doppler shifts and the flight time for every path are assumed to be fixed over a coherent transmission time Mt_b, where $t_b = Lt_c$ is the time taken to transmit one block of code, i.e., slow time. The transmit waveform takes the form

$$x_i(t) = \sum_{m=0}^{M-1}\sum_{l=0}^{L-1} a_m e^{j\zeta_l} s\left(t - lt_c - mt_b\right) e^{j2\pi f_c t} e^{j(i-1)kd\ \sin\beta}, \tag{1.1}$$

where $i \in [1, N_t]$ and $a_m = e^{j\varphi_m}$ denote differential PSK symbols (DPSK) over slow time (time for sending one code sequence). The DPSK modulation is robust to constant phase shifts. Further, $s(t)$ is the elementary baseband pulse shape, $\zeta_l \in \{0, \pi\}$ is the binary phase code, $e^{j(n-1)kd\sin\beta}$ is the beam-steering weight for the nth antenna, $k = \frac{2\pi}{\lambda}$ is the wave number, and β is the angle between the radiating beam and the perpendicular to the ULA (for simplicity, we consider only azimuth and ignore common elevation angles). The transmitter steers the beam in multiple transmission from $\left[\frac{-\pi}{2}, \frac{\pi}{2}\right]$, each time with angle β. As shown in Figure 1.3, the communications and radar waveform for PMCW-JRC are combined in analog hardware.

Let $\Delta V_q^{(1)}$ be the radial relative velocity between the transmitter and the qth path, where superscript $(\cdot)^{(1)}$ refers to the transmitter-target path, and the corresponding Doppler shift is $f_{D_q}^{(1)} = \frac{\Delta V_q^{(1)}}{c} f_c$, where $c = 3 \times 10^8$ m/s is the speed of light. The signal impinging on the qth scatterer is

$$z_{q,n}(t) = \sum_{m=0}^{M-1}\sum_{l=0}^{L-1} h_{q,n}^{(1)} a_m e^{j\zeta_l} s\left(t - lt_c - mt_b - \tau_q^{(1)}\right) e^{j2\pi f_c t - j2\pi f_{D_q}^{(1)} t - j2\pi f_c \tau_q^{(1)}}, \tag{1.2}$$

where $\tau_q^{(1)}$ and $h_{q,n}^{(1)}$ are the qth point scatterer time delay and propagation loss for each path, respectively. We exploit the standard narrowband assumption to express the received signal as a phase-Doppler-shifted version of the transmit signal. Assume $\tau_q = \tau_q^{(1)} + \tau_q^{(2)}$ to be the total flight time corresponding to a bi-static range $R_q = c\tau_q$, where the superscript $(\cdot)^{(2)}$ denotes variable dependency on the target-receiver path. Assume $f_{D_q} = f_{D_q}^{(1)} + f_{D_q}^{(2)}$ to be the bi-static Doppler shift, and ψ_q to be the angle between the qth scatterer and the perpendicular line to receive ULA. After TX/RX beamforming and frequency synchronization, the received signal at antenna p, obtained as a superposition of these reflections, takes the form

$$\begin{aligned}
\tilde{y}_p(t) &= \sum_{q=1}^{Q}\sum_{n=1}^{N_t} h_{q,p}^{(2)} z_{q,n}\left(t - \tau_q^{(2)}\right) e^{j2\pi f_{D_q}^{(2)} t} + \tilde{N}_p(t) \\
&= \sum_{q=1}^{Q}\sum_{n=1}^{N_t}\sum_{m=0}^{M-1}\sum_{l=0}^{L-1} h_{q,p}^{(2)} h_{q,n}^{(1)} a_m e^{j\zeta_l} s\left(t - lt_c - mt_b - \tau_q^{(1)} - \tau_q^{(2)}\right) \\
&\quad \times e^{j2\pi\left(f_c - f_{D_q}^{(1)} - f_{D_q}^{(2)}\right)t} e^{j\eta_q} e^{-jkd\sin(\psi_q)(p-1)} + \tilde{N}_p(t),
\end{aligned} \tag{1.3}$$

where $e^{j\eta_q} = e^{-j2\pi\left(f_c\left(\tau_q^{(1)} + \tau_q^{(2)}\right) + \tau_{D_q}^{(1)} \tau_q^{(2)}\right)}$ is a static phase shift, $h_{q,p}^{(2)}$ accumulates the effect of the qth transmitter-target-receiver point scatterer, path-loss, and RCS of the target, and $\tilde{N}_p(t)$ is complex circularly symmetric white Gaussian noise with variance σ^2. An extended target is modeled as a

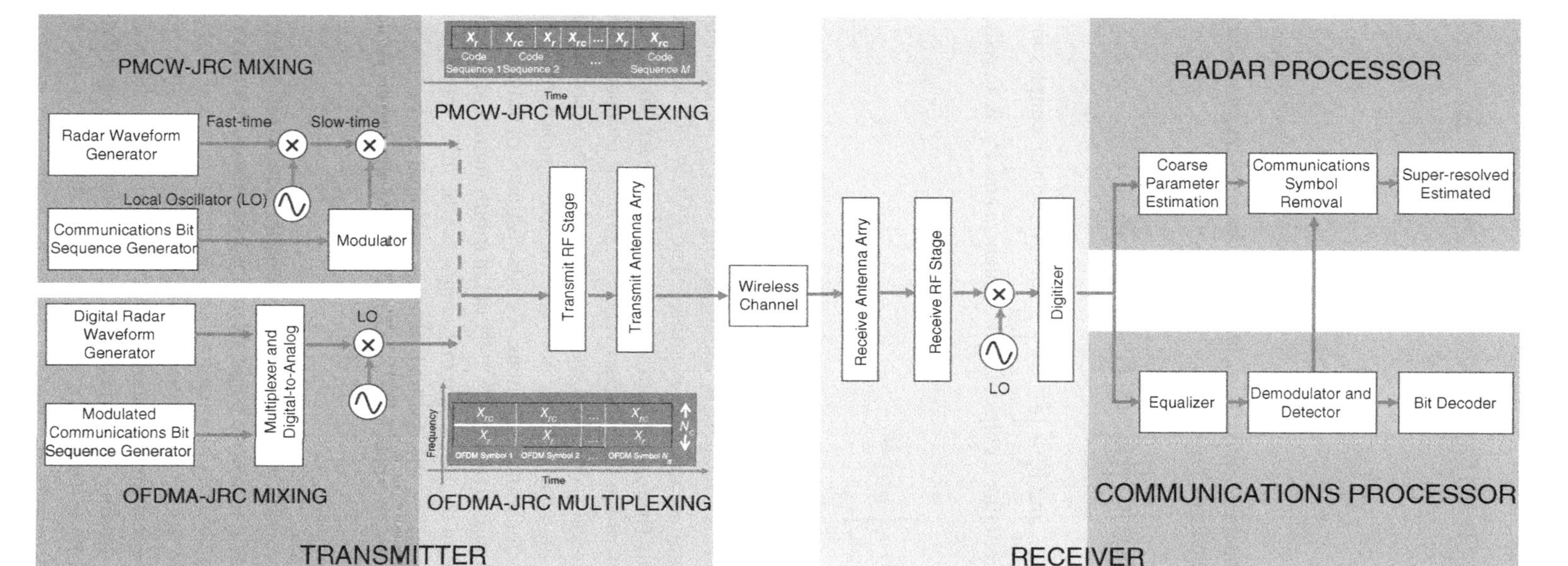

Figure 1.3 A simplified block diagram showing major steps of transmit and receive processing for a general mmWave JRC system. In the case of PMCW-JRC, the radar and communications waveforms are combined in the analog hardware before the RF stage. On the other hand, the information bits from these two subsystems are mixed digitally in OFDMA-JRC. The multiplexing of the radar-only and radar-communications frame for both PMCW- and OFDMA-JRC are depicted in the transmit portion. The receive processing for both systems is largely similar.

cluster of points. This combined with the superposition of reflections from the independent scatterer renders the model in (1.3) applicable for extended targets. After downconversion to baseband and ignoring RCS dependency on Tx and Rx antennas, i.e., $\sum_{n=1}^{N_t} h_{q,n}^{(1)} h_{q,p}^{(2)} e^{j\eta_q} = \sum_{n=1}^{N_t} d'_{q,p,n} = N_t d'_q = d_q$, the received signal is

$$y_p(t) = \sum_{q=1}^{Q} \sum_{m=0}^{M-1} \sum_{l=0}^{L-1} d_q a_m e^{-j2\pi f_{D_q} t} c_q^{p-1} e^{j\zeta_l} s\left(t - lt_c - mt_b - \tau_q\right) + N_p(t),$$

$$p \in \left[1, N_r\right], \tag{1.4}$$

where $c_q = e^{-jkd\sin(\psi_q)}$. Collecting the Nyquist time samples for the antenna p and rearranging them accordingly to slow/fast-time, we form a matrix,

$$\mathbf{Y}_p^{\text{PMCW-JRC}} = \sum_{q=1}^{Q} c_q^{p-1} d_q \text{Diag}\{\boldsymbol{a}\} \left[\left(\boldsymbol{b}_q^T \odot \boldsymbol{s}^T \boldsymbol{P}_{k_q}\right) \otimes \boldsymbol{e}_q\right] + \boldsymbol{N}_p \in \mathbb{C}^{M \times L}, \tag{1.5}$$

where vectors $\boldsymbol{e}_q = \left[e^{j2\pi f_{D_q} mLt_c}\right]_{m=1}^{M}$ and $\boldsymbol{b}_q = \left[e^{j2\pi f_{D_q} lt_c}\right]_{l=1}^{L}$ collect Doppler samples in slow and fast time, respectively, $\boldsymbol{s} = \left[e^{j\zeta_l}\right]_{l=0}^{L}$ contains L chips of code sequence, and $\boldsymbol{P}_{k_q}$ is a cyclic permutation matrix for a shift of k_q as

$$\boldsymbol{P}_{k_q} = \begin{bmatrix} \mathbf{0}_{K_q \times L-K_q} & \mathbf{I}_{K_q \times K_q} \\ \mathbf{I}_{L-K_q \times L-K_q} & \mathbf{0}_{L-K_q \times K_q} \end{bmatrix} \in \mathbb{C}^{L \times L}, \tag{1.6}$$

where $k_q \in \{0, \cdots, L-1\}$ is determined by range of the qth scatterer. If there is no delay between transmitter and receiver for all paths, then $k_q = 0$ for all q and $\boldsymbol{P}_{k_q}$ becomes an identity matrix.

In a PMCW-JRC, the communications symbols and Doppler parameters are coupled, thus leading to a non-identifiable model. This is resolved by a multiplexing strategy through which unknown parameters in the received signal are uniquely identified. The PMCW-JRC adopts time-division multiplexing between radar-only ($\mathbf{m}X_r$) and joint radar-communications ($\mathbf{m}X_{rc}$) frames that are transmitted for μ and $(1-\mu)$ % of the CPI, respectively. The value of μ depends on the amount of prior knowledge about the target scene. As a case in point, when the scene is stationary such as driving a straight path on a highway, we may not need full sensing capacity and can scale up the allocated time appropriately for communications. A coarse estimate of radar target parameters (range, angle, and Doppler) is obtained from $\mathbf{Y}_p^{\text{PMCW-JRC}}$ of radar-only frames $\mathbf{m}X_r$ while communications symbols are extracted from the received signal samples of the $\mathbf{m}X_{rc}$ frame. After extracting communications symbols from $\mathbf{m}X_{rc}$, the residual signal is exploited for further improving the radar target estimates through low-complexity JRC super-resolution algorithms [Dokhanchi et al., 2019a].

Multi-Carrier Waveforms Multi-carrier waveform radars provide additional DoFs to deal with dense spectral use and demanding mmWave target scenarios like drones, low-observable objects, and a large number of moving vehicles in an automotive scenario. Different DoFs can be used in an agile manner to achieve optimal performance depending on the radar task, nature of targets, and state of the radio spectrum. A general drawback of multi-carrier radar waveforms is their time-varying envelope leading to an increased Peak-to-Average-Power-Ratio (PAPR) or Peak-to-Mean-Envelope-Power-Ratio (PMEPR), which makes it difficult to use the amplifiers efficiently when high transmit powers are needed. However, in mmWave radars, the transmit powers tend to be small and surveillance ranges are short. The PAPR reduction is achieved by not allocating all subcarriers or by using appropriate coding/waveform design. Hence, the PAPR issue in mmWave may be less severe.

Multi-carrier Complementary Phase Coded (MCPC) waveform [Levanon, 2000], wherein each subcarrier is modulated by a pseudorandom code sequence of a specific length, is also a viable mmWave JRC candidate. The MCPC design exploits DoFs in the spectral and code domain. In a sense, it is related to OFDM because after each subcarrier is modulated by a code in the time-domain, the subcarriers remain orthogonal without ICI. If the subcarriers are uncoded, the waveform is exactly OFDM. The inter-carrier spacing in MCPC needs to accommodate the spreading of the signals in frequency due to phase codes such as Barker, P3, or P4 polyphase codes [Skolnik, 2008]. This is achieved by choosing the inter-carrier spacing to be the inverse of the chip duration. In OFDM, inter-carrier spacing is smaller. A Generalized Multi-carrier Radar (GMR) waveform devised in Bică and Koivunen [2014, 2016] subsumes most of the widely used radar waveforms such as pseudo random frequency hopping (FH), MCPC, OFDM, and linear step approximations of linear FM signals, as special cases. A matrix model of transmitter and receiver is developed for GMR that allows for defining the waveforms and codes, spreading in time and frequency domain, power allocations, and active subcarriers using a compact notation. Different waveforms are obtained by choosing the dimensions of the matrix model and filling in the entries appropriately. This approach allows for relaxing the perfect orthogonality requirement; this may lead to a better resolution of target delays and Doppler velocities at mmWave.

Spatial DoFs and Multiple Waveforms A few different solutions use the same waveform for both subsystems but make use of radar's spatial DoFs for communications symbols. For instance, in Hassanien et al. [2016], the radar array beampattern sidelobes are modulated by communications messages along user directions. In Hassanien et al. [2018a], the communications symbols are represented by different pairing of antennas and waveforms in a MIMO configuration. Spatial DoFs are also useful for adaptively canceling specific users. A joint beamforming method is suggested in Hassanien et al. [2018b] for a dual-function radar-communications (DFRC) that comprises MIMO radar and communications systems assuming full-duplex transmission. The downlink communications signal is embedded into the transmit radar waveform, and uplink communications takes place when the radar is in listening mode. This necessitates accurate synchronization among the subsystems. The technique utilizes spatial diversity by enforcing the spatial signature of the uplink signals to be orthogonal to the spatial steering vectors associated with the radar target returns. The receiver beamformer employs adaptive and non-adaptive strategies to separate the desired communications signal from echoes of targets, clutter, and noise even if they impinge the array from the same direction. Other solution paths consist of finding spatial filters to mitigate in-band MIMO communications interference through optimization of the sidelobe and cross-correlation levels in MIMO radar systems [Aittomäki and Koivunen, 2017; Li et al., 2016], exploiting co-array processing with multiple waveforms [Zhang et al., 2015] and designing precoders/decoders through interference alignment [Cui et al., 2017].

However, for mmWave JRC systems, the full-resolution ADCs at the baseband signal result in an unacceptably high power consumption. This makes it infeasible to utilize an RF chain for each antenna element, implying that the prevailing MIMO systems that employ fully digital beamforming are not practical for mmWave systems. Thus, the benefits of using multiple waveforms for spatial mitigation in mmWave JRC systems are yet to be carefully evaluated. Currently, a single data stream model that supports analog beamforming with frequency flat TX/RX beam steering vectors is more common [Kumari et al., 2018]. Use of large antenna arrays in mmWave suggests that a feasible JRC approach could be to simply partition the arrays for radar and communications functionalities [Mishra and Eldar, 2019].

1.7.3 Communications-Centric Waveform Design

The most popular communications signal for mmWave JRC is OFDM because it provides a stable performance in multipath fading and relatively simple synchronization [Donnet and Longstaff, 2006]. Also, frequency division in duplexing has an added advantage; unlike time-division duplexing, the former employs different bands for uplink and downlink so that the impact on the interference in radar systems is less severe. Some solutions [Donnet and Longstaff, 2006; Dokhanchi et al., 2019a] also employ the related Orthogonal Frequency Division Multiple Access (OFDMA) waveform for a JRC system. While the OFDM users are allocated on only the time domain, the OFDMA users can be differentiated by both time and frequency. The latter, therefore, provides DoFs in both temporal and spectral domains. Although OFDM-JRC offers high dynamic range and efficient receiver processing implementation based on fast Fourier transform (FFT), it requires additional processing to suppress high sidelobes in receiver processing and reduce PAPR. Further, the OFDM cyclic prefix (CP), used to transform the frequency selective channel to multiple frequency at channels leading to a simplified equalizer, may be a nuisance in the radar context. The CP may adversely affect the radar's ability to resolve ambiguities in radar ranging. Its length depends on the number of channels, particularly the maximum excess delay that the radar signal may experience (time difference between first and last received component of the signal). For radar applications, the CP duration should be equal to or longer than the total maximum signal travel time between the radar platform and target. Other communications waveforms proposed for mmWave automotive JRC include spread spectrum, noise-OFDM, and multiple encoded waveforms [Dokhanchi et al., 2019a]. We now examine mmWave OFDMA-JRC in detail.

OFDMA-JRC Consider the same bi-static scenario of Figure 1.2d that we earlier analyzed for the PMCW-JRC system. The OFDMA-JRC transmitter (Figure 1.3) sends N_s OFDM symbols from N_t transmit antennas, and reflections from Q targets impinge on N_r receive antennas. Assume that β is the angle of departure. The Doppler shift and flight time for the paths are assumed to be fixed over a CPI, i.e., $N_s T_{sym}$, where T_{sym} is the duration of one OFDM symbol and $a_{n,m}$ are multiplexed communications/radar DPSK on the nth carrier of the mth OFDM symbol. Let N_c be the number of subcarriers and Δf be the subcarriers spacing, then the joint transmit waveform in baseband neglecting the CP is

$$x_i(t) = \sum_{m=0}^{N_s-1} \sum_{n=0}^{N_c-1} a_{n,m} e^{j2\pi f_n t} e^{jk \sin(\beta)(i-1)\frac{\lambda}{2}} s\left(t - mT_{\mathrm{sym}}\right), \tag{1.7}$$

where $s(t)$ is a rectangular pulse of the width T_{sym}, $i \in [1, N_t]$, n and m are frequency and time indices, respectively, and $f_n = n\Delta f = \frac{n}{T_{sym}}$ [Dokhanchi et al., 2019a]. The received signal at the pth receiver over a CPI is

$$\begin{aligned} \tilde{y}_p(t) = & \sum_{m=0}^{N_s-1} \sum_{q=1}^{Q} \sum_{n=0}^{N_c-1} \sum_{i=1}^{N_t} d_{q,i,p} a_{n,m} e^{j2\pi f_n (t-\tau_q)} e^{j2\pi f_{D_q} t} e^{jk \sin(\psi_q)(p-1)\frac{\lambda}{2}} \\ & \times s\left(t - mT_{sym} - \tau_q\right) + \tilde{N}_p(t), \end{aligned} \tag{1.8}$$

where $\tilde{N}_p(t)$ is the additive noise on antenna p. Similar to PMCW-JRC, $d_{q,i,p}$ denotes path-loss, phase-shift caused by carrier frequency and RCS of the target; $d_{q,i,p}$ is independent of the subcarrier index due to narrowband assumption. Similarly, the Doppler is assumed to be identical for all subcarriers given a small inter-carrier spacing. For notational convenience, we omit the noise in the following. We sample (1.8) at intervals $t_s = \frac{1}{N_c \Delta f}$ as

$$\tilde{y}_p\left[t_s\right] = \sum_{m=0}^{N_s-1} \sum_{q=1}^{Q} \sum_{n=0}^{N_c-1} d_q s_{n,m} e^{j2\pi \frac{nl}{N_c}} s\left(lt_s - mT_{\mathrm{sym}} - \tau_q\right), \tag{1.9}$$

where $l \in [1, L]$, $n \in [1, N_c]$, and $L \leq N_c$, $d_q = \sum_{i=1}^{N_t} d_{q,i,p}$ as before, and $\tilde{s}_{n,m} = a_{n,m} e^{-j2\pi n \Delta f \frac{R_q}{c}} e^{j2\pi m T_{\mathrm{sym}} f_{D_q}} e^{j\pi \sin(\psi_q)(p-1)}$ $\tilde{s}_{n,m}$ contains information about range, Doppler, angle of arrival, and communications. We assume the number of inverse Fast Fourier Transform (IFFT) points N_c is equal to the number of fast-time samples L in each OFDM symbol. The received signal samples can be viewed as a radar data cube in spatial, spectral, and temporal domains with N_t antennas, N_c subcarriers, and N_s OFDM symbols. Let us stack the entire DPSK symbols into a matrix $\mathbf{m}A \in \mathbb{C}^{N_c \times N_s}$ and let $\mathbf{a}_m = [\mathbf{A}]_m$ be the communications symbols over all subcarriers at mth OFDM symbol time. For a given OFDM symbol, say m, collecting signals from all subcarriers across different antennas leads to the following slow-time slice of the data cube:

$$\mathbf{Y}_m^{\mathrm{OFDMA-JRC}} = \mathbf{F}_{N_c} Diag\left(\mathbf{a}_m\right) \Xi\left(\frac{-\Delta f R_q}{c}\right) Diag\left(\mathbf{d}\right) \boldsymbol{C} \in \mathbb{C}^{N_c \times N_r}, m \in \left[1, N_s\right], \tag{1.10}$$

where $\Xi\left(\frac{-\Delta f R_q}{c}\right) = \left[e^{-j2\pi n \Delta f \frac{R_q}{c}}\right]_{n=1,q=1}^{N_c,Q} \in \mathbb{C}^{N_c \times Q}$, $\mathbf{C} = \left[e^{jk\sin(\psi_q)(p-1)\frac{\lambda}{2}}\right]_{q=1,p=1}^{Q,N_r}$, $\mathbf{C} \in \mathbb{C}^{Q \times N_r}$, and $\mathbf{d} = [d_1 \cdots d_Q]$. Further, $\mathbf{F}_{N_c} = \left[e^{j2\pi \frac{nl}{N_c}}\right]_{l=0,n=0}^{N_c-1,N_c-1}$ denotes a N_c-point IFFT matrix. To estimate Doppler shifts, we consider a subcarrier slice of data cube (1.9):

$$\mathbf{Z}_n^{\mathrm{OFDMA-JRC}} = Diag\left(\mathbf{a}_n\right) \Xi\left(f_{D_q} T_{sym}\right) Diag\left(\mathbf{d}\right) \mathbf{C} \in \mathbb{C}^{N_s \times N_r}, \tag{1.11}$$

where $\mathbf{a}_n = [\mathbf{A}]_n \in \mathbb{C}^{N_s}$ are the DPSK symbols over slow time, $\Xi\left(f_{D_q} T_{sym}\right) = \left[e^{j2\pi m T_{sym} f_{D_q}}\right]_{m=1,q=1}^{N_s,Q}$.

As in PMCW-JRC, the receive processing of OFDMA-JRC is affected by coupling of communications symbols with a radar parameter (range in the case of OFDMA-JRC). To ensure that range estimation does not suffer by using all subcarriers, frequency-division multiplexing is employed in (1.2) such that μ% of the OFDMA subcarriers are allocated to radar (with known $a_{n,m}$ on these subcarriers) and the rest to JRC. The rest of the OFDMA-JRC receive processing is similar to PMCW-JRC (Figure 1.3) [Dokhanchi et al., 2019a].

Comparison of PMCW- and OFDMA-JRC While OFDMA encodes radar and communications simultaneously in the entire *time and space*, the PMCW does so in the entire *frequency and space*; hence, their DoFs and design spaces are in different domains. While it turns out that the receive system models of both waveforms are mathematically identical after matched filtering and retrieve all JRC parameters using similar super-resolution algorithms [Dokhanchi et al., 2017, 2019a], their individual performances mimic the respective communications and radar-centric properties. For example, the AF of the bi-static PMCW-JRC inherits the low sidelobes from its parent stand-alone PMCW radar waveform as shown in a comparison with the AF of OFDMA-JRC in Figure 1.4, given the same bandwidth. On the other hand, the PMCW-JRC is more sensitive to the number of users while the orthogonality of waveforms in OFDMA-JRC makes the latter robust to inter-channel interference. Finally, in a networked vehicle scenario, it requires less complex infrastructure and processing to apply PMCW with predefined or stored sequences rather than using OFDMA to adaptively allocate band to each user [Dokhanchi et al., 2019a; Donnet and Longstaff, 2006]. A comparison of estimation errors in the coupled parameter – range for OFDMA-JRC and Doppler for PMCW-JRC – using JRC super-resolution recovery [Dokhanchi et al., 2019a] is shown in Figure 1.5 for $\mu = 50\%$.

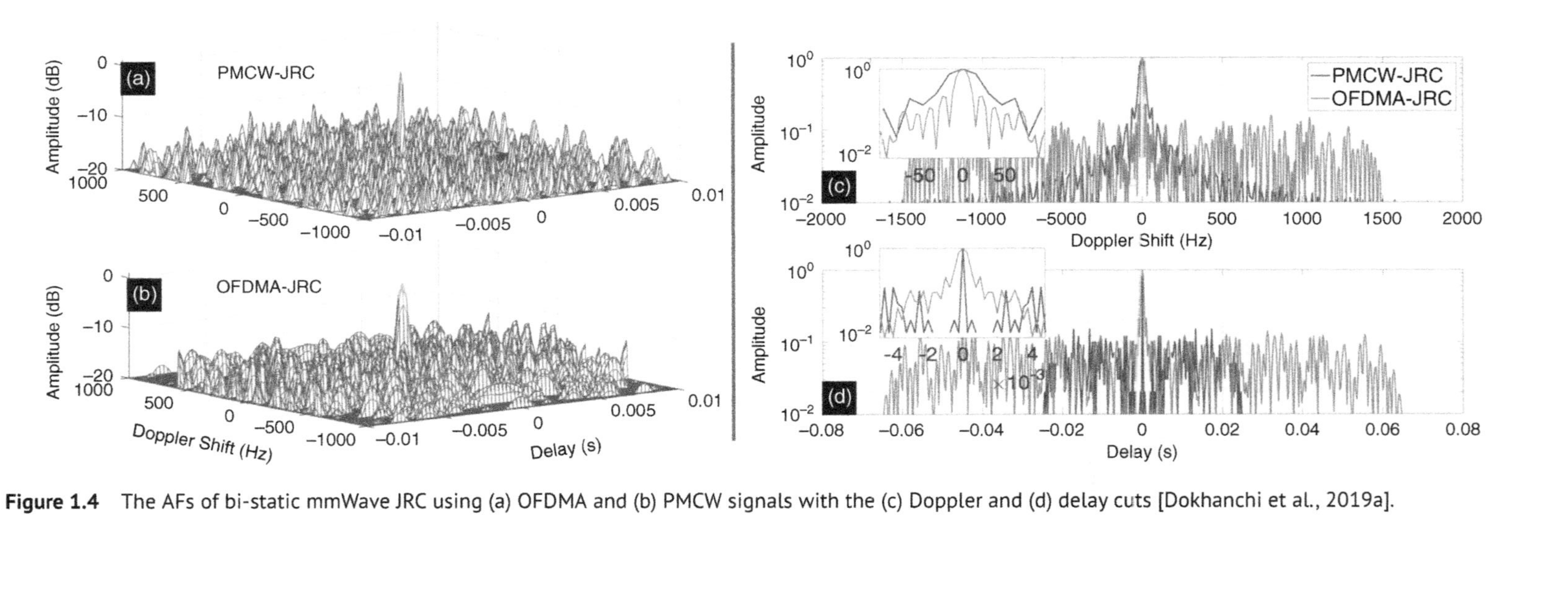

Figure 1.4 The AFs of bi-static mmWave JRC using (a) OFDMA and (b) PMCW signals with the (c) Doppler and (d) delay cuts [Dokhanchi et al., 2019a].

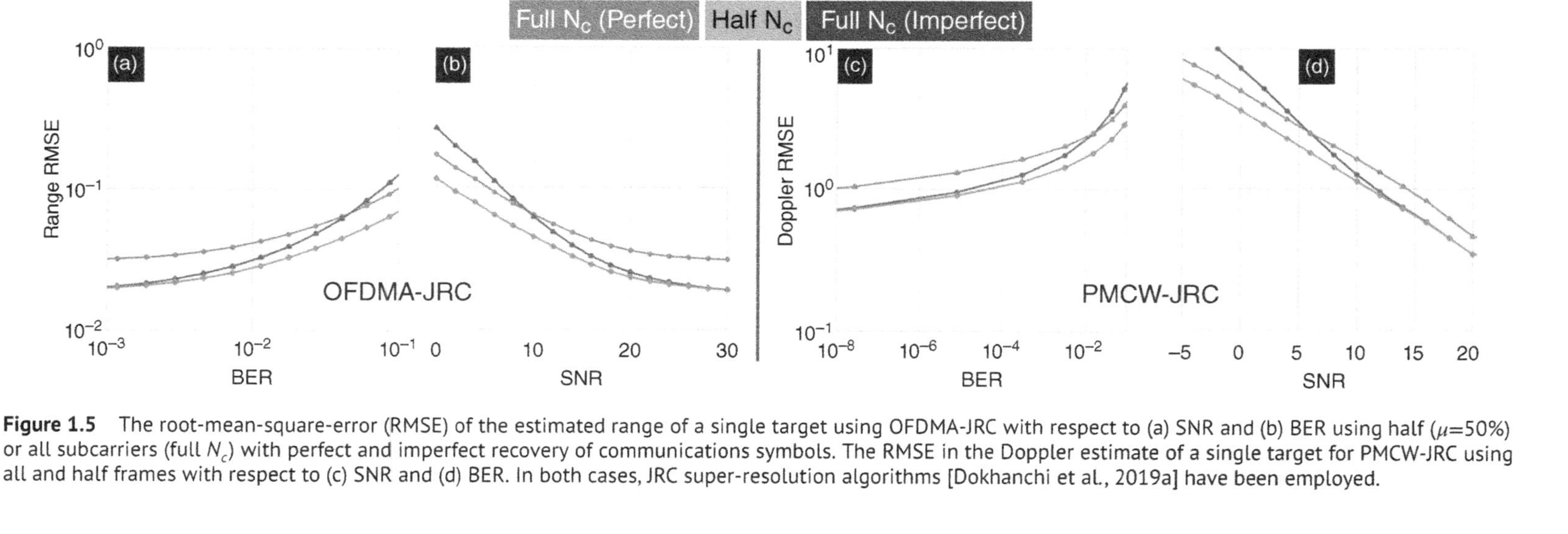

Figure 1.5 The root-mean-square-error (RMSE) of the estimated range of a single target using OFDMA-JRC with respect to (a) SNR and (b) BER using half (μ=50%) or all subcarriers (full N_c) with perfect and imperfect recovery of communications symbols. The RMSE in the Doppler estimate of a single target for PMCW-JRC using all and half frames with respect to (c) SNR and (d) BER. In both cases, JRC super-resolution algorithms [Dokhanchi et al., 2019a] have been employed.

1.7.4 Joint Coding

Recently, existing mmWave communications protocols that are embedded with codes that exhibit favorable radar ambiguity functions have been garnering much attention for JRC. In particular, the 60 GHz IEEE 802.11ad wireless protocol has been employed with time-division multiplexing of a radar-only and radar-communications frame. In general, these designs have temporal DoF (for a monostatic radar case). The IEEE 802.11ad single-carrier physical layer (SCPHY) frame consists of a short training field (STF), a channel estimation field (CEF), header, data, and beamforming training field. The STF and CEF together form the SCPHY preamble. CEF contains two 512-point sequences $Gu_{512}[n]$ and $Gv_{512}[n]$, each containing a *Golay complementary pair* of length 256, $\{Gau_{256}, Gbu_{256}\}$ and $\{Gav_{256}, Gbv_{256}\}$, respectively. A Golay pair has two sequences, Ga_N and Gb_N, each of the same length N with entries ± 1, such that the sum of their *aperiodic* autocorrelation functions has a peak of $2N$ and zero sidelobes:

$$Ga_N[n] * Ga_N[-n] + Gb_N[n] * Gb_N[-n] = 2N\delta[n], \tag{1.12}$$

where * denotes linear convolution. This property is useful for channel estimation and target detection.

By exploiting the preamble of a single SCPHY frame for radar, the existing mmWave 802.11ad waveform simultaneously achieves a cm-level range resolution and a Gbps data rate [Kumari et al., 2020]. The limited velocity estimation performance of this waveform can be improved by using multiple fixed-length frames in which preambles are reserved for radar [Kumari et al., 2018]. While this increases the radar integration duration leading to more accurate velocity estimation, the total preamble duration is also prolonged, causing a significant degradation in the communications data rate [Kumari et al., 2017]. A joint coding scheme based on the use of sparsity-based techniques in the time domain can minimize this trade-off between communications and radar [Kumari et al., 2020]. Here, the frame lengths are varied such that their preambles (exploited as radar pulses) are placed in non-uniformly. These non-uniform pulses in a CPI are then used to construct a virtual block of several pulses, increasing the radar pulse integration time and enabling an enhanced velocity estimation performance. If the channel is sparse, the same can be achieved in the frequency-domain using sub-Nyquist processing [Mishra and Eldar, 2017b]. In Duggal et al., [2020], the wide bandwidth of mmWave is exploited using a Doppler-resilient 802.11ad link to obtain very high resolution profiles in range and Doppler with the ability to distinguish various automotive targets. Figure 1.6 shows distinct, detailed movements of each wheel of a car and body parts of a pedestrian as detected by an 802.11ad-based Doppler-resilient short-range radar.

1.7.5 Carrier Exploitation

Selecting active subcarriers and controlling their power levels or PAPR in an adaptive manner is also useful for interference management. Radar systems generally utilize an entire bandwidth to achieve high resolution. On the other hand, communications systems often allocate resource blocks of a certain number of subcarriers to each user based on the channel quality indicator (CQI) to satisfy their rate and system QoS requirements. Through feedback from the receivers, spectrum sensing, databases, or other sources, the transmitters of both systems can have information about occupancy of different subcarriers, instantaneous or desired SINR levels, channel gains, and power

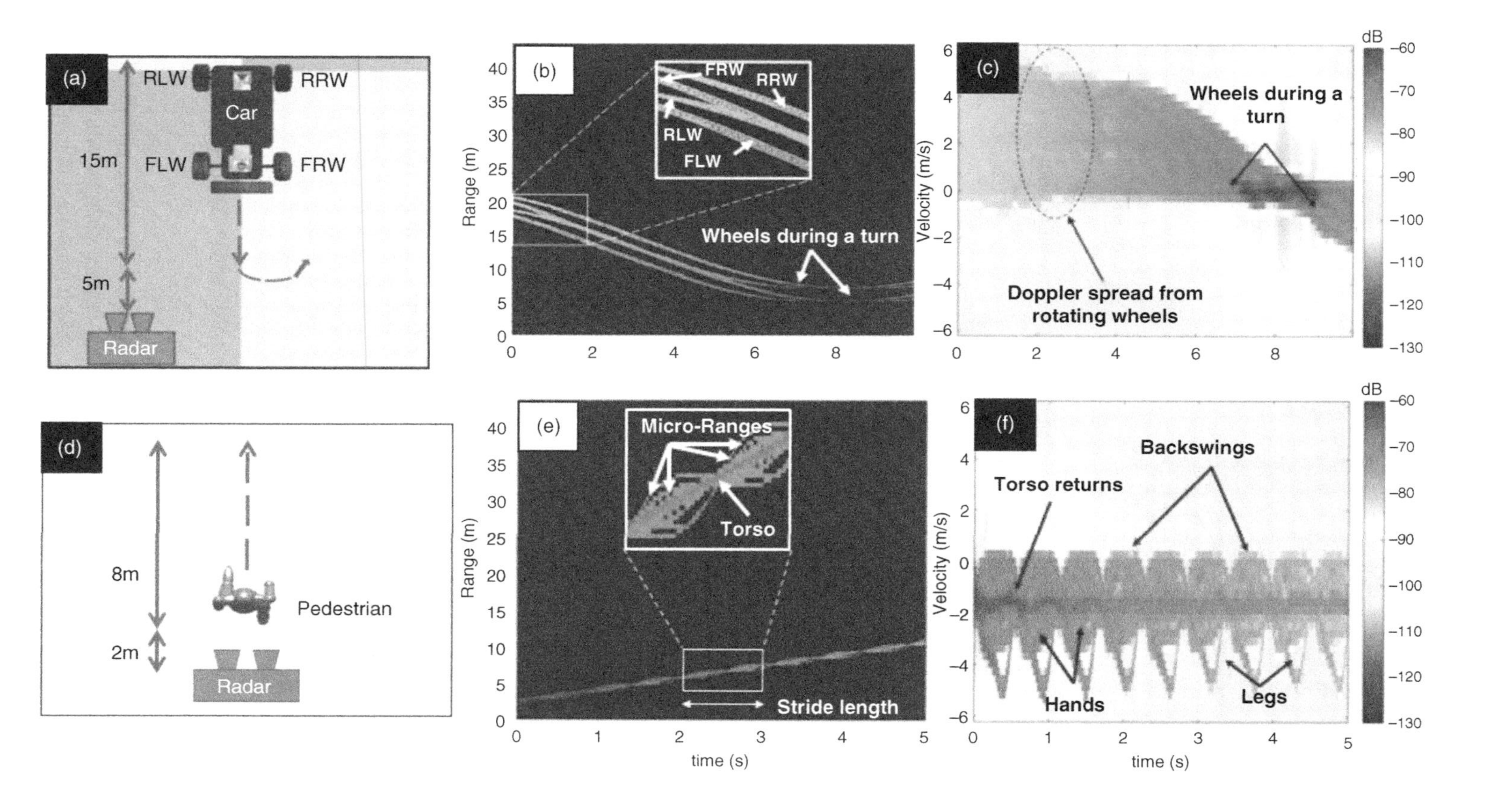

Figure 1.6 Radar signatures generated from animation models of (a) a small car and (d) a pedestrian using Doppler-resilient 802.11ad waveform [Duggal et al., 2019; Duggal et al., 2020]. As the targets move radially in front of the radar on the marked trajectories, the movements of the front right, front left, rear right, and rear left wheels (FRW, FLW, RRW, and LLW, respectively) of the car as well as the torso, arms, and legs of the pedestrian are individually observed in (b, e) range-time and (c, f) Doppler-time domains.

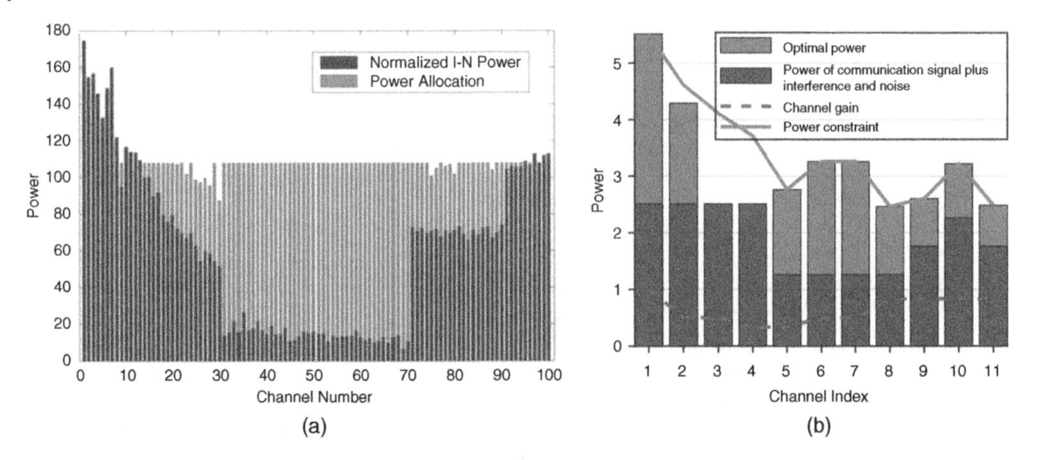

Figure 1.7 Power allocation solutions for JRC carrier exploitation via (a) water-filling and (b) Neyman-Pearson test [Bică et al., 2015].

constraints imposed by other coexisting subsystems. This awareness can be exploited in adaptively optimizing the power allocation among different subcarriers. An example of optimizing subcarrier power (P_k) allocations and imposing minimum desired rate constraints on wireless communications users and maximum power constraint P_{T} for the radar is as follows:

$$
\begin{aligned}
\underset{P_k,\eta}{\text{maximize}} \quad & p_{\mathrm{D}} \\
\text{subject to} \quad & p_{\mathrm{FA}} \leq \alpha, \\
& \log\left(1 + SINR_k\right) \geq t_k, \forall k, \\
& \sum_{k=0}^{N-1} P_k \leq P_{\mathrm{T}},
\end{aligned} \tag{1.13}
$$

where η is the detection threshold for a likelihood ratio test using Neyman-Pearson detection strategy with false alarm constraint α. Two example power allocations from the radar perspective are depicted in Figure 1.7. A water-filling solution (Figure 1.7a) obtained by maximizing mutual information between received data and target and channel response allocates the radar power to those parts of spectrum where the signal experiences the least attenuation and the interference level is low. The second approach (Figure 1.7b) takes into account channel gains, required SINR values at communications subsystems while maximizing the radar performance in the Neyman-Pearson sense for the target detection task.

1.8 Emerging JRC Applications

Some more recent enabling architectures and technologies for JRC where the system can sense, learn, and adapt to the changes in the channel are as follows.

Cognitive Systems Cognitive radars and radios sense the spectrum and exchange information to build and learn their radio environment state. This typically implies channel estimation and feedback on channel quality. Spectrum cartography methods, which generate a map of spectrum access in different locations and frequencies at different time instances, have been developed in this context [Kim et al., 2011]. Based on the obtained awareness, operational parameters of transmitters and receivers in each subsystem are adjusted to optimize their performance [Cohen et al., 2018]. Channel coherence times should be long enough for JRC to apply cognitive actions. Since this duration is in nanoseconds for mmWave environments, compressed sensing-based solutions aid in reducing the number of required samples for cognitive processing [Mishra and Eldar, 2017b,a].

Fast Waveforms Algorithms that develop cognitive waveforms should have low computational complexity in order to re-design waveforms on-the-fly, typically within a single CPI. This is especially important for mmWave systems where the fast-time radar waveform can easily have a length of tens of thousands of samples. In Li and Vorobyov [2018], waveform design in a spectrally dense environment does not exceed a quadratic complexity. In Mishra et al. [2017] and Mishra and Eldar [2017b], the mmWave radar based on sub-Nyquist sampling adaptively transmits in disjoint subbands, and the vacant slots are used by vehicular communications.

Machine Learning In order to facilitate fast configuration of mmWave JRC links with low latency and high efficiency, machine learning is useful to acquire situational awareness. This implies learning the evolution of spectrum state over time (including classifying radar target responses or other waveforms occupying the spectrum), acquiring the channel responses, identifying an underutilized spectrum, and exploiting it in an opportunistic manner. The deep learning methods are widely applied for tasks such as target classification, automatic waveform recognition, and determining optimal antennas and RF chains [Elbir and Mishra, 2019]. Optimal policies for coexisting systems may be learned using reinforcement learning approaches like the partially observable Markov decision process (POMDP) and the restless multi-arm bandit (RMAB) [Lunden et al., 2015].

Game Theoretic Solutions The interaction between radar and communications systems sharing a spectrum can be analyzed from a game theory perspective [Mishra et al., 2019]. The two systems or players form an adversarial, non-cooperative game because of conflicting interests in sharing the spectrum. The game is also dynamic due to continuously evolving spectral states over time. The utility function is designed to reflect the possible strategies based on the respective players' requirements. The solutions result in Nash or Stackelberg equilibrium, which are the game states with the property that none or one of the players can do better, respectively. In comparison to sub-6 GHz, the solution space for mmWave is several GHz wide with much lower maximum transmit power.

Resource Allocation A successful operation of spectrum-sharing systems relies on an efficient radio resource optimization [Wu et al., 2022]. In coexistence paradigms with a single antenna or fixed beamforming, the resources may include total transmit power, transmit signal bandwidth, and transmission time slots. Additionally, in phased array or MIMO systems, antennas need to be reserved for optimal sharing. In single-antenna JRC systems, the resource allocation objective is to optimize the transmit energy of the dual-purpose waveform based on the propagation channels of radar and communications users. In JRC systems that employ a transmit antenna array, highly directional beamforming toward the radar surveillance area while still ensuring sidelobes on the users offers an interesting solution. In JRC systems involving hybrid analog–digital transceiver architectures, antenna selection may also be viewed as a component of resource allocation objective.

Dual-Blind Deconvolution (DBD) In JRC systems, usually prior knowledge of the radar waveform and communications channel parameters is available. However, in applications such as covert and secure military systems, there is a need to decipher overlaid radar and communications signals with minimum prior information. Here, the primary objective shifts toward the simultaneous recovery of radar target parameters (e.g. delays, Doppler shifts), radar waveform, communications messages, and channel parameters. The radar transmit signal is also unknown in case of passive or multistatic sensing, where the transmitter and the receiver are separated platforms. Similarly, dynamic communications scenarios, like those found in mobile radio and vehicular networks, introduce a high level of channel dynamism, rendering prior channel estimates less accurate. In these cases of minimum prior information, recent works [Vargas et al., 2023; Monsalve et al., 2023; Jacome et al., 2023; Vargas et al., 2023] explored the low-rank structure of the waveforms and messages along with the sparsity of both radar and communications channels via atomic norm minimization. This framework facilitates the recovery of continuous-valued parameters, which is applicable to radar and communications channels. In Vargas et al. [2023], the recovery of both channels and signals is modeled as DBD, a highly ill-posed problem that is solved by minimizing the sum of multivariate atomic norms (SoMAN).

Tracking The presence of multiple targets in a distributed JRC scenario poses additional challenges. Compared to the well-studied single-target localization problems, one of the challenges involving multi-target localization using a widely distributed MIMO radar is as follows: since the relative distance of each target is different with respect to each receiver, the echoes from multiple targets are delayed by a different amount at each receiver. The result is that, after the detection procedure, each receiver ends up with a different ordering of targets in time. This makes it difficult to associate the detected echoes from all receiver uniquely to each target. To this end, the distributed JRC literature [Nayemuzzaman et al., 2024] suggests adapting various data association algorithms such as multiple hypothesis tracking, random finite sets, and joint probabilistic data association (JPDA) to assign detections correctly to each receiver. This technique paves way for integrating target tracking into distributed JRC scenarios.

1.9 Open Problems and Summary

In this chapter, we have explored the exciting field of JRC systems, which leverage the synergy between radar sensing and high-speed wireless communications. Throughout our discussion, we have identified several open problems that require further research and exploration to fully realize the potential of these integrated systems.

One of the key challenges is developing a correlation model between the sensing and communication channels. A quantitative description of the correlation between these channels is crucial for designing effective sensing-assisted communication schemes. Understanding the interplay between radar sensing and communication can enable the development of innovative techniques that exploit the shared information. These aspects need to be examined from both information theory and signal processing perspectives.

Coordination gain, which quantifies the benefits of JRC systems, requires further investigation. Developing accurate metrics and models to measure coordination gain will aid in evaluating the performance of these integrated systems and guide their design and optimization. The literature either lacks a common design/performance metric or employs a vaguely defined common metric for both systems.

Another important aspect is the design of the JRC frame protocol. Developing unified JRC waveforms or beams that seamlessly integrate sensing and communication functionalities is a complex task. Designing efficient frame protocols that enable simultaneous sensing and communication while minimizing interference is an open problem that warrants further research.

Joint resource allocation, waveform design, and deployment/trajectory planning are critical for efficient operation of radar-communication systems. Optimizing the allocation of resources such as frequency, power, and time, as well as designing suitable waveforms and trajectories, can enhance system performance and enable seamless integration of radar and communications functionalities.

Coordinated interference management is another challenge that needs to be addressed. As radar and communications systems share the same spectrum, interference can arise between different users or even within the same user. Developing interference management techniques that effectively mitigate interference and enhance system performance is an important area for future research.

Cooperative JRC, where multiple radar-communication systems collaborate to achieve common goals, is an interesting avenue that requires further exploration. Investigating cooperative strategies, such as joint beamforming, resource sharing, and coordinated sensing, can lead to significant performance improvements and enhanced system robustness.

Integrating intelligent algorithms and techniques, such as artificial intelligence (AI), into emerging platforms like unmanned aerial vehicles (UAVs) holds great promise for JRC research. AI-based algorithms for UAV-aided JRC can enable autonomous decision-making, adaptive resource allocation, and optimized system performance.

Security and privacy are critical concerns in JRC systems. Developing secure JRC protocols that protect against eavesdropping, jamming, and other malicious activities is crucial for ensuring the confidentiality and integrity of the transmitted information [Mishra et al., 2022].

In conclusion, JRC is a vibrant and rapidly evolving area of research. Addressing the open problems discussed in this chapter will pave the way for the development of efficient, robust, and secure integrated systems. Continued research and innovation in these areas will enable the realization of the full potential of JRC systems in various applications, ranging from autonomous vehicles and smart cities to IoT and beyond [Sedighi et al., 2021].

References

Report to the President: Realizing the full potential of government-held spectrum to spur economic growth, 2012. President's Council of Advisors on Science and Technology.

T. Aittomäki and V. Koivunen. Hybrid optimization method for cognitive and MIMO radar code design. In *European Signal Processing Conference*, pages 2226–2229, 2017.

M. Alaee-Kerahroodi, K. V. Mishra, M. R. Bhavani Shankar, and B. Ottersten. Discrete phase sequence design for coexistence of MIMO radar and MIMO communications. In *IEEE International Workshop on Signal Processing Advances in Wireless Communications*, pages 1–5, 2019.

A. Ayyar and K. V. Mishra. Robust communications-centric coexistence for turbo-coded OFDM with non-traditional radar interference models. In *2019 IEEE Radar Conference (RadarConf)*, pages 1–6, 2019.

J. Bernhard, J. Reed, J. M. Park, A. Clegg, A. Weisshaar, and A. Abouzeid. Final report of the National Science Foundation workshop on Enhancing Access to the Radio Spectrum (EARS). *National Science Foundation*, Arlington, Virginia, Tech. Rep., 2010.

M. Bică and V. Koivunen. Frequency agile generalized multicarrier radar. In *Annual Conference of Information Sciences Systems*, 2014.

M. Bică and V. Koivunen. Generalized multicarrier radar: Models and performance. *IEEE Transactions on Signal Processing*, 64(17), 2016.

M. Bică, K. Huang, U. Mitra, and V. Koivunen. Opportunistic radar waveform design in joint radar and cellular communication systems. In *IEEE Global Communications Conference*, pages 1–7, 2015.

M. Bică, K. Huang, V. Koivunen, and U. Mitra. Mutual information based radar waveform design for joint radar and cellular communication systems. In *IEEE International Conference on Acoustics Speech and Signal Processing*, pages 3671–3675, 2016.

D. W. Bliss. Cooperative radar and communications signaling: The estimation and information theory odd couple. In *IEEE Radar Conference*, pages 50–55, 2014.

D. Ciuonzo, A. De Maio, G. Foglia, and M. Piezzo. Pareto-theory for enabling covert intrapulse radar-embedded communications. In *IEEE Radar Conference*, pages 0292–0297, 2015.

D. Cohen, K. V. Mishra, and Y. C. Eldar. Spectrum sharing radar: Coexistence via Xampling. *IEEE Transactions on Aerospace and Electronic Systems*, 29:1279–1296, 2018.

Y. Cui, V. Koivunen, and X. Jing. Interference alignment based precoder-decoder design for radar-communication co-existence. In *Asilomar Conference on Signals Systems and Computing*, pages 1290–1295, 2017.

Y. Cui, V. Koivunen, and X. Jing. Interference alignment based spectrum sharing for mimo radar and communication systems. In *IEEE International Workshop on Signal Processing and Advanced Wireless Communications*, pages 1–5, 2018.

R. C. Daniels and R. W. Heath Jr. 60 GHz wireless communications: Emerging requirements and design recommendations. *IEEE Vehicular Technology Magazine*, 2(3), 2007.

N. Decarli, F. Guidi, and D. Dardari. A novel joint RFID and radar sensor network for passive localization: Design and performance bounds. *IEEE Journal of Selected Topics in Signal Processing*, 8(1):80–95, 2014.

S. H. Dokhanchi, M. R. Bhavani Shankar, Y. A. Nijsure, T. Stifter, S. Sedighi, and B. Ottersten. Joint automotive radar-communications waveform design. In *IEEE International Symposium on Personal, Indoor and Mobile Communications*, pages 1–7, 2017.

S. H. Dokhanchi, B. S. Mysore, K. V. Mishra, and B. Ottersten. A mmWave automotive joint radar-communications system. *IEEE Transactions on Aerospace and Electronic Systems*, 55(3):1241–1260, 2019a.

S. H. Dokhanchi, M. Alaee-Kerahroodi, B. S. Mysore R., and B. Ottersten. Mono-static automotive joint radar-communications system. In *2019 IEEE 30th Annual International Symposium on Personal, Indoor and Mobile Radio Communications (PIMRC)*, pages 1–6, 2019b.

S. H. Dokhanchi, M. R. Bhavani Shankar, K. V. Mishra, and B. Ottersten. Multi-constraint spectral co-design for colocated MIMO radar and MIMO communications. In *IEEE International Conference on Acoustics, Speech and Signal Processing*, pages 4567–4571, 2020.

B. J. Donnet and I. D. Longstaff. Combining MIMO radar with OFDM communications. In *IEEE Radar Conference*, 2006.

G. Duggal, S. S. Ram, and K. V. Mishra. Micro-Doppler and micro-range detection via Doppler-resilient 802.11ad-based vehicle-to-pedestrian radar. In *IEEE Radar Conference*, 2019. Prepress.

G. Duggal, S. Vishwakarma, K. V. Mishra, and S. S. Ram. Doppler-resilient 802.11ad-based ultrashort range automotive joint radar-communications system. *IEEE Transactions on Aerospace and Electronic Systems*, 56(5):4035–4048, Oct 2020. doi: 10.1109/TAES.2020.2990393.

A. M. Elbir and K. V. Mishra. Deep learning design for joint antenna selection and hybrid beamforming in massive MIMO. In *IEEE International Symposium on Antennas and Propagation*, 2019. Prepress.

A. M. Elbir, K. V. Mishra, and S. Chatzinotas. Terahertz-band joint ultra-massive MIMO radar-communications: model-based and model-free hybrid beamforming. *IEEE Journal of Selected Topics in Signal Processing*, 15(6):1468–1483, 2021.

A. M. Elbir, K. V. Mishra, S. Chatzinotas, and M. Bennis. Terahertz-band integrated sensing and communications: Challenges and opportunities. arXiv preprint arXiv:2208.01235, 2022.

A. M. Elbir, K. V. Mishra, M. R. Bhavani Shankar, and S. Chatzinotas. The rise of intelligent reflecting surfaces in integrated sensing and communications paradigms. *IEEE Network*, 1–8, 2023.

T. Esmaeilbeig, K. V. Mishra, and M. Soltanalian. Beyond diagonal RIS: Key to next-generation integrated sensing and communications? arXiv preprint arXiv:2402.14157, 21 Feb 2024.

G. Fortino, M. Pathan, and G. Di Fatta. BodyCloud: Integration of cloud computing and body sensor networks. In *IEEE International Conference on Cloud Computing Technology and Science*, pages 851–856, 2012.

C. Fragner, A. P. Weiss, F. P. Wenzl, and E. Leitgeb. Integrated sensing and communication in the visible spectral range: A novel closed loop controller. In *2022 International Conference on Broadband Communications for Next Generation Networks and Multimedia Applications (CoBCom)*, pages 1–7, 2022.

Z. Geng, R. Xu, H. Deng, and B. Himed. Fusion of radar sensing and wireless communications by embedding communication signals into the radar transmit waveform. *IET Radar, Sonar and Navigation*, 12(6):632–640, 2018.

A. Hassanien, M. G. Amin, Y. D. Zhang, and F. Ahmad. Dual-function radar communication: Information embedding using sidelobe control and waveform diversity. *IEEE Transactions on Signal Processing*, 64(8):2168–2181, 2016.

A. Hassanien, M. G. Amin, Y. D. Zhang, and F. Ahmad. Signaling strategies for dual-function radar communications: An overview. *IEEE Aerospace and Electronic Systems Magazine*, 83(10):36–45, 2017.

A. Hassanien, E. Aboutanios, M. G. Amin, and G. A. Fabrizio. A dual-function MIMO radar-communication system via waveform permulation. *Digital Signal Processing*, 83:118–128, 2018a.

A. Hassanien, C. Sahin, J. Metcalf, and B. Himed. Uplink signaling and receive beamforming for dual-function radar communications. In *IEEE International Workshop on Signal Processing and Advanced Wireless Communications*, pages 1–5, 2018b.

Q. He, Z. Wang, J. Hu, and R. S. Blum. Performance gains from cooperative MIMO radar and MIMO communication systems. *IEEE Signal Processing Letters*, 26(1):194–198, 2019.

J. A. Hodge, K. V. Mishra, B. M. Sadler, and A. I. Zaghloul. Index-modulated metasurface transceiver design using reconfigurable intelligent surfaces for 6G wireless networks. *IEEE Journal of Selected Topics in Signal Processing*, 17(6):1248–1263, 2023.

R. Jacome, E. Vargas, K. V. Mishra, B. M. Sadler, and H. Arguello. Factor graph processing for dual-blind deconvolution at ISAC receiver. In *2023 IEEE 9th International Workshop on Computational Advances in Multi-Sensor Adaptive Processing (CAMSAP)*, pages 276–280, 2023.

G. M. Jacyna, B. Fell, and D. McLemore. A high-level overview of fundamental limits studies for the DARPA SSPARC program. In *IEEE Radar Conference*, pages 1–6, 2016.

S.-J. Kim, E. Dall'Anese, and G. B. Giannakis. Cooperative spectrum sensing for cognitive radios using Kriged Kalman filtering. *IEEE Journal of Selected Topics in Signal Processing*, 5(1):24–36, 2011.

P. Kumari, D. H. N. Nguyen, and R. W. Heath. Performance trade-off in an adaptive IEEE 802.11ad waveform design for a joint automotive radar and communication system. In *IEEE International Conference on Acoustics Speech and Signal Processing*, 2017.

P. Kumari, J. Choi, N. Gonzalez-Prelcic, and R. W. Heath Jr. IEEE 802.11ad-based radar: An approach to joint vehicular communication-radar system. *IEEE Transactions on Vehicular Technology*, 67(4):3012–3027, 2018.

P. Kumari, S. A. Vorobyov, and R. W. Heath. Adaptive virtual waveform design for millimeter-wave joint communication–radar. *IEEE Transactions on Signal Processing*, 68:715–730, 2020. doi: 10.1109/TSP.2019.2956689.

N. Levanon. Multifrequency radar signals. In *IEEE International Radar Conference*, pages 683–688, 2000.

B. Li and A. Petropulu. Joint transmit designs for co-existence of MIMO wireless communications and sparse sensing radars in clutter. *IEEE Transactions on Aerospace and Electronic Systems*, 53(6):2846–2864, 2017.

Y. Li and S. A. Vorobyov. Fast algorithms for designing multiple unimodular waveform(s) with good correlation properties. *IEEE Transactions on Signal Processing*, 66(5):1197–1212, 2018.

B. Li, A. P. Petropulu, and W. Trappe. Optimum co-design for spectrum sharing between matrix completion based MIMO radars and a MIMO communication system. *IEEE Transactions on Signal Processing*, 64(17):4562–4575, 2016.

J. Lien, N. Gillian, M. E. Karagozler, P. Amihood, C. Schwesig, E. Olson, H. Raja, and I. Poupyrev. Soli: Ubiquitous gesture sensing with millimeter wave radar. *ACM Transactions on Graphics*, 35(4):142:4–142:19, 2016.

W.-C. Liu, F.-C. Yeh, T.-C. Wei, C.-D. Chan, and S.-J. Jou. A digital Golay-MPIC time domain equalizer for SC/OFDM dual-modes at 60 GHz band. *IEEE Transactions on Circuits and Systems I: Regular Papers*, 60(10):2730–2739, 2013.

J. Liu, K. V. Mishra, and M. Saquib. Co-designing statistical MIMO radar and in-band full-duplex multi-user MIMO communications. arXiv preprint arXiv:2006.14774, 2020.

J. Lunden, V. Koivunen, and H. V. Poor. Spectrum exploration and exploitation for cognitive radio: Recent advances. *IEEE Signal Processing Magazine*, 32(3):123–140, 2015.

G. R. MacCartney, Jr., S. Sun, T. S. Rappaport, Y. Xing, H. Yan, J. Koka, R. Wang, and D. Yu. Millimeter wave wireless communications: New results for rural connectivity. In *ACM Workshop on All Things Cellular*, pages 31–36, 2016.

J. A. Mahal, A. Khawar, A. Abdelhadi, and T. C. Clancy. Spectral coexistence of MIMO radar and MIMO cellular system. *IEEE Transactions on Aerospace and Electronic Systems*, 53(2):655–668, 2017.

R. Méndez-Rial, C. Rusu, N. González-Prelcic, A. Alkhateeb, and R. W. Heath. Hybrid MIMO architectures for millimeter wave communications: Phase shifters or switches? *IEEE Access*, 4:247–267, 2016.

K. V. Mishra and Y. C. Eldar. Performance of time delay estimation in a cognitive radar. In *IEEE International Conference on Acoustics Speech and Signal Processing*, pages 3141–3145, 2017a.

K. V. Mishra and Y. C. Eldar. Sub-Nyquist channel estimation over IEEE 802.11ad link. In *IEEE International Conference on Sampling Theory and Applications*, pages 355–359, 2017b.

K. V. Mishra and Y. C. Eldar. Sub-Nyquist radar: Principles and prototypes. In A. De Maio, Y. C. Eldar, and A. Haimovich, editors, *Compressed sensing in radar signal processing*. Cambridge University Press, 2019. Prepress.

K. V. Mishra, A. Gharanjik, M. R. Bhavani Shankar, and B. Ottersten. Deep learning framework for precipitation retrievals from communication satellites. In *European Conference on Radar in Meteorology and Hydrology*, page 023, 2018.

K. V. Mishra, A. Zhitnikov, and Y. C. Eldar. Spectrum sharing solution for automotive radar. In *IEEE Vehicular Technology Conference – Spring*, pages 1–5, 2017.

K. V. Mishra, A. F. Martone, and A. I. Zaghloul. Power allocation games for overlaid radar and communications. In *URSI Asia-Pacific Radio Science Conference*, 2019. Prepress.

K. V. Mishra, A. Chattopadhyay, S. S. Acharjee, and A. P. Petropulu. OptM3Sec: Optimizing multicast IRS-aided multiantenna DFRC secrecy channel with multiple eavesdroppers. In *IEEE International Conference on Acoustics, Speech and Signal Processing (ICASSP)*, pages 9037–9041, 2022.

J. Monsalve, E. Vargas, K. V. Mishra, B. M. Sadler, and H. Arguello. Beurling-Selberg extremization for dual-blind deconvolution recovery in joint radar-communications. In *2023 IEEE 9th International Workshop on Computational Advances in Multi-Sensor Adaptive Processing (CAMSAP)*, pages 246–250, 2023.

G. R. Muns, K. V. Mishra, C. B. Guerra, Y. C. Eldar, and K. R. Chowdhury. Beam alignment and tracking for autonomous vehicular communication using IEEE 802.11ad-based radar. In *IEEE Infocom Workshops – Hot Topics in Social and Mobile Connected Smart Objects*, 2019. Prepress.

S. Nayemuzzaman, J. Liu, K. V. Mishra, and M. Saquib. Co-designing statistical MIMO radar and in-band full-duplex multi-user MIMO communications – Part III: Multi-target tracking. arXiv preprint arXiv, 2024.

B. Paul, A. R. Chiriyath, and D. W. Bliss. Survey of RF communications and sensing convergence research. *IEEE Access*, 5:252–270, 2017.

T. S. Rappaport, G. R. MacCartney, M. K. Samimi, and S. Sun. Wideband millimeter-wave propagation measurements and channel models for future wireless communication system design. *IEEE Transactions on Communications*, 63(9), 2015.

S. Sedighi, K. V. Mishra, M. R. Bhavani Shankar, and B. Ottersten. Localization with one-bit passive radars in narrowband Internet-of-Things using multivariate polynomial optimization. *IEEE Transactions on Signal Processing*, 69:2525–2540, 2021.

T. Shi, S. Zhou, and Y. Yao. Capacity of single carrier systems with frequency-domain equalization. In *IEEE Symposium on Circuits and Systems. Emerging Technology: Frontiers of Mobile Wireless Communication*, volume 2, pages 429–432, 2004.

M. I. Skolnik. *Radar handbook*. McGraw-Hill, third edition, 2008.

K. Takizawa, M. Kyrö, K. Haneda, H. Hagiwara, and P. Vainikainen. Performance evaluation of 60 GHz radio systems in hospital environments. In *IEEE International Conference on Communications*, pages 3219–3295, 2012.

E. Vargas, K. V. Mishra, R. Jacome, B. M. Sadler, and H. Arguello. Dual-blind deconvolution for overlaid radar-communications systems. *IEEE Journal on Selected Areas in Information Theory*, 4:75–93, 2023. doi: 10.1109/JSAIT.2023.3287823.

S. A. Vorobyov. Adaptive and robust beamforming. In A. M. Zoubir, M. Viberg, R. Chellappa, and S. Theodoridis, editors, *Array and statistical signal processing*, volume 3 of *Academic Press Library in Signal Processing*, pages 503–552. Academic Press, 2014.

S.-Y. Wang, T. Erdoğan, U. Pereg, and M. R. Bloch. Joint quantum communication and sensing. In *2022 IEEE Information Theory Workshop (ITW)*, pages 506–511, 2022.

T. Wei, L. Wu, K. V. Mishra, and M. R. Bhavani Shankar. Multi-IRS-aided Doppler-tolerant wideband DFRC system. *IEEE Transactions on Communications*, 71(11):6561–6577, 2023a.

T. Wei, L. Wu, K. V. Mishra, and B. Shankar. RIS-aided wideband holographic DFRC. *IEEE Transactions on Aerospace and Electronic Systems*, 2023b. doi: 10.1109/TAES.2024.3374272.

L. Wu, K. V. Mishra, M. R. Bhavani Shankar, and B. Ottersten. Resource allocation in heterogeneously-distributed joint radar-communications under asynchronous Bayesian tracking framework. *IEEE Journal on Selected Areas in Communications*, 40(7):2026–2042, July 2022.

W. Zhang, S. A. Vorobyov, and L. Guo. DOA estimation in MIMO radar with broken sensors by difference co-array processing. In *IEEE International Workshop on Computational Advances in Multi-Sensor Adaptive Processes*, 2015.

2

Principles of Radar-Centric Dual-Function Radar-Communication Systems

Aboulnasr Hassanien[1] *and Moeness G. Amin*[2]

[1]*Aptiv PLC, Radar Systems Group, Agoura Hills, CA, USA*
[2]*Villanova University, Center for Advanced Communications, Villanova, PA, USA*

2.1 Background

During the last decade, the problems of radio frequency (RF) spectrum congestion and contention between disparate wireless systems has been the focus of intensive research. The explosive growth in commercial wireless communications services and the limited availability of the radio spectrum are putting other essential radio systems and services, such as radar, under immense pressure [Griffiths et al., 2015]. This has created fierce competition over bandwidth allocation, resulting in the problem of RF spectrum congestion. Particularly, radar and communications have become most in contention [Hayvaci and Tavli, 2014]. In this high-stakes game, some of frequency bands previously assigned solely to radar have been re-assigned for shared-access causing radar to yield and respond to commercial interests driving the communications revolution.

The response is to co-exist and to seek uncontested shared bandwidth between radar and communications, and generally among various services and users of the same RF spectrum [Hassanien et al., 2019]. This has propelled efforts to develop innovative techniques that simultaneously enable radar target illuminations and wireless information delivery. The emerging dual-function radar communication (DFRC) system is considered the least contentious approach. It does so by hosting both services on the same platform [Hassanien et al., 2016b,c; McCormick et al., 2019; Dokhanchi et al., 2019]. The overarching objective of the DFRC system approach is to use a common transmit platform to fulfill both radar and communication operations using the same frequency band, power, and transmit RF chains, while offering protection from interfering services [Chapin, 2013]. DFRC systems are also referred to as intentional modulation on a pulse (IMOP) [Nowak et al., 2016] and CoRadar [Gaglione et al., 2018]. In addition to eliminating competition over bandwidth, the use of DRFC system platforms leads to low size, weight, and power (SWaP) consumption requirements [Mishra et al., 2019; Martone and Amin, 2021]. DFRC is a promising technology that has potential to benefit several emerging applications such as automotive radar, intelligent transportation system, radar-based health monitoring systems, synthetic aperture radar (SAR) systems transmitting sensed data to ground stations, radar networks where scheduling and target information is communicated across the network. In defense applications, the use of DFRC systems allow for integrated command and control systems, and sensor management.

The DFRC system is one category of the cooperative coexistence strategies. From the dynamic spectrum allocation perspective, in cooperative coexistence, both radar and communication systems may operate from a common or separate platforms, where the two systems continuously

Signal Processing for Joint Radar Communications, First Edition.
Edited by Kumar Vijay Mishra, M. R. Bhavani Shankar, Björn Ottersten, and A. Lee Swindlehurst.

exchange information to achieve an efficient use of the spectrum as well as minimize mutual interference. (See [Martone and Amin, 2021] and references therein.) The other category of cooperative coexistence describes co-design, which assumes both systems play a role in the design at inception and follow a shared protocol for operations. A DFRC system can be viewed as a co-design strategy with a common platform if the modifications of the radar parameters to accommodate the communication operations are as such that the legacy system can no longer be attained or claimed. In essence, co-design assumes a more definite balance between radar and communication operations compared to the DFRC systems. Unlike DFRC systems and co-design approaches, the non-cooperative coexistence strategies are guided by whether radar is considered a primary or a secondary user of the spectrum, or it plays a dual role [Martone and Amin, 2021].

This chapter provides an overview of existing radar-centric DFRC techniques and discusses their potentials and future directions. The DFRC techniques presented benefit from strides made in waveform design and diversity, advances in software defined radio platforms, implementations of digital beamforming, and efficient solutions of constrained minimization problems based on convex optimization and relaxation [Hassanien et al., 2016b,c; Nowak et al., 2016; Euzière et al., 2014]. The techniques enabling DFRC systems discussed in this chapter assume that the radar is the host platform and the primary function that defines the system power, bandwidth, beamforming and pulse shape. The communication function can be carried out using the exact system resources as long as it is accomplished with minimum computational overhead and acceptable changes in radar parameters.

Radar-centric DFRC systems can implement different strategies to enable communication signals to use radar resources. These strategies have different levels of complexity and cause different modifications of the radar functionalities. One strategy may lead to clutter modulations while another can cause spectral leakage. These strategies are also viewed differently from the communication receiver perspectives in terms of achievable bit error rate (BER) and underlying demodulation assumptions. One can think of a cognitive DFRC where one strategy can be most preferred over others. For example, when the radar communicates its scheduling information, a slow data rate strategy can be accommodated, whereas when communicating the target range-Doppler map, a high data rate strategy is beneficial. Also, for strong clutter, the DFRC system can elect to avoid clutter modulations typically associated with the desire to send high data rate signals. This is more of an issue in air to ground rather than ground to air scene interrogations.

Based on this, the DFRC system strategies can be divided into the three categories.

1. **DFRC System Using Fixed Radar Pulse**: These techniques do not change the fast-time structure of the transmitted signal and do not change the radar waveforms from pulse to pulse except for a scaling factor of its amplitude and/or phase rotation. This strategy is suitable for scenarios when the radar has a conservative mode of operation and desires to keep its ambiguity function (AF) intact.
2. **DFRC System Using Modulated Radar Waveforms**: If the radar operation is flexible and willing to trade some performance degradation for increased communication data rate, then the strategy is to send communication information by changing the fast-time structure of the radar pulse. As a result, the waveform also changes from one pulse to another causing range-sidelobe modulation (RSM).
3. **DFRC System Using Index Modulation**: This DFRC system strategy sends the information by controlling the index of one or more of the DFRC system resources, such as waveform, transmit antenna, receive antenna, and frequency. This type of index modulation (IM) can be used with or without modulating the radar pulse. The radar under this strategy incurs additional signal processing complexity at the transmitter and/or receiver. Also, performance tradeoffs occur due to RSM and/or variations in the waveform spectral contents.

This chapter discusses the above strategies and present the different signal embedding techniques associated with each strategy. We consider both phased array and multiple-input multiple-output (MIMO) configurations. The chapter concludes by highlighting several challenges in DFRC systems and outlining some possible future research directions. It should be stressed that the DFRC system techniques overviewed in this chapter are developed in the physical layer and, therefore, can be integrated into legacy radar systems by modifying the signal processing block at the transmitter, receiver or both. In other words, these DFRC techniques can be used to modernize legacy radar systems as a software upgrade. Hence, this type of DFRC technology has a great economic impact. We also refer the reader to Chapter 10: Frequency-Hopping MIMO Radar-based Data Communications.

2.2 DFRC System Model

Consider a DFRC system hosted on a joint platform as shown in Figure 2.1. In this section, we provide the DFRC system configuration and its resources. We also highlight the achievable performances of the individual radar and communication functions and touch on the DFRC system tradeoff between the two individual performances. The section discusses the single function radar signal model as well as the generalized signal model of the DFRC system which lays foundations to communication signal embedding strategies that will be introduced in subsequent sections.

2.2.1 Resources and Performance Limits

The joint platform is equipped with a transmit array comprising M elements arranged in a linear shape. The total bandwidth assigned to the DFRC system is BW Hz and the total transmit power is P_t Watt.

2.2.1.1 Radar Performance

One essential feature of a radar system is its resolution which defines the radar's ability to distinguish between two or more closely spaced targets either in range or cross-range. The radar range-resolution is directly related to the bandwidth of its transmitted signal. For a given bandwidth, the radar range-resolution is defined as

$$\Delta_R = \frac{c}{2\text{BW}}, \tag{2.1}$$

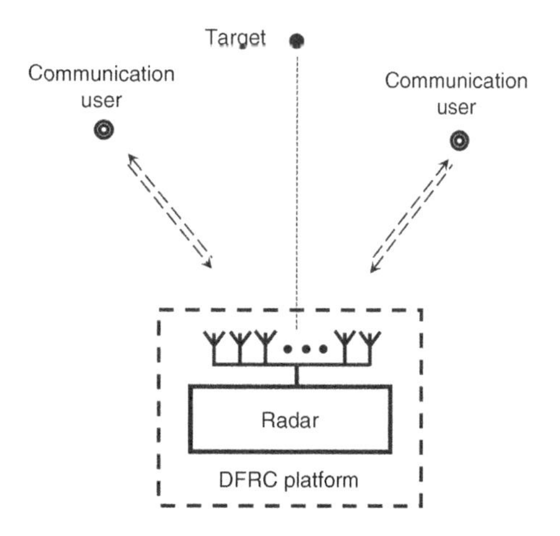

Figure 2.1 Illustrative diagram of an DFRC system.

where c is the speed of light. Therefore, from a radar system perspective, it is desirable that the spectral contents of the radar waveform occupy the entire available bandwidth. On the other hand, the radar's ability to distinguish between two moving targets with closely separated speeds defines its Doppler resolution which is inversely proportional to the duration of the radar's coherent processing interval (CPI).

2.2.1.2 Communications Performance

The efficacy of a communication system is commonly tied to its ability to transfer and reliably deliver information from one point to another. There is an increasingly growing demand for high data-rates in modern communication systems. Similar to range-resolution in radar, the data-rate of a communication system is also directly related to the system bandwidth. In presence of noise, the Shannon–Hartley theorem establishes the relationship between the maximum information transmission rate and the bandwidth, that is,

$$C = \mathrm{BW} \log_2 \left(1 + \frac{S_0}{N_{n+i}}\right), \tag{2.2}$$

where S_0 is the average received signal power over the bandwidth BW and N_{n+i} denotes the average power of the noise plus interference within the same bandwidth. It is worth noting that the noise power is also proportional to the bandwidth. Despite that, it can be observed from (2.2) that the information rate increases with the bandwidth. Therefore, from a communication system perspective, the larger the bandwidth is, the higher the achievable data rate becomes.

2.2.1.3 DFRC Performance Tradeoff

It can be observed from the radar resolution expression (2.1) and the communication information rate expression (2.2) that both operations require larger bandwidth to enhance their efficiency. Recent advances in DFRC system research show that joint operation using the same signals and the same bandwidth offers a win-win solution. Fundamental to DFRC system development is devising signaling strategies that enable the integrated operation and improved utilization of the finite bandwidth. This integrated operation presents a performance tradeoff between the radar and communication functions. This chapter focuses on techniques which prioritize the radar operation and allow communications to use the radar system as "system of opportunity," which includes radar spatio-temporal resources. The goal is to deliver information while minimally disturbing the radar functions. For related designs considering other perspectives, we refer the reader to Chapter 11 on "Optimized Resource Allocation for Joint Radar-Communications," and Chapter 4 on "Beamforming and Interference Management in Joint Radar-Communication System".

2.2.2 Radar Transmit Signal Model

The default operation of the joint transmit platform is to transmit waveforms designed to maximize the radar operation performance. Considering the MIMO radar signal model, let $\{\phi_m(t)\}$, $m = 1, \dots, M$ be a pre-designed set of orthogonal waveforms that satisfy the condition

$$\int_{T_\phi} \phi_m(t)\phi^*_{m'}(t)dt = \delta(m - m'), \tag{2.3}$$

where t denotes fast-time (i.e. time within the radar pulse), T_ϕ is the pulse duration, and $\delta(\cdot)$ stands for the Kronecker delta function. The spectral contents of the waveforms are fully overlapped and occupy the entire bandwidth BW. In practice, it is difficult to realize perfectly orthogonal waveforms. Instead, waveforms with low cross-correlations are commonly used (see [Blunt and Mokole, 2016], and references therein).

Typically, the radar necessitates its operating parameters to remain fixed within a CPI to ensure coherency and maintain high Doppler resolution. Therefore, the MIMO radar transmits consecutive train of identical pulses within the CPI. Within each pulse, the baseband transmit signal vector can be expressed as a linear combination of the individual orthogonal waveforms, that is,

$$\mathbf{s}_{\mathrm{MIMO}}(t) = \sqrt{\frac{P}{M}} \sum_{m=1}^{M} \tilde{\mathbf{w}}_m^* \phi_m(t) = \sqrt{\frac{P}{M}} \tilde{\mathbf{W}} \boldsymbol{\phi}(t), \tag{2.4}$$

where $\tilde{\mathbf{w}}_m$ is the $M \times 1$ transmit beamforming weight vector associated with the mth waveform, $(\cdot)^*$ is the conjugate operation, $\boldsymbol{\phi}(t) \triangleq [\phi_1(t), \ldots, \phi_M(t)]^T$ is the vector of orthogonal waveforms, and $(\cdot)^T$ stands for the transpose. The $M \times M$ transmit beamforming weight matrix $\tilde{\mathbf{W}} \triangleq [\tilde{\mathbf{w}}_1^*, \ldots, \tilde{\mathbf{w}}_M^*]$ is assumed to be normalized such that $\mathrm{tr}\{\tilde{\mathbf{W}}^H \tilde{\mathbf{W}}\} = M$, with $\mathrm{tr}\{\cdot\}$ being the trace of a square matrix and $(\cdot)^H$ denotes the Hermitian transpose. The weight matrix $\tilde{\mathbf{W}}$ and the vector of waveforms $\boldsymbol{\phi}(t)$ can be optimized to satisfy the requirements mandated by the MIMO radar operation.

Single-input multiple-output (SIMO) radar can be considered as a special case of the MIMO radar by restricting the number of waveforms to one. In this case, the $M \times 1$ baseband transmit signal vector can be defined as

$$\mathbf{s}_{\mathrm{SIMO}}(t) = \sqrt{P} \tilde{\mathbf{w}}^* \phi(t), \tag{2.5}$$

where $\tilde{\mathbf{w}}$ the unit-norm transmit beamforming weight vector and $\phi(t)$ is the SIMO radar waveform. The waveform $\phi(t)$ is assumed to have unit energy, i.e. $\int_{T_\phi} |\phi(t)|^2 dt = 1$. The transmit beamforming weight vector $\tilde{\mathbf{w}}$ and waveform $\phi(t)$ are assumed to satisfy the transmit beampattern and range-Doppler resolution requirements as mandated by the SIMO radar.

2.2.3 DFRC Transmit Signal Model

Communications can be integrated into the radar operation by modulating the radar transmit signals. The modulated radar signals are then used to simultaneously probe a common area of interest, where radar target(s) are possibly located, and to deliver information to one or more communication users (CUs). From a radar system perspective, the joint platform serves as a radar transmitter during transmit mode and functions as a receiver when listening to and intercepting the echo of the reflected signals. Form a communications perspective, the joint DFRC platform acts as a base station.

Different from the radar-only signal model, the DFRC system transmits different pulses at different times, depending on the information being embedded. Considering the MIMO DFRC system scenario, the transmit signals can be modeled as

$$\mathbf{s}_{\mathrm{DFRC}}(t; n) = \sqrt{\frac{P}{M}} \mathbf{W}(n) \mathcal{P}_n \left\{ \boldsymbol{\psi}(t; n) \odot \boldsymbol{\phi}(t) \right\}, \tag{2.6}$$

where t is the time index within the radar pulse, n is the slow-time index, i.e. the pulse number, $\mathbf{W}(n) \triangleq [\mathbf{w}_1^*(n), \ldots, \mathbf{w}_M^*(n)]$ is the DFRC transmit beamforming weight matrix associated with the nth pulse, $\boldsymbol{\psi}(t; n) \triangleq [\psi_1(t; n), \ldots, \psi_M(t; n)]^T$ is the vector of information bearing communication signals, $\odot$ is the Hadamard element-wise product, and $\mathcal{P}_n\{\cdot\}$ denotes a linear operator that can either permute the elements of its argument vector or select a subset of the elements of that vector. It can be observed from (2.6) that, during the nth DFRC pulse, information can be embedded in $\mathbf{W}(n)$, $\mathcal{P}_n$, $\psi_m(t; n)$, $m = 1, \ldots, M$, or any combination of these parameters. The generalized DFRC signal model (2.6) can represent most DFRC techniques reported in the literature which adopts the concept of information embedding as secondary to the primary radar function of the system.

In the following sections, we provide concise overviews of most of these techniques. Unless otherwise stated, the communication link between the DFRC and each CU is half-duplex, i.e. each user can either receive (downlink) or transmit (uplink) information at a time. In downlink communication, information is transmitted from the DFRC platform toward one or more CUs. The essence of downlink communication is to embed messages into the radar emissions preferably without any disturbance of the radar operation.

2.3 DFRC Using Fixed Radar Waveforms

This section presents an overview of DFRC information embedding strategies via modulation the transmit beampattern of the DFRC system without changing the radar waveforms. Unless otherwise stated, we assume that the communication symbol duration equals the pulse repetition interval (PRI) of the radar. Communications, being secondary to the primary radar function, can be incorporated by controlling the transmit weight matrix $\mathbf{W}(n)$ from one pulse to another depending on the information being embedded. In addition, information can be embedded by controlling the order of the orthogonal waveforms, i.e. by controlling which waveform is transmitted from which antenna.

2.3.1 DFRC Using Beampattern Modulation

For a communication receiver located at spatial direction θ_c, by appropriately designing the weight matrix $\mathbf{W}(n)$, information can be embedded using beampattern amplitude modulation (AM), beampattern phase modulation (PM), or beampattern quadrature amplitude modulation (QAM). Therefore, this type of DFRC fixes the radar waveforms, but changes the radar beampattern, possibly in amplitude and phase, depending on the transmitted information [Euzière et al., 2014; Hassanien et al., 2016a; Hassanien et al., 2016c; de Oliveira Ferreira et al., 2018]. In this case, the DFRC signal model simplifies to

$$\mathbf{s}_{\text{DFRC}}(t;n) = \sqrt{\frac{P}{M}}\mathbf{W}(n)\boldsymbol{\phi}(t). \tag{2.7}$$

Using the signal model (2.7), the covariance matrix of the DFRC transmit signal vector $\mathbf{s}_{\text{DFRC}}(t;n)$ is given by

$$\mathbf{R}_s \triangleq \mathbb{E}\left\{\mathbf{s}_{\text{DFRC}}(t;n)\mathbf{s}_{\text{DFRC}}^H(t;n)\right\} = \frac{P}{M}\mathbf{W}(n)\mathbf{W}^H(n), \tag{2.8}$$

where $\mathbb{E}\{\cdot\}$ denotes the expectation of a random variable. Then, the DFRC transmit beampattern is given by

$$\begin{aligned} G_{\text{DFRC}}(\theta, \mathbf{W}(n)) &= \mathbf{a}^T(\theta)\mathbf{R}_s\mathbf{a}(\theta) = \frac{P}{M}\mathbf{a}^T(\theta)\mathbf{W}(n)\mathbf{W}^H(n)\mathbf{a}(\theta) \\ &= \frac{P}{M}\sum_{m=1}^{M}\left|\mathbf{w}_m^H(n)\mathbf{a}(\theta)\right|^2 = \frac{P}{M}\sum_{m=1}^{M}\left|g_m(\theta, \mathbf{w}_m(n))\right|^2, \end{aligned} \tag{2.9}$$

where $g_m(\theta, \mathbf{w}_m(n)) \triangleq \mathbf{w}_m^H(n)\mathbf{a}(\theta)$ is the beamformer complex response associated with the mth waveform.

The essence of beampattern modulation is to keep the magnitude of the beampattern within the radar mainbeam unchanged while allowing for controlled variations in the sidelobe region. Hence, from a radar system perspective, the DFRC beampattern should be the same as that of the MIMO radar beampattern, that is,

$$\begin{aligned} G_{\text{DFRC}}(\theta, \mathbf{W}(n)) &= \frac{P}{M}\sum_{m=1}^{M}\left|g_m\left(\theta, \mathbf{w}_m(n)\right)\right|^2 \\ &\simeq \frac{P}{M}\sum_{m=1}^{M}\left|g_m\left(\theta, \tilde{\mathbf{w}}_m\right)\right|^2 \\ &= G_{\text{MIMO}}\left(\theta, \tilde{\mathbf{W}}\right), \qquad \forall\theta \in \Theta_{\text{MIMO}}, \end{aligned} \tag{2.10}$$

where Θ_{MIMO} is the mainbeam region of the MIMO radar. The MIMO radar operation may demand additional constraints in designing the overall modulated beampattern, such as sidelobe attenuation level, uniform virtual array structure, and/or transmit rotational invariance property (RIP) [Hassanien et al., 2016a]. It is worth noting that if the number of transmit beams is reduced to one, the DFRC beampattern in (2.10) simplifies to the SIMO radar case, that is,

$$\begin{aligned} G_{\text{DFRC}}(\theta, \mathbf{w}(n)) &= P\left|\mathbf{w}^H(n)\mathbf{a}(\theta)\right|^2 \\ &\simeq P\left|\tilde{\mathbf{w}}^H\mathbf{a}(\theta)\right|^2 = G_{\text{SIMO}}(\theta, \tilde{\mathbf{w}}), \quad \forall\theta \in \Theta_{\text{SIMO}}, \end{aligned} \tag{2.11}$$

where $\Theta_{\text{SIMO}} \triangleq \left[\theta_0 - \frac{1}{2}\Theta_{\text{bw}},\ \theta_0 + \frac{1}{2}\Theta_{\text{bw}}\right]$, θ_0 is the angle at the center of the SIMO radar mainbeam and Θ_{bw} denotes the width of the SIMO radar mainbeam.

From the communication system perspective, information can be embedded into the radar emissions via AM of the radar transmit gain in the sidelobe region and/or PM of the transmit beampattern in the entire spatial domain. For example, a communication symbol $S_m(n)$ can be transmitted toward a communication received in direction θ_c during the nth pulse over the mth beam by enforcing the condition

$$g_m(\theta_c, \mathbf{w}_m(n)) = \mathbf{w}_m^H(n)\mathbf{a}(\theta_c) = S_m(n). \tag{2.12}$$

Therefore, to achieve the radar and communication functions simultaneously, a DFRC system should design $\mathbf{w}_m(n)$ such that the conditions (2.10) and (2.12) are jointly satisfied.

The remainder of this section provides an overview of several techniques reported in the literature for the embedding of downlink communication symbols via beampattern modulation.

2.3.1.1 Beampattern Amplitude Modulation

The underlying principle of beampattern AM is to modulate the sidelobe level (SLL) of the radar beampattern toward the direction of the communication receiver θ_c such that a communication symbol is represented by a specific SLL. Let us define a dictionary of M-ary symbols of size N_{AM} as $\mathbb{D}_{\text{AM}} = \{\delta_1, \dots, \delta_{N_{\text{AM}}}\}$, where δ_l is a real positive number and $0 \le \delta_1 < \cdots < \delta_{N_{\text{AM}}}$. This dictionary of SLLs can be synthesized by designing N_{AM} transmit beamforming weight vector $\mathbf{q}_l$, $l = 1, \dots, N_{\text{AM}}$ such that

$$\left|\mathbf{q}_l^H\mathbf{a}(\theta_c)\right|^2 = \delta_l, \quad l = 1, \dots, N_{\text{AM}}. \tag{2.13}$$

One way to design the beamforming vectors $\mathbf{q}_l$, $l = 1, \dots, N_{\text{AM}}$ is to minimize the difference between the desired and actual transmit radiation patterns while satisfying the MIMO radar operation condition (2.10) and the beampattern SLLs (2.13). This can be formulated as the following optimization problem [Hassanien et al., 2016c]

$$\min_{\mathbf{q}_l}\max_{\theta} \left|d(\theta) - \left|\mathbf{q}_l^H\mathbf{a}(\theta)\right|\right|, \quad \theta \in \Theta_{\text{MIMO}}, \tag{2.14}$$

$$\text{subject to } \left|\mathbf{q}_l^H\mathbf{a}(\theta)\right| \le \varepsilon_{\text{SLL}}, \quad \theta \in \bar{\Theta}_{\text{MIMO}}, \tag{2.15}$$

$$\mathbf{q}_l^H\mathbf{a}(\theta_c) = \delta_l, \qquad l = 1, \dots, N_{\text{AM}}, \tag{2.16}$$

where $d(\theta)$ is the desired transmit beampattern, $\bar{\Theta}_{\text{MIMO}}$ denotes the sidelobe region, and $\epsilon_{\text{SLL}} > 0$ is a design parameter used to control the sidelobe attenuation level. Since ϵ_{SLL} is the highest SLL as mandated by the MIMO radar operation, the condition $\delta_l \leq \epsilon_{\text{SLL}},\ l = 1, \dots, N_{\text{AM}}$ should be satisfied. It is worth noting that the signal-to-noise ratio (SNR) at the communication receiver is directly related to the values $\delta_l \leq \epsilon_{\text{SLL}}$. Therefore, to maximize the average SNR at the communication receiver, the largest M-ary symbol should equal to ϵ_{SLL}.

Let $S_m(n) \in \mathbb{D}_{\text{AM}}$ be the communication symbol to be embedded during the nth pulse and transmitted over the mth beam. The corresponding transmit beamforming weight vector, $\mathbf{w}_m(n)$, is selected from the set of weight vectors $\{\mathbf{q}_1, \dots, \mathbf{q}_{N_{\text{AM}}}\}$ such that $\mathbf{w}_m^H(n)\mathbf{a}(\theta_c) = S_m(n)$. Therefore, the baseband signal at a single-antenna communication receiver located at direction θ_c is modeled as

$$y_{\text{com}}(t; n) = \alpha_{\text{ch}} \sqrt{\frac{P}{M}} \sum_{m=1}^{M} \underbrace{\mathbf{w}_m^H(n)\mathbf{a}(\theta_c)}_{S_m(n)} \phi_m(t) + z(t; n), \tag{2.17}$$

where α_{ch} is the channel coefficient which summarizes the propagation environment between the transmit array and the communication receiver and $z(t; n)$ is the additive white Gaussian noise with zero mean and power spectral density (PSD) N_0. Matched-filtering the communication received signal (2.17) to the transmitted waveforms yields

$$\begin{aligned} y_m^{(\text{AM})}(n) &= \int_{T_\phi} y_{\text{com}}(t; n)\phi_m^*(t)dt \\ &= \alpha_{\text{ch}} \sqrt{\frac{P}{M}} S_m(n) + z_m(n), \quad m = 1, \dots, M, \end{aligned} \tag{2.18}$$

where $z_m(n)$ is the additive noise term at the output of the matched-filter. Then, the embedded communication symbol can be detected by performing the comparison test

$$\hat{S}(n) = \delta_l, \quad \text{if} \quad T_{l-1} \leq \left| y_m^{(\text{AM})}(n) \right| < T_l, \quad l = 1, \dots, N_{\text{AM}}, \tag{2.19}$$

where $T_0 = 0 < T_1 < \cdots < T_{N_{\text{AM}}-1} < \delta_{N_{\text{AM}}}$ denote a set of appropriately selected thresholds.

For the SIMO radar case, a single symbol can be embedded during the nth pulse. In this case, the total transmit power is radiated over a single beam increasing the transmit processing gain by a factor M as compared to that of the MIMO radar case.

Beampattern Amplitude-Shift Keying (ASK) Beampattern beampattern amplitude-shift keying (ASK) employs only two SLLs in the communication direction to embed one bit per waveform per pulse. This reduces the symbol dictionary to $\mathbb{D}_{\text{ASK}} = \{\Delta_H, \Delta_L\}$, where $\Delta_H > \Delta_L$. Let $\Delta_m(n) \in \mathbb{D}_{\text{ASK}}$ be the binary symbol embedded in the mth transmit beam during nth pulse. Then, the signal at the output of the matched-filter of the communication receiver is

$$y_m^{(\text{ASK})}(n) = \begin{cases} \alpha_{\text{ch}} \sqrt{\frac{P}{M}} \Delta_{\text{H}} + z(n), & \Delta_m(n) \Leftrightarrow 1, \\ \alpha_{\text{ch}} \sqrt{\frac{P}{M}} \Delta_{\text{L}} + z(n), & \Delta_m(n) \Leftrightarrow 0, \quad m = 1, \dots, M. \end{cases} \tag{2.20}$$

The receiver detects the symbol $\Delta_m(n)$ by the performing the test $\left| y_{\text{ASK}}^{(m)}(n) \right| \lesssim T$, for an appropriately selected threshold T.

As an illustrative example, Figure 2.2 shows a realization of the optimum beampattern with $P = 1$, $\Delta_H^2 = 10^{-2}$, and $\Delta_L^2 = 10^{-5}$, giving SLLs of -20 and -50 dB respectively. The communication direction is $-50°$ and the radar mainbeam is $\Theta_{\text{SIMO}} = [-15°, 15°]$. It is evident that the two

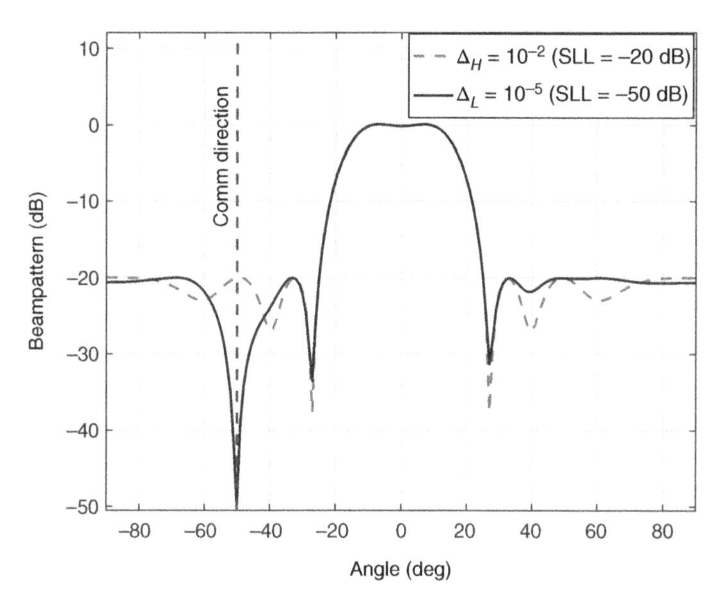

Figure 2.2 Transmit beampattern with two distinct SLLs toward the communication direction.

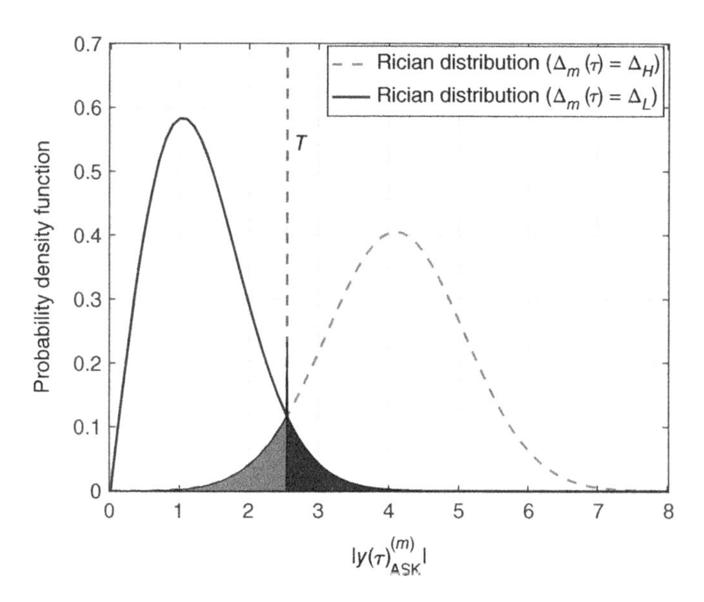

Figure 2.3 Probability distribution of the magnitude of signal at output of the matched filter.

beampatterns are almost identical within the mainbeam ensuring that the radar operation is not impacted by the information embedding. Figure 2.3 shows the probability density functions (pdfs) of $\left|y^{(m)}_{\text{ASK}}(n)\right|$. The figure shows that the pdf of the magnitude of the communication received signal is given by a Rician distribution. When symbol "0" is transmitted, i.e. $\Delta_m(n) = \Delta_L$, the corresponding distribution is given by the solid blue-colored curve. It is worth noting that the scale of the Rician distribution in this case is directly related to Δ_L. On the other hand, if symbol "1" is transmitted, the scale of the corresponding Rician pdf is proportional to Δ_H. This is shown as the dashed red-colored curve in the figure. The shaded areas represent the probability of bit error.

2.3.1.2 Beampattern Phase Modulation

This technique embeds information by controlling the phase of the transmit beampattern in the direction of the communication receiver. Information embedding using phase-shift keying (PSK) was recently reported in Hassanien et al. [2016a]. During each pulse, one PSK symbol from the dictionary $\mathbb{D}_{\text{PSK}} = \{\Omega_1, \dots, \Omega_{N_{\text{PSK}}}\}$ can be transmitted in each transmit beam. The PSK symbols can be chosen to be uniformly distributed on the unit circle. Each PSK symbol represents a unique sequence of $N_{\text{bit}} = \log_2(N_{\text{PSK}})$ binary bits. The beamforming weight vectors associated with the PSK modulated beampatterns can be designed using the formulation

$$\min_{\mathbf{q}_l} \max_{\theta} \left| d(\theta) - \left| \mathbf{q}_l^H \mathbf{a}(\theta) \right| \right|, \quad \theta \in \Theta_{\text{MIMO}}, \tag{2.21}$$

$$\text{subject to } \left| \mathbf{q}_l^H \mathbf{a}(\theta) \right| \leq \varepsilon_{\text{SLL}}, \quad \theta \in \bar{\Theta}_{\text{MIMO}}, \tag{2.22}$$

$$\mathbf{q}_l^H \mathbf{a}(\theta_c) = \left| d(\theta_c) \right| e^{j\Omega_l}, \qquad l = 1, \dots, N_{\text{PSK}}, \tag{2.23}$$

Here, Ω_n denotes the nth PSK symbol in the constellation $\mathbb{D}_{\text{PSK}}$, and $d(\theta_c)$ is the desired transmit gain in the communication direction. It is worth noting that the constraints in (2.23) dictate the phase of the beampattern, which then permits communications within the mainbeam of the radar. It is worth noting that the number cardinality of the PSK dictionary $\mathbb{D}_{\text{PSK}}$ can be much larger than the size of the AM dictionary $\mathbb{D}_{\text{AM}}$. For instance, if 256-PSK is used, then the formulation (2.21)–(2.23) has to be solved 256 times to obtain the require 256 transmit weight vectors $\{\mathbf{q}_1, \dots, \mathbf{q}_{N_{\text{PSK}}}\}$ required to implement the PSK-based information embedding scheme.

Assuming MIMO radar operation with M transmit beams, let $\Omega_m(n) \in \mathbb{D}_{\text{PSK}}$ be the symbol embedded during the nth pulse and transmitted over the mth beam. Then, the signal at the output of the matched filter of the communication receiver can be modeled as

$$y_m^{(\text{PSK})}(n) = \alpha_{\text{ch}} \sqrt{\frac{P}{M}} \left| d(\theta_c) \right| e^{j\Omega_m(n)} + z(n), \quad m = 1, \dots, M. \tag{2.24}$$

Using (2.24), the embedded PSK symbol can be extracted from the phase of $y_m^{(\text{PSK})}(n)$. For coherent communications, the channel coefficient α_{ch} should be known or accurately estimated. However, non-coherent PSK-based communications is also possible by transmitting two waveforms simultaneously. The PSK symbol is embedded as a phase rotation of one of the two waveforms with respect to the other [Hassanien et al., 2016a].

Illustrative Example We provide a simple example to illustrate the performances of the aforementioned beampattern modulation signaling strategies when 2 bits of information are embedded in a single beam during each pulse. Figure 2.4 shows the BER performance versus $E_b/N_0 \triangleq 10\log_{10} \frac{|\alpha_{\text{ch}}|^2 \Delta_H^2}{N_{\text{bit}} N_0}$, where E_b is the energy per bit. The figure compares the performances of the beampattern modulation techniques using AM, ASK, and PSK signaling for a SIMO radar with a 10-element array. The communication receiver is equipped with a single antenna located in the spatial direction $\theta_c = -50°$ and the radar mainbeam is focused toward the spatial direction $0°$. The number of bits per pulse is fixed to $N_{\text{bit}} = 2$ bits for all methods. To embed 2-bits per pulse, the AM method is implemented using a single waveform and four SLLs while the PSK method is implemented using four PSK symbols. However, since the ASK method employs two SLLs only, the 2 bits are embedded by simultaneously transmitting two orthogonal waveforms within the same pulse. To ensure that the total power constraint is satisfied, half the total power is assigned to each waveform. The figure shows that the ASK method using two SLLs and two orthogonal waveforms outperforms the AM method with four SLLs and a single waveform. The PSK approach offers the best performance of all three methods.

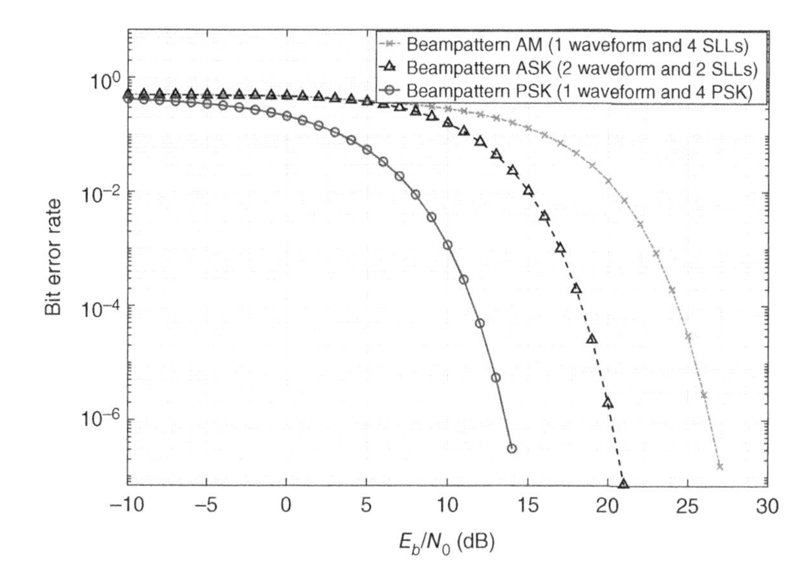

Figure 2.4 Bit error rate versus SNR for three beampattern modulation methods for a SIMO radar with 10-element transmit array; $N_{\text{bit}} = 2$ and $\theta_c = -50°$.

It is worth noting that beampattern modulation techniques may result in reduced transmit array efficiency of the radar transmit array due to non-constant modulus property of the designed beamforming weight vectors. As a result, the transmit coherent processing gain of the radar may be decreased leading to a reduction in SNR of the radar target signal. This can adversely affect the target detection performance and/or parameter estimation accuracy. However, the problem of constant-modulus beamforming weight design for beampattern modulation is open for future research.

2.3.1.3 Beampattern QAM Modulation

The main limitation of beampattern AM and PSK DRFC system strategies is the low data rate. An effort to increase the data rate based on beampattern QAM signaling has recently been reported in Ahmed et al. [2018]. This QAM-based information embedding technique permits supporting multiple communication receivers located in the sidelobe region. In addition, it enables multi-user access (MA) by allowing simultaneous transmission of distinct information streams to the communication receivers located in different directions. The essence of this scheme is to exploit amplitude as well as the PSK to enable the information embedding using multiple radar waveforms. The QAM symbol dictionary is larger than that of AM/PSK dictionaries. Hence, a larger number of transmit beamforming weight vectors are needed for implementing this QAM-based DFRC scheme which can be designed by solving the following optimization problem [Ahmed et al., 2018]

$$\min_{\mathbf{q}_{l_1 \dots l_Q, l'_1 \dots l'_Q}} \max_{\theta} \left| d(\theta) - \mathbf{q}^H_{l_1 \dots l_Q, l'_1 \dots l'_Q} \mathbf{a}(\theta) \right|, \quad \theta \in \Theta_{\text{MIMO}}, \tag{2.25}$$

$$\text{subject to } \left| \mathbf{q}^H_{l_1 \dots l_Q, l'_1 \dots l'_Q} \mathbf{a}(\theta) \right| \le \varepsilon_{\text{SLL}}, \quad \theta \in \bar{\Theta}_{\text{MIMO}}, \tag{2.26}$$

$$\mathbf{q}^H_{l_1 \dots l_Q, l'_1 \dots l'_Q} \mathbf{a}\left(\theta^c_1\right) = \Delta_{l_1}\left(\theta^c_1\right) e^{j\Omega_{l'_1}\left(\theta^c_1\right)}, \tag{2.27}$$

$$\vdots$$

$$\mathbf{q}^H_{l_1 \dots l_Q, l'_1 \dots l'_Q} \mathbf{a}\left(\theta^c_Q\right) = \Delta_{l_Q}\left(\theta^c_Q\right) e^{j\Omega_{l'_Q}\left(\theta^c_Q\right)}, \tag{2.28}$$

where Q is the number of communication receivers which are located at spatial directions $\theta_1^c, \dots, \theta_Q^c$, and $\Delta_{l_q}(\theta_q^c)$, $q = 1, \dots, Q$, $l_q = 1, \dots, N_{AM}$ and $\Omega_{l'_q}(\theta_q^c)$, $q = 1, \dots, Q$, $l'_q = 1, \dots, N_{PSK}$ denote the AM SLLs and PSK symbols associated with the qth communication receiver, respectively.

It is worth noting that the optimization problem (2.25)–(2.28) should be solved $(N_{AM}N_{PSK})^Q$ times to design the weight vectors $\mathbf{q}_{l_1 \dots l_Q, l'_1 \dots l'_Q}$, where $l_1 = 1, \dots, N_{AM}$, $l'_1 = 1, \dots, N_{PSK}$, $l_Q = 1, \dots, N_{AM}$, and $l'_Q = 1, \dots, N_{PSK}$. It should also be noted that QAM-based DFRC systems enables embedding of higher data rates to MA receivers at the cost of increased disturbance of the radar operation on the DFRC system.

2.3.2 DFRC Via Waveform Permutation

In DFRC systems employing a MIMO configuration, the antenna-waveform pairing can be used to convey information to a communications receiver at a known direction [BouDaher et al., 2016; Hassanien et al., 2018]. Although in a MIMO radar, the antenna-waveform pairing must be known, the association of a particular waveform to a specific antenna is arbitrary. Swapping waveforms between different antennas does not impact the operation of the radar provided this swap is reflected at the radar receiver. Thus, using the antenna-waveform pairing to embed information into the radar signal is a form of IM that is entirely transparent to the radar.

In a MIMO system with M antennas and a set of M orthogonal waveforms, shuffling the waveforms between the transmit antennas provides a constellation of $M!$ symbols, where $(\cdot)!$ denotes the factorial. This allows the transmission of $\lfloor \log_2(M!) \rfloor$ bits per PRI, where $\lfloor \cdot \rfloor$ stands for operator that gives the largest integer not greater than its argument. therefore, this information embedding scheme uses waveform permutation as means to transmit information. The signal model can be obtained from the general signal model (2.6) by substituting for $\mathbf{W}(n) = \mathbf{I}_{M \times M}$, $\boldsymbol{\psi}(t; n) = \mathbf{1}_M$, and replacing the operator $\mathcal{P}_n\{\cdot\}$ by a permutation matrix $\mathbf{P}$ of size $M \times M$. Hence, the signal model simplifies to

$$\mathbf{s}_{\mathrm{DFRC}}(t; n) = \sqrt{\frac{P}{M}} \mathbf{P} \boldsymbol{\phi}(t). \tag{2.29}$$

The communications receiver is assumed to have perfect knowledge of the radar waveforms, $\boldsymbol{\phi}(t)$. Upon correlation with the unshuffled set of waveforms $\boldsymbol{\phi}(t)$ the received signal becomes

$$\mathbf{r}_c(t) = \alpha_{\mathrm{ch}} \mathbf{P}^{\mathrm{T}} \mathbf{a}(\theta_{\mathbf{c}}) + \mathbf{n}(\mathbf{t}), \tag{2.30}$$

where $\mathbf{a}(\theta_{\mathbf{c}})$ is the steering vector in the direction of the communication receiver, θ_c, assumed known. Setting $\mathbf{a}_{\mathbf{s}}(\theta_{\mathbf{c}}) = \mathbf{P}^{\mathrm{T}} \mathbf{a}(\theta_{\mathbf{c}})$, we observe that the shuffling matrix $\mathbf{P}$ can be recovered from the received signal by comparing the received steering vector $\mathbf{a}_{\mathbf{s}}(\theta_{\mathbf{c}})$ to the un-shuffled vector $\mathbf{a}_{\mathbf{u}} = \mathbf{a}(\theta_{\mathbf{c}})$. Therefore, the transmitted symbol can be detected by searching for the permuted version of $\mathbf{a}_{\mathbf{u}}$ that best matches $\mathbf{a}_{\mathbf{s}}(\theta_{\mathbf{c}})$. The achievable bit rate of this scheme is $\lfloor \log_2(M!) \rfloor f_{\mathrm{PRF}}$ bps.

2.3.2.1 Illustrative Example

In order to demonstrate the ability of the waveform permutation based DFRC system for establishing communications over the MIMO configuration platform, we consider a joint platform with $M = 6$ antennas. The performance as a function of the receiver spatial angle for fixed SNR=10 dB is shown in Figure 2.5 for 1, 2, and 4 bits per symbol. The figure shows that the symbol error rate performance is poor at broadside where it has no capacity to transmit information. This is attributed to the fact that for $\theta_c = 0$, all elements of the array steering vector are equal and, therefore, any permutation of the vector yields the same vector. However, communications toward a receive located

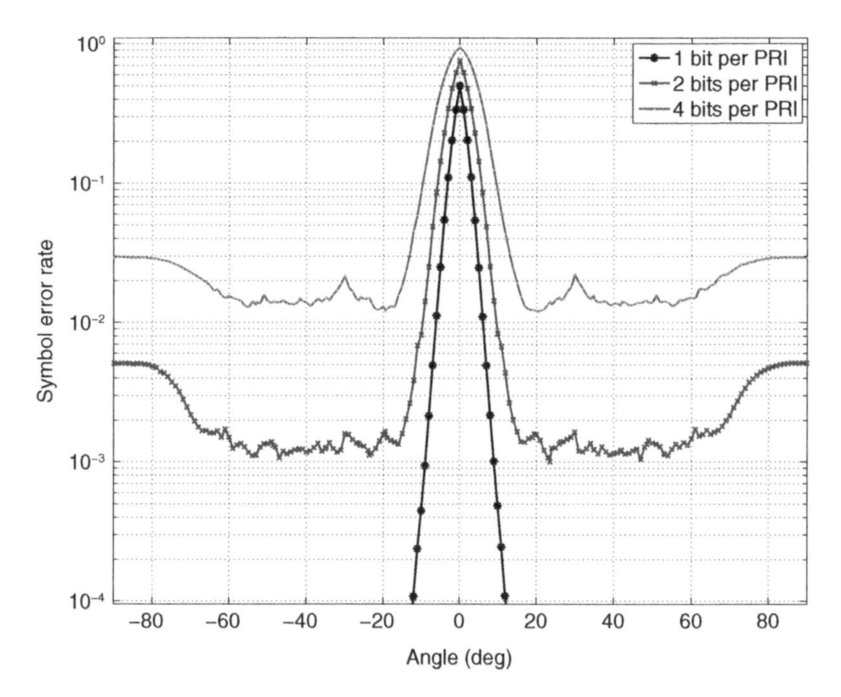

Figure 2.5 Symbol error rate versus spatial angle of the communication receiver.

at direction $\theta_c = 0$ can still be achieved by employing the phase induction, i.e. by multiplying each waveform by a pre-determined complex factor of unit magnitude.

2.4 DFRC Using Modulated Radar Waveforms

DFRC systems with fixed radar waveforms operate on a pulse-basis and their achievable bit rates are limited to the order of the pulse repetition frequency (PRF) of the system. To overcome this limitation, one approach is to incorporate information-bearing communication symbols into the emitted radar waveforms. Recent techniques achieve information embedding by modulating the radar waveform in fast-time [Nowak et al., 2016; Sahin et al., 2017a; Hassanien et al., 2017]. This leads to an increase in the communication data rate, However, the radar waveform changes from pulse to pulse, which can reduce radar performance.

2.4.1 DFRC Via Intended Modulation on Pulse (IMOP)

In Nowak et al. [2016], an approach to DFRC system is presented based on using mixed modulation through IMOP. Although the IMOP scheme is prone to cross interference, recent technological advances in digital electronics and signal processing have opened the door to re-exploring this approach in the context of joint radar-communication system design. This technique assumes that the waveforms are already designed to optimize the radar operation. Particularly, the basic linear frequency modulated (LFM) signal is adopted as the baseline radar waveform in Nowak et al. [2016]. Then, the radar pulse is divided into a number of sub-pulses and a communication symbol is embedded into each sub-pulse. To preserve the constant envelope feature of the original radar pulse, each sub-pulse is modulated by a binary phase shift keying (BPSK) communication symbol.

The data rate that can be achieved using the IMOP information embedding approach is proportional to the number of sub-pulses used within each radar pulse. The maximum data rate can be achieved when the sub-pulse duration equals the inverse of the total bandwidth of the DFRC system. However, in digital PM (e.g. BPSK) the carrier phase abruptly changes from one sub-pulse to another. This phase discontinuity results in high fractional out-of-band power causing poor spectral efficiency.

2.4.2 CPM-Embedded DFRC

Modulating the radar waveform during a CPI causes RSM resulting in increased residual clutter in the range-Doppler response, thus leading to reduced target visibility. To address this challenge and ensure spectral containment, continuous phase modulation (CPM) based approach to DFRC was recently developed [Sahin et al., 2017a]. In addition to the continuous phase feature of CPM signals, they also have constant envelope. These two features lead to tight spectral confinement as well as high transmit power efficiency. Motivated by the favorable features of CPM signals, communications embedding into radar emission is formulated in Sahin et al. [2017a]. According to this method, information-bearing sequences are modulated with CPM and phase-attached to the radar waveform. This enables preserving the constant modulus of the original radar waveform and enables frequency bandwidth confinement.

Considering the SIMO radar scenario (2.6), the CPM-based DFRC signal model can be stated as

$$\mathbf{s}_{\text{CPM}}(t;n) = \sqrt{P}\,(\psi(t;n)\cdot\phi(t)), \tag{2.31}$$

where $\phi(t)$ is the original SIMO radar waveform and $\psi(t;n)$ is the communication embedded signal during the nth pulse. The continuous phase of the communication signal $\psi(t;n)$ can be expressed as Sahin et al. [2017a]

$$\varphi(t;\boldsymbol{S}_{\text{CPM}}) = h\pi\int_0^t g_c(n) * \left[\sum_{n_s=1}^{N_s} S_{n_s}\delta(n-(n_s-1)T_s)\right] dn, \tag{2.32}$$

where N_s is the number of sub-pulses within a single radar pulse, $\boldsymbol{S}_{\text{CPM}} \triangleq [S_1, \dots, S_{N_s}]^T$ is the $N_s \times 1$ vector of embedded communication symbols, $g_c(t)$ is communication shaping filter, and h is the modulation index. It is worth noting that the modulation index h controls the magnitude of the total phase change due to a communication symbol embedding. It can be observed from (2.32) that the communication data rate can be increased by a factor N_s as compared to the data rates of DFRC systems with fixed radar waveforms.

As an example, the BER resulting from the use of binary CPM with a rectangular shaping filter that is phase-attached to a LFM radar waveform was examined in Sahin et al. [2017a]. The BER performance as a function of communication receiver SNR for modulation indices $h = 1/2, 1/4, 1/8$, and $1/16$ is shown in Figure 2.6. The figure shows that the required SNR to achieve a given BER increases with decreasing modulation index h. The impact of CPM-based information embedding on the spectral efficiency of the radar waveform is shown in Figure 2.7. The figure compares the PSD of the base radar waveform with the PSDs realized by communication-embedded radar waveforms (averaged over multiple independent information sequences) with and without guard symbols [Sahin et al., 2017a]. The figure shows that the use of guard symbols eliminates the spectral broadening caused by CPM signal embedding. It is worth noting that CPM modulation changes the AF characteristics of the waveform and leads to RSM. To address this problem, a mismatched filter design approach has been proposed in Sahin et al. [2017b].

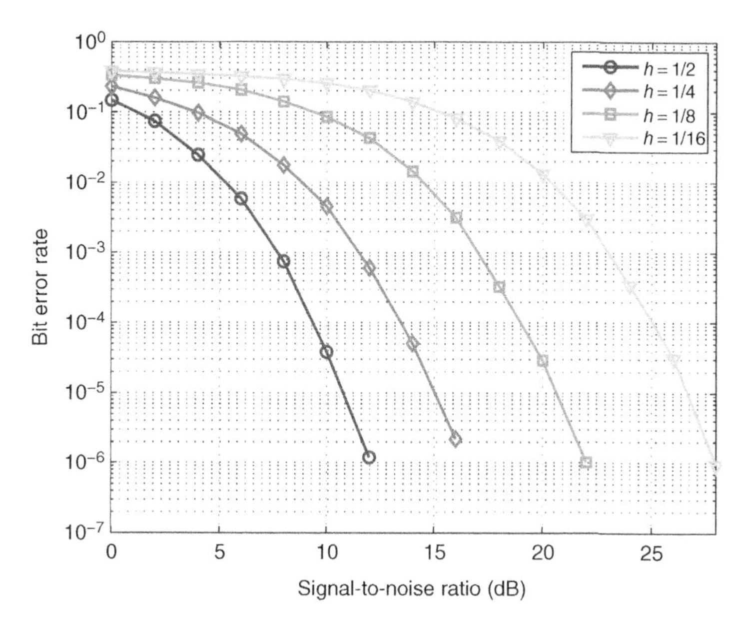

Figure 2.6 Bit error rate of binary CPM phase-attached to a LFM radar waveform versus SNR for various values of the modulation index h. Source: Courtesy Sahin et al. [2017a].

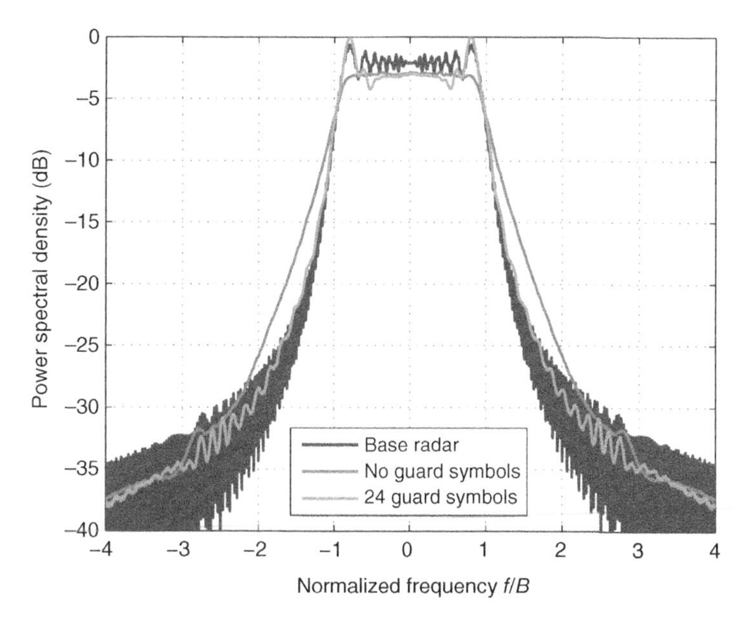

Figure 2.7 Power spectral density of the base radar waveform (dB) and CPM-based DFRC signals for cases with and without guard symbols. Source: Courtesy Sahin et al. [2017a].

2.4.3 DFRC Using FH Waveforms

The IMOP and CPM-based DFRC schemes discussed in Sections 2.4.1 and 2.4.2 are developed for SIMO radar configuration. To achieve even higher data rates, information embedding into the emission of MIMO radar using frequency-hopping (FH) waveforms, was first reported in Hassanien et al. [2017]. The basic idea behind this FH-DFRC scheme is to embed PSK communication symbols via phase-rotating the orthogonal transmit waveforms of the MIMO radar during each FH interval. Except for a phase discontinuity at transition from one FH interval to another, the embedded phase-rotations do not alter or compromise the offerings of the FH-based MIMO radar functionality. The achievable data rate under this signaling scheme is proportional to the number of transmit antennas and the PRF.

FH waveforms satisfy the high transmit power efficiency. requirements due to their constant-modulus feature. Moreover, they are simple to generate and immune to interference. The FH waveforms during one radar pulse can be expressed as

$$\phi_m(t) = \sum_{q=1}^{Q} e^{j2\pi c_{m,q}\Delta f t} u(t - \Delta t), \quad m = 1, \dots, M, \tag{2.33}$$

where $c_{m,q}$, $m = 1, \dots, M$, $q = 1, \dots, Q$ denote the FH code, Q is the FH code length, Δf and Δt stand for the frequency step and the hopping interval duration, respectively, and

$$u(t) \triangleq \begin{cases} 1, & 0 < t < \Delta t, \\ 0, & \text{otherwise.} \end{cases} \tag{2.34}$$

The radar pulse contains Q hopping intervals, i.e. $T_0 = Q\Delta t$. We assume that the FH code $c_{m,q} \in \{1, \dots, K\}$, where J is a pre-selected positive integer. Therefore, the pulse bandwidth is approximately given by $K\Delta f$. To achieve waveform orthogonality, $c_{m,q}$ should satisfy the following necessary constraint

$$c_{m,q} \neq c_{m',q}, \qquad \forall q, m \neq m'. \tag{2.35}$$

The problem of FH waveform optimization for MIMO radar has been discussed in a number of papers (see [Han and Nehorai, 2016]; and references therein). In the sequel, we assume that the orthogonal FH waveforms are already designed.

Let $\{\Omega_{m,q} \in [0, 2\pi]\}$, $m = 1, \dots, M$, $q = 1, \dots, Q$ be a set of MQ arbitrary phases. Define the phase-modulated waveforms $\psi_m(t)$, $m = 1, \dots, M$ as

$$\psi_m(t) = \sum_{q=1}^{Q} e^{j\Omega_{m,q}} e^{j2\pi c_{m,q}\Delta f t} u(t - \Delta t). \tag{2.36}$$

Assume that Δt is selected such that the following condition is satisfied

$$\int_0^{\Delta t} e^{j2\pi c_{m,q}\Delta f t} e^{-j2\pi c_{m',q'}\Delta f t} dt = 0, \quad m \neq m', \ q \neq q'. \tag{2.37}$$

The expression (2.37) implies that two hopped signals with different FH coefficients are orthogonal. This reduces the cross-correlation levels between the waveforms in (2.33). Making use of the condition (2.37) and the orthogonality between the waveforms given in (2.33), it can be easily verified that phase-modulated waveforms $\psi_m(t)$, $m = 1, \dots, M$ are also orthogonal. The phases $\Omega_{m,q}$ can be selected from the PSK constellation

$$\mathbb{D}_{\text{PSK}} = \left\{0, \frac{2\pi}{N_{\text{PSK}}}, \dots, \frac{(N_{\text{PSK}} - 1)2\pi}{N_{\text{PSK}}}\right\}. \tag{2.38}$$

During the nth pulse, the PSK-modulated set of orthogonal waveforms is defined as

$$\psi_m(t; n) = \sum_{q=1}^{Q} e^{j\Omega_{m,q}(n)} e^{j2\pi c_{m,q}\Delta f t} u(t - \Delta t), \tag{2.39}$$

where $\Omega_{m,q}(n) \in \mathbb{D}_{\text{PSK}}$, $m = 1, \dots, M$, $q = 1, \dots, Q$.

Assuming the communication receiver direction θ_c is known, the communication receiver signal model can be expressed as

$$r(t;n) = \alpha_{\text{ch}} \mathbf{a}^T(\theta_c)\boldsymbol{\psi}(t;n) + w(t;n), \tag{2.40}$$

where α_{ch} denotes the channel, $\boldsymbol{\psi}(t;n) \triangleq [\psi_1(t;n), \dots, \psi_M(t;n)]^T$ is the vector of transmitted FH-DFRC waveforms, and the communication receiver, and $w(t;n)$ represents interference plus noise, which is assumed to be white Gaussian with zero mean and variance σ_w^2. In (2.40), the channel coefficient α_{ch} is assumed to remain constant during the entire CPI. and an accurate estimate of the channel is assumed to be available.

Matched-filtering the communication received signal $r(t;n)$ to the FH waveforms yields

$$\begin{aligned} y_{m,q}(n) &= \int_0^{\Delta t} r(t;n)e^{-j2\pi c_{m,q}\Delta f t}u(t-\Delta t)dt \\ &= \alpha_{\text{ch}}\mathbf{a}_{[m]}e^{\Omega_{m,q}(n)} + w_{m,q}(n), \qquad m = 1,\dots,M,\ q = 1,\dots,Q, \end{aligned} \tag{2.41}$$

where $\mathbf{a}_{[m]} \triangleq e^{-j2\pi d_m \sin\theta_c}$ stands for the mth entry of $\mathbf{a}(\theta_c)$, d_m is the displacement between the first and the mth elements of the transmit array measured in wavelength, and $w_{m,q}(n) \triangleq \int_0^{\Delta t} w(t;n)e^{-j2\pi c_{m,q}\Delta f t}u(t-\Delta t)dt$ is the additive noise term at the output of the (m,q)th matched filter with zero mean and variance σ_w^2. Thus, the received communication signal at the output of the (m,q)th matched filter is a phase-shifted and noisy version of the mth entry of the steering vector $\mathbf{a}(\theta_c)$, meaning that the phase shift $\Omega_{m,q}(n)$ can be recovered from the received signal $y_{m,q}(n)$.

As an example, consider a MIMO radar system operating in the X-band with carrier frequency $f_c = 8.2$ GHz and bandwidth 500 MHz. The sampling frequency is taken as the Nyquist rate, i.e. $f_s = 10^9$ sample/s. The PRI is taken as $T_0 = 10$ μs, i.e. the PRF is 100 kHz. The transmit array is considered to be a uniform linear array (ULA) comprising $M = 16$ omni-directional transmit antennas spaced half a wavelength apart. A set of 16 FH waveforms with FH step $\Delta f = 10$ MHz is used. The FH code length $Q = 20$ is assumed and the FH interval duration $\Delta t = 0.1$ μs is used. The 16×20 FH code is generated randomly from the set $\{1,2,\dots,50\}$. The symbol error rate performance of the FH-DFRC technique is examined for BPSK, QPSK, and 16-PSK constellations. This corresponds to data rate of $R = 32$, 64, and 128 Mbps, respectively. Figure 2.8 shows the symbol error rate versus SNR for all constellation sizes considered. The figure shows that the smaller the constellation size is, the better the symbol error rate performance will be. The symbol error rate increases for denser constellations.

The impact of PSK symbol embedding on the AF of the FH waveforms was analyzed in Eedara et al. [2018]. Unlike the general expectation of adverse effect on radar operation brought about by embedding communication symbols, it was shown that modulation of the FH radar pulse can benefit both radar and communications. PSK symbol embedding analysis in terms of the AF of a MIMO radar revealed that symbol embedding yields a reduction in the SLLs of the AF of the original FH waveforms. A comparison between the AF with and without the PSK symbol embedding affirmed the benefit of communication embedding. Figures 2.9 and 2.10 show the zero Doppler cut of the AF for a series of 10 pulses of FH waveforms where each pulse comprises 16 FHs [Eedara et al., 2018]. The figures show that QPSK symbol embedding in the FH waveforms of the radar pulses significantly reduces the sidelobe peaks of the AF.

2.5 DFRC Using Index Modulation

IM refers to a wide class of modulation techniques where the information is represented by the index of some quantity from a set of available values for that quantity. For instance, the index of an

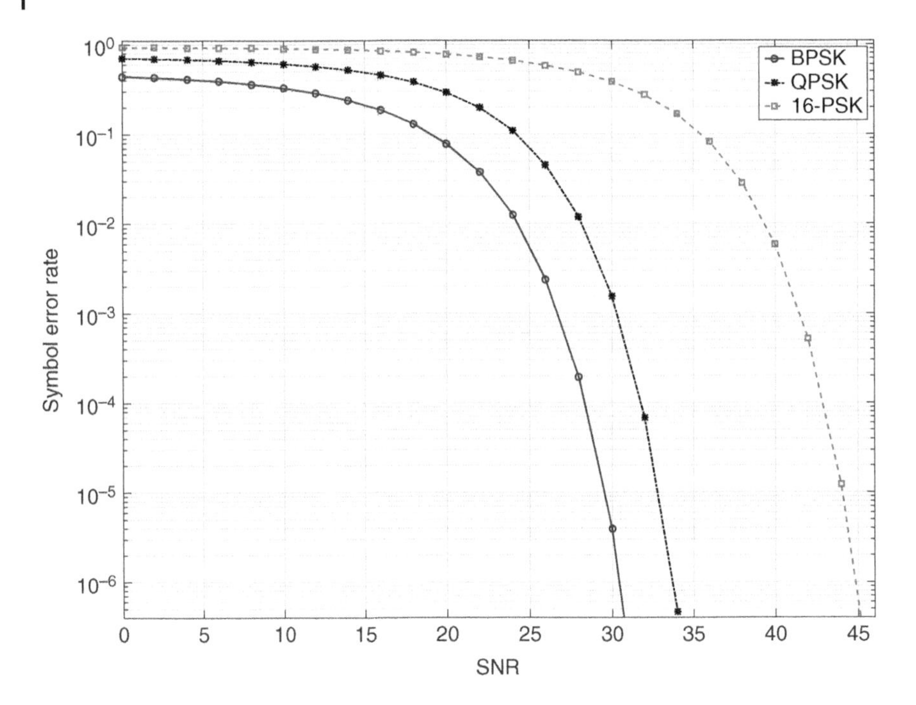

Figure 2.8 Symbol error rate versus SNR.

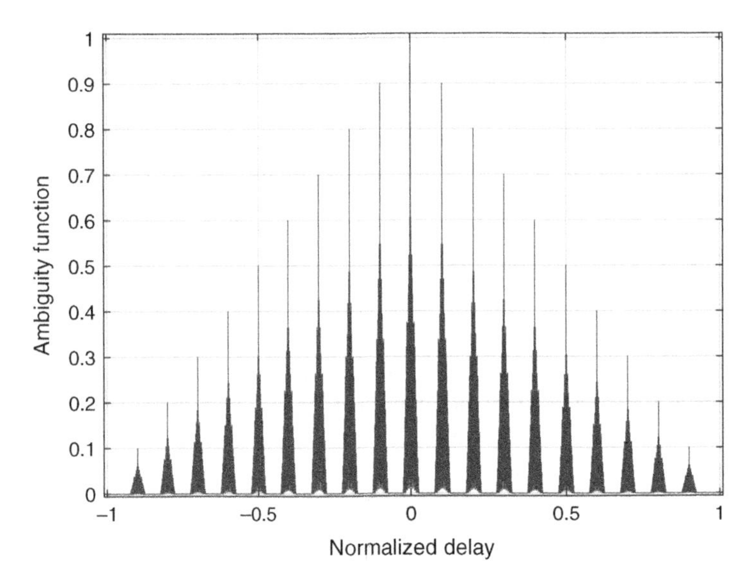

Figure 2.9 AF zero Doppler cut for a series of 10 pulses of FH waveforms without symbol embedding. Source: Courtesy Eedara et al. [2018].

antenna from a set of antennas can be used to represent information. A number of IM techniques have been recently developed in the context of DFRC. In this section, we overview four IM-based DFRC schemes namely, waveform selection [Blunt et al., 2010], antenna selection [Wang et al., 2019], FH code selection [Baxter et al., 2018], and OFDM-based carrier index selection [Huang et al., 2020].

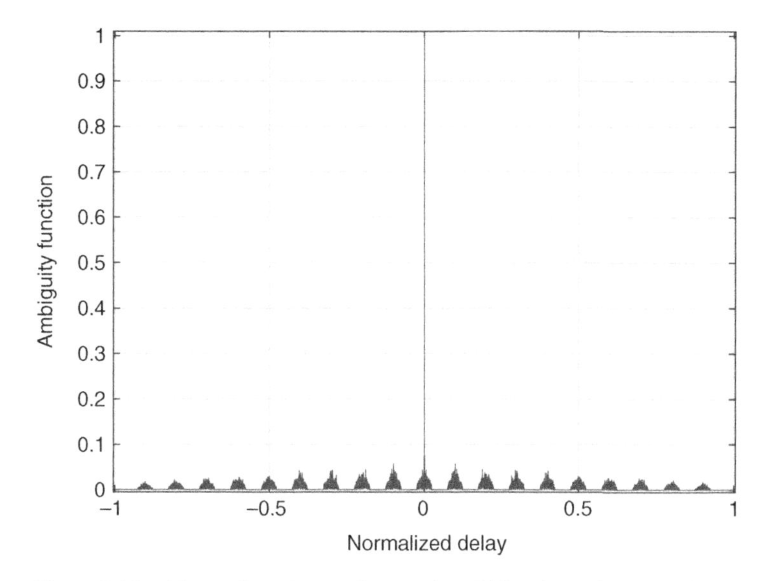

Figure 2.10 AF zero Doppler cut for a series of 10 pulses of FH waveforms with QPSK symbol embedding. Source: Courtesy Eedara et al. [2018].

2.5.1 DFRC by Waveform Selection

One of the early contributions on DFRC systems was developed in the context of SIMO radar [Blunt et al., 2010]. The key idea of that paper is to employ a bank of waveforms and to transmit one of those waveforms at a time depending on the information being embedded. Therefore, by selecting the radar waveform on a pulse-to-pulse basis, this scheme enables embedding $\log_2(L_w)$ bits of information per pulse, where L_w is the number of waveforms used [Blunt et al., 2010]. However, because different waveforms are utilized over a CPI, the range sidelobe response at the output of the radar receive filter can vary from pulse to pulse. This results in a loss of coherency within the CPI and, therefore, leads to a degradation of radar performance. Also, the data rate that can be achieved using this scheme is very low.

2.5.2 DFRC by Antenna Selection

The antenna index is used in Wang et al. [2019] to represent information in the context of MIMO radar. This scheme is specifically applicable when the number of orthogonal waveforms used is less than the number of antennas. Assuming that there are totally M antennas installed in the DFRC platform and an antenna selection network is deployed to select K out of M antennas. Then $K < M$ orthogonal waveforms, $\Psi_1(t), \dots, \Psi_K(t)$, can be deployed and transmitted via the selected K antennas.

The steering vector of the M-antenna full transmit array is denoted as $\mathbf{a}(\theta)$. Denote the $K \times M$ selection matrix during the nth radar pulse as $\mathbf{P}(n) \in \{0,1\}^{K\times M}$, where there is only one entry being "1" in each row and in the kth column corresponding to the kth selected antenna, $k \in \{1, \dots, M\}$. Applying the selection matrix $\mathbf{P}(n)$ to the steering vector $\mathbf{a}(\theta)$ of the full transmit array yields the $K \times 1$ steering vector of the selected subarray, that is,

$$\tilde{\mathbf{a}}(\theta; n) = \mathbf{P}(n)\mathbf{a}(\theta). \tag{2.42}$$

Let $\mathcal{P} = \{1, \dots, M\}$ label the full set of antennas installed in the transmit platform. During each radar pulse, a subset $\mathcal{S}$ of K antennas are selected from the full set $\mathcal{P}$ for waveform transmitting.

Such a selection is essentially a combinatorial problem. There are totally $L = C_M^K = \frac{M!}{K!(M-K)!}$ different subsets, $S_l \subset \mathcal{P}, l = 1, \dots, L$, and each subset S_l corresponds to a unique selection matrix $\mathbf{P}(n)$, which in turn corresponds to a unique steering vector $\tilde{\mathbf{a}}(\theta; n)$. For each subarray S_l, a communication symbol consisting of $\log_2 L$ bits of information can be defined. Assuming that the communication receiver knows its direction θ_c, the signal at the output of the communication receiver is expressed as,

$$x_c(t; n) = \alpha_{\text{ch}} \mathbf{a}^T(\theta_c) \mathbf{P}^T(n) \boldsymbol{\Psi}(t) + n_c(t; n). \tag{2.43}$$

Matched filtering the received data with the set of K orthogonal waveforms yields,

$$\begin{aligned}
\mathbf{y}_c(n) &= \text{vec}\left\{ \int_T x_c(t, n) \boldsymbol{\Psi}^H(t) dt \right\} \\
&= \alpha_{\text{ch}} \mathbf{P}(n) \mathbf{a}(\theta_c) + \mathbf{n}_c(n) \\
&= \alpha_{\text{ch}} \tilde{\mathbf{a}}(\theta_c; n) + \mathbf{n}_c(n),
\end{aligned} \tag{2.44}$$

where $\mathbf{y}_c(n) = [y_{c,1}(n), \dots, y_{c,K}(n)]^T$ and $\mathbf{n}_c(n) = [n_{c,1}(n), \dots, n_{c,K}(n)]^T$. Thus, the communication receiver signal at the output of the matched-filter is a scaled and noisy sparse selection of the full steering vector $\mathbf{a}(\theta_c)$, meaning that the selected sparse array S_l can be recovered from the received vector $\mathbf{y}_c(n)$. Hence, the steering vector of the sparse transmit array S_l is used as codes to embed communication symbols. It is worth noting that the use of sparse may result in higher SLLs in the beampattern of the array.

As an example, consider a radar with $M = 16$ antennas arranged in a ULA with inter-element spacing of 0.25 wavelength. Assume that $K = 8$ antennas are selected during each pulse. The performance of the system is evaluated by showing the symbol error rate as a function of SNR. Figure 2.11 shows the symbol error rate (SER) versus SNR for various numbers of bits per symbol. The figure

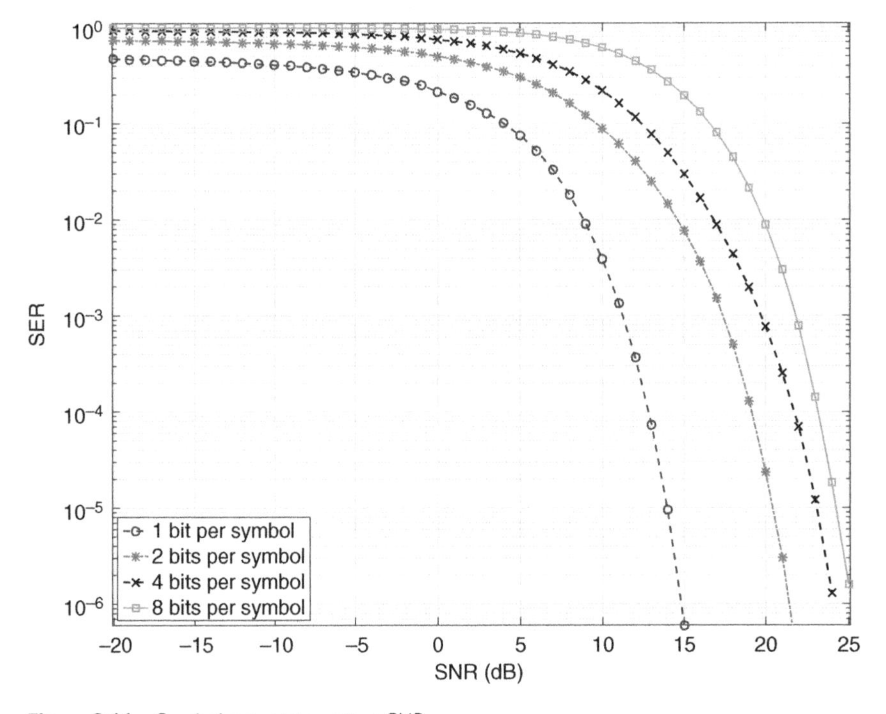

Figure 2.11 Symbol error rate versus SNR.

shows that the SER curves exhibit the expected standard behavior of a communication system, with the SER increasing with decreasing SNR and with increasing number of bits per symbol.

2.5.3 FH-based DFRC Via FH Index Modulation

A method of FH code selection as means to embed information into the emission of FH-MIMO radar is developed in Baxter et al. [2018]. This method adopts the FH-DFRC signal model (2.33). It is assumed that the number of antenna elements M is less than the number of available FH coefficients K. Therefore, during each FH sub-pulse, IM is used to represent symbols constructed by selecting M given K. Assuming $\Delta_f \Delta_t = 1$, the number of FH coefficients available is given by

$$\begin{aligned} K &= \frac{\mathrm{BW}}{\Delta_f} = \mathrm{BW}\Delta_t \\ &= \frac{1}{Q}\mathrm{BW}T_\phi, \end{aligned} \tag{2.45}$$

where the equality $\Delta_t \triangleq \frac{T_\phi}{Q}$ is used.

Define the FH code vectors $\mathbf{c}_q$, $q = 0, \dots, Q-1$. Each vector can represent a communication symbol. Therefore, Q communication symbols can be embedded; one symbol per FH sub-pulse. The total number of possible combinations that the entries of the vector $\mathbf{c}_q$ can be drawn from the set $\mathbb{F}_{\mathrm{FH}}$ is given by

$$L_{\mathrm{FH}} = \frac{K!}{M!(K-M)!} = \frac{(\mathrm{BW}T_\phi/Q)}{M!(\mathrm{BW}T_\phi/Q - M)!}. \tag{2.46}$$

Therefore, a dictionary of L_{FH} unique symbols can be constructed. The vector of the transmit signals during the qth chip of the nth pulse, can be defined as

$$\boldsymbol{\varphi}_q(t; n) = e^{j2\pi \mathbf{c}_q(n)\Delta_f t}, \quad q\Delta_t \leq t \leq (q+1)\Delta_t, \tag{2.47}$$

where $\mathbf{c}_q(n) \in \mathbb{S}$, $q = 1, \dots, Q$. Note that the FH code vector $\mathbf{c}_q(\tau)$ changes from pulse to another depending on the embedded binary information. Thus, the number of bits per pulse that can be embedded is

$$R_{\mathrm{FH}} = Q \log_2 L_{\mathrm{FH}}. \tag{2.48}$$

2.5.4 OFDM-based DFRC Using Index Modulation

A special case of the FH-DFRC model (2.33) is when the FH sub-pulse duration Δt is increased to occupy the entire pulse width. i.e. $\Delta t = T_\phi$. In this case, the FH step becomes $\Delta_f = 1/T_\phi$. Therefore, dividing the bandwidth BW into a number of frequencies separated by Δ_f (i.e. $Q = 1$) which corresponds to the conventional orthogonal frequency division multiplexing [OFDM]. This special case has recently been considered in Huang et al. [2020]. In that paper, the information is represented in the index of the OFDM frequencies. The total number of OFDM frequencies available is given by the time bandwidth product, that is,

$$K_{\mathrm{OFDM}} = \frac{\mathrm{BW}}{\Delta_f} = \mathrm{BW}T_\phi. \tag{2.49}$$

Therefore, the data rate that can be achieved using OFDM-DFRC is given by

$$R_{\mathrm{OFDM}} = \log_2 L_{\mathrm{OFDM}}, \tag{2.50}$$

where

$$L_{\text{OFDM}} = \frac{K_{\text{OFDM}}!}{M![K_{\text{OFDM}} - M]!}. \tag{2.51}$$

Comparing (2.48) and (2.50), the data rate that can be achieved using IM-based FH-DFRC is larger than the data rate that can be achieved using IM-based OFDM-DFRC.

2.6 Challenges and Future Trends

The main challenge in DFRC systems is how to achieve high data rates with minimum or no distortion to the radar spatio-temporal signal characteristics. To achieve higher data rates, signal embedding must be cognizant of wireless channel sensing, impairments, and equalization. For example, communications at sea faces different fading and multipath effects from those in land communications. The SNR needed for combating the channel must be examined in view of the radar equation and power requirement. Channel coding techniques would demand additional bits. This information redundancy may necessitate the use of additional radar pulses to communicate the intended message which may lead to reducing the information rate – a problem that was briefly addressed in Amin et al. [2019]. Channel estimation in a single function communication receiver typically requires a training sequence to be transmitted from the DFRC system. Avoiding the training sequence in FH radars was discussed in Wang et al. [2021]. If a separate link between the DFRC transmitter and single function communication receiver is to be avoided, this sequence must also be embedded into the radar waveforms. On the other hand, if the situation is reversed and the receiver is dual function, then separating the radar return from the received communication signal becomes vital for both signal demodulation and channel estimation. Without establishing timing and synchronizations, the communication receiver is compelled to perform radar pulse detection which in turn increases the overall probability of symbol error as delineated in Dong et al. [2019] and Chalise et al. [2020] for radar network operations. Further, joint carrier phase and delay estimation at the dual function receiver must be robust to the radar backscatterings. Notwithstanding the significant strides that have already been made, future DFRC systems stand to benefit greatly from the wealth of knowledge in the communications literature. For example, advances in symbol coding, symbol detection, channel estimation, multi-user interference mitigation, synchronization, and error detection and correction, should be informed by the state-of-the-art in the communications area. The future work in DFRC should apply the principle of metacognition recently introduced for the cognitive radar paradigm and dynamic spectrum allocations [Martone et al., 2020]. The DFRC Metacognition would decide on the best signal embedding strategy depending on the operating environment and task priorities

References

A. Ahmed, Y. D. Zhang, and Y. Gu. Dual-function radar-communications using QAM-based sidelobe modulation. *Digital Signal Processing*, 82:166–174, 2018.

M. G. Amin, Y. Dong, and G. A. Fabrizio. Scheduling data embedding in dual function radar networks. *IEEE 27th European Signal Processing Conference (EUSIPCO)*, Coruna, Spain, September 2019.

W. Baxter, E. Aboutanios, and A. Hassanien. Dual-function MIMO radar-communications via frequency-hopping code selection. In *Proceedings of the 52nd Asilomar Conference on Signals, Systems and Computers*, October 2018.

S. D. Blunt and E. L. Mokole. Overview of radar waveform diversity. *IEEE Aerospace and Electronic Systems Magazine*, 31(11):2–42, 2016.

S. D. Blunt, M. R. Cook, and J. Stiles. Embedding information into radar emissions via waveform implementation. In *Proceedings of International Waveform Diversity and Design Conference*, pages 195–199, August 2010.

E. BouDaher, A. Hassanien, E. Aboutanios, and M. G. Amin. Towards a dual-function MIMO radar-communication system. In *Proceedings of IEEE Radar Conference*, pages 1–6, Arlington, VA, May 2016.

B. K. Chalise, M. G. Amin, and G. A. Fabrizio. Multi-pulse processing of dual function radar waveforms without remodulation. In *IEEE Radar Conference*, pages 860–865, 2020.

J. M. Chapin. Shared spectrum access for radar and communications (SSPARC). *Technical Report DARPA BAA-13-24, Defense Advanced Research Projects Agency*, February 2013.

A. de Oliveira Ferreira, R. Sampaio-Neto, and J. M. Fortes. Robust sidelobe amplitude and phase modulation for dual-function radar-communications. *Transactions on Emerging Telecommunications Technologies*, e3314, 2018. https://doi.org/10.1002/ett.3314.

S. H. Dokhanchi, B. S. Mysore, K. V. Mishra, and B. Ottersten. A mmWave automotive joint radar-communications system. *EEE Transactions on Aerospace and Electronic Systems*, 55:1241–1260, 2019.

Y. Dong, G. A. Fabrizio, and M. G. Amin. Dual-functional radar waveforms without remodulation. In *IEEE Radar Conference*, pages 1–6, 2019.

I. P. Eedara, A. Hassanien, M. G. Amin, and B. D. Rigling. Ambiguity function analysis for dual-function radar communications using PSK signaling. In *Proceedings of the 52nd Asilomar Conference on Signals, Systems and Computers*, Pacific Grove, CA, October 2018.

J. Euzière, R. Guinvarc'h, M. Lesturgie, B. Uguen, and R. Gillard. Dual function radar communication time-modulated array. In *International Radar Conference*, pages 1–4, October 2014.

D. Gaglione, C. Clemente, C. V. Ilioudis, A. R. Persico, I. K. Proudler, J. J. Soraghan, and A. Farina. Waveform design for communicating radar systems using fractional Fourier transform. *Digital Signal Processing*, 80:57–69, 2018.

H. Griffiths, L. Cohen, S. Watts, E. Mokole, C. Baker, M. Wicks, and S. Blunt. Radar spectrum engineering and management: technical and regulatory issues. *Proceedings of the IEEE*, 103(1):85–102, 2015.

K. Han and A. Nehorai. Jointly optimal design for MIMO radar frequency-hopping waveforms using game theory. *IEEE Transactions on Aerospace and Electronic Systems*, 52(2):809–820, 2016. https://doi.org/10.1109/TAES.2015.140408.

A. Hassanien, M. G. Amin, Y. D. Zhang, and F. Ahmad. Phase-modulation based dual-function radar-communications. *IET Radar, Sonar and Navigation*, 10(8):1411 1421, 2016a.

A. Hassanien, M. G. Amin, Y. D. Zhang, and F. Ahmad. Signaling strategies for dual-function radar communications: an overview. *IEEE Aerospace and Electronic Systems Magazine*, 31(10):36–45, 2016b. https://doi.org/10.1109/MAES.2016.150225.

A. Hassanien, M. G. Amin, Y. D. Zhang, and F. Ahmad. Dual-function radar-communications: information embedding using sidelobe control and waveform diversity. *IEEE Transactions on Signal Processing*, 64(8):2168–2181, 2016c. https://doi.org/10.1109/TSP.2015.2505667.

A. Hassanien, B. Himed, and B. D. Rigling. A dual-function MIMO radar-communications system using frequency-hopping waveforms. In *IEEE Radar Conference*, pages 1721–1725, 2017. https://doi.org/10.1109/RADAR.2017.7944485.

A. Hassanien, E. Aboutanios, M. G. Amin, and G. A. Fabrizio. A dual-function MIMO radar-communication system via waveform permutation. *Digital Signal Processing*, 83:118–128, 2018.

A. Hassanien, M. G. Amin, E. Aboutanios, and B. Himed. Dual-function radar communication systems: a solution to the spectrum congestion problem. *IEEE Signal Processing Magazine*, 36(5):115–126, 2019. https://doi.org/10.1109/MSP.2019.2900571.

H. T. Hayvaci and B. Tavli. Spectrum sharing in radar and wireless communication systems: a review. In *Proceedings of the International Conference on Electromagnetics in Advanced Applications*, pages 810–813, August 2014.

T. Huang, N. Shlezinger, X. Xu, Y. Liu, and Y. C. Eldar. MAJoRCom: A dual-function radar communication system using index modulation. *IEEE Transactions on Signal Processing*, 68:3423–3438, 2020. https://doi.org/10.1109/TSP.2020.2994394.

P. M. McCormick, et al. FMCW implementation of phase-attached radar-communications (PARC). *IEEE Radar Conference*, Washington, DC, 2019.

A. Martone, and M. G. Amin. A view on radar and communication systems coexistence and dual functionality in the era of spectrum sensing. *Digital Signal Processing*, 119:103135, 2021.

A. Martone, et al. Metacognition for radar coexistence. In *IEEE International Radar Conference*, Washington, DC, 2020.

K. V. Mishra, M. B. Shankar, V. Koivunen, B. Ottersten, and S. A. Vorobyov. Toward millimeter-wave joint radar communications: a signal processing perspective. In *IEEE Signal Processing Magazine*, 36(5):100–114, 2019.

M. Nowak, M. Wicks, Z. Zhang, and Z. Wu. Co-designed radar-communication using linear frequency modulation waveform. *IEEE Aerospace and Electronic Systems Magazine*, 31(10):28–35, 2016.

C. Sahin, J. Jakabosky, P. M. McCormick, J. G. Metcalf, and S. D. Blunt. A novel approach for embedding communication symbols into physical radar waveforms. In *Proceedings of the IEEE Radar Conference*, pages 1498–1503, May 2017a.

C. Sahin, J. G. Metcalf, and S. D. Blunt. Filter design to address range sidelobe modulation in transmit-encoded radar-embedded communications. In *IEEE Radar Conference*, pages 1509–1514, 2017b. https://doi.org/10.1109/RADAR.2017.7944446.

X. Wang, A. Hassanien, and M. G. Amin. Dual-function MIMO radar communications system design via sparse array optimization. *IEEE Transactions on Aerospace and Electronic Systems*, 55(3):1213–1226, 2019. https://doi.org/10.1109/TAES.2018.2866038.

D. M. Wang, B. K. Chalise, J. Metcalf, and M. G. Amin. Information decoding and SDR implementation of DFRC systems without training signals. *IIEEE International Conference on Acoustics, Speech and Signal Processing (ICASSP)*, 55(3):8218–8222, 2021.

3

Interference, Clutter, and Jamming Suppression in Joint Radar–Communications Systems – Coordinated and Uncoordinated Designs

Jeremy Johnston[1], *Junhui Qian*[2], *and Xiaodong Wang*[1]

[1] *Columbia University, Electrical Engineering Department, New York, NY, USA*
[2] *Chongqing University, School of Microelectronic and Communication Engineering, Chongqing, China*

3.1 Introduction

The explosive development of services exploiting the terrestrial radio channel has brought up the problem of designing overlaid architectures, consisting of systems with different functions that share bandwidth while guaranteeing prescribed Quality of Service (QoS) [Griffiths and Blunt, 2014; Griffiths et al., 2014]. A scenario of major importance relates to areas that should be, on the one hand, covered by wireless communications and, on the other hand, kept under surveillance through one or more radars – a problem of radar/communication coexistence, i.e. of efficient exploitation of the assigned spectrum and cross-interference minimization [Aubry et al., 2016]. Some early results concerning the preservation of radar detection capabilities in the presence of coexisting, possibly unlicensed wireless users have been established: A combined approach based on mutual information (MI) and signal-to-interference-plus-noise ratio (SINR) has been developed in Turlapaty and Jin [2014]. The philosophy of dual-function radar–communication instead relies on considering the radar as a primary function and the communication as a secondary one, whose data can be embedded in the radar waveform [Blunt et al., 2010; Hassanien et al., 2016a]. An information theoretic approach to the radar waveform design is taken in Bica et al. [2016] whereby the MI is optimized under the constraint that the radar does not produce excessive interference on the coexisting communications system. Attention is steered back to the performance of the communications system in Li et al. [2016], wherein matrix-completion-based multiple-input multiple-output (MIMO) radars are made to coexist with wireless systems by constraining the average capacity of the latter, while minimizing the measure of the interference induced on the former. Communications and radar functions are likewise given equal weight in Chiriyath et al. [2016] and Paul et al. [2016], which aimed at investigating the interplay between the estimation accuracy in target localization on the delay-Doppler plane and the performance of a multiple-access coexisting system, characterized through rate achievability regions of active users.

The open literature is now rich with a number of studies, mainly assuming single-input single-output (SISO) radars and communications systems, focused on appropriate waveform design to allow coexistence with overlaid communications systems and/or determining the fundamental limits achievable in the presence of spectrum sharing (see, e.g., [Zheng et al., 2018b]). In general, available studies so far focus on interference mitigation to guarantee the performance of the communication link through projection onto the null space of the interference channel from the radar to the communications system [Khawar et al., 2014], or through beamforming

Signal Processing for Joint Radar Communications, First Edition.
Edited by Kumar Vijay Mishra, M. R. Bhavani Shankar, Björn Ottersten, and A. Lee Swindlehurst.

and nulling [Puglielli et al., 2016]. The performance of the radar system is typically guaranteed through techniques such as spatial filtering [Deng and Himed, 2013] and careful waveform design for SINR maximization, possibly exploiting some form of cognition (see [Aubry et al., 2013, 2016]). On the other hand, the widespread use of MIMO architectures for communications [Tse and Viswanath, 2005] and the potential of MIMO radars [Li and Stoica, 2007] are naturally conducive to investigate the feasibility of coexistence MIMO structures. An interesting new framework, proposed in Li and Petropulu [2016] and Qian et al. [2018a], is the philosophy of co-design: the radar–communications system pair is seen as a holistic structure whose constituent subsystems share information and cooperate to jointly design the transmitted signals in order to take advantage of the increased number of design degrees of freedoms (DOFs). In radar system design, to apply the matrix completion techniques in colocated MIMO radars, the data matrices need to be low rank and satisfy certain coherence conditions [Sun et al., 2015]. More generally, waveform diversity [Gini et al., 2012] may be regarded as a key tool to guarantee the dual-function radar–communication, and in this context space–time processing is key to improve the performance of coexisting systems [Hassanien et al., 2016b].

When such cooperation and co-design are not feasible – perhaps because of security concerns or financial burden – interference suppression methods must process the received and transmitted signals by exploiting a priori interference models and built-in environment sensing and cognition. For tasks such as image recovery and data demodulation, estimating the signal of interest is inherently coupled with estimation of the interference signal; a poor estimate of one will likely degrade the estimate of the other. And even though methods such as filtering and thresholding may be effective if the interference is much stronger than the desired signal, they run the risk of inadvertently distorting the desired signal. Hence, joint estimation of the signal and interference may be preferable. To do so, parametric methods estimate the parameters of a statistical signal model via either subspace methods or optimization. Greedy methods, e.g. CLEAN Blunt et al. [2008] and matching pursuit Zhu et al. [2020], iteratively remove the most dominant interference component, which is found by projecting the recording onto an interference dictionary, until a stopping criterion is met. If the received interference is concentrated in narrow regions along some dimension, e.g. time, frequency, or space, and hence is sparse in a known dictionary, convex relaxation methods such as ℓ_1-minimization can be effective [Li et al., 2019b; Li et al., 2019a]. In addition to suppression measures implemented at the receiver, the transmitted radar waveform or communication signal constellation can be periodically adjusted so as to adapt to the interference environment. They can be obtained through optimizing a suitable objective subject to constraints on, for example, the interference caused to other systems [Aubry et al., 2014]. Another approach is to force the transmitted signals to lie in the null space of the interference channel, as enabled by blind null space estimation [Manolakos et al., 2012].

In this chapter, we describe coordinated and uncoordinated approaches to interference suppression from both radar- and communication-centric perspectives. In Section 3.2, we consider a scenario wherein a MIMO radar with colocated antennas [Li and Stoica, 2007] shares bandwidth with a MIMO communications system and the transmitted signals can be jointly designed, for both the static target case and moving target case. In particular, we first extend to this new scenario to the channel and clutter models, mainly based on a geometrical approach, presented in Sayeed and Member [2002], Aubry et al. [2013], and Karbasi et al. [2015], assuming cognitive paradigms. The objects of interest in system design are the radar transmit space–time code, its receive filter, and the communications system codebook. We state the design problem as a constrained maximization of the SINR subject to a number of constraints, guaranteeing performance of the radar through a similarity constraint with some standard waveform [De Maio et al., 2009], and the communications system via the communication rate [Qian et al., 2018b]. The resulting problem is nonconvex, and we devise an iterative algorithm based on alternating maximization applied to three different subproblems. Finally, thorough performance assessment is undertaken so as to illustrate the merits of the proposed algorithm. In Section 3.3, we explore an approach whereby interference suppression is integrated into the system's primary function: prior knowledge of the interference structure is embedded in an unconstrained optimization problem that forms the basis for interference-robust processing. A detailed application to super-resolution MIMO radar imaging is presented along with two algorithms – an iterative optimization method and a corresponding model-based deep learning method. The section concludes with a vignette of how this uncoordinated approach can be employed for radar interference suppression at a communication receiver performing data symbol demodulation.

3.1.1 Notations

$\mathcal{CN}(\mu, \boldsymbol{\Sigma})$ denotes the circularly symmetric complex Gaussian distribution with mean μ and covariance matrix $\boldsymbol{\Sigma}$. We denote vectors by boldface lowercase letters, e.g. $\mathbf{a}$, and matrices by boldface uppercase letters, e.g. $\mathbf{A}$. The Hermitian, transpose, and conjugate operators are denoted by $(\cdot)^H$, $(\cdot)^T$, and $(\cdot)^*$, respectively. $\otimes$ and $\odot$ denote the Kronecker product and Hadamard product, respectively. $\mathrm{Tr}(\mathbf{A})$ denotes the trace of $\mathbf{A}$. $\mathrm{diag}(\cdot)$ denotes the diagonal matrix formed by the entries of the vector argument. $\det(\mathbf{A})$ stands for the determinant of the matrix $\mathbf{A}$. $\|\mathbf{a}\|$ denotes the Euclidean norm of the vector $\mathbf{a}$. $\mathbf{I}_m$ stands for the $m \times m$ identity matrix. $\mathbb{C}^m$ denotes the sets of m-dimensional vectors of complex numbers. The letter j represents the imaginary unit. $\mathbb{E}[\cdot]$ denotes statistical expectation. For an $m \times n$ matrix $\mathbf{A} = [\mathbf{a}_1\ \mathbf{a}_2\ \dots\ \mathbf{a}_n]$, we define the $mn \times 1$ vector $\mathrm{vec}(\mathbf{A}) = [\mathbf{a}_1^T\ \mathbf{a}_2^T\ \dots\ \mathbf{a}_n^T]^T$.

3.2 Joint Design of Coordinated Joint Radar–Communications Systems

In this section, we explore the joint design of a MIMO radar with colocated antennas and a MIMO communications system for shared spectrum access based on coordinated designs.

3.2.1 System Description

We consider a communications system and a mono-static radar operating on the same frequency band, as illustrated in Figure 3.1 (see Table 3.1). The two systems are assumed to use narrowband waveforms with the same number of symbols K in one transmit period. The radar is equipped with a linear array with $M_{t,R}$ transmit and $M_{r,R}$ receive antennas and is pointing toward the azimuth direction θ_0, while the communication link employs a linear array with $M_{t,C}$ antennas at the transmitter (TX) and a linear array with $M_{r,C}$ antennas at the receiver (RX). Baseband continuous-time waveforms are sampled at rate B, where B is the two-sided shared bandwidth. The communications system operates in a local rich scattering environment (e.g. an urban area). The radar monitors a wide region that includes the smaller area where the communications system is placed. We assume that the two systems are synchronized at the pulse/symbol level, and, therefore, in what follows, we focus on the signal models in one transmit period that consists of K pulses/symbols.

For simplicity, we assume that the radar system waveform symbol duration matches that of the communications system. For the mismatched case, the mathematical expressions have the same form as those in the matched case except the delay factor. To calculate the corresponding delay factor, the communications system only needs to know the sampling time of the radar system.

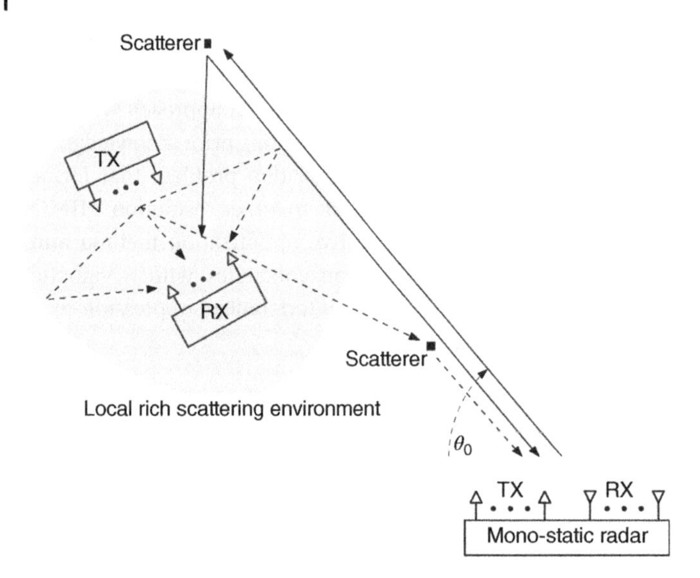

Figure 3.1 Example of a MIMO communications system coexisting with a mono-static MIMO radar.

Table 3.1 Symbol definitions for joint design.

Symbol	Definition	Symbol	Definition
$\mathbf{s}$	Radar TX waveform	K	Number of radar epochs
$\mathbf{x}$	Com. TX space–time code	M	Number of rad-com interference paths
$\mathbf{w}$	Radar RX filter	P	Number of com-rad interference paths
$\mathbf{H}$	Com. channel	Q	Number of radar clutters
$\mathbf{V}$	Radar RX array response	$\mathbf{G}_1$	Com-rad interference channel
$\mathbf{p}$	Doppler steering vector	$\mathbf{G}_2$	Rad-com interference channel
$M_{t,R}$	Number of radar TX elements	$M_{t,C}$	Number of com. TX elements
$M_{r,R}$	Number of radar RX elements	$M_{r,C}$	Number of com. RX elements

Therefore, the spectrum sharing problems in such cases can still be solved using the proposed algorithms [Li et al., 2016].

3.2.1.1 Radar Signal Model

The radar emits a coherent train of K pulses from each transmit antenna. Let $\mathbf{s}(k) = \left[s_1(k), \dots, s_{M_{t,R}}(k)\right]^T \in \mathbb{C}^{M_{t,R}\times 1}$ denote the vector of the amplitudes transmitted at the kth epoch by the $M_{t,R}$ transmit elements, so that the transmitted waveform can be equivalently represented by the vector $\mathbf{s} = \text{vec}([\mathbf{s}(1), \dots, \mathbf{s}(K)]) \in \mathbb{C}^{M_{t,R}K\times 1}$. In the considered coexistence scenario, a far-field moving target at the azimuth angle θ_0 generates at epoch k an observation $\mathbf{y}(k) \in \mathbb{C}^{M_{r,R}\times 1}$ modeled as (see, e.g. [Karbasi et al., 2015; Aubry et al., 2013; Gini et al., 2012])

$$\mathbf{y}(k) = \alpha_0 e^{j2\pi(k-1)f_{d,0}}\mathbf{v}_r(\theta_0)\mathbf{v}_t^T(\theta_0)\mathbf{s}(k) + \mathbf{y}_s(k) + \mathbf{y}_c(k) + \mathbf{y}_n(k), \tag{3.1}$$

where α_0 represents the complex path loss including propagation loss and reflection coefficient, $f_{d,0}$ is the target normalized Doppler frequency, $\mathbf{y}_s(k)$ represents the clutter (signal-dependent interference) at the receive side, $\mathbf{y}_c(k)$ denotes interference resulting from the communications system (signal-independent interference), and $\mathbf{y}_n(k) \sim \mathcal{CN}\left(0, \sigma_R^2 \mathbf{I}_{M_{r,R}}\right)$ is the additive white Gaussian noise (AWGN) at the radar receive array. For a uniform linear array (ULA) with $\tilde{M}$ antenna elements with inter-antenna spacing d, we define the array response vector to the received signal from angle θ as

$$\mathbf{u}(\theta; \tilde{M}, d) = \frac{1}{\sqrt{\tilde{M}}}\left[1 \;\; e^{j2\pi d \sin\theta/\lambda} \;\; \ldots \;\; e^{j2\pi(\tilde{M}-1)d\sin\theta/\lambda}\right]^T, \tag{3.2}$$

where λ is the carrier wavelength. Then, the transmit and receive steering vectors for the radar and communications systems are, respectively,

$$\begin{aligned} \mathbf{v}_t(\theta) &= \mathbf{u}(\theta; M_{t,R}, d_{t1}), \\ \mathbf{v}_r(\theta) &= \mathbf{u}(\theta; M_{r,R}, d_{r1}), \\ \mathbf{a}_t(\theta) &= \mathbf{u}(\theta; M_{t,C}, d_{t2}), \\ \mathbf{a}_r(\theta) &= \mathbf{u}(\theta; M_{r,C}, d_{r2}), \end{aligned} \tag{3.3}$$

where d_{t1} and d_{r1} are the array inter-element spacing of the radar TX and the radar RX, respectively, and d_{t2} and d_{r2} are the array inter-element spacing of the communication TX and the communication RX, respectively. For notational simplicity, we introduce the transmit–receive spatial steering vector $\mathbf{V}(\theta) = \mathbf{v}_r(\theta)\mathbf{v}_t^T(\theta) \in \mathbb{C}^{M_{r,R} \times M_{t,R}}$ for the azimuth angle θ. For a signal with Doppler shift f_d, the temporal steering vector can be expressed as

$$\mathbf{p}\left(f_d\right) = \left[1, e^{j2\pi f_d}, \ldots, e^{j2\pi f_d(K-1)}\right]^T. \tag{3.4}$$

Then the corresponding space–time steering vector can be expressed as

$$\tilde{\mathbf{V}}\left(f_d, \theta\right) = \operatorname{diag}\left(\mathbf{p}\left(f_d\right)\right) \otimes \mathbf{V}(\theta). \tag{3.5}$$

To proceed with the signal model, we need a statistical characterization of both interfering contributions. We start with the signal-dependent contribution $\mathbf{y}_s$. Assume that a target is located in a given range-azimuth cell, indexed as (0,0) for simplicity, where the first coordinate corresponds to range and the second to azimuth. The cell under test is subject to its own clutter return, due to the entire transmitted signal, plus the fraction of the responses from possible range cells. The signal-dependent interference is defined by a number, say Q, of range rings causing interference to the cell under test, in keeping with the model outlined in Aubry et al. [2013] and Cui et al. [2017]. Since we assume point-like scatterers, the qth being located at distance $r_q \in \left\{0, \ldots, \hat{K}-1\right\}, \hat{K} \leq K$,[1] we have that the signal-dependent interference obeys the model

$$\mathbf{y}_s(k) = \sum_{q=1}^{Q} \alpha_q e^{j2\pi f_{d,q}(k-1)} V(\theta_q) s(k - r_q), \quad 0 \leq r_q \leq k-1, \tag{3.6}$$

where α_q, $f_{d,q}$, and θ_q denote the complex amplitude, normalized Doppler frequency, and look angle of the qth clutter, respectively. Similar to (3.6), we also adopt a physical propagation path channel model for the radar–communication channel, extending to the case of coexisting systems outlined in Sayeed and Member [2002] with reference to MIMO communications. For the signal-independent interference, we assume the presence of P uncorrelated scatterers reflecting the communication signal toward the radar. Given the channel

$$\mathbf{G}_1 = \sum_{p=1}^{P} \beta_1(p)\mathbf{v}_r(\theta_p)\mathbf{a}_t^T(\varphi_p) = \sum_{p=1}^{P} \beta_1(p)\mathbf{G}_{1,p}, \tag{3.7}$$

1 Here, the time basis is the considered one transmit period itself.

where $\mathbf{a}_t(\varphi_p)$ is the communications system transmit steering vector of the pth propagation path and $\beta_1(p)$ is a zero-mean complex random variable with variance $\sigma^2_{\beta,p}$ accounting for the response of the pth scatterer, the interference produced by the communications system simplifies to

$$\mathbf{y}_c(k) = \sum_{p=1}^{P_1} \beta_1(p)\mathbf{G}_{1,p}\mathbf{x}(l - r'_p)e^{j2\pi f'_{d,p}(k-1)}, \quad 0 \leq r'_p \leq k-1, \tag{3.8}$$

where $f'_{d,p}$ and r'_p denote the communication Doppler frequency and the delay for the pth path, respectively Foerster [2002] and Wang et al. [2014].

3.2.1.2 Communication Signal Model

In the considered narrowband, flat-fading environment, the signal at the communication RX can be written as

$$\mathbf{z}(k) = \mathbf{H}\mathbf{x}(k)e^{j2\pi\mu_0(k-1)} + \mathbf{z}_r(k) + \mathbf{z}_n(k), \tag{3.9}$$

where $\mathbf{H} \in \mathbb{C}^{M_{r,c} \times M_{t,c}}$ is modeled as Rayleigh fading, accounting for the channel matrix between the communication TX and RX, μ_0 denotes the Doppler shift of the communication signal, $\mathbf{z}_r(k)$ represents the interference from the radar system, and $\mathbf{z}_n(k) \sim \mathcal{CN}\left(0, \sigma^2_C \mathbf{I}_{M_{r,c}}\right)$ is the AWGN vector at the communication receiving array. To model the interference caused by the radar, we borrow the same arguments used to model the effect of the communication system on the radar. Denoting M the number of paths impinging on the communication RX, $\mu_{d,m}$ the Doppler shift of the mth path, φ_m the direction under which the path is seen, and θ_m the direction of transmission of the ray originating it, we have, under nondispersive channels,

$$\mathbf{z}_r(k) = e^{j2\pi\mu_{d,m}(k-1)}\mathbf{G}_2\mathbf{s}(k). \tag{3.10}$$

The matrix $\mathbf{G}_2$ is the exact counterpart of the matrix $\mathbf{G}_1$ defined previously (3.7), and thus reads

$$\mathbf{G}_2 = \sum_{m=1}^{M} \beta_2(m)\sqrt{\eta_m}\mathbf{a}_r^*(\varphi_m)\mathbf{v}_t^H(\theta_m) = \sum_{m=1}^{M} \beta_2(m)\mathbf{G}_{2,m}. \tag{3.11}$$

Here, $\beta_2(m)$ denotes the random, complex-valued unit-mean-square-value path gain, and η_m denotes the average intensity of the mth propagation path in the interference channel between the radar TX and the communication RX.

Because all the channels are assumed to be flat-fading and remain the same over K symbol intervals, (3.9) can be expressed in a compact space–time form as

$$\mathbf{z} = \tilde{\mathbf{H}}\mathbf{x} + \sum_{m=1}^{M} \beta_2(m)\tilde{\mathbf{G}}_{2,m}\left(\mu_{d,m}\right)\mathbf{s} + \mathbf{z}_n, \tag{3.12}$$

where $\tilde{\mathbf{H}} = \left(\mathrm{diag}\left(\mathbf{p}\left(\mu_0\right)\right) \otimes \mathbf{H}\right)$, $\tilde{\mathbf{G}}_{2,m}\left(\mu_{d,m}\right) = \mathbf{G}_{2,m} \otimes \mathrm{diag}\left(\mathbf{p}\left(\mu_{d,m}\right)\right)$, and $\mathbf{z}_n = \mathrm{vec}\left(\mathbf{z}_n(1), \ldots, \mathbf{z}_n(K)\right) \in \mathbb{C}^{M_{r,c}K \times 1}$. Once again, cognition of the matrices $\{\mathbf{G}_{2,m}\}_{m=1}^{M}$ is assumed, implying that the channel state information can be periodically communicated among the communication RX, TX and the radar system through a pilot channel [Filo et al., 2009]. (See Chapters 4 and 8.) In what follows, we treat the cases of moving targets and static targets separately and then provide the corresponding problem formulations.

3.2.1.3 Moving Targets

Most of the aforementioned techniques focus on stationary targets. While this assumption is reasonable during the detection process confirmation or for an already tracked target, it cannot be met during the standard search radar operation. We first consider the case of moving targets.

Following Cui et al. [2017], each $f_{d,q}$ is assumed uniformly distributed around mean value $\bar{f}_{d,q}$, i.e. $f_{d,q} \sim U\left(\bar{f}_{d,q} - \frac{\varepsilon_q}{2}, \bar{f}_{d,q} + \frac{\varepsilon_q}{2}\right), q \in 1,2,\dots,Q$, where ε_q accounts for the uncertainty on $\bar{f}_{d,q}$. Then the signal-dependent interference vector covariance matrix can be expressed as

$$\boldsymbol{\Sigma}_{\mathbf{y}_s}(\mathbf{s}) = \mathbb{E}\left[\mathbf{y}_s\mathbf{y}_s^H\right] = \sum_{q=1}^{Q} \left(\mathbf{J}^{r_q} \otimes \mathbf{V}(\theta_q)\right) \left[\left(\mathbf{s}\mathbf{s}^H\right) \odot \boldsymbol{\Xi}_q\right] \left(\mathbf{J}^{r_q} \otimes \mathbf{V}(\theta_q)\right)^H, \tag{3.13}$$

where $\mathbf{J}^r \in \mathbb{C}^{K\times K}$ denotes the shift matrix and $\boldsymbol{\Xi}_q = \sigma_q^2 \boldsymbol{\Phi}_{\varepsilon_q}^{\bar{f}_{d,q}} \otimes \boldsymbol{\Upsilon}_{M_{t,R}}$, with

$$\boldsymbol{\Phi}_{\varepsilon_q}^{\bar{f}_{d,q}}(k_1,k_2) = e^{j2\pi\bar{f}_{d,q}(k_1-k_2)} \frac{\sin\left[\pi\varepsilon_q(k_1-k_2)\right]}{\pi\varepsilon_q(k_1-k_2)}, \quad \forall (k_1,k_2) \in \{1,\dots,K\}^2 \tag{3.14}$$

and $\boldsymbol{\Upsilon}_{M_{t,R}} = \mathbf{1}_{M_{t,R}}\mathbf{1}_{M_{t,R}}^T$, $\mathbf{1}_{M_{t,R}} = [1,1,\dots,1]^T \in \mathbb{C}^{M_{t,R}\times 1}$.

For the signal-independent interference generated by the communications system, we assume that Gaussian space–time random coding is undertaken with the code block length K: we assume $\left[\mathbf{x}^T(1),\dots,\mathbf{x}^T(K)\right]^T \sim \mathcal{CN}\left(\mathbf{0},\tilde{\mathbf{X}}\right)$ [Zhang et al., 2016]; hence, the covariance matrix of the signal-independent interference vector $\mathbf{x}_C$ has a structure very similar to that of the signal-dependent component (3.13) and (3.14). Here, we assume that the Doppler shift of the communication signal has been previously estimated through pilot symbols and is known. Denoting $f'_{d,p}$ such a shift, we can define $\boldsymbol{\Xi}_C = \boldsymbol{\Psi}^{f'_{d,p}} \otimes \boldsymbol{\Upsilon}_{M_{t,C}}$, in which

$$\boldsymbol{\Psi}^{f'_{d,p}}(k_1,k_2) = e^{j2\pi f'_{d,p}(k_1-k_2)}, \quad \forall (k_1,k_2) \in \{1,\dots,K\}^2, \tag{3.15}$$

and $\boldsymbol{\Upsilon}_{M_{t,C}} = \mathbf{1}_{M_{t,C}}\mathbf{1}_{M_{t,C}}^T$ with $\mathbf{1}_{M_{t,C}} = [1,1,\dots,1]^T \in \mathbb{C}^{M_{t,C}\times 1}$. As a result, the communication interference covariance matrix has a form similar to the matrix in (3.13) – namely,

$$\boldsymbol{\Sigma}_{\mathbf{y}_c}(\tilde{\mathbf{X}}) = \sum_{p=1}^{P} \left(\mathbf{J}^{r_p} \otimes \mathbf{G}_{1,p}\right) \left[\tilde{\mathbf{X}} \odot \boldsymbol{\Xi}_C\right] \left(\mathbf{J}^{r_p} \otimes \mathbf{G}_{1,p}\right)^H. \tag{3.16}$$

According to (3.13) and (3.16), once again, the second-order statistical characterizations of the interference require the knowledge of the interference power and the Doppler parameters for any interference path. Such information can be obtained via a cognitive paradigm [Haykin, 2006; De Maio et al., 2009; Mahal et al., 2017; Filo et al., 2009]. Specifically, we assume that the system exploits a dynamic environmental database, including a geographical information system (GIS), meteorological information, tracking files, and interference models. As to the communications system, if the channel matrices $\tilde{\mathbf{G}}_{2,m}\left(\mu_{d,m}\right)$ in (3.12) are available, the conditional covariance given $\tilde{\mathbf{G}}_{2,m}\left(\mu_{d,m}\right)$ of interference plus noise can be expressed as

$$\mathbf{R}_{\text{Cin}} = \sum_{m=1}^{M} \tilde{\mathbf{G}}_{2,m}\left(\mu_{d,m}\right)\left(\mathbf{s}\mathbf{s}^H\right)\left(\tilde{\mathbf{G}}_{2,m}\left(\mu_{d,m}\right)\right)^H + \sigma_C^2\mathbf{I}_{M_{r,C}K}. \tag{3.17}$$

Thus, we obtain the covariance matrices of the radar clutter, of the interference due to the communications system, and of the interference due to the radar system. In Section 3.2.2, we propose a joint design of the communication and radar transceiver.

3.2.1.4 Static Targets

In this subsection, we account for possible inequality in the number of pulses, and, for simplicity, we consider the case of static targets: all Doppler shifts are zero, i.e. $f_{d,0} = 0$ in (3.1) and $f_{d,q} = 0$ in (3.5). The radar waveform in one transmit period is assumed such that the first segment is nonzero and the rest is zero (see also Figure 3.2), i.e. $\mathbf{s}(k) \neq 0$ for $k \leq \tilde{K}$ and $\mathbf{s}(k) = 0$ for $\tilde{K}+1 \leq k \leq K$. Assuming that the radar is blind when the probing signal is transmitted, we can define $K - 2\tilde{K} + 1$ range gates, one for each resolvable delay $\tau \in \{\tilde{K},\dots,K-\tilde{K}\}$, as shown in Figure 3.2; each

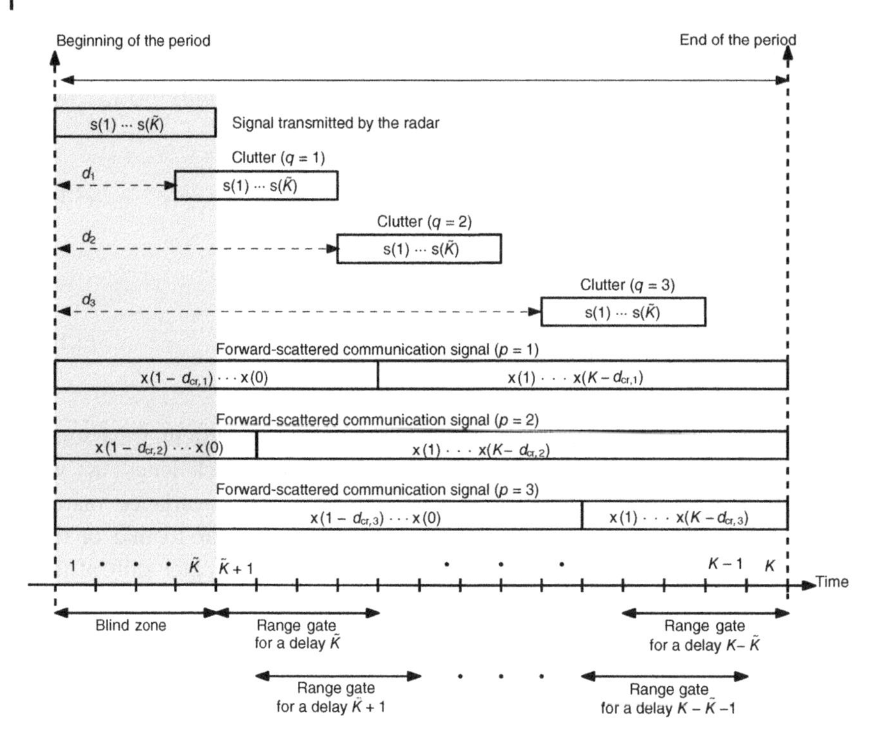

Figure 3.2 Signal received by the radar over the first transmit period.

delay τ corresponds to a unambiguous radar range of $c\tau/(2B)$. We denote by $\mathcal{T} \subseteq \{\tilde{K}, \dots, K-\tilde{K}\}$ the set of range gates inspected by the radar. Typically, $\mathcal{T} = \{\tilde{K}, \dots, K-\tilde{K}\}$ in a search mode, while $\mathcal{T}$ only contains few delays in a confirmation or tracking mode. As to the communications system, let

$$\tilde{\mathbf{x}}(p) = \text{vec}\left(\left[\mathbf{x}((p-1)K+1+\tau_{\text{off}}), \dots, \mathbf{x}(pK+\tau_{\text{off}})\right]\right) \tag{3.18}$$

be the data symbols spanning the pth period, where $\tau_{\text{off}} \in \{0,1,\dots,K-1\}$ is the offset between the beginning of one considered period at the radar and communication TX. For simplicity, we assume $\tau_{\text{off}} = 0$ and that $\tilde{\mathbf{x}}(p)$ is a sequence of independent identically-distributed vectors with covariance matrix $\tilde{\mathbf{X}} = \mathbb{E}\left[\tilde{\mathbf{x}}(p)\tilde{\mathbf{x}}^H(p)\right]$; otherwise stated, space–time codewords over different periods are independent and drawn from the same codebook.

To simplify the exposition and without loss of generality, we analyze here the radar operation during the first period spanned by the time indexes $\{1,\dots,K\}$. We assume the presence of Q uncorrelated scatterers reflecting back the radar signal; the qth scatterer is at the azimuth $\theta_q \in \Theta$ and presents a round-trip delay $d_q \in \{1,\dots,K-1\}$, for $q = 1,\dots,Q$. Hence, the clutter at the range gate $\tau \in \mathcal{T}$, corresponding to $\{f_{d,q}\} = 0$ in (3.6), can be written as

$$\mathbf{y}_s(\tau) = \sum_{q=1}^{Q} \alpha_q \left(\mathbf{J}_K^T(\tau-d_q) \otimes \mathbf{V}(\theta_q)\right)\mathbf{s}. \tag{3.19}$$

Also, for the presence of P uncorrelated scatterers reflecting the communication signal toward the radar, we define the delay $d_{\text{cr},p} \in \{0,\dots,\tilde{K}-1\}$ with respect to the beginning of the period, for

$p = 1, \dots, P$. Hence, for the communication interference, corresponding to $\{f'_{d,p}\} = 0$ in (3.8), we have[2]

$$\mathbf{y}_c(\tau) = \sum_{p=1}^{P} \beta_1(p) \left(\mathbf{I}_{\tilde{K}} \otimes \mathbf{G}_{1,p}\right) \mathrm{vec}([\mathbf{x}(\tau - d_{\mathrm{cr},p} + 1) \cdots \mathbf{x}(\tau - d_{\mathrm{cr},p} + \tilde{K})]). \tag{3.20}$$

For the same range gate, a different segment of the communication signal is observed along reflected paths with different delays; likewise, for the same reflected path, a different segment of the communication signal is observed in different range gates. Also, notice $\tau - d_{\mathrm{cr},p} \in \{\tilde{K} - (K-1), \dots, K - \tilde{K}\}$, so that the data segment $\left[\mathbf{x}(\tau - d_{\mathrm{cr},p} + 1), \dots, \mathbf{x}(\tau - d_{\mathrm{cr},p} + \tilde{K})\right]$ spans the first (current) codeword if $\tau - d_{\mathrm{cr},p} \in \{0, \dots, K - \tilde{K}\}$, the zeroth (previous) codeword if $\tau - d_{\mathrm{cr},p} \in \{\tilde{K} - (K-1), \dots, -\tilde{K}\}$, and two consecutive codewords (the current and previous ones) if $\tau - d_{\mathrm{cr},p} \in \{-\tilde{K} + 1, \dots, -1\}$. The radar RX performs joint spatial beamforming and temporal pulse compression by processing $\mathbf{y}(\tau)$ with a filter $\mathbf{w}(\tau) = \mathrm{vec}(\mathbf{W}(\tau))$, with $\mathbf{W}(\tau) = [\mathbf{w}_1(\tau), \dots, \mathbf{w}_{\tilde{K}}(\tau)] \in \mathbb{C}^{M_{r,R} \times \tilde{K}}$; the corresponding output SINR is

$$\mathrm{SINR}(\tilde{\mathbf{X}}, \mathbf{s}, \mathbf{w}(\tau), \tau) = \frac{\sigma_{\alpha,0}^2 \left|\mathbf{w}(\tau)^H \mathbf{V}_0 \mathbf{s}\right|^2}{\mathbf{w}(\tau)^H \boldsymbol{\Sigma}(\tilde{\mathbf{X}}, \mathbf{s}, \tau) \mathbf{w}(\tau)}, \tag{3.21}$$

where $\boldsymbol{\Sigma}(\tilde{\mathbf{X}}, \mathbf{s}, \tau) = \boldsymbol{\Sigma}_s(\mathbf{s}, \tau) + \boldsymbol{\Sigma}_c(\tilde{\mathbf{X}}, \tau) + \sigma_R^2 \mathbf{I}_{M_{r,R}K}$ is the interference-plus-noise covariance matrix. $\boldsymbol{\Sigma}_s(\mathbf{s}, \tau) = \mathbb{E}\left[\mathbf{y}_s(\tau)\mathbf{y}_s^H(\tau)\right]$ and $\boldsymbol{\Sigma}_c(\tilde{\mathbf{X}}, \tau) = \mathbb{E}\left[\mathbf{y}_c(\tau)\mathbf{y}_c^H(\tau)\right]$ are the covariance matrices of the clutter and of the communication interference, respectively, which can be derived similar to (3.13) and (3.16). Moreover, the covariance matrix of the data segment $\mathrm{vec}\left(\left[\mathbf{x}(\ell + 1) \cdots \mathbf{x}(\ell + \tilde{K})\right]\right)$ can be expressed as

$$\tilde{\mathbf{X}}_c(\ell) = \begin{cases} [\tilde{\mathbf{X}}]_{(K+\ell)M_{t,C}+1:(K+\ell+\tilde{K})M_{t,C}}, & \ell = \tilde{K} - (K-1), \dots, -\tilde{K}, \\ \begin{bmatrix} [\tilde{\mathbf{X}}]_{(K+\ell)M_{t,C}+1:KM_{t,C}} & \mathbf{O}_{(K+\ell)M_{t,C}+1,(\ell+\tilde{K})M_{t,C}}, \\ \mathbf{O}_{(\ell+\tilde{K})M_{t,C},-\ell M_{t,C}} & [\tilde{\mathbf{X}}]_{1:(\ell+\tilde{K})M_{t,C}}, \end{bmatrix}, & \ell = -\tilde{K} + 1, \dots, -1, \\ [\tilde{\mathbf{X}}]_{\ell M_{t,C}+1:(\ell+\tilde{K})M_{t,C}}, & \ell = 0, \dots, K - \tilde{K}, \end{cases} \tag{3.22}$$

where $[\tilde{\mathbf{X}}]_{a:b}$ is the submatrix of $\tilde{\mathbf{X}}$ obtained by maintaining only the rows from a to b and the columns from a to b, and $\mathbf{O}_{a,b}$ is the all-zero matrix with a rows and b columns [Grossi et al., 2020].

In this chapter, we assume that σ_R^2 is known; also, the radar optimization is undertaken for some nominal values of the target response $\sigma_{\alpha,0}^2$ and of the clutter parameters $\{\theta_q, d_q, \sigma_{\alpha,q}^2\}_{q=1}^{Q}$, which typically specify a worst-case operating scenario. Finally, we assume that the interference information change over a time scale much longer than one transmit period; hence, they can be estimated through pilot signals [Haykin, 2006; De Maio et al., 2009; Mahal et al., 2017; Filo et al., 2009; Liu et al., 2020].

The signal at the communication RX over one transmit period is (see also Figure 3.3)

$$\mathbf{z}(k) = \mathbf{H}\mathbf{x}(k) + \mathbf{z}_r(k) + \mathbf{z}_n(k). \tag{3.23}$$

Assuming the presence of M uncorrelated scatterers reflecting the radar signal toward the communication RX, the radar interference can be expressed as $\mathbf{z}_r(k) = \sum_{m=1}^{M} \gamma_m \mathbf{a}_r(\phi_{\mathrm{rc},m}) \mathbf{v}_t^T(\varphi_{\mathrm{rc},m}) \mathbf{s}(k - d_{\mathrm{rc},m})$, where γ_m accounts for the response of the m-th scatterer and is a zero-mean complex random variable with variance $\sigma_{\gamma,m}^2$, and $\varphi_{\mathrm{rc},m}$, $\phi_{\mathrm{rc},m}$, and $d_{\mathrm{rc},m} \in \{0, \dots, K-1\}$ are the angles of departure and arrival and the delay offset (with respect to the beginning of the transmit period), respectively, of the ray hitting the mth scatterer. Notice here that the radar interference is time-varying over the processed period, so that the K spatial symbols composing

---

2 The direct link between the communication TX and the radar RX, if present, is also included in (3.20). Scatterers reflecting both the radar and the communication signals are included in both (3.19) and (3.20).

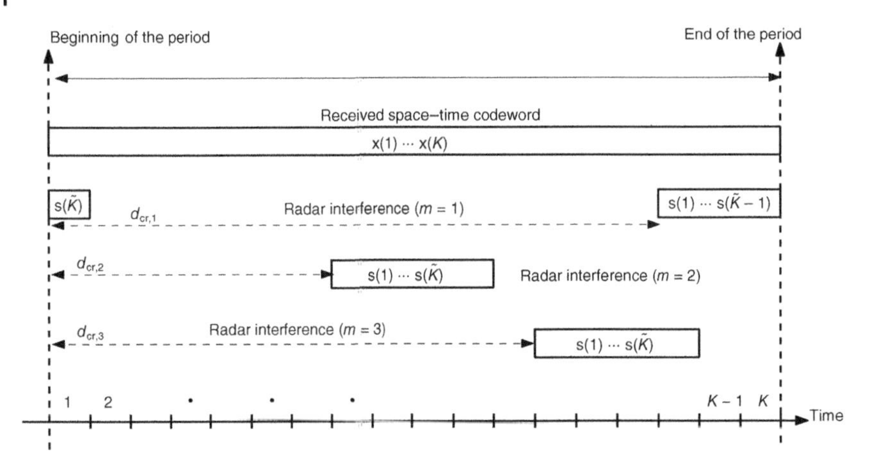

Figure 3.3 Signal received by the communication RX over the first PRI.

the communication codeword will experience a different channel quality. Let $\mathbf{e}(k) \in \mathbb{C}^K$ be a vector whose kth entry is 1, while all the other entries are zero, for $k = 1, \ldots, K$; also, let $\boldsymbol{\Delta} = [\mathbf{I}_{\tilde{K}}\ \mathbf{0}_{\tilde{K},K-\tilde{K}}] \in \mathbb{C}^{\tilde{K}\times K}$ and $\boldsymbol{\Delta}_m = \boldsymbol{\Delta}\left(\mathbf{J}_K(-d_{\text{rc},m}) + \mathbf{J}_K(K - d_{\text{rc},m})\right) \in \mathbb{C}^{\tilde{K}\times K}$. Then $\mathbf{s}(k - d_{\text{rc},m}) = \mathbf{S}\boldsymbol{\Delta}_m\mathbf{e}(k) = \left(\mathbf{e}(k)^T\boldsymbol{\Delta}_m^T \otimes \mathbf{I}_{M_{t,R}}\right)\mathbf{s}$ and the radar interference can be rewritten as $\mathbf{z}_r(k) = \sum_{m=1}^{M}\gamma_m\mathbf{L}_m(k)\mathbf{s}$, where $\mathbf{L}_m(k) = \mathbf{e}(k)^T\boldsymbol{\Delta}_m^T \otimes \mathbf{a}_r(\phi_{\text{rc},m})\mathbf{v}_t^T(\varphi_{\text{rc},m}) \in \mathbb{C}^{M_{r,C}\times\tilde{K}M_{t,R}}$. Finally, we model $\mathbf{z}_n(1), \ldots, \mathbf{z}_n(K)$ as independent identically distributed, circularly symmetric Gaussian vectors with covariance matrix $\sigma_C^2\mathbf{I}_{M_{r,C}}$. Upon defining $\mathbf{z} = \text{vec}([\mathbf{z}(1), \ldots, \mathbf{z}(K)])$, $\mathbf{z}_r = \text{vec}\left([\mathbf{z}_r(1), \ldots, \mathbf{z}_r(K)]\right)$, $\mathbf{z}_n = \text{vec}\left([\mathbf{z}_n(1), \ldots, \mathbf{z}_n(K)]\right)$, we can write

$$\mathbf{z} = \tilde{\mathbf{H}}\tilde{\mathbf{x}}(1) + \underbrace{\sum_{m=1}^{M}\gamma_m\mathbf{L}_m\mathbf{s}}_{\mathbf{z}_r} + \mathbf{z}_n \in \mathbb{C}^{\tilde{K}M_{r,C}}. \tag{3.24}$$

3.2.2 Joint Design Formulations

In this subsection, we investigate the spectrum sharing system under the framework of joint optimization. We briefly discuss the spectrum sharing formulations for moving targets and static targets. The corresponding optimization algorithms are proposed based on alternating optimization.

3.2.2.1 Joint Design for Moving Targets

The optimization entails the constrained maximization of the average SINR of the radar. At the communication side, we define a measure of the achievable rate as the system rate per channel use and per DOF were the interference Gaussian, i.e.

$$\text{MI}\left(\tilde{\mathbf{X}}, \mathbf{s}\right) = \frac{1}{K}\log\det\left(\mathbf{I}_{KM_{r,C}} + \mathbf{R}_{\text{Cin}}^{-1}\tilde{\mathbf{H}}\tilde{\mathbf{X}}\tilde{\mathbf{H}}^H\right) \tag{3.25}$$

nats per channel use[3] ($\mathbf{R}_{\text{Cin}}$ can be calculated by (3.17)), on which we impose two constraints: (i) on the maximum transmit power $\mathcal{P}_C$ whereby $\text{Tr}\left(\tilde{\mathbf{X}}\right) \leq \mathcal{P}_C$ and (ii) on the guaranteed communication performance whereby the MI is no less than a given threshold MI_0. The radar operates under two constraints: (i) on the maximum transmit power $\mathcal{P}_R$ whereby $\|\mathbf{s}\|^2 \leq \mathcal{P}_R$ and (ii) on the similarity

3 Throughout this chapter, nats are used as the information units and log denotes natural logarithm.

of the transmit code to a reference signal $\mathbf{s}_0$ whereby $\|\mathbf{s}-\mathbf{s}_0\|^2 \leq \xi$, where ξ rules the extent of the similarity. The reference signal is normalized such that $\|\mathbf{s}_0\|^2 = \mathcal{P}_R$ and possesses some desired properties in terms of sidelobe levels and envelope constancy [De Maio et al., 2009]. Summing up, the full constrained optimization problem can be stated as

$$\begin{aligned} \max_{\mathbf{s},\mathbf{w},\tilde{\mathbf{X}}\succeq \mathbf{0}} \quad & \frac{|\mathbf{w}^H \tilde{\mathbf{V}}(f_{d,0},\theta_0)\mathbf{s}|^2}{\mathbf{w}^H \boldsymbol{\Sigma}_{\mathbf{y}_i}(\mathbf{s})\mathbf{w} + \mathbf{w}^H \boldsymbol{\Sigma}_{\mathbf{x}_C}(\tilde{\mathbf{X}})\mathbf{w} + \sigma_R^2 \mathbf{w}^H\mathbf{w}} \\ \text{s.t.} \quad & \frac{1}{K}\log\det\left(\mathbf{I}_{KM_{r,C}} + \mathbf{R}_{\text{Cin}}^{-1}\tilde{\mathbf{H}}\tilde{\mathbf{X}}\tilde{\mathbf{H}}^H\right) \geq \text{MI}_0, \\ & \|\mathbf{s}-\mathbf{s}_0\|^2 \leq \xi, \\ & \text{Tr}\left(\tilde{\mathbf{X}}\right) \leq \mathcal{P}_C, \\ & \|\mathbf{s}\|^2 \leq \mathcal{P}_R. \end{aligned} \tag{3.26}$$

Problem (3.26) is nonconvex with respect to the triplet $(\mathbf{s},\mathbf{w},\tilde{\mathbf{X}})$, and its solution entails a prohibitive complexity. In Qian et al. [2018b], an alternating-optimization-based iterative method is developed for solving (3.26), whereby each variable is solved for with the other two variables fixed. Here, we show some joint design results through a simulation example.

In the simulation, the moving MIMO radar system consists of $M_{t,R} = 3$ transmit and $M_{r,R} = 8$ receive antennas with half wavelength element separation. The MIMO communications system consists of $M_{t,C} = 3$ transmit elements and $M_{r,C} = 3$ receive elements with half wavelength element separation. We consider the Barker code of length $K = 4$ as the radar reference waveform $\mathbf{s}_0$. We assume that the signal-to-noise ratio (SNR) is 10 dB, with the normalized Doppler frequency $f_{d,0} = 0.35$. For each range-azimuth clutter bin, we consider a clutter-to-noise ratio (CNR) of 20 dB, and the Doppler uncertainty $\epsilon_q = 0.04$. The clutters are uniformly generated in the azimuth angular sector $\left[-\frac{\pi}{2},\frac{\pi}{2}\right]$, and the number of azimuth cells in each ring is 90. We consider the range ring $r_q = 0$ for all range-azimuth ground clutter bins. For the interferences between radar and communications system, we set $P = M = 6$ and the communication interference range bin is zero. For the channel matrix $\mathbf{G}_1$, we select angle parameters $\theta_p = \varphi_p$ uniformly at random in the range $\left[-30^\circ,-10^\circ\right]$, the corresponding Doppler shift belongs to the interval $[-0.35,-0.2]$, and the INR equals 15 dB. For the channel matrix $\mathbf{G}_2$, we select angle parameters $\varphi_m = \theta_m$ uniformly at random in the range $\left[10^\circ,30^\circ\right]$, the corresponding Doppler shift belongs to the interval $[0.15,0.3]$, and the INR equals 25 dB. For simplicity, we assume that $\mathcal{P}_R = \mathcal{P}_C = 1$, $\xi = 1.8$ and $\sigma_R^2 = \sigma_C^2 = 0.001$, respectively. The entries of $\mathbf{H}$ are independently generated following the distribution of $\mathcal{CN}(0,1)$. We set the minimum achievable MI for the communications system $\text{MI}_0 = 6$.

Figure 3.4a,b depicts the cross-ambiguity function (CAF) [Karbasi et al., 2015] of the optimized waveforms in the 3rd and 30th iteration, respectively. Specifically, we plot the normalized CAF with

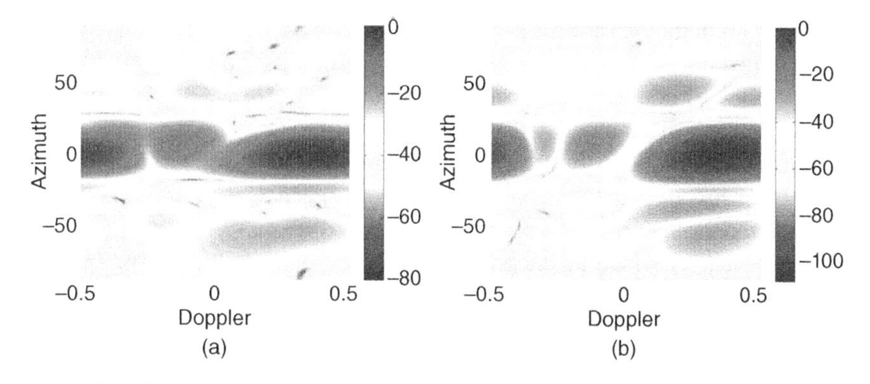

Figure 3.4 Doppler-azimuth plane of CAF at $r_q = 0$ for $\xi = 1.8$ for iteration number $n = \{3,30\}$.

respect to the target response, that is, $g\left(\mathbf{s},\mathbf{w},\theta,f_d\right) = \left|\mathbf{w}^H\left(\mathbf{J}^r \otimes \mathbf{I}_{M_{r,R}}\right)\tilde{\mathbf{V}}\left(f_d,\theta\right)\mathbf{s}\right|^2$, for $r = 0$ (the target ring) [Qian et al., 2017]. In Figure 3.4, we can observe the mainlobes at the target location. In addition, it can be seen that the clutter is coupled in spatial and temporal domains, and its clutter ridge is diagonally distributed in the sidelooking geometry of the radar. The $g\left(\mathbf{s},\mathbf{w},\theta,f_d\right)$ exhibits smaller values in the clutter region as the iteration number increases. This trend highlights that the co-design is able to suitably shape the CAF in order to suppress the clutter. Note that the $\tilde{\mathbf{X}}$ design can effectively reduce the interference energy at the radar receive antennas such that the notch is not guaranteed at the communication interference regions; that is, the $g\left(\mathbf{s},\mathbf{w},\theta,f_d\right)$ is not guaranteed to exhibit smaller values in the communication interference region as the iteration number increases.

3.2.2.2 Joint Design for Static Targets

This formulation exploits constraints similar to those in (3.26). At the communications system side, assuming that the data symbols $\tilde{\mathbf{x}}(1)$ and the channel coefficients $\{\gamma_m\}_{m=1}^M$ are circularly symmetric Gaussian, the MI between $\mathbf{z}$ and $\tilde{\mathbf{x}}(1)$ is (3.25) with $\mathbf{R}_{\text{Cin}} = \sum_{m=1}^M \sigma_{\gamma,m}^2 \mathbf{L}_m \mathbf{s}\mathbf{s}^H \mathbf{L}_m^H + \sigma_C^2 \mathbf{I}_{KM_{r,C}}$ [Biglieri et al., 2007; Qian et al., 2020]. At the radar side, we consider the weighted harmonic mean of the observed SINRs since the harmonic mean of a list of numbers is strongly biased toward the smallest element. Assuming $|\mathbf{w}(\tau)^H \mathbf{V}_0 \mathbf{s}| \neq 0$ for $\tau \in \mathcal{T}$, the considered figure of merit is

$$\text{SINR}_h(\tilde{\mathbf{X}}, \mathbf{s}, \mathbf{w}(\mathcal{T})) = \left(\sum_{\tau \in \mathcal{T}} \frac{\omega(\tau)}{\text{SINR}(\tilde{\mathbf{X}}, \mathbf{s}, \mathbf{w}(\tau), \tau)} \right)^{-1} \tag{3.27}$$

where $\{\omega(\tau)\}_{\tau\in\mathcal{T}}$ are positive weights, normalized to have $\sum_{\tau\in\mathcal{T}}\omega(\tau) = 1$, and $\mathbf{w}(\mathcal{T})$ is a shorthand notation to indicate $\{\mathbf{w}(\tau)\}_{\tau\in\mathcal{T}}$. The design variables are $\mathbf{s}$, $\mathbf{w}(\tau)$, and $\tilde{\mathbf{X}}$. We aim at maximizing SINR_h under a constraint on the minimum required MI, say MI_0; in particular,

$$\begin{aligned} \max_{\tilde{\mathbf{X}}\succeq 0, \mathbf{s}, \mathbf{w}(\mathcal{T})} \quad & \left(\sum_{\tau\in\mathcal{T}} \frac{\omega(\tau)}{\text{SINR}(\tilde{\mathbf{X}}, \mathbf{s}, \mathbf{w}(\tau), \tau)} \right)^{-1}, \\ \text{s.t.} \quad & \text{MI}(\tilde{\mathbf{X}}, \mathbf{s}) \geq \text{MI}_0, \\ & \text{Tr}(\tilde{\mathbf{X}}) \leq \mathcal{P}_C, \\ & \|\mathbf{s} - \mathbf{s}_0\|^2 \leq \xi, \\ & \|\mathbf{s}\|^2 \leq \mathcal{P}_R, \end{aligned} \tag{3.28}$$

where $\{\omega(\tau)\}_{\tau\in\mathcal{T}}$ are positive weights, normalized to have $\sum_{\tau\in\mathcal{T}}\omega(\tau) = 1$, and $\mathbf{w}(\mathcal{T})$ is a shorthand notation to indicate $\{\mathbf{w}(\tau)\}_{\tau\in\mathcal{T}}$. As in Section 3.2.2.1, to solve this nonconvex problem we may resort to an alternating optimization approach, whereby the variables $\tilde{\mathbf{X}}$, $\mathbf{w}(\mathcal{T})$, and $\mathbf{s}$ are optimized one at a time in an iterative fashion [Razaviyayn et al., 2013].

The simulation parameters remain the same as in Section 3.2.2.1, unless explicitly stated otherwise, as follows. For the static targets, the radar operates with a transmit period of $K = 40$, $\tilde{K} = 4$, and six consecutive range gates are controlled, as specified by set $\mathcal{T} = \{15, 16, \dots, 20\}$. We assume that the target has equal range weights, namely, $\omega(\tau) = 1/|\mathcal{T}|$. At the design stage, we employ the nonuniform interference and unequal transmitted power environment. We assume the presence of $Q = 6$ scatterers reflecting back the probing signal, with $d_q = 14 + q$ and θ_q randomly chosen in $\left(25^\circ 50^\circ\right)$. Also, we assume the average transmit power $\mathcal{P}_R = 100\mathcal{P}_C$. Figure 3.5 reports the value of the objective function in (3.26) provided by the proposed solution versus $\text{INR}_R = \text{INR}_C$ (panel a) and the index of the inspected range cell in $\mathcal{T}$ (panel b). As a benchmark, we consider a *disjoint design* whereby the communication code $\tilde{\mathbf{X}}$ is matched to the channel matrix $\tilde{\mathbf{H}}$ by standard water-filling and the radar selects $\mathbf{s} = \mathbf{s}_0$ and computes the receive filter ignoring the presence of the communication interference. Also, we consider a solution where $\tilde{\mathbf{X}}$ is chosen in the disjoint

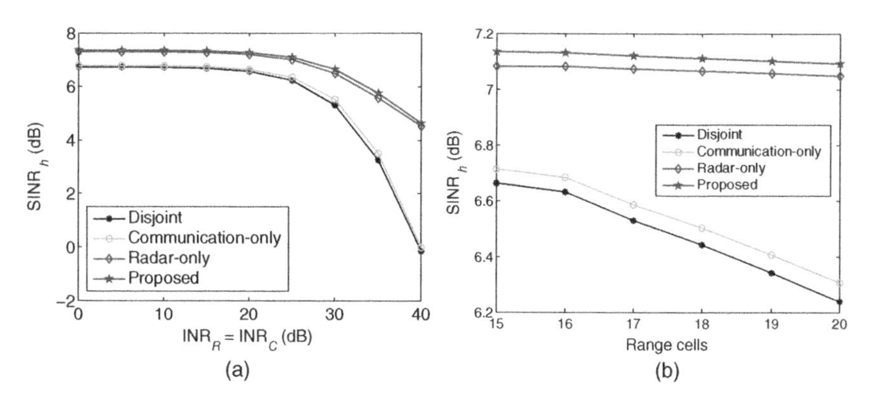

Figure 3.5 (a) $SINR_h$ versus $INR_R = INR_C$ when $CNR_R = 30$ dB; (b)$SINR_h$ versus the index of the inspected range cell in $\mathcal{T}$ when $INR_R = INR_C = 25$ dB and $CNR_R = 30$ dB.

design and the problem in (3.26) is only solved over $\mathbf{s}$ and $\mathbf{w}(\mathcal{T})$ (referred to as *radar-only design*) and a solution where the $\mathbf{s}$ and $\mathbf{w}(\mathcal{T})$ are chosen in the disjoint design and the problem in (3.26) is only solved over $\tilde{\mathbf{X}}$ (referred to as *communication-only design*). It should be noted that the MI constraint should be active at the optimal point. This means that the achieved MI is always MI_0, as we can always shrink $\tilde{\mathbf{X}}$ until the MI reduces to MI_0. It is seen that the proposed solution is robust to the external interference, incurring only a small $SINR_h$ degradation in the presence of severe clutters and a strong coupling among the systems; it safeguards the worst-case radar SINR over the inspected region.

3.3 Interference Suppression in Uncoordinated Joint Radar–Communications Systems

In this section, we turn to radar target parameter estimation in an uncoordinated spectrum sharing scenario. Here, unlike the coordinated setup of Section 3.2, the radar and communication TX waveforms and the radar RX filter are fixed. Such a scenario arises, for example, when regulation or the use of legacy systems prohibits changes to the transmit waveforms, thus rendering the joint design approach of Section 3.2 infeasible. Instead, uncoordinated spectrum sharing may be enabled through interference-robust processing algorithms that exploit prior knowledge of the interference structure.

3.3.1 System Description

We consider an uncooperative scenario where a stepped-frequency radar tasked with sparse angle-range-Doppler imaging operates amid communications that overlap with portions of the radar spectrum. Employing a multi-frame processing architecture, the radar transmits a series of simple pulse trains to obtain a set of low-resolution radar measurements with which to synthesize an image. Although the total radar bandwidth is large ($\sim$ 1 GHz is typical of stepped-frequency radars), by virtue of the pulse-by-pulse processing only the communication signals that spectrally overlap with a given pulse interfere in the recording of the pulse echo. Moreover, since communication signals tend to be sparse in the frequency domain, owing to periods of low activity or otherwise underutilized spectrum, the interference manifests as sparse noise in the radar measurements.

Applicable scenarios lie between two extremes. At one (Figure 3.6a), the total radar bandwidth overlaps with multiple communication carriers and the radar frequency step is on the order of

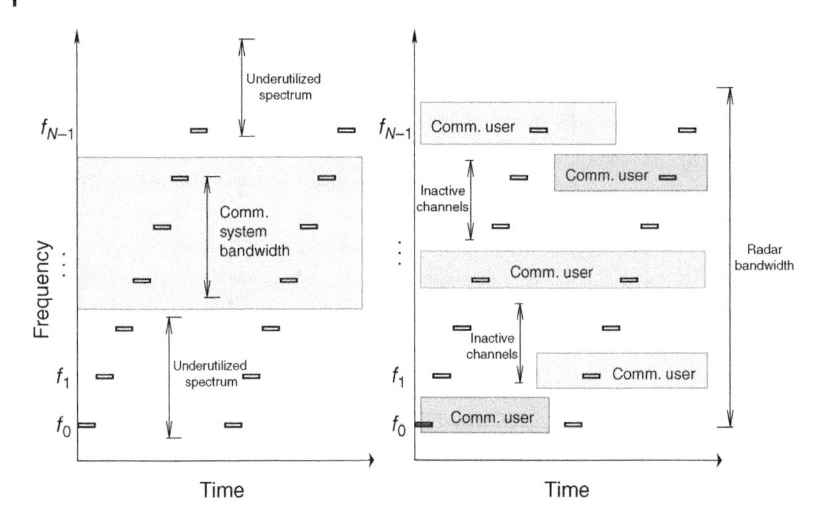

Figure 3.6 Frequency occupation versus time for two representative spectrum sharing scenarios. The black strips indicate the spectrum occupied by the radar system over time; the colored strips indicate the spectrum occupied by the communications system. Only the overlapping regions cause interference to the radar.

the communication carrier bandwidth; for example, stepped-frequency radars may have a step size of 20 MHz [Counts et al., 2007], while the maximum long-term evolution (LTE) bandwidth is 20 MHz [3rd Generation Partnership Project, 2017] and in sub-6GHz 5G the maximum channel bandwidth is 100 MHz [Kim et al., 2019]. At the other (Figure 3.6b), the radar overlaps with a single communication carrier; the carrier comprises subchannels sized on the order of the radar frequency step size that are assigned to opportunistic communications users. For example, 5G employs channels with bandwidths in the hundreds of megahertz to a few gigahertz [Shafi et al., 2017], and stepped-frequency radars often have a sweep bandwidth on that order. In either case, the key property that enables the radar to coexist is that significant portions of spectrum tend to be underutilized [Haykin, 2005; Kim et al., 2019]; in light of this, the interference induced by the active portions can be mitigated.

3.3.1.1 Radar Signal Model

Consider a frequency-stepped pulsed MIMO radar with $M_{t,R}$ transmitters and $M_{r,R}$ receivers (see Table 3.2). Each of the transmitted waveforms u^p, $p = 1, \dots, M_{t,R}$, has duration T seconds and the waveforms are assumed to be approximately mutually incoherent (see (3.45)). The scene is illuminated by N_d trains of N pulses: within the mth train, the nth pulse emitted by the pth transmit antenna is given by

$$s^p(m, n, t) = u^p(t - nT_r - mNT_r)\exp(j2\pi f_n t), \tag{3.29}$$

where t is continuous time, $0 \leq m \leq N_d - 1, 0 \leq n \leq N - 1, 1 \leq p \leq M_{t,R}; f_n = n\Delta f + f_0$, where f_0 is the lowest carrier frequency, and $N\Delta f$ is the overall bandwidth. Each pulse echo recording length is $T_r \gg T$ seconds, the pulse repetition interval (PRI). A complete observation consists of N_dN PRIs.

We consider a scene of L scatterers with scattering coefficients x_i and radial velocities v_i. The signal received by the qth receiver, $q = 1, \dots, M_{r,R}$, is

$$r^q(n, t) = \sum_{m=0}^{N_d-1}\sum_{p=1}^{M_{t,R}}\sum_{i=1}^{L} x_i s^p(m, n, t - \tau_i^{pq}(t)), \tag{3.30}$$

where

$$\tau_i^{pq}(t) = \frac{2v_i}{c}t + \tau_i + \delta_i^p + \varepsilon_i^q \tag{3.31}$$

Table 3.2 Symbol definitions for stepped-frequency MIMO radar.

Symbol	Definition	Symbol	Definition
N	Number of frequency steps	t	Continuous fast-time, absolute
N_d	Number of sweeps	m	Sweep index
$M_{t,R}$	Number of transmitters	n	Pulse index within sweep
$M_{r,R}$	Number of receivers	i	Scatterer index
f_0	Start frequency	L	Number of scatterers
Δf	Frequency step size	x_i	Scattering coeff.
f_n	$f_0 + n\Delta f,\ 0 \leq n \leq N-1$	τ_i^{pq}	Absolute delay, (p, q) TX/RX pair
u^p	Transmitter p's pulse envelope	δ_i^p	Marginal delay, pth TX
s^p	Transmitter p's waveform	ε_i^q	Marginal delay, qth RX
r^q	Radar return at receiver q	τ_i	Absolute delay, reference TX/RX pair
T	Pulse duration (all transmitters)	$\overline{\tau}_i(k)$	Delay offset, kth range cell
T_r	Pulse-repetition interval	v_i	Radial velocity
		θ_i	Direction coordinates

is the ith scatterer's delay; δ_i^p and ε_i^q are the marginal delays due to array geometry associated with antenna pair (p, q); and τ_i is the absolute round-trip delay observed by a reference antenna pair during the first PRI. We assume the velocities are constant throughout the series of sweeps.

We further make the following assumptions:

- The range variation throughout the series of sweeps is negligible with respect to the range resolution of each pulse:

$$2v_i N_d N T_r / c \ll T.$$

- The array element spacing is much smaller than the range resolution granted by the overall transmitted bandwidth:

$$|\delta_i^p + \varepsilon_i^q| \ll \frac{1}{N\Delta f}. \tag{3.32}$$

  This, in turn, constraints f_0 since the spacing is typically a function of wavelength.

  Since the pulse is unsophisticated, $T\Delta f \simeq 1$; hence, (3.32) implies $|\delta_i^p + \varepsilon_i^q| \ll T$, whereby

$$u^p(t - \tau_i^{pq}(t)) \simeq u^p(t - \tau_i). \tag{3.33}$$

In (3.30), the phase of $s^p(m, n, t - \tau_i^{pq}(t))$ is

$$\exp\left(j2\pi(f_0 + n\Delta f)(t - (2v_i t/c + \tau_i + \delta_i^p + \varepsilon_i^q))\right). \tag{3.34}$$

In (3.34), the term $\exp(-j2\pi n\Delta f(\delta_i^p + \varepsilon_i^q))$ may be neglected since, by (3.32), $|n\Delta f(\delta_i^p + \varepsilon_i^q)| \ll 1$, $n = 0, 1, \ldots, N-1$. Under these assumptions, (3.30) becomes

$$r^q(n, t) = \sum_{m=0}^{N_d-1} \sum_{p=1}^{M_{t,R}} \sum_{i=1}^{L} x_i \exp(-j2\pi f_0(\delta_i^p + \varepsilon_i^q)) \times u^p(t - nT_r - mNT_r - \tau_i) \exp\left(j2\pi f_n\left(t - \tau_i - \frac{2v_i}{c}t\right)\right). \tag{3.35}$$

3.3.1.2 Communication Signal Model

Suppose there are N_c carriers that spectrally overlap with the radar band, with center frequencies f_i^C and bandwidths B_i, $i = 0, 1, \ldots, N_c - 1$. Here, "carrier" refers to any communication transmission within the radar band, e.g. a particular block of subcarriers within a communication band, the aggregate transmission over a communication band, etc. Thus, the received communication signal at receiver q has the form

$$s_c^q(t) = \sum_{i=0}^{N_c-1} g_i^q(t) \exp(j2\pi f_i^C t), \tag{3.36}$$

where g_i^q represents the information signal, including the path gain, on carrier i and is a zero-mean random process whose power spectral density G_i^q satisfies

$$G_i^q(f) = 0 \text{ if } |f| > \frac{B_i}{2}. \tag{3.37}$$

SC-FDMA As a concrete example, consider an uplink SC-FDMA system, as specified in the 5G New Radio standard released by 3GPP in December 2017. Suppose the system bandwidth consists of N_s subcarriers with uniform spacing Δf^C and every $K_s \in \mathbb{Z}^+$ consecutive subcarriers are grouped into channels with center frequencies $f_i^C = f_0^C + iK_s\Delta f^C,\ 0 \le i \le N_c - 1$, where f_0^C is the start frequency, each channel has bandwidth $K_s\Delta f^C$, for a total of $N_c = \lfloor N_s/K_s \rfloor$ channels. Users are assigned one or more channels over which to transmit. The signal transmitted over channel i has the form

$$\begin{aligned} g_i^q(t) = \sqrt{\gamma_i} h_i^q \sum_{n_c=-\infty}^{\infty} \sum_{k=0}^{K_s-1} a_{ik}(n_c) u_C(t - n_c T_c) \\ \times \exp\left[j2\pi(f_i^C + k\Delta f^C)t\right], \end{aligned} \tag{3.38}$$

where

- γ_i is the power level assigned to channel i.
- $h_i^q \sim \mathcal{CN}(0, \beta)$ is an i.i.d. channel fading coefficient. A block fading channel model is assumed and K_s is chosen such that $K_s\Delta f^C$ equals the coherence bandwidth ($\sim$ 0.5 MHz) [D'Andrea et al., 2020]. Therefore, each channel i is characterized by a single fading coefficient h_i^q that is statistically independent of all other channels. The variance β accounts for additional user-dependent effects (e.g. path loss and log-normal shadowing) [D'Andrea et al., 2020]. For simplicity, we assume β is the same for all users.
- $\{a_{ik}(n_c) \in \mathbb{C} : 0 \le k \le K_s - 1,\ 0 \le i \le N_c - 1,\ n_c \in \mathbb{Z}\}$ are random variables representing the transmitted symbol sequence, comprising the data and cyclic prefix, with $a_{ik}(n_c)$ transmitted on subcarrier k of channel i during the n_cth data block. In SC-FDMA, the transmitted symbols $a_{ik}(n_c)$, $k = 0, \ldots, K_s - 1$ are the isometric discrete Fourier transform (DFT) coefficients of the original data symbol sequence. We assume the original data symbols adhere to a memoryless modulation format.
- T_c is the block duration (cyclic prefix plus data); for example, in 5G $\Delta f^C \sim 15$ kHz, so $T_c \sim 1/(15\ \text{kHz}) = 66\ \mu\text{s}$.
- $$u_C(t) \triangleq \begin{cases} \sqrt{\frac{1}{T_c}} & 0 \le t \le T_c \\ 0 & \text{otherwise} \end{cases} \tag{3.39}$$

  is the normalized pulse envelope.

3.3.1.3 Signal Processing at Radar Receiver

Receiver q's recording of the nth pulse return, denoted $\tilde{y}^q(n, \cdot)$, takes the form

$$\tilde{y}^q(n,t) = r^q(n,t) + s_c^q(t) + e(t), \quad 0 < t < N_d N T_r, \tag{3.40}$$

where $e(t)$ is AWGN. The recording is processed as follows. Each pulse return is divided into $\lfloor T_r/T \rfloor$ range gates of size T seconds, a range interval of $\frac{cT}{2}$ meters, centered at times $t_k = kT + \frac{T}{2}$, $k = 0, \dots, \lfloor T_r/T \rfloor - 1$. The qth receiver's recordings are projected onto the pth transmit waveform shifted to range cell k, i.e. onto the functions $\{s^p(m,n,t-t_k) : 0 \le m \le N_d - 1, 0 \le n \le N-1, 1 \le p \le M_{t,R}, 0 \le k \le \lfloor T_r/T \rfloor - 1\}$, to obtain the output sequence $y^q(m,n,p,k)$, given by

$$y^q(m,n,p,k) = \langle \tilde{y}^q(n,t), s^p(m,n,t-t_k) \rangle \tag{3.41}$$

$$\triangleq y_R^q(m,n,p,k) + y_C^q(m,n,p,k) + \bar{e}(m,n,p,k), \tag{3.42}$$

where $\langle y_1(t), y_2(t) \rangle \triangleq \int_{-\infty}^{\infty} y_1(t) y_2^*(t) dt$ and the terms y_R^q, y_C^q, and $\bar{e}$ are the projections of the radar echoes, the communication signal, and AWGN, respectively. This operation is equivalent to matched filtering each of the N echo recordings and sampling the output at times t_k [Zheng et al., 2018c]. Next, we derive each of the terms in (3.42).

3.3.1.4 Radar Signal Component

We have

$$\begin{aligned} y_R^q(m,n,p,k) &= \langle r^q(n,t), s^p(m,n,t-t_k) \rangle \\ &\simeq \sum_{p'=1}^{M_{t,R}} \sum_{i=1}^{L} x_i \exp(-j2\pi f_0(\delta_i^{p'} + \epsilon_i^q)) R_{u^{p'}u^p}(t_k - \tau_i) \\ &\quad \times \exp\left(-j2\pi f_n \left(\tau_i + \frac{2v_i}{c}(nT_r + mNT_r) - t_k\right)\right), \end{aligned} \tag{3.43}$$

where $R_{uv}(\tau) \triangleq \langle u(t), v(t-\tau) \rangle$, and we have used the fact that $\{u^p(t - nT_r - mNT_r - t_k)\}_{m=0}^{N_d-1}$ is orthogonal along t. The approximation in (3.43) assumes the target velocities are small such that each target's position is constant within a single PRI. Since each u^p has duration T, the autocorrelation $R_{u^pu^p}$ has a duration of approximately $2T$; therefore, we assume

$$R_{u^pu^p}(\tau) \simeq \begin{cases} 1 & |\tau| < T/2, \\ 0 & |\tau| > T/2. \end{cases} \tag{3.44}$$

We also assume the waveforms are incoherent, i.e.

$$R_{u^{p'}u^p}(\tau) \simeq \begin{cases} R_{u^pu^p}[\tau] & \text{if } p = p', \\ 0 & \text{if } p \neq p', \end{cases} \quad \tau \in \left[-\frac{T}{2}, \frac{T}{2}\right]. \tag{3.45}$$

This could be achieved, for example, through time-domain multiplexing (which would require increasing the illumination period in order to maintain a given maximum unambiguous range) or MIMO radar waveform design. Define $\mathcal{I}_k \triangleq \{i : |\tau_i - t_k| < T/2\}$, the indices of the scatterers that belong to range cell k. Applying (3.44) and (3.45), (3.43) becomes

$$\begin{aligned} y_R^q[m,n,p,k] &= \sum_{i \in \mathcal{I}_k} x_i \exp[-j2\pi f_0[\delta_i^p + \epsilon_i^q]] \\ &\quad \times \exp\left[-j2\pi \left(n\Delta f[\tau_i - t_k] + f_n \frac{2v_i}{c} nT_r\right)\right] \\ &\quad \times \exp\left[-j2\pi \left(f_0 \frac{2v_i}{c} mNT_r + n\Delta f \frac{2v_i}{c} mNT_r\right)\right], \end{aligned} \tag{3.46}$$

where x_i now includes the term $\exp(-j2\pi f_0(\tau_i - t_k))$.

The TX/RX array elements, in general, are distributed on a plane and the delays $\delta_i^p = \delta_i^p(\boldsymbol{\theta})$ and $\epsilon_i^q = \epsilon_i^q(\boldsymbol{\theta})$ are functions of the scatterer's angular coordinates $\boldsymbol{\theta} \in \mathbb{R}^2$, e.g. azimuth and elevation, relative to the array plane. We consider a generic array response matrix $\mathbf{V} \in \mathbb{C}^{M_{t,R} \times M_{r,R}}$ where

$$(\mathbf{V}(\boldsymbol{\theta}))_{pq} \triangleq \exp\left(-j2\pi f_0(\delta_i^p(\boldsymbol{\theta}) + \epsilon_i^q(\boldsymbol{\theta}))\right) \tag{3.47}$$

and let $\mathbf{v} \triangleq \text{vec}\,(\mathbf{V}) \in \mathbb{C}^{M_{t,R}M_{r,R}}$.

We define steering vectors for the intra- and inter-frame time scales: for intra-frame, the range steering vector $\mathbf{r}[\tau, \upsilon] \in \mathbb{C}^N$ where

$$[\mathbf{r}(\tau, \upsilon)]_n \triangleq \exp\left[-j2\pi\left(n\Delta f \tau + f_n \frac{2\upsilon}{c} nT_r\right)\right], \tag{3.48}$$

for inter-frame, the velocity steering vector $\mathbf{d}[\upsilon] \in \mathbb{C}^{N_d}$ where

$$[\mathbf{d}(\upsilon)]_m \triangleq \exp\left[-j2\pi f_0 \frac{2\upsilon}{c} mNT_r\right]. \tag{3.49}$$

Additionally, define the vector of "distortion terms" $\mathbf{c}(\upsilon) \in \mathbb{C}^{NN_d}$ where

$$[\mathbf{c}(\upsilon)]_{n+mN} \triangleq \exp\left[-j2\pi n\Delta f \frac{2\upsilon}{c} mNT_r\right]. \tag{3.50}$$

Now let

$$\phi(\boldsymbol{\theta}, \tau, \upsilon) \triangleq \mathbf{v}(\boldsymbol{\theta}) \otimes [(\mathbf{d}(\upsilon) \otimes \mathbf{r}(\tau, \upsilon)) \odot \mathbf{c}(\upsilon)] \in \mathbb{C}^{M_{t,R}M_{r,R}NN_d}, \tag{3.51}$$

where $\odot$ is the Hadamard product. Hence, the radar signal component can be expressed in vector form as

$$\mathbf{y}_R(k) = \sum_{i \in \mathcal{I}_k} x_i \phi(\boldsymbol{\theta}_i, \overline{\tau}_i(k), \upsilon_i), \tag{3.52}$$

where the coordinate

$$\overline{\tau}_i(k) \triangleq \tau_i - t_k \in \left[-\frac{T}{2}, \frac{T}{2}\right] \tag{3.53}$$

is the ith scatterer's offset from the center of the kth range cell.

3.3.1.5 Communication Signal Component

The interference component in the projection of receiver q's recording is

$$y_C^q(m, n, p, k) = \langle s_C^q(t), s^p(m, n, t - t_k)\rangle. \tag{3.54}$$

The power spectral density of y_C^q for any q is

$$S_C^q(f) = \sum_{i \in C_n} G_i^q(f - f_i^C)|U^p(f - f_n)|^2, \tag{3.55}$$

where

$$C_n \triangleq \left\{ i \mid |f_n - f_i^C| \leq \frac{\Delta f}{2} + \frac{B_i}{2} \right\} \tag{3.56}$$

is the set of carriers that overlap with radar pulse n. Any communication carrier spectrally overlaps with at least one radar pulse; but, in general, a radar pulse may or may not overlap with any carriers, in which case C_n would be empty. We have

$$\mathbb{E}[|y_C^q[m, n, p, k]|^2] = \int_{-\infty}^{\infty} \sum_{i \in C_n} G_i^q(f - f_i^C)|U^p(f - f_n)|^2 df, \tag{3.57}$$

where the expectation is over the G_i^q, implying that only the carriers C_n may interfere with the radar. Moreover, only a subset of the carriers C_n actually interferes because G_i^q implicitly depends on whether carrier i is in use. Therefore, $\mathbb{E}[\left|y_C^q(m,n,p,k)\right|^2] = 0$ whenever (i) $C_n = \emptyset$ or (ii) none of the carriers C_n are in use.

Define $\mathbf{B}(k) \in \mathbb{C}^{M_{t,R}\times M_{r,R}\times N_d\times N}$ such that $[\mathbf{B}(k)]_{pqmn} \triangleq y_C^q(m,n,p,k)$ and let $\mathbf{b}(k) \triangleq \text{vec}(\mathbf{B}(k)) \in \mathbb{C}^{M_{t,R}M_{r,R}N_dN}$, such that element i of $\mathbf{b}[k]$ is consistent with element i of $\mathbf{y}_R[k]$. Then, the number of nonzero entries in $\mathbf{b}[k]$ is equal to $M_{t,R}M_{r,R}N_d$ times the number of occurrences of spectral overlap. Roughly speaking, if the probability of spectrum overlap with an active carrier is small, then $\mathbf{b}[k]$ will be sparse. For now, we assume that $\mathbf{b}[k]$ contains mostly zeros.

Finally, the projection onto range cell k can be written as

$$\mathbf{y}(k) = \sum_{i\in\mathcal{I}_k} x_i\boldsymbol{\phi}(\theta_i, \overline{\tau}_i(k), \upsilon_i) + \mathbf{b}(k) + \mathbf{e}(k), \tag{3.58}$$

where $\mathbf{e}(k) \sim \mathcal{CN}(0, \sigma^2\mathbf{I})$.

3.3.2 Optimization and Neural Network Approaches

The radar's task is to reconstruct the angle-range-Doppler image from the measurements (3.58) and suppress the interference component $\mathbf{b}[k]$. To that end, we grid the image domain and formulate a convex optimization problem whereby the image and the interference signal can be jointly recovered and hence separated. The following approach images the contents of a single-coarse range cell k; in practice, the following would be applied separately to each desired cell.

Radar images may exhibit considerable sparsity, in that the number of dominant scatterers in the illuminated area may be relatively small Potter et al. [2010, Sec. 3]. Sparsity-inducing image reconstruction algorithms exploit the relatively low information content to enable "super-resolution" finer than conventional processing methods. Compressed sensing theory has provided theoretical underpinnings for ℓ_1-minimization techniques, in particular. Such methods have been applied to angle-range-Doppler imaging [Strohmer and Friedlander, 2012; Yu et al., 2012; Herman and Strohmer, 2009], and are the basis for the following imaging methods.

3.3.2.1 Problem Formulation

The radar data consists of a coherent batch of echo returns from N_d sweeps, given by (3.40). The projection operation in (3.41) isolates the returns of all scatterers located in range cell k, yielding a measurement vector of length $M_{t,R}M_{r,R}N_dN$, given by (3.58). We assume the scatterers' coordinates in angle-range-velocity space lie on the grid $\mathcal{G} \subset \mathbb{R}^4$, where $|\mathcal{G}| \triangleq P \gg M_{t,R}M_{r,R}N_dN$. Define $\boldsymbol{\Phi} \in \mathbb{C}^{M_{t,R}M_{r,R}N_dN\times P}$ whose columns form the dictionary $D \triangleq \{\boldsymbol{\phi}(\theta,\overline{\tau},\upsilon) \mid (\theta,\overline{\tau},\upsilon) \in \mathcal{G}\}$, where $\boldsymbol{\phi}$ is given by (3.51). By the on-grid assumption, we have $\{\boldsymbol{\phi}(\theta_i,\overline{\tau}_i(k),\upsilon_i) \mid i \in \mathcal{I}_k\} \subseteq D$. Thus, the radar signal component (3.52) can be expressed as

$$\mathbf{y}_R(k) = \boldsymbol{\Phi}\mathbf{w}(k), \tag{3.59}$$

where $\mathbf{w}(k) \in \mathbb{C}^P$ is the vectorized angle-range-Doppler image. The nonzero entries of $\mathbf{w}[k]$ form $\{x_i \mid i \in \mathcal{I}_k\}$ and are positioned such that x_i weights $\boldsymbol{\phi}(\theta_i,\overline{\tau}_i(k),\upsilon_i)$. Substituting (3.59) in (3.58), we obtain

$$\mathbf{y} = \boldsymbol{\Phi}\mathbf{w} + \mathbf{b} + \mathbf{e}, \tag{3.60}$$

with dependence on k hereafter implied.

Sparsity manifests in two forms: $\mathbf{b}$ is sparse because of the frequency-domain sparsity of the communication signals; $\mathbf{w}$ is sparse because the radar scene is sparsely populated. In light of these two properties, we formulate the following optimization problem to jointly recover $\mathbf{w}$ and $\mathbf{b}$:

$$\min_{\mathbf{w},\mathbf{b}} \ \| \mathbf{y} - \boldsymbol{\Phi}\mathbf{w} - \mathbf{b} \|_2^2 + \lambda_1 \| \mathbf{w}\|_1 + \lambda_2 \| \mathbf{b}\|_1. \tag{3.61}$$

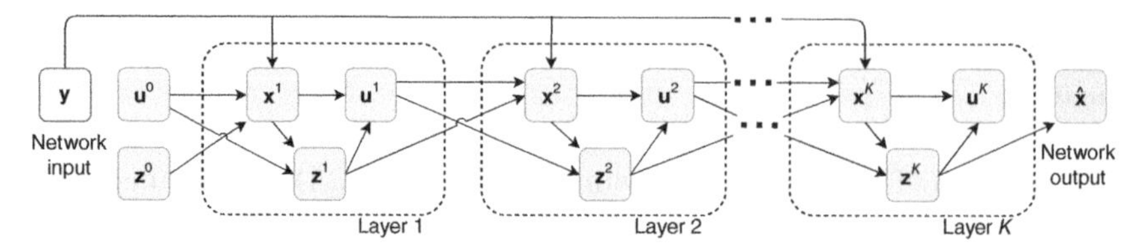

Figure 3.7 Data flow graph for ADMM-Net.

Given the measurement $\mathbf{y}$, (3.61) seeks sparse $\mathbf{w}$ and $\mathbf{b}$ that fit (3.60), where the hyperparameters $\lambda_1, \lambda_2 > 0$ control the sparsity levels. The optimal $\mathbf{w}$ is the recovered image.

The objective in (3.61) is convex; hence, off-the-shelf convex solvers can be used, although the complexity of such solvers scales poorly with problem size. Iterative algorithms, on the other hand, recursively apply a fixed set of simple operations, such as constant matrix multiplication, and are thus well suited for high-dimensional problems where interior point methods are computationally prohibitive. We present two efficient algorithms in this vein. The first is a first-order iterative optimization method, and the other is a model-based neural network inspired by the former.

3.3.2.2 Alternating Direction Method of Multipliers

Consider the equivalent constrained problem

$$\begin{aligned} \min_{\mathbf{x},\mathbf{z}} \quad & \| \mathbf{y} - \mathbf{A}\mathbf{x} \|_2^2 + \lambda_1 \| \mathbf{D}_1 \mathbf{z} \|_1 + \lambda_2 \| \mathbf{D}_2 \mathbf{z} \|_1, \\ \text{s.t.} \quad & \mathbf{x} - \mathbf{z} = 0, \end{aligned} \tag{3.62}$$

where $\mathbf{A} := [\boldsymbol{\Phi}\ \mathbf{I}_N] \in \mathbb{C}^{N\times(P+N)}$, $\mathbf{D}_1 := [\mathbf{I}_P\ 0] \in \mathbb{R}^{P\times(P+N)}$, $\mathbf{D}_2 := [0\ \mathbf{I}_N] \in \mathbb{R}^{N\times(P+N)}$, $\mathbf{x} := [\mathbf{w}^T\ \mathbf{b}^T]^T \in \mathbb{C}^{P+N}$. The *method of multipliers* seeks to maximize the augmented Lagrange dual function $f_\rho(\mathbf{u}) \triangleq \min_{\mathbf{x},\mathbf{z}} L_\rho(\mathbf{x}, \mathbf{z}, \mathbf{u})$, where

$$L_\rho[\mathbf{x}, \mathbf{z}, \mathbf{u}] = \| \mathbf{y} - \mathbf{A}\mathbf{x} \|_2^2 + \lambda_1 \| \mathbf{D}_1 \mathbf{z} \|_1 + \lambda_2 \| \mathbf{D}_2 \mathbf{z} \|_1 + \mathbf{u}^H[\mathbf{x} - \mathbf{z}] + \frac{\rho}{2} \| \mathbf{x} - \mathbf{z} \|_2^2$$

is the augmented Lagrangian with parameter $\rho > 0$ and $\mathbf{u}$ is the dual variable. In each iteration, the primal variables are updated via $\operatorname{argmin}_{\mathbf{x},\mathbf{z}} L_\rho(\mathbf{x}, \mathbf{z}, \mathbf{u})$, and then $\mathbf{u}$ is updated via the gradient of L_ρ with respect to $\mathbf{u}$ evaluated at the updated $\mathbf{x}, \mathbf{z}$.

Evaluating f_ρ is intractable in general because of the joint minimization over $\mathbf{x}$ and $\mathbf{z}$. The *alternating direction method of multipliers* [ADMM] replaces joint minimization with block coordinate minimization through a sequence of subproblems, and thus we obtain the ADMM iteration

$$\begin{aligned} \mathbf{x}^{k+1} &= \operatorname*{argmin}_{\mathbf{x}} L_\rho(\mathbf{x}, \mathbf{z}^k, \mathbf{u}^k), \\ \mathbf{z}^{k+1} &= \operatorname*{argmin}_{\mathbf{z}} L_\rho(\mathbf{x}^{k+1}, \mathbf{z}, \mathbf{u}^k), \\ \mathbf{u}^{k+1} &= \mathbf{u}^k + \rho \nabla_{\mathbf{u}} L_\rho(\mathbf{x}^{k+1}, \mathbf{z}^{k+1}, \mathbf{u}). \end{aligned}$$

The $\mathbf{u}$-update is a gradient step with step size ρ for the objective $L_\rho(\mathbf{x}^{k+1}, \mathbf{z}^{k+1}, \mathbf{u})$, instead of $f_\rho(\mathbf{u})$ as in the method of multipliers. Each subproblem here admits a closed-form solution that can be computed efficiently; see Algorithm 1.

3.3.2.3 Deep Unfolded Neural Network

Deep Unfolding Hershey et al. [2014] is an approach to neural network design that draws inspiration from a chosen iterative algorithm such as ADMM. Unfolded networks have been shown to offer dramatically reduced complexity and improved accuracy compared to the algorithms on which they are based. The idea is to view a single algorithm iteration as a function with learnable

Algorithm 1 ADMM for super-resolution angle-range-Doppler imaging with interference suppression

Input: measurement $\mathbf{y} \in \mathbb{C}^N$, measurement matrix $\mathbf{A} \in \mathbb{C}^{N\times(P+N)}$, parameters λ_1, λ_2, $\rho \in \mathbb{R}$, and stopping criterion ϵ

Output: reconstructed vectorized angle-range-doppler image $\hat{\mathbf{w}}$

Initialize $\mathbf{z}^0 = \mathbf{u}^0 = 0$.

Define $S_\kappa : \mathbb{C}^{P+N} \to \mathbb{C}^{P+N}$ such that $[S_\kappa(\mathbf{a})]_i = \frac{a_i}{|a_i|}\max(|a_i| - \kappa, 0)$, $\kappa > 0$,

$\mathbf{a} = [a_1 \; \cdots \; a_{P+N}]^T \in \mathbb{C}^{P+N}$, $\mathbf{D}_1 := [\mathbf{I}_P \; 0] \in \mathbb{R}^{P\times(P+N)}$, and $\mathbf{D}_2 := [0 \; \mathbf{I}_N] \in \mathbb{R}^{N\times(P+N)}$

repeat for $k = 0, 1, \ldots$

$\mathbf{x}^{k+1} = (\mathbf{A}^H\mathbf{A} + \rho\mathbf{I}_{P+N})^{-1}(\mathbf{A}^H\mathbf{y} + \rho(\mathbf{z}^k - \mathbf{u}^k))$

$\mathbf{z}_1^{k+1} = S_{\lambda_1/\rho}(\mathbf{D}_1\mathbf{x}^{k+1} + \mathbf{u}_1^k)$ (radar image update)

$\mathbf{z}_2^{k+1} = S_{\lambda_2/\rho}(\mathbf{D}_2\mathbf{x}^{k+1} + \mathbf{u}_2^k)$ (interference update)

$\mathbf{z}^{k+1} = \left[(\mathbf{z}_1^{k+1})^T \; (\mathbf{z}_2^{k+1})^T\right]^T$

$\mathbf{u}^{k+1} = \mathbf{u}^k + \mathbf{x}^{k+1} - \mathbf{z}^{k+1}$

break if $\|\mathbf{z}^{k+1} - \mathbf{z}^k\|_2/\|\mathbf{z}^k\|_2 < \epsilon$

return $\hat{\mathbf{w}} := \mathbf{z}_1^{k+1}$

parameters and compose the function with itself several times, say 5 or 10, thus forming a deep learning model. Then, using real-world training data gathered from the intended system, stochastic gradient descent is employed to learn all parameters. The network, thus, inherits domain knowledge, as encapsulated by the signal model, and suitable operations, as informed by the unfolded algorithm – similar to how convolutional layers exploit invariant structures found in images, but instead draws insight from the signal model and, implicitly, the algorithm's theoretical underpinning. The final network, however, is ultimately determined by real-world measurement data.

For example, the ADMM iteration given by Algorithm 1 may be recast as

$$\mathbf{x}^{k+1} = \mathbf{M}_1^{k+1}\mathbf{y} + \mathbf{M}_2^{k+1}(\mathbf{z}^k - \mathbf{u}^k)), \tag{3.63}$$

$$\mathbf{z}_1^{k+1} = S_{\lambda_1^{k+1}}(\mathbf{D}_1\mathbf{x}^{k+1} + \mathbf{u}_1^k), \tag{3.64}$$

$$\mathbf{z}_2^{k+1} = S_{\lambda_2^{k+1}}(\mathbf{D}_2\mathbf{x}^{k+1} + \mathbf{u}_2^k), \tag{3.65}$$

$$\mathbf{z}^{k+1} = \left[(\mathbf{z}_1^{k+1})^T \quad (\mathbf{z}_2^{k+1})^T\right]^T, \tag{3.66}$$

$$\mathbf{u}^{k+1} = \mathbf{u}^k + \mathbf{x}^{k+1} - \mathbf{z}^{k+1}, \tag{3.67}$$

where we have introduced iteration-dependent parameters $\mathbf{M}_1^{k+1} \in \mathbb{C}^{(P+N)\times(P+N)}$, $\mathbf{M}_2^{k+1} \in \mathbb{C}^{(P+N)\times N}$, $\lambda_1^{k+1} \in \mathbb{R}$, and $\lambda_2^{k+1} \in \mathbb{R}$. Note these equations are equivalent to one ADMM iteration if, given values for λ_1, λ_2, and ρ, we assign

$$\mathbf{M}_1^{k+1} \leftarrow (\mathbf{A}^H\mathbf{A} + \rho\mathbf{I}_{P+N})^{-1}\mathbf{A}^H, \qquad \mathbf{M}_2^{k+1} \leftarrow \rho\mathbf{A}^H,$$

$$\lambda_1^{k+1} \leftarrow \lambda_1/\rho, \qquad \lambda_2^{k+1} \leftarrow \lambda_2/\rho,$$

for all k. Equations (3.63)–(3.67) form a computational unit, a *layer*, with inputs $\{\mathbf{z}^k, \mathbf{u}^k\}$, outputs $\{\mathbf{z}^{k+1}, \mathbf{u}^{k+1}\}$, and parameter set $\Omega_{k+1} = \{\mathbf{M}_1^{k+1}, \mathbf{M}_2^{k+1}, \lambda_1^{k+1}, \lambda_2^{k+1}\}$. By chaining together K layers, we obtain a model with parameter set $\cup_{k=1}^K \Omega_k$ that can be learned via gradient descent. This example illustrates the core idea of deep unfolding: parameterize an iterative algorithm so that it can be improved through deep learning.

The process of converting Algorithm 1 into the layer defined by (3.63)–(3.67) is not clear cut, but we propose the following guidelines. Again, the main goal is to map the desired algorithm's iteration onto a parameterized function that defines the network's layer, the basic building block that is sequenced to form the network. The algorithm's data flow graph provides a schematic for the layer's internal connections (see inside the dashed lines in Figure 3.7), where each node is associated with an algorithm iterate and the connections indicate functional dependence as defined by the algorithm. How to designate the learnable parameters, however, is less straightforward. Most iterative algorithms have inherent parameters, e.g. step sizes, penalty parameters, that must be set; one approach is to learn only such parameters, in effect automating a typical hyperparameter cross-validation procedure for the original algorithm [He et al., 2018; Barratt and Boyd, 2021]. While such a parameterization is readily interpretable, there may be too few parameters per layer and hence insufficient learning capacity; dozens of layers might still be needed to surpass the accuracy of the base algorithm. Network learning capacity can be increased by parameterizing the algorithm operations themselves, especially by replacing a constant matrix with an entry-wise learnable matrix, as in (3.63). A network comprising just a few – in some cases an order of magnitude smaller than the number of iterations required for the base algorithm to converge – highly parameterized layers, each computationally equivalent to a single iteration, may outperform the base algorithm. Initializing the parameters can be based on the algorithm's prescribed operations [Yang et al., 2016; Borgerding and Schniter, 2016], so that the initialized network's forward operation is equivalent to a number of algorithm iterations. Algorithms whose iteration comprises a nonlinearity composed with matrix operations, as exemplified by ADMM, are suitable for deep unfolding.

While the network improves the original algorithm, it also inherits some of the algorithm's shortcomings. Chief among them perhaps is the off-grid assumption, and ADMM and CVX are significantly more robust in this regard.

3.3.3 Discussion

3.3.3.1 Simulations

We compare the performance of ADMM-Net, ADMM, and the CVX semi-definite program solver in a simulated interference environment in which a MIMO stepped-frequency radar shares spectrum with the aforementioned SC-FDMA communications system. Simulated radar measurements are generated according to the on-grid model (3.60). The TX and RX arrays are coplanar ULAs arranged in a cross-shaped geometry [Ungan et al., 2020]. The scattering coefficients x_i are independently sampled from $\mathcal{CN}[0,1]$. Without loss of generality, we consider the radar processing for the range cell $k = 0$. The ADMM-Net was trained on data randomly generated via the signal model (3.58) with specified sparsities, SNR, and signal-to-interference ratio [SIR]. Training data sets contained $N_{\text{train}} = 4.5 \times 10^6$ samples for problem size $N = 64$, $P = 150$.

Figure 3.8a illustrates the potential improvement in computational complexity offered by the unfolded network relative to the base algorithm. Each point on the gray curve represents a K-layer network, with K given by the abscissa. Evidently, the 10-layer ADMM-Net (computationally equivalent to 10 ADMM iterations) attains the same accuracy as 40 ADMM iterations. Figure 3.8b plots estimation accuracy versus test set SNR for ADMM, the CVX semidefinite program solver for (3.62), and ADMM-Net. The diamond and solid-circle curves report the accuracy for two networks, each trained on a data set containing samples with a certain SNR (diamond was trained on low SNR data, solid-circle was trained on high SNR data), evaluated over the range of SNRs. Each point on the triangle curve corresponds to a network trained with SNR equal to the abscissa. As expected, a network is most accurate when the test set is drawn from the training distribution, and accuracy worsens as the test set SNR deviates from the training set SNR. The square curve corresponds to a single network trained on data comprising samples with a variety of SNRs drawn uniformly at random; this network outperforms diamond and is overall more robust

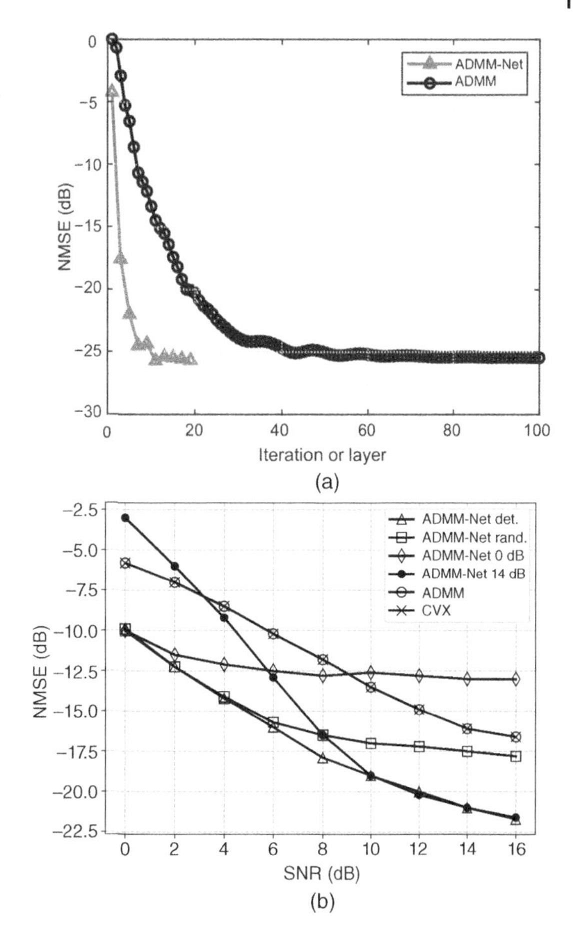

Figure 3.8 (a) Test set normalized mean squared error (NMSE) versus iteration/layers of ADMM/ADMM-Net. The per layer computational cost of ADMM-Net equals the per iteration cost of ADMM. (b) Test set NMSE versus SNR.

than solid-circle, although less accurate than solid-circle at higher SNRs. ADMM and CVX yield practically identical performance, and even though they are more computationally complex, in most cases they are significantly less accurate than ADMM-Net.

To assess super-resolution capability, we simulate two scatterers located at adjacent range grid points with the same velocity and angle coordinates. Figure 3.9 shows each method's output range-velocity image slice. The errors of the full images are −12.0 dB for ADMM and −18.4 dB for ADMM-Net, approximately their corresponding 1000-sample averages for this scenario.

3.3.3.2 Communication-centric Example

We conclude with a vignette, based on Zheng et al. [2018a], from the perspective of a communications system and show how an analogous joint optimization approach can be employed to suppress radar interference. To jointly demodulate the communication data symbols and extract the radar signal, sparsity is imposed on [i] the vector of symbol demodulation errors so as to maximize the number of correctly demodulated symbols and [ii] the radar signal's representation in a known dictionary. In the same vein as the previous example, an ℓ_1 minimization problem is formulated and solved, except this time it is one subproblem inside an outer iteration.

Consider data symbol demodulation performed at a communication receiver in the presence of interference due to multiple spectrally overlapping radars. The received radar signal is assumed to be a sparse linear combination $\Upsilon\boldsymbol{\alpha}$ with a known dictionary matrix Υ and unknown weights $\boldsymbol{\alpha}$ corresponding to the radar target locations. The received communication signal is modeled by $\mathbf{Ab}_0$,

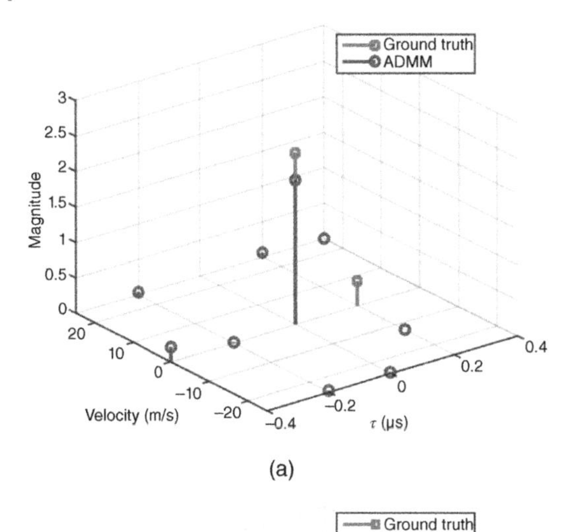

(a)

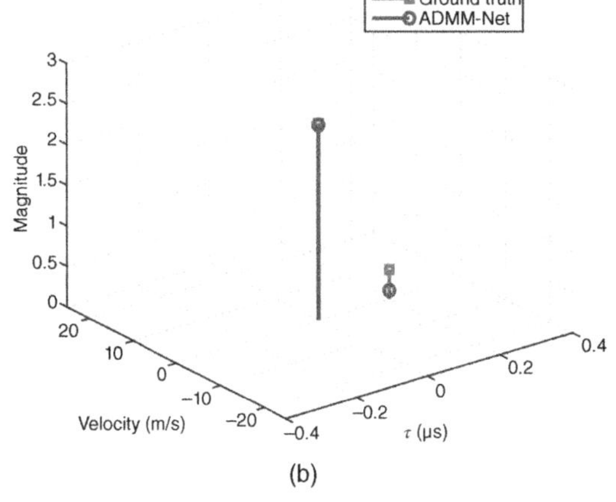

(b)

Figure 3.9 Recovered range-velocity image slice for ADMM (a) and ADMM-Net (b). The two scatterers have magnitudes 2.4 and 0.3 and the same angular position.

where $\mathbf{A}$ is the composite linear transformation due to the communication channel, transmission scheme, and preprocessing, and $\mathbf{b}_0 \in \mathcal{B}$ is the vector of transmitted data symbols to be recovered, where $\mathcal{B}$ is the set of all possible such vectors. The communication receiver measurement is denoted $\mathbf{y}$ and is modeled by $\mathbf{y} = \mathbf{A}\mathbf{b}_0 + \Upsilon\boldsymbol{\alpha} + \mathbf{n}$, where $\mathbf{n}$ is Gaussian noise.

The data demodulation process proceeds as follows. An initial estimate of the data $\mathbf{b}^{(0)}$ is obtained via a generic demodulation scheme assumed given. Then this estimate is iteratively refined via the iteration

$$\mathbf{z}^{(l)} = \mathbf{y} - \mathbf{A}\mathbf{b}^{(l-1)} \tag{3.68}$$

$$(\boldsymbol{\alpha}^{(l)}, \mathbf{v}^{(l)}) = \underset{\boldsymbol{\alpha},\mathbf{v}}{\text{argmin}} \; \| \mathbf{z}^{(l)} - \Upsilon\boldsymbol{\alpha} - \mathbf{A}\mathbf{v} \|_2^2 + \lambda_1 \, \|\mathbf{v}\|_1 + \lambda_2 \, \|\boldsymbol{\alpha}\|_1 \tag{3.69}$$

$$\mathbf{b}^{(l)} = \underset{\mathbf{b}\in\mathcal{B}}{\text{argmin}} \; \|\mathbf{b} - \mathbf{b}^{(l-1)} - \mathbf{v}^{(l)}\|_2 \tag{3.70}$$

for $l = 0, 1, \ldots,$ where $\mathbf{v}$ represents the vector of demodulation errors. The objective in (3.69) enforces sparsity in both the error $\mathbf{v}$ and the radar coefficients $\boldsymbol{\alpha}$, while (3.70) uses $\mathbf{v}^{(l)}$ to obtain a symbol vector that belongs to $\mathcal{B}$. (According to the model, $\mathbf{z}^{(l)} = \mathbf{A}(\mathbf{b}_0 - \mathbf{b}^{(l-1)}) + \Upsilon\boldsymbol{\alpha} + \mathbf{n}$, so that $\mathbf{v}^{(l)}$ is an estimate of $\mathbf{b}_0 - \mathbf{b}^{(l-1)}$, the current error vector.) Thus, the radar interference is extracted and the data symbols are recovered simultaneously. See Zheng et al. [2018a] for more details regarding the signal model and more sophisticated algorithms; (3.69) can be enhanced by instead leveraging the atomic norm to promote sparsity.

3.4 Conclusion

We have addressed the problem of joint design of the radar transmit code, the radar receive filter, and the communications system codebook for coexistence of MIMO radar and MIMO communications. At the design stage, the figure of merit at the radar side, the harmonic mean of the SINR across the inspected range gates, accounts for the presence of a range-dependent interference, while at the communication side, a minimum MI is enforced. The simulation results show that co-design may produce significant SINR improvements although it is important to underline that all of the above conclusions hold true under perfect information on the structure of the interference produced by the coexisting architectures. The sensitivity of such an assumption can be probed by introducing some (random) deviations of the actual matrices from the nominal ones; as shown in Qian et al. [2018b], all of the algorithms appear somewhat sensitive to channel state estimation errors, showing that further investigation is needed to include in the design suitable constraints guaranteeing robustness to unreliable or uncertain channel information.

For uncoordinated interference suppression, we have illustrated an unconstrained joint-optimization approach that incorporates interference suppression into the processing method used for the system's primary function, exploiting prior knowledge of the interference structure. In particular, we have seen, from a radar perspective, that the stepped-frequency radar processing sparsifies the communication interference to the extent that the communications are underutilized, paving the way for sparse reconstruction; and from a communication perspective, if the radar signals admit a sparse representation in a known dictionary, further sparsity can be imposed on the vector of demodulation errors so as to enable interference-robust demodulation. For such problems, deep learning, via the deep unfolding framework, may be employed to significantly improve upon iterative optimization algorithms such as ADMM, although more investigation is needed to determine the extent to which learning is able to compensate for model mismatch and whether deep unfolding can accommodate off-grid recovery methods.

References

3rd Generation Partnership Project. LTE; Evolved Universal Terrestrial Radio Access (E-UTRA); Physical Channels and Modulation. *3GPP TS 36.211 Version 14.2.0 Release*, 14 2017.

A. Aubry, A. De Maio, A. Farina, and M. Wicks. Knowledge-aided (potentially cognitive) transmit signal and receive filter design in signal-dependent clutter. *IEEE Transactions on Aerospace and Electronic Systems*, 49(1):93–117, 2013.

A. Aubry, A. De Maio, M. Piezzo, and A. Farina. Radar waveform design in a spectrally crowded environment via nonconvex quadratic optimization. *IEEE Transactions on Aerospace and Electronic Systems*, 50(2):1138–1152, 2014.

A. Aubry, V. Carotenuto, and A. De Maio. Forcing multiple spectral compatibility constraints in radar waveforms. *IEEE Signal Processing Letters*, 23(4):483–487, 2016.

S. T. Barratt and S. P. Boyd. Least squares auto-tuning. *Engineering Optimization*, 53(5):789–810 2021.

M. Bica, K. Huang, V. Koivunen, and U. Mitra. Mutual information based radar waveform design for joint radar and cellular communication systems. In *2016 IEEE International Conference on Acoustics, Speech and Signal Processing (ICASSP)*, pages 3671–3675, 2016. https://doi.org/10.1109/ICASSP.2016.7472362.

E. Biglieri, R. Calderbank, A. Constantinides, A. Goldsmith, A. Paulraj, and H. V. Poor. *MIMO wireless communications*. Cambridge, U.K.: Cambridge University Press, 2007.

S. D. Blunt, W. Dower, and K. Gerlach. Hybrid interference suppression for multistatic radar. *IET Radar, Sonar and Navigation*, 2(5):232–333, 2008.

S. D. Blunt, M. R. Cook, and J. Stiles. Embedding information into radar emissions via waveform implementation. In *2010 International Waveform Diversity and Design Conference*, pages 195–199, 2010.

M. Borgerding and P. Schniter. Onsager-corrected deep learning for sparse linear inverse problems. *https://arxiv.org/abs/1607.05966v1*, 2016.

A. R. Chiriyath, B. Paul, G. M. Jacyna, and D. W. Bliss. Inner bounds on performance of radar and communications co-existence. *IEEE Transactions on Signal Processing*, 64(2):464–474, 2016. https://doi.org/10.1109/TSP.2015.2483485.

T. Counts, A. C. Gurbuz, W. R. Scott, J. H. McClellan, and K. Kim. Multistatic ground-penetrating radar experiments. *IEEE Transactions on Geoscience and Remote Sensing*, 45(8):2544–2553, 2007.

G. Cui, X. Yu, V. Carotenuto, and L. Kong. Space–time transmit code and receive filter design for colocated MIMO radar. *IEEE Transactions on Signal Processing*, 65(5):1116–1129, 2017.

C. D'Andrea, S. Buzzi, and M. Lops. Communications and radar coexistence in the massive MIMO regime: uplink analysis. *IEEE Transactions on Wireless Communications*, 19(1):19–33, 2020.

A. De Maio, S. De Nicola, Y. Huang, and Z. Q. Luo. Design of phase codes for radar performance optimization with a similarity constraint. *IEEE Transactions on Signal Processing*, 57(2):610–621, 2009.

H. Deng and B. Himed. Interference mitigation processing for spectrum-sharing between radar and wireless communications systems. *IEEE Transactions on Aerospace and Electronic Systems*, 49(49):1911–1919, 2013.

M. Filo, A. Hossain, A. R. Biswas, and R. Piesiewicz. Cognitive pilot channel: enabler for radio systems coexistence. In *Proceedings of the 2nd International Workshop on Cognitive Radio and Advanced Spectrum Management*, pages 17–23. IEEE, 2009.

J. R. Foerster. The performance of a direct-sequence spread ultrawideband system in the presence of multipath, narrowband interference, and multiuser interference. In *2002 IEEE Conference on Ultra Wideband Systems and Technologies*, 2002. Digest of Papers, pages 87–91. IEEE, 2002.

F. Gini, A. De Maio, and L. Patton. *Waveform design and diversity for advanced radar systems*. Radar, Sonar and Navigation. IET - Institution of Engineering and Technology, 2012.

H. Griffiths and S. Blunt. T09 – Spectrum engineering and waveform diversity. In *Radar Conference*, page 36, 2014.

H. Griffiths, L. Cohen, S. Watts, E. Mokole, C. Baker, M. Wicks, and S. Blunt. Radar spectrum engineering and management: technical and regulatory issues. *Proceedings of the IEEE*, 103(1):85–102, 2014.

E. Grossi, M. Lops, and L. Venturino. Joint design of surveillance radar and MIMO communication in cluttered environments. *IEEE Transactions on Signal Processing*, 68(1):1544–1557, 2020. ISSN 1941-0476. https://doi.org/10.1109/TSP.2020.2974708.

A. Hassanien, M. G. Amin, Y. D. Zhang, F. Ahmad, and B. Himed. Non-coherent PSK-based dual-function radar-communication systems. In *2016 IEEE Radar Conference (RadarConf)*, pages 1–6, 2016a. https://doi.org/10.1109/RADAR.2016.7485066.

A. Hassanien, M. G. Amin, Y. D. Zhang, and F. Ahmad. Dual-function radar-communications: information embedding using sidelobe control and waveform diversity. *IEEE Transactions on Signal Processing*, 64(8):2168–2181, 2016b.

S. Haykin. Cognitive radio: brain-empowered wireless communications. *IEEE Journal on Selected Areas in Communications*, 23(2):201–220, 2005.

S. Haykin. Cognitive radar: a way of the future. *IEEE Signal Processing Magazine*, 23(1):30–40, 2006.

H. He, C. Wen, S. Jin, and G. Y. Li. A model-driven deep learning network for MIMO detection. In *2018 IEEE Global Conference on Signal and Information Processing (GlobalSIP)*, pages 584–588, 2018. https://doi.org/10.1109/GlobalSIP.2018.8646357.

M. A. Herman and T. Strohmer. High-resolution radar via compressed sensing. *IEEE Transactions on Signal Processing*, 57(6):2275–2284, 2009.

J. R. Hershey, J. L. Roux, and F. Weninger. Deep unfolding: model-based inspiration of novel deep architectures. *https://arxiv.org/abs/1409.2574*, 2014.

S. M. Karbasi, A. Aubry, V. Carotenuto, M. M. Naghsh, and M. H. Bastani. Knowledge-based design of space–time transmit code and receive filter for a multiple-input–multiple-output radar in signal-dependent interference. *IET Radar, Sonar and Navigation*, 9(8):1124–1135, 2015.

A. Khawar, A. Abdel-Hadi, and T. C. Clancy. Spectrum sharing between S-band radar and LTE cellular system: a spatial approach. In *2014 IEEE International Symposium on Dynamic Spectrum Access Networks (DYSPAN)*, pages 7–14. IEEE, 2014.

Y. Kim, Y. Kim, J. Oh, H. Ji, J. Yeo, S. Choi, H. Ryu, H. Noh, T. Kim, F. Sun, Y. Wang, Y. Qi, and J. Lee. New radio (NR) and its evolution toward 5G-advanced. *IEEE Wireless Communications*, 26(3):2–7, 2019.

B. Li and A. Petropulu. MIMO radar and communication spectrum sharing with clutter mitigation. In *Radar Conference (RadarConf), 2016 IEEE*, pages 1–6. IEEE, 2016.

J. Li and P. Stoica. MIMO radar with colocated antennas. *IEEE Signal Processing Magazine*, 24(5):106–114, 2007.

B. Li, A. P. Petropulu, and W. Trappe. Optimum co-design for spectrum sharing between matrix completion based MIMO radars and a MIMO communication system. *IEEE Transactions on Signal Processing*, 64(17):4562–4575, 2016.

Y. Li, L. Zheng, M. Lops, and X. Wang. Interference removal for radar/communication co-existence: the random scattering case. *IEEE Transactions on Wireless Communications*, 18(10):4831–4845, 2019a.

Y. Li, X. Wang, and Z. Ding. Multi-target position and velocity estimation using OFDM communication signals. *IEEE Transactions on Communications*, 68(2):1160–1174, 2019b.

F. Liu, C. Masouros, A. Petropulu, H. Griffiths, and L. Hanzo. Joint radar and communication design: applications, state-of-the-art, and the road ahead. *IEEE Transactions on Communications*, 68(6):3834–3862, 2020.

J. A. Mahal, A. Khawar, A. Abdelhadi, and T. C. Clancy. Spectral coexistence of MIMO radar and MIMO cellular system. *IEEE Transactions on Aerospace and Electronic Systems*, 53(2):655–668, 2017.

A. Manolakos, Y. Noam, K. Dimou, and A. J. Goldsmith. Blind null-space tracking for MIMO underlay cognitive radio networks. In *Global Communications Conference (GLOBECOM), 2012 IEEE*, pages 1223–1229. IEEE, 2012.

B. Paul, A. R. Chiriyath, and D. W. Bliss. Joint communications and radar performance bounds under continuous waveform optimization: the waveform awakens. In *2016 IEEE Radar Conference (RadarConf)*, pages 1–6, 2016. https://doi.org/10.1109/RADAR.2016.7485103.

L. C. Potter, E. Ertin, J. T. Parker, and M. Cetin. Sparsity and compressed sensing in radar imaging. *Proceedings of the IEEE*, 98(6):1006–1020, 2010.

A. Puglielli, A. Townley, G. LaCaille, V. Milovanović, P. Lu, K. Trotskovsky, A. Whitcombe, N. Narevsky, G. Wright, T. Courtade, et al. Design of energy-and cost-efficient massive MIMO arrays. *Proceedings of the IEEE*, 104(3):586–606, 2016.

J. Qian, M. Lops, L. Zheng, and X. Wang. Joint design for co-existence of MIMO radar and MIMO communication system. *51st Asilomar Conference on Signals, Systems, and Computers*, pages 568–572, 2017 (Invited Paper).

J. Qian, Z. He, N. Huang, and B. Li. Transmit designs for spectral coexistence of MIMO radar and MIMO communication systems. *IEEE Transactions on Circuits and Systems II: Express Briefs*, 65(12):2072–2076, 2018a.

J. Qian, M. Lops, L. Zheng, X. Wang, and Z. He. Joint system design for coexistence of MIMO radar and MIMO communication. *IEEE Transactions on Signal Processing*, 66(13):3504–3519, 2018b.

J. Qian, M. Lu, and N. Huang. Radar and communication co-existence design based on mutual information optimization. *IEEE Transactions on Circuits and Systems II: Express Briefs*, 67(12):3577–3581, 2020.

M. Razaviyayn, M. Hong, and Z. Q. Luo. A unified convergence analysis of block successive minimization methods for nonsmooth optimization. *SIAM Journal on Optimization*, 23(2):1126–1153, 2013.

A. M. Sayeed and S. Member. Deconstructing multi-antenna fading channels. *IEEE Transactions on Signal Processing*, 50(10):2563–2579, 2002.

M. Shafi, A. F. Molisch, P. J. Smith, T. Haustein, P. Zhu, P. De Silva, F. Tufvesson, A. Benjebbour, and G. Wunder. 5G: A tutorial overview of standards, trials, challenges, deployment, and practice. *IEEE Journal on Selected Areas in Communications*, 35(6):1201–1221, 2017.

T. Strohmer and B. Friedlander. Analysis of sparse MIMO radar. *Applied and Computational Harmonic Analysis*, 37(3):361–388 2014.

S. Sun, W. U. Bajwa, and A. P. Petropulu. MIMO-MC radar: a MIMO radar approach based on matrix completion. *IEEE Transactions on Aerospace and Electronic Systems*, 51(3):1839–1852, 2015.

D. Tse and P. Viswanath. *Fundamentals of wireless communication*. Cambridge, U.K.: Cambridge University Press, 2005.

A. Turlapaty and Y. Jin. A joint design of transmit waveforms for radar and communications systems in coexistence. In *2014 IEEE Radar Conference*, pages 0315–0319, 2014. https://doi.org/10.1109/RADAR.2014.6875606.

C. U. Ungan, Ç. Candan, and T. Ciloglu. A space–time coded mills cross MIMO architecture to improve DOA estimation and its performance evaluation by field experiments. *IEEE Transactions on Aerospace and Electronic Systems*, 56(3):1807–1818, 2020.

T. Wang, Y. Shen, S. Mazuelas, H. Shin, and M. Z. Win. On OFDM ranging accuracy in multipath channels. *IEEE Systems Journal*, 8(1):104–114, 2014.

Y. Yang, J. Sun, H. Li, and Z. Xu. Deep ADMM-NET for compressive sensing MRI. In D. D. Lee, M. Sugiyama, U. V. Luxburg, I. Guyon, and R. Garnett, editors, *Advances in neural information processing systems 29*, pages 10–18. Curran Associates, Inc., 2016. http://papers.nips.cc/paper/6406-deep-admm-net-for-compressive-sensing-mri.pdf.

Y. Yu, A. P. Petropulu, and H. V. Poor. Cssf MIMO radar: compressive-sensing and step-frequency based MIMO radar. *IEEE Transactions on Aerospace and Electronic Systems*, 48(2):1490–1504, 2012.

X. Zhang, H. Li, J. Liu, and B. Himed. Joint delay and doppler estimation for passive sensing with direct-path interference. *IEEE Transactions on Signal Processing*, 64(3):630–640, 2016.

L. Zheng, M. Lops, and X. Wang. Adaptive interference removal for uncoordinated radar/communication coexistence. *IEEE Journal on Selected Topics in Signal Processing*, 12(1):45–60, 2018a.

L. Zheng, M. Lops, X. Wang, and E. Grossi. Joint design of overlaid communication systems and pulsed radars. *IEEE Transactions on Signal Processing*, 66(1):139–154, 2018b.

L. Zheng, M. Lops, X. Wang, and E. Grossi. Joint design of overlaid communication systems and pulsed radars. *IEEE Transactions on Signal Processing*, 66(1):139–154, 2018c.

D. Zhu, J. Li, and G. Li. RFI source localization in microwave interferometric radiometry: a sparse signal reconstruction perspective. *IEEE Transactions on Geoscience and Remote Sensing*, 58(6):4006–4017, 2020.

4

Beamforming and Interference Management in Joint Radar–Communications Systems

Tuomas Aittomäki[1], Yuanhao Cui[1,2], and Visa Koivunen[1]

[1] *Aalto University, Department of Signal Processing and Acoustics, Espoo, Finland*
[2] *Beijing University of Posts and Telecommunications, School of Information and Communication Engineering, Beijing, China*

4.1 Introduction

Joint radar–communications (JRC) systems operate in shared and often congested – possibly even contested – spectrum with the goal of providing both reliable communication and radar capabilities. JRC systems consist of communications and sensing subsystems that may be colocated or in distinct locations, with the source and destination of communication being obviously in different locations. Sensing and communications systems may coexist, cooperate or they may be even co-designed for mutual benefits. Co-designed systems that share waveforms, hardware, and antenna resources are commonly called integrated sensing and communications (ISAC) systems. Multiuser communications systems that share spectrum among multiple different radio systems performing communications or sensing are of particular interest to us since they are the key components in JRC systems.

There are many difficulties in designing JRC systems. The state of the spectrum may vary rapidly as a function of time, location, and frequency. Moreover, radio wave propagation phenomena such as attenuation and multipath are very different in different frequency bands, for example in 77 GHz band and 3.5 GHz band. In multiuser systems, interference is the main factor limiting the performance. There are a variety of techniques for communications and sensing systems to enhance signal quality and manage interference. These systems have a number of degrees of freedom (DoF) and operational parameters that can be selected or adjusted to optimize their performance. Examples of such parameters are frequency band, beampattern, antenna selection, the modulation method, precoder–decoder design, and power allocation. A variety of systems models, configurations, and applications of JRC and ISAC systems have been considered for many joint communication and sensing applications in different frequency bands; see for example Sturm and Wiesbeck [2011], Hassanien et al. [2016a], Paul et al. [2017], Mishra et al. [2019], and Liu et al. [2020a]. A JRC system with co-designed system in the presence of both cooperative and noncooperative users and different types of interference are illustrated in Figure 4.1.

Depending on the type of the JRC system, different configurations and different duplexing and multiple access (MA) techniques are possible. Three different categories of systems can be distinguished: coexistence, cooperation, and co-design [Paul et al., 2017]. In the case of plain coexistence, radar and communications systems will be running on different hardware platforms and in distinct locations. They will not share awareness about the state of the radio environment such as channel state information (CSI). Cooperative systems will be either colocated or in different locations

Signal Processing for Joint Radar Communications, First Edition.
Edited by Kumar Vijay Mishra, M. R. Bhavani Shankar, Björn Ottersten, and A. Lee Swindlehurst.

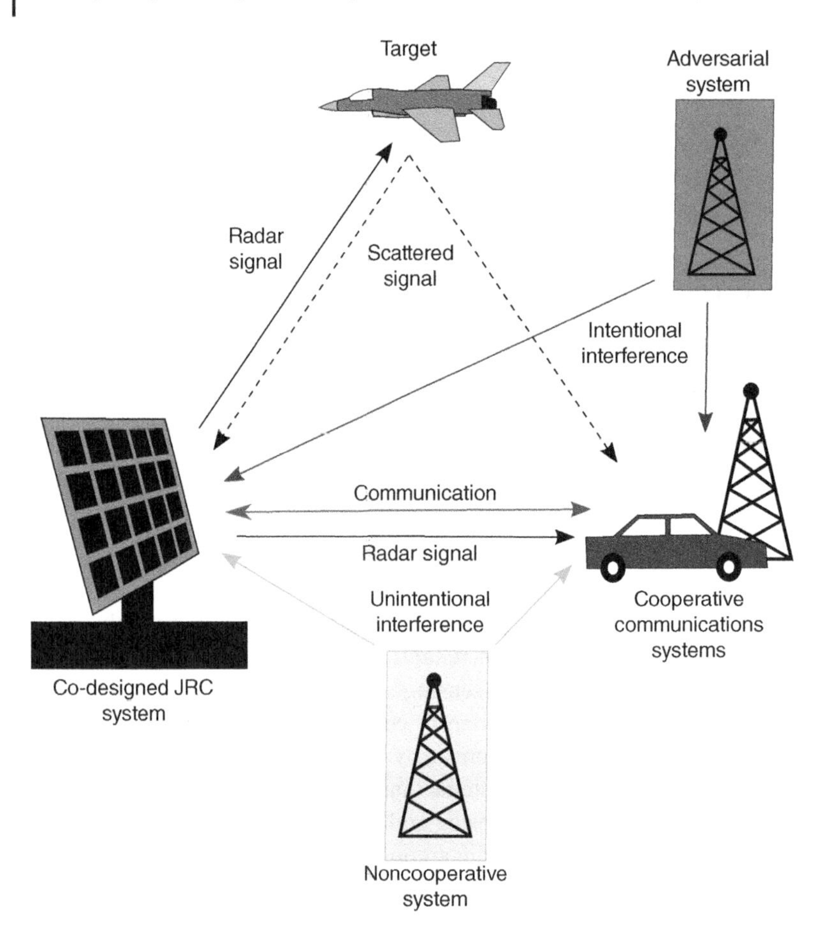

Figure 4.1 Co-design, cooperation, and noncooperative users in a JRC system. The system should be able to deal with different sensing and communication signals and sources of interference.

and share information for improved performance. In a co-designed JRC system, the radar and communication will typically be colocated, use dual-function waveforms, and share transceiver hardware and antenna resources. Radar systems in question may have monostatic, bistatic, or multistatic configurations. Multiple-input multiple-output (MIMO) radars that may be colocated or distributed can also be employed.

The duplexing approaches are coupled with different sensing and communication waveform designs. In the case of pulsed radar, the maximum unambiguous range is determined by pulse repetition time (PRT) and pulse duration. The radar transmitter needs to be switched off in order to listen to the returning pulses and determine the range. A radar may be subject to self-interference unless transmit and receive antennas are isolated well. Very efficient decoupling may be achieved using emerging full-duplex transceiver structures. If the achieved isolation of transmitters and receivers is sufficient, for example about 100 dB overall isolation, the JRC system could transmit continuously and receive signals simultaneously. This would allow for continuous transmission of multicarrier signals such as orthogonal frequency-division multiplexing (OFDM) for communications while performing sensing. A radar could also have a bistatic configuration such that the transmitter and receiver are in distinct locations, in which case the required isolation is achieved.

In communications systems, either time-division duplexing (TDD) or frequency-division duplexing (FDD) is commonly used between downlink and uplink transmissions. In TDD, there are different time slots for downlink and uplink communications taking place at the same frequency band, whereas in FDD, uplink and downlink communications use different frequencies. The duplexing method impacts the interference experienced at radar receiver and its signal processing. In radar, the signal-to-noise ratio (SNR) is inversely proportional to the distance as R^4 instead of R^2 in communications, so radar systems may need a significantly higher transmit power than communications to achieve a similar SNR. The communications subsystems in ISAC may be subject to strong interference from radar due to this power imbalance. For example, doubling the range of a radar would mean that 16 times higher transmit power is needed. Alternatively, the applications of ISAC systems may be limited to shorter distances.

Spatial processing of signals using sensor arrays is a key enabling technology in wireless communications, radar, medical imaging and acoustics, for example. Spatial filtering of signals using multiple active or passive sensors is called beamforming. It allows for radiating transmissions in desired directions, i.e. focusing transmitted signal energy to desired users or targets at certain angles. In the case of beamforming at the receiver, signals arriving from desired directions are enhanced while suppressing interfering signals from other directions. Beamforming improves the SNR or signal to interference and noise ratio (SINR) of the received signals; as a consequence, a higher data rate and a larger coverage can be achieved in communications as well as more reliable target detection, recognition, and tracking in sensing tasks.

Transmit beamforming enables directional transmission that can reduce unintended interference caused to other systems. Angular resolution of a beamformer depends on the aperture of the sensor array, i.e. the size of the sensor array compared to the wavelength. In modern millimeter-wave sensing and communications systems, large antenna systems with high angular resolution may be achieved even if the physical array occupies a reasonably small space. In order to do beamforming in both azimuth and elevation angles, a two-dimensional sensor array is needed. Sparse array techniques can be used to reduce the number of antennas and their expensive front-end electronics without sacrificing the resolution. In sub-6 GHz frequencies, digital beamforming is typically used, whereas hybrid and analog beamformers are common in millimeter-wave (mm-Wave) bands, even though there has been significant effort in developing digital beamformers for such higher frequencies.

Interference management is a key task in both radio frequency (RF) sensing and communication scenarios with shared spectrum and multiple users. Systems can be designed to be orthogonal such that signals can be separated from each other and they cause little or no interference to other users or systems. In communications systems, the objective is typically to maximize the sum rate of data transmission. The most common and the simplest approach is to use orthogonal design in the time–frequency plane, as is done in time-division multiple access (TDMA) and frequency-division multiple access (FDMA) systems. Cellular communications systems can accommodate a large number of users by reusing frequencies in sufficiently far away locations. This design produces co-channel interference. Another orthogonal design approach for managing interference is based on orthogonal codes. In code-division multiple-access (CDMA), each user is allocated a code sequence that is orthogonal to other users' sequences so that interference within a cell can be avoided in ideal conditions. Intercell interference can then be avoided by using different scrambling codes for close-by cells or sectors a base station is serving. This approach facilitates a frequency reuse factor of 1 since intercell interference turns into randomized additive noise.

OFDM uses multiple narrowband subcarriers with intercarrier spacing that makes them orthogonal. OFDM-based MA method called orthogonal frequency-division multiple access (OFDMA) allocates different subsets of subcarriers to each user so that interference caused to neighboring cells remains tolerable even with a reuse factor of 1. Spatial domain interference management is based on multiantenna technologies – in particular on phased-arrays or MIMO systems. Interference may be managed spatially using either beamforming or launching multiple parallel data streams using spatial multiplexing or by designing precoders and decoders based on the CSI. Prime examples include massive MIMO techniques forming a large number of pencil beams in 5G systems and MIMO with spatial multiplexing in 4G. Interference management is a thoroughly studied topic in wireless communications [Veeravalli and El Gamal, 2018] and more recently in congested spectrum in vehicular sensing and communication [Aydogdu et al., 2020].

Radar systems use phased-array technologies, waveform diversity and more recently distributed or colocated MIMO transceiver structures to deal with intentional and unintentional interference. Both transmitters and receivers may be adapted. Radars have a variety of tasks including target detection, recognition, tracking, and parameter estimation and may have to perform multiple tasks simultaneously. Hence, different task-dependent objective functions have to be employed in interference management.

To fully take advantage of beamforming and various interference management techniques, radio receivers should acquire situational awareness about the state of the radio spectrum, especially interference and channel awareness. This typically requires the use of statistical parameter estimation and detection tools or machine learning. Awareness can be represented in the form of CSI; directions of the desired users, targets, or interferers; SINR values at the receiver in different frequencies; and directions and channels between different transmitter and receiver pairs. The obtained awareness needs to be shared among different systems and their users. Based on the awareness, transmitters and receivers can optimize and adjust their operational parameters such as beampatterns, power allocation, waveforms, frequency, bandwidth, and precoder and decoder designs.

There are many ways to share awareness depending on the level of cooperation among the systems and radios. The radar and communications system can just *coexist* without exchanging awareness information with other systems. In such cases, other users or systems are competing for the same spectrum resources and are sources of unintentional interference to each other. In such noncooperative settings with no sharing of awareness among different systems, each receiver needs to estimate the parameters describing channel states and interference individually. Most commercial wireless communication standards have an option to provide quantized feedback about instantaneous channel state or its longer term statistics to transmitters in the same system. Alternatively, if TDD is used, the channel is reciprocal and the channel estimate is the same in both directions over the coherence time of the channel.

In JRC systems, it is highly beneficial if communications and sensing subsystems cooperate and share relevant information. In such cases, a protocol is needed for exchanging awareness information. In existing communications systems, this would require changes in standards. If communications and sensing subsystems are co-designed and colocated, the exchange of information between them is simple to arrange. Different subsystems may have access to shared memory containing information on CSI, interference and spectrum occupancy. Moreover, timing information needed for synchronization and estimating delays may be shared. Given the awareness about the state of the spectrum, transmitters and receivers can adjust and optimize the use of their DoF for the communication and sensing task at hand. To reduce the overhead, the feedback provided to transmitters is typically quantized with only few bits. Adjusting operational parameters is commonly modeled as a constrained optimization problem where minimum desired performance levels for other subsystems impose the constraints. For example, one may maximize the probability of detection in the radar subsystem while ensuring that the communication users are guaranteed to achieve some minimum required data rate. Alternatively, in a communication-centric case, the data rate of a user could be maximized while ensuring minimum tolerable performance for a radar in target detection and parameter estimation.

Orthogonal designs and multicarrier systems such as OFDM, spatial multiplexing MIMO, orthogonal codes in space–time coding, and precoder and decoder designs that project desired signals and interference to separate subspaces are used mainly in current and emerging wireless communications systems. Even if the system is designed to be orthogonal at the transmitter, that may not be the case at the receiver as various propagation effects and the mobility of the subsystems and scatterers may cause a loss of orthogonality, resulting in an increased level of interference. The loss of orthogonality and increased interference may also be caused by transmitter and receiver nonidealities, signals leaking from adjacent bands and reuse of frequencies. It is important to remember that *a receiver is always the victim of interference*. Orthogonal designs in time, frequency, and spatial domains employing beamforming, MIMO, spread spectrum, and waveform diversity techniques are commonly used to manage interference in modern radar systems. JRC systems can use a variety of approaches from both fields to enhance signal quality and manage interference.

Seamless integration of communication and sensing in ISAC may be achieved by co-designing radar and communications subsystems. This approach facilitates maximizing the joint performance of ISAC instead of optimizing the performance of each subsystem individually that may lead to deterioration in the performance of other subsystems. Co-design allows for using the same transceiver hardware and algorithms, waveforms, multiantenna systems and their front-ends, timing circuitry for time and frequency synchronization, and delay estimation for sensing and communication. Moreover, communications and sensing subsystems can jointly build awareness about the state of the radio spectrum and then exploit it in optimizing their operational parameters. The waveforms used by co-designed JRC systems are commonly called *dual-functional radar communication* (*DFRC*) waveforms. Co-design is a key component of the emerging RF convergence paradigm in which different RF systems share their resources including their key hardware in a mutually beneficial manner. The outcome of co-design is a multifunction RF system.

This chapter is organized as follows. In Section 4.2, the basic JRC system model is described with a focus on spatial processing using beamforming and interference management techniques. Section 4.3 discusses beamformer designs for JRC systems. In Section 4.4, multicarrier waveforms and related interference management in JRC are considered. Two different examples are provided. First, a design where subcarrier allocation to radar and communication tasks is performed using OFDM model and mutual information (MI) criterion is described. A second example design uses multicarrier spread spectrum model with orthogonal variable spreading factor (OVSF) codes where radar and communication signals are allocated to different subcarriers and separated by long scrambling codes. In Section 4.5, different precoding and decoding approaches for mitigating interference in JRC systems are described. Finally, some concluding remarks are provided and emerging topics in JRC are discussed in Section 4.6.

4.2 System Overview

JRC system designs can be radar-centric or communication-centric so that the other task is done on the side with the requirement of satisfying a constraint on a minimum performance level. In a co-designed ISAC system, the performance of both sensing and communications can be jointly optimized. These three scenarios have different signal models both at the transmitter and receiver. Hence, there does not exist any commonly used signal model that would be applicable in all three cases. A step to that direction was taken in the Generalized Multicarrier model in Bică and Koivunen [2016].

JRC systems impose quantitative performance criteria to both radar and communications. The goal of communications systems is to use the spectrum efficiently while maintaining tolerable error levels. Commonly used quantitative criteria include data rate, quality-of-service (QoS), spectral efficiency (in bits/s/HZ), mutual information (MI), SINR, as well as bit error rate (BER), symbol error rate (SER), and block error rate (BLER). Theoretical performance bounds stemming from the classical Shannon formula have been established in a variety of scenarios.

Radars have a variety of tasks, and the employed performance criteria are task-dependent. Moreover, modern multifunction radars perform multiple tasks at the same time and the tasks may have different priorities. Target-detection performance is characterized by probabilities of false alarm and detection, whereas target tracking and parameter estimation use criteria stemming from estimation theory associated with errors in distance, velocity, and direction, for example. Typical quantitative criteria are the mean square error (MSE) and the variance compared to the Cramer–Rao bound (CRB) There are also several radar design parameters that determine the performance, for example the resolution in range, Doppler or angular domains, antenna gains, coverage as well as the DoF for resolving multiple targets or cancelling multiple interference sources simultaneously. The ambiguity function (AF) of the employed radar waveform in particular characterizes the ability to discriminate targets in both range and Doppler domains.

Communication and sensing tasks may also have different priorities that are reflected in an objective function of waveform designs or imposed constraints. The system model depends on the employed system configuration, including whether sensing and communication are colocated or distributed; the employed antenna systems and transceiver structures; the amount of information exchanged among subsystems; and the MA method and the duplexing scheme employed in the communications subsystem. Consequently, it is difficult to define a general performance criterion or signal model for JRC systems. Instead, we describe a variety of system models through design examples. The examples describe how beamforming methods can be applied in JRC and different dual-function waveforms used in spatial processing. We describe an orthogonal OFDM-based multicarrier waveform design for JRC systems that may be optimized to take into account desired performance criteria for sensing and communication. Moreover, a multicarrier direct-sequence CDMA (MC-DS-CDMA) design based on OVSF codes and long scrambling codes for separating the sensing and communications subsystems is described briefly. In a cooperative system, interference can be managed by designing precoders and decoders that project the desired signals and interfering signals into different orthogonal subspaces, hence enhancing the signal quality. Examples of such precoding and decoding techniques are provided for joint sensing and communication.

4.2.1 Radio Channel Properties

JRC systems have been considered in sub-6 GHz and mm-Wave frequencies. The mm-Wave communication and sensing (30–300 GHz) suffer from high attenuation caused by physical barriers, weather phenomena, and atmospheric absorption. The RF channels may have very short coherence times or narrow coherence bandwidths because of mobility, target responses, and scattering environment. Channels may be even doubly selective and vary rapidly over time and frequency. Some JRC applications have a large Doppler range to cover because some targets may be stationary, move very slowly or extremely fast – vehicular applications are a prime example. At the same time, mm-Wave channels provide a high-range resolution and high data rates due to the availability of broad bandwidth. Multipath is a less severe issue compared to sub-6 GHz frequencies. On the other hand, blockage caused by obstacles or walls may limit the coverage severely. A very high spatial resolution may also be achieved by multiantenna systems since large aperture antenna arrays require only a small physical space. Consequently, JRC in mm-Wave bands is suitable for short-range applications such as automotive systems, autonomous driving, human gesture recognition, vital sign and well-being monitoring, and RF identification.

In sub-6 GHz frequencies, there are many existing systems that need to share spectrum, for example, S-band and C-band radars, 4G and 5G cellular, WiFi, Bluetooth, navigation systems, and Citizen's Broadband. Many of the existing and emerging communications systems use multicarrier OFDM modulation. Therefore, it is interesting to study the feasibility of multicarrier signals as DFRC waveforms. Techniques for coexistence and in particular cooperation may require changes in standardization of existing wireless systems. Propagation phenomena are different in sub-6 GHz, too. Therefore, transceiver signal processing including channel estimation and dealing with multipath effects, frequency offsets, and interference are more complicated and important tasks. Hence, more advanced coding schemes and frequent pilot signals are required. Multiantenna systems with high spatial resolution such as massive MIMO arrays occupy a large physical space. On the other hand, managing mobility is easier in sub-6 GHz regime.

4.2.2 Cooperation in JRC

JRC systems may have different configurations and use different duplexing and MA techniques to distinguish between downlink and uplink traffic in communications. They may accommodate multiple simultaneous users in both sensing and communication tasks and, thus, need to manage different types of interference.

Communications and sensing systems can coexist while competing for the same spectral resources or collaborate for mutual benefits. In the case of noncooperative coexistence, radar and communications systems are running on different hardware platforms and in distinct locations and access the spectrum in an opportunistic manner. Each participating subsystem would need to estimate the channel and interference information individually and adapt its own transmitter and receiver parameters in a selfish manner while taking into account the spectrum regulation in the employed frequency band.

Cooperative JRC requires that information such as the experienced SNR or SINR levels and the channel responses are exchanged among the users. For cooperation, protocols for sharing the awareness among the users are needed. If existing communications systems are involved in the cooperation, changes in the standards may be necessary.

The most advanced mode of a JRC system is achieved via co-design. Co-designed ISAC systems share the same transceiver hardware, timing circuitry, multiantenna configurations, and signal processing algorithms. DFRC waveforms are used for sensing and communications tasks jointly. Co-design is a part of RF convergence research and development effort where multifunction transceivers are being developed for sensing and communication.

In multiuser JRC systems, duplexing and MA methods also play an important role in accessing the shared spectrum. Downlink and uplink traffic and different users can be separated in frequency, time, spatial, or code domains. Radars have to listen to their own transmitted signals scattered off the targets. If the radar is transmitting continuously, sufficient isolation between transmitters and receivers have to be ensured. In pulsed radars, pulse repetition rate and pulse width need to be designed such that ambiguities in determining the target range can be avoided. Synchronization is also important in obtaining high-resolution performance in sensing, in particular localizing targets and estimating their velocity. Shared timing circuitry also facilitates saving energy and compensating for clock offsets and skews.

4.3 JRC Beamforming

The beamforming capability of a radar can be used for transmitting data to communication users in addition to the main radar task. It is possible to steer beams separately toward the communication receiver, but this would take resources away from the operation of the radar. More efficient approaches maintain the performance of the radar while transmitting the communication signal on the side, or jointly optimize the beampattern for both radar and communication use. A JRC system using beamforming is illustrated in Figure 4.2.

In the former radar-centric approach, the beampattern is designed so that the communication signal can be transmitted while incurring minimal losses to the radar performance. The communication data can be transmitted using amplitude, phase, or both [Hassanien et al., 2019]. In this approach, the communication capabilities, including the data rate, depend on the operational parameters of the radar.

If the communication capabilities are not sufficient in the radar-centric approach, joint beamforming approaches can be used. The beampattern is optimized so that quality constraints for the communication signals are satisfied. This, however, requires diverting some resources from the radar operation to the communication. Common design criteria for the beampattern include the squared error between the desired beampattern and the actual beampattern as well as the probability of detection, whereas receiver SINR and data rate are commonly used criteria for the communication [Hassanien et al., 2019, Liu et al., 2017a, 2020b].

The implementation of beamformers depends on the employed frequency band. In sub-6 GHz bands, beamforming is typically digital, whereas in mm-wave frequencies such as 77 GHz band, analog and hybrid beamformers are common.

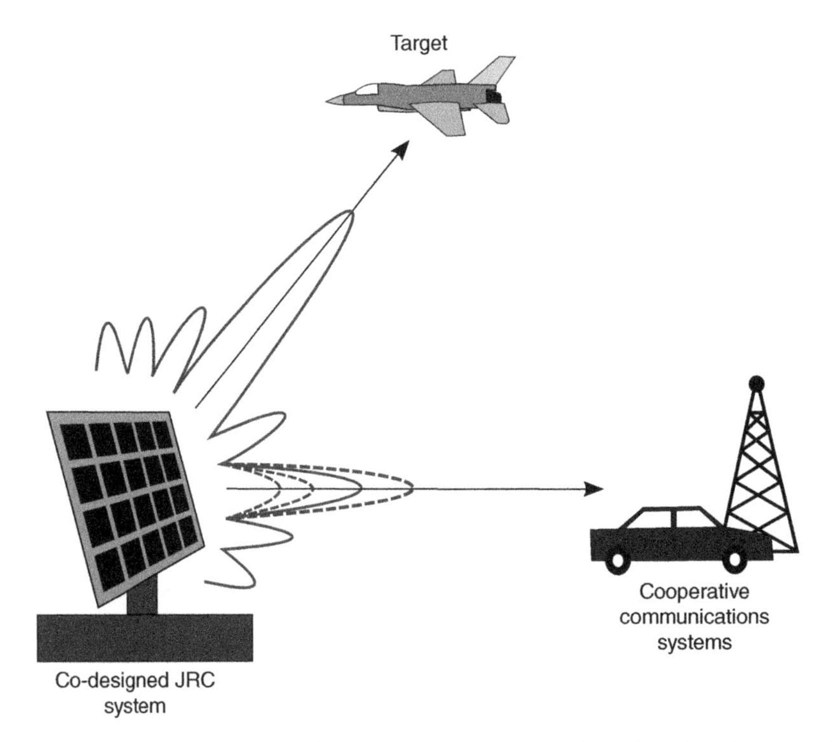

Figure 4.2 Beamforming in JRC system. The beampattern is designed so that the radar operation is not affected while communication is performed simultaneously.

4.3.1 Radar-centric Approach

Beamforming can be used in JRC systems for simultaneous communication and sensing tasks. In the radar-centric approach, the beampattern is adjusted to transmit communication signals with minimal effect on the main operation of the radar.

The transmit beampattern of the radar can be written as [Fuhrmann and Antonio, 2004, Stoica et al., 2007]

$$B(\theta) = \mathbf{a}_T^H(\theta)\mathbf{\Gamma}_\mathbf{x}\mathbf{a}_T(\theta), \tag{4.1}$$

where $\mathbf{a}_T(\theta)$ is the transmit steering vector and $\mathbf{\Gamma}_\mathbf{x}$ is the covariance matrix of the transmitted signal. In a typical beampattern design, either the maximum square error or the integrated square error between the desired beampattern B_0 and the actual beampattern is minimized on the angular region of interest Θ_M while constraining the sidelobe level to a tolerable threshold. This design problem can be written, e.g. as

$$\underset{\alpha,\mathbf{\Gamma}_\mathbf{x}}{\text{minimize}} \; \max \left|\alpha B_0(\theta) - \mathbf{a}_T^H(\theta)\mathbf{\Gamma}_\mathbf{x}\mathbf{a}_T(\theta)\right|^2 \qquad \theta \in \Theta_M, \tag{4.2a}$$

$$\text{subject to } \mathbf{a}_T^H(\theta)\mathbf{\Gamma}_\mathbf{x}\mathbf{a}_T(\theta) \leq c_{\mathrm{SL}} \qquad \theta \in \Theta_{\mathrm{SL}}, \tag{4.2b}$$

$$\mathrm{tr}(\mathbf{\Gamma}_\mathbf{x}) \leq P_{\mathrm{tot}}, \tag{4.2c}$$

$$(\mathbf{\Gamma}_\mathbf{x})_{ii} \leq P_{\mathrm{el}}, \tag{4.2d}$$

$$\mathbf{\Gamma}_\mathbf{x} \succeq 0, \tag{4.2e}$$

where α is a scaling constant to match the desired shape of the beampattern with the available power [Stoica et al., 2007], c_{SL} is the allowed sidelobe level, Θ_{SL} is the sidelobe region, P_{tot} is the total transmit power, and P_{el} is the maximum transmit power of each element. Equation (4.2e) constrains $\mathbf{\Gamma}_\mathbf{x}$ to be positive-semidefinite.

Let the coding matrix of the transmitted waveforms be $\mathbf{W}$. If the waveforms are initially uncorrelated, the transmit covariance matrix is $\mathbf{\Gamma}_\mathbf{x} = \mathbf{W}\mathbf{W}^H$. For a full-rank $\mathbf{\Gamma}_\mathbf{x}$, the beampattern design problem can be solved initially for $\mathbf{\Gamma}_\mathbf{x}$, and then the coding matrix $\mathbf{W}$ can be obtained with a matrix square root or Cholesky factorization.

If there are fewer waveforms than the number of antenna elements, the covariance matrix becomes rank-deficient. For a single waveform in particular, $\mathbf{\Gamma}_\mathbf{x} = \mathbf{w}\mathbf{w}^H$. Solving (4.2) with a rank constraint is difficult. Common approaches either try to solve $\mathbf{w}$ directly or to obtain it using the full-rank solution. One such method is semidefinite relaxation in which the full rank $\mathbf{\Gamma}_\mathbf{x}$ is solved first, and the low-rank or rank-one solution is obtained then by randomization [Luo et al., 2010], for example.

The communication data can be transmitted by modulating the beampattern in a suitable way. The modulation can be done on the power or the phase.

4.3.1.1 Power Modulation

The idea in the power modulation is to change the power radiated toward the communication receiver while maintaining the other properties of the transmit beampattern. In order for this not to disrupt the main operation of the radar, the communication receiver cannot be in the mainlobe of the radar. This approach is also called amplitude-shift keying [Hassanien et al., 2019], although the data is coded in the change of the power from pulse to pulse.

A power level of the beampattern corresponds to each communication symbol, so the number of symbols should be chosen based on the permitted power level and the channel quality. Denote the communication symbols by $\{\gamma_1, \gamma_2, \dots, \gamma_{N_\gamma}\}$ and let the direction of the communication receiver be θ_c. The idea is to maintain the beampattern shape elsewhere, but transmit the symbol to the communication receiver so that $\mathbf{a}_T^H(\theta_c)\mathbf{\Gamma}_\mathbf{x}\mathbf{a}_T(\theta_c) = \gamma_i$. The beampattern design problem needs thus to be solved separately for each communication symbol γ_i.

In order to solve the single-waveform beampattern, Hassanien et al. [2019] proposed minimizing the absolute error between the ideal array response b_0, where $|b_0(\theta)|^2 = B_0(\theta)$, and the actual one $\mathbf{w}^H\mathbf{a}_T(\theta)$. This optimization problem can then be written as

$$\underset{\alpha,\mathbf{w}}{\text{minimize}} \;\max \left|\alpha b_0(\theta) - \mathbf{w}^H\mathbf{a}_T(\theta)\right| \qquad \theta \in \Theta_M, \tag{4.3a}$$

$$\text{subject to } |\mathbf{a}_T^H(\theta)\mathbf{w}|^2 \leq c_{\text{SL}} \qquad \theta \in \Theta_{\text{SL}}, \tag{4.3b}$$

$$||\mathbf{w}||^2 \leq P_{\text{tot}}, \tag{4.3c}$$

$$|(\mathbf{w})_{ii}|^2 \leq P_{\text{el}}, \tag{4.3d}$$

$$\mathbf{w}^H\mathbf{a}_T(\theta_c) = \gamma_i^{1/2}, \tag{4.3e}$$

where the linear constraint (4.3e) is used to transmit the desired symbol. This problem is convex, but it is generally necessary to solve a nonconvex problem to obtain $b_0(\theta)$ as the desired beampattern $B_0(\theta)$ does not contain the phase information.

Using a linear constraint for the communication symbol has some drawbacks. Firstly, it unnecessarily fixes the phase. Secondly, small error in the assumed direction of the communication receiver θ_c could lead to large changes in the power of the transmit beampattern and thus result in a high error rate. In de Oliveira Ferreira et al. [2018], a constraint on the derivative of the beampattern,

$$\mathbf{w}^H\frac{d\,\mathbf{a}_T(\theta_c)}{d\,\theta} = 0, \tag{4.4}$$

is used to control how fast the beampattern can change in the vicinity of θ_c. By setting the derivative to zero, a local extremum point can be generated in the beampattern. However, this does not control how fast the beampattern changes in directions other than θ_c.

It is possible to formulate the beampattern design problem so that the phase is not fixed for the communication direction while also limiting the change of the beampattern in the vicinity of the direction of the communication receiver. The problem can be written as

$$\underset{\alpha,\mathbf{\Gamma}_\mathbf{x}}{\text{minimize}} \;\epsilon_M, \tag{4.5a}$$

$$\text{subject to } \max \left|\alpha B_0(\theta) - \mathbf{a}_T^H(\theta)\mathbf{\Gamma}_\mathbf{x}\mathbf{a}_T(\theta)\right|^2 < \epsilon_M \qquad \theta \in \Theta_M, \tag{4.5b}$$

$$\mathbf{a}_T^H(\theta)\mathbf{\Gamma}_\mathbf{x}\mathbf{a}_T(\theta) \leq c_{\text{SL}} \qquad \theta \in \Theta_{\text{SL}}, \tag{4.5c}$$

$$\text{tr}(\mathbf{\Gamma}_\mathbf{x}) \leq P_{\text{tot}}, \tag{4.5d}$$

$$(\mathbf{\Gamma}_\mathbf{x})_{ii} \leq P_{\text{el}}, \tag{4.5e}$$

$$\left|\gamma_i - \mathbf{a}_T^H(\theta)\mathbf{\Gamma}_\mathbf{x}\mathbf{a}_T(\theta)\right| < \epsilon_\gamma \qquad \theta \in \Theta_c, \tag{4.5f}$$

$$\mathbf{\Gamma}_\mathbf{x} \succeq 0, \tag{4.5g}$$

where Θ_c is the set of possible directions of the communication receiver and ϵ_γ controls the error in the vicinity of the communication direction. In a MIMO radar with full-rank covariance matrix $\mathbf{\Gamma}_\mathbf{x}$, the direction θ can be discretized to obtain a convex optimization problem. Alternatively, it is possible to fix the mainlobe error ϵ_M and obtain a convex feasibility problem that is solvable via the sum-of-squares representation [Aittomäki and Koivunen, 2017].

When there are multiple communication receivers, an additional constraint

$$|\gamma_{i_k} - \mathbf{a}_T^H(\theta)\mathbf{\Gamma}_\mathbf{x}\mathbf{a}_T(\theta)| < \epsilon_{\gamma,k}, \quad \theta \in \Theta_{c,k}$$

is needed for each user k. It should be noted that in this case, the beampattern design problem has to be solved for each combination of the transmitted symbols $\{\gamma_{i_1}, \gamma_{i_2}, \dots\}$.

The benefit of the power modulation in JRC is that the communication can be carried out with almost no extra cost. The drawback, however, is that the data rate is largely determined by the PRT of the radar, and that can also change during the operation. Furthermore, communication is not possible in the direction of the mainlobe.

Figure 4.3 shows an example of transmit beampattern optimization with communication using power modulation. The transmit array is a 10-element uniform linear MIMO array with 10 waveforms. The mainlobe is centered at −15° and its width is 30°. The direction of the communication receiver is 45° and two symbols are used at $\gamma_1 = -20$ dB and $\gamma_2 = -40$ dB relative to the mainlobe power, which is typically very high. The sidelobe constraint is $c_{\mathrm{SL}} = -15$ dB.

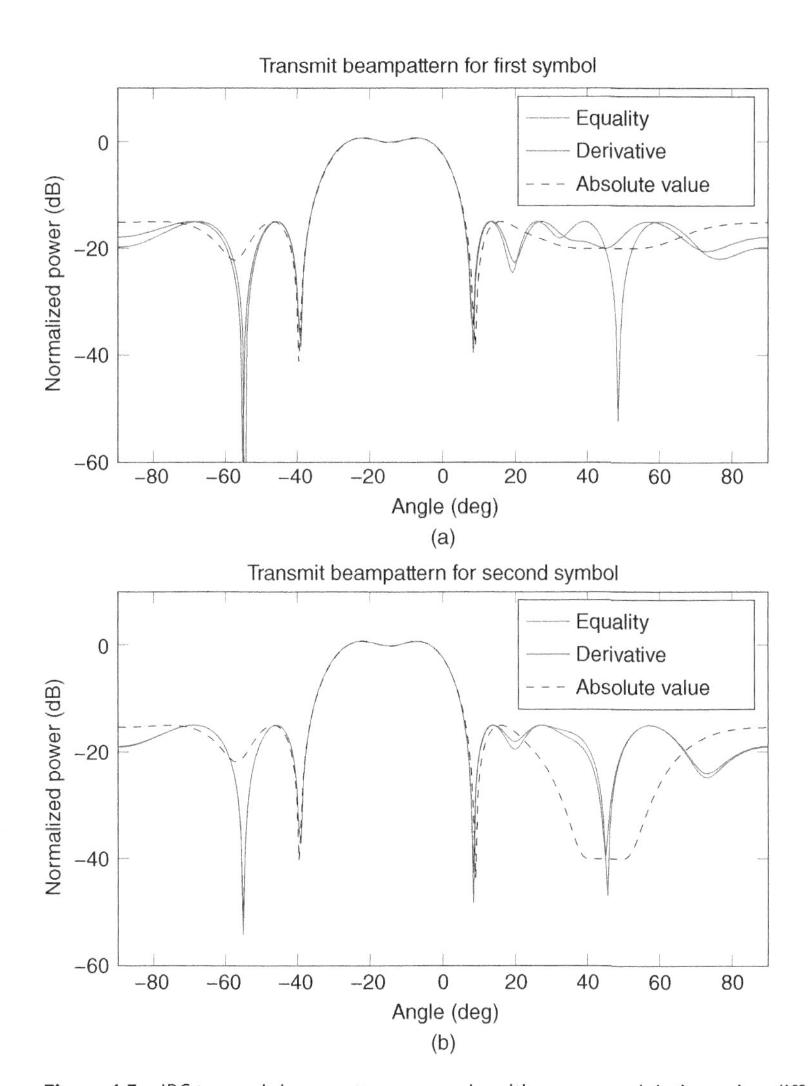

Figure 4.3 JRC transmit beampattern example with power modulation using different constraints. The equality constraint can set the power only for some angles. The derivative constraint can be used in addition to limit the rate of change of the beampattern near the communication direction, but this approach also lacks precise control. Using the constraint for the absolute value, a sufficient control of the transmit power over a range of angles can be achieved. (a) $\gamma_1 = -20$ dB and (b) $\gamma_2 = -40$ dB.

Three different design approaches are used, the first one using an equality constraint similar to (4.3e), whereas the second method also uses (4.4) so that the derivative of the beampattern with respect to θ is constrained to zero. The third approach uses the absolute error constraint in (4.3e) where the angle range Θ_c for the communication symbol is 10° and the error $\epsilon_\gamma = 10^{-4}$.

All the three approaches have a similar mainlobe. However, it can be seen in Figure 4.3 that while the equality constraint can be used for constraining the power levels, there can be large changes in the power for small changes in the angle, which could lead to increased BER. The derivative constraint is more successful with the first symbol but suffers from the same problem with the second symbol when using only the equality constraint. On the other hand, the absolute value constraint can be used to achieve a relatively flat beampattern over the desired range of angles, leading to robust communication.

4.3.1.2 Phase Modulation

In phase modulation, the data is coded into the phase of the transmission. The set of symbols is typically assumed to be $\gamma_i = e^{j2\pi i/N_\gamma}, i = 0, \dots, N_\gamma - 1$, corresponding phase-shift keying. In JRC systems, using phase modulation is simpler than power modulation as the phase of the transmitted signal has no significance in the operation of the radar; only the phase difference of the transmitted and the received signals is significant, and, thus, the phase can be selected for communication purposes. As the transmit power does not need to be adjusted according to the transmitted communication symbols, communicating to a user within the mainlobe of the radar is also possible.

Let the M transmit waveforms be denoted by $\mathbf{s}_r(t)$. The received signal at the communication receiver is then

$$y_c(t) = h_c \mathbf{a}_T^H(\theta_c) \sqrt{\frac{P_{\text{tot}}}{M}} \mathbf{W} \mathbf{s}_r(t - \tau) + n_c(t), \tag{4.6}$$

where h_c is the communication channel coefficient, and $n_c(t)$ is the communication receiver noise. We see that shifting the phase of the transmit precoding matrix $\tilde{\mathbf{W}} = \gamma_i \mathbf{W}$ introduces an equal phase shift to the signal y_c of the communication receiver. In order to take into account the communication channel coefficient h_c, the receiver needs to do channel estimation by using, e.g. pilot signals.

In case the channel coherence time is short or phase synchronization is difficult, the phase may be coded in the phase of an additional waveform, provided that such is available [Hassanien et al., 2019]. A complex scheme was proposed in Hassanien et al. [2016b] with different weights for each of the waveforms. A simpler approach is to add the extra waveform with a phase shift to one of the previously used waveforms. Let $s_{r,0}(t)$ be the additional waveform with sufficient orthogonality. Then using

$$\tilde{s}_{r,k}(t) = \frac{1}{\sqrt{2}} s_{r,k}(t) + \frac{\gamma_i}{\sqrt{2}} s_{r,0}(t) \tag{4.7}$$

will result in the same beampattern, but the communication receiver can retrieve the symbol γ_i as the phase difference of the output of the filters matched to $s_{r,0}$ and $s_{r,k}$. This scheme can be readily extended for multiple communication receivers. If an additional waveform is not available, differential phase-shift keying can be used to code the symbol in the phase difference of two subsequent pulses.

If the communication receiver is within the mainlobe of the radar, no extra constraints are required in the beampattern design problem. In case the receiver is in the sidelobe region, the constraint (4.5f) can be modified to

$$|P_c - \mathbf{a}_T^H(\theta) \mathbf{\Gamma}_{\mathbf{x}} \mathbf{a}_T(\theta)| < \epsilon_\gamma, \quad \theta \in \Theta_c, \tag{4.8}$$

where P_c is the desired power level for the communication. This constraint guarantees that sufficient power is transmitted toward the communication receiver.

When there are multiple communication receivers using phase-shift keying, the beampattern design becomes more complicated. For the first receiver, the phase can be coded as outlined before. For the second receiver, however, it is necessary to fix the phase of the transmitted signal to the *phase difference* of the transmitted symbols. For a single-waveform case, this can be done either using a linear constraint

$$\mathbf{a}_T^H(\theta_{c,2})\mathbf{w} = \gamma_{i_1}^{-1}\gamma_{i_2}\sqrt{P_c} \tag{4.9}$$

or a quadratic constraint

$$\left|\gamma_{i_1}^{-1}\gamma_{i_2}\sqrt{P_c} - \mathbf{a}_T^H(\theta_{c,2})\mathbf{w}\right|^2 < \epsilon_{\gamma,2}, \quad \theta \in \Theta_{c,2}. \tag{4.10}$$

These constraints are convex, but it is necessary obtain the desired response $b_0(\theta)$ beforehand as in (4.3). For the MIMO case with multiple waveforms, the phase depends on the coding matrix $\mathbf{W}$, the direction θ, and the waveforms $\mathbf{s}_r(t)$. The phase can be constrained to some point in time, by e.g. a constraint $\mathbf{a}_T^H(\theta_{c,2})\mathbf{W}\mathbf{s}_r(0) = \gamma_{i_1}^{-1}\gamma_{i_2}$, but this makes the retrieval of the communication symbol prone to errors.

It is possible to use quadrature amplitude modulation by using both phase and magnitude constraints. Alternatively, different modulation types can be used for communicating with different users. The benefits and drawbacks of the phase-shift keying are largely the same as for the amplitude-shift keying, namely, the communication comes with little extra cost but is largely constrained by the operation of the radar.

4.3.2 Joint Transmit Beamforming

The beamforming methods in the previous sections transmit the communication data alongside the main radar operation. To perform the communication in a manner not tied to the operation of the radar, different approaches are needed. A beamforming method for achieving this was proposed in Liu et al. [2020b].

Let $\mathbf{s}_r(t)$ be the radar waveforms and $\mathbf{s}_c(t)$ be the communication signal. Furthermore, assume that these are uncorrelated and have identity covariance matrices, i.e.

$$\mathrm{E}[\mathbf{s}_r(t)\mathbf{s}_r^H(t)] = \mathbf{I},$$
$$\mathrm{E}[\mathbf{s}_c(t)\mathbf{s}_c^H(t)] = \mathbf{I}.$$

The transmitted signal consists of both the radar and communication signals encoded by their respective precoding matrices $\mathbf{W}_r$ and $\mathbf{W}_c$. The transmitted signal can thus be written as

$$\mathbf{x}(t) = \mathbf{W}_r\mathbf{s}_r(t) + \mathbf{W}_c\mathbf{s}_c(t) \tag{4.11}$$

and the covariance matrix of the transmitted signal is

$$\mathbf{\Gamma}_\mathbf{x} = \mathbf{W}_r\mathbf{W}_r^H + \mathbf{W}_c\mathbf{W}_c^H \tag{4.12}$$

as the $\mathbf{s}_r(t)$ and $\mathbf{s}_c(t)$ are uncorrelated.

The signal received by the communication receiver is

$$\mathbf{y}_c(t) = \mathbf{H}\mathbf{x}(t) + \mathbf{n}_c(t) = \mathbf{H}\mathbf{W}_r\mathbf{s}_r(t) + \mathbf{H}\mathbf{W}_c\mathbf{s}_c(t) + \mathbf{n}_c(t), \tag{4.13}$$

where $\mathbf{H}$ is the channel matrix for the communication and $\mathbf{n}_c(t)$ is the communication receiver noise.

The signal power of the kth receiver is

$$\mathrm{E}\left[|(\mathbf{HW}_c)_{kk}(\mathbf{s}_c(t))_k)|^2\right] = |(\mathbf{HW}_c)_{kk}|^2 = |\mathbf{h}_k^H \mathbf{w}_{c,k}|^2, \tag{4.14}$$

where $\mathbf{h}_k^H$ is the kth *row* of $\mathbf{H}$ and $\mathbf{w}_{c,k}$ is the kth *column* of $\mathbf{W}_c$. Correspondingly, the SINR can be written as

$$\mathrm{SINR}_k = \frac{|\mathbf{h}_k^H \mathbf{w}_{c,k}|^2}{\sum_{n\neq k}|\mathbf{h}_n^H \mathbf{w}_{c,n}|^2 + \sum_m |\mathbf{h}_m^H \mathbf{w}_{r,m}|^2 + \sigma_{n,c}^2}, \tag{4.15}$$

where $\sigma_{n,c}^2$ is the noise power of the communication receivers. One now wishes to solve the beampattern optimization problem while guaranteeing that the SINR of each receiver is above a threshold c_{SINR}, that is

$$\underset{\mathbf{W}_r, \mathbf{W}_c}{\text{minimize}} \; \max\, |\alpha B_0(\theta) - \mathbf{a}_T^H(\theta)\boldsymbol{\Gamma}_{\mathbf{x}}\mathbf{a}_T(\theta)|^2 \qquad \theta \in \Theta_M, \tag{4.16a}$$

$$\text{subject to } \mathbf{a}_T^H(\theta)\boldsymbol{\Gamma}_{\mathbf{x}}\mathbf{a}_T(\theta) \leq c_{\mathrm{SL}} \qquad \theta \in \Theta_{\mathrm{SL}}, \tag{4.16b}$$

$$\mathrm{tr}(\boldsymbol{\Gamma}_{\mathbf{x}}) \leq P_{\mathrm{tot}}, \tag{4.16c}$$

$$(\boldsymbol{\Gamma}_{\mathbf{x}})_{ii} \leq P_{\mathrm{el}}, \tag{4.16d}$$

$$\mathrm{SINR}_k \geq c_{\mathrm{SINR}}. \tag{4.16e}$$

This problem is not convex because of the SINR constraints. However, a convex problem can be obtained by using semidefinite relaxation [Luo et al., 2010]. To this end, let

$$\mathbf{U}_m = \mathbf{w}_{r,m}\mathbf{w}_{r,m}^H, \quad m = 1, \ldots, M \tag{4.17}$$

and

$$\mathbf{U}_{M+k} = \mathbf{w}_{c,k}\mathbf{w}_{c,k}^H, \quad k = 1, \ldots, K \tag{4.18}$$

such that

$$\boldsymbol{\Gamma}_{\mathbf{x}} = \sum_{n=1}^{M+K} \mathbf{U}_n. \tag{4.19}$$

The SINR is thus

$$\begin{aligned} \mathrm{SINR}_k &= \frac{\mathbf{h}_k^H \mathbf{U}_{M+k}\mathbf{h}_k}{\sum_{n\neq M+k}\mathbf{h}_k^H \mathbf{U}_n \mathbf{h}_k + \sigma_{n,c}^2} \\ &= \frac{\mathbf{h}_k^H \mathbf{U}_{M+k}\mathbf{h}_k}{\mathbf{h}_k^H \boldsymbol{\Gamma}_{\mathbf{x}}\mathbf{h}_k - \mathbf{h}_k^H \mathbf{U}_{M+k}\mathbf{h}_k + \sigma_{n,c}^2}. \end{aligned} \tag{4.20}$$

Finally, the joint beampattern optimization problem can be written as [Liu et al., 2020b]

$$\underset{\mathbf{U}_1, \ldots, \mathbf{U}_{M+K}}{\text{minimize}} \; \max\, \left|\alpha B_0(\theta) - \mathbf{a}_T^H(\theta)\boldsymbol{\Gamma}_{\mathbf{x}}\mathbf{a}_T(\theta)\right|^2 \qquad \theta \in \Theta_M, \tag{4.21a}$$

$$\text{subject to } \mathbf{a}_T^H(\theta)\boldsymbol{\Gamma}_{\mathbf{x}}\mathbf{a}_T(\theta) \leq c_{\mathrm{SL}} \qquad \theta \in \Theta_{\mathrm{SL}}, \tag{4.21b}$$

$$\mathrm{tr}(\boldsymbol{\Gamma}_{\mathbf{x}}) \leq P_{\mathrm{tot}}, \tag{4.21c}$$

$$(\boldsymbol{\Gamma}_{\mathbf{x}})_{ii} \leq P_{\mathrm{el}}, \tag{4.21d}$$

$$\boldsymbol{\Gamma}_{\mathbf{x}} = \sum_{n=1}^{M+K} \mathbf{U}_n \geq 0, \tag{4.21e}$$

$$\mathbf{U}_n \succeq 0 \qquad n = 1, \ldots, M+K, \tag{4.21f}$$

$$(1 + c_{\text{SINR}}^{-1})\mathbf{h}_k^H \mathbf{U}_{M+k} \mathbf{h}_k > \mathbf{h}_k^H \mathbf{\Gamma}_{\mathbf{x}} \mathbf{h}_k + \sigma_{n,c}^2 \qquad k = 1, \ldots, K, \tag{4.21g}$$

where the SINR constraints are now linear in $\mathbf{U}_n$. This relaxed problem was shown in Liu et al. [2020b] to be tight, meaning that its solution is the solution of the original, nonrelaxed problem.

4.3.3 Receiver Processing

To facilitate both the target detection and the reception of the communication data, it is necessary for the JRC system to minimize the interference that the radar and communication signals cause to each other. At the receiver, the interference can be attenuated by beamforming or temporal filtering.

The signal received by the radar can be written as

$$\mathbf{y}_r(t) = \sum_{i=1}^{N_b} \beta_i \mathbf{a}_R(\theta_{b,i}) \mathbf{a}_T^H(\theta_{b,i}) \mathbf{x}(t) + \mathbf{H}\mathbf{W}_d \mathbf{s}_d(t) + \mathbf{n}_r(t), \tag{4.22}$$

where N_b is the number of targets, β_i is the path loss and scattering coefficient, $\mathbf{a}_R(\theta_{b,i})$ is the receiver steering vector, and $\mathbf{a}_T(\theta_{b,i})$ is the transmit steering vector for the ith target. Matrix $\mathbf{H}$ is the channel matrix and $\mathbf{n}_r(t)$ is the receiver noise at the radar. Here, $\mathbf{W}_d$ is the communication transmit coding matrix and $\mathbf{s}_d(t)$ denotes the signal transmitted by the communications subsystem. It is necessary for the radar to check for the presence of targets in the assumed directions $\theta_{b,i}$. Given that the limiting factor in the receiver is the computational complexity and not the transmit power, it makes sense to form multiple narrow beams for each $\theta_{b,i}$ unlike in the transmit beampattern design where the power is limited.

A sensible design criterion for the receiver beamformer weights is to minimize the interference and noise power. If the receiver beamforming weights are given by $\mathbf{u}$, the receiver beamformer optimization can be written as

$$\underset{\mathbf{u}}{\text{minimize}} \ \mathbf{u}^H \mathbf{H}\mathbf{W}_d \mathbf{W}_d^H \mathbf{H}^H \mathbf{u}, \tag{4.23a}$$

$$\text{subject to } \mathbf{u}^H \mathbf{a}(\theta_{b,n}) = 1, \tag{4.23b}$$

where $\mathrm{E}[\mathbf{H}\mathbf{W}_d \mathbf{s}_d(t) \mathbf{s}_d^H(t) \mathbf{W}_d \mathbf{H}^H] = \mathbf{H}\mathbf{W}_d \mathbf{W}_d^H \mathbf{H}^H$ is the interference covariance matrix. The solution in this case will be the well-known minimum variance distortionless response (MVDR) beamformer. However, if the communications system can design its transmit encoding matrix including the off-diagonal elements (e.g. in the case of a MIMO transmitter), an alternative problem to minimize the interference caused to the radar by the communication signal can be written as

$$\underset{\mathbf{W}_d}{\text{minimize}} \ \text{tr}(\mathbf{A}_T^H \mathbf{H}\mathbf{W}_d \mathbf{W}_d^H \mathbf{H}^H \mathbf{A}_T) = \text{tr}(\mathbf{W}_d^H \mathbf{H}^H \mathbf{A}_T \mathbf{A}_T^H \mathbf{H}\mathbf{W}_d), \tag{4.24a}$$

$$\text{subject to } (\mathbf{W}_d^H \mathbf{W}_d)_{kk} = 1, \quad k = 1, \ldots, K, \tag{4.24b}$$

where $\mathbf{A}_T = \begin{bmatrix} \mathbf{a}_T(\theta_{b,1}) & \mathbf{a}_T(\theta_{b,2}) & \ldots \end{bmatrix}$, K is the number of communication transmitters, and it was assumed without loss of generality that the communication transmitters transmit at unit power. By denoting each column of $\mathbf{W}_d$ as $\mathbf{w}_{d,k}$, the code design problem can be alternatively written as

$$\underset{\mathbf{w}_{d,1},\ldots,\mathbf{w}_{d,K}}{\text{minimize}} \sum_{k=1}^{K} \mathbf{w}_{d,k}^H \mathbf{H}^H \mathbf{A}_T \mathbf{A}_T^H \mathbf{H} \mathbf{w}_{d,k}, \tag{4.25a}$$

$$\text{subject to } \mathbf{w}_{d,k}^H \mathbf{w}_{d,k} = 1, \quad k = 1, \ldots, K. \tag{4.25b}$$

The solution of this problem is that $\mathbf{w}_k$ are the eigenvectors corresponding to the smallest eigenvalues of $\mathbf{H}^H \mathbf{A}_T \mathbf{A}_T^H \mathbf{H}$, or equally the right singular vectors corresponding to the smallest singular values of $\mathbf{A}_T^H\mathbf{H}$. Naturally, this solution requires that the communication transmitter has the full knowledge of the CSI and also the target directions $\theta_{b,i}$, which can be problematic in a dynamic scenario.

In Hassanien et al. [2018], a particular solution preventing interference such that $\mathbf{W}_c^H\mathbf{a}_T(\theta_{b,i}) = 0$ is proposed for a MIMO radar employing a virtual aperture. More general methods for reducing the interference between the radar and multiple communication transmitters are presented in Section 4.5.

When using a virtual array aperture with the MIMO radar, it is typically necessary to filter the received signal with matched filters before the beamforming or any other spatial processing is applied [Forsythe et al., 2004, Li et al., 2008]. However, matched filtering is not optimal in the presence of nonwhite interference, so the communication signal could hinder the operation of the radar.

In the operation of the radar, it is crucial to control the sidelobe levels in addition to the interference. Therefore, mismatched filters [Ackroyd and Ghani, 1973] can be used for detection and also separation of the different waveforms of MIMO radar. Using mismatched filters, it is possible to control sidelobes, the interference power as well as the cross-correlation with the used waveforms.

Define $L \times 1$ vectors

$$\boldsymbol{\psi} = \begin{bmatrix}\psi(0) & \psi(1) & \dots & \psi(L-1)\end{bmatrix}^T \tag{4.26}$$

and

$$\mathbf{s}_{r,k}(n) = \begin{bmatrix}s_{r,k}^*(n-0) & s_{r,k}^*(n-1) & \dots & s_{r,k}^*(n-L+1)\end{bmatrix}^H, \tag{4.27}$$

where $\boldsymbol{\psi}$ contains the filter coefficients and $\mathbf{s}_{r,k}(n)$ is the kth transmitted radar waveform. The mismatched filter design to minimize the interference power for fast-time codes (insignificant Doppler) can now be written as [Aittomäki and Koivunen, 2014]

$$\underset{\boldsymbol{\psi}}{\text{minimize}}\ \boldsymbol{\psi}^H\boldsymbol{\Gamma}_{\text{i+n}}\boldsymbol{\psi}, \tag{4.28a}$$

$$\text{subject to } \mathbf{s}_{r,k}^H(0)\boldsymbol{\psi} = 1, \tag{4.28b}$$

$$|\mathbf{s}_{r,k}^H(n)\boldsymbol{\psi}|^2 \leq c_{\text{PSL}}, \qquad n = -L+1\dots -1, 1, \dots, L-1, \tag{4.28c}$$

$$|\mathbf{s}_{r,k}^H(n)\boldsymbol{\psi}|^2 \leq c_{\text{PCC}}, \qquad m \neq k,\ n = -L+1, \dots, L-1, \tag{4.28d}$$

where c_{PSL} is the peak sidelobe level constraint and c_{PCC} is the peak cross-correlation constraint between the different waveforms, and $\boldsymbol{\Gamma}_{\text{i+n}}$ is the interference-plus-noise covariance matrix.

If the Doppler shift can be significant (slow-time codes), the filter design problem needs to be adjusted accordingly. Define the Doppler-shifted transmit waveform as

$$\tilde{\mathbf{s}}_{r,k}(n,f) = \begin{bmatrix}s_{r,k} * (n-0) & s_{r,k} * (n-1)e^{-j2\pi f} & \dots & s_{r,k}^*(n-L+1)e^{-j2\pi(L-1)f}\end{bmatrix}^H. \tag{4.29}$$

The mismatched filter optimization can then be formulated as

$$\underset{\boldsymbol{\psi}}{\text{minimize}}\ \boldsymbol{\psi}^H\boldsymbol{\Gamma}_{\text{i+n}}\boldsymbol{\psi}, \tag{4.30a}$$

$$\text{subject to } \max_{f,n} \left|\boldsymbol{\psi}^H\tilde{\mathbf{s}}_{r,k}(n,f)\right|^2 \leq c_{\text{PSL}}, \qquad |f - f_b| \geq \delta_{0n}f_w, \tag{4.30b}$$

$$\max_{f,n,i} \left|\boldsymbol{\psi}^H\tilde{\mathbf{s}}_{r,i}(n,f)\right|^2 \leq c_{\text{PCC}}, \qquad i \neq k, \tag{4.30c}$$

$$\boldsymbol{\psi}^H\tilde{\mathbf{s}}_{r,k}(0,f_b) = 1, \tag{4.30d}$$

where f_b is the normalized frequency of Doppler bin of interest, δ_{ij} is the Kronecker delta, and f_w is the width of the mainlobe of the AF in frequency. Since the Doppler shift is a continuous variable, it

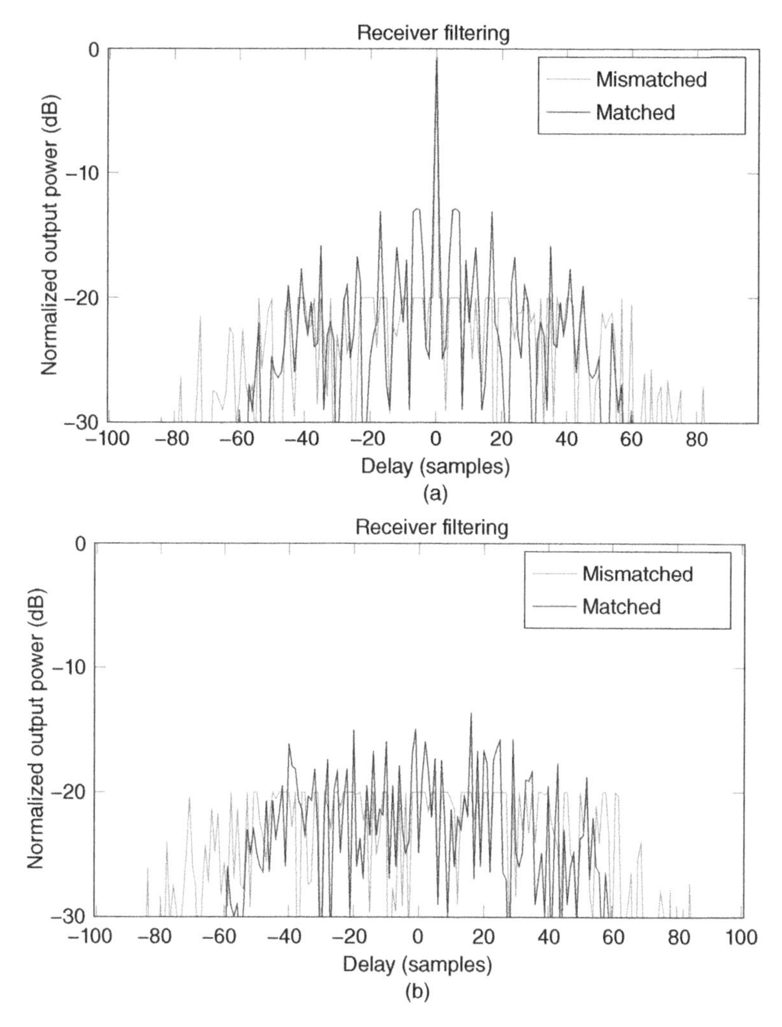

Figure 4.4 Receiver filter example using matched and mismatched filtering. The mismatched filter can be used to reduce the peak sidelobe and cross-correlation levels while improving the SINR – in this example by 6.1 dB. (a) Code 1 and (b) Code 2.

is necessary to either discretize it to obtain a quadratically constrained problem or use the bounded real lemma to obtain a convex optimization problem [Aittomäki and Koivunen, 2020].

Figure 4.4 shows a numerical example of the mismatched filtering. The radar uses two pseudo-random quadrature phase-shift keying (QPSK) signals with 63 symbols. There is an interfering 8-PSK signal with the interference-to-noise ratio of 10 dB and a frequency offset relative to the radar baseband of 0.3. The symbol rate for the interference is 5% of that of the radar (if the radar signal and the interference are on the same frequency band, the interference has to have a smaller bandwidth in order to be attenuated with temporal filtering). The mismatched filter with 131 coefficients is designed to have PSL and PCC of −20 dB while minimizing the interference power.

Figure 4.4a shows the filter output power for Code 1 that the filter is intended to receive, while Figure 4.4b shows the output power for Code 2, which in this case interferes with the reception of Code 1. It can be seen that the PSL and PCC were reduced to the required −20 dB level. Additionally, the SINR was improved by 6.1 dB compared to the matched filter.

4.4 Multicarrier Waveforms for JRC

Multicarrier waveforms, OFDM in particular, have been used in most present and emerging wireless communications systems. Multicarrier waveforms have been considered for radar applications as well, especially in passive radar systems. Consequently, multicarrier systems provide a very promising basis for co-designing RF sensing and communication as well as a common hardware platform for RF convergence. A variety of multicarrier waveform designs for radar and JRC tasks have been proposed, for example multicarrier phase-coded (MCPC) waveforms [Braun et al., 2013, Donnet and Longstaff, 2006, Levanon, 2000, Sturm et al., 2009] and generalized multicarrier waveforms in Bică and Koivunen [2016], see Dokhanchi et al. [2019], Liu et al. [2017b], and Mishra et al. [2019] for additional examples. The block diagram of OFDM JRC system is shown in Figure 4.5.

If no coding or spreading operation along subcarriers takes place, then a signal model and subcarrier spacing is similar to the conventional OFDM. If a signal is coded or spread over subcarriers or time, the signal model is typically similar to the ones used in a variety of multicarrier spread spectrum systems, including multicarrier CDMA (MC-CDMA) and MC-DS-CDMA. A prime example is the MCPC waveform using complementary sets of P3 or P4 polyphase codes [Levanon, 2000]. If spreading is used, it typically impacts the intercarrier spacing and the number of subcarriers employed, too. In OFDM, the intercarrier spacing is typically selected to be the reciprocal of the symbol duration; however, if spreading is used, the intercarrier spacing is commonly selected to be the reciprocal of the chip duration. Interference may be managed by appropriately allocating distinct subcarriers to radar tasks and communication users as illustrated in Figure 4.6. The transmitted waveform can be optimized by adjusting the power allocation to subcarriers based on the channel quality or interference levels or by optimizing the use of multiple antennas and subcarriers jointly. Radar range resolution depends on the bandwidth; hence, it may make sense to allocate radar signal subcarriers to the band edges.

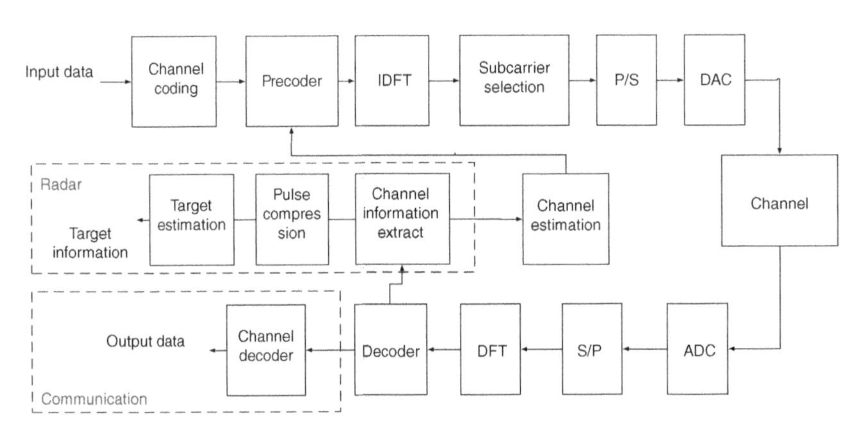

Figure 4.5 Block diagram of OFDM joint radar and communications system.

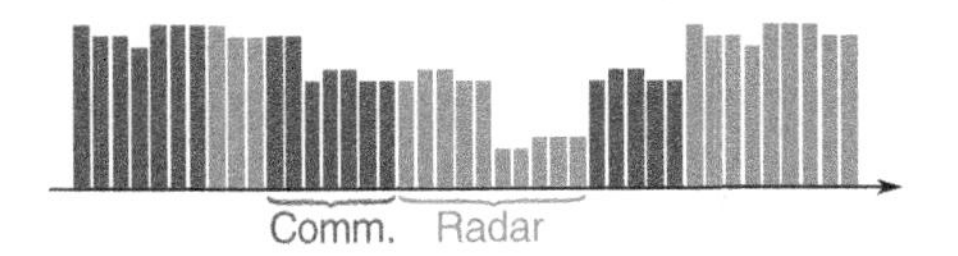

Figure 4.6 Subcarrier assignment for radar or communications users. Communication users could be allocated resource blocks of 32 subcarriers, for example. Allocating radar signals to band edges would help in achieving higher range resolution. Source: Bică and Koivunen 2019/IEEE.

Ideally, an orthogonal design allows for interference-free operation. In reality, however, the orthogonality is degraded at the receiver because of transceiver nonidealities, mobility, and a variety of propagation effects. This degradation causes unintentional interference. However, multicarrier techniques provide waveform diversity and frequency diversity as well as facilitate short time on target. Such systems also provide many DoFs to optimize the transmitted waveforms depending on the state of the spectrum and sensing or communication task at hand, such as power and time–frequency profile of subcarriers. For example, the observed SINR values at the receiver can be used to optimize the transmission parameters and the use of DoFs for a specific radar task and communication requirements. This would require feedback from the receiver if the transmitter and receiver are not colocated. Alternatively, the channels are assumed to be reciprocal and may be estimated sufficiently accurately. The transmission power can be allocated to subcarriers with high SINR or high channel gain using the water-filling principle, see for example Bică and Koivunen [2019]. Subcarriers that are subject to jamming or high attenuation have a low SINR; hence, they may be allocated zero transmit power; see for example Bică and Koivunen [2016].

Multicarrier systems can be used as a platform for RF convergence between sensing and communication both in terms of shared hardware and signal processing algorithms. If waveforms are used both for communication and sensing, then interference cancelling receivers such as serial interference canceller (SIC) may be used at the receiver. Multicarrier systems typically transmit continuously. Therefore, there needs to be highly effective isolation between the transmitter and receiver using a full-duplex transceiver structure. Alternatively, a bistatic configuration could be used for radar subsystems. The duplexing principle used in communications subsystems also plays a role. In TDD, downlink and uplink transmissions take place in an alternating manner in the same frequency band, whereas in FDD different frequencies are used for downlink and uplink. TDD systems can exploit the channel reciprocity when adjusting their transmitting waveforms. Co-designed JRC can be also used for covert sensing or communication.

Multicarrier waveforms have also some drawbacks in radar and JRC. They have a nonconstant signal amplitude, which may lead to a large peak-to-mean envelope power ratio (PMEPR). Consequently, the use of power amplifiers can become inefficient or the waveforms may be suitable mainly to shorter range applications to avoid distortions caused by amplifiers. Furthermore, multicarrier waveforms typically contain some periodicities that may degrade processing based on correlation or matched filtering in radars, for example by masking some target range-Doppler cells. A long cyclic prefix (CP) may decrease the unambiguous radial velocity. Correlations are induced by modulation, coding, guard intervals, CP, and pilot signals. Both intra-symbol and inter-symbol correlations take place.

Doppler sensitivity is often considered a drawback of multicarrier communications. However, in multicarrier radar, it may be beneficial since velocity is a key target parameter of interest and it can be estimated fast. In fact, Doppler may be resolved using only one pulse because of the effects of Doppler on several subcarrier frequencies in one time instant.

In the following, we present two examples of multicarrier waveform design for JRC in which subcarriers or subsets of subcarriers can be assigned to the radar or the communication tasks. First, an OFDM-type system model is considered. We consider two approaches: a radar-centric approach where the performance in a radar task is maximized while imposing constraints on minimum required performance levels for wireless communication users, for example in terms of data rate or QoS. On the other hand, in a cooperative formulation, a joint objective function comprised of terms related to the radar and communication performance is maximized. The approach chosen here is to jointly maximize the sum of two MI terms. The second design uses MC-DS-CDMA model with OVSF codes and long scrambling codes to separate the subsystems.

4.4.1 OFDM Design Example

We consider OFDM waveform as a special case of the generalized multicarrier radar (GMR) [Bică and Koivunen, 2016]. The approach is applicable to a variety of models where the subcarrier signals are spread. This example gives an overview of the method described in detail in Bică and Koivunen [2019]. In the case of N subcarriers, the dual-use waveform of N subcarriers is defined as

$$\mathbf{x} = \mathbf{F}^H[\mathbf{D}\mathbf{s}_r + (\mathbf{I} - \mathbf{D})\mathbf{s}_c], \tag{4.31}$$

where vector $\mathbf{s}_r$ of size N contains the frequency-domain transmitted radar symbols and vector $\mathbf{s}_c$ of size N contains the frequency-domain transmitted communication symbols. The diagonal matrix $\mathbf{D}$ of size $N \times N$ is a subcarrier selection matrix whose elements indicate if the subcarrier is modulated and take the values $\{0,1\}$. Matrix $\mathbf{F}^H$ is the inverse discrete Fourier transform (IDFT) matrix. The transmitted waveform passes through one radar channel $\mathbf{h}_r$, which contains the target, and one communication channel $\mathbf{h}_c$ between the JRC transceiver and the communication receiver.

The observed data at the radar and the communication receivers may be written in compact matrix notation as follows:

$$\begin{cases} \mathbf{y}_r = \mathbf{X}_r\mathbf{h}_r + \mathbf{X}_c\mathbf{h}_r + \mathbf{n}_r, \\ \mathbf{y}_c = \mathbf{X}_r\mathbf{h}_c + \mathbf{X}_c\mathbf{h}_c + \mathbf{n}_c, \end{cases} \tag{4.32}$$

where $\mathbf{X}_r$ and $\mathbf{X}_c$ are Toeplitz matrices constructed from the corresponding signal. Toeplitz matrices of sufficiently large dimensions, as is the case here, can be approximated by circulant matrices. Circulant matrices can be diagonalized by unitary discrete Fourier transform (DFT) matrices. It is assumed that the noise vectors $\mathbf{n}_r$ and $\mathbf{n}_c$, the radar channel $\mathbf{h}_r$ and communication channel coefficients $\mathbf{h}_c$ are all zero-mean complex Gaussian random vectors with known covariance matrices.

MI-based criteria are used for allocating the subcarriers for radar and communications. The radar subsystem wishes to maximize the MI between the received signal reflected off the target and the impulse response of the target. The communications subsystem wishes to maximize the MI between the received signal at the communication receiver and the transmitted signal. The MI-based joint objective function may be formulated as [Bică and Koivunen, 2019]

$$\frac{1}{2}\left[\sum_{k=0}^{N-1}\log\left(1+\frac{w[k]p_r[k]\eta_r[k]}{u[k]p_c[k]\eta_r[k]+\sigma_{n_c}^2}\right)+\sum_{k=0}^{N-1}\log\left(1+\frac{u[k]p_c[k]\eta_c[k]}{w[k]p_r[k]\eta_c[k]+\sigma_{n_r}^2}\right)\right], \tag{4.33}$$

where $p_r[k]$ and $p_c[k]$ denote the power of the kth subcarrier of the radar and the communication signal, whereas $\eta_r[k]$ and $\eta_c[k]$ denote the channel gains of the radar and communication channels on kth subcarrier, respectively. The selection coefficients $w[k]$ and $u[k]$ denote whether or not the kth subcarrier is available for radar or communication use, respectively, and take only values of 0 or 1. The noise power is denoted by $\sigma_{n_r}^2$ at the radar and by $\sigma_{n_c}^2$ at the communications subsystems.

A radar-centric design that optimizes the radar performance is obtained by solving the two following optimization problems. Initially, the radar system is optimized by solving

$$\begin{aligned} &\underset{\{p_r[k]\}}{\text{maximize}}\sum_{k=0}^{N-1}\log\left(1+\frac{w[k]p_r[k]\eta_r[k]}{\sigma_n^2}\right), \\ &\text{subject to}\sum_{k=0}^{N-1}w[k]p_r[k]\leq P_r, \end{aligned} \tag{4.34}$$

where P_r is the transmitted maximum radar power. Based on the obtained optimal power allocation on high-quality subcarriers for the radar subsystem, the remaining subcarriers can be assigned

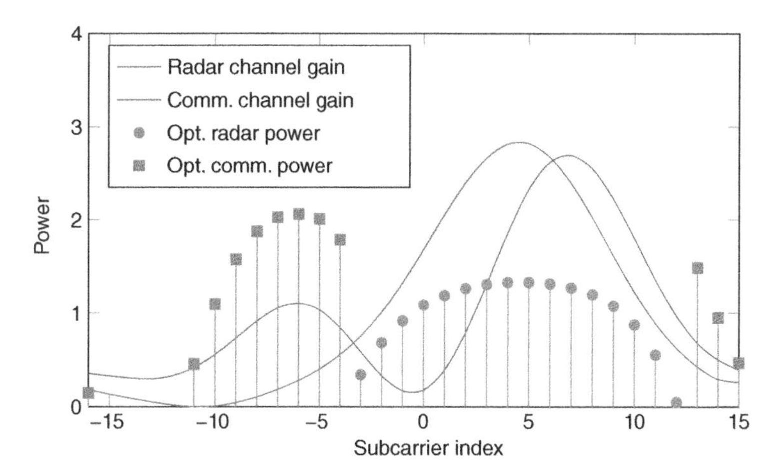

Figure 4.7 An example of radar-centric design and obtained water-filling power allocation for radar and communications subsystems. More power is allocated to subcarriers with higher channel gain and low noise and interference levels. Source: Bică and Koivunen 2019/IEEE

to the communications subsystem. The communications subsystem is thus optimized using the following formulation:

$$\begin{aligned}&\underset{\{p_c[k]\}}{\text{maximize}} \sum_{k=0}^{N-1} \log\left(1+\frac{u[k]p_c[k]\eta_c[k]}{\sigma_m^2}\right),\\&\text{subject to} \sum_{k=0}^{N-1} u[k]p_c[k] \leq P_c,\end{aligned} \tag{4.35}$$

where P_c is the power available for the communication. To prevent the communications system from using the subcarriers allocated for the radar, $u[k]$ is zero when $p_r[k] > 0$. Both Problems (4.34) and (4.35) can be solved exactly as in Bică and Koivunen [2016], for example, by using the Lagrangian form and the Karush–Kuhn–Tucker (KKT) conditions [Boyd and Vandenberghe, 2004]. The resulting solutions are water-filling solutions, where more power is allocated to subcarriers with higher channel gain and lower noise and interference power.

An example of obtained power allocation based on water-filling principle is depicted in Figure 4.7. An example of optimization problem formulation where the power allocation and subcarrier selection for both radar and communications system are done jointly without priorities or preferences between the subsystems can be found in Bică and Koivunen [2019].

4.4.2 MC-DS-CDMA Design Example

Spread spectrum waveforms are widely used in ranging and localization applications such as GPS, multiuser wireless communications (3G, WCDMA) and in radar. They facilitate high-resolution range estimation and have desirable low probability of intercept (LPI)/low probability of detection (LPD) and narrowband interference mitigation properties, making them useful in JRC systems as well. This example gives a condensed version of the multicarrier spread spectrum waveform design described in detail in Sharma et al. [2020].

We focus on MC-DS-CDMA waveforms with OVSFs since they allow for different data rates for different users in multiuser communications, powerful isolation of radar and communication signals through code design, desirable AF shapes as well as a convenient way to obtain LPI/LPD waveforms. OVSFs are well-known CDMA channelization codes known for their code tree construction. Spreading is done in the time domain. Long scrambling codes are used to reduce mutual correlations between the radar and communications subsystems as well as among different communication subcarriers, users, or sectors. Consequently, both sensing and communication signals will experience less unintentional interference.

In the following, we describe a JRC multicarrier waveform that uses a total of $N = N_c + N_r$ subcarriers, where N_c subcarriers are dedicated for communications subsystem and N_r for radar subsystem. An example case for such waveform systems could be cellular downlink communication scenario where multiple users are served by a base station. Each user is assigned multiple spreading codes (from an OVSF tree) similarly to Wideband CDMA in 3G. Scrambling sequences are used to separate the radar and communication signals, and if necessary, base stations, angular sectors, or users. A JRC transmitter using MC-DS-CDMA model is illustrated in Figure 4.8. The radar and communication subcarriers can be arranged differently, not necessarily in only two clusters as shown in the figure. Subcarriers for radar can also be allocated to band edges to obtain wider bandwidths and consequently better range resolution.

We first describe the transmitter structure for a communications subsystem using a joint waveform, shown in Figure 4.8. Let b_{k,M_k} denote a stream of M_k bits on the kth subcarrier, and c_k denotes the OVSF code (with Spreading Factor = SF_k) used on the kth subcarrier. To maintain the same sampling rate on all subcarriers, the following relation needs to hold:

$$\mathrm{SF}_s = M_1\mathrm{SF}_1 = M_2\mathrm{SF}_2 = \cdots = M_{N_c}\mathrm{SF}_{N_c} \tag{4.36}$$

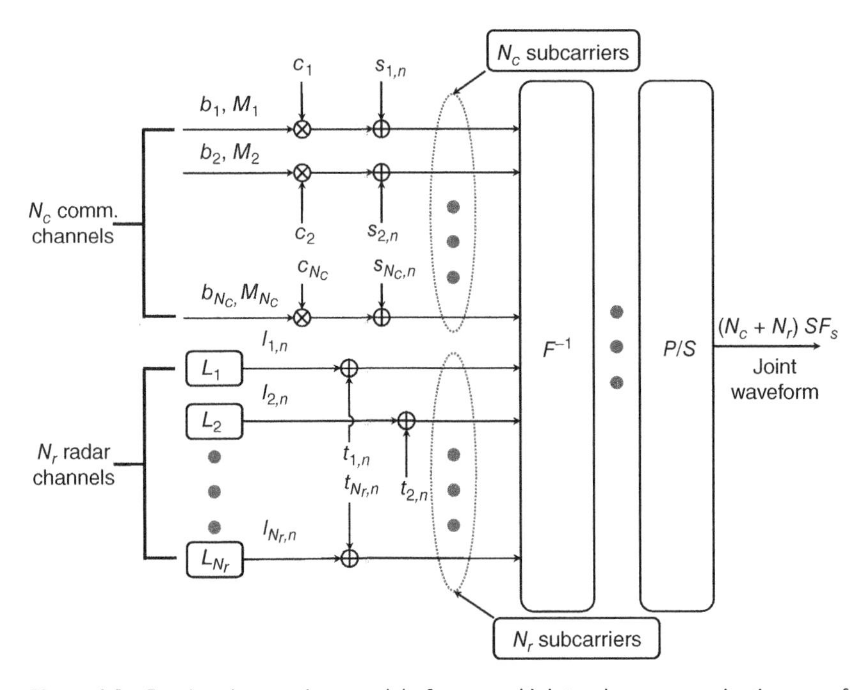

Figure 4.8 Baseband transmitter model of proposed joint radar–communication waveform for $N = N_c + N_r$ subcarriers. Communication signals are DS-CDMA modulated on N_c subcarriers using OVSF codes. Radar signals are modulated on N_r subcarriers using spreading codes. Radar and communications subsystems are isolated and users separated by scrambling using long codes. Source: Sharma et al. 2020/IEE

The chip matrix for communication signals can be written in matrix notation as

$$\mathbf{B} = \begin{bmatrix} \begin{bmatrix} b_{1,1} & \cdots & b_{1,M_1} \end{bmatrix} \otimes c_1 \\ \begin{bmatrix} b_{2,1} & \cdots & b_{2,M_2} \end{bmatrix} \otimes c_2 \\ \vdots \\ \begin{bmatrix} b_{N_c,1} & \cdots & b_{N_c,M_{N_c}} \end{bmatrix} \otimes c_{N_c} \\ \end{bmatrix}_{N_c \times \mathrm{SF}_s}, \tag{4.37}$$

where $\otimes$ is Kronecker product. The commonly used OVSF codes do not have ideal auto- and cross-correlation properties; hence, long scrambling codes should be used. The long scrambling code sequences are denoted by $s_{k,n}$, which refers to the nth chip of the sequence on kth subcarrier. Since scrambling is a modulo-addition operation, it can be written as

$$\mathbf{X} = [\mathbf{B}]_{N_c \times \mathrm{SF}_s} \oplus \begin{bmatrix} s_{1,1} & \cdots & s_{1,\mathrm{SF}_s} \\ s_{2,1} & \cdots & s_{2,\mathrm{SF}_s} \\ \vdots & \ddots & \vdots \\ s_{N_c,1} & \cdots & s_{N_c,\mathrm{SF}_s} \end{bmatrix}_{N_c \times \mathrm{SF}_s}$$

$$\implies \mathbf{X} = [\mathbf{B} \oplus \mathbf{S}_c]_{N_c \times \mathrm{SF}_s},$$

where $\oplus$ denotes modulo-addition operation and $\mathbf{S}_c$ is the scrambling chip matrix. The scrambling is followed by the inverse fast Fourier transform (IFFT) stage, which is described later.

Next, we describe the radar part of the joint waveform. The radar code sequence deployed on the kth subcarrier with the nth chip is given by $l_{k,n}$. Scrambling used on radar subcarriers is denoted by $t_{k,n}$, which defines the nth chip on the kth subcarrier. Note that the chip rate for the radar sequences is the same as the chip rate for the communication streams. Suitable choices for radar and scrambling sequences include well-known pulse compression codes such as polyphase codes, Gold codes, Zadoff–Chu codes or Kasami codes – see Dinan and Jabbari [1998]. After the scrambling operation, the radar chip matrix is

$$\mathbf{Y} = \mathbf{L} \oplus \mathbf{T}, \tag{4.38}$$

where $\mathbf{L}$ and $\mathbf{T}$ are the radar sequence chip matrix and the radar scrambling chip matrix, respectively, such that

$$\mathbf{L} = \begin{bmatrix} l_{1,1} & \cdots & l_{1,\mathrm{SF}_s} \\ l_{2,1} & \cdots & l_{2,\mathrm{SF}_s} \\ \vdots & \ddots & \vdots \\ l_{N_r,1} & \cdots & l_{N_r,\mathrm{SF}_s} \end{bmatrix}_{N_r \times \mathrm{SF}_s}$$

and

$$\mathbf{T} = \begin{bmatrix} t_{1,1} & \cdots & t_{1,\mathrm{SF}_s} \\ t_{2,1} & \cdots & t_{2,\mathrm{SF}_s} \\ \vdots & \ddots & \vdots \\ t_{N_r,1} & \cdots & t_{N_r,\mathrm{SF}_s} \end{bmatrix}_{N_r \times \mathrm{SF}_s}.$$

Some remarks about the JRC waveform are in place. The joint waveforms may be generated by modulating chips on N subcarriers and combining them using the MC-DS-CDMA principles. For

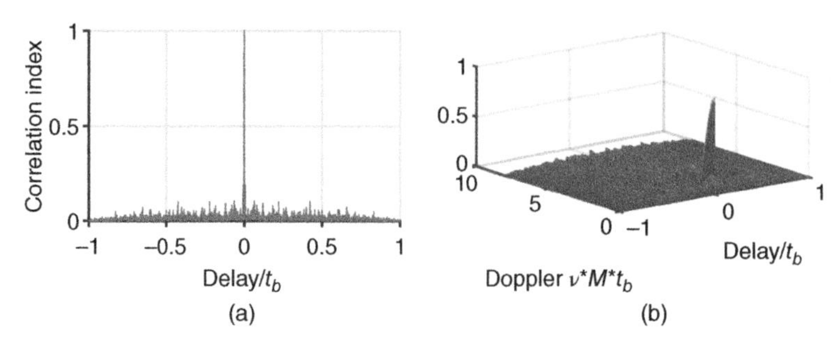

Figure 4.9 Zero-Doppler cut and AF plots for MC-DS-CDMA joint radar–communication waveform for SF= 128. It can be observed that the communications subsystem introduces no sidelobes or distortions to the AF. (a) $M = 1$ and SF= 128 and (b) $M = 1$ and SF= 128. Source: Sharma et al. 2020/IEEE

the IFFT block, using $c = \exp(j2\pi \frac{\Delta f T_c}{N})$, the IDFT matrix can be written as

$$\mathbf{F}^{-1} = \begin{bmatrix} 1 & 1 & \cdots & 1 \\ 1 & c & \cdots & c^{N-1} \\ \vdots & \vdots & \ddots & \vdots \\ 1 & c^{(N-1)} & \cdots & c^{(N-1)(N-1)} \end{bmatrix}_{N\times N}.$$

For $\mathbf{Z} \triangleq \left[\mathbf{X}^H \;\; \mathbf{Y}^H\right]^H$, the generated JRC waveform can be written in a compact matrix form as

$$\mathbf{s}_{\text{joint}} = \text{vec}(\mathbf{F}^{-1}\mathbf{Z}). \tag{4.39}$$

Signal processing at the receiver follows the conventional MC-DS-CDMA model. Separate bit streams are available for further processing after a parallel-to-serial conversion. If the radar receiver is colocated with the JRC transmitter, the radar can also take advantage of the communication signals since the transmitted signals are known to it.

An example of the AF of the generated waveform is depicted in Figure 4.9. This AF is very close to the ideal thumbtack shape in the delay-Doppler domain. The parameters in this simulation were $N = 16$, $N_r = N_c = 8$, and spreading factor SF = 128. Gold codes were used for radar and scrambling.

4.5 Precoder Design for Multiple JRC Users

In this section, we focus on the benefits of co-designing multiple radio systems. These systems can function as RF sensing or communications systems either jointly or individually. An example of a possible scenario is widely deployed communication transceivers forming a cooperative passive sensing network (e.g. existing cellular network) that assists a radar to improve detection performance [He et al., 2019]. In vehicular applications, another interesting approach is a network of roadside units (RSU) that could extend the sensing range of a passing-by vehicle beyond the line-of-sight and the field-of-view. Furthermore, multiple sensors can also be connected (e.g. as a wireless sensing network) or deployed in the same device (e.g. an autonomous vehicle) to perform high-precision day-and-night environment sensing cooperatively. In such scenarios, the sensing and the communications systems should share their spectrum awareness and optimize their signaling strategies jointly for mutual benefits.

A lot of research has been carried out on modeling and managing mutual interference in different shared spectrum scenarios [Aydogdu et al., 2020, Cheng et al., 2019, Dokhanchi et al., 2020,

Liu et al., 2017a, 2020b, Su et al., 2019, Tian et al., 2019, Veeravalli and El Gamal, 2018, Zhang et al., 2020, Zheng et al., 2019]. The impact of interference in such joint systems was investigated in EU project MOSARIM [Kunert, 2012]. It was discovered that interfering signals (particularly for linear frequency modulation (LFM) waveforms) received by an automotive radar are unlikely to cause ghost targets but rather they will create noise-like combined interference. For OFDM interference at the radar receiver, ghost targets and interference are observed simultaneously after radar postprocessing [Kumari et al., 2018]. Interference is the main factor limiting the performance of the JRC systems; hence, interference management is essential for enabling reliable sensing and communication.

4.5.1 Cooperative Active/Passive Sensing for Multiple Users

In emerging 5G and 6G wireless communications systems, radio access networks will operate on higher frequency bands, in particular on mm-Wave and terahertz bands. High attenuation of such frequencies will lead to ultra-dense and cell-free architectures. Consequently, the communication infrastructure, such as cellular base stations and RSU for vehicular networks, will be deployed in small and crowded areas. The densely deployed radio transceivers naturally form a passive sensing network.

In the presence of radar targets, the communication network with passive sensing capability can detect the targets using the scattered communication signals. If the transmitted communication signal is a known pilot signal (e.g. the channel estimation field (CEF) defined by IEEE 802.11ad standards) or is shared among the users, communication receivers are able to measure the frequency-shift and the time-delay, which are typically treated as channel distortion in communication literature. Delay and frequency shift are, however, key parameters of interest to the radar in the form of range and Doppler shift caused by the radial component of the target motion. Angle of arrival (AoA) estimation of target echoes is also available because mm-Wave communications systems typically employ large-aperture antenna arrays capable of high-resolution spatial processing. Therefore, the joint application of distance, Doppler, and AoA estimation can be exploited for target detection and localization.

When a radar shares information on its location and transmitted waveforms, the radar and the communication network can be considered a *cooperative JRC system* performing both active and passive sensing. The target information can be extracted from target echoes at the radar receiver as well as from the scattered signals reaching the communication receivers, as shown in Figure 4.10.

In this subsection, we assume that both communication and radar systems transmit multicarrier waveforms used in most current and emerging communications systems. Therefore, the presented algorithms can be applied to existing communication networks with minimal modifications. Let us start from a simplified scenario in which a monostatic radar cooperating with multiple communication base stations is tracking a target. The radar and the base stations are all employing OFDM signals with N subcarriers. It is assumed that the channel impulse responses for the radar and the communication are wide-sense stationary (WSS) Gaussian processes. Following Bică and Koivunen [2016] and Bica et al. [2016], we extend the signal model (4.32) to a scenario where multiple radar systems cooperate with a communication network. The received signal at the radar receiver can be written as

$$\mathbf{y}_r = \mathbf{X}_r\mathbf{h}_r + \mathbf{X}_s\mathbf{h}_s + \mathbf{X}_s\mathbf{h}_d + \mathbf{n}, \tag{4.40}$$

where $\mathbf{y}$ denotes an N-dimensional received observation vector at the radar receiver, $\mathbf{h}_r$ and $\mathbf{h}_s$ are N-dimensional channel coefficients of the propagation path radar transmitter → target → radar receiver for active sensing and of the propagation path communication transmitter →

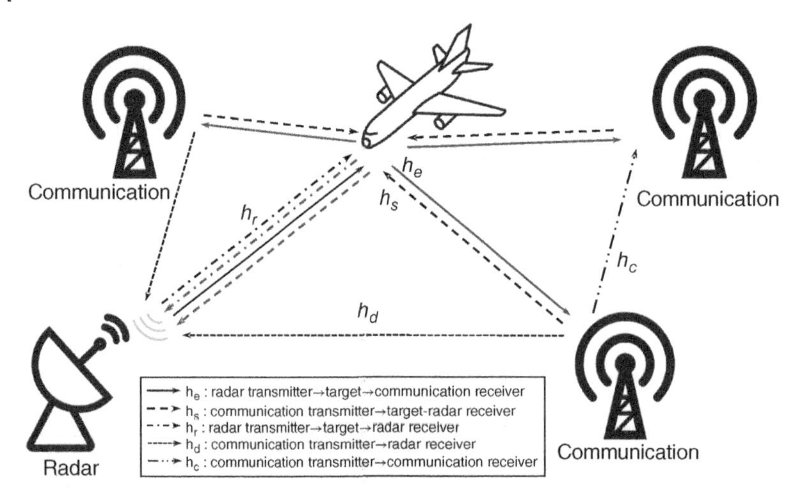

Figure 4.10 System model for cooperative JRC system. The regular lines and the dashed lines denote the transmitted and the reflected signals, respectively.

target → radar receiver for passive sensing, respectively. Vector $\mathbf{h}_d$ denotes channel coefficients of the direct path between the communication transmitter and the radar receiver. Additionally, $\mathbf{h}_e$ denotes the L-dimensional channel coefficients of the propagation path radar transmitter → target → communication receiver for hybrid active–passive sensing. Matrices $\mathbf{X}_r$ and $\mathbf{X}_s$ are $N \times N$ Toeplitz matrices that can be approximated by circulant matrices since their dimensions are sufficiently large [Bică and Koivunen, 2016]. Vector $\mathbf{n}$ is modeled as a Gaussian white noise vector with variance σ_n^2.

We are interested in radar-centric waveform design for JRC systems. This design aims to optimize the radar performance while guaranteeing a lower bound for communication capacity and limiting the total transmitted power. Three kinds of MI-based waveform design criteria are considered:

- Treat communication signal as useful energy maximizing $I(\mathbf{y}; \mathbf{h}_r, \mathbf{h}_s)$.
- Ignore communication signal maximizing $I(\mathbf{y}; \mathbf{h}_r)$.
- Treat communication signal as inference maximizing $I(\mathbf{y}; \mathbf{h}_r|\mathbf{h}_s)$.

4.5.1.1 Communication Signal as Useful Energy

In this case, the objective function of

$$\begin{aligned} I(\mathbf{y}; \mathbf{h}_r,\ \mathbf{h}_s) &= H(\mathbf{y}) - H(\mathbf{y}|\mathbf{h}_r,\ \mathbf{h}_s) \\ &= \sum_{k=0}^{N-1} \log\left(1 + \frac{p_r[k]\eta_r[k] + p_s[k]\eta_s[k]}{p_s[k]\eta_d[k] + \sigma_n^2[k]}\right). \end{aligned} \tag{4.41}$$

Therefore, we obtain the optimization problem as

$$\underset{p_r[k]}{\text{maximize}} \sum_{k=0}^{N-1} \log\left(1 + \frac{p_r[k]\eta_r[k] + p_s[k]\eta_s[k]}{p_s[k]\eta_d[k] + \sigma_n^2[k]}\right), \tag{4.42a}$$

$$\text{subject to } \log\left(1 + \frac{|p_s[k]|^3\sigma_{h_c}^2[k]}{p_r[k]\eta_e[k] + \sigma_n^2}\right) \geq t_k, \tag{4.42b}$$

$$0 \leq \sum_{k=0}^{N-1} p_r[k] \leq P_r, \tag{4.42c}$$

where $p_r[k]$ and $p_s[k]$ are the power of the transmitted radar and communication signals for the kth subcarrier, respectively, while $\eta_r[k], \eta_s[k], \eta_d[k], \eta_e[k]$ denote the power of kth subcarrier corresponding to channel $\mathbf{h}_r$, $\mathbf{h}_s$, $\mathbf{h}_d$, and $\mathbf{h}_e$, respectively. The constraint (4.42b) states that the channel capacity of lth subcarrier in the communications system should be better than t_k, and (4.42c) ensures the transmit power is not higher than the limit P_r.

The optimization problem (4.42) is convex. Furthermore, the global optimum can be found by using KKT conditions [Bica et al., 2016], and the corresponding Lagrange dual variable can be efficiently found by bisection search [Fox and Landi, 1970].

4.5.1.2 Ignoring Communication Signal

If the passive sensing signal on the channel $\mathbf{h}_s$ (the communication transmitter → target → radar receiver) is not of sufficient quality for radar use, it can be ignored and (4.41) can be written as

$$I(\mathbf{y}; \mathbf{h}_r, \mathbf{h}_s) = H(\mathbf{y}) - H(\mathbf{y}|\mathbf{h}_r, \mathbf{h}_s)$$

$$= \max_{p_r[k]} \sum_{k=0}^{N-1} \log\left(1 + \frac{p_r[k]\eta_r[k]}{p_s[k]\eta_d[k] + \sigma_n^2[k]}\right). \tag{4.43}$$

In this case, the waveform optimization problem (4.44a) can be formulated as

$$\underset{p_r[k]}{\text{maximize}} \sum_{k=0}^{N-1} \log\left(1 + \frac{p_r[k]\eta_r[k]}{p_s[k]\eta_d[k] + \sigma_n^2[k]}\right), \tag{4.44a}$$

$$\text{subject to } \log\left(1 + \frac{p_s[k]\eta_s[k]}{p_r[k]\eta_e[k] + \sigma_n^2}\right) \geq t_k, \tag{4.44b}$$

$$\sum_{l=0}^{L-1} p_r[k] \leq P_r. \tag{4.44c}$$

The optimal solution can be found following the same procedure with (4.42), see Bica et al. [2016].

4.5.1.3 Communication Signal as Interference

In this case, the optimization objective can be given by

$$I(\mathbf{y}; \mathbf{h}_r) = H(\mathbf{y}) - H(\mathbf{y}|\mathbf{h}_r) = H(\mathbf{y}) - H(\mathbf{r}_s + \mathbf{s} + \mathbf{n}), \tag{4.45}$$

where $\mathbf{r}_s$ is an L-dimensional signal vector received by the radar that follows the passive sensing path (the communication transmitter → target → radar receiver). We then get an optimization problem

$$\underset{p_r[k]}{\text{maximize}} \sum_{k=0}^{N-1} \log\left(1 + \frac{p_r[k]\eta_r[k]}{p_s[k]\eta_s[k] + p_s[k]\eta_d[k] + \sigma_n^2[k]}\right), \tag{4.46a}$$

$$\text{subject to } \log\left(1 + \frac{p_s[k]\sigma_{h_c}^2[k]}{p_r[k]\eta_e[k] + \sigma_n^2}\right) \geq t_l, \tag{4.46b}$$

$$\sum_{k=0}^{N-1} p_r[k] \leq P_r. \tag{4.46c}$$

The optimal waveform can be solved by KKT conditions as well; for more details, please refer to Bica et al. [2016].

4.5.2 Interference Management for Multiple JRC Users

Most of the prior radar–communication spectrum sharing approaches address interference management by employing orthogonal designs. The employed waveforms are optimized while guaranteeing acceptable performance for other subsystems. Orthogonality may be used in time, frequency, code, or spatial domains. However, perfect orthogonality is commonly lost at the receiver end. Cognitive radio systems commonly exploit temporal occupancy information via opportunistic spectrum sharing [Bica et al., 2015], in which communications or radar systems emit their signal whenever the spectrum is detected to be idle.

Given the multiantenna technology used in modern radar and communications systems, interference management by designing spatial or spectral precoders leads to better spectrum efficiency. The prior knowledge required by precoder designs is the interference channel state information (ICSI). In the spatial domain, the precoding strategy can be designed to project either the radar or the communication signal into the null space (orthogonal to signal space) of the other system, which is called *null-space projection* (NSP) in Babaei et al. [2013]. The solution of the NSP-based precoding design can be obtained with the singular value decomposition (SVD), so its computational complexity is typically lower compared to optimization-based precoding designs [Zheng et al., 2018]. However, NSP may not be feasible when the ICSI matrix does not have any zero singular values [Babaei et al., 2013], as in noisy scenarios. The solution of the projecting matrix depends on the channel state matrix, which is in turn determined by the propagation and the target environments [Khawar et al., 2016]. Furthermore, this type of precoding design allows avoiding interference on either subsystem but not both.

Also convex optimization methods can be utilized in designing the precoding matrix for interference management; see other chapters of this book for details. The precoder of a multiple-input multiple-output matrix completion (MIMO-MC) radar is jointly designed with a point-to-point MIMO communications system to maximize the radar SINR while guaranteeing desired data rate [Li et al., 2016]. A MIMO-MC radar reconstructs the target data matrix from sampled target returns via matrix completion and a pseudo-random sub-Nyquist scheme. The advantage of this system comes from the way that sampling scheme modifies ICSI originating from the communication transmitters, and, therefore, the radar signal subspaces may be utilized more efficiently.

Another robust precoding design that minimizes interference power is proposed for coexistence between MIMO radar and downlink multiuser MIMO communications system in Liu et al. [2018]. In Zheng et al. [2018], a radar waveform and a communication codebook are jointly designed by maximizing the compound capacity and the Kullback–Leibler divergence between the densities of target detection observations under two alternative hypotheses via Neyman–Pearson criterion. Furthermore, the transmit power, linear receive filters, and the space-time communication codebook are jointly designed for cooperating surveillance radar and communications systems, using a block coordinate ascent method [Grossi et al., 2020].

Moreover, optimizing the transmitted waveforms and the receiver processing for the radar and the communications systems can be done in a joint fashion. For example, precoders and decoders in radar and communications systems may be jointly designed to construct signal spaces and orthogonal interference spaces in order to achieve effective and flexible interference management. In order to figure out the number of DoFs that can be achieved when radar and communications systems coexist, information theoretic analysis has been conducted using high-dimensional statistics. It has been proved that the upper bound for the DoFs for radar–communication spectrum sharing in $K + 1$ users interference channel can be achieved via joint precoder–decoder designs [Cui et al., 2018b].

In the following subsections, we first describe some of the considered interference management techniques. A particularly powerful method based on interference alignment (IA) is presented in more detail [Cadambe and Jafar, 2008, Cui et al., 2017]. Finally, we discuss the type of information that should be shared between radar and communications systems.

4.5.2.1 Null Space-based Precoder Design

To take fully advantage of beamforming and precoding, one of the most common requirements in interference management techniques is the ICSI awareness. Another approach is embedding the same pilot signal in the radar coherent intervals as well as the communication frames. It is also noteworthy that if the radar and the communications subsystems are colocated, the ICSI exchange can take place efficiently with shared memory. In this subsection, the ICSI is considered to be perfectly known, obtained through feedback or estimation at the communication and the radar transceivers. One feasible way to achieve this is to treat the known radar waveforms as shared pilots to estimate channels among the communication and the radar users or subsystems. After the ICSI awareness is acquired, null space–based precoding strategy can be used to cancel interference either at the radar or at the communications system.

Consider a colocated MIMO radar that is interfering K communication base stations. The NSP method is applied at the radar transmitter for interference mitigation. The received radar interference at ith communication receiver can be written as

$$\Upsilon_i = \mathbf{H}_i \mathbf{W}_r \mathbf{s}_r(t), \tag{4.47}$$

where $\mathbf{H}_i$ is the ICSI matrix of the radar signal. To ensure that the received signals at K communication receivers are interference-free, the criteria $\Upsilon_i = \mathbf{0},\ \forall i = \{1, \dots, K\}$ should be satisfied, which is equivalent to the following condition:

$$\mathbf{W}_r \mathbf{s}_r(t) \in \mathcal{N}\left(\mathbf{H}_i\right), \quad \forall i \in \{1, \dots, K\}, \tag{4.48}$$

where $\mathcal{N}(\mathbf{H}_i)$ denotes the null space of $\mathbf{H}_i$. Therefore, the transmitted radar signal must lie in the intersection of null spaces of the K communications systems, that is

$$\mathbf{W}_r \mathbf{s}_r(t) \in \mathcal{N}\left(\mathbf{H}_1\right) \cap \mathcal{N}\left(\mathbf{H}_2\right) \cdots \cap \mathcal{N}\left(\mathbf{H}_K\right) = \mathcal{N}\left(\overline{\mathbf{H}}\right), \tag{4.49}$$

where $\overline{\mathbf{H}} = \left[\mathbf{H}_1^H \ \mathbf{H}_2^H \ \cdots \ \mathbf{H}_K^H\right]^H$, where the intersections of the null spaces are obtained using the identity $\mathcal{N}(\mathbf{A}) \cap \mathcal{N}(\mathbf{B}) = \mathcal{N}\left(\left[\mathbf{A}^H \ \mathbf{B}^H\right]^H\right)$. One can obtain the NSP precoder by applying SVD on $\overline{\mathbf{H}}$ [Babaei et al., 2013]. Therefore, the NSP precoder is given as

$$\mathbf{W}_r = \overline{\mathbf{V}}_0 (\overline{\mathbf{V}}_0^H \overline{\mathbf{V}}_0)^{-1} \overline{\mathbf{V}}_0^H, \tag{4.50}$$

where the columns of $\overline{\mathbf{V}}_0$ are the right singular vectors of $\overline{\mathbf{H}}$ that correspond to the zero singular values. It's worth noting that $\mathbf{W}_r$ exists if and only if $\mathcal{N}(\overline{\mathbf{H}})$ is not empty, which happens when the number of the spatial DoFs of the radar (e.g. number of radar antennas) is larger than the number of DoFs in the communications systems. Moreover, when the number of communication users increases, it is more difficult to find the intersection $\mathcal{N}(\overline{\mathbf{H}})$ and therefore also the NSP precoder.

To overcome this issue, matrix $\overline{\mathbf{V}}_0$ can be replaced with the singular vectors associated with singular values less than threshold $\delta_{\min}$, the resulting matrix denoted by $\overline{\mathbf{V}}_{\delta_{min}}$. One reasonable option of $\delta_{\min}$ is the noise threshold. Therefore, in addition to the null space, the projection space is expanded to small singular value space, which is named *switched small singular value space projection* (*SSSVSP*). The resulting precoder can be given by [Mahal et al., 2017],

$$\mathbf{W}_r = \overline{\mathbf{V}}_{\delta_{\min}} (\overline{\mathbf{V}}_{\delta_{\min}}^H \overline{\mathbf{V}}_{\delta_{\min}})^{-1} \overline{\mathbf{V}}_{\delta_{\min}}^H. \tag{4.51}$$

In general, threshold $\delta_{\min}$ implies the maximum tolerable interference level of communications systems. Even though the SSSVSP precoding strategy allows more flexible usage of signal spaces in radar–communication spectrum sharing, it still has the drawback that the resulting projector can cancel the mutual interference either in the radar or in the communications system but not in both.

4.5.2.2 Precoder–Decoder Co-design

As mentioned in Section 4.5.2.1, the precoder-only design projects interference into the signal subspace spanned by the columns of channel state matrices. The basis of the signal subspaces is determined by the radio propagation environment that cannot be controlled by the designer, unless intelligent reflective surfaces are used. To overcome this limitation, interference mitigation can be further applied at the signal decoding stage of the radar and the communication receivers. Motivated by canceling mutual interference as well as achieving better performance, a precoding matrix and a decoding matrix can be co-designed for both radar and communications systems to achieve flexible signal space usage. Moreover, both the interference between radar and communication and the interference between multiple communications systems can be managed efficiently. Compared to previously described interference management techniques, joint precoder–decoder design is able to achieve the upper bound of the DoFs [Cadambe and Jafar, 2008], i.e. $K + 1/2$, according to IA theory; see Cui et al. [2018b]. This leads to improved detection performance in the radar subsystems and a higher rate in the communications subsystems.

Let us assume a signal model of a radar coexisting with K communications systems such that the radar interferes with the communications systems and also suffers communication interference simultaneously, known as the $K + 1$-user radar–communication interference channel [Cui et al., 2017]. After decoding, the received signal at the radar receiver can be written as

$$\mathbf{y}_r(t) = \underbrace{\mathbf{Q}_r^H\mathbf{G}(\theta)\mathbf{W}_r\mathbf{s}_r(t)}_{\text{desired signal}} + \underbrace{\mathbf{Q}_r^H\sum_{i=1}^{K}\mathbf{H}_{r,i}\mathbf{W}_{c,i}\mathbf{s}_{c,i}(t)}_{\text{comm. interference}} + \underbrace{\mathbf{Q}_r^H\mathbf{n}_r(t)}_{\text{clutter + noise}}, \tag{4.52}$$

where $\mathbf{Q}_r \in \mathbb{C}^{M\times d}$ is the linear decoding matrix employed in the radar, $\mathbf{W}_r \in \mathbb{C}^{M\times d}$ is the precoding matrix for the radar, and d denotes the corresponding desired DoFs. Matrix $\mathbf{G}$ is the target response matrix for the target of interest to the radar, and it can be expressed with the transmit steering vector $\mathbf{a}_T(\theta)$, the receive steering vector $\mathbf{a}_R$, and the complex scattering coefficient β as $\mathbf{G}(\theta) = \beta\mathbf{a}_R(\theta)\mathbf{a}_T^H(\theta)$.

Matrix $\mathbf{H}_{r,i}$ is the ICSI from the ith communication transmitter to the radar receiver. Conversely, in the following subsection, $\mathbf{H}_{i,r}$ denotes the ICSI from the radar transmitter to the ith communication receiver (the transmitter(s) and receiver of each subsystem are not necessarily in the same place). In addition, we define $\mathbf{H}_{i,j}$ as the ICSI from the jth communication transmitter to the ith communication receiver. Also, $\mathbf{H}_{i,i}$ denotes the communication channel from the ith communication transmitter to its designated receiver.

Following the signal model (4.52), the interference-free criterion at the radar receiver is given by

$$\Upsilon_r = \mathbf{Q}_r^H\sum_{i=1}^{K}\mathbf{H}_{r,i}\mathbf{W}_{c,i}\mathbf{s}_i(t) = \mathbf{0}, \tag{4.53}$$

or equivalently,

$$\mathbf{W}_{c,i}\mathbf{s}_{c,i}(t) \in \mathcal{N}\left(\mathbf{Q}_r^H\mathbf{H}_{r,i}\right), \quad i = 1, \dots, K. \tag{4.54}$$

Equation (4.54) implies that, for each of the K communication users, the interfering signal from the communication transmitter to the radar receiver is projected onto the null subspace that is jointly formed by the decoder matrix and the ICSI. The transmitted signal is projected onto the null space formed jointly by the decoder and the channel state matrix. If $\mathbf{J}_r$ and $\mathbf{J}_i$ are the matrices spanning the interference subspace corresponding to the radar and ith communications system, and $\mathbf{S}_r$ and $\mathbf{S}_i$ are matrices spanning the respective signal subspaces, we have

$$\mathbf{J}_r \triangleq \mathbf{Q}_r^H[\{\mathbf{H}_{r,i}\mathbf{W}_{c,i}\mathbf{s}_i(t)\}_{i=1}^K], \tag{4.55a}$$

$$\mathbf{J}_i \triangleq \mathbf{Q}_i^H[\{\mathbf{H}_{i,r}\mathbf{W}_r\mathbf{s}_r(t)\}\{\mathbf{H}_{i,j}\mathbf{W}_{c,j}\mathbf{s}_j(t)\}_{j\neq i,j=1}^K], \tag{4.55b}$$

$$\mathbf{S}_r \triangleq \mathbf{Q}_r^H\mathbf{G}\mathbf{W}_r, \tag{4.55c}$$

$$\mathbf{S}_i \triangleq \mathbf{Q}_i^H\mathbf{H}_{i,i}\mathbf{W}_i, \quad i = 1, \dots, K, \tag{4.55d}$$

where $\mathbf{Q}_i$ denotes the decoder matrix of the ith communication receiver, and $\mathbf{H}_{i,j}$ is the channel between the ith and the jth communications systems. Notation $\{\mathbf{A}_i\}_{i=1}^K$ means the horizontal concatenation of matrices $\mathbf{A}_1, \dots, \mathbf{A}_K$.

In order to have zero interference for all the systems regardless of the used signals, we would like to have the matrices spanning the interference subspaces to be all zeroes. This ideal IA condition can be written as

$$\mathbf{J}_r = \mathbf{0}, \quad \mathbf{J}_i = \mathbf{0}, \tag{4.56a}$$

$$\text{rank}(\mathbf{S}_r) = d, \quad \text{rank}(\mathbf{S}_i) = d, \quad i = 1, \dots, K. \tag{4.56b}$$

The DoFs of the signal subspaces of the radar systems and the communications systems are constrained by (4.56b), where d is the number of required spatial DoFs in a certain radar or communications system. For communications, the number of DoFs implies the number of independent streams that can be sent at a high SNR, denoted as

$$d = \lim_{SNR\to\infty} \log \frac{C(\text{SNR})}{\text{SNR}},$$

where $C(\text{SNR})$ is the channel capacity at the given SNR value. From an information theoretic perspective [Cui et al., 2020], the goal of the joint precoder–decoder design in a communications system is to enable interference-free communication. On the other hand, the joint precoder–decoder design in the radar system aims to find an interference-aware waveform–filter pair such that information extraction and target characteristics are not hindered by the interference.

Since $\mathbf{J}_i$, the interference subspace matrix for communication, depends on all the other precoder and decoder matrices $\mathbf{W}_{c,j}$ and $\mathbf{Q}_j$, achieving the ideal IA condition is an NP-hard and nonconvex problem. Therefore, relaxation methods and iterative approaches are commonly employed to find an IA solution [Zhao et al., 2016] that consists of K communication precoder–decoder pairs and a radar waveform–filter pair. Here, we introduce three different design criteria, namely, the SINR maximization, rank-constrained rank minimization (RCRM), and MI maximization.

SINR Maximization Criterion Let us start with the radar system. The SINR of the radar is defined as the power of the returned target signals with respect to the power of the noise and interference signals, and it largely determines the target detection probability. Assuming the powers of the transmitted signals satisfy $\mathrm{E}[||\mathbf{s}_r||^2] = 1$ and $\mathrm{E}[||\mathbf{s}_c||^2] = 1$ and that the signals are mutually independent, the interference-plus-noise covariance matrix $\mathbf{\Gamma}_{\text{i+n}}$ at the radar receiver can be written as

$$\mathbf{\Gamma}_{\text{i+n}} = \mathbf{Q}_r^H\left(\sum_{i=1}^K \mathbf{H}_{r,i}\mathbf{W}_{c,i}\mathbf{W}_{c,i}^H\mathbf{H}_{r,i}^H + \sigma_{n,r}^2\mathbf{I}\right)\mathbf{Q}_r, \tag{4.57}$$

where $\sigma_{n,r}^2$ is the noise power at the radar receiver. We can the express the radar-centric SINR maximization subproblem as

$$\underset{\mathbf{W}_r,\mathbf{Q}_r}{\text{maximize}}\ \text{tr}(\boldsymbol{\Gamma}_{\text{i+n}}^{-1}\mathbf{Q}_r^H\mathbf{G}(\theta)\mathbf{W}_r\mathbf{W}_r^H\mathbf{G}(\theta)^H\mathbf{Q}_r), \quad (5.58a)$$

$$\text{subject to } \mathbf{Q}_r^H\mathbf{Q}_r = \mathbf{I}_d, \quad (5.58b)$$

$$\text{tr}(\mathbf{W}_r^H\mathbf{W}_r) \leq P_r, \quad (5.58c)$$

where $\mathbf{I}_d$ is a $d \times d$ identity matrix. The constraint (4.58b) guarantees that the rank of the signal space is not reduced below d by the decoder.

Similarly, the interference-plus-noise covariance matrix $\boldsymbol{\Phi}_i$ at the ith communication receiver is

$$\boldsymbol{\Phi}_i = \mathbf{Q}_i^H\left(\sum_{\substack{j=1\\j\neq i}}^{K}\mathbf{H}_{i,j}\mathbf{W}_{c,j}\mathbf{W}_{c,j}^H\mathbf{H}_{i,j}^H + \mathbf{H}_{i,r}\mathbf{W}_r\boldsymbol{\Sigma}_r\mathbf{W}_r^H\mathbf{H}_{i,r}^H + \sigma_{n,c}^2\mathbf{I}\right)\mathbf{Q}_i. \quad (4.59)$$

Therefore, for the ith communication user, the communication-centric SINR maximization subproblem is

$$\begin{aligned}&\underset{\mathbf{W}_{c,i},\mathbf{Q}_i}{\text{maximize}}\ \text{tr}(\boldsymbol{\Phi}_i^{-1}\mathbf{Q}_i^H\mathbf{H}_{i,i}\mathbf{W}_{c,i}\mathbf{W}_{c,i}^H\mathbf{H}_{i,i}^H\mathbf{Q}_i),\\ &\text{subject to } \mathbf{Q}_i^H\mathbf{Q}_i = \mathbf{I}_d,\\ &\qquad\text{tr}(\mathbf{W}_{c,i}^H\mathbf{W}_{c,i}) \leq P_i.\end{aligned} \quad (4.60)$$

Problems (4.58) and (4.60) can be solved in an alternating manner, by finding $\mathbf{Q}_r$ and $\mathbf{Q}_i$ solutions first with fixed $\mathbf{W}_r$ and $\mathbf{W}_{c,i}$, and vice versa. For more details, please refer to Cui et al. [2017].

Rank-constrained Rank Minimization (RCRM) If the rank of each of the matrices spanning the interference subspaces were zero, the ideal IA condition would be satisfied. Thus, in order to minimize the amount of interference, one can recast (4.56) as a rank minimization problem with the rank constraint (4.56b):

$$\underset{\mathbf{Q}_i,\mathbf{W}_{c,i},\mathbf{Q}_r,\mathbf{W}_r}{\text{minimize}}\ \text{rank}(\mathbf{J}_r) + \sum_{i=1}^{K}\text{rank}(\mathbf{J}_i), \quad (4.61a)$$

$$\text{subject to } \text{rank}(\mathbf{S}_r) = d, \quad (4.61b)$$

$$\text{rank}(\mathbf{S}_i) = d, \quad (4.61c)$$

$$\text{tr}(\mathbf{W}_r^H\mathbf{W}_r) \leq P_r, \quad (4.61d)$$

$$\text{tr}(\mathbf{W}_{c,i}^H\mathbf{W}_{c,i}) \leq P_{c,i}, \quad i = 1, \dots, K. \quad (4.61e)$$

Finding the global optimum of this RCRM problem is difficult, so a locally optimal solution should be accepted to mitigate the interference. In addition to the rank constraint (4.61b), a constraint for the diversity gain of the radar may be imposed by requiring $\mathbf{S}_r$ to have d^2 nonzero elements [Cui et al., 2018a]. For the communication side, there is no similar benefit in diversity gain.

Problem (4.61) is a highly nonconvex problem. However, it can be relaxed by using the nuclear norm [Recht et al., 2010] to obtain a tight convex approximation as

$$\begin{aligned}\text{CE}\left[\text{rank}(\mathbf{J}_r) + \sum_{i=1}^{K}\text{rank }(\mathbf{J})_i\right] &= \text{CE}\left[\text{rank}(\text{blkdiag}(\mathbf{J}_r, \mathbf{J}_1, \dots, \mathbf{J}_i)\right]\\ &= \|\mathbf{J}_r\|_* + \sum_{i=1}^{K}\|\mathbf{J}_i\|_*,\end{aligned} \quad (4.62)$$

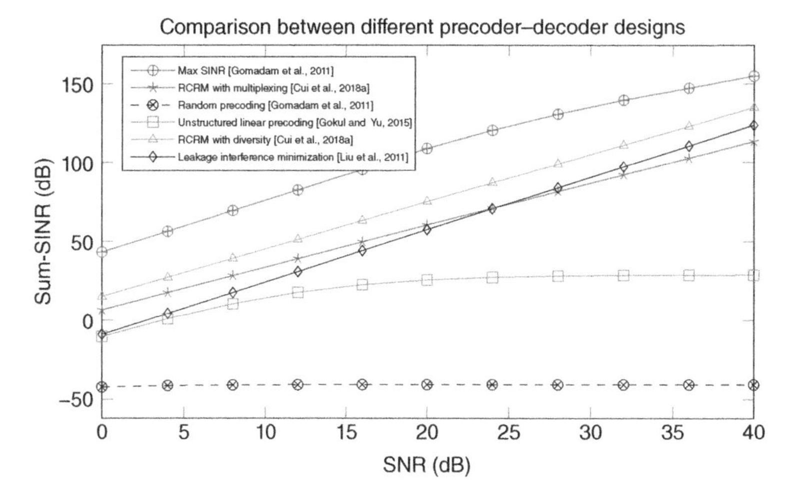

Figure 4.11 Comparison between rank-constrained rank minimization (RCRM) and various other interference management techniques. The RCRM design of Cui et al. [2018a] performs better than most of other tested precoder–decoder designs except the Max-SINR. Source: Based on Cui et al. 2018.

where CE[f] denotes the convex envelope of function f, blkdiag($\mathbf{J}_r, \mathbf{J}_1, \dots, \mathbf{J}_i$) denotes the block diagonal matrix that has the diagonal blocks $\mathbf{J}_r, \mathbf{J}_1, \dots, \mathbf{J}_i$. The nuclear norm $||\mathbf{J}_i||_*$ is the sum of the singular values of $\mathbf{J}_i$. Following Theorem 2.2 in Recht et al. [2010], when used as the objective function in optimization, the sum of nuclear norms in (4.62) is equivalent to the convex envelope of the sum of interference ranks in (4.62). The rank constraint (4.61b) can be relaxed by matrix inequality constraints to get a convex problem; for more details, please refer to Cui et al. [2018a].

The performance of different precoder–decoder designs is illustrated in Figure 4.11, which shows the sum of the SINR of all the systems as a function of the SNR. All receivers are assumed to have equal SNR in this case. The Max-SINR approach [Gomadam et al., 2011] performs the best, while the RCRM [Cui et al., 2018a] comes next performing better than the random precoding [Gomadam et al., 2011], unstructured linear precoding [Gokul and Yu, 2015] and leakage interference minimization [Liu et al., 2011]. When a radar or a communications system increases its transmit power, the joint precoder–decoder designs perform more efficiently than the precoder-based designs. For example, the system using unstructured linear precoder strategy is overwhelmed by interference at high SNRs.

Mutual Information Maximization Aiming to find the maximum information transmission for communication and the information extraction waveform for radar, the sum-rate of the communications system and the MI between the target impulse response and the target echoes on radar system can be jointly maximized to find the precoder–decoder pairs. The sum-rate of the K-user communication interference channel is given by the sum of the maximum MI $I(\mathbf{y}_i; \mathbf{s}_i|\mathbf{H}_{i,i})$ between $\mathbf{y}_i$ and $\mathbf{s}_i$,

$$\sum_{k=1}^{K} \mathrm{I}(\mathbf{y}_i; \mathbf{s}_i|\mathbf{H}_{i,i}) = \sum_{k=1}^{K} \log\det(\mathbf{I}_d + \mathbf{\Phi}_i^{-1}\mathbf{Q}_i^H\mathbf{H}_{i,i}\mathbf{W}_{c,i}\mathbf{W}_{c,i}^H\mathbf{H}_{i,i}^H\mathbf{Q}_i). \tag{4.63}$$

In the radar, the received signal is typically processed with matched (or mismatched) filtering. We assume that the filter outputs are stacked into a $dM \times 1$ vector. Assuming that the transmitted radar signals are orthogonal, the interference-plus-noise covariance matrix of the filter output is given by

$$\boldsymbol{\Psi}_r = \sum_{i=1}^{K} \bar{\mathbf{Q}}_r^H \overline{\mathbf{H}}_{r,i} \bar{\mathbf{W}}_{c,i} \boldsymbol{\Sigma}_{\mathbf{s}_i} \bar{\mathbf{W}}_{c,i}^H \overline{\mathbf{H}}_{r,i}^H \bar{\mathbf{Q}}_r + \sigma_{n,r}^2 \bar{\mathbf{Q}}_r^H \bar{\mathbf{Q}}_r, \tag{4.64}$$

where $\text{vec}(\mathbf{S}_i)\,\text{vec}(\mathbf{S}_i)^H = \boldsymbol{\Sigma}_{\mathbf{s}_i}, \forall i \in \{1, \dots, K\}$ is the temporal auto-correlation of the vectorized communication signal, and $\bar{\mathbf{Q}}_r = \mathbf{I}_L \otimes \mathbf{Q}_r$, $\overline{\mathbf{H}}_{r,i} = \mathbf{I}_L \otimes \mathbf{H}_{r,i}$,$\bar{\mathbf{W}}_{c,i} = \mathbf{I}_L \otimes \mathbf{W}_{c,i}$ are applied to simplify the formulation. The corresponding radar MI waveform design is given by

$$\text{I}(\mathbf{y}_r; \mathbf{g}|\bar{\mathbf{S}}_r) = \log\det(\mathbf{I}_{d^2} + \boldsymbol{\Psi}_r^{-1} \bar{\mathbf{S}}_r^H \boldsymbol{\Gamma}_{\mathbf{g}} \bar{\mathbf{S}}_r), \tag{4.65}$$

where $\boldsymbol{\Gamma}_{\mathbf{g}} = \text{E}[\text{vec}(\mathbf{G}(\theta))\text{vec}\,(\mathbf{G}(\theta))^H]$ is the covariance matrix of the stacked target response vector $\text{vec}(\mathbf{G}(\theta))$, and $\bar{\mathbf{S}}_r = \mathbf{W}_r \otimes \mathbf{Q}_r$ denotes the equivalent radar signal.

Now, the MI maximization problem can be formulated as

$$\underset{\substack{\{\mathbf{W}_{c,i}\}_{i=1}^K, \{\mathbf{Q}_i\}_{i=1}^K \\ \mathbf{W}_r, \mathbf{Q}_r}}{\text{maximize}} \sum_{i=1}^{K} \text{I}(\mathbf{y}_i; \mathbf{s}_i|\mathbf{H}_{i,i}) + \text{I}(\mathbf{y}_r; \mathbf{g}|\bar{\mathbf{S}}_r), \tag{4.66a}$$

$$\text{subject to } \text{tr}(\mathbf{W}_r^H \mathbf{W}_r) \le P_r, \tag{4.66b}$$

$$\text{tr}(\mathbf{W}_{c,i}^H \mathbf{W}_{c,i}) \le P_{c,i}, \quad i = 1, \dots, K, \tag{4.66c}$$

where P_r and $P_{c,i}$ denote the maximal transmit power of the radar and the communications system, respectively. A solution to this problem can be found efficiently by alternating primal–dual subgradient ascent, as shown in Cui et al. [2020].

4.5.2.3 Awareness, Cognition, and Cooperation

In both cooperative and coexistence JRC systems, communications and radar subsystems benefit from the awareness information shared among subsystems. Such information could be employed in JRC signal processing strategies to achieve performance gain, even if the radar and communications systems are not fully integrated. Therefore, one needs to consider what kind of information awareness needs to be shared from the signal processing perspective.

To address this issue, we summarize the shared/unshared information awareness of cognitive communication/radar, joint-precoder–based interference managements, joint precoder–decoder–based interference managements, and fully cooperative JRCs into the following formulations:

Cognition-based designs
$$\underbrace{\mathbf{G}(\theta)\mathbf{W}_r\mathbf{s}_r(t) + \sum_{i=1}^{K} \mathbf{H}_{r,i}\mathbf{W}_{c,i}\mathbf{s}_i(t)}_{\text{All unshared}}, \tag{4.67a}$$

Precoder-based designs
$$\underbrace{\mathbf{G}(\theta)\mathbf{W}_r\mathbf{s}_r(t)}_{\text{Unshared}} + \sum_{i=1}^{K} \underbrace{\mathbf{H}_{r,i}}_{\text{Shared}} \underbrace{\mathbf{W}_{c,i}\mathbf{s}_i(t)}_{\text{Unshared}}, \tag{4.67b}$$

Joint precoder–decoder designs
$$\underbrace{\mathbf{Q}_r^H}_{\text{Shared}} \underbrace{\mathbf{G}(\theta)\mathbf{W}_r\mathbf{s}_r(t)}_{\text{Unshared}} + \sum_{i=1}^{K} \underbrace{\mathbf{Q}_r^H \mathbf{H}_{r,i}}_{\text{Shared}} \underbrace{\mathbf{W}_{c,i}\mathbf{s}_i(t)}_{\text{Unshared}}, \tag{4.67c}$$

Fully cooperative JRCs
$$\underbrace{\mathbf{G}(\theta)\mathbf{W}_r\mathbf{s}_r(t) + \sum_{i=1}^{K} \mathbf{H}_{r,i}(\theta)\mathbf{W}_{c,i}\mathbf{s}_i(t)}_{\text{Fully shared}}. \tag{4.67d}$$

It should be also noted that noise is ignored in the above four formulations.

- **Cognition-based Interference Management**: In general, cognitive communication/radar exploits the spare spectrum of the other system to achieve radar–communication coexistence. The common procedure first perceives the spectrum holes and then performs radio access when time–frequency resources are underutilized. The channel information, the knowledge about the radar waveforms, and the communication symbols are not necessarily shared externally. The cognitive radar/communications systems cannot benefit from co-designed radar–communication signal processing strategies.
- **Precoder-based Interference Management**: Precoder-based information management designs require the ICSI shared between the radar and the communications subsystems, i.e. the shared term indicated in (4.67b). Unlike the cognition-based interference management, the radar and the communications subsystems periodically cooperate to emit cross-subsystem pilots for ICSI estimation and then share the estimated ICSI around the interference-governed radio systems. However, the periodic ICSI estimation and transmission leads to additional consumption of the frequency and time resources. If the communications and the radar subsystems are colocated, the estimation and exchange can be efficiently addressed by shared memory and the existing communication infrastructure, e.g. optical fiber.
- **Joint Precoder–Decoder Interference Management**: Compared to the precoder-based information management designs, the joint precoder–decoder information management schemes further exploit the signal spaces by adjusting the basis that spans the signal subspace and the interference subspace. Therefore, the mutual interference between the radar and the communications subsystems can be aligned into a designed null subspace, which results in gain in DoFs compared to signal strategies based on spatial orthogonality. The cross-system information requirement is the designed decoders and corresponding ICSI, i.e. the shared factors in (4.67c).
- **Cooperative or Co-designed JRC systems**: If the waveform and transmitted communication signals are fully shared (4.67d), both the radar and the communication receivers can resolve the scattered signals to extract target parameters in which the radar is interested. In this scenario, an improvement in the detection performance is achieved from jointly designed radar–communication signaling strategies [He et al., 2019].

4.6 Summary

In this chapter, beamforming and interference management in JRC systems and shared spectrum scenarios were considered. We focused on cooperative settings where information about the state of the radio spectrum and awareness of the environment may be exchanged among the users and subsystems for mutual benefits. Consequently, the radar and communications subsystems can adjust their operational parameters to achieve improved performance in the sensing task at hand or improved data rate in the communications subsystem.

There is no general model describing all JRC systems because of the wide variety of conditions, including frequency bands, target scenarios, sensing tasks and their objectives, waveform designs, multiantenna system models, distributed and colocated configurations, MA schemes, duplexing methods in communications, and isolation approaches between transmitters and receivers, to name but a few. Therefore, we presented a few illustrative example designs and system models developed for JRC systems. The examples include the DFRC waveforms used for both sensing and communication purposes, spatial processing and beamforming methods in cooperative settings, orthogonal multicarrier designs facilitating simultaneous communication and sensing and efficient Doppler estimation. Different frequency bands have different propagation properties, including attenuation, blocking, and multipath, which influences the impact of the interference on the receiver and also the sensing and communication transceiver design.

We considered a variety of precoding and decoding methods that allow for managing interference and projecting desired signals and interference into different subspaces to avoid significant deterioration in sensing or communication performance. To optimize the JRC system performance, information about the state of the radio spectrum and interference needs to be shared among the users. If sensing and communication are co-designed, sharing of hardware resources (transceivers, antenna systems and their front-ends, timing circuitry) and awareness of spectrum and employed waveforms will be easier. Such RF convergence will lead to multifunction hardware suitable for both sensing and communication at radio frequencies for mutual benefits. Future work may include the spatial DoFs analysis, novel waveform designs for balancing the performance between radar and communications subsystems, and beamforming designs under the constraints of the existing 5G communication network architecture.

List of Symbols

The following list describes the symbols that are used within this chapter.

$\mathbf{h}_d$	Channel coefficients of the direct path between the communication transmitter and the radar receiver
$\mathbf{h}_e$	Channel coefficients of the path radar transmitter → target → communication receiver
Υ_i	Interference signal from ith communication transmitter to radar receiver
α	Scaling constant to match the desired shape of the beampattern with the available power
β	Target scattering coefficient
ψ	Radar receiver filter coefficients
$\mathbf{a}_R(\theta)$	Receiver steering vector
$\mathbf{a}_T(\theta)$	Transmit steering vector
$\mathbf{G}$	Target response matrix
$\mathbf{H}$	Channel matrix
$\mathbf{h}_k^H$	kth row of $\mathbf{H}$
$\mathbf{n}_c(t)$	Communication receiver noise
$\mathbf{n}_r(t)$	Receiver noise at the radar
$\mathbf{Q}_i$	Decoder matrix of ith communication receiver
$\mathbf{s}_r(t)$	Transmitted radar signals at time t
$\mathbf{s}_r(t)$	Transmitted radar waveforms
$\mathbf{S}_c$	Communication scrambling chip matrix
$\mathbf{W}$	Transmitted waveform coding matrix
$\mathbf{W}_c$	Communication precoding matrix
$\mathbf{W}_d$	Communication transmit coding matrix
$\mathbf{W}_r$	Radar precoding matrix
$\mathbf{w}_{c,k}$	kth *column* of $\mathbf{W}_c$
$\mathbf{x}(t)$	Joint transmitted signal
$\tilde{\mathbf{W}}$	Phase-shifted transmit precoding matrix
$\mathbf{\Gamma}_\mathbf{g}$	Covariance matrix of the stacked target response vector vec($\mathbf{G}(\boldsymbol{\theta})$)
$\mathbf{\Gamma}_\mathbf{x}$	Covariance matrix of the transmitted signal
$\mathbf{\Gamma}_{\mathrm{i+n}}$	Interference-plus-noise covariance matrix

δ_{ij}	Kronecker delta
$\eta_c[k]$	Channel gain of communication channel on kth subcarrier
$\eta_r[k]$	Channel gain of the radar channel on kth subcarrier
γ_i	Symbol for communication signal
$\mathbf{\Phi}_i$	Interference-plus-noise covariance matrix at ith communication receiver
$\mathbf{B}$	Chip matrix for communication signal
$\mathbf{D}$	Diagonal subcarrier selection matrix
$\mathbf{F}^H$	Inverse discrete Fourier transform (IDFT) matrix
$\mathbf{h}_c$	Communication channel
$\mathbf{h}_r$	Radar channel
$\mathbf{h}_r$	Channel coefficients of the propagation path: radar transmitter → target → radar receiver for active sensing
$\mathbf{h}_s$	Channel coefficients of the propagation path: communication transmitter → target → radar receiver for passive sensing
$\mathbf{J}_i$	Interference subspace matrix corresponding to ith communications system
$\mathbf{J}_r$	Interference subspace matrix corresponding to radar
$\mathbf{L}$	MC-DS-CDMA radar sequence chip matrix
$\mathbf{s}_c$	Frequency-domain transmitted communication symbols
$\mathbf{s}_r$	Transmitted radar symbols
$\mathbf{T}$	MC-DS-CDMA radar scrambling chip matrix
$\mathbf{X}$	Transmitted MC-DS-CDMA communication chip matrix
$\mathbf{X}_r$	Communication Toeplitz matrix
$\mathbf{X}_r$	Radar Toeplitz matrix
$\mathbf{Y}$	Transmitted MC-DS-CDMA radar chip matrix
$\mathbf{Z}$	Transmitted MC-DS-CDMA joint radar–communication chip matrix
$\mathcal{N}(\mathbf{H}_i)$	Null subspace of $\mathbf{H}_i$
SF_k	Spreading factor of the kth communication user
$\overline{\mathbf{V}}_0$	Right singular vectors of $\overline{\mathbf{H}}$ corresponding to the zero singular values
$\overline{\mathbf{V}}_{\delta_{\min}}$	Right singular vectors of $\overline{\mathbf{H}}$ corresponding to the singular values less than threshold $\delta_{\min}$
$\overline{\mathbf{H}}$	Stacking of ICSI matrices
$\sigma^2_{n,c}$	Noise power of the communication receivers
Θ_c	Set of possible directions of the communication receiver
θ_c	Direction of the communication receiver
Θ_M	Angular range of the radar mainlobe
Θ_{SL}	Sidelobe region
B_0	Desired beampattern
b_0	Ideal array response
c_{PCC}	Peak cross-correlation constraint between the different waveforms
c_{PSL}	Peak sidelobe level constraint
c_{SL}	Allowed sidelobe level
f_b	Normalized frequency of Doppler bin of interest
f_w	Ambiguity function mainlobe width in frequency
h_c	Communication channel coefficient
$l_{k,n}$	Radar code sequence deployed on the kth subcarrier with nth chip in MC-DS-CDMA signal
M_k	Number of communication bits
N	Number of subcarriers

N_c	Number of subcarriers allocated to communications subsystem
$n_c(t)$	Communication receiver noise
N_r	Number of subcarriers allocated to radar subsystem
N_γ	Number of symbols for communication signal
P_c	Communication power level
$p_c[k]$	Communication signal power of the kth subcarrier
$p_r[k]$	Radar signal power of the kth subcarrier
P_{el}	Maximum transmit power of each antenna element
P_r	Transmitted maximum radar power
P_{tot}	Total transmit power constraint
$t_{k,n}$	Scrambling used on nth chip on kth subcarrier of radar in MC-DS-CDMA signal
$u[k]$	Communications subsystem subcarrier selection coefficient
$w[k]$	Radar subsystem subcarrier selection coefficient
c_k	OVSF code
$s_{k,n}$	Long scrambling code sequence

References

M. H. Ackroyd and F. Ghani. Optimum mismatched filters for sidelobe suppression. *IEEE Transactions on Aerospace and Electronic Systems*, 9(2):214–218, 1973.

T. Aittomäki and V. Koivunen. MIMO radar filterbank design for interference mitigation. In *IEEE International Conference on Acoustics, Speech and Signal Processing*, pages 5297–5301, May 2014.

T. Aittomäki and V. Koivunen. MIMO radar beampattern optimization with ripple control using sum-of-squares representation. In *51st Asilomar Conference on Signals, Systems, and Computers*, pages 578–583, 2017.

T. Aittomäki and V. Koivunen. Optimal mismatched filter design for radar using bounded real lemma. In *IEEE Radar Conference (RadarConf20)*, pages 1–5, 2020.

C. Aydogdu, M. F. Keskin, G. K. Carvajal, O. Eriksson, H. Hellsten, H. Herbertsson, E. Nilsson, M. Rydstrom, K. Vanas, and H. Wymeersch. Radar interference mitigation for automated driving: exploring proactive strategies. *IEEE Signal Processing Magazine*, 37(4):72–84, 2020.

A. Babaei, W. H. Tranter, and T. Bose. A nullspace-based precoder with subspace expansion for radar/communications coexistence. In *IEEE Global Communications Conference (GLOBECOM)*, pages 3487–3492, 2013.

M. Bică and V. Koivunen. Generalized multicarrier radar: models and performance. *IEEE Transactions on Signal Processing*, 64(17):4389–4402, 2016.

M. Bică and V. Koivunen. Multicarrier radar-communications waveform design for RF convergence and coexistence. In *IEEE International Conference on Acoustics, Speech and Signal Processing (ICASSP)*, pages 7780–7784, 2019.

M. Bica, K. Huang, U. Mitra, and V. Koivunen. Opportunistic radar waveform design in joint radar and cellular communication systems. In *IEEE Global Communications Conference (GLOBECOM)*, pages 1–7, December 2015.

M. Bica, K. Huang, V. Koivunen, and U. Mitra. Mutual information based radar waveform design for joint radar and cellular communication systems. In *IEEE International Conference on Acoustics, Speech and Signal Processing (ICASSP)*, pages 3671–3675, March 2016.

S. Boyd and L. Vandenberghe, editors. *Convex optimization*. Cambridge University Press, 2004.

M. Braun, R. Tanbourgi, and F. K. Jondral. Co-channel interference limitations of OFDM communication-radar networks. *IEEE Transactions on Signal Processing*, 2013(1):207–222, 2013.

V. R. Cadambe and S. A. Jafar. Interference alignment and degrees of freedom of the k-user interference channel. *IEEE Transactions on Information Theory*, 54(8):3425–3441, 2008.

Z. Cheng, B. Liao, S. Shi, Z. He, and J. Li. Co-design for overlaid MIMO radar and downlink MISO communication systems via Cramér–Rao bound minimization. *IEEE Transactions on Signal Processing*, 67(24):6227–6240, 2019.

Y. Cui, V. Koivunen, and X. Jing. Interference alignment based precoder-decoder design for radar communication co-existence. In *51st Asilomar Conference on Signals, Systems, and Computers*, pages 1290–1295, October 2017.

Y. Cui, V. Koivunen, and X. Jing. Interference alignment based spectrum sharing for MIMO radar and communication systems. In *IEEE 19th International Workshop on Signal Processing Advances in Wireless Communications (SPAWC)*, pages 1–5, 2018a.

Y. Cui, V. Koivunen, and X. Jing. A perspective on degrees of freedom for radar in radar-communication interference channel. In *52nd Asilomar Conference on Signals, Systems, and Computers*, pages 403–408, 2018b.

Y. Cui, V. Koivunen, and X. Jing. Mutual information based co-design for coexisting MIMO radar and communication systems. In *IEEE International Conference on Communications Workshops (ICC Workshops)*, pages 1–6, 2020.

A. de Oliveira Ferreira, R. Sampaio-Neto, and J. M. Fortes. Robust sidelobe amplitude and phase modulation for dual-function radar-communications. *Transactions on Emerging Telecommunications Technologies*, 29(9):e3314, 2018. e3314 ett.3314.

E. H. Dinan and B. Jabbari. Spreading codes for direct sequence CDMA and wideband CDMA cellular networks. *IEEE Communications Magazine*, 36(9):48–54, 1998.

S. H. Dokhanchi, B. S. Mysore, K. V. Mishra, and B. Ottersten. A mmWave automotive joint radar-communications system. *IEEE Transactions on Aerospace and Electronic Systems*, 55(3):1241–1260, 2019.

S. H. Dokhanchi, M. R. B. Shankar, K. V. Mishra, and B. Ottersten. Multi-constraint spectral co-design for colocated MIMO radar and MIMO communications. In *ICASSP 2020 - 2020 IEEE International Conference on Acoustics, Speech and Signal Processing (ICASSP)*, pages 4567–4571, May 2020.

B. J. Donnet and I. D. Longstaff. Combining MIMO radar with OFDM communications. In *European Radar Conference*, pages 37–40, 2006.

K. W. Forsythe, D. W. Bliss, and G. S. Fawcett. Multiple-input multiple-output (MIMO) radar: performance issues. In *Conference Record of the Thirty-Eighth Asilomar Conference on Signals, Systems and Computers*, Volume 1, pages 310–315, 2004.

B. L. Fox and D. M. Landi. Searching for the multiplier in one-constraint optimization problems. *Operations Research*, 18(2):253–262, 1970.

D. R. Fuhrmann and G. S. Antonio. Transmit beamforming for MIMO radar systems using partial signal correlation. In *Conference Record of the Thirty-Eighth Asilomar Conference on Signals, Systems and Computers*, Volume 1, pages 295–299, November 2004.

S. Gokul and W. Yu. Degrees of freedom of MIMO cellular networks: decomposition and linear beamforming design. *IEEE Transactions on Information Theory*, 61(6):3339–3364, 2015.

K. Gomadam, V. R. Cadambe, and S. A. Jafar. A distributed numerical approach to interference alignment and applications to wireless interference networks. *IEEE Transactions on Information Theory*, 57(6):3309–3322, 2011.

E. Grossi, M. Lops, and L. Venturino. Joint design of surveillance radar and MIMO communication in cluttered environments. *IEEE Transactions on Signal Processing*, 68:1544–1557, 2020.

A. Hassanien, M. G. Amin, Y. D. Zhang, and F. Ahmad. Signaling strategies for dual-function radar communications: an overview. *IEEE Aerospace and Electronic Systems Magazine*, 31(10):36–45, 2016a.

A. Hassanien, M. G. Amin, Y. D. Zhang, and F. Ahmad. Phase-modulation based dual-function radar-communications. *IET Radar, Sonar and Navigation*, 10(8):1411–1421, 2016b.

A. Hassanien, C. Sahin, J. Metcalf, and B. Himed. Uplink signaling and receive beamforming for dual-function radar communications. In *IEEE 19th International Workshop on Signal Processing Advances in Wireless Communications (SPAWC)*, pages 1–5, 2018.

A. Hassanien, M. G. Amin, E. Aboutanios, and B. Himed. Dual-function radar communication systems: a solution to the spectrum congestion problem. *IEEE Signal Processing Magazine*, 36(5):115–126, 2019b.

Q. He, Z. Wang, J. Hu, and R. S. Blum. Performance gains from cooperative MIMO radar and MIMO communication systems. *IEEE Signal Processing Letters*, 26(1):194–198, 2019.

A. Khawar, A. Abdelhadi, and T. C. Clancy. Coexistence analysis between radar and cellular system in los channel. *IEEE Antennas and Wireless Propagation Letters*, 15:972–975, 2016.

P. Kumari, J. Choi, N. González-Prelcic, and R. W. Heath. IEEE 802.11ad-based radar: an approach to joint vehicular communication-radar system. *IEEE Transactions on Vehicular Technology*, 67(4):3012–3027, 2018.

M. Kunert. The EU project Mosarim: a general overview of project objectives and conducted work. In *9th European Radar Conference*, pages 1–5, 2012.

N. Levanon. Multifrequency complementary phase-coded radar signal. *IEE Proceeedings Radar, Sonar and Navigation*, 147(6):276–284, 2000.

J. Li, P. Stoica, and X. Zheng. Signal synthesis and receiver design for MIMO radar imaging. *IEEE Transactions on Signal Processing*, 56(8):3959–3968, 2008.

B. Li, A. P. Petropulu, and W. Trappe. Optimum co-design for spectrum sharing between matrix completion based MIMO radars and a MIMO communication system. *IEEE Transactions on Signal Processing*, 64(17):4562–4575, 2016.

Y. F. Liu, Y. H. Dai, and Z. Q. Luo. On the complexity of leakage interference minimization for interference alignment. In *2011 IEEE 12th International Workshop on Signal Processing Advances in Wireless Communications*, pages 471–475, 2011.

Y. Liu, G. Liao, J. Xu, Z. Yang, and Y. Zhang. Adaptive OFDM integrated radar and communications waveform design based on information theory. *IEEE Communications Letters*, 21(10):2174–2177, 2017b.

F. Liu, C. Masouros, A. Li, and T. Ratnarajah. Robust MIMO beamforming for cellular and radar coexistence. *IEEE Wireless Communications Letters*, 6(3):374–377, 2017a.

F. Liu, C. Masouros, A. Li, H. Sun, and L. Hanzo. MU-MIMO communications with MIMO radar: from co-existence to joint transmission. *IEEE Transactions on Wireless Communications*, 17(4):2755–2770, 2018.

X. Liu, T. Huang, N. Shlezinger, Y. Liu, J. Zhou, and Y. C. Eldar. Joint transmit beamforming for multiuser MIMO communications and MIMO radar. *IEEE Transactions on Signal Processing*, 68:3929–3944, 2020b.

F. Liu, C. Masouros, A. P. Petropulu, H. Griffiths, and L. Hanzo. Joint radar and communication design: applications, state-of-the-art, and the road ahead. *IEEE Transactions on Communications*, 68(6):3834–3862, 2020a.

Z.-Q. Luo, W.-K. Ma, A. M.-C. So, Y. Ye, and S. Zhang. Semidefinite relaxation of quadratic optimization problems. *IEEE Signal Processing Magazine*, 27(3):20–34, 2010.

J. A. Mahal, A. Khawar, A. Abdelhadi, and T. C. Clancy. Spectral coexistence of MIMO radar and MIMO cellular system. *IEEE Transactions on Aerospace and Electronic Systems*, 53(2):655–668, 2017.

K. V. Mishra, M. R. B. Shankar, V. Koivunen, B. Ottersten, and S. A. Vorobyov. Toward millimeter-wave joint radar communications: a signal processing perspective. *IEEE Signal Processing Magazine*, 36(5):100–114, 2019.

B. Paul, A. R. Chiriyath, and D. W. Bliss. Survey of RF communications and sensing convergence research. *IEEE Access*, 5:252–270, 2017.

B. Recht, M. Fazel, and P. A. Parrilo. Guaranteed minimum-rank solutions of linear matrix equations via nuclear norm minimization. *SIAM Review*, 52(3):471–501, 2010.

S. Sharma, M. Melvasalo, and V. Koivunen. Multicarrier DS-CDMA waveforms for joint radar-communication system. In *IEEE Radar Conference*, pages 1–6, 2020.

P. Stoica, J. Li, and Y. Xie. On probing signal design for MIMO radar. *IEEE Transactions on Signal Processing*, 55(8):4151–4161, 2007.

C. Sturm and W. Wiesbeck. Waveform design and signal processing aspects for fusion of wireless communications and radar sensing. *Proceedings of the IEEE*, 99(7):1236–1259, 2011.

C. Sturm, T. Zwick, and W. Wiesbeck. An OFDM system concept for joint radar and communications operations. In *IEEE 69th Vehicular Technology Conference*, pages 1–5, 2009.

N. Su, F. Liu, and C. Masouros. Enhancing the physical layer security of dual-functional radar communication systems. In *IEEE Global Communications Conference (GLOBECOM)*, pages 1–6, December 2019.

T. Tian, T. Zhang, G. Li, and T. Zhou. Mutual information-based power allocation and co-design for multicarrier radar and communication systems in coexistence. *IEEE Access*, 7:159300–159312, 2019.

V. V. Veeravalli and A. El Gamal. *Interference management in wireless networks: fundamental bounds and the role of cooperation*. Cambridge University Press, 2018.

S. Zhang, A. Ahmed, Y. D. Zhang, and S. Sun. DOA estimation exploiting interpolated multi-frequency sparse array. In *2020 IEEE 11th Sensor Array and Multichannel Signal Processing Workshop (SAM)*, pages 1–5, June 2020.

N. Zhao, F. R. Yu, M. Jin, Q. Yan, and V. C. M. Leung. Interference alignment and its applications: a survey, research issues, and challenges. *IEEE Communication Surveys and Tutorials*, 18(3):1779–1803, 2016.

L. Zheng, M. Lops, X. Wang, and E. Grossi. Joint design of overlaid communication systems and pulsed radars. *IEEE Transactions on Signal Processing*, 66(1):139–154, 2018.

L. Zheng, M. Lops, Y. C. Eldar, and X. Wang. Radar and communication coexistence: an overview: a review of recent methods. *IEEE Signal Processing Magazine*, 36(5):85–99, 2019.

5

Information Theoretic Aspects of Joint Sensing and Communications

Mari Kobayashi[1] *and Giuseppe Caire*[2]

[1] *Technical University of Munich, Chair of Communications Engineering, Munich, Germany*
[2] *Technical University of Berlin, Chair of Communications and Information Theory, Berlin, Germany*

5.1 Introduction

Future-generation wireless networks are expected to support a number of autonomous and intelligent applications that rely strongly on accurate sensing and localization techniques [Flagship, 2020; Heath, 2020]. One of the typical and relevant examples is vehicular-to-everything (V2X), whose goal is to ensure the safety and to further offer a wide range of real-time autonomous services based on accurate sensing capabilities in the presence of high mobility. The key enabler of these challenging applications is the ability to continuously track the dynamically changing environment, hereafter called *state*, and react accordingly by exchanging information with each other in the network.

The high cost of spectrum and hardware will inevitably encourage that both radar sensing and communications shall be operated by sharing the same radio frequency bands. Such an observation has motivated the joint radar and communications (JRC) paradigm (see e.g. [Sturm and Wiesbeck, 2011; Patole et al., 2017; Ma et al., 2020; Mishra et al., 2019; Hassanien et al., 2019; Liu et al., 2020] and references therein). Assuming that both radar and data communication are performed over the common resources and at the same device[1], existing works on JRC can be roughly classified into two classes. The first class considers *resource-splitting approach*, such that time, frequency, or space resources are split into either radar or data communication and two functions are operated separately (e.g. see [Ma et al., 2020] and references therein). The second class considers a common waveform for both radar and communication tasks, sometimes referred to as *co-design*. A number of possible waveforms have been considered such as information-embedded radar waveforms (e.g. see [Ma et al., 2020; Zheng et al., 2019; Dokhanchi et al., 2019] and references therein) and standard communication waveforms applied to radar detection (e.g. see [Zheng et al., 2019; Sturm and Wiesbeck, 2011; Liu et al., 2017; Nguyen and Heath, 2017; Raviteja et al., 2019b]).

The main objective of the second class is to design the common waveform that jointly optimizes radar and communications through appropriate performance metrics capturing some tension or trade-off between two functions. Different trade-offs have been considered in the literature, including communication rate versus estimation rate[2] [Bliss, 2014; Chiriyath et al., 2016; Paul et al.,

1 This is relevant to the scenario where communications devices are equipped with mono-static radar.
2 The estimation rate in Bell [1993] is defined as the mutual information between the state and the observation at the radar receiver.

Signal Processing for Joint Radar Communications, First Edition.
Edited by Kumar Vijay Mishra, M. R. Bhavani Shankar, Björn Ottersten, and A. Lee Swindlehurst.

2017], communication rate versus estimation mean square error (MSE) [Kumari et al., 2018a], communication MSE versus estimation MSE [Kumari et al., 2017], as well as communication rate versus target detection probability [Kumari et al., 2018b]. Although these works provide system guidelines or propose waveforms suitable to some specific scenarios, they do not address the fundamental performance limit above which a joint sensing and communications system cannot exceed irrespectively of computational complexities, choices of parameters to be estimated, or further assumptions.

This chapter precisely addresses the fundamental limit of a joint sensing and communications system in a simplified albeit representative point-to-point communication scenario. Namely, we assume that a transmitter, equipped with on-board sensors, wishes to convey a message to its intended receiver and simultaneously estimate the corresponding parameters of interest from the backscattered signals. Such a scenario is relevant to the joint radar parameter estimation and communication, once communication receivers are detected via suitable target detection methods in a previous (initial acquisition) phase. We model the backscattered signals as so-called *generalized feedback*, that is, an additional strictly causal channel output depending on the channel input and the states [Carleial, 1982; Willems et al., 1983; Shayevitz and Wigger, 2013]. Generalized feedback is necessary for state sensing at the transmitter and may be simultaneously useful to improve the communication performance as demonstrated in many communication scenarios without state sensing. Although C. Shannon proved that feedback does not increase the capacity of memoryless point-to-point channels [Shannon, 1956], feedback improves the reliability of memoryless point-to-point channels [Schalkwijk and Kailath, 1966]) and increases the capacity of a number of multiuser channels [Ozarow, 1984; Ozarow and Leung-Yan-Cheong, 1984; Dueck, 1980; Kramer, 2002; Wang, 2012; Gatzianas et al., 2013; Shayevitz and Wigger, 2013]. We can expect these gains similarly in joint sensing and communications systems.

We consider the capacity and the distortion to characterize the fundamental trade-off between communications and state sensing. On the one hand, the capacity is the fundamental communication metric introduced by C. Shannon in Shannon [1948], which presents the largest communication rate above which a message cannot be conveyed reliably with an arbitrarily long codeword. On the other hand, the distortion is a general metric that measures the difference between the source (parameters to be estimated) and its reconstruction (the estimated parameters). Depending on whether the source is continuous or discrete, there are different distortion measures. The most relevant measure in the context of parameter estimation is minimum mean square error (MMSE). The trade-off between the rate and the distortion is a metric widely used in information theory. For example, the rate distortion function is used to characterize the quality of the reconstruction for a given compression rate in lossy source coding. The capacity–distortion trade-off has been also considered in the context of joint state estimation and communication, where the transmitter knows the state and wishes to communicate it together with some data to the receiver [Zhang et al., 2011; Kim et al., 2008; Choudhuri et al., 2013; Sutivong et al., 2005].

5.2 Information Theoretic Model

We consider a simple point-to-point communication setup depicted in Figure 5.1, where the transmitter wishes to send a message W to its receiver and simultaneously estimate states from generalized feedback. We let X, Y, Z, and S denote the channel input, output, feedback output, and state random variables that take values in the sets $\mathcal{X}$, $\mathcal{Y}$, $\mathcal{Z}$, and $\mathcal{S}$, respectively. We let x denote a realization of X. The relation between these random variables is characterized by a memoryless channel with independently and identically distributed (i.i.d.) states. By letting n denote the transmission

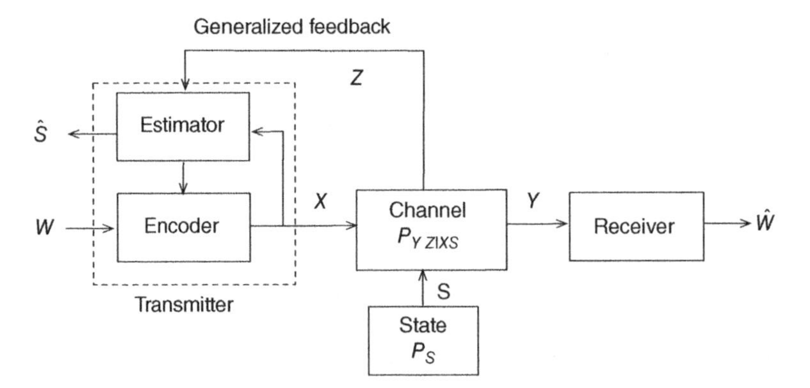

Figure 5.1 Information theoretic model for joint state sensing and communications.

duration expressed in channel uses, the joint probability distribution of the considered model is given by

$$P_{WX^nS^nY^nZ^n}(w, \mathbf{x}, \mathbf{s}, \mathbf{y}, \mathbf{z}) = P(w)\prod_{i=1}^{n} P_S(s_i)\prod_{i=1}^{n} P(x_i|wz^{i-1})P_{YZ|XS}(y_iz_i|x_is_i), \tag{5.1}$$

where we repeatedly use the notation $X^n = (X_1, \dots, X_n)$ for all variables. We remark that the state $P_S(\cdot)$ and the state-dependent channel $P_{YZ|XS}(\cdot)$ are both time-invariant.[3] In order to focus on state sensing at the transmitter, we assume ideally that the state is known perfectly at the receiver. This assumption can be reasonably justified by dedicating a fraction of resources to send pilot symbols so that the receiver estimates the channel state. Under this setting, we need to further introduce a strategy, parameterized by its length n and its rate R in bits per channel use, denoted by $(2^{nR}, n)$, where 2^{nR} denotes the size of the message set $\mathcal{W}$. The strategy consists of

- an encoder that sends a symbol $x_i = \phi_i(w, z^{i-1})$ at channel use i for each message $w \in \mathcal{W}$ and each delayed feedback output $z^{i-1} \in \mathcal{Z}^{i-1}$;
- a decoder that assigns a message estimate $\hat{w} = g(y^n, s^n) \in \mathcal{W}$ for the observation sequence y^n and the state sequence s^n;
- a state estimator that assigns an estimation sequence $\hat{s}^n \in \hat{\mathcal{S}}^n$ to each feedback output sequence $z^n \in \mathcal{Z}^n$ and the channel input sequence $x^n \in \mathcal{X}^n$. The set $\hat{\mathcal{S}}$ denotes the reconstruction alphabet.

The state estimate is measured by the expected distortion given by

$$\mathbb{E}[d(S^n, \hat{S}^n)] = \frac{1}{n}\sum_{i=1}^{n} \mathbb{E}[d(S_i, \hat{S}_i)], \tag{5.2}$$

where $d : \mathcal{S} \times \hat{\mathcal{S}} \mapsto [0, \infty)$ is a distortion function. In practical communications systems, we typically impose a cost constraint such as average or peak power constraints on the channel input. The expected cost is given by

$$\mathbb{E}[b(X^n)] = \frac{1}{n}\sum_{i=1}^{n} \mathbb{E}[b(X_i)], \tag{5.3}$$

where $b : \mathcal{X} \mapsto \mathbb{R}^+$ denotes a cost function. A rate distortion triple (R, D, B) is said to be achievable if there exist $(2^{nR}, n)$ strategies that satisfy the reliability condition, the distortion constraint, and

3 The generalization to the state-dependent channels with memory remains an interesting yet challenging open problem.

the cost constraint, i.e.

$$\lim_{n\to\infty} P(\hat{W} \neq W) = 0, \tag{5.4a}$$

$$\limsup_{n\to\infty} \mathbb{E}[d(S^n, \hat{S}^n)] \leq D, \tag{5.4b}$$

$$\mathbb{E}[b(X^n)] \leq B. \tag{5.4c}$$

The capacity–distortion–cost trade-off $C(D, B)$ is defined as the supremum of R such that (R, D, B) is achievable. From the well-known result on a point-to-point memoryless channel with i.i.d. varying states where the state is available only at the decoder [Gamal and Kim, 2011], the standard capacity without the distortion constraint is

$$C(D = \infty, B) = \max_{P_X} I(X; Y, S) = \max_{P_X} I(X; Y|S), \tag{5.5}$$

where the maximum is over the input distribution P_X satisfying the cost constraint (5.4c). This capacity is achieved by ignoring the feedback as demonstrated by C. Shannon Shannon [1956].

5.3 Fundamental Trade-off Between Sensing and Communications

This section characterizes the capacity–distortion–cost trade-off of the point-to-point joint state sensing and communication model presented in Section 5.2.

5.3.1 Capacity-Distortion-Cost Trade-off

First, we provide a useful lemma justifying a simple deterministic estimator.

Lemma 5.1 *We can choose without loss of generality a deterministic estimator given by*

$$\hat{s}(x, z) \triangleq \arg\min_{s' \in S} \sum_{s \in S} P_{S|XZ}(s|x, z) d(s, s') \tag{5.6}$$

for all x, z.

Proof:

$$\begin{aligned}
\mathbb{E}[d(S, \hat{S})] &= \mathbb{E}\left[\mathbb{E}[d(S, \hat{S})|X, Z]\right] \\
&\overset{(a)}{=} \sum_{x,z} P_{XZ}(xz) \sum_{\hat{s} \in S} P_{\hat{S}|XZ}(\hat{s}|xz) \sum_{s} P_{S|XZ}(s|xz) d(s, \hat{s}) \\
&\geq \sum_{x,z} P_{XZ}(xz) \min_{\hat{s} \in S} \sum_{s} P_{S|XZ}(s|xz) d(s, \hat{s}) \\
&\overset{(b)}{=} \sum_{x,z} P_{XZ}(xz) \sum_{s} P_{S|XZ}(s|xz) d(s, \hat{s}(x, z)) \\
&= \mathbb{E}[d(S, \hat{s}(X, Z))],
\end{aligned} \tag{5.7}$$

where (a) follows from the Markov chain $\hat{S} - XZ - S$ and (b) follows by choosing (5.6). ■

Using the deterministic estimator, we define the estimation cost $c(x)$ as

$$c(x) = \sum_{z \in Z} P_{Z|X}(z|x) \sum_{s \in S} P_{S|XZ}(s|xz) d(s, \hat{s}(x, z)). \tag{5.8}$$

Now we are ready to present the capacity–distortion–trade-off result.

Theorem 5.1 *The capacity–distortion–cost trade-off of the state-dependent memoryless channel with the i.i.d. states is given by*

$$C(D,B) = \max_{P_X \in \mathcal{P}_D \cap \mathcal{P}_B} I(X;Y|S), \tag{5.9}$$

where we define

$$\mathcal{P}_D = \left\{ P_X \mid \sum_{x \in \mathcal{X}} P_X(x)c(x) \le D \right\}, \tag{5.10a}$$

$$\mathcal{P}_B = \left\{ P_X \mid \sum_{x \in \mathcal{X}} P_X(x)b(x) \le B \right\}, \tag{5.10b}$$

and the joint distribution of $SXYZ\hat{S}$ is given by $P_X(x)P_S(s)P_{YZ|XS}(yz|xs)P_{\hat{S}|XZ}(\hat{s}|xz)$.

Proof: Please refer to Appendix 5.A. ■

Some remarks are in order. First, Theorem 5.1 is a simple result that extends the capacity of the state-dependent channel with states known at the receiver to the case with an additional distortion constraint. On the one hand, by ignoring the sensing part of the problem, we recover the standard ergodic capacity of the state-dependent channels in (5.5). On the other hand, by ignoring the communication part of the problem, the trade-off reduces to the distortion minimization under the cost constraint. In some simple cases, the solutions to these individual problems are well known as shown in Section 5.3.3. Second, the optimal input distribution that achieves the capacity–distortion–cost trade-off for discrete variables can be solved efficiently by modifying the Blahut–Arimoto algorithm, i.e. an iterative algorithm developed by Arimoto Arimoto [1972] and Blahut Blahut [1972], as shown in Section 5.3.2.

We provide the useful property of the function $C(D, B)$.

Lemma 5.2 *$C(D, B)$ is a nondecreasing concave function of (D, B) for $D \ge D_{\min} \triangleq \min_{x \in \mathcal{X}} c(x)$ and for any $B \ge 0$.*

Proof: The proof is a straightforward extension of Zhang et al. [2011, Corollary 1] to the case of two cost functions and the state-dependent channel. The nondecreasing property follows immediately from the definition (5.9) because we have $\mathcal{P}_{D_1} \subseteq \mathcal{P}_{D_2}$ and $\mathcal{P}_{B_1} \subseteq \mathcal{P}_{B_2}$ for any $D_1 \le D_2$ and $B_1 \le B_2$. In order to verify the concavity of $C(D, B)$ with respect to (D, B), we consider time-sharing between two input distributions, denoted by $P_X^{(1)}$ and $P_X^{(2)}$, that achieve $C(D_1, B_1)$ *and* $C(D_2, B_2)$, respectively. For any $\theta \in (0,1)$, by dedicating a fraction θ of time to $P_X^{(1)}$ and another fraction $(1-\theta)$ of time to $P_X^{(2)}$, we obtain

$$\begin{aligned}\theta C(D_1,B_1) + (1-\theta)C(D_2,B_2) &\le \max_{P_X \in \mathcal{P}_{\theta D_1 + (1-\theta)D_2} \cap \mathcal{P}_{\theta B_1 + (1-\theta)B_2}} I(X;Y|S) \\ &= C\left(\theta D_1 + (1-\theta)D_2, \theta B_1 + (1-\theta)B_2\right).\end{aligned}$$

This establishes the concavity of $C(D, B)$. ■

5.3.2 Numerical Method for Optimization

This section presents a numerical method to solve the optimization problem stated in Theorem 5.1 for discrete variables $(X, Y, S) \in \mathcal{X} \times \mathcal{Y} \times \mathcal{S}$. The mutual information term $I(X;Y|S)$ can be explicitly rewritten as a function of the input distribution P_X to be optimized as well as given channel and state distributions, i.e. $P_{Y|XS}, P_S$.

$$I(P_X, P_{Y|XS}|P_S) = \sum_{s\in\mathcal{S}} P_S(s) \sum_{x\in\mathcal{X}} \sum_{y\in\mathcal{Y}} P_X(x) P_{Y|XS}(y|xs) \log \frac{P_{Y|XS}(y|xs)}{P_{Y|S}(y|s)}. \tag{5.11}$$

The optimization problem can be stated as

$$\text{maximize} \quad I(P_X, P_{Y|XS}|P_S), \tag{5.12a}$$

$$\text{subject to} \quad \sum_x b(x) P_X(x) \leq B, \tag{5.12b}$$

$$\sum_x c(x) P_X(x) \leq D, \tag{5.12c}$$

$$\sum_x P_X(x) = 1, \quad P_X(x) \geq 0, \ \forall x, \tag{5.12d}$$

where (5.12d) reflects that $\{P_X(x)\}_{x\in\mathcal{X}}$ is the probability simplex. The problem (5.12) is convex as the objective functional is concave in P_X and the constraints are linear in P_X. By recalling (5.10b) and (5.10a), we let $\mathcal{F}(B, D) = \mathcal{P}_B \cap \mathcal{P}_D$ denote a feasible probability simplex set satisfying constraints (5.12b) and (5.12c). Assuming for simplicity that $\mathcal{F}(B, D)$ is not empty, $\mathcal{F}(B, D)$ is a convex compact set. In order to compute the capacity–distortion trade-off for a given cost, it is convenient to consider a parametric form of the optimization problem by incorporating one cost function as a penalty term in the objective function and focusing on the other cost function. The new problem is to maximize

$$I(P_X, P_{Y|XS}|P_S) - \mu \sum_x c(x) P_X(x) \tag{5.13}$$

subject to the constraints (5.12b) and (5.12d), where $\mu \geq 0$ is a fixed parameter. We proceed as in the derivation of the standard Blahut–Arimoto algorithm [Arimoto, 1972; Blahut, 1972] that computes the capacity-cost function. In particular, let $Q_{X|YS}$ denote a general conditional distribution of X given Y, S and consider the function

$$J(P_X, P_{Y|XS}, Q_{X|YS}|P_S) = \sum_s P_S(s) \sum_x \sum_y P_X(x) P_{Y|XS}(y|xs) \log \frac{Q_{X|YS}(x|ys)}{P_X(x)}.$$

We generalize the standard Blahut–Arimoto algorithm that calculates the capacity of discrete channels Cover and Thomas [2006, Chapter 10.8] to the capacity–distortion trade-off characterization of discrete state-dependent channels. First, we prove the important properties of the alternating optimization procedure as stated below.

Theorem 5.2 *The following statements hold:*

(a) By letting $L_\mu(B, D)$ denote the optimal value of (5.13), we have

$$L_\mu(B, D) = \max_{P_X\in\mathcal{P}(B)} \min_{Q_{X|YS}} J(P_X, P_{Y|XS}, Q_{X|YS}|P_S) - \mu \sum_x c(x) P_X(x). \tag{5.14}$$

(b) For fixed $P_X \in \mathcal{P}(B)$, the function $J(\cdot) - \mu \sum_x c(x) P_X(x)$ is maximized over $Q_{X|YS}$ by

$$Q^\star_{X|YS}(x|ys) = \frac{P_X(x) P_{Y|XS}(y|xs)}{\sum_{x'} P_X(x') P_{Y|XS}(y|x's)}. \tag{5.15}$$

(c) For fixed $Q_{X|YS}$, the function $J(\cdot) - \mu \sum_x c(x) P_X(x)$ is maximized over $P_X \in \mathcal{P}(B)$ by

$$P^\star_X(x) = \frac{e^{g(x)}}{\sum_{x'} e^{g(x')}}, \tag{5.16}$$

where

$$g(x) = \sum_{s}\sum_{y} P_S(s)P_{Y|XS}(y|xs)\log Q_{X|YS}(x|ys) - \lambda b(x) - \mu c(x) \tag{5.17}$$

and $\lambda \geq 0$ *is chosen so that* (5.12b) *holds with equality.*

Proof: Please refer to Appendix 5.B. ∎

Based on a general result on alternating optimization Cover and Thomas [2006, Chapter 10.8], we propose the following algorithm summarized in Algorithm 1. The algorithm yields a pair of capacity–distortion values $(C_\mu(B), D_\mu(B))$ for any fixed input cost B and μ. By letting $P^{\infty}_{X,\mu}$ denote the convergent input distribution produced by the algorithm for given B and μ, we have

$$C_\mu(B) = \mathcal{I}(P^{(\infty)}_{X,\mu}, P_{Y|XS}|P_S), \tag{5.18a}$$

$$D_\mu(B) = \sum_{x} c(x)P^{(\infty)}_{X,\mu}(x). \tag{5.18b}$$

By varying μ, we obtain the capacity–distortion trade-off for fixed input cost B. Note that for $\mu = 0$ we obtain the standard capacity-cost function of the channel (disregarding the distortion), while as $\mu \to \infty$ the problem becomes a distortion minimization for a given input cost B. Moreover, by varying the input cost B, we obtain the family of all such trade-offs, and therefore the whole boundary of the achievable capacity–distortion–cost trade-off region.

Algorithm 1 Modified Blahut–Arimoto algorithm

Input: B, $\mu \geq 0$, $P^{(0)}_X(x) = \frac{1}{|\mathcal{X}|}, \forall x \in \mathcal{X}$.

Output: $P^{(\infty)}_{X,\mu}(x), \forall x \in \mathcal{X}$.

For $k = 1, 2, 3, \ldots$ do:

1. Let

$$Q^{(k)}_{X|YS}(x|ys) = \frac{P^{(k-1)}_X(x)P_{Y|XS}(y|xs)}{\sum_{x'} P^{(k-1)}_X(x')P_{Y|XS}(y|x's)}. \tag{5.19}$$

2. Choose $\lambda^{(0)} > 0$ and, for $\ell = 1, 2, \ldots$, repeat:
   a) compute primal variables: $p^{(\ell)}(x) = \frac{e^{g^{(\ell)}(x)}}{\sum_{x'} e^{g^{(\ell)}(x')}}$ with

$$g^{(\ell)}(x) = \sum_{s,y} P_S(s)P_{Y|XS}(y|xs)\log Q^{(k)}_{X|YS}(x|ys) - \lambda^{(\ell-1)}b(x) - \mu c(x) \tag{5.20}$$

   b) update dual variables:

$$\lambda^{(\ell)} = \left[\lambda^{(\ell-1)} + \alpha_\ell\left(\sum_{x} b(x)p^{(\ell)}(x) - B\right)\right]_+ \tag{5.21}$$

   where α_ℓ is the gradient adaptation step. Let $P^{(k)}_X(x) = \lim_{\ell\to\infty} p^{(\ell)}(x)$.

5.3.3 Numerical Examples

We provide two examples to illustrate the gain of our proposed joint scheme with respect to the resource-splitting approach.

Definition 1 A resource-splitting approach refers to a scheme whose resources are divided into either state sensing with generalized feedback or data communication without feedback.

For simplicity, generalized feedback is assumed to be perfect $Z = Y$.

5.3.3.1 Binary Channels with Multiplicative Bernoulli State

Consider binary channels with multiplicative Bernoulli states given by

$$Y = SX, \tag{5.22}$$

where the state S is Bernoulli distributed such that $P_S(1) \triangleq q \in [0,1/2]$, and the multiplication is binary such that $y = 1$ if $x = s = 1$ and $y = 0$ otherwise. We use the Hamming distortion measure $d(s, \hat{s}) = s \oplus \hat{s}$. Notice that the input cost constraint is not active.

We characterize the input distribution $P_X(0) \triangleq p \in [0,1]$ that maximizes the capacity–distortion trade-off $C(D)$. The two extreme points on the capacity–distortion trade-off are as follows.

- If the transmitter chooses $p = 0$, i.e. sends always $x = 1$, the state is perfectly estimated from the output feedback $s = y$. This choice achieves the minimum distortion $D_{\min} = 0$, while letting the transmission rate zero $C(D_{\min}) = 0$.
- If the transmitter chooses $p = 1/2$, i.e. sends $x = 1$ and $x = 0$ with the equal probability, then the unconstrained capacity of $C(D_{\max}) = q$ is achieved. By noticing that the Hamming distortion is zero when $y = 1$ with probability $P_S(1)P_X(1)$, the corresponding distortion is given by $D_{\max} = q/2$.

More generally, we have the following result.

Proposition 5.1 The capacity–distortion trade-off of the binary channel with the Bernoulli state in (5.22) is given by

$$C(p) = qH_2(p), \quad D(p) = qp, \tag{5.23}$$

where $H_2(p) = -p \log_2 p - (1 - p) \log_2 (1 - p)$ for $p \in [0,1]$ denotes the binary entropy function.

Proof: For a given state distribution q and a given input distribution p, the capacity of the binary channel (5.22) can be expressed as follows:

$$\begin{aligned} C(p) &= I(X; Y|S) = H(Y|S) - H(Y|X, S) \\ &\overset{(a)}{=} H(Y|S) \\ &= -\sum_s P_S(s) \sum_y P_{Y|S}(y|s) \log P_{Y|S}(y|s) \\ &\overset{(b)}{=} qH_2(p), \end{aligned}$$

where (a) follows from the state-dependent channel assumption; (b) follows by noticing $P_{Y|S}(0|0) = 1$ and $P_{Y|S}(1|1) = p$. To compute the distortion, we first determine the deterministic estimator $\hat{s}(x, y)$ and the resulting cost function $c(x)$. From Lemma 5.1, we have

$$\hat{s}(x, 0) = 0, \ \ \forall x, \ \ \hat{s}(x, 1) = 1, \ \ \forall x. \tag{5.24}$$

In fact, since $y = 1$ cannot be generated from the input $x = 0$, the value of $\hat{s}(0,1)$ is irrelevant. Using (5.8) and the conditional probability mass function (pmf), we have

$$P_{S|XY}(s|xy) = \begin{cases} 1 - q & \text{if } (x, y, s) = (0,0,0) \\ q & \text{if } (x, y, s) = (0,0,1) \\ 1 & \text{if } (x, y, s) = (1,0,0) \text{ or } (1,1,1) \\ 0 & \text{else} \end{cases} \tag{5.25}$$

yielding the estimation cost function $c(1) = 0$ and $c(0) = q$. By plugging these into $\sum_{x\in\{0,1\}} P_X(x)c(x)$, the desired distortion function $D(p)$ is obtained. ■

Figure 5.2 evaluates the capacity–distortion trade-off for the case $q = 0.4$. The resource-splitting approach achieves a time-sharing between the following two corner points: (i) in sensing mode, the distortion-rate pair of $(D, C) = (0,0)$ is achieved by dedicating the full resources; (ii) in communications mode, the distortion-rate pair of $(D, C) = (q, q)$ is achieved by choosing the input distribution $p = 1/2$ and considering a fixed estimator $\hat{s} = 0$ independent of feedback. Observe that our proposed joint scheme yields a significant gain over the resource-splitting approach.

5.3.3.2 Rayleigh Fading Channels

Next, we consider the real Gaussian channel with Rayleigh fading. The channel input and output relation is given by

$$Y_i = S_i X_i + N_i, \tag{5.26}$$

where X_i is the channel input satisfying the cost constraint $\frac{1}{n}\sum_i \mathbb{E}[|X_i|^2] \leq B$, and both N_i and S_i are i.i.d. Gaussian distributed with zero mean and unit variance. Focusing on the quadratic distortion measure or MMSE such that $d(s, \hat{s}) = |s - \hat{s}|^2$, we first remark two extreme cases:

- If we relax the distortion constraint, the Gaussian input maximizes the capacity and yields the unconstrained capacity $C_{\max} = \frac{1}{2}\mathbb{E}[\log(1 + |S|^2 P)]$ (bit/channel use) by averaging overall possible fading states. The corresponding expected distortion is $\mathbb{E}\left[\frac{1}{1+|X|^2}\right]$ where the expectation is with respect to the Gaussian distributed X.
- The minimum distortion $D_{\min}$ is achieved by 2-ary pulse amplitude modulation (PAM) and is equal to $\frac{1}{1+B}$. The corresponding achievable rate of PAM modulation is found in the literature Ungerboeck [1982].

Figure 5.3 shows the capacity–distortion trade-off for the average power constraint of 10 dB ($B = 10$), calculated by applying the modified Blahut–Arimoto algorithm presented in Section 5.3.2 to the quantized real additive white Gaussian noise (AWGN) channel and M-ary PAM. The resource-splitting approach achieves two corner points; (i) in sensing mode, the distortion-rate

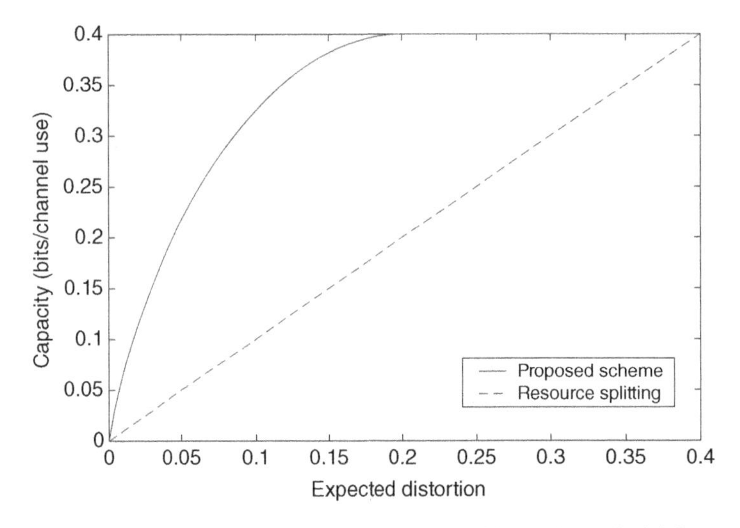

Figure 5.2 Capacity–distortion trade-off of the binary channel with Bernoulli states of $q = 0.4$.

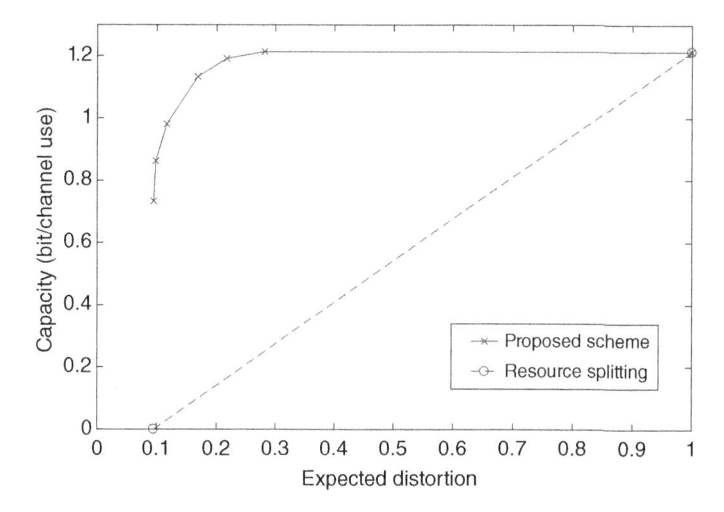

Figure 5.3 Capacity–distortion trade-off of the Rayleigh fading channels with 10 dB.

pair $(D_{\min}, 0)$ is achieved by dedicating full resources to state estimation; (ii) in communications mode, the distortion-rate pair of $(D_{\max}, C_{\max})$ with $D_{\max} \triangleq var[S] = 1$ is achieved by ignoring feedback and sending data with Gaussian distribution. It is worth noticing that the ideal operating point $(D_{\min}, C_{\max}) = (0.091, 1.213)$ is not far from the optimal trade-off.

These two examples illustrate the potential gains of the proposed joint scheme compared to the resource-splitting approach that operates sensing and communications separately. Although our framework builds on simplified point-to-point memoryless channels, the capacity–distortion–cost trade-off is the first rigorous characterization of the fundamental limit of a joint sensing and communications system.

5.4 Application to Joint Radar and Communications

Motivated by the potential gains of the proposed joint sensing and communications scheme observed in Section 5.3.3, this section addresses a more practical scenario of joint radar parameter estimation and communication using orthogonal frequency-division multiplexing (OFDM). OFDM is currently the most popular choice for communications standards such as LTE [Sesia et al., 2011] due to its robustness and its ability to deal with time–invariant frequency selective channels. Moreover, since it is also considered in the wireless LAN standards for vehicular networks IEEE 802.11p and IEEE 802.11ad, OFDM has been extensively studied as one of the promising waveforms for JRC (see e.g. [Sturm and Wiesbeck, 2011; Braun, 2014; Nguyen and Heath, 2017; Gaudio et al., 2019; Carvajal et al., 2020] and references therein). A nice property of OFDM-based radar is that the delay and Doppler estimation can be decoupled under some mild condition as explained in the following.

5.4.1 System Model

We consider a JRC system operating over a total bandwidth of W_b (Hz) and at the carrier frequency f_c (Hz). We assume that a transmit vehicle, equipped with a mono-static full-duplex radar, wishes to convey a message modulated by OFDM to its target receiver, while estimating the relative range and the relative velocity related to the same receiver. Full-duplex operations can be achieved with

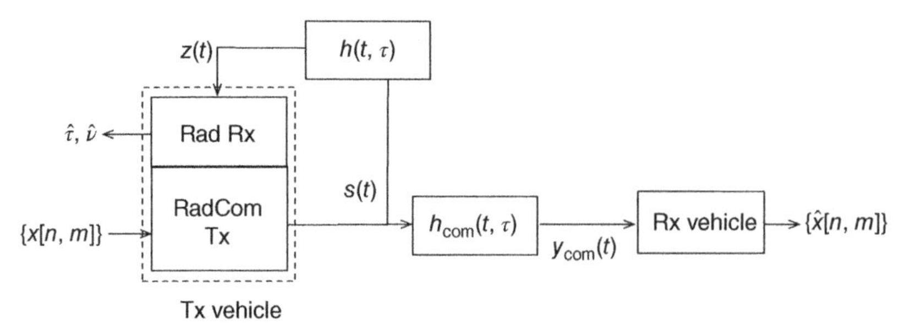

Figure 5.4 Joint radar sensing and communications model.

sufficient isolation and cancellation [Sabharwal et al., 2014]. The model at hand is illustrated in Figure 5.4.

In the OFDM system, the total bandwidth is divided into M subcarriers such that $W_b = M\Delta f$, where Δf (Hz) denotes the subcarrier spacing. For a given maximum Doppler shift $\nu_{\max}$, the subcarrier spacing is typically chosen to satisfy[4]

$$\nu_{\max} \ll \Delta f. \tag{5.27}$$

For each OFDM data symbol of duration $T = 1/\Delta f$ (seconds), we typically append cyclic prefix (CP) to avoid the inter-symbol interference (ISI). Provided the maximum delay spread T_d, the length of CP is chosen as $T_{\rm cp} = C\frac{T}{M}$, with $C = \lceil \frac{T_d}{T/M} \rceil$, where $\lceil \cdot \rceil$ is the rounding-up operation. The resulting OFDM symbol duration including CP is given by $T_o = T_{\rm cp} + T$. By considering N symbols, the OFDM frame duration is $T_f^{\rm ofdm} = NT_o$.

For simplicity, we assume that the radar performs beam sweeping such that there is a single target within a narrow angular sector possibly surrounded by other scatterers. We model the radar channel as a P-tap time–frequency selective channel given by

$$h(t,\tau) = \sum_{p=0}^{P-1} h_p \delta(\tau - \tau_p) e^{j2\pi\nu_p t}, \tag{5.28}$$

where P denotes the number of scattering reflections or paths and the 0th path corresponds to the target (communication receiver), h_p is the complex channel gain, $\nu_p = \frac{2v_p f_c}{c}$ and $\tau_p = \frac{2r_p}{c}$ denotes a round-trip Doppler shift and delay, respectively. By taking into account the one-way Doppler shift and delay, the forward communication channel is given by

$$h_{\rm com}(t,\tau) = g_0 e^{j\pi\nu_0 t} \delta\left(\tau - \frac{\tau_0}{2}\right), \tag{5.29}$$

where g_0 denotes the complex channel gain. The data symbols $x[n,m]$, for $n = 0, \dots, N-1$ and $m = 0, \dots, M-1$, are arranged in an $N \times M$ grid. We consider the average power constraint given by

$$\frac{1}{NM} \sum_{n=0}^{N-1} \sum_{m=0}^{M-1} \mathbb{E}\left[|x[n,m]|^2\right] \leq B. \tag{5.30}$$

The continuous-time OFDM transmitted signal with CP is given by

$$s(t) = \sum_{n=0}^{N-1} \sum_{m=0}^{M-1} x[n,m] \mathrm{rect}(t - nT_o) e^{j2\pi m\Delta f(t - T_{\rm cp} - nT_o)}, \tag{5.31}$$

4 Note that this approximation can be justified in a number of scenarios. For example, consider a scenario inspired by IEEE 802.11p with $f_c = 5.89$ GHz and the subcarrier spacing $\Delta f = 156.25$ KHz. This yields $v_{\max} \ll 14325$ km/h, which is reasonable even for a relative speed of 400 km/h. The same holds for IEEE 802.11ad with $f_c = 60$ GHz and $\Delta f = 5.156\ 25$ MHz [Cordeiro et al., 2010].

where rect(t) is 1 for $t \in [0, T_o]$ and zero otherwise. The signal received after the time–frequency selective channel (5.28) is

$$\begin{aligned} z(t) &= \int h(t,\tau)s(t-\tau)d\tau + \zeta(t) \\ &= \sum_{p=0}^{P-1} h_p s(t-\tau_p)e^{j2\pi\nu_p t} + \zeta(t), \end{aligned} \tag{5.32}$$

where $\zeta(t) \sim \mathcal{N}_C(0,1)$ is the noise assumed to be i.i.d. complex additive Gaussian random process. By sampling every $\frac{T}{M}$ and removing the CP in each OFDM symbol, we obtain

$$\begin{aligned} z'[n,m] &= z(t)|_{t=nT_o+T_{cp}+mT/M} \\ &= \sum_{p=0}^{P-1} h_p e^{j2\pi nT_o\nu_p} \sum_{m'=0}^{M-1} x[n,m']e^{j2\pi\frac{m}{M}\left(\frac{\nu_p}{\Delta f}+m'\right)} e^{-j2\pi m'\Delta f\tau_p} + \zeta'[n,m], \end{aligned} \tag{5.33}$$

where $\zeta'[n,m] = \zeta(nT_o + T_{cp} + mT/M) \sim \mathcal{N}_C(0,1)$ are i.i.d. across n and m. Applying the discrete Fourier transform (DFT) and using the orthogonal property, the output is given by

$$\begin{aligned} z'[n,m] &= \frac{1}{M}\sum_{i=0}^{M-1} z'[n,i]e^{-j2\pi\frac{mi}{M}} + \zeta[n,m] \\ &\approx \frac{1}{M}\sum_{p=0}^{P-1} h_p e^{j2\pi nT_o\nu_p} \sum_{m'=0}^{M-1} x[n,m']e^{-j2\pi m'\Delta f\tau_p} \sum_{i=0}^{M-1} e^{j2\pi\frac{i(m'-m)}{M}} + \zeta[n,m] \\ &= \sum_{p=0}^{P-1} h_p e^{j2\pi nT_o\nu_p} e^{-j2\pi m\Delta f\tau_p} x[n,m] + \zeta[n,m], \end{aligned} \tag{5.34}$$

where $\zeta[n,m] = \frac{1}{M}\sum_{i=0}^{M-1}\zeta[n,i]e^{-j2\pi\frac{mi}{M}}$ has the same statistics as $\zeta'[n,m]$ and the approximation follows from (5.27). Under the approximated channel input–output relation (5.34), it readily follows that the Doppler shift and the delay are decoupled, which makes joint range and velocity estimation simple (see e.g. [Braun, 2014; Sturm and Wiesbeck, 2011]).

Following similar steps, we obtain the communication channel output. Under the assumption that the communication receiver knows the channel state and can shift the phase, the input–output relation is given by

$$y[n,m] \approx g_0 x[n,m] + \eta[n,m], \tag{5.35}$$

where $\eta[n,m] \sim \mathcal{N}_C(0,1)$ and the approximation follows from (5.27).

5.4.2 Maximum Likelihood Estimator

This section derives the maximum likelihood (ML) estimator to extract the range and the velocity from NM observations. By letting $\mathbf{x} = [x[0,0], \dots, x[N-1, M-1]]$, $\mathbf{z} = [z[0,0], \dots, z[N-1, M-1]]$, $\boldsymbol{\zeta} = [\zeta[0,0], \dots, \zeta[N-1, M-1]]$, and defining a NM-dimensional diagonal matrix $\mathbf{U}_p = diag(\phi_p[0,0], \dots, \phi_p[N-1, M-1])$ with $\phi_p[n,m] = e^{j2\pi(nT_o\nu_p - m\Delta f\tau_p)}$, we write a vector observation model[5] as

$$\mathbf{z} = \sum_{p=0}^{P-1} h_p \mathbf{U}_p(\tau_p, \nu_p)\mathbf{x} + \boldsymbol{\zeta}. \tag{5.36}$$

5 Notice that the model (5.36) applies to any general channel beyond the parallel channels of (5.33) by modifying the pth channel matrix $\mathbf{U}_p$ accordingly.

We derive the ML estimator to estimate jointly channel gain, range, and velocity, denoted by a $3P$-dimensional vector $\theta = (\mathbf{h}, \boldsymbol{\tau}, \boldsymbol{v})$, where $\mathbf{h} = (h_0, \ldots, h_{P-1})$ and $\boldsymbol{\tau}, \boldsymbol{v}$ are defined similarly. Since the radar receiver is colocated at the communication transmitter and knows $\mathbf{x}$, the likelihood function conditioned on $\mathbf{x}$ is given by

$$p(\mathbf{z}|\theta, \mathbf{x}) = \frac{1}{(\pi)^{NM}} \exp\left(-l(\mathbf{z}|\theta, \mathbf{x})\right) \tag{5.37}$$

with

$$l(\mathbf{z}|\theta, \mathbf{x}) = \left\| \mathbf{z} - \sum_{p=0}^{P-1} h_p \mathbf{U}_p \mathbf{x} \right\|^2. \tag{5.38}$$

By taking logarithm of (5.37), the ML estimator is given as the solution to the minimization of the log-likelihood function (5.38), i.e.

$$\hat{\theta} = \min_{\theta \in \mathbb{C}^P \times \mathbb{R}^P \times \mathbb{R}^P} l(\mathbf{z}|\theta, \mathbf{x}). \tag{5.39}$$

As an extension of the well-known sinusoid parameter estimation or Doppler signal estimation Richards [2014, Chapter 7.2.2] to the multipath and multicarrier channel, we proceed the derivation in two steps. First, assuming that $(\boldsymbol{\tau}, \boldsymbol{v})$ is known, we remark that the log-likelihood function (5.38) is quadratic in the complex channel gain h_p. Hence, the minimization of (5.38) w.r.t. $\mathbf{h}$ for fixed $(\boldsymbol{\tau}, \boldsymbol{v})$ is readily given as a solution to the linear equation system:

$$\sum_{q=0}^{P-1} h_q \mathbf{x}^{\mathsf{H}} \mathbf{U}_p^{\mathsf{H}} \mathbf{U}_q \mathbf{x} = \mathbf{x}^{\mathsf{H}} \mathbf{U}_p^{\mathsf{H}} \mathbf{z}, \quad p = 0, \ldots, P-1. \tag{5.40}$$

We can alternatively rewrite (5.40) by focusing on the pth path as

$$h_p = \frac{\mathbf{x}^{\mathsf{H}} \mathbf{U}_p^{\mathsf{H}} \mathbf{z} - \sum_{q \neq p}^{P-1} h_q \mathbf{x}^{\mathsf{H}} \mathbf{U}_p^{\mathsf{H}} \mathbf{U}_q \mathbf{x}}{\|\mathbf{x}\|^2}. \tag{5.41}$$

Plugging the equality (5.40) into (5.38), we readily see that

$$\begin{aligned} l(\mathbf{z}|\theta, \mathbf{x}) &= \|\mathbf{z}\|^2 - \sum_{p=0}^{P-1} h_p \mathbf{z}^{\mathsf{H}} \mathbf{U}_p \mathbf{x} \\ &= \|\mathbf{z}\|^2 - l_2(\mathbf{z}|\theta, \mathbf{x}), \end{aligned} \tag{5.42}$$

where the last equality follows by using (5.41) and defining

$$l_2(\mathbf{z}|\theta, \mathbf{x}) = \sum_{p=0}^{P-1} \left[S_p(\tau_p, v_p) - I_p(\tau_p, v_p, \mathbf{h}_{\bar{p}}, \boldsymbol{\tau}_{\bar{p}}, \boldsymbol{v}_{\bar{p}}) \right], \tag{5.43}$$

where $S_p(\cdot), I_p(\cdot)$ denote the desired term and the interference term, which are given, respectively, by

$$S_p(\tau_p, v_p) = \frac{|\mathbf{x}^{\mathsf{H}} \mathbf{U}_p^{\mathsf{H}} \mathbf{z}|^2}{\|\mathbf{x}\|^2}, \tag{5.44a}$$

$$I_p(\tau_p, v_p, \mathbf{h}_{\bar{p}}, \boldsymbol{\tau}_{\bar{p}}, \boldsymbol{v}_{\bar{p}}) = \frac{\left(\sum_{q \neq p}^{P-1} h_q^* \mathbf{x}^{\mathsf{H}} \mathbf{U}_q^{\mathsf{H}} \mathbf{U}_p \mathbf{x} \right) \mathbf{x}^{\mathsf{H}} \mathbf{U}_p^{\mathsf{H}} \mathbf{z}}{\|\mathbf{x}\|^2}, \tag{5.44b}$$

where $\bar{p} = \{0, \ldots, P-1\} \backslash \{p\}$. From (5.42), it readily follows that the minimization of $l(\mathbf{z}|\theta, \mathbf{x})$ reduces to maximizing the term $l_2(\mathbf{z}|\theta, \mathbf{x})$. A few remarks are in order. First, for a special case of a single-path channel ($P = 1$), the ML estimator is the one maximizing $S_0(\tau_0, v_0)$ with respect to (τ_0, v_0) for a given h_p. This can be solved efficiently by a two-dimensional search that generalizes the periodogram for the case of the known data symbols. Second, the function $l_2(\mathbf{z};\theta|\mathbf{x})$ is not separable in the pairs of parameters (τ_p, ν_p) for different values of p because of the dependency of the interference terms I_p on all (τ_q, ν_q) for $q \neq p$. Nevertheless, this dependency through the "cross-term" coefficients, given by $\frac{\mathbf{x}^H\mathbf{U}_p^H\mathbf{U}_q\mathbf{x}}{\|\mathbf{x}\|^2} \approx \sum_{n,m} e^{-j2\pi m\Delta f(\tau_q-\tau_p)}e^{j2\pi nT_o(\nu_q-\nu_p)}$ for the uniform power allocation, tends to be weak for parallel (OFDM) channels as well as for typically sparse multipath channels.

Given the above observations, we propose an iterative block-wise optimization that alternates between the optimization of each pair (τ_p, ν_p) by fixing $\mathbf{h}_{\bar{p}}, \boldsymbol{\tau}_{\bar{p}}, \boldsymbol{\nu}_{\bar{p}}$ and the optimization of $\mathbf{h}$ by fixing $\boldsymbol{\tau}, \boldsymbol{\nu}$ as summarized in Algorithm 2.

Algorithm 2 Approximate ML algorithm for range, Doppler shift, and channel estimation over P-path OFDM channels

---

Input: $h_p^{(0)} = 0, \forall p$

Output: $\hat{\mathbf{h}}, \hat{\boldsymbol{\tau}}, \hat{\boldsymbol{\nu}}$

For $k = 1, 2, 3, \dots$ do:

1. Delay and Doppler shift update:
   For $p = 0, \dots, P-1$, solve

$$(\hat{\tau}_p^{(k)}, \hat{\nu}_p^{(k)}) = \arg\max_{\tau,\nu}\left[S_p(\tau,\nu) - I_p(\tau,\nu,\hat{\mathbf{h}}_{\bar{p}}^{(k-1)}, \hat{\boldsymbol{\tau}}_{\bar{q}}^{(k-1)}, \hat{\boldsymbol{\nu}}_{\bar{p}}^{(k-1)})\right]$$

2. Complex channel gain update: solve $\mathbf{h}^{(k)}$ satisfying the linear equation

$$\sum_{q=0}^{P-1} h_q^{(k)}\mathbf{x}^{\mathsf{H}}\mathbf{U}_p(\hat{\tau}_p^{(k)}, \hat{\nu}_p^{(k)})^{\mathsf{H}}\mathbf{U}_q(\hat{\tau}_q^{(k)}, \hat{\nu}_q^{(k)})\mathbf{x} = \mathbf{x}^{\mathsf{H}}\mathbf{U}_p^{\mathsf{H}}(\hat{\tau}_p^{(k)}, \hat{\nu}_p^{(k)})\mathbf{z}.$$

5.4.3 Cramér–Rao Lower Bound

To evaluate the performance of our proposed iterative algorithm for joint range and velocity estimation, this subsection derives the Cramer–Rao lower bound (CRLB) of the corresponding error variance. The CRLB is a well-known lower bound on the variance of any unbiased estimator [Kay, 1993]. Let $\boldsymbol{\theta}$ denote the parameter set and $\mathbf{y}$ be the observation vector whose conditional pdf is given by $p(\mathbf{y}|\boldsymbol{\theta})$. Under the mild condition, the variance of the ith element of $\boldsymbol{\theta}$, denoted by $\sigma_{\hat{\theta}_i}^2$, is lower bounded by Kay [1993]

$$\sigma_{\hat{\theta}_i}^2 \geq \left[\mathbf{I}^{-1}(\boldsymbol{\theta})\right]_{i,i}, \tag{5.45}$$

where $\mathbf{I}(\boldsymbol{\theta})$ denotes the Fisher information matrix and its (i,j)th element is given by

$$[\mathbf{I}(\boldsymbol{\theta})]_{i,j} = -\mathbb{E}\left[\frac{\partial^2 \ln(\mathbf{y}|\boldsymbol{\theta})}{\partial\theta_i\partial\theta_j}\right], \tag{5.46}$$

where the expectation is with respect to $p(\mathbf{y}|\boldsymbol{\theta})$.

By focusing on the case of a single-path channel ($P = 1$) and ignoring the path index, we apply the above general formula to compute the error variance of (h, τ, ν). Consider the vector of four real parameters $\boldsymbol{\theta} = (\alpha, \varphi, \nu, \tau) \in \mathbb{R}^4$ to be estimated, where $\alpha = |h|, \varphi = \angle(h)$, we can rewrite (5.36) under a more convenient form:

$$z[n,m] = \sqrt{B}\alpha e^{j\varphi}e^{j2\pi nT_o\nu}e^{-j2\pi m\Delta f\tau} + \nu[n,m], \tag{5.47}$$

where we consider uniform power allocation $|x[n,m]|^2 = B$ for any pair n, m and the phase is compensated by the radar receiver. By letting $s[n,m] = x[n,m]\alpha e^{j\varphi} e^{j2\pi n T_o \nu} e^{-j2\pi m \Delta f \tau}$, we derive the 4×4 Fisher information matrix.

$$
\begin{aligned}
[\mathbf{I}(\boldsymbol{\theta}|\mathbf{x})]_{i,j} &= -\mathbb{E}\left[\frac{\partial^2 l(\mathbf{z}|\boldsymbol{\theta},\mathbf{x})}{\partial\theta_i\partial\theta_j}\right] \\
&= \mathbb{E}\left[\frac{\partial}{\partial\theta_j\partial\theta_i}\sum_{n,m}(z[n,m]-s[n,m])^*(z[n,m]-s[n,m])\right] \\
&= -2B\,\mathbb{E}\left[\frac{\partial}{\partial\theta_j}\sum_{n,m}\mathrm{Re}\left\{(z[n,m]-s[n,m])\frac{\partial s^*[n,m]}{\partial\theta_i}\right\}\right] \\
&= 2B\,\mathrm{Re}\left\{\sum_{n,m}\left[\frac{\partial s[n,m]}{\partial\theta_i}\right]^*\left[\frac{\partial s[n,m]}{\partial\theta_j}\right]\right\},
\end{aligned}
\tag{5.48}
$$

where the last equality follows because $\mathbb{E}[z[m,n]-s[n,m]] = 0$ for any n, m and we used $\frac{\partial s^*[n,m]}{\partial\theta_i} = \left[\frac{\partial s[n,m]}{\partial\theta_i}\right]^*$ by noticing that four parameters are all real. After straightforward algebra, it readily follows that the CRLB of range $\hat{r}$ and that of velocity $\hat{v}$ are given, respectively, by

$$
\sigma_{\hat{r}}^2 \geq \frac{6}{MN(M^2-1)|h|^2 B}\left(\frac{c}{4\pi\Delta f}\right)^2, \tag{5.49a}
$$

$$
\sigma_{\hat{v}}^2 \geq \frac{6}{MN(N^2-1)|h|^2 B}\left(\frac{c}{4\pi f_c T_o}\right)^2. \tag{5.49b}
$$

These expressions, or variations using the attenuation coefficient given in (5.51), are well known in the literature (see e.g. Braun [2014, Section 3.3]). We can see immediately that the range estimation depends on the bandwidth $W_b = M\Delta f$ while the velocity estimation depends on the OFDM frame length $T_o N$.

5.4.4 Simulation Results

We evaluate the JRC performance over time-varying channels using OFDM. The performance of the proposed iterative estimation algorithm for joint range and velocity estimation is compared with frequency-modulated continuous waveform (FMCW), one of the popular automotive waveforms [Patole et al., 2017]. For simplicity, focusing on a single-path case ($P = 1$) and recalling the radar and communication channel input–output relations in (5.34) and (5.35), we let the received radar and communication SNR be

$$
\mathrm{SNR}_{\mathrm{rad}} = |h|^2 B, \quad \mathrm{SNR}_{\mathrm{com}} = |g|^2 B, \tag{5.50}
$$

with

$$
|h| = \sqrt{\frac{\lambda^2\sigma_{\mathrm{rcs}}G^2}{(4\pi)^3 r^4}}, \quad |g| = \sqrt{\frac{\lambda^2 G^2}{(4\pi)^2 r^2}}, \tag{5.51}
$$

where $\lambda = c/f_c$ is the wavelength, σ_{rcs} is the radar cross section in m^2, G is the antenna gain, and r is the distance between the transmitter and the receiver. To characterize the JRC performance, we provide the communication rate achieved by the Gaussian input signals.

$$
C_{\mathrm{OFDM}} = \left(\frac{T}{T+T_{\mathrm{cp}}}\right)\log\left(1+\mathrm{SNR}_{\mathrm{com}}\right). \tag{5.52}
$$

Table 5.1 Simulation parameters.

Carrier frequency	$f_c = 5.89\,\text{GHz}$
Bandwidth	$W_b = 10\,\text{MHz}$
Number of subcarriers	$M = 64$
Number of OFDM symbols	$N = 50$
Subcarrier spacing	$\Delta f = \frac{W_b}{M} = 156.25\,\text{kHz}$
Data symbol length	$T = \frac{1}{\Delta f} = 6.4\,\mu\text{s}$
Cyclic prefix length	$T_{\text{cp}} = \frac{1}{4}T = 1.6\,\mu\text{s}$
Radar cross section	$\sigma_{\text{rcs}} = 1\,\text{m}^2$
Antenna gain	$G = 100$
Target range	$r = 20\,\text{m}$
Target velocity	$v = 80\,\text{km/h}$

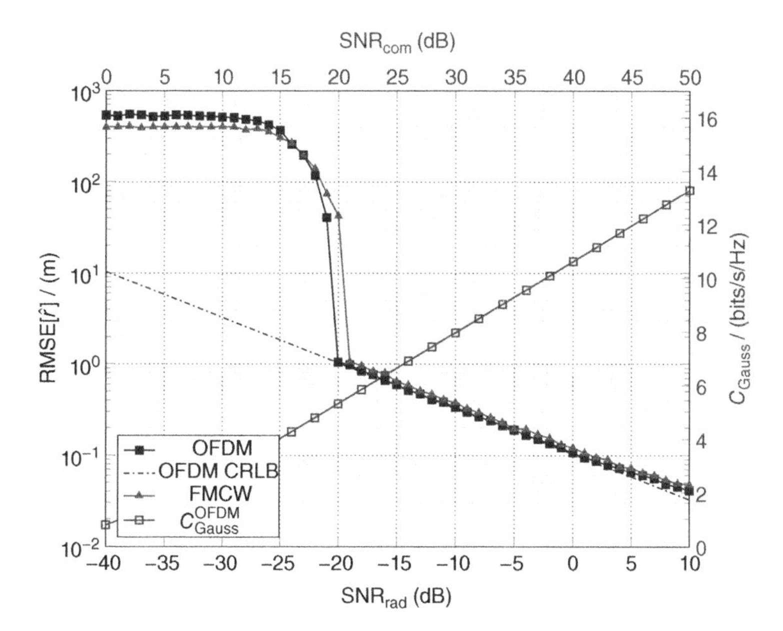

Figure 5.5 RMSE of the target range estimation $\hat{r}$ versus SNR_{rad}.

Using the parameters of Table 5.1 inspired by IEEE 802.11p [Nguyen and Heath, 2017], we show the range and velocity estimation in terms of root mean square error (RMSE) in Figures 5.5 and 5.6, respectively. As a reference, we show the estimation performance of FMCW, using the same bandwidth as well as the CRLB. It is remarkable that OFDM provides as accurate parameter estimation as FMCW by simultaneously sending data symbols.

5.5 Concluding Remarks

In this chapter, we introduced the first information theoretic model for a joint state sensing and communications system such that a transmitter, equipped with on-board sensors, operates both

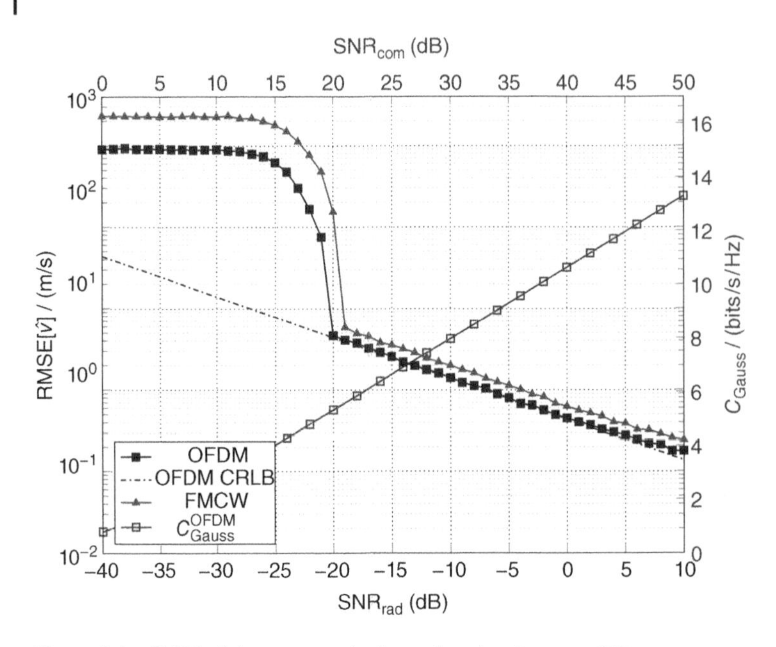

Figure 5.6 RMSE of the target velocity estimation $\hat{v}$ versus SNR$_{rad}$.

functions simultaneously. For point-to-point memoryless channels, the optimal trade-off between state sensing and communications in terms of the capacity–distortion–cost trade-off has been fully characterized. We also proposed a modified Blahut–Arimoto algorithm to find iteratively the optimal input distribution achieving the capacity–distortion trade-off for a given cost. The simple toy examples over binary channels and Gaussian channels illustrate the benefits of the proposed joint scheme compared to the resource-splitting approach that divides the resources in an orthogonal fashion. Moreover, our Gaussian example reveals also that the gap between the optimal trade-off and ideal operating point, where the minimum distortion and the capacity are simultaneously achieved, is rather small. We expect that the gap between the ideal operating point and the trade-off will further decrease under a more realistic setup such as block-fading channels and channels with memory.

Motivated by the potential gains of the proposed joint scheme, we further considered a more practical scenario of joint radar parameter estimation and communication based on OFDM, i.e. one of the promising waveforms for a JRC system. In this application example, rather than optimizing the waveform, we evaluated the joint parameter estimation and communication performance by varying the communication cost (power). We proposed an approximated ML algorithm that estimates range, velocity, and channel coefficient of P paths over the time-varying parallel channels. Finally, the numerical examples inspired by the IEEE 802.11p setup demonstrate that the proposed approximate ML algorithm offers as accurate parameter estimation as FMCW, one of the most popular automotive radar waveforms, while achieving very high data rates. These results suggest that joint radar parameter estimation and communication can be effectively operated without compromising neither the achievable data rates nor the radar performance.

Some follow-up results of this chapter are available in our recent publications. The capacity–distortion trade-off presented in Section 5.3 based on Kobayashi et al. [2018] has been extended to two-user multiuser memoryless channels, i.e. broadcast (downlink) channels [Ahmadipour et al., 2020] and multiple access (uplink) channels [Kobayashi et al., 2019]. In both multiuser channels, the resources shall be properly used to achieve different operating points in the capacity–distortion trade-off region for a given cost constraint according to the system requirement. A carefully designed multiuser scheme can deal with heterogenous sensing or communication qualities across users. Moreover, in analogy to the standard multiuser communications systems, feedback is demonstrated to be beneficial to improve the joint sensing and communication performance. This is contrasted to the single-user channels presented in Section 5.3, where feedback is useful only for state sensing. The capacity–distortion–cost trade-off for more general channels such as multiple-input multiple-output (MIMO) channels or temporally correlated channels remains an interesting open problem. The joint parameter estimation and communication presented in Section 5.4 was extended to orthogonal time frequency space (OTFS) modulation in Gaudio et al. [2020a]. OTFS is a multicarrier modulation robust to Doppler shifts and is naturally suited to sparse channels in the delay-Doppler domain [Hadani et al., 2017; Raviteja et al., 2018; Shen et al., 2019; Raviteja et al., 2019a]. OFDM provides similar parameter estimation performance and better communication rate than OFDM at the cost of higher decoding complexity. A further extension to the MIMO radar with OTFS, where the transmitter detects targets and then estimates the angle of arrival in addition to the range and velocity, is found in Gaudio et al. [2020b].

5.A Proof of Theorem 5.1

The proof of Theorem 5.1 consists of two parts: the converse to prove that any strategy cannot exceed the rate–distortion–cost trade-off and the achievability to prove that there exists a strategy achieving the trade-off.

We start with the converse part. From Fano's inequality that guarantees the decodability of the message given the observation and state sequences, i.e. $H(W|Y^n, S^n) \leq n\epsilon_n$ for an arbitrarily small ϵ_n, we have

$$\begin{aligned}
nR &\leq I(W; Y^n, S^n) + n\epsilon_n \\
&= I(W; Y^n|S^n) + n\epsilon_n \\
&= \sum_{i=1}^{n} H(Y_i|Y^{i-1}S^n) - H(Y_i|W, Y^{i-1}S^n) + n\epsilon_n \\
&\overset{(a)}{\leq} \sum_{i=1}^{n} H(Y_i|S_i) - H(Y_i|X_i, Y^{i-1}, W, S^n) + n\epsilon_n \\
&\overset{(b)}{=} \sum_{i=1}^{n} H(Y_i|S_i) - H(Y_i|X_i, S_i) + n\epsilon_n \\
&= \sum_{i=1}^{n} I(X_i; Y_i|S_i) + n\epsilon_n,
\end{aligned} \tag{5.A.1}$$

where (a) follows by removing the conditioning on $\{S_l\}_{l\neq i}, Y^{i-1}$ in the first term and adding the conditioning on X_i in the second term; (b) follows because $(W, Y^{i-1}, \{S_l\}_{l\neq i}) - (S_i, X_i) - Y_i$ forms a Markov chain. We also have

$$\begin{aligned}
R &\leq \frac{1}{n}\sum_{i=1}^{n} I(X_i; Y_i|S_i) + \epsilon_n \\
&\overset{(a)}{\leq} \frac{1}{n}\sum_{i=1}^{n} C\left(\sum_x P_{X_i}(x)c(x), \sum_x P_{X_i}(x)b(x)\right) + \epsilon_n
\end{aligned}$$

$$\overset{(b)}{\leq} C\left(\frac{1}{n}\sum_{i=1}^{n}\sum_{x} P_{X_i}(x)c(x), \frac{1}{n}\sum_{i=1}^{n}\sum_{x} P_{X_i}(x)b(x)\right) + \epsilon_n$$

$$\overset{(c)}{\leq} C(D, B) \tag{5.A.2}$$

where (a) follows from the definition of $C(D, B)$; (b) follows from the concavity property of Lemma 5.2; (c) follows from the nondecreasing property of Lemma 5.2.

Next, we provide the achievability. The decoding and the error analysis build on the following notion of joint typical sequences Gamal and Kim [2011, Chapter 2].

Definition 5.A.1 Let (x^n, y^n) be a pair of sequences of length n with elements drawn from a pair of finite alphabets $\mathcal{X}, \mathcal{Y}$. A joint empirical pdf of (x^n, y^n) is defined as

$$\pi(x, y|x^n, y^n) = \frac{|\{i : (x_i, y_i) = (x, y)\}|}{n}, \quad (x, y) \in \mathcal{X} \times \mathcal{Y} \tag{5.A.3}$$

Let $(X, Y) \sim P_{XY}(x, y)$. The set of ϵ-typical n-sequences is defined as

$$\mathcal{T}_\epsilon^{(n)}(X, Y) = \left\{(x^n, y^n) : |\pi(x, y|x^n, y^n) - P_{XY}(x, y)| \leq \epsilon P_{XY}(x, y), \forall (x, y) \in \mathcal{X} \times \mathcal{Y}\right\} \tag{5.A.4}$$

We prove Theorem 5.1 when the distortion function $d(\cdot)$ is bounded by $D_{\max} = \max_{x\in\mathcal{X}} c(x) < \infty$. The proof can be extended, as usual, to $d(\cdot)$ for which there is a letter s^* such that $\mathbb{E}[d(S, s^*)] \leq D_{\max}$.

Codebook generation: Fix $P_X(\cdot)$ and functions $\hat{s}(x, z)$ that achieve $C(D/(1+\epsilon), B)$, where D is the desired distortion and B is the target cost. Randomly and independently generate 2^{nR} sequences $x^n(w)$ for each $w \in [1 : 2^{nR}]$. This defines the codebook C that is revealed to the encoder and the decoder.

Encoding: To send a message $w \in \mathcal{W}$, the encoder transmits $x^n(w)$.

Decoding: The decoder finds a unique index $\hat{w}$ such that $(y^n, s^n, x^n(\hat{w}))$ is jointly typical Gamal and Kim [2011, Chapter 2], i.e.

$$(y^n, s^n, x^n(\hat{w})) \in \mathcal{T}_\epsilon^{(n)}(Y, S, X) \tag{5.A.5}$$

Estimation: The estimator computes the reconstruction sequence from the transmitted codeword $x^n(w)$ and the feedback sequence z^n as $\hat{s}^n = \hat{s}(x^n(w), z^n)$.

Analysis of expected distortion: In order to bound the distortion averaged over a random choice of the codebooks C, we define the decoding error event. The decoder makes an error if and only if one or both of the following events occur.

$$\mathcal{E}_1 = \{(y^n, s^n, x^n(1)) \notin \mathcal{T}_\epsilon^{(n)}(Y, S, X)\}, \tag{5.A.6}$$

$$\mathcal{E}_2 = \{(y^n, s^n, x^n(\hat{w})) \in \mathcal{T}_\epsilon^{(n)}(Y, S, X) \text{ for some } w \neq 1\}, \tag{5.A.7}$$

where we assume without loss of generality that $w = 1$ is sent. By the union bound, we have

$$P_e = P(\mathcal{E}_1 \cup \mathcal{E}_2) \leq P(\mathcal{E}_1) + P(\mathcal{E}_2).$$

The first term goes to zero as $n \to \infty$ by the law of large numbers. The second term also tends to zero as $n \to \infty$ if $R < I(X; Y|S)$ by the independence of the codebooks and the packing lemma Gamal and Kim [2011, Lemma 3.1]. Therefore, P_e tends to zero as $n \to \infty$ if $R < I(X; Y|S)$.

If there is no decoding error, we have $(Y^n, S^n, X^n(1)) \in \mathcal{T}_\epsilon^{(n)}(Y, S, X)$. The expected distortion averaged over the random codebook, encoding and decoding, is upper bounded as

$$\limsup_{n\to\infty} \mathbb{E}[d(S^n, \hat{S}^n)] = \limsup_{n\to\infty} \frac{1}{n}\sum_{i=1}^{n}\sum_{x\in\mathcal{X}} P_{X_i}(x)c(x)$$

$$\overset{(a)}{\leq} \limsup_{n\to\infty} \left(P_e D_{\max} + (1+\epsilon)(1-P_e)\frac{1}{n}\sum_{i=1}^{n}\sum_{x\in\mathcal{X}} P_{X_i}(x)c(x) \right)$$
$$\overset{(b)}{\leq} \limsup_{n\to\infty} \left(P_e D_{\max} + (1+\epsilon)(1-P_e)D \right)$$
$$\overset{(c)}{=} (1+\epsilon)D, \tag{5.A.8}$$

where (a) follows by applying the upper bound on the distortion function to the decoding error event and the typical average lemma Gamal and Kim [2011, Ch. 2.4] to the successful decoding event; (b) follows from the assumption on P_X that satisfies $\sum_{x\in\mathcal{X}} P_X(x)c(x) \leq D$; (c) follows because $\limsup_{n\to\infty} P_e \to 0$ if $R < I(X;Y|S) = C(\frac{D}{1+\epsilon}, B)$, which proves the achievability of the pair $(C(\frac{D}{1+\epsilon}, B), D)$. Finally, by the continuity of $C(D,B)$ in D, the desired $C(D,B)$ is achieved as $\epsilon \to 0$.

5.B Proof of Theorem 5.2

In order to prove statement (a), notice first that by construction we have

$$P_{X,Y|S} = P_X P_{Y|XS} = P_{Y|S} Q^{\star}_{X|YS}, \tag{5.B.1}$$

where we define the conditional marginal $P_{Y|S} = \sum_x P_X(x)P_{Y|XS}(y|xs)$. From (5.B.1) and (5.11), the conditional mutual information can be written as

$$I(P_X, P_{Y|XS}|P_S) = J(P_X, P_{Y|XS}, Q^{\star}_{X|YS}|P_S). \tag{5.B.2}$$

Then, for fixed $P_X \in \mathcal{P}(B)$ and a general $Q_{X|YS}$, we have

$$\begin{aligned}
& I(P_X, P_{Y|XS}|P_S) - J(P_X, P_{Y|XS}, Q_{X|YS}|P_S) \\
&= \sum_s P_S(s)\sum_x\sum_y P_X(x)P_{Y|XS}(y|xs)\log\frac{Q^{\star}_{X|YS}(x|ys)}{P_X(x)} \\
&\quad - \sum_s P_S(s)\sum_x\sum_y P_X(x)P_{Y|XS}(y|xs)\log\frac{Q_{X|YS}(x|ys)}{P_X(x)} \\
&= \sum_s P_S(s)\sum_x\sum_y P_{Y|S}(y|s)Q^{\star}_{X|YS}(x|ys)\log\frac{Q^{\star}_{X|YS}(x|ys)}{Q_{X|YS}(x|ys)} \\
&= \sum_s\sum_y P_S(s)P_{Y|S}(y|s)\sum_x Q^{\star}_{X|YS}(x|ys)\log\frac{Q^{\star}_{X|YS}(x|ys)}{Q_{X|YS}(x|ys)} \\
&= D_{\mathrm{KL}}\left(Q^{\star}_{X|YS}\|Q_{X|YS}|P_{SY}\right) \geq 0,
\end{aligned} \tag{5.B.3}$$

where $D_{\mathrm{KL}}(p\|q)$ denotes the Kullback–Leiber divergence. This proves statement (b) since $J(P_X, P_{Y|XS}, Q_{X|YS}|P_S)$ is maximized by $Q^{\star}_{X|YS}$.

To prove statement (c), we notice that the constrained maximization $J - \mu\sum_x c(x)P_X(x)$ over $P_X \in \mathcal{P}(B)$ is equivalent to

$$\min_{\lambda\geq 0,\nu\geq 0}\max_{P_X} \mathcal{L}(P_X, \lambda, \nu), \tag{5.B.4}$$

with

$$\mathcal{L}(P_X, \lambda, \nu) = J(P_X, P_{Y|XS}, Q_{X|YS}|P_S) - \mu\sum_x c(x)P_X(x)$$

$$-\lambda\left(\sum_{x} b(x)P_X(x) - B\right) - \nu\left(\sum_{x} P_X(s) - 1\right). \tag{5.B.5}$$

Applying the KKT conditions together with the non-negativity constraint, the optimal pmf is given by

$$P_X(x) = \kappa e^{\sum_{s}\sum_{y} P_S(s)P_{Y|XS}(y|xs)\log Q_{X|YS}(x|ys) - \lambda b(x) - \mu c(x)} \tag{5.B.6}$$

with $\kappa = \exp(-(\nu + 1))$, yielding (5.B.16). Since the problem is strictly convex and, by assumption, feasible, there exists a unique solution that satisfies (5.12b) with equality.

Acknowledgment

The works of both authors are supported by the DFG Grant KR 3517/11-1/CA 1340/11-1. The authors wish to thank Lorenzo Gaudio, Giulio Colavolpe, University of Parma, as well as Gerhard Kramer, Technical University of Munich for the collaboration.

References

M. Ahmadipour, M. Wigger, and M. Kobayashi. Joint sensing and communication over memoryless broadcast channels. *arXiv preprint arXiv:2011.03379 (to appear in Proceedings IEEE Information on Theory Workshop (ITW))*, 2020.

S. Arimoto. An algorithm for computing the capacity of arbitrary discrete memoryless channels. *IEEE Transactions on Information Theory*, 18(1):14–20, 1972.

M. R. Bell. Information theory and radar waveform design. *IEEE Transactions on Information Theory*, 39(5):1578–1597, 1993.

R. Blahut. Computation of channel capacity and rate-distortion functions. *IEEE Transactions on Information Theory*, 18(4):460–473, 1972.

D. W. Bliss. Cooperative radar and communications signaling: the estimation and information theory odd couple. In *Radar Conference, 2014 IEEE*, pages 0050–0055. IEEE, 2014.

M. Braun. OFDM Radar Algorithms in Mobile Communication Networks. *Ph.D. Thesis at Karlsruhe Institute of Technology*, 2014.

A. Carleial. Multiple-access channels with different generalized feedback signals. *IEEE Transactions on Information Theory*, 28(6):841–850, 1982.

G. K. Carvajal, M. F. Keskin, C. Aydogdu, O. Eriksson, H. Herbertsson, H. Hellsten, E. Nilsson, M. Rydström, K. Vänas, and H. Wymeersch. Comparison of automotive FMCW and OFDM radar under interference. In *2020 IEEE Radar Conference (RadarConf20)*, pages 1–6, 2020. https://doi.org/10.1109/RadarConf2043947.2020.9266449.

A. R. Chiriyath, B. Paul, G. M. Jacyna, and D. W. Bliss. Inner bounds on performance of radar and communications co-existence. *IEEE Transactions on Signal Processing*, 64(2):464–474, 2016.

C. Choudhuri, Y.-H. Kim, and U. Mitra. Causal state communication. *IEEE Transactions on Information Theory*, 59(6):3709–3719, 2013.

C. Cordeiro, D. Akhmetov, and M. Park. IEEE 802.11 ad: introduction and performance evaluation of the first multi-Gbps WiFi technology. In *Proceedings of the 2010 ACM International Workshop on mmWave Communications: From Circuits to Networks*, pages 3–8. ACM, 2010.

T. M. Cover and J. A. Thomas. *Elements of information theory, 2nd edition (wiley series in telecommunications and signal processing)*. Wiley-Interscience, 2006.

S. H. Dokhanchi, M. B. Shankar, K. V. Mishra, and B. Ottersten. A mm Wave automotive joint radar-communications system. *IEEE Transactions on Aerospace and Electronic Systems*, 55(3):1241–1260, 2019.

G. Dueck. Partial feedback for two-way and broadcast channels. *Information and Control*, 46(1):1–15, 1980.

A. El Gamal and Y.-H. Kim. *Network information theory*. Cambridge University Press, 2011.

6G Flagship. 6G white paper on localization and sensing. *University of Oulu, Finland*, June 2020.

M. Gatzianas, L. Georgiadis, and L. Tassiulas. Multiuser broadcast erasure channel with feedback: capacity and algorithms. *IEEE Transactions on Information Theory*, 59(9):5779–5804, 2013.

L. Gaudio, M. Kobayashi, B. Bissinger, and G. Caire. Performance analysis of joint radar and communication using OFDM and OTFS. In *2019 IEEE International Conference on Communications Workshops (ICC Workshops)*, pages 1–6. IEEE, 2019.

L. Gaudio, M. Kobayashi, C. Caire, and G. Colavolpe. On the effectiveness of OTFS for joint radar parameter estimation and communication. *IEEE Transactions on Wireless Communications*, 19(9):5951–5965, 2020a.

L. Gaudio, M. Kobayashi, G. Caire, and G. Colavolpe. Joint radar target detection and parameter estimation with MIMO OTFS. In *2020 IEEE Radar Conference (RadarConf20)*, pages 1–6. IEEE, 2020b.

R. Hadani, S. Rakib, A. F. Molisch, C. Ibars, A. Monk, M. Tsatsanis, J. Delfeld, A. Goldsmith, and R. Calderbank. Orthogonal time frequency space (OTFS) modulation for millimeter-wave communications systems. In *Microwave Symposium (IMS), 2017 IEEE MTT-S Internationl*, pages 681–683. IEEE, 2017.

A. Hassanien,M. G. Amin, E. Aboutanios, and B. Himed. Dual-function radar communication systems: a solution to the spectrum congestion problem. *IEEE Signal Processing Magazine*, 36(5):115–126, 2019.

R. W. Heath. Communications and sensing: an opportunity for automotive systems [from the editor]. *IEEE Signal Processing Magazine*, 37(4):3–13, 2020.

S. M. Kay. *Fundamentals of statistical signal processing*. Prentice Hall PTR, 1993.

Y.-H. Kim, A. Sutivong, and T. M. Cover. State amplification. *IEEE Transactions on Information Theory*, 54(5):1850–1859, 2008.

M. Kobayashi, G. Caire, and G. Kramer. Joint state sensing and communication: optimal tradeoff for a memoryless case. In *Proceedings of IEEE International Symposium on Information Theory*, pages 111–115, 2018.

M. Kobayashi, H. Hamad, G. Kramer, and G. Caire. Joint state sensing and communication over memoryless multiple access channels. In *Proceedings of IEEE International Symposium on Information Theory*, pages 270–274. IEEE, 2019.

G. Kramer. Feedback strategies for white Gaussian interference networks. *IEEE Transactions on Information Theory*, 48(6):1423–1438, 2002.

P. Kumari, D. H. N. Nguyen, and R. W. Heath. Performance trade-off in an adaptive IEEE 802.11 ad waveform design for a joint automotive radar and communication system. In *IEEE International Conference on Acoustics, Speech, and Signal Processing (ICASSP)*, pages 4281–4285. IEEE, 2017.

P. Kumari, J. Choi, N. González-Prelcic, and R. W. Heath. IEEE 802.11ad-based radar: an approach to joint vehicular communication-radar system. *IEEE Transactions on Vehicular Technology*, 67(4):3012–3027, 2018a.

P. Kumari, M. E. Eltayeb, and R. W. Heath. Sparsity-aware adaptive beamforming design for IEEE 802.11 ad-based joint communication-radar. In *Radar Conference (RadarConf18), 2018 IEEE*, pages 0923–0928. IEEE, 2018b.

Y. Liu, G. Liao, J. Xu, Z. Yang, and Y. Zhang. Adaptive OFDM integrated radar and communications waveform design based on information theory. *IEEE Communications Letters*, 21(10):2174–2177, 2017.

F. Liu, C. Masouros, A. P. Petropulu, H. Griffiths, and L. Hanzo. Joint radar and communication design: applications, state-of-the-art, and the road ahead. *IEEE Transactions on Communications*, 68(6):3834–3862, 2020.

D. Ma, N. Shlezinger, T. Huang, Y. Liu, and Y. C. Eldar. Joint radar-communication strategies for autonomous vehicles: combining two key automotive technologies. *IEEE Signal Processing Magazine*, 37(4):85–97, 2020. https://doi.org/10.1109/MSP.2020.2983832.

K. V. Mishra, M. B. Shankar, V. Koivunen, B. Ottersten, and S. A. Vorobyov. Towards millimeter wave joint radar-communications: a signal processing perspective. *IEEE Signal Processing Magazine*, 36(5):100–114, 2019.

D. H. N. Nguyen and R. W. Heath. Delay and Doppler processing for multi-target detection with IEEE 802.11 OFDM signaling. In *Proceedings of IEEE International Conference on Acoustics, Speech, and Signal Processing (ICASSP)*, pages 3414–3418, March 2017.

L. Ozarow. The capacity of the white Gaussian multiple access channel with feedback. *IEEE Transactions on Information Theory*, 30(4):623–629, 1984.

L. Ozarow and S. Leung-Yan-Cheong. An achievable region and outer bound for the Gaussian broadcast channel with feedback (corresp.). *IEEE Transactions on Information Theory*, 30(4):667–671, 1984.

S. M. Patole, M. Torlak, D. Wang, and M. Ali. Automotive radars: a review of signal processing techniques. *IEEE Signal Processing Magazine*, 34(2):22–35, 2017. ISSN 1053-5888. https://doi.org/10.1109/MSP.2016.2628914.

B. Paul, A. R. Chiriyath, and D. W. Bliss. Survey of RF communications and sensing convergence research. *IEEE Access*, 5: 252–270, 2017.

P. Raviteja, K. T. Phan, Y. Hong, and E. Viterbo. Interference cancellation and iterative detection for orthogonal time frequency space modulation. *IEEE Transactions on Wireless Communications*, 17(10):6501–6515, 2018. ISSN 1536-1276. https://doi.org/10.1109/TWC.2018.2860011.

P. Raviteja, K. T. Phan, and Y. Hong. Embedded pilot-aided channel estimation for OTFS in delay-Doppler channels. *IEEE Transactions on Vehicular Technology*, 68(5):4906–4917, 2019a.

P. Raviteja, K. T. Phan, Y. Hong, and E. Viterbo. Orthogonal time frequency space (OTFS) modulation based radar system. In *2019 IEEE Radar Conference (RadarConf)*, pages 1–6, 2019b. https://doi.org/10.1109/RADAR.2019.8835764.

M. A. Richards. *Fundamentals of radar signal processing, 2nd edition*. McGraw-Hill Education, 2014.

A. Sabharwal, P. Schniter, D. Guo, D. W. Bliss, S. Rangarajan, and R. Wichman. In-band full-duplex wireless: challenges and opportunities. *IEEE Journal on Selected Areas in Communications*, 32(9):1637–1652, 2014.

J. Schalkwijk and T. Kailath. A coding scheme for additive noise channels with feedback–I: no bandwidth constraint. *IEEE Transactions on Information Theory*, 12(2):172–182, 1966.

S. Sesia, M. Baker, and I. Toufik. *LTE-the UMTS long term evolution: from theory to practice*. John Wiley & Sons, 2011.

C. E. Shannon. A mathematical theory of communication. *The Bell System Technical Journal*, 27: 379–423, 1948.

C. Shannon. The zero error capacity of a noisy channel. *IRE Transactions on Information Theory*, 2(3):8–19, 1956.

O. Shayevitz and M. Wigger. On the capacity of the discrete memoryless broadcast channel with feedback. *IEEE Transactions on Information Theory*, 59(3):1329–1345, 2013.

W. Shen, L. Dai, J.-P. An, P. Fan, and R. W. Heath. Channel estimation for orthogonal time frequency space (OTFS) massive MIMO. *IEEE Transactions on Signal Processing*, 67(16):4204–4217, 2019.

C. Sturm and W. Wiesbeck. Waveform design and signal processing aspects for fusion of wireless communications and radar sensing. *Proceedings of the IEEE*, 99(7):1236–1259, 2011. ISSN 0018-9219. https://doi.org/10.1109/JPROC.2011.2131110.

A. Sutivong, M. Chiang, T. M. Cover, and Y.-H. Kim. Channel capacity and state estimation for state-dependent Gaussian channels. *IEEE Transactions on Information Theory*, 51(4):1486–1495, 2005.

G. Ungerboeck. Channel coding with multilevel/phase signals. *IEEE Transactions on Information Theory*, 28(1):55–67, 1982.

C. C. Wang. On the capacity of 1-to-k broadcast packet erasure channels with channel output feedback. *IEEE Transactions on Information Theory*, 58(2):931–956, 2012.

F. Willems, E. van der Meulen, and J. Schalkwijk. Achievable rate region for the multiple access channel with generalized feedback. In *Proceedings of Annual Allerton Conferene on Communication, Control and Computing*, pages 284–292, 1983.

W. Zhang, S. Vedantam, and U. Mitra. Joint transmission and state estimation: a constrained channel coding approach. *IEEE Transactions on Information Theory*, 57(10):7084–7095, 2011.

L. Zheng, M. Lops, Y. C. Eldar, and X. Wang. Radar and communication co-existence: an overview: a review of recent methods. *IEEE Signal Processing Magazine*, 36(5):85–99, 2019.

Part II

Physical-Layer Signal Processing

6

Radar-aided Millimeter Wave Communication

Nuria González-Prelcic[1], Anum Ali[2], and Yun Chen[1]

[1] *Department of Electrical and Computer Engineering, North Carolina State University, Raleigh, NC, USA*
[2] *Standards and Mobility Innovation Laboratory, Samsung Research America, Plano, TX, USA*

Them bats is smart. They use radar!

Source: David Letterman

6.1 Motivation for Radar-aided Communication

Millimeter wave (mmWave) multiple-input multiple-output (MIMO) communication has become a key technology for wireless communication and sensing, including cellular communication, personal and local area networks, vehicular networks, communications in virtual reality scenarios, wearable networks, satellite communications, automotive radar, satellite-based remote sensing or security screening to name a few. MIMO communication systems operating at mmWave rely on antenna arrays with a large number of elements that provide enough antenna aperture, array gain, and narrow beam patterns that enable highly directional communication and high data rates. Using large arrays is challenging, however, due to the high overhead for link establishment and operation, either when using beam training protocols or channel estimation algorithms for initial access or array reconfiguration due to blockage.

Out-of-band (OOB) information coming from sensors or other communication systems operating at sub-6 GHz frequencies can assist mmWave link configuration to reduce the associated overhead [González-Prelcic et al., 2017]. The limitation of using channel information from sub-6 GHz systems to aid mmWave communication is that it requires the state of the sub-6 GHz and the mmWave channel to be the same, either line-of-sight (LOS) or non-line-of-sight (NLOS) [Ali et al., 2019b]. The main motivation for using sensors is that many devices are already equipped with them and can also be easily mounted in a base station (BS) [Fuller and Waters, 2019; Ali et al., 2020b]. Sensors extract features of the propagation environment that also impact the communication channel. Some sensors, radar in particular, can provide useful information even in NLOS scenarios. Unlike other sensors, radar can operate in the same band as the communication system. In addition, the manufacture of miniature and cost-effective radar sensors is now a reality. Radar is especially useful to obtain information about the wireless communication channel, for example directions of arrival. Furthermore, radar channels and communication channels are closely related if the transceiver and the radar sensor are colocated, as we will show later in this chapter. In the next sections, we describe strategies that exploit the similarity between radar and communication channels to define an alternative paradigm for radar and communication coexistence based on the idea of collaboration, rather than a joint design and operation as in previous chapters.

Signal Processing for Joint Radar Communications, First Edition.
Edited by Kumar Vijay Mishra, M. R. Bhavani Shankar, Björn Ottersten, and A. Lee Swindlehurst.

6.1.1 Sensing on the Wireless Infrastructure

Sensors in general, and radar in particular, can provide information about the propagation environment that determines the communication channel. When mounted at the cellular infrastructure, sensors will have a birds-eye-view of the propagation scenario, and the information they provide can be used to assist MIMO communication in different ways [Ali et al., 2020b].

First, radar mounted at the cellular infrastructure can provide position information about the vehicles, which can be used to reduce the overhead associated to the beam training process employed to configure the phased arrays. The beam training protocol consists of testing all possible beampatterns that can be generated with a given codebook at both sides of the communication link, to decide later which pair provides the highest signal-to-noise ratio (SNR). The duration of the beam training phase is long when large arrays and codebooks are considered, but can be dramatically reduced if information about the position of the vehicle is available. The details of a beam training strategy aided by position information obtained with radar are discussed in Section 6.2.

Second, a BS equipped with sensors can detect with high accuracy the state of the communication channel, i.e. whether the channel is LOS or NLOS. Information about the channel state is relevant for the operation of the MIMO communication system. For example, it can be used to choose the appropriate model for the temporal evolution of the channel when performing channel tracking. Another application of channel state information (CSI) is positioning based on communication signals, since the specific mapping between channel parameters to position and orientation depends on the channel state [Shahmansoori et al., 2018]. As a final example, channel state can be used as prior information about the maximum number of channel paths in compressive channel estimation algorithms for mmWave and massive MIMO systems [Lee et al., 2016; Venugopal et al., 2017; González-Coma et al., 2018; Rodríguez-Fernández and González-Prelcic, 2019; Xie and González-Prelcic, 2020].

Third, sensor information can also help to mitigate blockage in mmWave communication systems. To this aim, sensors mounted at the BS can be used to track mobile elements in the environment (vehicles or pedestrians for example), so the potential blockages they will introduce can be predicted in advance to avoid link failures [Simić et al., 2016]. The information provided by sensors can also be combined with that extracted from the received signal to increase the accuracy of blockage detection and prediction algorithms [Xie and González-Prelcic, 2021].

Finally, long-term statistical CSI, in particular the spatial covariance, can also be obtained from a radar mounted at the BS. The spatial covariance of the radar channel can be used directly, or can be further refined, to provide an estimate of the spatial covariance of the communication channel. Previous work has shown the effectiveness of the spatial covariance to configure analog or hybrid architectures in MIMO systems with large arrays [Park et al., 2020], though there is high overhead associated with estimating the spatial covariance through the transmission of training pilots. An estimation based on the radar covariance can significantly reduce, or even avoid, the overhead associated to the transmission of training pilots. Section 6.3 introduces the details of several approaches that estimate the communication spatial covariance from the output of a radar sensor.

Sensing at the infrastructure has additional benefits besides assisting communication. Vehicles on the road have sensing capabilities limited by their LOS view and the obstructions created by traffic, but a sensing BS can provide augmented sensing capabilities to the vehicle by sharing the data collected by its sensors. This information can also be used to increase the safety level of vulnerable road users such as pedestrians or bicyclists, or to extract data for urban planning. These extra benefits of sensing, targeting applications that directly impact people, make it attractive for cities to deploy infrastructure with a sensing component.

6.1.2 Sensing at the User Equipment (UE)

Automotive radar has become a standard sensor even in mid-class vehicles. Its role in active safety functions and automated driving and its robustness to weather and illumination conditions makes it an irreplaceable sensor in the vehicle. Recent work proposes a new function for automotive radar: aid the high data rate mmWave vehicular communication systems that will enable many different applications, from raw sensor data sharing to infotainment. An interesting example of using radar to establish mmWave vehicle-to-vehicle (V2V) links is described in Aydogdu et al. [2020]. The ego-vehicle is assumed to be equipped with a multi-band communication system so that the position information from neighboring vehicles obtained by global positioning system (GPS) is received through the low frequency system with low data rates. The position of the intended receiver could be used by the ego-vehicle to steer the beams in the appropriate direction, but the large uncertainty in the information provided by GPS limits the interest of this approach. To overcome this limitation, the automotive radar in the ego-vehicle can detect with small uncertainty the positions of moving objects, including the intended vehicle to communicate with. A data association algorithm can be used to match the GPS information about neighboring vehicles with the objects detected by the automotive radar, to finally identify, with low uncertainty, the position of the desired receiver and the best beam for communication.

Radar is now also becoming more prevalent in consumer electronics. The primary function of the radar on the consumer electronics today remains sensing. A prime example is Google Soli project [Lien et al., 2016], in which a high bandwidth radar is used for high precision dynamic micro-gesture recognition. The Soli radar platform was commercially launched in Pixel 4. Recently, however, there is interest in using the smart-phone radar to help the communication. To this end, note that the federal communications commission (FCC) regulates maximum permissible exposure (MPE) for human safety. The purpose of this regulation is to ensure that a human tissue is not exposed to a high level of electromagnetic radiation for a long period of time. One naive strategy is to always operate with the transmission power that will ensure that MPE is not violated. An intelligent strategy would be to detect the presence of a human tissue near the radar, and limit the transmit power only if a human tissue near the radar is detected. If the human tissue is not detected, the smartphone can transmit with high power to achieve a high uplink data rate. Such strategy is part of the patent application by Apple Cetinoneri et al. [2020], in which the transmit power is adjusted depending on whether an object near the smartphone is detected or not. As the objective of the radar is to detect the presence of a nearby object, for this application, the radar can operate with considerably lower power compared to the communication system. This is an example of a radar at the user equipment (UE) enabling high rate mmWave communication (when human presence is not detected). In addition, for multipanel UEs, the radar can be used to detect blockage, and hence help with identifying the panels that are more suitable for transmission/reception of communication signals.

6.2 Radar-aided Communication Exploiting Position Information

Position information of the terminal can help to reduce the overhead associated to configuring/reconfiguring the large arrays in massive MIMO or mmWave MIMO communication when the link is LOS. This is especially critical in vehicular scenarios and when operating at mmWave frequencies, since the channel is highly dynamic in this case and the antenna arrays need to be configured frequently. The accuracy of position information provided by radar is higher than that extracted from GPS signals and can be updated more frequently. This leads to a larger reduction in communication overhead when exploiting position information provided by a radar sensor than when leveraging GPS-based position, as experimentally shown in Graff et al. [2019]. In Section 1.2.1 we describe a particular strategy for location-based beam training in a vehicle-to-infrastructure (V2I) communication setting as described in Ali et al. [2019a], and provide results of the overhead reduction that can be obtained when the position is estimated by a radar mounted at a BS.

6.2.1 MmWave Beamtraining Using a BS Mounted Radar

To illustrate the effectiveness of radar-aided communication exploiting position information, we consider a vehicular communication scenario where the BS is equipped with a mmWave transceiver with an analog beamforming architecture, and with an active radar that detects the position of the neighboring vehicles, as illustrated in Figure 6.1. The vehicle also has a phased array and employs an analog combiner. The BS phased array has N_{BS} antennas and the vehicle's one has N_v antennas. The system uses OFDM with K subcarriers over a frequency selective MIMO channel represented by $\mathbf{H}[k]$, where k is the subcarrier index. The details of the channel model are provided in Ali et al. [2019a]. The analog beamformer is denoted as $\mathbf{f}$, while the analog combiner is denoted as $\mathbf{q}$. The received signal model can be written as

$$\mathrm{y}[k] = \mathbf{q}^*\mathbf{H}[k]\mathbf{f}\mathrm{s}[k] + \mathbf{q}^*\mathbf{v}[k], \tag{6.1}$$

where $\mathbf{v}[k] \sim \mathcal{CN}(\mathbf{0}, \sigma_{\mathbf{v}}^2\mathbf{I})$ models the noise and $\mathrm{s}[k]$ is the symbol sequence satisfying $\mathbb{E}[\mathrm{s}[k]\mathrm{s}^*[k]] = \frac{P_c}{K}$, where P_c is the average transmit power.

When considering a codebook-based beam training, the beamformer and the combiner are chosen from a set of possible beamforming/combing vectors as the pair that maximizes a given metric, usually the SNR. To construct the codebooks for the beamformer and the combiner, denoted as $\mathbf{W}$ and $\mathbf{Z}$ respectively, we consider a grid of beams (GoB) approach. In this case, the vectors in the beamforming (combining) codebook are constructed by sampling the array response vector of the BS (vehicle) at the angular values specified in a predefined grid. If uniform linear arrays (ULAs) are considered, the array steering vector for the BS array is

$$\mathbf{a}_{\text{BS}}(\phi) = [1, e^{\mathrm{j}\pi\Delta\sin(\phi)}, \ldots, e^{\mathrm{j}\pi(N_{\text{BS}}-1)\Delta\sin(\phi)}]^T, \tag{6.2}$$

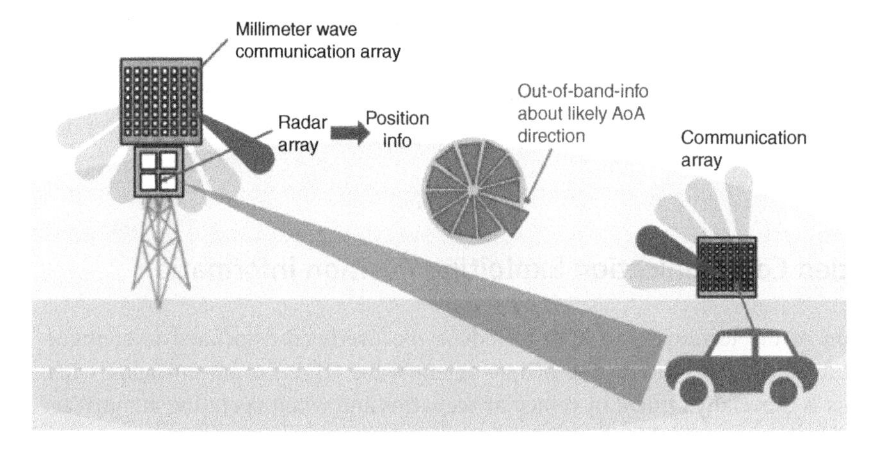

Figure 6.1 MIMO V2I communication set up with an active BS mounted radar that tracks the position of neighboring vehicles.

where Δ is the inter-element spacing in half-wavelengths, and ϕ is the angle of departure. The array response vector for the vehicle is defined in the same way by substituting N_{BS} by N_v and ϕ by the angle of arrival (AoA), denoted as θ. Setting the size of the codebooks to the size of the corresponding phased array, the resulting codebooks $\mathbf{W}$ and $\mathbf{Z}$ are discrete Fourier transform (DFT) matrices. Thus, the nth beamformer in the codebook $\mathbf{W} = [\mathbf{w}_1, \mathbf{w}_2, \dots, \mathbf{w}_{N_{\text{BS}}}]$ is defined as $\mathbf{w}_n = \frac{1}{\sqrt{N_{\text{BS}}}}\mathbf{a}_{\text{BS}}(\arcsin(\frac{2n-N_{\text{BS}}-1}{N_{\text{BS}}}))$, $n = 1, \dots, N_{\text{BS}}$. The combiners in $\mathbf{Z}$ can be defined in a similar way from the array steering vector corresponding to the vehicle array, as detailed in Ali et al. [2019a].

The GoB beam training protocol based on exhaustive search and these codebooks generates the received signal for all possible combinations of beamformers and combiners, and then chooses the pair that maximizes SNR. The associated training overhead is $N_{\text{BS}}N_v$ OFDM symbols, which is high when operating with large arrays. The position of the vehicle can be used to significantly reduce the size of the codebook $\mathbf{W}$ and consequently the communication overhead. To this aim, position information can be easily projected first into AoD information in a LOS scenario. Then, the beamforming codebook can be reduced to the set of beams whose direction of maximum directivity is around the estimated AoD, denoted as $\hat{\phi}$. A set of beams is considered in the reduced codebook instead of only the closest one to $\hat{\phi}$ to account for the error in the vehicle angular position, denoted as $\Delta\phi$. This way, the reduced codebook $\mathbf{W}_{\mathcal{L}}$ will contain the codewords indexed by the index set $\mathcal{L}$ such that $n \in \mathcal{L}$ if

$$\sin(\hat{\phi} - \Delta\phi) + 1 \leq \frac{2n}{N_{\text{BS}}} \leq \sin(\hat{\phi} + \Delta\phi) + 1 + 2/N_{\text{BS}}. \tag{6.3}$$

Operating with the reduced beamforming codebook leads to a reduction in training overhead of $(1 - |\mathcal{L}|/N_{\text{BS}})$. Note that the reduction that can be achieved depends on the accuracy of the position information. Position information obtained by GPS can have an error in the order of meters, which translates to a large angular error (in the order of tens of degrees). The accuracy in the angular position of the vehicle estimated by a MIMO radar mounted at the BS depends on the specific system parameters. As an example, it was shown in Ali et al. [2019a] that under some assumptions and using some realistic system parameters and an off-the-shelf radar, the angular error is in the order of one-tenth of a degree. Given the relationship between the angular error and the overhead reduction, a BS mounted radar is clearly preferred as the mechanism to provide position information to assist beam training.

To further illustrate the performance of position-aided beamtraining when using GPS and a BS mounted radar to obtain localization information, we consider the practical radar-aided communication setup with parameters defined in Table 6.1. Note that realistic mmWave channels have been generated for these numerical experiments based on urban macrocell (Uma) LOS scenario in the 5G channel model [3GPP, 2017] using QuaDRiGa [Jaeckel et al., 2014, 2017]. We define T_{train} as the duration of the beam training process, and T_{coh} as the angle coherence time, i.e. the time in which the beamformer does not change. The selected metric to evaluate performance is the spectral efficiency (SE) defined as

$$\text{SE} = \mathbb{E}\left[\left(1 - \frac{T_{\text{train}}}{T_{\text{coh}}}\right)\frac{1}{K}\sum_{k=1}^{K}\log_2\left(1 + \frac{P_c|\underline{\mathbf{q}}^*\mathsf{H}[k]\underline{\mathbf{f}}|^2}{KN_{0,c}}\right)\right], \tag{6.4}$$

where $N_{0,c}$ is the noise spectral density in the communication receiver. Note that it heavily depends on $\frac{T_{\text{train}}}{T_{\text{coh}}}$, which is the fraction of coherence time spent in training. The system initiates the beam training protocol once in every coherence interval. The SE for this system when averaging the results for 1000 channels is shown in Figure 6.2. Radar location-aided beam training clearly outperforms GPS-aided beam training. SE for GoB (exhaustive search) or GPS-aided beam training decreases as the vehicle speed increases because a larger fraction of T_{coh} is used by the beam

Table 6.1 System parameters for the simulation of a radar-aided mmWave vehicular communication system exploiting position information.

Parameter	Symbol	Value	Units
Communication system			
Carrier frequency	f_c	73	GHz
BS height	h_{BS}	5	m
Inter-site distance	ISD	100	m
Distance BS to closest point on the road	d	10	m
Number of antennas at the BS	N_{BS}	64	
Number of antennas at the vehicle	N_v	16	
Number of phase shifter bits	D	2	bits
Number of subcarriers	K	512	
Subcarrier spacing		240	kHz
Cyclic prefix length	L_c	0.6	μs
Radar system			
Center frequency	f_c	24	GHz
Bandwidth	B_r	1	GHz
Transmit power	P_r	10	dBm
Virtual array size	M_r	29	
Number of snapshots	N_s	1024	
Nonfluctuating target radar cross section	σ	10	m^2
Signal-to-noise ratio	SNR_r	−21.88	dB

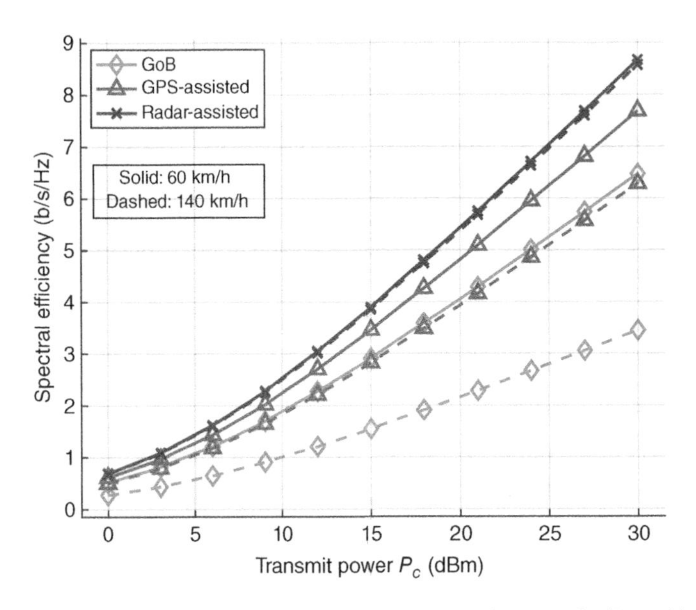

Figure 6.2 The spectral efficiency versus transmit power P_c: $N_{BS} = 64$ and $N_v = 16$ with vehicle speeds of 60 km/h and 140 km/h.

training protocol. Radar aided beam training is not sensitive to the specific value of T_{coh} because $T_{\mathrm{train}} \ll T_{\mathrm{coh}}$ independently of the vehicle speed. These simulations show the effectiveness of a BS mounted radar to significantly reduce the beam training overhead in a LOS vehicular communication system operating at mmWave frequencies.

6.3 Radar-aided Communication Exploiting Covariance Information

Position information obtained from a BS mounted radar is useful to reduce communication overhead in a LOS scenario. When operating in a NLOS link and with a hybrid MIMO architecture, additional information about the channel is needed to configure the hybrid precoders and combiners. These are typically designed based on either the full instantaneous CSI or the spatial channel covariance information (also referred to as statistical CSI, partial CSI, or imperfect CSI). On the one hand, accurate estimation of the frequency selective MIMO channel matrices is challenging because of the reduced number of radio-frequency (RF) chains in a hybrid architecture. It also introduces a non-negligible overhead when operating with large arrays. On the other hand, estimating the channel covariance is not free of difficulties and additional overhead [Park et al., 2020]. Nevertheless, the dimension of the problem to be solved is smaller, since the spatial covariance matrix will be used to configure only the frequency flat analog precoders and combiners, and only an averaged covariance over all subcarriers needs to be estimated. Regarding the performance of hybrid designs based on full or partial CSI, it has been shown that they provide similar SEs [Méndez-Rial et al., 2015; Park et al., 2017].

A BS mounted radar can assist in the estimation of the spatial channel covariance matrix to reduce the overhead associated to the array configuration when establishing the link (initial access). Specific strategies that exploit the radar covariance in the context of a V2I link have been proposed in González-Prelcic et al. [2016] and Ali et al. [2020a]. The pioneering work in González-Prelcic et al. [2016] proposes a BS mounted radar that estimates the covariance with an active radar. However, we only focus here on the system model and the approach designed in Ali et al. [2020a], which leverages radar information available at the BS obtained with a passive radar receiver. To understand the usefulness of radar information to estimate the spatial communication covariance, it is necessary to define first a metric of similarity between radar and communication. In Section 1.3.1 we introduce this metric and present some similarity measurements that show that the spatial features of the radar and communication channels are congruent when considering a mmWave link and a mmWave radar. The subsequent section describes a specific way to leverage radar covariance information for mmWave NLOS link configuration and presents the results that show the overhead reduction that can be achieved.

6.3.1 Measuring Radar and Communication Congruence

To guarantee that the radar covariance matrix provides useful information for the configuration of the beampatterns, we need to show that there is a similarity between the angular information extracted from the radar and the communication spatial covariances. To measure this congruence in the angular domain, we can define a metric that quantifies the similarity between the angular power spectrum (APS) of the received radar signals and the APS of communication signals. Comparing the over-the-air APS is one way of evaluating similarity. For example, using the toy example in Figure 6.3, we can analyze the similarity between APS1 obtained from the radar signal, and APS2, obtained from the communication signal, both shown in Figure 6.3a. The usefulness of APS1 to design the beamformers when the actual communication spectrum is APS2 depends on

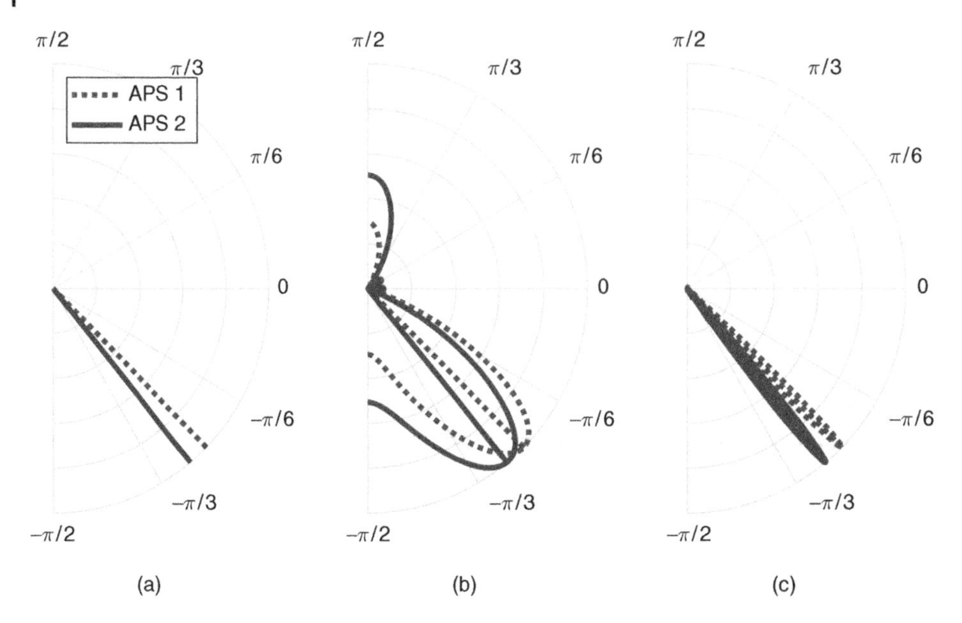

Figure 6.3 (a) Over the air azimuth power spectra. (b) Observed power spectra through arrays of four antenna elements. (c) Observed power spectra through arrays of 32 antenna elements.

the system parameters. To illustrate this impact, Figure 6.3b,c shows the beam-patterns of a ULA pointing in the directions suggested by the two possible spectra when the number of antennas for communication is 4 and 32. While the angular information in APS1 is useful for beamforming when a four element array is used for communication, it cannot be used to assist communication if the number of antennas is 32. In this case, the beams are much narrower and will not provide an appropriate gain in the main direction for communication if steered using the radar APS.

The mathematical definition of similarity between radar and communication APS was proposed in Ali et al. [2020a]. Given two N point spectra $\mathbf{d}_1$ and $\mathbf{d}_2$ obtained from radar and communication respectively, and defining the index set $\mathcal{I}_1$ ($\mathcal{I}_2$) of cardinality $L \leq N$ that contains indices of L largest entries of $\mathbf{d}_1$ ($\mathbf{d}_2$), the similarity metric is defined as

$$\mathrm{S}_{1\to 2}(L, N) = \frac{\sum_{i\in\mathcal{I}_1} \mathbf{d}_2[i]}{\sum_{i\in\mathcal{I}_2} \mathbf{d}_2[i]}. \tag{6.5}$$

The intuitive meaning of the metric is illustrated in Figure 6.4. The denominator in (6.5) is the power of the L largest spectral components of the communication APS in $\mathbf{d}_2$, indexed by $\mathcal{I}_2$ and marked in light gray in Figure 6.4, while the numerator is the power of the spectral components of the communication APS suggested as dominant by the radar APS in $\mathbf{d}_1$, indexed by $\mathcal{I}_1$ and marked in dark gray in Figure 6.4. Note that when the radar and communication APS are congruent, $\mathrm{S}_{1\to 2} \to 1$ for reasonable values of the window length L.

This similarity metric is also related to the relative precoding efficiency (RPE) defined in Park et al. [2020] as

$$\mathrm{RPE} = \frac{\mathrm{tr}(\mathbf{U}_1^* \mathbf{R}_2 \mathbf{U}_1)}{\mathrm{tr}(\mathbf{U}_2^* \mathbf{R}_2 \mathbf{U}_2)}, \tag{6.6}$$

where $\mathbf{R}_2$ and $\mathbf{R}_1$ are the communication spatial covariance and its estimate (obtained from radar in our radar assisted communication system), and $\mathbf{U}_1$ ($\mathbf{U}_2$) are the L singular vectors of $\mathbf{R}_1$ ($\mathbf{R}_2$) corresponding to the L largest singular values. The connection between the similarity metric and

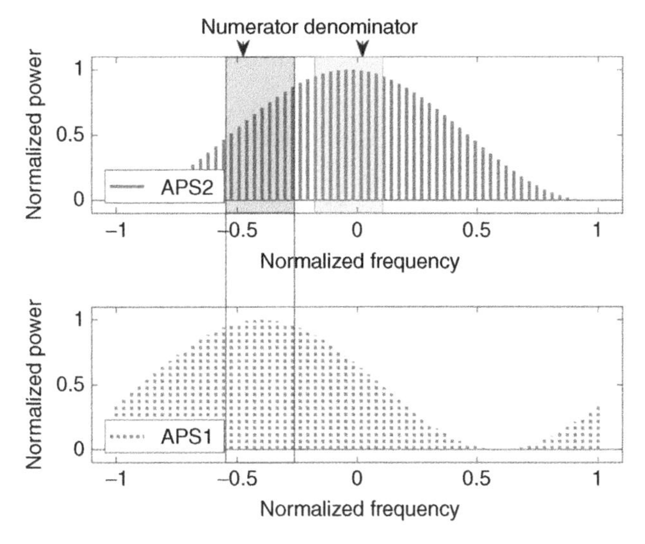

Figure 6.4 Intuitive explanation of the similarity metric (6.5).

the RPE is clear when writing a general APS **d** as

$$\mathbf{d} = \mathrm{diag}(\mathbf{F}^*\mathbf{R}\mathbf{F}), \tag{6.7}$$

where **R** is a covariance matrix and **F** contains the L columns of the DFT matrix corresponding to the L largest spectral components. Using the APS definition in terms of the spatial covariance matrix, the similarity metric can be written as

$$\mathrm{S} = \frac{\mathrm{tr}(\mathbf{F}_1^*\mathbf{R}_2\mathbf{F}_1)}{\mathrm{tr}(\mathbf{F}_2^*\mathbf{R}_2\mathbf{F}_2)}. \tag{6.8}$$

Note the equivalence between the similarity metric and the RPE when using linear arrays and all the AoAs fall on the grid created by the DFT matrix. This is because in this case the singular vectors of any covariance matrix **R** are elements of the DFT matrix, and S and RPE are the same. The connection between the RPE and the achievable rate was further established in Park et al. [2020], creating also a bridge between the similarity metric and the rate in on-grid scenarios.

The similarity metric was also experimentally evaluated in Graff et al. [2019] for a prototype radar aided V2I link operating at mmWave frequencies tested in 20 different outdoor environments. Figure 6.5 shows a picture of the prototype, based on two INRAS Radarbooks (a software-defined radar evaluation platform [http://www.inras.at/en/products/radarbook.html]). One of the Radarbooks is operating as a radar and the other one as a MIMO communication transceiver. An example of the similarity measurements obtained in one of these urban environments is provided in Figure 6.6, together with the beam patterns created from the radar and communications spectra when using 29 antenna elements and $N = 64$. Even for small values of L there is a high similarity between radar and communication APS.

6.3.2 Radar Covariance Estimation with a Passive Radar at the BS

After showing the usefulness of radar APS to configure the communication link, we consider a V2I set up where the BS is equipped with a passive mmWave radar receiver with a ULA of N_r antennas. The communication system operates at mmWave frequencies with OFDM, a hybrid MIMO architecture and ULAs at both the vehicle and the BS. The ego-vehicle on the road is also equipped

Figure 6.5 Prototype of a radar assisted V2I link based on two INRAS Radarbooks.

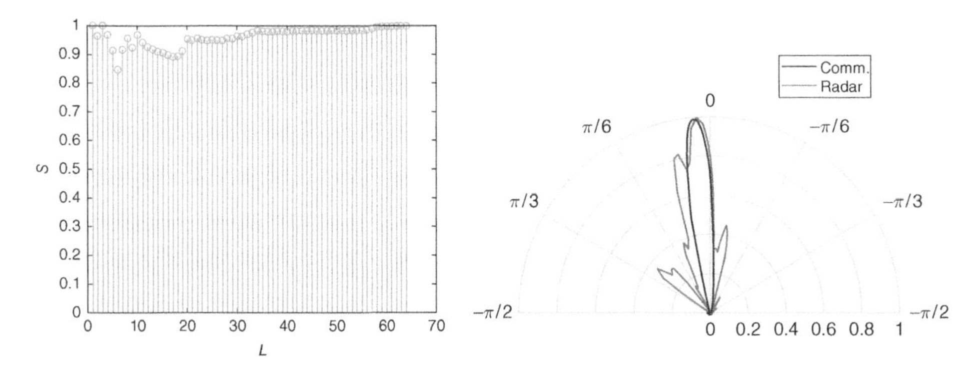

Figure 6.6 Similarity metric as a function of L evaluated in an outdoor urban scenario and corresponding beam patterns generated from the communication and radar APS.

with multiple medium range frequency-modulated continuous-wave (FMCW) radars placed at the four corners of the car. The vehicle has several communication arrays located on the roof. To assist communication, the passive radar at the BS will tap the signals coming from the automotive radars in the ego-vehicle, as illustrated in Figure 6.7. Note the relationship between the multipath components of the automotive radar signal and the communication signal. The spatial covariance of these radar signals will be estimated at the BS to aid in the design of the frequency flat analog precoder and combiner in the hybrid mmWave MIMO system as in Ali et al. [2020a]. The digital precoder and combiner can be designed independently for every subcarrier from an estimate of the low dimensional equivalent channel once the analog stage is fixed. Note that the low dimensional equivalent channel represents the combined response of the analog precoder, analog combiner, and the MIMO channel matrix $\mathbf{H}[k]$ between the highest SNR antenna array in the vehicle and the BS array.

There are several challenges in using FMCW radar information for mmWave link configuration based on spatial covariance. The first one is that the conventional estimate of the radar covariance $\mathbf{R}_r \in \mathbb{C}^{N_r \times N_r}$ is built from the signal received at the BS radar receiver as

$$\mathbf{R} = \frac{1}{I}\mathbf{Y}\mathbf{Y}^*, \tag{6.9}$$

where $\mathbf{Y} \in \mathbb{C}^{N_r \times I}$ contains I samples of each one of the received signal coming from the different antennas of the radar receiver. The mathematical expression for this signal is provided in

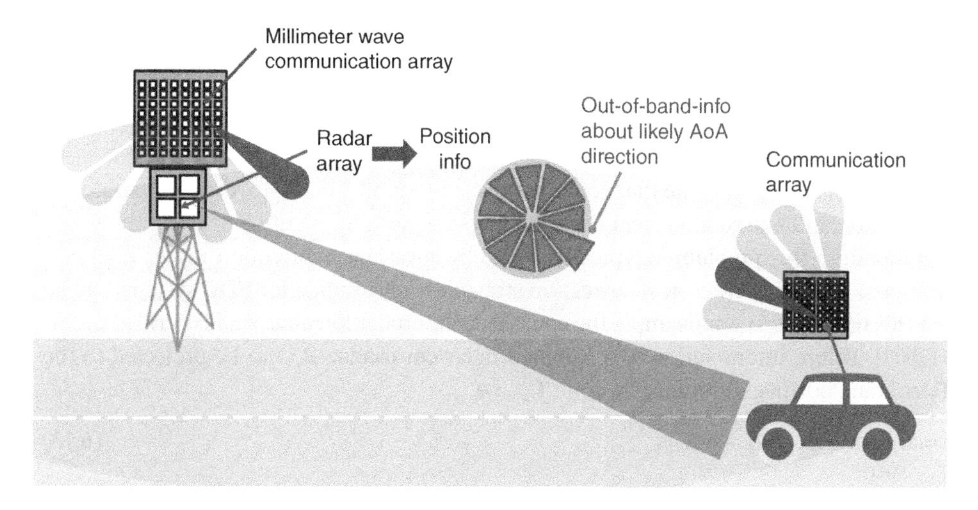

Figure 6.7 MIMO V2I communication set up with a BS mounted passive radar receiver.

Ali et al. [2020a], although it is not essential to understand the key idea behind the proposed approach, so we refer the reader to this publication for further details. These signals are obtained in a conventional FMCW radar receiver after mixing the received echoes with a chirp reference signal of initial frequency f_r and chirp rate β, and then low pass filtering. But the reference signal is unknown for the passive radar at the BS. Only the radar system at the vehicle knows the reference signal used to generate the transmit radar waveform. To circumvent this issue, an alternative processing chain that provides the same spatial covariance without knowledge of the reference signal was proposed in Ali et al. [2020a]. It is based on the idea of mixing with a sinusoidal reference signal of frequency $f_r + \Delta$, where Δ is an unknown frequency offset, to generate the receive samples $\check{\mathbf{Y}}$ instead of $\mathbf{Y}$. The radar spatial covariance that can be estimated in this case is

$$\check{\mathbf{R}}_r = \frac{1}{I}\check{\mathbf{Y}}\check{\mathbf{Y}}^*. \tag{6.10}$$

It has been shown in Ali et al. [2020a] that, because the spatial covariance depends only on the phase difference across antennas, the specific frequency of the resulting signal after mixing is not relevant, and $\mathbf{R} = \check{\mathbf{R}}$. Note, however, that this processing chain cannot provide the information needed for range and Doppler estimation, and only the spatial covariance can be estimated.

The second issue is that there is a bias in angle estimation through FMCW radar. The true and the estimated angles for a single point target, denoted by θ and $\hat{\theta}$, respectively, are related by

$$\sin\hat{\theta} = \left(1 + \frac{B_r}{2f_r}\right)\sin\theta. \tag{6.11}$$

This bias is critical in large antenna systems as small pointing errors can have a significant impact on the system performance. As an example, for 256 antennas, 76 GHz frequency, and 1.2 GHz bandwidth, it can be shown that if we did not correct the bias, the beam pointing error will create a null in the direction of the true angle. Therefore, it is essential to correct the bias. To solve this problem, a connection with the mismatch that appears in frequency division duplexing (FDD) systems between the uplink and downlink channels was also found in Ali et al. [2020a].

In FDD systems the same array is used in the uplink and the downlink. This array can be configured either according to the uplink half wavelength, λ_{UL}, or downlink half wavelength, λ_{DL}. If the arrays are set using half wave length, and the observations are made in the uplink, the true and estimated angles will be related by

$$\sin \hat{\theta} = \frac{\lambda_{\mathrm{DL}}}{\lambda_{\mathrm{UL}}} \sin \theta. \tag{6.12}$$

This relationship is similar to the one that generates bias in FMCW. Therefore, the same strategies developed for FDD can be used to correct the spatial covariance and rectify bias. Note, however, that in FDD literature this problem is typically solved by covariance correction rather than correcting the angles. Different covariance correction strategies well studied for FDD systems can be employed in this new context, for example the covariance interpolation approach described in Jordan et al. [2009]. Before interpolation, the sampled radar covariance $\check{\mathbf{R}}_r$ can be projected to the Toeplitz, Hermitian, positive semi-definite cone $\mathbf{T}_+^{N_r}$, i.e.

$$\hat{\mathbf{R}}_r = \arg\min_{\mathbf{X}\in\mathbf{T}_+^{N_r}} \|\mathbf{X} - (\check{\mathbf{R}}_r - \sigma_n^2\mathbf{I})\|_F. \tag{6.13}$$

to obtain a Toeplitz Hermitian matrix. In this case, $\hat{\mathbf{R}}_r$ is described by its first column, denoted as $\hat{\mathbf{r}}$. After covariance interpolation, the corrected covariance vector $\hat{\mathbf{r}}_c$ can be used to obtain the final estimate of the corrected covariance matrix $\hat{\mathbf{R}}_c$ by Toeplitz completion.

Figure 6.8 shows the performance of this covariance estimation approach for $T_{\mathrm{coh}} \to \infty$, $T_{\mathrm{coh}} = 4N_{\mathrm{BS}}N_v$, and the system parameters defined in Table 6.2. The performance is measured as the achieved rate as a function of the number of beams tested during the training process, which is smaller than the number of total beams in radar-assisted and position-assisted strategies. Note that for exhaustive search, all the beams are always tested, so the rate does not depend on this parameter. For testing, the radar and communication channels have been simulated using a ray tracing

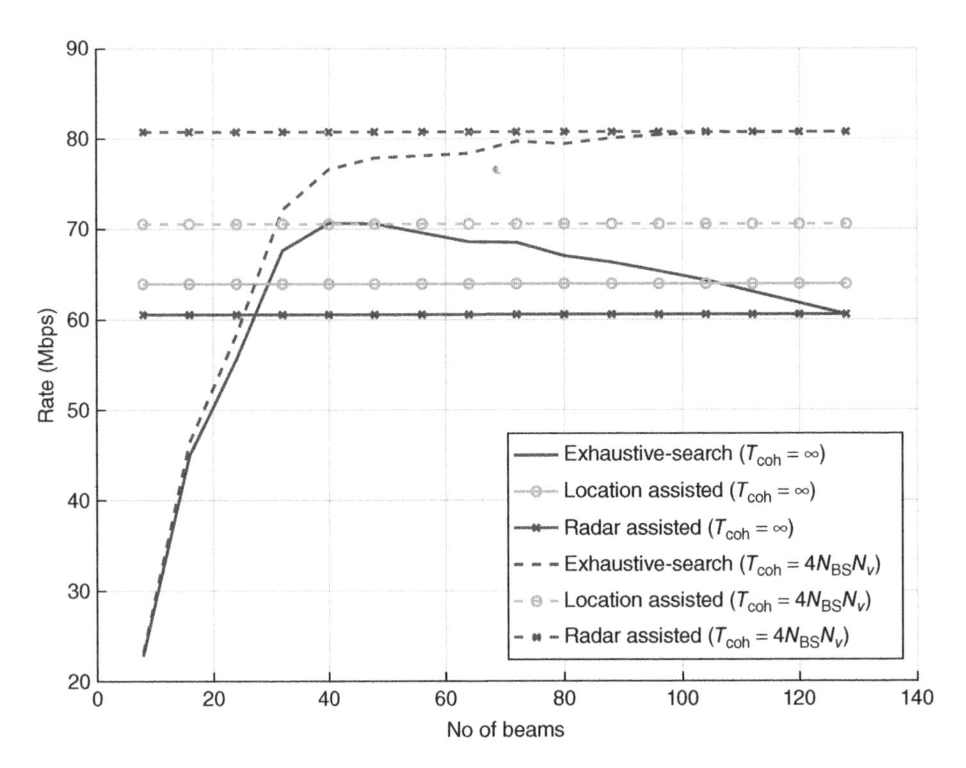

Figure 6.8 Rate versus the number of beams for $T_{\mathrm{coh}} \to \infty$ and $T_{\mathrm{coh}} = 4N_{\mathrm{BS}}N_v$ in NLOS channel.

Table 6.2 System parameters for the simulation of a radar-aided mmWave vehicular communication system based on radar covariance estimation.

Parameter	Symbol	Value	Units
Communication system			
Transmit power	P_c	30	dBm
Carrier frequency	f_c	73	GHz
Bandwidth	B_c	1	GHz
BS height	h_{BS}	5	m
Vertical separation of arrays at the BS		10	cm
Distance BS to closest point on the road	d	10	m
Number of antennas at the BS	N_{BS}	128	
Number of RF chains at the BS		1	
Number of antennas at the vehicle	N_v	16	
Number of arrays at the vehicle		4	
Number of RF chains at the vehicle		1	
Height of the communication arrays at the vehicle		1.6	m
Number of phase shifter bits	D	2	bits
Number of subcarriers	K	2048	
Subcarrier spacing		240	kHz
Cyclic prefix length	L_c	511	samples
Radar system			
Center frequency	f_c	76	GHz
Bandwidth	B_r	1	GHz
Transmit power	P_r	30	dBm
Number of antennas	N_r	128	
Chirp period	T_p	500	μs
Samples per chirp	I	1024	samples
Height of the vehicle radars		0.75	m

approach as described in Ali et al. [2020a]. For $T_{\text{coh.}} \to \infty$, the radar assisted beam configuration strategy based on the estimation of the covariance matrix clearly outperforms the location-assisted approach with fewer measurements. For $T_{\text{coh.}} = 4N_{BS}N_v$, only 30 measurements are needed to achieve the exhaustive-search rate, implying an overhead reduction of around 77%. The spatial covariance information obtained from radar and the proposed estimation strategy can effectively reduce the training overhead associated to the configuration of a V2I mmWave link.

6.3.3 Learning Mismatches Between Radar and Communication Channels

The approach described in Section 1.3.2 exploits the relationship between the radar and communication covariances to reduce mmWave link configuration overhead. In particular, the radar estimated covariance is used in Section 1.3.2 as a direct estimate of the communication spatial covariance, without any additional refinement step to compensate for the mismatch between radar and communication channels. Note that in the scenario depicted in Figure 6.7, there is, however, an inherent mismatch between the radar and communication spatial covariances. It is due to several factors:

- Different operation frequencies, which may experience different propagation conditions even if we consider mmWave radar and mmWave communication.
- Different size and antenna geometry, which impact the observed angular parameters.
- Different location of the radar and the communication transceiver, which introduces differences in the delay and angle domains.

If these mismatches could be considered somehow in the estimation strategy, it would be possible to further reduce the error in the communication covariance estimate. Due to the high number of parameters involved and the uncertainty about some of them (for example, the BS will not know the exact location of the communication module and the radars in the car), it is difficult to develop an analytical model for the received radar signal including mismatches, especially in a NLOS scenario. Deep learning architectures as proposed in Chen et al. [2021], are an interesting alternative to translating the radar covariance into the communication covariance if an appropriate training data set of matching pairs can be constructed. Current trends in the development of artificial intelligence solutions to manage and optimize cellular networks [Shafin et al., 2020] suggest that the BS will be able to obtain and handle huge amounts of data, justifying the use of machine learning approaches also in the context of radar aided mmWave communication systems.

When translating the radar information into the communication space, different objective functions can be considered. The approaches proposed in Chen et al. [2021] and Graff et al. [2021] include three different possible functions to be estimated/translated: APS, covariance vector, and dominant eigenvector of the covariance matrix. Different functions require tailored deep neural network (DNN) architectures to maximize the radar to communication (R2C) translation performance. In the following sections we review the DNN architectures proposed in Chen et al. [2021] and Graff et al. [2021] to map the information from the radar channel to the communication channel.

Given the interest in using the communication APS to configure the mmWave arrays, the mean squared error (MSE) between the estimated APS, denoted as $\hat{\mathbf{d}}_c$, and the true communication APS, denoted as $\mathbf{d}_c$, is considered as the loss function to train the networks,

$$\mathcal{L}(\mathbf{w}^{\mathrm{NN}}) = ||\hat{\mathbf{d}}_c - \mathbf{d}_c||^2, \tag{6.14}$$

where $\mathbf{w}^{\mathrm{NN}}_{\{\cdot\}}$ represents the network parameters to be trained for each method. The APS translation method directly approximates the communication APS from the measured radar APS, while when translating the covariance vector or the eigenvector, the estimated APS will be computed using the DFT matrix as in (6.7).

6.3.3.1 Learning the Covariance Vector

The most intuitive way of R2C translation is to map the radar covariance directly to the communication covariance. But mapping the entire covariance matrix is inefficient. As we previously discussed, since the covariance matrix is Toeplitz, Hermitian and positive semi-definite, it is defined by the first column $\tilde{\mathbf{r}}$, also called covariance vector. The R2C translation problem is reduced to predicting the communication covariance vector $\hat{\mathbf{r}}_c$ from the input radar covariance vector $\mathbf{r}_r$ using a DNN denoted as $\mathcal{N}_{\mathrm{cov}}(\cdot)$ such that

$$\hat{\mathbf{r}}_c = \mathcal{N}_{\mathrm{cov}}(\mathbf{r}_r; \mathbf{w}^{\mathrm{NN}}_{\mathrm{cov}}). \tag{6.15}$$

The components of the covariance vector are complex numbers, so it is reasonable to split the real and imaginary parts as a 2-channel input for the DNN. As the covariance column does not have

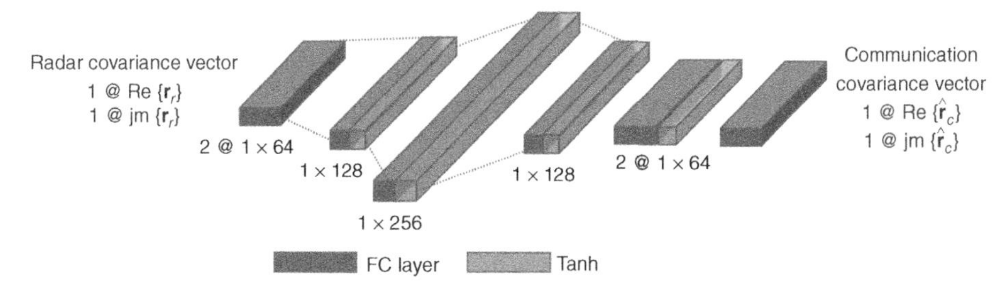

Figure 6.9 DNN architecture for learning the communication covariance vector.

specific structural patterns, e.g. it does not have specific patterns in the magnitude and phase, fully connected (FC) layers are selected as the dominant parts of the DNN for feature extractions (note that FC layers do not require prior information of the input distribution), as shown in Figure 6.9. The hyperbolic tangent activation function (Tanh), which constrains the value to $[-1, 1]$, is selected, since the input has both positive and non-positive values. There are four hidden layers inside the DNN, which are sufficient to parse the input information while saving computational resources (FC layers usually require a larger amount of parameters to train compared with convolutional layers with the same depth, so fewer hidden layers can help reduce both the time and spatial complexity). Finally, the 2-channel output from the DNN contains the real and imaginary parts of the estimated communication covariance vector. After getting $\hat{\mathbf{r}}_c$, the APS can be derived by $\hat{\mathbf{d}}_c = \mathbf{F}^* \mathbf{R}(\tilde{\mathbf{r}}_c)\mathbf{F}$.

6.3.3.2 Learning the Dominant Eigenvectors

The dominant eigenvector of the covariance matrix contains the information about the main channel direction. A DNN can also be designed for R2C eigenvector translation. Figure 6.10 shows the architecture proposed in Graff et al. [2021] to solve the eigenvector translation problem. The approximated communication eigenvector $\hat{\mathbf{e}}_c$ can be obtained from the dominant eigenvector of the radar covariance $\mathbf{e}_r$ as

$$\hat{\mathbf{e}}_c = \mathcal{N}_{\text{eig}}(\mathbf{e}_r; \mathbf{w}_{\text{eig}}^{\text{NN}}), \tag{6.16}$$

where $\mathcal{N}_{\text{eig}}(\cdot)$ denotes the particular DNN used to solve this problem. Similar to the covariance vector translation, the eigenvector is complex, and needs to be split into real and imaginary components as a 2-channel input. As eigenvectors do not have recognizable spatial features either, FC layers are adopted for general feature extractions. LeakyReLU [Xu et al., 2015] is selected as the activation function to avoid the dying ReLU [Maas et al., 2013] and vanishing gradient [Hochreiter, 1998] problems. There are five hidden layers to make the DNN deep enough to process the input and extract the information in the eigenvector, and the dropout layer is added to avoid overfitting. The 2-channel output from the DNN similarly contains the real and imaginary parts of

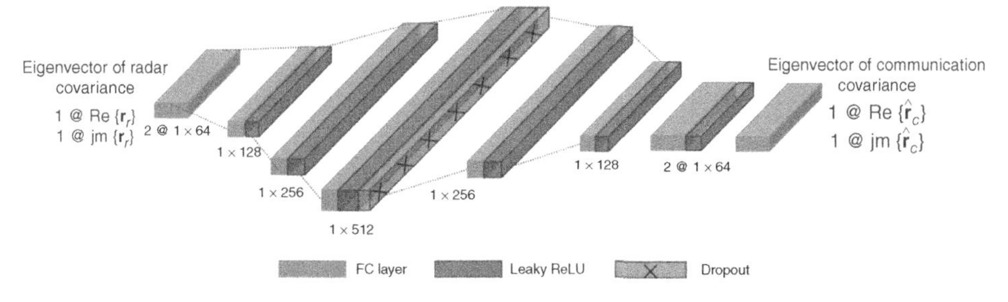

Figure 6.10 DNN architecture for learning the dominant eigenvector of the communication covariance. Source: Based on Graff et al. [2021].

the estimated communication eigenvector $\hat{\mathbf{e}}_c$. Then, the estimated APS can be derived from $\hat{\mathbf{e}}_c$ as $\hat{\mathbf{d}}_c = \mathbf{F}^* \cdot \hat{\mathbf{e}}_c$.

6.3.3.3 Learning the APS

Unlike the covariance column and the dominant eigenvectors, the APS has apparent local features such as sharp peaks that can be learned by the DNN. Convolutional layers are well suited for extracting local features [Jogin et al., 2018], and become the main components of the DNN $\mathcal{N}_{\text{APS}}(\cdot)$ for learning the APS. The DNN is inspired by the Encoder–Decoder architecture [Badrinarayanan et al., 2017], which reduces the input dimension to extract the key features and then goes back to the original dimension as the output. Since the APS is non-negative and real, the DNN takes the radar APS as a single channel input, and outputs the approximated communication APS

$$\hat{\mathbf{d}}_c = \mathcal{N}_{\text{APS}}(\mathbf{d}_r; \mathbf{w}_{\text{APS}}^{\text{NN}}). \tag{6.17}$$

The input goes through 1-D convolutional layers, and then MaxPooling is used for dimension reduction instead of AvgPooling. This is because MaxPooling can select the components with the highest magnitude from the APS, while AvgPooling will smooth out the APS and the sharp peak may not be identified. LeakyReLU is used as the activation function with the negative slope of 0.01. It has a similar performance as ReLU but can avoid the dying ReLU problem and speed up the training as well [Clevert et al., 2015]. The DNN architecture is shown in Figure 6.11.

An example of the estimated communication APS from the translation of the covariance vector using the network in Figure 6.9 is shown in Figure 6.12. In the DNN training phase, for this and the other described DNNs, a training set of 9600 samples and a testing set of 2400 samples was used. The Adam optimizer [Kingma and Ba, 2014] was used with a learning rate of 0.0001. A learning rate decay of 0.9 was adopted, which allows for changing the learning rate over time for refined training convergence.

The similarity metric [Ali et al., 2020a] is used for evaluating the similarity between the true and the estimated communication APS after deep learning-based translation. Higher similarity implies good APS alignment, which will lead to a significantly reduced overhead. The CDF of the similarity values ($L = 5$) based on the test dataset are shown in Figure 6.13. The similarity between the radar and the communication APS is treated as the benchmark, where only a small portion (30%) have high similarity (0.9). All of the three proposed networks generate APS translations that increase similarity. Though the APS method has the lowest rate [70%] of high similarity values [$\geq$0.9], generally the values exceed 0.75 [fifth-percentile value]. Covariance vector predictions provide a higher similarity than direct APS predictions, with the 5th, 50th, and 95th-percentile of the values being 0.85, 0.95, 0.99. The eigenvector translation brings the largest similarity among all the methods. The fifth-percentile achieves 0.9 and only 8% of the values are smaller than 1.

To test the beam training results based on the estimated APS, we simulate a MIMO OFDM communication system operating at mmWave with a single stream transmission and a transmit power of 24 dBm. The bandwidth is 491.52 MHz and 2048 subcarriers are used for the transmission. A DFT codebook as in Section 6.2 with 2 bits quantization for the phase shifters is assumed,

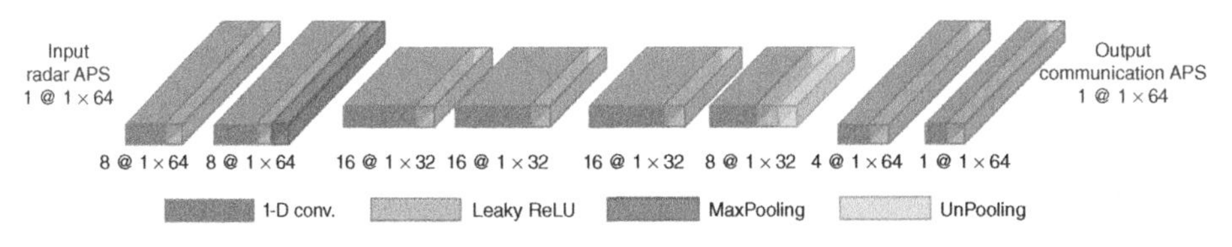

Figure 6.11 The DNN architecture for learning the communication APS.

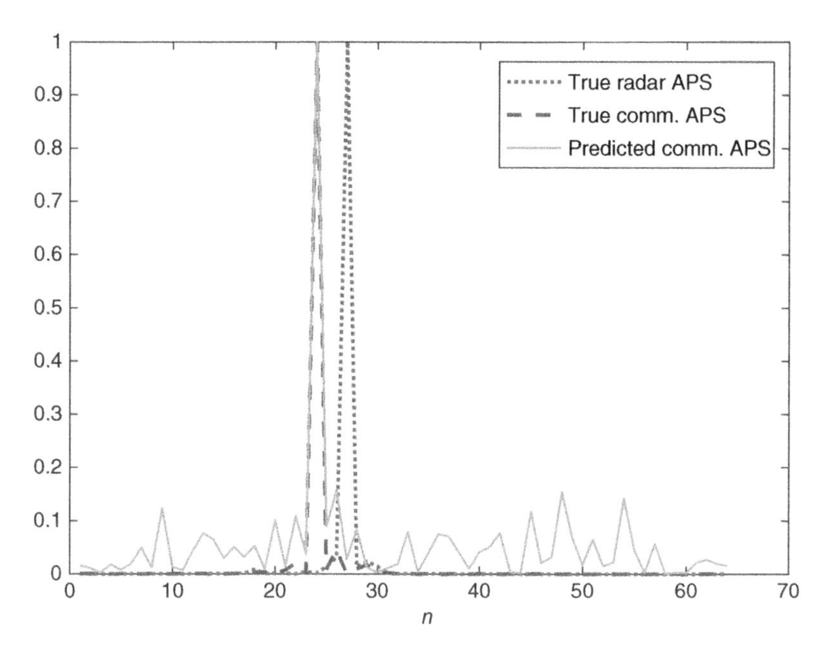

Figure 6.12 Example of the estimated communication APS from the translation of the radar covariance vector.

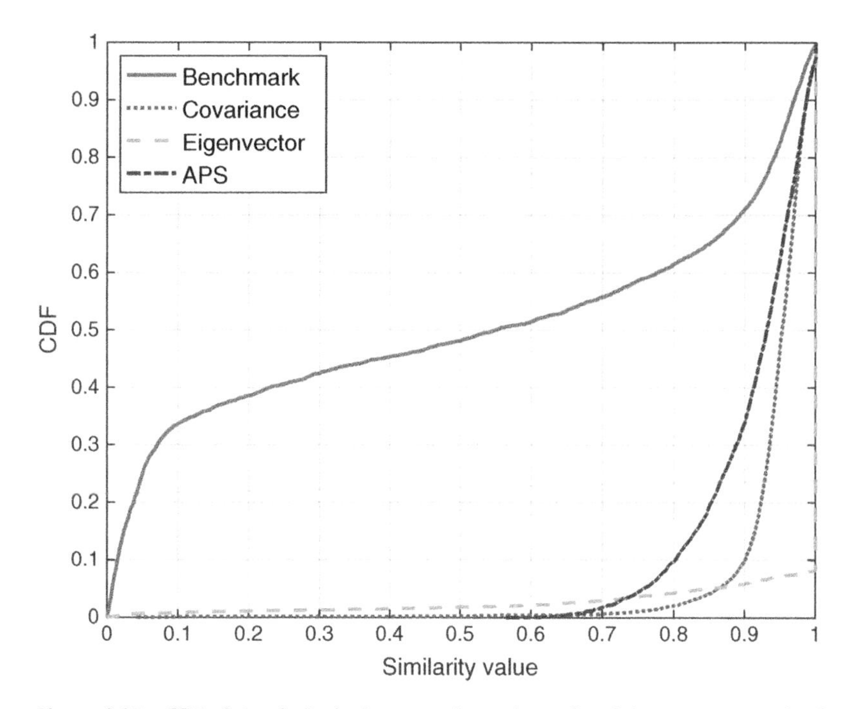

Figure 6.13 CDF of the similarity between the estimated and the true communication APS for the proposed covariance column, eigenvector and APS prediction methods.

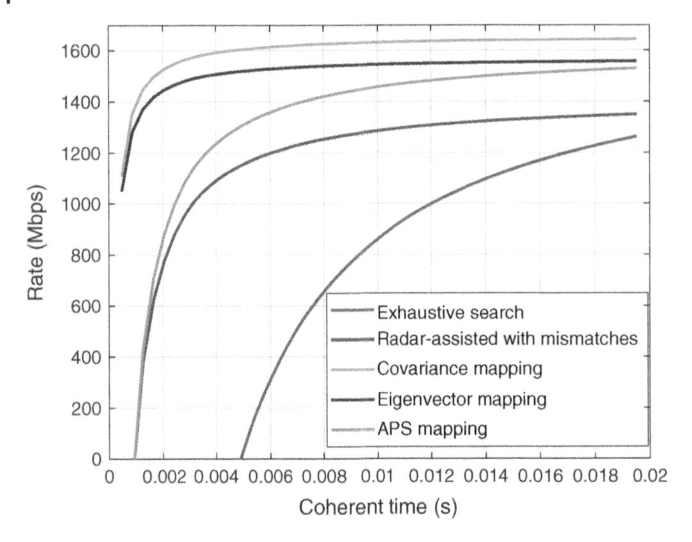

Figure 6.14 Rate results with the beam search size of 64 for exhaustive search, 12 for radar APS based search and search based on the APS prediction, and 2 for search based on the covariance column and eigenvector prediction.

considering a 180° sector spanning from $[-\frac{\pi}{2}, \frac{\pi}{2}]$. The coherent time of the V2I channel is again set to $4N_{\text{BS}}N_v$ blocks.

Exhaustive and radar APS-based search are treated as the benchmarks. In exhaustive search, the best combiner is selected by searching all over the codebook, while for radar APS aided search [without a R2C stage], the beam search size is reduced to 12 beams in the codebook, i.e. searching among the top 12 beam directions that are closer to the angle corresponding to the maximum in the radar APS. The beam search size remains to be 12 when using a R2C translation strategy based on learning the APS, while it is further reduced to 2 when leveraging a prediction of the covariance vector or the eigenvector. The rate results are shown in Figure 6.14. Exhaustive search offers the lowest rate as it allocates most of the resources to training. The APS prediction method requires similar time for training as radar APS aided search, due to the same search size, but then it leads to a higher achievable rate because more accurate APS information is available. The translations of the covariance vector and eigenvector lead to a much smaller training overhead, with achievable rates of 1.644 and 1.557 GHz, respectively. In summary, compared with exhaustive search, learning the covariance vector provides the best beam training performance with a rate increase of 30.26%, while learning the dominant eigenvector and the communication APS increase the achievable rate by 23.38% and 21.16%.

These results show the effectiveness of DNNs to learn the mismatch between the radar and communication channel and further reduce the training overhead associated to mmWave link configuration in vehicular channels.

6.4 Challenges and Opportunities

Leveraging different types of information extracted from radar sensors is an effective strategy to significantly reduce MIMO communication overhead, both in LOS and NLOS scenarios. This area of research is in its infancy though. Approaches that go beyond exploiting position information or the spatial covariance need to be investigated. For example, it would be interesting to understand how the radar signals can be leveraged in combination with the received communication signals to estimate the full instantaneous CSI with low overhead. Understanding the error bounds for CSI estimation in a MIMO system with large arrays when the available information includes a limited number of collocated radar and communication measurements is also a challenging and interesting problem that has not been addressed so far.

A passive radar receiver mounted in the BS can detect and track the automotive radar signals coming from a neighboring vehicle to generate an estimate of the channel spatial covariance. To extend this idea to a multi-vehicle scenario, additional signal processing techniques that handle the multi-vehicle interference and/or separate the radar signals coming from the automotive radars in the different vehicles are needed [Graff et al., 2021]. This problem is challenging due to the lack of information about the reference waveforms employed by the different automotive radars.

The mismatches between the radar and communication channels have an impact on the similarity of the APS. Machine learning techniques have proven to be effective to learn this mismatch in particular scenarios, but a complete theoretical analysis of the impact of frequency separation, degree of collocation or array size differences is needed. In this context, new techniques that enable covariance translation between radar and communication when assuming different antenna array sizes and geometries also need to be developed.

Finally, machine learning techniques can play a more significant role in the radar aided MIMO communication system than merely learning mismatches. In site-specific problems, for example a specific cellular cell, where a large number of measurements are possible, radar and communication signals coming from the different vehicles can be stored as matching pairs together with the channel estimates and/or the states of the channel to aid array reconfiguration and blockage prediction.

References

3GPP. Study on channel model for frequencies from 0.5 to 100 GHz. Tr 38.901, 3rd Generation Partnership Project (3GPP), December 2017. http://www.3gpp.org/DynaReport/38901.htm. Version 14.3.0.

A. Ali, N. González-Prelcic, and A. Ghosh. Millimeter wave V2I beam-training using base-station mounted radar. In *2019 IEEE Radar Conference (RadarConf)*, pages 1–5, 2019a.

A. Ali, N. González-Prelcic, and R. W. Heath. Spatial covariance estimation for millimeter wave hybrid systems using out-of-band information. *IEEE Transactions on Wireless Communications*, 18(12):5471–5485, 2019b.

A. Ali, N. González-Prelcic, and A. Ghosh. Passive radar at the roadside unit to configure millimeter wave vehicle-to-infrastructure links. *IEEE Transactions on Vehicular Technology*, 69(12): 14903–14917, 2020a.

A. Ali, N. González-Prelcic, R. W. Heath, and A. Ghosh. Leveraging sensing at the infrastructure for mmWave communication. *IEEE Communications Magazine*, 58(7):84–89, 2020b.

C. Aydogdu, F. Liu, C. Masouros, H. Wymeersch, and M. Rydström. Distributed radar-aided vehicle-to-vehicle communication. In *2020 IEEE Radar Conference (RadarConf20)*, pages 1–6, 2020.

V. Badrinarayanan, A. Kendall, and R. Cipolla. SegNet: A deep convolutional encoder-decoder architecture for image segmentation. *IEEE Transactions on Pattern Analysis and Machine Intelligence*, 39(12):2481–2495, 2017.

B. Cetinoneri, I. Sarkas, and Q. Yu. Systems and methods for object detection by radio frequency systems. United States Patent Application, September 2020.

Y. Chen, A. Graff, N. González-Prelcic, and T. Shimizu. Radar aided mmWave vehicle-to-infrastructure link configuration using deep learning. In *IEEE Global Communications Conference*, 2021.

D.-A. Clevert, T. Unterthiner, and S. Hochreiter. Fast and accurate deep network learning by exponential linear units (ELUs). *arXiv preprint arXiv:1511.07289*, 2015.

S. Fuller and G. Waters. Smarter Infrastructure for a Smarter World. March 2019. https://www.nxp.com/docs/en/white-paper/SMRTRINFRASTRWP.pdf.

J. P. González-Coma, J. Rodríguez-Fernández, N. González-Prelcic, L. Castedo, and R. W. Heath. Channel estimation and hybrid precoding for frequency selective multiuser mmWave MIMO systems. *IEEE Journal of Selected Topics in Signal Processing*, 12(2):353–367, 2018.

N. González-Prelcic, R. Méndez-Rial, and R. W. Heath Jr. Radar aided beam alignment in mmWave V2I communications supporting antenna diversity. In *Proceedings Information Theory and Applications (ITA) WKSP*, pages 1–5, February 2016.

N. González-Prelcic, A. Ali, V. Va, and R. W. Heath. Millimeter-wave communication with out-of-band information. *IEEE Communications Magazine*, 55(12):140–146, 2017.

A. Graff, A. Ali, and N. González-Prelcic. Measuring radar and communication congruence at millimeter wave frequencies. In *53rd Asilomar Conference on Signals, Systems, and Computers*, pages 925–929, 2019.

A. Graff, Y. Chen, N. González-Prelcic, and T. Shimizu. Deep-learning based link configuration for radar-aided multiuser mmWave vehicle-to-infrastructure communication. *arXiv preprint*, 2021.

S. Hochreiter. The vanishing gradient problem during learning recurrent neural nets and problem solutions. *International Journal of Uncertainty, Fuzziness and Knowledge-Based Systems*, 6(02):107–116, 1998.

INRAS Radarbook. http://www.inras.at/en/products/radarbook.html.

S. Jaeckel, L. Raschkowski, K. Borner, and L. Thiele. QuaDRiGa: A 3-D multi-cell channel model with time evolution for enabling virtual field trials. *IEEE Transactions on Antennas and Propagation*, 62(6): 3242–3256, 2014.

S. Jaeckel, L. Raschkowski, K. Borner, L. Thiele, F. Burkhardt, and E. Eberlein. QuaDRiGa-Quasi deterministic radio channel generator, user manual and documentation. *Fraunhofer Heinrich Hertz Institute, Tech. Rep. v2.0.0*, 2017.

M. Jogin, M. S. Madhulika, G. D. Divya, R. K. Meghana, S. Apoorva, et al. Feature extraction using convolution neural networks (CNN) and deep learning. In *2018 3rd IEEE International Conference on Recent Trends in Electronics, Information & Communication Technology (RTEICT)*, pages 2319–2323. IEEE, 2018.

M. Jordan, X. Gong, and G. Ascheid. Conversion of the spatio-temporal correlation from uplink to downlink in FDD systems. In *Proceedings IEEE Wireless Communications and Networking Conference (WCNC)*, pages 1–6, April 2009.

D. P. Kingma and J. Ba. Adam: A method for stochastic optimization. *arXiv preprint arXiv:1412.6980*, 2014.

J. Lee, G.-T. Gil, and Y. H. Lee. Channel estimation via orthogonal matching pursuit for hybrid MIMO systems in millimeter wave communications. *IEEE Transactions on Communications*, 64(6): 2370–2386, 2016. https://doi.org/10.1109/TCOMM.2016.2557791.

J. Lien, N. Gillian, M. E. Karagozler, P. Amihood, C. Schwesig, E. Olson, H. Raja, and I. Poupyrev. Soli: Ubiquitous gesture sensing with millimeter wave radar. *ACM Transactions on Graphics*, 35(4):1–19, 2016. ISSN 0730-0301. https://doi.org/10.1145/2897824.2925953.

A. L. Maas, A. Y. Hannun, and A. Y. Ng. Rectifier nonlinearities improve neural network acoustic models. In *Proceedings of the 30th International Conference on International Conference on Machine Learning ICML*, volume 30, page 3. CiteSeer, 2013.

R. Méndez-Rial, N. González-Prelcic, and R. W. Heath. Adaptive hybrid precoding and combining in MmWave multiuser MIMO systems based on compressed covariance estimation. In *IEEE 6th*

International Workshop on Computational Advances in Multi-Sensor Adaptive Processing (CAMSAP), pages 213–216, 2015.

S. Park, J. Park, A. Yazdan, and R. W. Heath. Exploiting spatial channel covariance for hybrid precoding in massive MIMO systems. *IEEE Transactions on Signal Processing*, 65(14):3818–3832, 2017.

S. Park, A. Ali, N. González-Prelcic, and R. W. Heath. Spatial channel covariance estimation for hybrid architectures based on tensor decompositions. *IEEE Transactions on Wireless Communications*, 19(2):1084–1097, 2020.

J. Rodríguez-Fernández and N. González-Prelcic. Channel estimation for hybrid mmWave MIMO systems with CFO uncertainties. *IEEE Transactions on Wireless Communications*, 18(10):4636–4652, 2019.

R. Shafin, L. Liu, V. Chandrasekhar, H. Chen, J. Reed, and J. C. Zhang. Artificial intelligence-enabled cellular networks: a critical path to beyond-5G and 6G. *IEEE Wireless Communications*, 27(2):212–217, 2020.

A. Shahmansoori, G. E. Garcia, G. Destino, G. Seco-Granados, and H. Wymeersch. Position and orientation estimation through millimeter-wave MIMO in 5G systems. *IEEE Transactions on Wireless Communications*, 17(3):1822–1835, 2018.

L. Simić, J. Arnold, M. Petrova, and P. Mähänen. RadMAC: Radar-enabled link obstruction avoidance for agile mm-Wave beamsteering. In *Proceedings of the 3rd WKSP on Hot Topics Wireless*, pages 61–65, 2016.

K. Venugopal, A. Alkhateeb, N. González Prelcic, and R. W. Heath. Channel estimation for hybrid architecture-based wideband millimeter wave systems. *IEEE Journal on Selected Areas in Communications*, 35(9):1996–2009, 2017.

H. Xie and N. González-Prelcic. Dictionary learning for channel estimation in hybrid frequency-selective mmWave MIMO systems. *IEEE Transactions on Wireless Communications*, 19(11):7407–7422, 2020.

H. Xie and N. González-Prelcic. Blockage detection and channel tracking in wideband mmWave MIMO systems. In *Proceedings of IEEE International Conference on Communications (ICC)*, 2021.

B. Xu, N. Wang, T. Chen, and M. Li. Empirical evaluation of rectified activations in convolutional network. *arXiv preprint arXiv:1505.00853*, 2015.

7

Design of Constant-Envelope Radar Signals Under Multiple Spectral Constraints

Augusto Aubry[1], Jing Yang[2], Antonio DeMaio[1], Guolong Cui[2], and Xianxiang Yu[2]

[1] *University of Naples, Federico II, Department of Electrical and Information Technology Engineering, Napoli, Italy*
[2] *University of Electronic Science and Technology of China, School of Information and Communication Engineering, Chengdu, China*

7.1 Introduction

The increasing demand for radio frequency (RF) spectrum access mainly driven by the mobile communications industry has raised, in recent years, serious concerns on spectral cohabitation among wireless systems [Wicks, 2010; Govoni, 2016]. Not surprisingly, the design of effective spectrum sharing strategies, capitalizing on environmental cognition and waveform diversity for a smart and dynamic RF spectrum usage to enable joint radar and communication (JRC) activities, has attracted attention of many scientists and engineers during the last decade [Wicks, 2010; Griffiths et al., 2015; Blunt and Mokole, 2016; Cui et al., 2020].

JRC problem has been approached according to different spectral sharing paradigms, either with a communication-centric perspective or following a radar-centric philosophy. The former mainly involves the co-designing of radar and communication signals assuming a full-cooperation among the RF nodes, with interesting and technically sound strategies available in the open literature (the interested reader may refer to Paul et al. [2016], Hu et al. [2021], Liu et al. [2018, 2020], Hassanien et al. [2015], Dokhanchi et al. [2020], Li et al. [2016], Li and Petropulu [2017], Qian et al. [2018], and Zheng et al. [2018] where communication-specific constraints are considered too). The latter, which is the core of this chapter, instead refers to a partial cooperation among the RF systems, with communications systems that do not alter their behavior while radar systems dynamically devise their probing signals to guarantee spectral cohabitation while boosting their performance.

Many studies in the open literature have dealt with the problem of devising radar signals with an appropriate frequency allocation [Nunn and Moyer, 2012; Govoni, 2016], namely according to a radar-centric approach, so as to induce acceptable interference levels on frequency-overlaid systems, while optimizing radar performance in terms of range-Doppler resolution, low range/Doppler sidelobes, detection/tracking capabilities. In Gerlach [1998], a waveform design technique is introduced to confer some desired spectral nulls to the radar signal. The idea is to perturb a stepped-frequency-modulated waveform forcing an additional fast-time polyphase code. The approach is extended in Gerlach et al. [2011] to the case of continuous-phase waveforms with nulls at some specific frequencies. The effectiveness of both mentioned strategies is assessed in Cook et al. [2010], via an experimental analysis. An alternating projection algorithm for the construction of chirp-like constant-modulus signals with a single spectral null is proposed in Selesnick et al. [2010], whereas in Selesnick and Pillai [2011] its extension, addressing the synthesis of multiple notches, is established. Some iterative algorithms are introduced in Lindenfeld [2004]

Signal Processing for Joint Radar Communications, First Edition.
Edited by Kumar Vijay Mishra, M. R. Bhavani Shankar, Björn Ottersten, and A. Lee Swindlehurst.

for the joint design of the transmit signal and the receive filter achieving frequency stopband suppression and range sidelobes minimization. Besides, in Tang and Liang [2018], two algorithms are introduced to synthesize waveforms with desired spectral behavior according to a weighted least-squares (WLS) fitting approach.

In He et al. [2010] and Wang and Lu [2011], sparse frequency constant modulus radar signals with a low integrated sidelobe level (ISL) are built optimizing a suitable combination of the ISL metric and a penalty function accounting for the waveform frequency allocation. In Patton et al. [2012], a spectrum-centric signal design is developed based on the minimization of the transmitted energy over a set of disjoint stopband frequencies under a unimodularity constraint and autocorrelation function (ACF) masking. In Romero and Shepherd [2015], a friendly spectral-shaped radar waveform design is considered to allow the coexistence among the radar and one or more communications systems.

In Aubry et al. [2014a], the authors focus on the synthesis of radar waveforms that optimize the signal-to-interference-plus-noise-ratio (SINR) while controlling the total interference radiated on the overlaid systems together with the similarity between the transmitted signal and a reference waveform. This framework is extended to incorporate multiple spectral compatibility constraints in Aubry et al. [2016a]. To comply with the current amplifier technology, extensions to address optimized synthesis of constant-envelop waveforms with continuous phase [Aubry et al., 2020] and the finite alphabet [Yang et al., 2020] are developed. This design framework is further investigated in Aubry et al. [2014b] and Aubry et al. [2016b] where signal-dependent interference is accounted for at the transceiver design stage, while in Tang et al. [2018] the alternating direction method of multipliers (ADMM) approach is exploited to efficiently synthesize the waveform sought [Aubry et al., 2014a]. Still forcing a constraint on the global spectral interference, Wu et al. [2018], Yu et al. [2020], and Cheng et al. [2018] propose waveform design procedures in the context of multiple-input multiple-output (MIMO) radar systems operating in highly reverberating environments. Moreover, Wu et al. [2018] also extends the developed method considering multiple spectral constraints, but just an energy constraint is forced on each transmitted signal. Furthermore, in Tang and Li [2019] an iterative algorithm to synthesize waveforms for MIMO radars is developed considering the mutual information between the target response and the received echo as a figure of merit.

Additional radar waveform design techniques for spectral cohabitation and their performance assessment can be found in Jing et al. [2018], Zhao et al. [2018], Jakabosky et al. [2015], Ravenscroft et al. [2018], and Bica and Koivunen [2018].

In this chapter, some techniques for constant- modulus waveform synthesis to face spectrally dense environments are reviewed [Aubry et al., 2020; Yang et al., 2020], where the SINR serves as the optimized performance metric. Specifically, a local control on the interference energy radiated by radar on each reserved frequency bandwidth is performed, so as to enable JRC activities. To comply with the current amplifiers technology (operating in saturation regime), constant-envelope waveforms are considered, with either arbitrary or discrete phases. Besides, to fulfill basic radar requirements, in addition to an upper bound to the maximum radiated energy, a similarity constraint is enforced to bestow relevant waveform hallmarks, i.e. a well-shaped ambiguity function. The optimization process is restricted to constant-modulus waveforms for compatibility with current amplifier technology. The resulting problem is NP-hard, and an iterative algorithm based on the coordinate descent (CD) method is presented for code design. In each iteration, non-convex optimization problems are globally solved in closed form by computing elementary functions. The overall computational burden of the algorithm is linear with respect to the code length and the number of iterations, and is less than quadratic with respect to the number of licensed emitters. At the analysis stage, some interesting case studies are illustrated to assess the capability of the devised algorithm to improve the radar detection performance while guaranteeing spectral compatibility.

This chapter is organized as follows. The system model is introduced in Section 7.2, followed by the definition and the description of the key performance metrics and the constraints involved in the formulation of radar waveform design problems under investigation. Hence, in Section 7.3, solution methods are developed to handle the resulting NP-hard optimization problems. Section 7.4 is devoted to the performance analysis of the developed algorithms. Finally, conclusions and possible future research avenues are drawn in Section 7.5.

7.1.1 Notations

We adopt the notation of boldface for vectors $\mathbf{a}$ (lowercase) and matrices $\mathbf{A}$ (uppercase). The (m, n)th entry of $\mathbf{A}$ is denoted by $\mathbf{A}_{m,n}$ or $\mathbf{A}(m, n)$. The transpose, conjugate, and conjugate transpose operators are denoted by $(\cdot)^T$, $(\cdot)^*$, and $(\cdot)^\dagger$, respectively. $\mathbf{I}$ and $\mathbf{0}$ denote, respectively, the identity matrix and the matrix with zero entries (their size is determined from the context). $\mathbb{R}^N$, $\mathbb{C}^N$, and $\mathbb{H}^N$ are, respectively, the sets of N-dimensional vectors of real numbers, of N-dimensional vectors of complex numbers, and of $N \times N$ Hermitian matrices. The curled inequality symbol $\succeq$ (and its strict form $\succ$) is used to denote generalized matrix inequality: for any $\mathbf{A} \in \mathbb{H}^N$, $\mathbf{A} \succeq \mathbf{0}$ means that $\mathbf{A}$ is a positive semidefinite matrix ($\mathbf{A} \succ \mathbf{0}$ for positive definiteness). The letter j represents the imaginary unit (i.e. $j = \sqrt{-1}$). For any complex number x, $|x|$ represents the modulus of x. $\arg(x) \in [-\pi, \pi)$, $\overline{\arg}(x) \in [0, 2\pi)$ are the argument of the complex number x. Moreover, for any $\mathbf{x} \in \mathbb{C}^N$, $\|\mathbf{x}\|$ and $\|\mathbf{x}\|_\infty$ denote the Euclidean norm and l-infinity norm, respectively. $\mathbf{diag}(\mathbf{x})$ indicates the diagonal matrix whose ith diagonal element is the ith entry of $\mathbf{x}$. Given two integer numbers x and y, $\mathrm{mod}(x, y)$ provides the remainder of the integer division of x by y. $\lfloor a \rfloor$ and $\lceil a \rceil$ ($a \in \mathbb{R}$) provide the greatest integer not larger than a and the lowest integer not smaller than a, respectively. $\lceil a \rfloor, a \in \mathbb{R}$ denotes the integer closest to a. If the decimal part of a is 0.5, $\lceil a \rfloor$ is the smallest integer larger than a if $a < 0$; otherwise, it is the largest integer smaller than a. $\odot$ is the Hadamard element-wise product. Finally, $\mathbb{E}[\cdot]$ denotes statistical expectation and for any optimization problem $\mathcal{P}$, $v(\mathcal{P})$ is its optimal value.

7.2 System Model and Problem Formulation

This section focuses on the introduction of the radar system and signal models accounting for RF congested environment, as well as on the definition of the constrained optimization problem that formalizes in mathematical terms the radar transmit resource management process. A pictorial representation of the spectrally crowded scenario under consideration in this chapter is shown in Figure 7.1.

7.2.1 System Model

Let $\mathbf{c} = [c_1, \dots, c_N]^T \in \mathbb{C}^N$ be the radar fast-time (range) code sequence with length N (see Figure 7.2 as an example). The waveform at the receiver end is down-converted to baseband, undergoes a sub-pulse matched filtering operation, and then is sampled. As a result, the N-dimensional column vector $\mathbf{v} = [v_1, \dots, v_N]^T \in \mathbb{C}^N$ of the fast-time observations from the range-azimuth cell under test can be expressed as

$$\mathbf{v} = \alpha_T \mathbf{c} + \mathbf{n}, \tag{7.1}$$

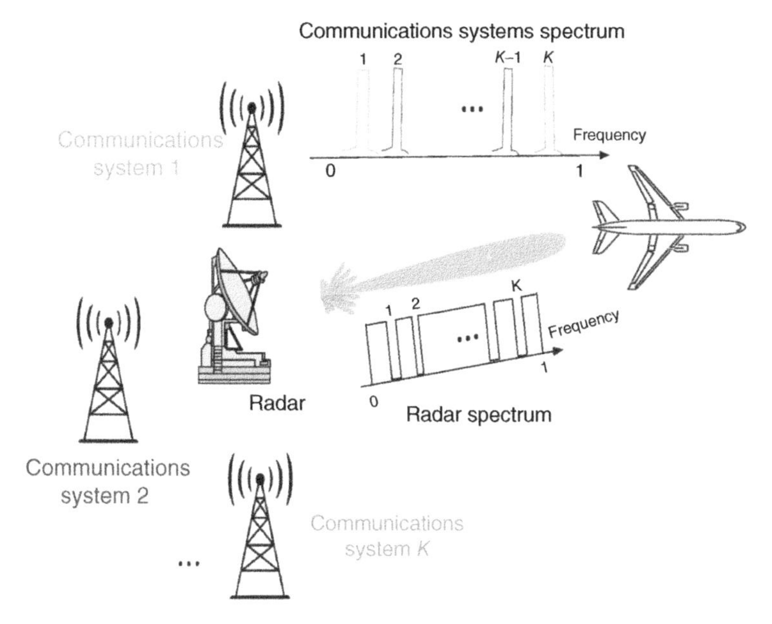

Figure 7.1 A pictorial representation of the spectrally crowded scenario under consideration.

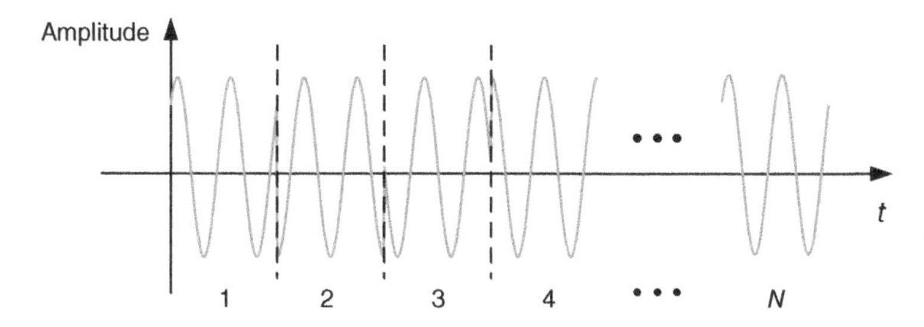

Figure 7.2 Radar probing waveform with fast-time phase coding.

where α_T is a complex parameter accounting for channel propagation and backscattering effects from the target within the range-azimuth bin of interest, and $\mathbf{n}$ is the N-dimensional column vector containing the filtered disturbance echo samples. Specifically, the vector $\mathbf{n}$ accounts for both white internal thermal noise and interfering signals due to unknown and unlicensed, possibly hostile jammers, as well as licensed overlaid telecommunications networks sharing the same frequencies as the radar of interest and possible hot-clutters. Moreover, $\mathbf{n}$ is modeled as a complex, zero-mean, circularly symmetric Gaussian random vector with covariance matrix $\mathbb{E}[\mathbf{n}\mathbf{n}^\dagger] = \mathbf{M}$.

As to the description of the congested RF environment, it is supposed the presence of K licensed emitters coexisting with the radar of interest, each of them operating over a frequency band $\Upsilon_k = [f_1^k, f_2^k]$, $k = 1, \ldots, K$, where f_1^k and f_2^k denote the lower and upper normalized frequencies (with respect to the underlying radar sampling frequency) for the kth system, respectively.

Now, to ensure spectral compatibility with overlaid RF systems, the radar has to shape properly its transmit waveform to manage the amount of interfering energy produced on each of the K shared frequency bandwidths. In this respect, to note that the disturbance energy injected on the kth band is

$$\int_{f_1^k}^{f_2^k} S_c(f)\, df = \mathbf{c}^\dagger \mathbf{R}_I^k \mathbf{c}, \tag{7.2}$$

where $S_c(f) = \left|\sum_{n=1}^{N} c_n e^{-j2\pi fn}\right|^2$ is the energy spectral density (ESD) of the fast-time code $\mathbf{c}$ and for $(m, l) \in \mathcal{N}^2$ with $\mathcal{N} = \{1, \dots, N\}$,

$$\mathbf{R}_I^k(m, l) = (f_2^k - f_1^k)e^{j\pi(f_2^k + f_1^k)(m-l)} \times \text{sinc}(\pi(f_2^k - f_1^k)(m - l)).$$

Thus, denoting by $E_I^k, k = 1, \dots, K$, the acceptable level of disturbance on the kth bandwidth, that is related to the quality of service (QoS) required by the kth telecommunications networks as well as its relative position with respect to the radar, the transmitted waveform has to comply with the constraints

$$\mathbf{c}^\dagger \mathbf{R}_I^k \mathbf{c} \le E_I^k, \quad k = 1, \dots, K. \tag{7.3}$$

From the radar perspective, the coexisting wireless networks exhibit a different importance, based, for instance, on their distance from the radar and their tactical relevance (i.e. navigation systems, defense communications, and public services). Evidently, for a smart and efficient spectrum allocation, it is of paramount importance for the sensing radar to have an accurate, reliable, and comprehensive radio environment awareness, to glean knowledge about the number K of coexisting licensed radiators, emitters' bandwidths (referred to as stopband in the following and defined by the lower and upper normalized frequencies f_1^k and f_2^k for Υ_k) and their location. Radio environment map (REM) [Zhao et al., 2006] represents the key to gain the aforementioned spectrum awareness, which is at the base of an intelligent and agile spectrum management. It can be defined as an integrated database, digitizing and indexing the available electromagnetic information. Figure. 7.3 illustrates some REM multi-domain knowledge sources represented by geographical features, available services and spectral regulations, locations and activities of telecommunications networks, and previous sensing acquisitions (radio experiences and measurements) [Zhao et al., 2007; Yucek and Arslan, 2009]. Indeed, the idea behind REM is to store and process a variety of data, so as to make it possible the inference of a multitude of environmental characteristics, such as locations of transmitters, prevailing propagation conditions, and estimates of spectrum usage over time and space. To populate and update the REM, both a priori knowledge and spectral sensing techniques (like feature-based signal detectors) can be used [Yucek and

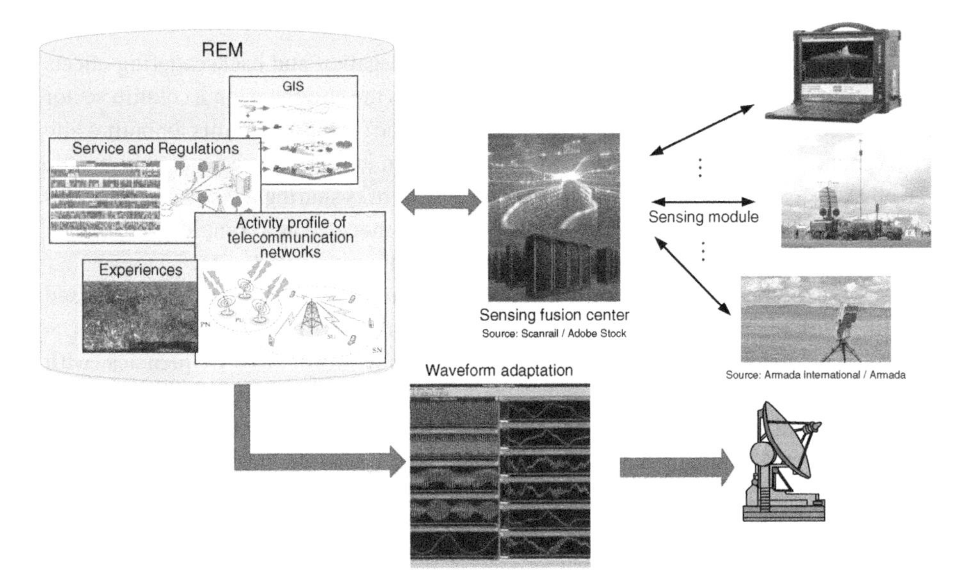

Figure 7.3 A pictorial representation of the REM and its usage in radar. Source: ktsdesign / Adobe Stock.

Arslan, 2009]. In this respect, as depicted in Figure. 7.3, a dedicated sensor network could be also employed to improve the quality of the scenario monitoring. Hence, exploiting the REM, the radar can become aware of the surrounding electromagnetic environment; hence, it can intelligently use the available knowledge sources to adapt its transmission to the actual operative scenario suitably diversifying the probing waveform [Baker et al., 2012; Cohen et al., 2017]. Remarkably, a prediction of the overlaid networks coverage can be obtained and used together with the radar antenna pointing direction, to select a suitable upper bound E_I^k for each stopband.

7.2.2 Code Design Optimization Problem

In this subsection, a radar waveform design approach is introduced aimed at optimizing target detection probability while controlling both the amount of interfering energy produced in each shared band and some hallmarks of the transmitted waveform.

According to the signal model in (7.1), the problem of establishing the presence or absence of a target in the cell under test can be formulated as the following binary hypothesis test:

$$\begin{cases} H_0 : \mathbf{v} = \mathbf{n}, \\ H_1 : \mathbf{v} = \alpha_T \mathbf{c} + \mathbf{n}. \end{cases} \tag{7.4}$$

Under the assumption of Gaussian interference with known positive-definite covariance matrix $\mathbf{M}$, the generalized likelihood ratio test detector over α_T for model (7.4), which coincides with the optimum test (according to the Neyman–Pearson criterion) if the phase of α_T is uniformly distributed in $[0, 2\pi)$, is given by De Maio et al. [2010]

$$\left|\mathbf{v}^\dagger \mathbf{M}^{-1} \mathbf{c}\right|^2 \underset{H_0}{\overset{H_1}{\gtrless}} \eta, \tag{7.5}$$

where η is the detection threshold set according to a desired value of the false alarm probability (P_{fa}). In the case of nonfluctuating target, the detection probability (P_d), for a given value of P_{fa}, can be analytically expressed as De Maio et al. [2010]

$$P_d = Q\left(\sqrt{2|\alpha_T|^2 \mathbf{c}^\dagger \mathbf{M}^{-1} \mathbf{c}}, \sqrt{-2 \ln P_{fa}}\right), \tag{7.6}$$

where $Q(\cdot, \cdot)$ denotes the Marcum Q function of order 1. Equation (7.6) shows that, given P_{fa}, P_d depends on the radar code and the disturbance covariance matrix only through the SINR, defined as

$$\text{SINR}(\mathbf{c}) = |\alpha_T|^2 \mathbf{c}^\dagger \mathbf{R} \mathbf{c}. \tag{7.7}$$

which is considered as a figure of merit, where $\mathbf{R} = \mathbf{M}^{-1}$ is the inverse of the interference covariance matrix.

To assure compatibility with current amplifier technology, the optimization process is restricted to phase-only waveforms, which is tantamount to requiring that

$$|c_i| = |c_j|, \quad i \neq j. \tag{7.8}$$

Furthermore, as a limited number of bits are available in digital waveform generators, the finite alphabet constraint is possibly forced, requiring $\arg(c_i) \in \Omega_M$, where $\Omega_M = \frac{2\pi}{M}\left\{-\lfloor\frac{M}{2}\rfloor, \dots, -\lfloor\frac{M}{2}\rfloor + M - 1\right\}$ denotes the set of discrete phases with the cardinality M. Furthermore, to control some peculiarities of the transmitted waveform, other than an energy requirement, namely $\|\mathbf{c}\|^2 \leq 1$, a similarity constraint is imposed on the probing code, i.e.

$$\left\|\frac{\mathbf{c}}{\|\mathbf{c}\|} - \mathbf{c}_0\right\|_\infty \leq \frac{\epsilon}{\sqrt{N}}, \tag{7.9}$$

where the parameter $0 \leq \epsilon \leq 2$ rules the size of the trust hypervolume, and $\mathbf{c}_0$ is a specific phase-only reference code, with $\|\mathbf{c}_0\|^2 = 1$. Indeed, an optimization without any explicit control on the code structure can lead to signals with high peak sidelobe level (PSL)/ISL values and more generally with an undesired ambiguity function behavior. These complications can be partially mitigated, forcing the solution to be similar to a known code $\mathbf{c}_0$ that shares some advantageous properties.

In summary, leveraging the aforementioned guidelines, the design of constant-envelope signals complying with spectral restrictions can be formulated as the following non-convex optimization problem:

$$\mathcal{P}_p \begin{cases} \underset{\mathbf{c}\in\mathbb{C}^N}{\text{maximize}} & \overline{\text{SINR}}(\mathbf{c}), \\ \text{subject to} & \mathbf{c}^\dagger \boldsymbol{R}_I^k \mathbf{c} \leq E_I^k, \quad k = 1, \dots, K, \\ & \|\mathbf{c}\|^2 \leq 1, \\ & |c_i| = |c_j|, \quad i \neq j, \\ & \left\| \frac{\mathbf{c}}{\|\mathbf{c}\|} - \mathbf{c}_0 \right\|_\infty \leq \frac{\epsilon}{\sqrt{N}}, \\ & \arg(c_i) \in \Omega_p, \quad i = 1, \dots, N, \end{cases} \tag{7.10}$$

where $p \in \{\infty, M\}$ is either an integer number or ∞ and specifies the code alphabet size, $\Omega_\infty = [-\pi, \pi)$, and

$$\overline{\text{SINR}}(\mathbf{c}) = \mathbf{c}^\dagger \mathbf{R} \mathbf{c} \tag{7.11}$$

is the normalized SINR.

7.3 Radar Waveform Design Procedure

In this section, an iterative design procedure is developed to devise optimized constant-envelope sequences leveraging the CD paradigm [Wright, 2015]. By invoking elementary function, each non-convex optimization subproblem is solved in closed form, and the monotonic improvement of the SINR is ensured along the iterations. As a first step toward this goal, let us re-parameterize the optimization vector $\mathbf{c}$ as

$$\mathbf{c} = \bar{\mathbf{c}} \odot \mathbf{c}_0, \tag{7.12}$$

where $\bar{\mathbf{c}} = \sqrt{x_{N+1}}\left[e^{jx_1}, \dots, e^{jx_N}\right]^T \in \mathbb{C}^N$ with $x_i, i \in \mathcal{N}$ and x_{N+1} denoting the code phases and amplitude level, respectively, ($0 \leq x_{N+1} \leq 1$ for constant envelope and energy constraints). The similarity constraint is equivalent to $\Re\left\{e^{j\arg(x_i)}\right\} \geq 1 - \epsilon^2/2, i \in \mathcal{N}$. Hence, denoting by $\delta = \arccos(1 - \epsilon^2/2)$,

- $x_i \in \Psi_\infty = [-\delta, \delta], i \in \mathcal{N}$ for the continuous case [Aubry et al., 2020].
- $x_i \in \Psi_M = \frac{2\pi}{M}\left\{\alpha_\epsilon, \alpha_\epsilon + 1, \dots, \alpha_\epsilon + \omega_\epsilon - 1\right\}, i \in \mathcal{N}$ with

$$\alpha_\epsilon = -\left\lfloor \frac{M\delta}{2\pi} \right\rfloor, \tag{7.13}$$

$$\omega_\epsilon = \begin{cases} 1 - 2\alpha_\epsilon, & \text{if } \epsilon \in [0, 2), \\ M, & \text{if } \epsilon = 2. \end{cases} \tag{7.14}$$

for the finite alphabet case [Yang et al., 2020]. The interested reader may also refer to Section 7.A.1.

Additionally, the spectral constraints can be recast as

$$\bar{\mathbf{c}}^{\dagger}\bar{\mathbf{R}}_I^k\bar{\mathbf{c}} \leq E_I^k, \quad k = 1, \dots, K, \tag{7.15}$$

where $\bar{\mathbf{R}}_I^k = \text{diag}\left(\mathbf{c}_0^*\right)\mathbf{R}_I^k\text{diag}\left(\mathbf{c}_0\right)$. Furthermore, the objective function in (7.11) can be expressed as

$$\overline{\text{SINR}}(\bar{\mathbf{c}}) = \bar{\mathbf{c}}^{\dagger}\bar{\mathbf{R}}\bar{\mathbf{c}} \tag{7.16}$$

with

$$\bar{\mathbf{R}} = \mathbf{diag}(\mathbf{c}_0^*)\mathbf{R}\mathbf{diag}(\mathbf{c}_0).$$

Hence, the radar waveform design problem with p either M or ∞ boils down to

$$\bar{\mathcal{P}}_p \begin{cases} \underset{\mathbf{x}}{\text{maximize}} & \overline{\text{SINR}}(\mathbf{x}), \\ \text{subject to} & \bar{\mathbf{c}} = \sqrt{x_{N+1}}\left[e^{jx_1}, e^{jx_2}, \dots, e^{jx_N}\right]^T, \\ & \bar{\mathbf{c}}^{\dagger}\bar{\mathbf{R}}_I^k\bar{\mathbf{c}} \leq E_I^k, \quad k = 1, \dots, K, \\ & 0 \leq x_{N+1} \leq 1, \\ & x_i \in \Psi_p, \quad i \in \mathcal{N}, \end{cases} \tag{7.17}$$

where $\mathbf{x} = \left[x_1, \dots, x_{N+1}\right]^T \in \mathbb{R}^{N+1}$ is the re-parameterized optimization vector. Last, with a slight abuse of notation, $\overline{\text{SINR}}\,(\mathbf{x})$ indicates the function (7.16) evaluated at $\bar{\mathbf{c}} = \sqrt{x_{N+1}}\left[e^{jx_1}, \dots, e^{jx_N}\right]^T$, i.e. $\overline{\text{SINR}}\,(\mathbf{x}) = \overline{\text{SINR}}(\mathbf{c})|\mathbf{c} = \frac{\sqrt{x_{N+1}}}{N}[e^{j(x_1+\arg(c_{0,1}))}, \dots, e^{j(x_N+\arg(c_{0,N}))}]^T$.

7.3.1 Code Phase Optimization

In this section, iterative algorithms that exploit the CD maximization paradigm are introduced to get optimized solutions with some quality guarantee to the NP-hard Problem $\mathcal{P}_p$ [Aubry et al., 2020; Yang et al., 2020]. The idea is to alternate the maximization of the objective $\overline{\text{SINR}}(\mathbf{x})$ among the entries of $\mathbf{x}$, namely optimizing one variable at time while keeping others fixed. This is tantamount to solving appropriate univariate optimization problems in a loop [Wright, 2015]. With reference to $\bar{\mathcal{P}}_p$, at each iteration a specific entry of $\mathbf{x}$ is selected as the optimization variable. This leads to the following problem at step $n+1$:

$$\bar{\mathcal{P}}_p^{d,\mathbf{x}^{(n)}} \begin{cases} \text{maximize}_{x_d} & \widehat{\text{SINR}}\left(x_d; \mathbf{x}_{-d}^{(n)}\right), \\ \text{subject to} & f_k\left(x_d; \mathbf{x}_{-d}^{(n)}\right) \leq E_I^k, \quad k = 1, \dots, K, \\ & x_d \in \Theta_d. \end{cases} \tag{7.18}$$

where

- $\mathbf{x}^{(n)}$ is the optimized vector at step n, x_d is the variable to optimize at step $n+1$, and $\mathbf{x}_{-d}^{(n)} = \left[x_1^{(n)}, \dots, x_{d-1}^{(n)}, x_{d+1}^{(n)}, \dots, x_{N+1}^{(n)}\right]^T \in \mathbb{R}^N$;

  $$\widehat{\text{SINR}}\left(x_d; \mathbf{x}_{-d}^{(n)}\right) = \overline{\text{SINR}}\left(x_1^{(n)}, \dots, x_{d-1}^{(n)}, x_d, x_{d+1}^{(n)}, \dots, x_{N+1}^{(n)}\right)$$

  is the normalized SINR objective function restricted to x_d only with other parameters fixed to the value of the previous step;

-

$$f_k\left(x_d;\mathbf{x}_{-d}^{(n)}\right)=\begin{cases}\left(\bar{z}_{k,d}^{(n)}+2\Re\left\{z_{k,d}^{(n)}e^{jx_d}\right\}\right)x_{N+1}^{(n)}, & d\in\mathcal{N}\\ x_d p_k^{(n)}, & d=N+1\end{cases} \tag{7.19}$$

is the restriction of the kth spectral constraint (7.15) to x_d only while keeping the other variables fixed (see Section 7.A.2 for the formal definition of the parameters $\bar{z}_{k,d}^{(n)}$, $z_{k,d}^{(n)}$, $p_k^{(n)}$ and technical details);

-

$$\Theta_d=\begin{cases}\Psi_p, & d\in\mathcal{N},\\ [0,1], & d=N+1.\end{cases} \tag{7.20}$$

Thus, denoting by $x_{d,n+1}^{\star}$ the optimal solution to $\bar{\mathcal{P}}_p^{d,\mathbf{x}^{(n)}}$, the optimized radar code parameters vector at step $n+1$ is

$$\mathbf{x}^{(n+1)}=[x_1^{(n)},\dots,x_{d-1}^{(n)},x_{d,n+1}^{\star},x_{d+1}^{(n)},\dots,x_{N+1}^{(n)}]^T.$$

As a result, starting from a feasible solution $\mathbf{x}^{(0)}$, a sequence $\mathbf{x}^{(1)},\mathbf{x}^{(2)},\mathbf{x}^{(3)},\dots$ of radar codes can be obtained iteratively. A summary of the proposed approach is given in Algorithm 1.

Remarkably, being $\mathbf{x}^{(n)}$ feasible to $\bar{\mathcal{P}}$ for any n,

$$\overline{\text{SINR}}^{(n)}=\widehat{\text{SINR}}\left(x_d^{(n)};\mathbf{x}_{-d}^{(n)}\right)\leq\widehat{\text{SINR}}\left(x_{d,n+1}^{\star};\mathbf{x}_{-d}^{(n)}\right)=\overline{\text{SINR}}^{(n+1)} \tag{7.21}$$

implying that the objective function monotonically increases along the iterations; hence, because the objective function is bounded (from above), it is possible to claim the convergence of the sequence of objective values. It is also worth pointing out that the maximum block improvement (MBI) updating rule [Chen et al., 2012] can be used in place of the cyclic one.

Algorithm 1 Phase code design with spectral compatibility requirements

Input: Initial feasible code parameters $\mathbf{x}^{(0)}$, $p\in\{\infty,M\}$, and the minimum required improvement $\bar{\epsilon}$.

Output: Optimized solution $\mathbf{x}^{\star}$;

1. **Initialization**.
   - Set $d:=1$ and $n:=0$ as well as compute the initial objective value $\overline{\text{SINR}}^{(n)}=\overline{\text{SINR}}\left(\mathbf{x}^{(0)}\right)$.
2. **Improvement**.
   - Solve $\bar{\mathcal{P}}_p^{d,\mathbf{x}^{(n)}}$ obtaining $x_{d,n+1}^{\star}$.
   - Set $n:=n+1$,
     $\mathbf{x}^{(n)}=\left[x_1^{(n-1)},\dots,x_{d-1}^{(n-1)},x_{d,n}^{\star},x_{d+1}^{(n-1)},\dots,x_{N+1}^{(n-1)}\right]^T$,
     $\overline{\text{SINR}}^{(n)}=\overline{\text{SINR}}(\mathbf{x}^{(n)})$.
3. **Stopping Criterion**.
   - If mod$(n,N+1)=0$ and $\left|\overline{\text{SINR}}^{(n)}-\overline{\text{SINR}}^{(n-N-1)}\right|<\bar{\epsilon}$, stop. Otherwise, update d, i.e. $d=$ mod$(d,N+1)+1$, and go to the step 2.
4. **Output**.
   - Set $\mathbf{x}^{\star}=\mathbf{x}^{(n)}$.

7.3.2 Solution Technique for $\bar{\mathcal{P}}_p^{d,\mathbf{x}^{(n)}}$

In the following, efficient procedures to tackle $\bar{\mathcal{P}}_p^{d,\mathbf{x}^{(n)}}, d \in \{1, \dots, N+1\}$ are developed. From an optimization theory point of view, this is the main technical contribution of this chapter. Besides, some heuristic procedures are proposed, to perform a bespoke initialization.

7.3.2.1 Code Amplitude Optimization

Let us start with the study of Problem $\bar{\mathcal{P}}_p^{d,\mathbf{x}^{(n)}}$ as $d = N+1$. In this case, it is easy to show that (7.18), regardless of p, reduces to the following linear programming problem with a positive objective slope

$$\begin{cases} \underset{x_{N+1}}{\text{maximize}} & x_{N+1} P^{(n)}, \\ \text{subject to} & x_{N+1}\, p_k^{(n)} \leq E_I^k, \quad k = 1, \dots, K, \\ & 0 \leq x_{N+1} \leq 1, \end{cases} \tag{7.22}$$

where $P^{(n)} = \tilde{\mathbf{c}}^{(n)\dagger}\, \mathbf{diag}(\mathbf{c}_0^*) \mathbf{R} \mathbf{diag}(\mathbf{c}_0) \tilde{\mathbf{c}}^{(n)}$ with

$$\tilde{\mathbf{c}}^{(n)} = \left[e^{jx_1^{(n)}}, e^{jx_2^{(n)}}, \dots, e^{jx_N^{(n)}} \right]^T.$$

Hence, the optimal solution is just the highest feasible value, which is given by

$$x_{N+1,n+1}^{\star} = \min\left(\min_{k=1,\dots,K} \left(\frac{E_I^k}{p_k^{(n)}} \right), 1 \right)$$

with $p_k^{(n)}$ the parameter involved in the definition of $f_k(x_{N+1}; \mathbf{x}_{-(N+1)}^{(n)})$, specified in Section 7.A.2.

7.3.2.2 Code Phase Optimization

With reference to $1 \leq d \leq N$, $\bar{\mathcal{P}}_p^{d,\mathbf{x}^{(n+1)}}$ is equivalent to

$$\mathcal{P}_p^{x_d^{(n)}} \begin{cases} \underset{x_d}{\text{maximize}} & \chi\left(x_d\right), \\ \text{subject to} & \Re\{z_{k,d}^{(n)} e^{jx_d}\} \leq \bar{c}_{k,d}^{(n)}, \quad k = 1, \dots, K \\ & x_d \in \Psi_p \end{cases} \tag{7.23}$$

with $\chi\left(x_d\right) = \Re\{\bar{b}_d^{(n)} e^{jx_d}\}$, $\bar{b}_d^{(n)} = \sum_{i \neq d}^{N} e^{-jx_i^{(n)}} \bar{\mathbf{R}}(i,d)$, and $\bar{c}_{k,d}^{(n)} = \frac{E_I^k}{2x_{N+1}^{(n)}} - \frac{\bar{z}_{k,d}^{(n)}}{2} \in \mathbb{R}$.

The monotonicity properties of the objective function are analyzed in Section 7.A.3 along with the derivation of global maximizer and minimizer $\phi_{g,d}^{(n)}$ and $\phi_{s,d}^{(n)}$, respectively. Besides, as shown in Section 7.A.4, the feasible set[1] $\bar{\mathcal{F}}_p$ of Problem $\bar{\mathcal{P}}_p^{d,\mathbf{x}^{(n)}}, p \in \{\infty, M\}, d \in \mathcal{N}$, can be expressed as union of appropriate disjoint compact sets, i.e.

$$\bar{\mathcal{F}}_p = \begin{cases} \bigcup_{i=1}^{K_\infty} \left[\hat{l}_i^\infty, \hat{u}_i^\infty \right], & \text{if } p = \infty, \\ \bigcup_{t=1}^{K_M} \{ \hat{l}_t^M, \hat{l}_t^M + \frac{2\pi}{M}, \dots, \hat{u}_t^M \}, & \text{if } p = M, \end{cases} \tag{7.24}$$

where (see Section 7.A.4 for details)

1 To avoid unnecessary complications, $\epsilon < 2$ is assumed.

- K_∞ is the number of disjoint intervals that compose $\bar{\mathcal{F}}_\infty$, with $\hat{l}_i^\infty, \hat{u}_i^\infty \in \Psi_M, i = 1, \dots, K_\infty$ depending on the specific instance of Problem $\mathcal{P}_\infty^{x_d}$ and such that $\hat{l}_i^\infty \le \hat{u}_i^\infty < \hat{l}_{i+1}^\infty, i = 1, \dots, K_\infty - 1$.
- $K_M \le K_\infty$ is the number of disjoint discrete sets that form $\bar{\mathcal{F}}_M$, with $\hat{l}_t^M, \hat{u}_t^M \in \Psi_M, t = 1, \dots, K_M$, and $\hat{l}_t^M \le \hat{u}_t^M < \hat{l}_{t+1}^M, t = 1, \dots, K_M - 1$.
- $\hat{l}_i^\infty, \hat{u}_i^\infty, i = 1, \dots, K_\infty$ can be derived relying on De Morgan law and "union-find" algorithm (see Section 7.A.4).
- $\hat{l}_t^M, \hat{u}_t^M, t = 1, \dots, K_M$ can be derived performing a bespoke intersection between $\bar{\mathcal{F}}_\infty$ and Ψ_M.

The following proposition paves the way to the solution of the non-convex optimization problem $\mathcal{P}_\infty^{x_d^{(n)}}$.

Proposition 7.1 An optimal solution $x_d^\star$ to $\mathcal{P}_\infty^{x_d^{(n)}}$ can be evaluated in closed form via the computation of elementary function as follows:

- if the unconstrained global optimal solution $\phi_{g,d}^{(n)}$ is feasible, i.e. $\phi_{g,d}^{(n)} \in \bar{\mathcal{F}}_\infty$, the optimal solution is $x_d^\star = \phi_{g,d}^{(n)}$;
- elseif $\phi_{g,d}^{(n)} \notin \left[\hat{l}_1^\infty, \hat{u}_{K_\infty}^\infty\right]$, the optimal solution is among $\{\hat{l}_1^\infty, \hat{u}_{K_\infty}^\infty\}$, i.e.

$$x_d^\star = \arg \underset{x_d \in \{\hat{l}_1^\infty, \hat{u}_{K_\infty}^\infty\}}{\text{maximize}} \chi\left(x_d\right); \tag{7.25}$$

- else, denoting by $r^\star = \arg \text{maximize}_{\hat{u}_r^\infty < \phi_{g,d}^{(n)}} r$, the optimal solution is

$$x_d^\star = \arg \underset{x_d \in \{\hat{l}_1^\infty, \hat{u}_{r^\star}^\infty, \hat{l}_{r^\star+1}^\infty, \hat{u}_{K_\infty}^\infty\}}{\text{maximize}} \chi\left(x_d\right). \tag{7.26}$$

Proof: See the Section 7.A.5. ■

According to Section 7.A.5, as long as $\phi_{g,d}^{(n)}$ does not belong to one of the closed and bounded interval $[\hat{l}_i^\infty, \hat{u}_i^\infty], i = 1, \dots, K_\infty$, the global optimal solution is one of the intervals extremes. As a consequence, embedding $\bar{\mathcal{F}}_M$ in an appropriate union of closed intervals, the following corollary holds true.

Corollary 7.1 *An optimal solution $x_d^\star$ to $\mathcal{P}_M^{x_d}$ can be derived as follows:*

- *if there exists an index $h^\star \in \{1, \dots, K_M\}$ satisfying $\hat{l}_{h^\star}^M \le \phi_{g,d}^{(n)} \le \hat{u}_{h^\star}^M$, then*

$$x_d^\star = \frac{2\pi}{M} \left\lfloor \frac{M\phi_{g,d}^{(n)}}{2\pi} \right\rceil; \tag{7.27}$$

- *elseif $\phi_{g,d}^{(n)} \notin \left[\hat{l}_1^M, \hat{u}_{K_M}^M\right]$, $x_d^\star = \arg$ maximize$_{x_d \in \{\hat{l}_1^M, \hat{u}_{K_M}^M\}} \chi\left(x_d\right)$;*
- *else, $x_d^\star = \arg$ maximize$_{x_d \in \{\hat{l}_1^M, \hat{u}_{r^\star}^M, \hat{l}_{r^\star+1}^M, \hat{u}_{K_M}^M\}} \chi\left(x_d\right)$, where $r^\star = \arg$ maximize$_{\hat{u}_r^M < \phi_{g,d}^{(n)}} r$.*

Proof: See the Section 7.A.6. ■

A summary of the complete procedure to optimize x_d, $d \in \mathcal{N}$, is provided in Algorithm 2.

Algorithm 2 Phase code entry optimization

Input: Initial code vector $\mathbf{x}^{(n)}$, code entry d, phase cardinality p;
Output: Optimal solution $x_d^\star$;

1. Compute the feasible set $\bar{\mathcal{F}}_p$ as given in (7.24)
2. **If** $\phi_{g,d}^{(n)} \in [\hat{l}_i^p, \hat{u}_i^p], i \in \{1, \dots, K_p\}$,
   - **if** $p = \infty, x_d^\star = \phi_{g,d}^{(n)}$;
   - **else**, $x_d^\star = \frac{2\pi}{M} \lceil \frac{M\phi_{g,d}^{(n)}}{2\pi} \rfloor$.
3. **Elseif** $\phi_{g,d}^{(n)} \notin \left[\hat{l}_1^p, \hat{u}_{K_p}^p\right]$,
$$x_d^\star = \arg \underset{x_d \in \{\hat{l}_1^p, \hat{u}_{K_p}^p\}}{\text{maximize}} \, \chi\left(x_d\right).$$
4. **Else**,
$$x_d^\star = \arg \underset{x_d \in \{\hat{l}_1^p, \hat{u}_{r^\star}^p, \hat{l}_{r^\star+1}^p, \hat{u}_{K_p}^p\}}{\text{maximize}} \, \chi\left(x_d\right), \tag{7.28}$$
   where $r^\star = \arg \underset{\hat{u}_r^p < \phi_{g,d}^{(n)}}{\text{maximize}} \, r$.
5. **Output** $x_d^\star$.

The main idea behind Algorithm 2 is the joint exploitation of both the objective function monotonicities and the feasible set structure, to come up with a closed-form optimal solution to the optimization problem at hand.

Remark 7.1 To establish the computational complexity of Algorithm 2 (computed the problem parameters, i.e. $\bar{b}_d^{(n)}, \bar{z}_{k,d}^{(n)}, z_{k,d}^{(n)}, p_k^{(n)}$), it is necessary to observe that the main actions in its implementation are:

1. evaluation of $\bar{\mathcal{F}}_p$;
2. computation of the optimal phase.

As to the former step, note that $\bar{\mathcal{F}}_\infty$ determination can be evaluated with a computational complexity of $O(K \log(K))$ leveraging [Aubry et al., 2020]. Besides, according to Section 7.A.4, $\bar{\mathcal{F}}_M$ can be accomplished by the discretization of $\hat{l}_i^\infty$ and $\hat{u}_i^\infty, i = 1, \dots, K_\infty$ and discarding the empty sets with an overall computational complexity of $O(K)$. With reference to the second item, the unconstrained optimal solution $\phi_{g,d}^{(n)}$ can be evaluated via elementary function using $\bar{b}_d^{(n)}$ with the computational complexity of $O(1)$, while the optimal solution to the Problem $\mathcal{P}_\infty^{x_d^{(n)}}$ can be accomplished with a complexity at most of $O(K)$ placing $\phi_{g,d}^{(n)}$ within $\{\hat{l}_1^p, \hat{u}_1^p, \dots, \hat{l}_{K_p}^p, \hat{u}_{K_p}^p\}$ ($\hat{l}_i^p, \hat{u}_i^p, i = 1, \dots, K_p, p \in \{\infty, M\}$are obtained by the output of the previous step) in the correct sorted location. Since Problem $\mathcal{P}_p^{x_d^{(n)}}$ parameters can be evaluated with a complexity proportional to KN, the overall computational burden of Algorithm 1 is $O((\log(K) + N)K\bar{N})$, where $\bar{N}$ is the number of iterations. Indeed, the parameters of Problem $\mathcal{P}_p^{x_d^{(n)}}$, $2 \le d \le N$ can be efficiently computed updating those of Problem $\hat{\mathcal{P}}_{d-1,\mathbf{x}^{(n-1)}}^p$. In this respect, note that it is enough to cancel out from $\bar{z}_{k,d-1}^{(n-1)}$, $k = 1, \dots, K$ the terms related to the variable $x_d^{(n-1)}$ and add those involving $x_{d-1}^{(n)}$, which can be accomplished with a complexity of $O(N)$. Similar considerations hold true with respect to $d = 1$ and $d = N + 1$.

Before concluding this section, note that Algorithm 1 requires an appropriate initialization; to this end, two heuristic procedures are proposed in the Section 7.3.3.

7.3.3 Heuristic Methods for Algorithm Initialization

The solution provided by Algorithm 1 depends on the initial sequence. Thus, the development of some heuristic procedures ensuring high-quality starting points is valuable. To this end, in Section 7.3.3.1 an algorithm based on the semidefinite relaxation (SDR) and randomization paradigm is proposed to synthesize optimized and feasible phase-only codes. Moreover, in Subsection 7.3.3.2, a radar code design aimed at minimizing the energy transmitted on both licensed and jammed bandwidths, so as to reduce both produced and perceived interference, while fulfilling the constant modulus and similarity constraints, is introduced. In particular, an iterative procedure based on alternating optimization is devised to handle the resulting optimization problem.

7.3.3.1 Relaxation and Randomization-based Approach

Problem $\mathcal{P}_p$ can be equivalently expressed in matrix form as

$$\mathcal{P}_1^p \begin{cases} \underset{\mathbf{C},\mathbf{c}}{\text{maximize}} & \text{tr}\,(\mathbf{CR}), \\ \text{subject to} & \text{tr}\left(\mathbf{C}\bar{\mathbf{R}}_k\right) \le 1, \quad k = 1, \dots, K, \\ & \text{tr}\,(\mathbf{C}) \le 1, \\ & \mathbf{C}_{i,i} = \mathbf{C}_{h,h}, \quad (i,h) \in \mathcal{N}^2, \\ & \left\| \frac{\mathbf{c}}{\|\mathbf{c}\|} - \mathbf{c}_0 \right\|_\infty \le \frac{\epsilon}{\sqrt{N}}, \\ & \mathbf{C} = \mathbf{c}\mathbf{c}^\dagger \succeq \mathbf{0}, \\ & \arg(c_i) \in \Omega_p, \quad i = 1, \dots, N, \end{cases} \tag{7.29}$$

where $\bar{\mathbf{R}}_k = \frac{\mathbf{R}_I^k}{E_I^k}$ and the non-convexity is completely confined in the last constraint. The SDR of Problem $\bar{\mathcal{P}}_1^p$, obtained dropping the rank-one, discrete phase, and the similarity constraints, is

$$\bar{\mathcal{P}}_2^p \begin{cases} \underset{\mathbf{C}}{\text{maximize}} & \text{tr}\,(\mathbf{CR}), \\ \text{subject to} & \text{tr}\left(\mathbf{C}\bar{\mathbf{R}}_k\right) \le 1, \quad k = 1, \dots, K, \\ & \text{tr}\,(\mathbf{C}) \le 1, \\ & \mathbf{C}_{i,i} = \mathbf{C}_{h,h}, \quad (i,h) \in \mathcal{N}^2, \\ & \mathbf{C} \succeq \mathbf{0}. \end{cases} \tag{7.30}$$

Problem $\bar{\mathcal{P}}_2^p$ is a semidefinite programming (SDP) convex problem, and its optimal solution $\mathbf{C}^\star$ can be obtained in polynomial time with arbitrary precision. Nevertheless, $\mathbf{C}^\star$ could not be feasible to $\bar{\mathcal{P}}_1^p$. Hence, to synthesize good-quality approximate solutions to $\mathcal{P}_p$, a randomization procedure (inspired to [De Maio et al., 2009]) is employed, which capitalizes on $\mathbf{C}^\star$ to build a feasible solution to $\bar{\mathcal{P}}_1^p$. In Algorithm 3, the relaxation and randomization procedure is described. Therein, H represents the number of randomizations, i.e. random trials involved in the procedure. Specifically, the parameter H appearing in steps 4 and 5 allows to improve the approximation quality. In fact, the randomized feasible solution yielding the largest objective value will be chosen as the approximate solution. Interestingly, Algorithm 3 provides waveforms with quality guarantee as the following approximation bound holds true:

$$v(\bar{\mathcal{P}}_1^p) \ge \mathbb{E}\left[\frac{1}{\text{maximize}\left(N, \text{maximize}_{k=1,\dots,K}\left(\tilde{\mathbf{c}}^\dagger \bar{\mathbf{R}}_k \tilde{\mathbf{c}}\right)\right)} \tilde{\mathbf{c}}^\dagger \mathbf{R} \tilde{\mathbf{c}} \right]$$

$$\geq \frac{1}{N \text{ maximize}\left(1, \text{maximize}_{k=1,\ldots,K} \lambda_{\max}\left(\bar{\mathbf{R}}_k\right)\right)} \mathbb{E}\left[\tilde{\mathbf{c}}^\dagger \mathbf{R} \tilde{\mathbf{c}}\right]$$
$$\geq \frac{\text{tr}\left(N\mathbf{R}\mathbf{C}^\star\right) R_c^p}{N \text{ maximize}\left(1, \text{maximize}_{k=1,\ldots,K} \lambda_{\max}\left(\bar{\mathbf{R}}_k\right)\right)}$$
$$= \frac{v(\bar{\mathcal{P}}_2^p) R_c^p}{\text{maximize}\left(1, \text{maximize}_{k=1,\ldots,K} \lambda_{\max}\left(\bar{\mathbf{R}}_k\right)\right)},$$

where the first inequality is due to the sub-optimality of the generated codes, the second inequality stems from the definition of maximum eigenvalue, and the last inequality follows from De Maio et al. [2009, Theorem 4.1] (therein the energy code is constrained to be N) with

$$R_c^p = \begin{cases} \frac{\pi(1-\cos(\delta))}{2(2\pi-\delta)^2}, & \text{if } p = \infty, \\ \frac{\sin^2\left(\pi \frac{1}{\omega_\epsilon}\right)\sin^2\left(\pi \frac{\omega_\epsilon}{M}\right)}{4\pi \sin^2\left(\pi\left(\frac{1}{\omega_\epsilon} - \frac{1}{M}\right)\right)}, & \text{else.} \end{cases}$$

Algorithm 3 Randomization algorithm to devise radar waveforms fulfilling spectrum requirements

---

Input: $\mathbf{R}$, $\mathbf{c}_0$, $\{\mathbf{R}_I^k\}_{k=1}^K$, $\{E_I^k\}_{k=1}^K$, p;
Output: A randomized approximate solution $\mathbf{c}_1^\star$ to $\bar{\mathcal{P}}_1^p$.

1. Solve the SDP problem $\bar{\mathcal{P}}_2^p$ and denote by $\mathbf{C}^\star$ an optimal solution.
2. Generate H random vectors $\mathbf{y}_i \in \mathbb{C}^N$, $i = 1, \ldots, H$ from the zero-mean circularly symmetric Gaussian distribution with covariance $\mathbf{C}^\star \odot \bar{\mathbf{c}}_0^p \bar{\mathbf{c}}_0^{p\dagger}$, $\bar{c}_0^p(n) = \begin{cases} e^{-j(\overline{\arg}(c_0(n))-\delta)}, & \text{if } p = \infty, \\ e^{-j\frac{2\pi}{M}\left(\frac{M\overline{\arg}(c_0(n))}{2\pi}+\alpha_\epsilon\right)}, & \text{if } p = M; \end{cases}$
3. Assign to each $\tilde{c}_i^p(n) = \bar{c}_0^{p*}(n)\sigma^p\left(y_i(n)\right)$, $n = 1, \ldots, N$, $i = 1, \ldots, H$, where for $x \in \mathbb{C}$, $\sigma^p(x) = \begin{cases} \exp\left(j\frac{\overline{\arg}(x)}{\pi}\delta\right), & \text{if } p = \infty, \\ \exp\left(j\lfloor\frac{\omega_\epsilon \overline{\arg}(x)}{2\pi}\rfloor\frac{2\pi}{M}\right), & \text{if } p = M; \end{cases}$
4. Define $\mathbf{c}_i = \frac{\tilde{\mathbf{c}}_i}{\sqrt{\text{maximize}\left(N, \text{maximize}_{k=1,\ldots,K}\left(\tilde{\mathbf{c}}_i^\dagger \bar{\mathbf{R}}_k \tilde{\mathbf{c}}_i\right)\right)}}$, $i = 1, \ldots, H$.
5. Evaluate $\text{SINR}_i = \mathbf{c}_i^\dagger \mathbf{M}^{-1} \mathbf{c}_i$, $i = 1, \ldots, H$, and pick up the maximal value over $\{\text{SINR}_1, \ldots, \text{SINR}_H\}$, and the index achieving this value say $i^\star$.
6. **Output.**
   - Set $\mathbf{c}_1^\star = \mathbf{c}_{i^\star}$.

7.3.3.2 Free-Band Capitalization Code Design

The idea pursued in this subsection is to synthesize a radar signal that reduces as much as possible the transmitted energy in both the shared and the jammed bands while fulfilling the constant modulus and the similarity constraints. Specifically, the goal is to devise waveforms that lie in the subspace "orthogonal" to both the shared and jammed bands so as to maximize the energy radiated in the "free-bands." To this end, let

$$\boldsymbol{\Sigma} = \sum_{k=1}^{K} \frac{\mathbf{R}_I^k}{\Delta f_k} + \sum_{h=1}^{K_J} \mathbf{R}_{J,h} = \bar{\mathbf{U}}\boldsymbol{\Lambda}\bar{\mathbf{U}}^\dagger, \tag{7.31}$$

with K_J the number of jammers, $\mathbf{R}_{J,h}$ the normalized covariance matrix associated with the hth unlicensed source,[2] and $\bar{\mathbf{U}}$ a unitary matrix containing the eigenvectors of $\boldsymbol{\Sigma}$. To proceed further, let $\mathbf{U} \in \mathbb{C}^{N,m_1}$ be the sub-matrix of $\bar{\mathbf{U}}$ obtained extracting the eigenvectors whose eigenvalues are lower than or equal to a certain fraction, β say, of the maximum eigenvalue of $\boldsymbol{\Sigma}$. Hence, aimed at minimizing the code energy spill-over outside the subspace $\mathbf{U}$ the code design problem can be formulated as

$$\bar{\mathcal{P}}_3^p \begin{cases} \min\limits_{\mathbf{c},\mathbf{y}} & \|\mathbf{c} - \mathbf{U}\mathbf{y}\|^2, \\ \text{subject to} & \left\| \frac{\mathbf{c}}{\|\mathbf{c}\|} - \mathbf{c}_0 \right\|_\infty \leq \frac{\epsilon}{\sqrt{N}}, \\ & |c_i|^2 = \frac{1}{N}, \quad i = 1, \dots, N, \\ & \arg(c_i) \in \Omega_p \quad i =, \dots, N, \end{cases} \tag{7.32}$$

where the auxiliary variable $\mathbf{y} \in \mathbb{C}^{m_1}$ controls the representation of $\mathbf{c}$ in the subspace $\mathbf{U}$. To handle Problem $\bar{\mathcal{P}}_3^p$ the alternating optimization paradigm is exploited. The resulting optimization procedure is summarized in Algorithm 4.

Finally, the initial code to Algorithm 1 is obtained as

$$\mathbf{c}_2^\star = \frac{\bar{\mathbf{c}}^\star}{\sqrt{\text{maximize}\left(1, \text{maximize}\limits_{k=1,\dots,K}\left(\bar{\mathbf{c}}^{\star\dagger}\bar{\mathbf{R}}^k\bar{\mathbf{c}}^\star\right)\right)}}.$$

Algorithm 4 Free-band capitalization code design

Input: $\mathbf{c}_0$, $\mathbf{c}^{(0)}$, $\mathbf{U}$, $\bar{\epsilon}_1$, P;
Output: Optimized solution $\bar{\mathbf{c}}^\star$ to $\bar{\mathcal{P}}_3^p$.

1. **Initialization**.
   - Set $\mathbf{y}^{(0)} = \mathbf{U}^\dagger\mathbf{c}^{(0)}$, $\mathbf{z}^{(0)} = \mathbf{U}\mathbf{y}^{(0)}$.
   - Compute $\text{obj}^{(0)} = \|\mathbf{c}^{(0)} - \mathbf{U}\mathbf{y}^{(0)}\|^2$.
   - $n := 1$;
2. **Improvement**.
   - Compute $\mathbf{c}^{(n)}$, with $c_i^{(n)} = \frac{1}{\sqrt{N}} e^{j(\arg(c_{0,i}) + \psi_{p,i}^{(n)})}$, $i = 1, \dots, N$, where

   $$\psi_{p,i}^{(n)} = \begin{cases} \text{maximize}(\min(\arg(z_i^{(n-1)}c_{0,i}^*), \delta), -\delta), & \text{if } p = \infty, \\ \frac{2\pi}{M}\,\text{maximize}\left(\min\left(\left\lceil \frac{\arg(z_i^{(n-1)}c_{0,i}^*)M}{2\pi} \right\rfloor, m_{\epsilon,M}\right), \alpha_\epsilon\right), & \text{if } p = M; \end{cases}$$

   with $m_{\epsilon,M} = \begin{cases} M/2, & \text{if } M \text{ is even, and } \epsilon = 2 \\ \alpha_\epsilon + \omega_\epsilon - 1, & \text{else} \end{cases}$.
   - Compute $\mathbf{y}^{(n)} = \mathbf{U}^\dagger\mathbf{c}^{(n)}$ and $\mathbf{z}^{(n)} = \mathbf{U}\mathbf{y}^{(n)}$.
   - Set $\text{obj}^{(n)} = \|\mathbf{c}^{(n)} - \mathbf{z}^{(n)}\|^2$.
3. **Stopping Criterion**.
   - If $|\text{obj}^{(n-1)} - \text{obj}^{(n)}| < \bar{\epsilon}_1$, stop. Otherwise, $n := n + 1$ and go to the step 2.
4. **Output**.
   - Set $\bar{\mathbf{c}}^\star = \mathbf{c}^{(n)}$.

---

2 Exploiting electronic support measures (ESMs), a rough prediction of the jammers' attributes (such as their bandwidth and power) can be obtained.

7.4 Performance Analysis

In this section, the performance assessment of Algorithm 1 is conducted. Specifically, achievable detection performance, spectral shape, and autocorrelation features associated with the devised waveforms are analyzed. To this end, a radar whose baseband equivalent transmitted signal has a two-sided bandwidth of 900 kHz is considered. The disturbance (interference-plus-thermal noise) covariance matrix is described as

$$\mathbf{M} = \sigma_0 \mathbf{I} + \sum_{k=1}^{K} \frac{\sigma_{I,k}}{\Delta f_k} \mathbf{R}_I^k + \sum_{k=1}^{K_J} \sigma_{J,k} \mathbf{R}_{J,k}, \tag{7.33}$$

where $\sigma_0 = 0\,\text{dB}$ is the thermal noise level; $K = 6$ is the number of licensed emitters; for $K = 1, \ldots, 6$, $\sigma_{I,k}$ accounts for the energy of the kth coexisting telecommunications network operating over the normalized frequency interval $[f_2^k, f_1^k]$ with $\Delta f_k = f_2^k - f_1^k$ the related bandwidth extent; $K_J = 2$ is the number of active and unlicensed narrowband interference sources; $\sigma_{J,k}$, $k = 1, \ldots, K_J$, accounts for the energy of the kth active source ($\sigma_{J,i\ \text{dB}} = 40\,\text{dB}$, $i = 1,2$); $\mathbf{R}_{J,k}$, $k = 1, \ldots, K_J$, is the normalized disturbance covariance matrix of the kth active source, whose normalized carrier frequency and bandwidth are $f_{J,k}$ and $\Delta_{J,k}$ ($f_{J,1} = 0.60, f_{J,2} = 0.73$, and $\Delta_{J,k} = 4\ \ 10^{-3}$, $k = 1,2$).

As to the overlaid telecommunications systems spectrally coexisting with the radar of interest, the following normalized baseband equivalent radar stopbands are considered:

$$\Upsilon_1 = [0.060, 0.091], \Upsilon_2 = [0.121, 0.239], \Upsilon_3 = [0.2810.302],$$
$$\Upsilon_4 = [0.315, 0.375], \Upsilon_5 = [0.402, 0.465], \Upsilon_6 = [0.919, 0.969].$$

Hence, it is required that the radar probing waveform fulfills the spectral compatibility constraints corresponding to $E_{I\ \text{dB}}^k = 10\log_{10}(E_I^k) = -30\,\text{dB}$ for $k = 3, 6$, and $E_{I\ \text{dB}}^k = -20\,\text{dB}$ for the other frequency bands, assuming that the provided interference control guarantees a desired worst-case communication rate on each bandwidth. As to the reference code $\mathbf{c}_0$, a unitary energy linear frequency modulated (LFM) pulse of 200 μs and a chirp rate $K_s = (850 \times 10^3)/(200 \times 10^{-6})$ Hz/s is employed for the continuous phase. The M-quantized version of the above chirp is instead considered for the finite alphabet case, with cardinality M.

In Figure 7.4, the probability of detection (P_d) versus $|\alpha_T|^2$ (assuming a false alarm probability (P_{fa}) of 10^{-6}) is shown considering different similarity parameter values.[3] For comparison, it is

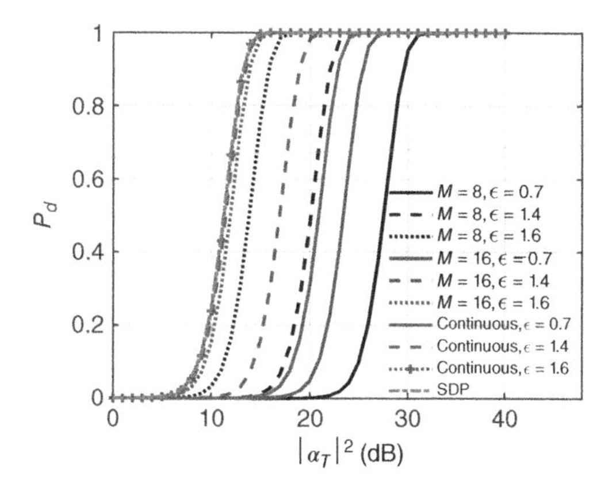

Figure 7.4 P_d versus $|\alpha_T|^2$ (in dB) of codes synthesized for different ϵ and phase code cardinalities, assuming $P_{fa} = 10^{-6}$.

3 Note that for any ϵ, Algorithm 1 (using $\bar{\epsilon} = 10^{-4}$) is run with three different initializations: the first is obtained via Algorithm 3 (with $H = 1000$), the second according to Algorithm 4 (using $\bar{\epsilon}_1 = 10^{-4}$, $\mathbf{c}^{(0)} = \mathbf{c}_1^\star$, and $\beta = 10^6$), and the third corresponds to the optimized code at the previous ϵ value (a grid of ϵ values, i.e. $\epsilon_1 \leq \epsilon_2 \leq \cdots \leq \epsilon_M$, is supposed). Hence, among the three synthesized codes, the one providing the higher $\overline{\text{SINR}}$ is picked up.

also plotted the P_d associated with the upper bound to the normalized SINR provided by the SDR of $\mathcal{P}_\infty$, i.e. $\bar{\mathcal{P}}_2^\infty$. The curves show that increasing the similarity parameter, a better performance is obtained as a result of the larger degrees of freedom available at the design stage. Interestingly, as ϵ is large enough, the upper bound performance is substantially achieved, highlighting the effectiveness of Algorithm 1. Besides, better detection probabilities can be achieved increasing the size of the alphabet code.

To further assess the impact of the similarity parameter on the radar detection performance, the achieved normalized SINR and the energy modulation required to comply with spectral compatibility requirements versus ϵ is reported for continuous and discrete phase codes ($M = 2, 4, 8, 16, 64$) in Figure 7.5. According to the results of Figure 7.4, the higher ϵ, the better the achieved SINR and the larger the energy radiated by radar without impairing licensed emitters regardless of phase cardinality. Besides, in agreement with the results of Figure 7.4, the finer the phase discretization, still the better the performance, with $\overline{\text{SINR}}$ and energy curves of the synthesized discrete phase codes closer and closer to that of the continuous-phase benchmark. Finally, it is worth pointing out that if $\epsilon = 0$ for the continuous phase or $\alpha_\epsilon = 0$ for the finite alphabet, i.e. $\varphi_i = 0, i \in \mathcal{N}$, the designed sequence coincides with a scaled version of reference code, with an energy modulation implemented to comply with the forced spectral constraints. Consequently, the similarity parameter should be carefully selected to balance the detection performance and waveform characteristics.

In Figure 7.6, the spectral behavior of the signals synthesized through Algorithm 1 (in terms of ESD versus the normalized frequency) is provided as a function of the similarity parameter. Specifically, Figure 7.6a refers to the continuous phase design, whereas Figure 7.6b is related to $M = 16$. Therein, the stopbands $\Upsilon_i, i = 1, \dots, K$, are shaded in light gray. Inspection of the figures reveals that, regardless of phase code cardinality, the ESD curve corresponding to $\epsilon = 0.1$ almost coincides with $\epsilon = 0$, which is a scaled version of the considered reference code, as a result of the energy modulation performed to comply with the forced spectral constraints along with the demanding similarity requirement. An improvement in the "useful" energy distribution is achieved as ϵ increases. Note also that possible slight differences in the curves can be experienced if different initialization sequences/schemes are considered.

The normalized ACFs for both the continuous phase and $M = 16$ cases are depicted in Figure 7.7 with $\epsilon \in \{0.1, 1, 1.9\}$. The sequence designed with $\epsilon = 0.1$, whose ACF is almost aligned with that of $\mathbf{c}_0$, experiences an energy modulation to satisfy the spectral constraints. By increasing ϵ, a better energy allocation within the spectrum can be derived at the cost of a possible degradation in ACF behavior, e.g. ISL or PSL.

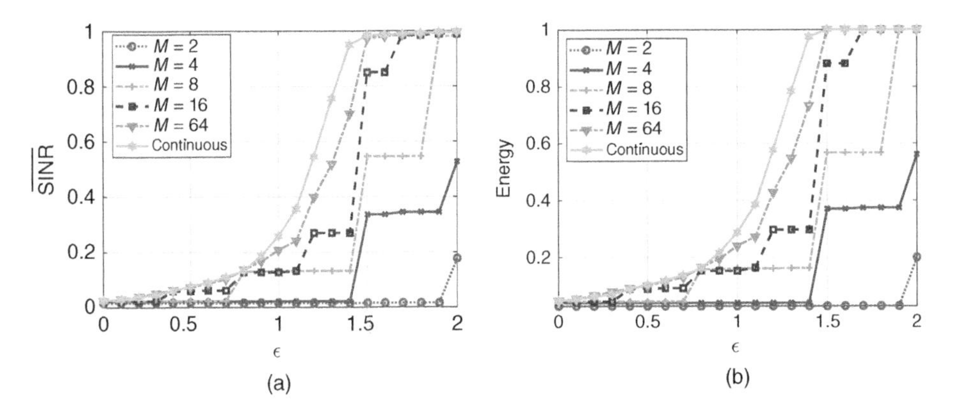

Figure 7.5 Achieved $\overline{\text{SINR}}$ and the transmitted energy versus ϵ of the continuous and discrete phase codes. (a) $\overline{\text{SINR}}$; (b) Transmitted energy.

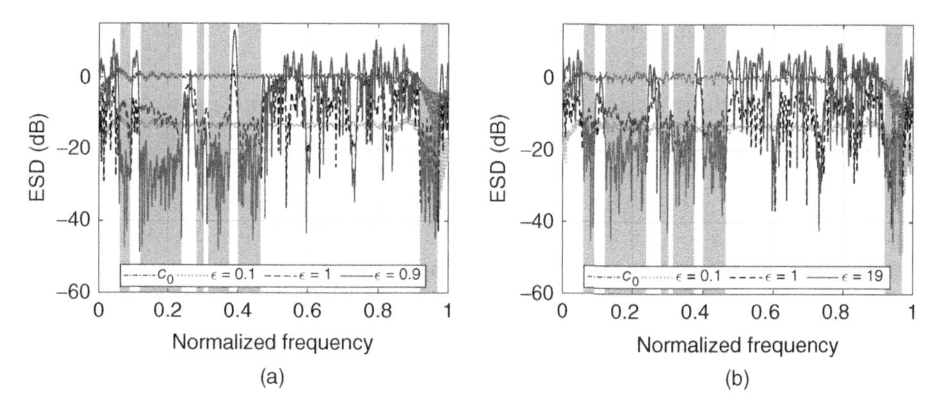

Figure 7.6 ESD versus normalized frequency of the phase codes designed for $\epsilon = 0, 0.1, 1, 1.9$. (a) Continuous phase; (b) $M = 16$.

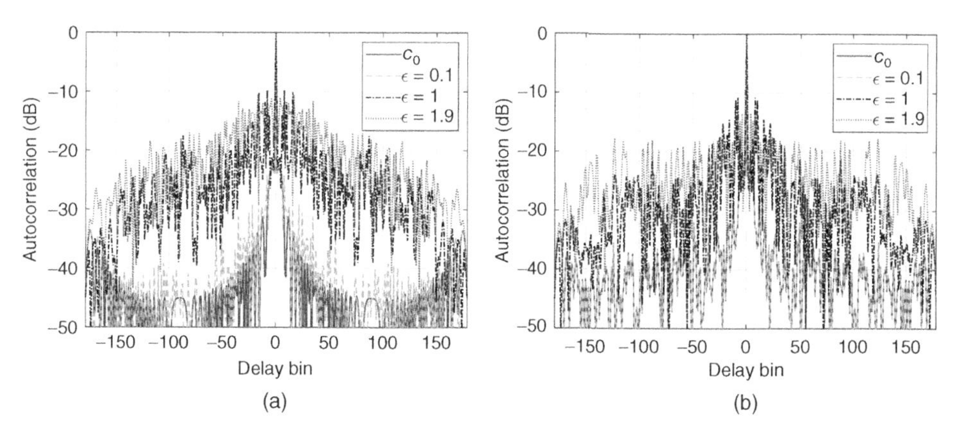

Figure 7.7 Normalized ACF of the phase codes designed for $\epsilon = 0, 0.1, 1, 1.9$. (a) Continuous phase; (b) $M = 16$.

To assess the convergence property and computational complexity of the developed waveform design scheme, Figure 7.8 depicts $\overline{\text{SINR}}$ versus the iteration number with $\epsilon = 1.5$, for the continuous phase and $M = 16$. Algorithm 3 is used in this case study to initialize Algorithm 1. As expected, the $\overline{\text{SINR}}$ monotonically increases along the iterations. Moreover, a larger number of iterations is required for the continuous phase code synthesis, where more degrees of freedom are available at the design stage that allow to achieve better performance.

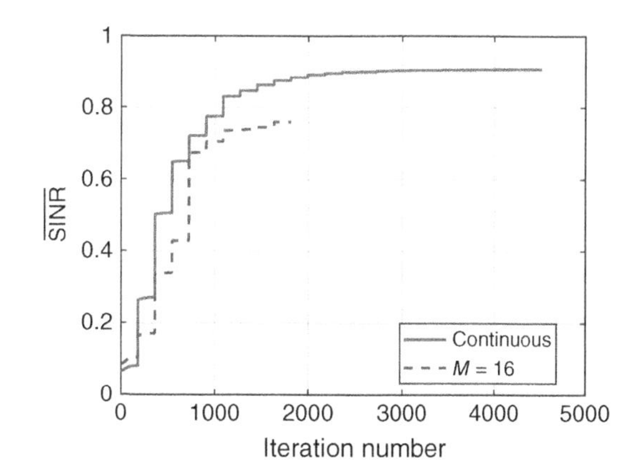

Figure 7.8 Achieved $\overline{\text{SINR}}$ versus the iteration number with $\epsilon = 1.5$ for the continuous phase and $M = 16$.

7.5 Conclusion

Constant-modulus radar waveform design in a spectrally crowded environment was explored, with reference to either continuous or finite alphabet phase codes. Specifically, multiple spectral compatibility constraints were enforced at the design stage to ensure a tight and local control on the interference energy induced on each shared/reserved bandwidth. Furthermore, to comply with the amplifier and digital-to-analog converter technologies, the design was focused on constant-envelope waveforms with a possible restriction on the phase code alphabet. Hence, by leveraging the CD framework, an iterative solution technique with ensured convergence properties was illustrated to synthesize optimized radar waveforms. Substantially, each step of the procedure requires the solution of a non-convex optimization problem whose global optimal point is provided in closed form. From an optimization theory point of view, this represents the main technical contribution of the chapter. Notably, the overall computational complexity of the algorithm is polynomial with the code length and the number of licensed emitters.

At the analysis stage, the performance of the synthesized waveforms was assessed by studying the trade-off among the achievable detection probability, spectral shape, and ACF features. The results highlighted that the proposed algorithm is capable of ensuring both spectral coexistence with the overlaid RF emitters and reasonable detection performance at the cost of other desirable radar signal features, such as sidelobe levels and/or range resolution.

Future research tracks might concern the extension of the framework to account for signal-dependent interference and, possibly, slow-time coding to perform optimized Doppler processing. Besides, it could be worth of consideration the possible exploitation of a MIMO architecture where spatial diversity may induce other degrees of freedom, which can be optimized to improve the performance of the overall method [Yu et al., 2018; Aubry et al., 2016c; Alhujaili et al., 2019].

7.A Appendix

7.A.1 Derivation of Ψ_M

Proof: In this appendix, some details about the derivation of Ψ_M are provided. Let us now re-parameterize the optimization vector $\mathbf{c}$ as in (7.12), i.e. $\mathbf{c} = \bar{\mathbf{c}} \odot \mathbf{c}_0$, with $\bar{\mathbf{c}} = \sqrt{x_{N+1}}\left[e^{jx_1}, \dots, e^{jx_N}\right]^T \in \mathbb{C}^N$, and with

$$x_i \in H_M = \begin{cases} \frac{2\pi}{M}\left\{-\frac{M}{2}, 1-\frac{M}{2}, \dots, \frac{M}{2}-1\right\}, & \text{if } M \text{ is even,} \\ \frac{2\pi}{M}\left\{-\frac{M-1}{2}, \left(1-\frac{M-1}{2}\right) \dots, \frac{M-1}{2}\right\}, & \text{if } M \text{ is odd,} \end{cases} \quad i \in \mathcal{N}, \tag{7.A.1}$$

where[4] M is the cardinality of the finite alphabet and $\arg(c_i), \arg(c_{0,i}) \in H_M$. As a consequence, the similarity constraint can be recast as $\Re\{e^{jx_i}\} \geq 1 - \epsilon^2/2, i \in \mathcal{N}$, namely

$$-\delta \leq x_i \leq \delta \tag{7.A.3}$$

with $\delta = \arccos(1 - \epsilon^2/2)$.

---

4 Note that H_M can be expressed in compact form as

$$H_M = \frac{2\pi}{M}\left\{-\left\lfloor\frac{M}{2}\right\rfloor, 1-\left\lfloor\frac{M}{2}\right\rfloor, \dots, -\left\lfloor\frac{M}{2}\right\rfloor + M - 1\right\}. \tag{7.A.2}$$

If $\epsilon < 2$, being $x_i = 2\pi i/M$, $i \in \left\{-\left\lfloor\frac{M}{2}\right\rfloor, 1-\left\lfloor\frac{M}{2}\right\rfloor, \dots, -\left\lfloor\frac{M}{2}\right\rfloor + M - 1\right\}$, Eq. (7.A.3) implies that the similarity constraint is tantamount to considering all the discrete phases in (7.A.2) belonging to $\left[\frac{2\pi}{M}\alpha_\epsilon, -\frac{2\pi}{M}\alpha_\epsilon\right]$ with $\alpha_\epsilon = -\left\lfloor\frac{M\delta}{2\pi}\right\rfloor$. Otherwise stated, the intersection between (7.A.1) and the set induced by (7.A.3) leads to $x_i \in \frac{2\pi}{M}\left\{\alpha_\epsilon, \alpha_\epsilon + 1, \cdots, -\alpha_\epsilon\right\}$, $i \in \mathcal{N}$.

If $\epsilon = 2$, the intersection between (7.A.1) and (7.A.3) coincides with (7.A.1) (indeed, the similarity constraint is inactive). In particular, it can be rewritten as

$$x_i \in \frac{2\pi}{M}\left\{\alpha_2, 1+\alpha_2, \dots, \alpha_2 + M - 1\right\}, \quad i \in \mathcal{N}, \tag{7.A.4}$$

Now, denoting by

$$\omega_\epsilon = \begin{cases} 1 - 2\alpha_\epsilon, & \text{if } \epsilon \in [0,2) , \\ M, & \text{if } \epsilon = 2, \end{cases} \tag{7.A.5}$$

it follows that the feasible set can be written compactly, i.e. regardless of ϵ and M, as $x_i \in \Psi_M$, $i \in \mathcal{N}$, where

$$\Psi_M = \frac{2\pi}{M}\left\{\alpha_\epsilon, \alpha_\epsilon + 1, \dots, \alpha_\epsilon + \omega_\epsilon - 1\right\}. \tag{7.A.6}$$

■

7.A.2 Derivation of Equation (7.19)

Proof: Let

$$\bar{\mathbf{c}}_d^{(n)} = \sqrt{x_{N+1}^{(n)}}\left[e^{jx_1^{(n)}}, \dots, e^{jx_d}, \dots, e^{jx_N^{(n)}}\right]^T \in \mathbb{C}^N, d \in \mathcal{N}, \tag{7.A.7}$$

it follows that

$$\begin{aligned} f_d^k(x_d; \mathbf{x}_{-d}^{(n)}) &= \bar{\mathbf{c}}_d^{(n)\dagger}\bar{\mathbf{R}}_I^k\bar{\mathbf{c}}_d^{(n)} \\ &= x_{N+1}^{(n)}\left(\sum_{i\neq d}^{N}\sum_{h\neq d}^{N} e^{j(x_h^{(n)} - x_i^{(n)})}\bar{\mathbf{R}}_{Ii,h}^k \right. \\ &\quad \left. + \bar{\mathbf{R}}_{Id,d}^k + 2\Re\left\{e^{jx_d}\sum_{i\neq d}^{N} e^{-jx_i^{(n)}}\bar{\mathbf{R}}_{Ii,d}^k\right\}\right). \end{aligned} \tag{7.A.8}$$

Hence, Eq. (7.19) is obtained defining

$$\bar{z}_{k,d}^{(n)} = \left(\sum_{i\neq d}^{N}\sum_{h\neq d}^{N} e^{j(x_h^{(n)} - x_i^{(n)})}\bar{\mathbf{R}}_{Ii,h}^k\right) + \bar{\mathbf{R}}_{Id,d}^k, \tag{7.A.9}$$

$$z_{k,d}^{(n)} = \sum_{i\neq d}^{N} e^{-jx_i^{(n)}}\bar{\mathbf{R}}_{Ii,d}^k. \tag{7.A.10}$$

Now, denoting by

$$\tilde{\mathbf{c}}^{(n)} = \left[e^{jx_1^{(n)}}, \dots, e^{jx_d^{(n)}}, \dots, e^{jx_N^{(n)}}\right]^T \in \mathbb{C}^N,$$

the case $d = N+1$ follows from

$$f_d^k(x_d; \mathbf{x}^{(n)}) = x_d\tilde{\mathbf{c}}^{(n)\dagger}\bar{\mathbf{R}}_I^k\tilde{\mathbf{c}}^{(n)} \tag{7.A.11}$$

with $p_k^{(n)} = \tilde{\mathbf{c}}^{(n)\dagger}\bar{\mathbf{R}}_I^k\tilde{\mathbf{c}}^{(n)}$. ■

7.A.3 Characterization of the Objective in $\mathcal{P}_p^{x_d^{(n)}}$

Following the same approach as in (7.A.8), it is easy to show that, if $d \in \mathcal{N}$,

$$\widehat{\text{SINR}}\left(x_d; \mathbf{x}_{-d}^{(n)}\right) = x_{N+1}^{(n)}\left(\bar{a}_d^{(n)} + 2\Re\{e^{jx_d}\bar{b}_d^{(n)}\}\right)$$

with

$$\bar{a}_d^{(n)} = \sum_{i\neq d}^{N}\sum_{h\neq d}^{N} e^{j(x_h^{(n)}-x_i^{(n)})}\bar{\mathbf{R}}_{i,h} + \bar{\mathbf{R}}_{d,d},$$

$$\bar{b}_d^{(n)} = \sum_{i\neq d}^{N} e^{-jx_i^{(n)}}\bar{\mathbf{R}}_{i,d}.$$

Hence, $\bar{\mathcal{P}}_p^{d,\mathbf{x}^{(n)}}$ can be recast as

$$\mathcal{P}_p^{x_d^{(n)}}\begin{cases}\underset{x_d}{\text{maximize}} & \chi\left(x_d\right),\\ \text{subject to} & \Re\{z_{k,d}^{(n)}e^{jx_d}\} \leq \bar{c}_{k,d}^{(n)}, \quad k = 1,\ldots,K\\ & x_d \in \Psi_p\end{cases} \tag{7.A.12}$$

with $\chi\left(x_d\right) = \Re\{\bar{b}_d^{(n)}e^{jx_d}\}$, and $\bar{c}_{k,d}^{(n)} = \frac{E_I^k}{2x_{N+1}^{(n)}} - \frac{\bar{z}_{k,d}^{(n)}}{2} \in \mathbb{R}$.

Before proceeding further, let us observe that both the characterization of the objective function monotonicities and the feasible set derivation can be performed by means of a change of variables, which defines a one-to-one monotonically increasing mapping. To this end, let us consider $t = \tan(x_d/2), x_d \in (-\pi, \pi)$; thus, according to the trigonometric relationships [Abramowitz and Stegun, 1964; Ben-Tal and Nemirovski, 2001],

$$\sin(\phi_d) = \frac{2\tan\left(\frac{\phi_d}{2}\right)}{1+\tan^2\left(\frac{\phi_d}{2}\right)}, \tag{7.A.13}$$

and

$$\cos(\phi_d) = \frac{1-\tan^2\left(\frac{\phi_d}{2}\right)}{1+\tan^2\left(\frac{\phi_d}{2}\right)}, \tag{7.A.14}$$

after straightforward algebraic manipulations, the objective function of Problem (7.23) can be rewritten as

$$R_d^{(n)}(t) = \frac{a_d^{(n)}t^2 + b_d^{(n)}t - a_d^{(n)}}{1+t^2}, \tag{7.A.15}$$

where $a_d^{(n)} = -\Re\left(\bar{b}_d^{(n)}\right)$ and $b_d^{(n)} = -2\Im\left(\bar{b}_d^{(n)}\right)$.

Let us first characterize the behavior of the function $R_d^{(n)}(t)$. Note that the first-order derivative of (7.A.15) is

$$R_d^{(n)\prime}(t) = \frac{-b_d^{(n)}t^2 + 4a_d^{(n)}t + b_d^{(n)}}{(1+t^2)^2}, \quad t \in \mathcal{R}. \tag{7.A.16}$$

As a consequence, the behavior of $R_d^{(n)}(t)$ is listed as follows:

- if $b_d^{(n)} \neq 0$, (7.A.15) presents the minimum point at $t_{1,d}^{(n)} = \frac{2a_d^{(n)} - \sqrt{(2a_d^{(n)})^2 + b_d^{(n)2}}}{b_d^{(n)}}$ and the maximum point at $t_{2,d}^{(n)} = \frac{2a_d^{(n)} + \sqrt{(2a_d^{(n)})^2 + b_d^{(n)2}}}{b_d^{(n)}}$,
  - if $b_d^{(n)} > 0$, $R_d^{(n)}(t)$ is strictly decreasing over $t < t_{1,d}^{(n)}$, strictly increasing within $t_{1,d}^{(n)} < t < t_{2,d}^{(n)}$, and strictly decreasing over $t > t_{2,d}^{(n)}$;
  - if $b_d^{(n)} < 0$, $R_d^{(n)}(t)$ is strictly increasing within $t < t_{2,d}^{(n)}$, strictly decreasing over $t_{2,d}^{(n)} < t < t_{1,d}^{(n)}$, and strictly increasing over $t > t_{1,d}^{(n)}$;
- if $b_d^{(n)} = 0$, it follows that
  - if $a_d^{(n)} = 0$, thus, implying $R_d^{(n)\prime}(t) = 0$ and $R_d^{(n)}(t) = 0$;
  - if $a_d^{(n)} < 0$, $R_d^{(n)}(t)$ monotonically increases over $t \leq 0$ and monotonically decreases for $t > 0$ with $t_{1,d}^{(n)} = 0$ the maximum point, and the finite infimum, i.e. $a_d^{(n)}$, is achieved at $\pm\infty$;
  - if $a_d^{(n)} > 0$, $R_d^{(n)}(t)$ monotonically decreases over $t \leq 0$ and monotonically increases for $t > 0$ with $t_{2,d}^{(n)} = 0$ the minimum point, and the finite supremum, i.e. $a_d^{(n)}$, is achieved at $\pm\infty$.

Since $y_h \in (-\pi, \pi)$ is mapped to $t \in (-\infty, \infty)$ by a strictly increasing function, the monotonicity of $\chi(x_d), x_d \in (-\pi, \pi)$ can be derived leveraging the behavior of $R_d^{(n)}(t)$. For instance, supposing $b_d^{(n)} < 0$ without loss of generality, $\chi(x_d)$ is strictly increasing over $-\pi < x_d < \phi_{g,d}^{(n)}$ with $\phi_{g,d}^{(n)} = 2\arctan t_{2,h}^{(n)}$, then decreasing over $\phi_{g,d}^{(n)} < x_d < \phi_{s,d}^{(n)}$ with $\phi_{s,d}^{(n)} = 2\arctan(t_{1,h}^{(n)})$, and increasing over $\phi_{s,d}^{(n)} < y_h < \pi$. As a result, $\phi_{g,d}^{(n)}$ and $\phi_{s,d}^{(n)}$ can be assigned as

$$\phi_{g,d}^{(n)} = \begin{cases} \pi, & \text{If } a_d^{(n)} \geq 0, b_d^{(n)} = 0, \\ 0, & \text{If } a_d^{(n)} < 0, b_d^{(n)} = 0, \\ 2\arctan\left(\dfrac{2a_d^{(n)} + \sqrt{(2a_d^{(n)})^2 + b_d^{(n)2}}}{b_d^{(n)}}\right), & \text{If } b_d^{(n)} \neq 0, \end{cases} \tag{7.A.17}$$

and

$$\phi_{s,d}^{(n)} = \begin{cases} 0, & \text{If } a_d^{(n)} \geq 0, b_d^{(n)} = 0, \\ -\pi, & \text{If } a_d^{(n)} < 0, b_d^{(n)} = 0, \\ 2\arctan\left(\dfrac{2a_d^{(n)} - \sqrt{(2a_d^{(n)})^2 + b_d^{(n)2}}}{b_d^{(n)}}\right), & \text{If } b_d^{(n)} \neq 0. \end{cases} \tag{7.A.18}$$

7.A.4 Derivation of the Feasible Set of $\mathcal{P}_p^{x_d^{(n)}}$

Since $\mathbf{x}^{(n)}$ is a feasible solution to $\bar{\mathcal{P}}_p$, the feasible set of Problem $\mathcal{P}_p^{x_d^{(n)}}$ is not empty. Precisely, omitting the dependence on d and n for notational simplicity, the feasible set in the transformed domain $t = \tan(x_d/2)$ is given by

$$\mathcal{F}_p = \left(\cap_{k=1}^{K} \mathcal{S}_k\right) \cap \overline{\Psi}_p, \tag{7.A.19}$$

where

$$\mathcal{S}_k = \left\{ t : a_{k,d}^{(n)} t^2 + b_{k,d}^{(n)} t + c_{k,d}^{(n)} \leq 0 \right\}, \quad k = 1, \ldots, K, \tag{7.A.20}$$

$$\overline{\Psi}_{\infty} = \left[-\tan\left(\frac{\delta}{2}\right), \tan\left(\frac{\delta}{2}\right)\right], \tag{7.A.21}$$

$$\overline{\Psi}_{M} = \left\{\tan\left(\frac{\pi\alpha_{\epsilon}}{M}\right), \dots, \tan\left(\frac{\pi(\alpha_{\epsilon}+\omega_{\epsilon}-1)}{M}\right)\right\}, \tag{7.A.22}$$

with $a_{k,d}^{(n)} = -\Re\left(z_{k,d}^{(n)}\right) - \bar{c}_{k,d}^{(n)}$, $b_{k,d}^{(n)} = -2\Im\left(z_{k,d}^{(n)}\right)$, and $c_{k,d}^{(n)} = \Re\left(z_{k,d}^{(n)}\right) - \bar{c}_{k,d}^{(n)}$.

As to the set $\mathcal{S}_k$, $k = 1, \dots, K$, denoting by

$$\Delta_{k,d}^{(n)} = b_{k,d}^{(n)2} - 4a_{k,d}^{(n)}c_{k,d}^{(n)},$$

$$e_{k,d}^{(1,n)} = \frac{-b_{k,d}^{(n)} - \sqrt{\Delta_{k,d}^{(n)}}}{2a_{k,d}^{(n)}}, \text{ assuming } \Delta_{k,d}^{(n)} \geq 0 \text{ and } a_{k,d}^{(n)} \neq 0,$$

$$e_{k,d}^{(2,n)} = \frac{-b_{k,d}^{(n)} + \sqrt{\Delta_{k,d}^{(n)}}}{2a_{k,d}^{(n)}}, \text{ assuming } \Delta_{k,d}^{(n)} \geq 0 \text{ and } a_{k,d}^{(n)} \neq 0,$$

it follows that

$$\mathcal{S}_k = \begin{cases} \mathcal{R}, & \text{if } \Delta_{k,d}^{(n)} < 0 \text{ or } a_{k,d}^{(n)} = 0 \text{ and } b_{k,d}^{(n)} = 0, \\ \left[e_{k,d}^{(1,n)}, e_{k,d}^{(2,n)}\right], & \text{if } a_{k,d}^{(n)} > 0 \text{ and } \Delta_{k,d}^{(n)} \geq 0, \\ \left(-\infty, e_{k,d}^{(2,n)}\right] \cup \left[e_{k,d}^{(1,n)}, \infty\right), & \text{if } a_{k,d}^{(n)} < 0 \text{ and } \Delta_{k,d}^{(n)} \geq 0, \\ \left(-\infty, \frac{-c_{k,d}^{(n)}}{b_{k,d}^{(n)}}\right], & \text{if } a_{k,d}^{(n)} = 0 \text{ and } b_{k,d}^{(n)} > 0, \\ \left[\frac{-c_{k,d}^{(n)}}{b_{k,d}^{(n)}}, \infty\right), & \text{if } a_{k,d}^{(n)} = 0 \text{ and } b_{k,d}^{(n)} < 0. \end{cases}$$

Based on the associative property of the intersection, the sets $\mathcal{S}_k$, $k = 1, \dots, K$, can be grouped in two classes, corresponding to: (i) closed, possibly unbounded, intervals, i.e. $[y_k^l, y_k^u]$; (ii) sets given by the union of two unbounded intervals, i.e. $]-\infty, y_k^l] \cup [y_k^u, \infty[$.

To proceed further, let us assume without loss of generality that type (1) sets correspond to the indexes $k \in \{1, \dots, I_1\}$, whereas $k \in \{I_1 + 1, \dots, I_2\}$, with $I_2 = K$, refers to type (2) sets. As a consequence,

$$\mathcal{F}_{\infty} = \left(\cap_{k=I_1+1}^{I_2} \mathcal{S}_k\right) \cap [y^l, y^u],$$

where

$$[y^l, y^u] = \cap_{k=1}^{I_1} \mathcal{S}_k \bigcap \overline{\Psi}_{\infty},$$

with

$$y^l = \text{maximize}\{y_k^l, k = 1, \dots, I_1, -\tan(\delta/2)\}, \tag{7.A.23}$$

$$y^u = \min\{y_k^u, k = 1, \dots, I_1, \tan(\delta/2)\}. \tag{7.A.24}$$

Furthermore, as shown in Aubry et al. [2020], $\mathcal{F}_\infty$ can be expressed as union of disjoint closed intervals, i.e.

$$\mathcal{F}_\infty = \cup_{i=1}^{K_\infty} [l_i, u_i], \tag{7.A.25}$$

with $l_i \le u_i < l_{i+1}, i = 1, \dots, K_\infty - 1$ and $K_\infty \le K + 1$ depending on the specific parameter values of $a_{k,d}^{(n)}, b_{k,d}^{(n)}, c_{k,d}^{(n)}$, and δ.

Evidently, the feasible set of Problem $\mathcal{P}_p^{x_d^{(n)}}$ is given by

$$\bar{\mathcal{F}}_\infty = \cup_{i=1}^{K_\infty} \left[\hat{l}_i^\infty, \hat{u}_i^\infty\right], \tag{7.A.26}$$

where $\hat{l}_i^\infty = 2\arctan(l_i)$, $\hat{u}_i^\infty = 2\arctan(u_i)$.

As to the discrete phase case, leveraging the above results, it follows that

$$\mathcal{F}_M = \mathcal{F}_\infty \cap \overline{\Psi}_M. \tag{7.A.27}$$

As a consequence,

$$\bar{\mathcal{F}}_M = \bigcup_{\substack{i=1 \\ i:\ \hat{l}_i \le \hat{u}_i}}^{K_\infty} \Gamma_i, \tag{7.A.28}$$

where

$$\Gamma_i = \left\{\hat{l}_i, \hat{l}_i + \frac{2\pi}{M}, \dots, \hat{u}_i\right\},$$

and

$$\hat{l}_i = \left\lceil \frac{\hat{l}_i^\infty M}{2\pi} \right\rceil \frac{2\pi}{M}, \tag{7.A.29}$$

$$\hat{u}_i = \left\lfloor \frac{\hat{u}_i^\infty M}{2\pi} \right\rfloor \frac{2\pi}{M}, \tag{7.A.30}$$

with the condition $\hat{l}_i \le \hat{u}_i$ ensuring that Γ_i is not empty. Now denoting by $K_M \le K_\infty$ the number of the actual Γ_i involved in the union operation in $\bar{\mathcal{F}}_M$ (i.e. the number of the actual disjoint closed sets), there exists an increasing mapping i_t: $t \in \{1, \dots, K_M\} \to i_t \in \{1, \dots, K_\infty\}$, such that $\bar{\mathcal{F}}_M = \bigcup_{t=1}^{K_M} \Gamma_{i_t}$. Without loss of generality, denoting by $\Lambda_t = \Gamma_{i_t} = \left\{\hat{l}_t^M, \hat{l}_t^M + \frac{2\pi}{M}, \dots, \hat{u}_t^M\right\}$, $\bar{\mathcal{F}}_M$ can be rewritten as

$$\bar{\mathcal{F}}_M = \bigcup_{t=1}^{K_M} \Lambda_t \tag{7.A.31}$$

with $-\pi < \hat{l}_t^M \le \hat{u}_t^M < \hat{l}_{i+1}^M < \pi, i = 1, \dots, K_M - 1$ and $\hat{l}_t^M, \hat{u}_t^M \in \Psi_M$.

7.A.5 Proof of Proposition 7.1

Proof: The optimal solution $x_d^\star$ to Problem $\mathcal{P}_\infty^{x_d}$ can be derived as follows:

- if $\phi_{g,d}^{(n)} \in \bar{\mathcal{F}}_\infty$, then $x_d^\star = \phi_{g,d}^{(n)}$;
- if $\phi_{g,d}^{(n)} \le \hat{l}_1^\infty$, the optimal solution depends on the actual monotonicity of the objective function
  - if $b_d^{(n)} = 0$ and $a_d^{(n)} = 0$, $\chi(x_d) = 0\ \forall x_d$, thus $x_d^\star = \hat{l}_1^\infty$;
  - if $b_d^{(n)} > 0$, or $b_d^{(n)} = 0$ and $a_d^{(n)} < 0$, $\chi(x_d)$ monotonically decreases over $x_d \ge \phi_{g,d}^{(n)}$, thus $x_d^\star = \hat{l}_1^\infty$;
  - if $b_d^{(n)} < 0$, or $b_d^{(n)} = 0$ and $a_d^{(n)} > 0$, $\chi(x_d)$ monotonically decreases over $\phi_{g,d}^{(n)} \le x_d < \phi_{s,d}^{(n)}$, and increases over $\phi_{s,d}^{(n)} \le x_d < \pi$, thus $x_d^\star \in \{\hat{l}_1^\infty, \hat{u}_{K_\infty}^\infty\}$;

- if $\phi_{g,d}^{(n)} \geq \hat{u}_{K_\infty}^{\infty}$, a situation dual to $\phi_{g,d}^{(n)} < \hat{l}_1^{\infty}$ occurs
  - if $b_d^{(n)} = 0$ and $a_d^{(n)} = 0$, $\chi(x_d) = 0\ \forall x_d$, thus $x_d^\star = \hat{l}_1^{\infty}$;
  - if $b_d^{(n)} > 0$, or $b_d^{(n)} = 0$ and $a_d^{(n)} < 0$, $\chi(x_d)$ monotonically increases over $x_d \leq \phi_{g,d}^{(n)}$, thus $x_d^\star = \hat{u}_{K_\infty}^{\infty}$;
  - if $b_d^{(n)} < 0$, or $b_d^{(n)} = 0$ and $a_d^{(n)} > 0$, $\chi(x_d)$ monotonically decreases over $-\pi \leq x_d < \phi_{s,d}^{(n)}$, and then increases over $\phi_{s,d}^{(n)} \leq x_d < \pi$, thus $x_d^\star \in \{\hat{l}_1^{\infty}, \hat{u}_{K_\infty}^{\infty}\}$;
- otherwise, denoting by $r^\star = \arg\text{maximize}_{\hat{u}_r^{\infty} < \phi_{g,d}^{(n)}} r$, it follows that $\hat{u}_{r^\star}^{\infty} < \phi_{g,d}^{(n)} < \hat{l}_{r^\star+1}^{\infty}$,
  - if $b_d^{(n)} = 0$ and $a_d^{(n)} = 0$, $\chi(x_d) = 0\ \forall x_d$, thus $x_d^\star = \hat{l}_1^{\infty}$;
  - if $b_d^{(n)} > 0$, $\chi(x_d)$ monotonically decreases over $\phi_{g,d}^{(n)} \leq x_d \leq \hat{u}_{K_\infty}^{\infty}$, thus $x_d^\star \in \{\hat{l}_1^{\infty}, \hat{u}_{r^\star}^{\infty}, \hat{l}_{r^\star+1}^{\infty}\}$;
  - if $b_d^{(n)} < 0$, $\chi(x_d)$ monotonically increases over $\hat{l}_1^{\infty} \leq x_d \leq \phi_{g,d}^{(n)}$, thus $x_d^\star \in \{\hat{u}_{K_\infty}^{\infty}, \hat{u}_{r^\star}^{\infty}, \hat{l}_{r^\star+1}^{\infty}\}$;
  - if $b_d^{(n)} = 0$ and $a_d^{(n)} < 0$, $\chi(x_d)$ monotonically increases over $-\pi \leq x_d \leq \phi_{g,d}^{(n)}$ then decreases over $x_d \geq \phi_{g,d}^{(n)}$, thus $x_d^\star \in \{\hat{u}_{r^\star}^{\infty}, \hat{l}_{r^\star+1}^{\infty}\}$;

  As a result, $x_d^\star \in \{\hat{l}_1^{\infty}, \hat{u}_{r^\star}^{\infty}, \hat{l}_{r^\star+1}^{\infty}, \hat{u}_{K_\infty}^{\infty}\}$.

■

7.A.6 Proof of Corollary 7.1

Proof: Two situations may occur: there exists an index $h^\star \in \{1, \dots, K_M\}$ satisfying $\hat{l}_{h^\star}^{M} \leq \phi_{g,d}^{(n)} \leq \hat{u}_{h^\star}^{M}$ or such an index does not exist. As to the former case, $\phi_l^M = \left\lfloor \frac{\phi_{g,d}^{(n)} M}{2\pi} \right\rfloor \frac{2\pi}{M}$ and $\phi_u^M = \left\lceil \frac{\phi_{g,d}^{(n)} M}{2\pi} \right\rceil \frac{2\pi}{M}$ represent the feasible points closest from the bottom and from the up, respectively, to $\phi_{g,d}^{(n)}$. To proceed further, let us consider the optimal solution to the relaxed problem:

$$\begin{cases} \underset{x_d}{\text{maximize}} & \chi(x_d), \\ \text{subject to} & x_d \in [\hat{l}_1^{M}, \phi_l^M] \bigcup [\phi_u^M, \hat{u}_{K_M}^{M}], \end{cases} \tag{7.A.32}$$

where $\chi(x_d) = |\bar{b}_d^{(n)}| \cos(x_d - \phi_{g,d}^{(n)})$. It follows that

- if $\phi_{g,d}^{(n)} > 0$, $\chi(x_d)$ monotonically decreases over $\phi_{g,d}^{(n)} \leq x_d \leq \pi$, thus $\chi(\hat{u}_{K_M}^{M}) \leq \chi(\phi_u^M)$ as $\phi_{g,d}^{(n)} \leq \phi_u^M \leq \hat{u}_{K_M}^{M} < \pi$. Moreover, being $\phi_{s,d}^{(n)} = \phi_{g,d}^{(n)} - \pi$,
  - if $\phi_{s,d}^{(n)} \leq \hat{l}_1^{M} \leq \phi_{g,d}^{(n)}$, $\chi(x_d)$ monotonically increases over $\hat{l}_1^{M} \leq x_d \leq \phi_{g,d}^{(n)}$, thus $\chi(\hat{l}_1^{M}) \leq \chi(\phi_l^M)$ as $\hat{l}_1^{M} \leq \phi_l^M \leq \phi_{g,d}^{(n)}$. Then, owing to the symmetry of $\chi(x_d)$ and the existence of a single local minimum within $[-\pi, \pi)$, the optimal solution is the point between ϕ_l^M and ϕ_u^M closest to $\phi_{g,d}^{(n)}$, i.e. $x_d^\star = \frac{2\pi}{M} \left\lceil \frac{M\phi_{g,d}^{(n)}}{2\pi} \right\rfloor$;
  - if $-\pi \leq \hat{l}_1^{M} \leq \phi_{s,d}^{(n)}$, $\chi(x_d)$ monotonically decreases over $-\pi \leq x_d \leq \hat{l}_1^{M}$, thus $\chi(\hat{l}_1^{M}) \leq \chi(-\pi) = \chi(\pi) \leq \chi(\phi_u^M)$, and again the optimal solution is $x_d^\star = \frac{2\pi}{M} \left\lceil \frac{M\phi_{g,d}^{(n)}}{2\pi} \right\rfloor$;
- if $\phi_{g,d}^{(n)} < 0$, following a line of reasoning similar to that for $\phi_{g,d}^{(n)} > 0$, it can be concluded that $x_d^\star = \frac{2\pi}{M} \left\lceil \frac{M\phi_{g,d}^{(n)}}{2\pi} \right\rfloor$;
- if $\phi_{g,d}^{(n)} = 0$, $\chi(x_d)$ monotonically increases over $-\pi \leq x_d \leq 0$, and decreases over $0 < x_d < \pi$, thus $x_d^\star = \frac{2\pi}{M} \left\lceil \frac{M\phi_{g,d}^{(n)}}{2\pi} \right\rfloor = 0$.

Let us now focus on the latter situation. In this case let us consider the following relaxed problem,

$$\begin{cases} \underset{x_d}{\text{maximize}} & \chi(x_d), \\ \text{subject to} & x_d \in \cup_{i=1}^{K_M} \left[\hat{l}_i^M, \hat{u}_i^M\right]. \end{cases} \tag{7.A.33}$$

Based on Proposition 7.1, it follows that

- if $\phi_{g,d}^{(n)} \notin [\hat{l}_1^M, \hat{u}_1^M]$, the optimal solution to Problem (7.A.33) is $x_d^\star = \arg\text{maximize}_{x_d \in \{\hat{l}_1^M, \hat{u}_{K_M}^M\}} \chi(x_d)$;
- otherwise, denoting by $r^\star = \arg\text{maximize}_{\hat{u}_r^M < \phi_{g,d}^{(n)}} r$, it follows that $\hat{u}_{r^\star}^M < \phi_{g,d}^{(n)} < \hat{l}_{r^\star+1}^M$, thus $x_d^\star = \arg\text{maximize}_{x_d \in \{\hat{l}_1^M, \hat{u}_{r^\star}^M, \hat{l}_{r^\star+1}^M, \hat{u}_{K_M}^M\}} \chi(x_d)$.

Being the optimal solution to (7.A.32) or (7.A.33) feasible to Problem $\mathcal{P}_M^{x_d^{(n)}}$, the proof is concluded. ■

References

M. Abramowitz and I. A. Stegun. *Handbook of mathematical functions: with formulas, graphs, and mathematical tables*, volume 55. Courier Corporation, 1964.

K. Alhujaili, V. Monga, and M. Rangaswamy. Transmit MIMO radar beampattern design via optimization on the complex circle manifold. *IEEE Transactions on Signal Processing*, 67(13):3561–3575, 2019.

A. Aubry, A. De Maio, M. Piezzo, and A. Farina. Radar waveform design in a spectrally crowded environment via nonconvex quadratic optimization. *IEEE Transactions on Aerospace and Electronic Systems*, 50(2):1138–1152, 2014a.

A. Aubry, A. De Maio, M. Piezzo, M. M. Naghsh, M. Soltanalian, and P. Stoica. Cognitive radar waveform design for spectral coexistence in signal-dependent interference. In *2014 IEEE Radar Conference*, Cincinnati (OH), USA, May 2014b.

A. Aubry, V. Carotenuto, and A. De Maio. Forcing multiple spectral compatibility constraints in radar waveforms. *IEEE Signal Processing Letters*, 23(4):483–487, 2016a.

A. Aubry, V. Carotenuto, A. De Maio, A. Farina, and L. Pallotta. Optimization theory-based radar waveform design for spectrally dense environments. *IEEE Aerospace and Electronic Systems Magazine*, 31(12):14–25, 2016b.

A. Aubry, A. De Maio, and Y. Huang. MIMO radar beampattern design via PSL/ISL optimization. *IEEE Transactions on Signal Processing*, 64(15):3955–3967, 2016c.

A. Aubry, A. De Maio, M. A. Govoni, and L. Martino. On the design of multi-spectrally constrained constant modulus radar signals. *IEEE Transactions on Signal Processing*, 68:2231–2243, 2020. https://doi.org/10.1109/TSP.2020.2983642.

C. Baker, H. Griffiths, and A. Balleri. *Waveform design and diversity for advanced radar systems.* Institution of Engineering and Technology London, UK 2012.

A. Ben-Tal and A. Nemirovski. *Lectures on modern convex optimization: analysis, algorithms, and engineering applications*, volume 2. SIAM, 2001.

M. Bica and V. Koivunen. Radar waveform optimization for target parameter estimation in cooperative radar-communications systems. *IEEE Transactions on Aerospace and Electronic Systems*, 2018. ISSN 0018-9251. https://doi.org/10.1109/TAES.2018.2884806.

S. D. Blunt and E. L. Mokole. Overview of radar waveform diversity. *IEEE Aerospace and Electronic Systems Magazine*, 31(11):2–42, 2016. ISSN 0885-8985. https://doi.org/10.1109/MAES.2016.160071.

B. Chen, S. He, Z. Li, and S. Zhang. Maximum block improvement and polynomial optimization. *SIAM Journal on Optimization*, 22(1):87–107, 2012.

Z. Cheng, B. Liao, Z. He, Y. Li, and J. Li. Spectrally compatible waveform design for MIMO radar in the presence of multiple targets. *IEEE Transactions on Signal Processing*, 66(13):3543–3555, 2018.

D. Cohen, K. V. Mishra, K. K. Bae, and Y. C. Eldar. *Spectral coexistence in radar using xampling*. In *2017 IEEE Radar Conference (RadarConf)*, 2017.

M. R. Cook, T. Higgins, and A. K. Shackelford. *Thinned spectrum radar waveforms*. August 2010. https://doi.org/10.1109/WDD.2010.5592622.

G. Cui, A. De Maio, A. Farina, and J. Li. *Radar waveform design based on optimization theory*. SciTech Publishing, 2020.

A. De Maio, S. De Nicola, Y. Huang, Z. Luo, and S. Zhang. Design of phase codes for radar performance optimization with a similarity constraint. *IEEE Transactions on Signal Processing*, 57(2):610–621, 2009. ISSN 1053-587X. https://doi.org/10.1109/TSP.2008.2008247.

A. De Maio, S. De Nicola, Y. Huang, D. P. Palomar, S. Zhang, and A. Farina. Code design for radar STAP via optimization theory, *IEEE Transactions on Signal Processing*, 58(2):679–694, 2010.

S. H. Dokhanchi, M. R. B. Shankar, K. V. Mishra, and B. Ottersten. Multi-constraint spectral co-design for colocated MIMO radar and MIMO communications. *ICASSP 2020–2020 IEEE International Conference on Acoustics, Speech and Signal Processing (ICASSP)*, 4567–4571, 2020. https://doi.org/10.1109/ICASSP40776.2020.9054680.

K. Gerlach. Thinned spectrum ultrawideband waveforms using stepped-frequency polyphase codes. *IEEE Transactions on Aerospace and Electronic Systems*, 34(4):1356–1361, 1998.

K. Gerlach, M. R. Frey, M. J. Steiner, and A. Shackelford. Spectral nulling on transmit via nonlinear FM radar waveforms. *IEEE Transactions on Aerospace and Electronic Systems*, 47(2):1507–1515, 2011. ISSN 0018-9251. https://doi.org/10.1109/TAES.2011.5751276.

M. A. Govoni. Enhancing spectrum coexistence using radar waveform diversity. In *2016 IEEE Radar Conference*, Philadelphia (PA), USA, May 2016.

H. Griffiths, L. Cohen, S. Watts, E. Mokole, C. Baker, M. Wicks, and S. Blunt. Radar spectrum engineering and management: technical and regulatory issues. *Proceedings of the IEEE*, 103(1):85–102, 2015.

A. Hassanien, M. G. Amin, Y. D. Zhang, and F. Ahmad. Dual-function radar-communications: information embedding using sidelobe control and waveform diversity. *IEEE Transactions on Signal Processing*, 64(8):2168–2181, 2015. https://doi.org/10.1109/TSP.2015.2505667.

H. He, P. Stoica, and J. Li. *Waveform design with stopband and correlation constraints for cognitive radar*. Elba Island, IT, June 2010.

X. Hu, C. Masouros, F. Liu, and R. Nissel. MIMO-OFDM dual-functional radar-communication systems: low-PAPR waveform design. *arXivpreprint arXiv:2109.13148*, 2021.

J. Jakabosky, S. D. Blunt, and A. Martone. *Incorporating hopped spectral gaps into nonrecurrent nonlinear FMCW radar emissions*. Cancun, Mexico, December 2015. https://doi.org/10.1109/CAMSAP.2015.7383791.

Y. Jing, J. Liang, D. Zhou, and H. C. So. Spectrally constrained unimodular sequence design without spectral level mask. *IEEE Signal Processing Letters*, 25(7):1004–1008, 2018. ISSN 1070-9908. https://doi.org/10.1109/LSP.2018.2836219.

B. Li and A. P. Petropulu. Joint transmit designs for coexistence of MIMO wireless communications and sparse sensing radars in clutter. *IEEE Transactions on Aerospace and Electronic Systems*, 53(6):2846–2864, 2017. ISSN 0018-9251. https://doi.org/10.1109/TAES.2017.2717518.

B. Li, A. P. Petropulu, and W. Trappe. Optimum co-design for spectrum sharing between matrix completion based MIMO radars and a MIMO communication system. *IEEE Transactions on Signal Processing*, 64(17):4562–4575, 2016. ISSN 1053-587X. https://doi.org/10.1109/TSP.2016.2569479.

M. J. Lindenfeld. Sparse frequency transmit-and-receive waveform design. *IEEE Transactions on Aerospace and Electronic Systems*, 40(3):851–861, 2004. ISSN 0018-9251. https://doi.org/10.1109/TAES.2004.1337459.

F. Liu, L. Zhou, M. Christos, A. Li, W. Luo, and A. Petropulu. Toward dual-functional radar-communication systems: optimal waveform design. *IEEE Transactions on Signal Processing*, 66(16):4264–4279, 2018. https://doi.org/10.1109/TSP.2018.2847648.

J. Liu, K. V. Mishra, and M. Saquib. Co-designing statistical MIMO radar and in-band full-duplex multi-user MIMO communications. *arXiv preprint arXiv:2006.14774*, 2020

C. Nunn and L. R. Moyer. Spectrally-compliant waveforms for wideband radar. *IEEE Aerospace and Electronic Systems Magazine*, 27(8):11–15, 2012. ISSN 0885-8985. https://doi.org/10.1109/MAES.2012.6329156.

L. K. Patton, C. A. Bryant, and B. Himed. *Radar-centric design of waveforms with disjoint spectral support*. Atlanta (GA), USA, May 2012. https://doi.org/10.1109/RADAR.2012.6212149.

B. Paul, A. R. Chiriyath, and D. W. Bliss. Survey of RF communications and sensing convergence research. *IEEE Access*, 5:252–270, 2016. ISSN 2169-3536. https://doi.org/10.1109/ACCESS.2016.2639038.

J. Qian, M. Lops, X. Wang, and Z. He. Joint system design for coexistence of MIMO radar and MIMO communication. *IEEE Transactions on Signal Processing*, 66(13):3504–3519, 2018. ISSN 1053-587X. https://doi.org/10.1109/TSP.2018.2831624.

B. Ravenscroft, J. W. Owen, J. Jakabosky, S. D. Blunt, A. F. Martone, and K. D. Sherbondy. Experimental demonstration and analysis of cognitive spectrum sensing and notching for radar. *IET Radar, Sonar and Navigation*, 12(12):1466–1475, 2018. ISSN 1751-8784. https://doi.org/10.1049/iet-rsn.2018.5379.

R. A. Romero and K. D. Shepherd. Friendly spectrally shaped radar waveform with legacy communication systems for shared access and spectrum management. *IEEE Access*, 3:1541–1554, 2015.

I. W. Selesnick and S. U. Pillai. *Chirp-like transmit waveforms with multiple frequency-notches*. Kansas City (MO), USA, May 2011. https://doi.org/10.1109/RADAR.2011.5960706.

I. W. Selesnick, S. U. Pillai, and R. Zheng. *An iterative algorithm for the construction of notched chirp signals*. Washington (DC), USA, May 2010. https://doi.org/10.1109/RADAR.2010.5494625.

B. Tang and J. Li. Spectrally constrained MIMO radar waveform design based on mutual information. *IEEE Transactions on Signal Processing*, 67(3):821–834, 2019. ISSN 1053-587X. https://doi.org/10.1109/TSP.2018.2887186.

B. Tang and J. Liang. Efficient algorithms for synthesizing probing waveforms with desired spectral shapes. *IEEE Transactions on Aerospace and Electronic Systems*, 2018. ISSN 0018-9251. https://doi.org/10.1109/TAES.2018.2876585.

B. Tang, J. Li, and J. Liang. Alternating direction method of multipliers for radar waveform design in spectrally crowded environments. *Signal Processing*, 142:398–402, 2018.

G. Wang and Y. Lu. Designing single/multiple sparse frequency waveforms with sidelobe constraint. *IET Radar, Sonar and Navigation*, 5(1):32–38, 2011. ISSN 1751-8784. https://doi.org/10.1049/iet-rsn.2009.0255.

M. Wicks. *Spectrum crowding and Cognitive Radar*. Elba Island, IT, June 2010.

S. J. Wright. Coordinate descent algorithms. *Mathematical Programming*, 151(1):3–34, 2015.

L. Wu, P. Babu, and D. P. Palomar. Transmit waveform/receive filter design for MIMO radar with multiple waveform constraints. *IEEE Transactions on Signal Processing*, 66(6):1526–1540, 2018.

J. Yang, A. Aubry, A. De Maio, X. Yu, and G. Cui. Design of constant modulus discrete phase radar waveforms subject to multi-spectral constraints. *IEEE Signal Processing Letters*, 27:875–879, 2020.

X. Yu, G. Cui, T. Zhang, and L. Kong. Constrained transmit beampattern design for colocated MIMO radar. *Signal Processing*, 144:145–154, 2018.

X. Yu, K. Alhujaili, G. Cui, and V. Monga. MIMO radar waveform design in the presence of multiple targets and practical constraints. *IEEE Transactions on Signal Processing*, 68:1974–1989, 2020.

T. Yucek and H. Arslan. A survey of spectrum sensing algorithms for cognitive radio applications. *IEEE Communication Surveys and Tutorials*, 11(1):116–130, 2009.

Y. Zhao, J. Gaeddert, K. K. Bae, and J. H. Reed. *Radio environment map-enabled situation-aware cognitive radio learning algorithms*. Orlando, FL, USA, November 2006.

Y. Zhao, L. Morales, J. Gaeddert, K. K. Bae, and J. H. Reed. Applying radio environment maps to cognitive wireless regional area networks. In *2nd IEEE International Symposium on New Frontiers in Dynamic Spectrum Access Networks, 2007. DySPAN 2007*, 2007.

D. Zhao, Y. Wei, and Y. Liu. Hopped-frequency waveform design for range sidelobe suppression in spectral congestion. *IET Radar, Sonar and Navigation*, 12(1):87–94, 2018. ISSN 1751-8784. https://doi.org/10.1049/iet-rsn.2017.0232.

L. Zheng, M. Lops, X. Wang, and E. Grossi. Joint design of overlaid communication systems and pulsed radars. *IEEE Transactions on Signal Processing*, 66(1):139–154, 2018. ISSN 1053-587X. https://doi.org/10.1109/TSP.2017.2755603.

8

Spectrum Sharing Between MIMO Radar and MIMO Communication Systems*

Bo Li[1] *and Athina P. Petropulu*[2]

[1]*Aurora Innovation, Inc., Research and Development Department, Pittsburgh, PA, USA*
[2]*Rutgers, the State University of New Jersey, Department of Electrical and Computer Engineering, Piscataway, NJ, USA*

8.1 Introduction

The radio frequency (RF) wireless spectrum supports a wide array of technologies, including broadcasting, wireless communications, position and navigation systems, and military and civilian radars. The "doctrine of spectrum scarcity" [Rossini, 1985] makes radio spectrum among one of the most tightly regulated resources. For example, US spectrum allocations regulate spectrum from 9 kHz up to 300 GHz. Spectrum scarcity is an ever-growing problem due to continuously emerging wireless communications applications such as LTE, 5G, Wi-Fi, Internet of Things (IoT), Citizens Broadband Radio Services (CBRS), and low Earth orbit (LEO) broadband satellite network. The Federal Communications Commission (FCC) C-band spectrum auction for 5G deployment, which started in December 2020 and closed in January 2021, raised $80.9 billion in bids, which makes it the biggest US spectrum auction to date [Web, 2020].

The regulated spectrum allocation reduces interference from unlicensed transmitters, whereas the fixed allocation introduces unbalanced spectrum utilization. Until recently, radar and communications systems jointly consumed most of the spectrum below 6 GHz, with each being assigned on different bands. However, recent studies have shown that large chunks of spectrum designated for radar applications are underutilized [PCAST, 2012], while there is spectrum congestion in bands devoted to commercial wireless communications. To address the need for more efficient use of spectrum, US Government agencies have been examining the possibility of allowing wireless broadband communications systems to operate in the 3500–3650 MHz C-band, which was previously used exclusively by high-powered shipborne, airborne, and ground-based radars operated by the Department of Defense (DoD) [FCC, 2012, Locke and Strickling, 2012]. The regulation changes in the C-band have prompted significant interest among government agencies, industry, and academia, on how to efficiently use this new spectrum by minimizing interference effects.

Techniques for efficient spectrum sharing would be important in a large number of applications. For example, consider a crowded emergency room scenario, in which Doppler radar, used for physical state monitoring, operates in the vicinity of 5G communications devices. Indeed, there is a growing interest in 24 GHz Doppler radar sensors [Droitcour et al., 2004], or ultra-wideband (UWB) radar sensors [Immoreev et al., 2005] for detection of vital signs, considering their abil-

* This work was supported by NSF under Grants ECCS-1408437 and ECCS-2033433.

Signal Processing for Joint Radar Communications, First Edition.
Edited by Kumar Vijay Mishra, M. R. Bhavani Shankar, Björn Ottersten, and A. Lee Swindlehurst.

ity of nonintrusive, privacy-preserving wireless monitoring, and their sensitivity to small body movements, such as respiration and heartbeat monitoring, and larger movements such as fall detection [Mercuri et al., 2013, Li et al., 2017, Liang et al., 2018, Quaiyum et al., 2018, Liu et al., 2018b]. Moreover, the UWB radar sensors perform robustly and reliably under different sound, dirt, and light conditions and are superior to ultrasound, infrared, and optical sensors. Radar sensors can easily penetrate clothing, curtains, glass, and wooden obstacles to provide continuous monitoring. Multiple-input multiple-output (MIMO) radars have been used for human localization and vital sign detection for multiple humans [Liu et al., 2018b, Shang et al., 2020, Cardillo and Caddemi, 2020]. However, the 24 GHz mm-wave band has been licensed for 5G deployment in most major US cities [Web, 2019]. It is therefore inevitable that there will be cases in which UWB radar vital sign detectors will operate in the vicinity of 5G communications devices. In such scenarios, achieving the high reliability requirement of vital sign monitoring and also the high data rate requirement of 5G communications systems requires careful mutual interference management. Further, the low latency requirement of 5G communications requires an efficient implementation of the interference management mechanism. All of these challenging factors set up a perfect application scenario and motivate research on spectrum sharing techniques.

Another example of spectrum sharing is the coexistence of automotive radars and vehicle-to-vehicle (V2V)/vehicle-to-infrastructure (V2I) (or V2X) communications. Autonomous driving has become the most anticipated and heavily invested-upon technology by the automotive industry. Along with cameras and LIDAR, high-resolution automotive imaging radars are essential sensors, enabling high-performance long-range environment perception, while unlike cameras and LIDAR, they function reliably under all lighting and weather conditions. Next-generation radars operate in 76–81 GHz W-band spectrum, benefiting from wide bandwidths for high-range resolution. Current V2X communications operate at the 5.9 GHz frequency band and using around 30 MHz bandwidth, for both ITS-G5 (also known as IEEE 802.11p) [Sjoberg et al., 2017] and LTE-V2X [LTE, 2018]. Connected vehicles [Uhlemann, 2018] is an area currently attracting huge attention and envisions enabling vehicles to share their sensor measurements with each other and with a roadside infrastructure or the cloud. Sensing information from multiple cars and infrastructure nodes can be combined to extend the visual horizon of each vehicle; this would significantly improve scene analysis and prediction, thus enabling proactive rather than reactive subsequent control actions, improving safety and efficiency. However, sharing sensor information would involve large amounts of data and thus would require a lot of bandwidth. Latency need to be kept low in order to support the tight control of vehicles. As it is impossible to achieve the required Gigabits per second (Gbps) rate via the narrow bandwidth channels in the 5.9 GHz band, researchers are looking into the possibilities of sharing the W-band spectrum, used by the automotive radars [Kumari et al., 2018]. This is an ideal scenario for spectrum sharing, as managing interference would be very important in meeting sensing and communication rate requirements. Automotive imaging radars usually have large MIMO arrays for high angular resolution [Sun et al., 2020]. Massive MIMO techniques are also key technology for 5G mm-wave communications, with similar expectation for the next-generation V2X. The large spatial degree of freedom of both systems will greatly facilitate the sharing of the spectrum. Accompanying challenges include reduction of the hardware cost of large arrays and design of effective and efficient spectrum sharing algorithms that control mutual interference.

In this chapter we study spectrum sharing between two *coexisting systems*, namely, a sparse sensing based MIMO radar system and a MIMO communications system, operating in the vicinity of each other. We formulate spectrum sharing as a constrained optimization problem and present efficient joint design methods for controlling interference, which push the envelope of addressing the aforementioned challenges.

8.1.1 Literature Review

Spectrum sharing methods for radar and communications systems are focused on enabling the two systems to share the spectrum efficiently by minimizing interference effects. There are two lines of work on this topic, namely, sharing between coexisting radar and communications systems, where the two systems are spatially distributed and separately operated by different entities but share the same spectrum, and sharing in systems that use a single platform, also referred to as dual-functional radar–communications (DFRC) systems. DFRC systems achieve sensing and communication using the same transmitter or receiver resources. Readers can refer to Zheng et al. [2019], Hassanien et al. [2016], Liu et al. [2020], and Mishra et al. [2019] for a comprehensive overview of spectrum sharing systems. This chapter focuses on coexisting systems and presents a brief overview on related works.

8.1.1.1 Noncooperative Spectrum Sharing Methods

Early works on coexistence of radar and communication systems operating in the 3500–3600 MHz band have explored large physical separation [Lackpour et al., 2011, Sanders et al., 2012, Bell et al., 2014], or the usage of exclusion zones [Drocella et al., 2015]. The latter approach was proposed by US National Telecommunications and Information Administration (NTIA) to protect base stations (BSs) from interference due to radar operating nearby. However, those zones cover large US metropolitan areas, and such an approach is not spectrally efficient. During the 5G rollout in the United States, it was delayed near 88 airports across the country due to possible interference with aircraft radar altimeters, which are critical for navigation in low-visibility situations [Sweeney et al., 2022]. To mitigate the issue, the Federal Aviation Administration (FAA) and wireless carriers temporarily agreed to implement "buffer zones" around airports across the country, which would lead to inefficient spectral utilization near many major airports. Dynamic spectrum access methods [Zhao and Sadler, 2007, Hossain et al., 2009, Wang et al., 2008, Bhat et al., 2012, Saruthirathanaworakun et al., 2012] using orthogonal frequency-division multiplexing (OFDM) signals have also been proposed, that optimally allocate subcarriers between the two systems [Surender et al., 2010, Gogineni et al., 2013, Turlapaty and Jin, 2014]. Methods that use specially designed radar waveforms have also been considered [Aubry et al., 2014, 2015, Huang et al., 2015, Bica et al., 2016]. In Zheng et al. [2018a], the authors consider a noncooperative spectrum sharing scenario for a communications system coexisting with a set of radar systems, where the communications system knows that the interfering radar waveforms fall in the subspace of a known dictionary. Based on the sparse property of the radar interference, compressed sensing techniques are used to simultaneously estimate the radar interference and demodulate the communication symbols. There are also works that explore the spatial degrees of freedom enabled by the use of multiple antennas at both systems that allow coexistence in time and space of the two systems. MIMO radars offer large spatial degrees of freedom and thus have been considered for spectrum sharing with communications systems [Sodagari et al., 2012, Babaei et al., 2013, Amuru et al., 2013, Khawar et al., 2014, 2016, Shahriar et al., 2015, Deng and Himed, 2013]. However, in these works, the interference mitigation is addressed either for the communications system [Sodagari et al., 2012, Babaei et al., 2013, Khawar et al., 2014, Amuru et al., 2013] or for the radar [Deng and Himed, 2013], but not for both. For example, in the well-studied null space projection (NSP) based scheme of Sodagari et al. [2012], Babaei et al. [2013], Amuru et al. [2013], Khawar et al. [2014, 2016], and Shahriar et al. [2015], the radar eliminates its interference to the communications system by projecting its waveforms onto the null space of the interference channel. However, the interference generated by the communications system to the radar is not considered. By applying spatial multiplexing to each system in isolation, however, cannot achieve the gain a coordinated operation of the two systems could.

8.1.1.2 Cooperative Spectrum Sharing Methods

Joint design of the signaling schemes of coexisting radar and communications systems enjoys more design degrees of freedom and thus has the potential to better suppress the mutual interference. That argument was first advocated and verified in Li et al. [2016b] in the context of cooperative spectrum sharing between MIMO radars employing sparse sensing and matrix completion (MC) (MIMO-MC) and point-to-point MIMO communications systems. In Li et al. [2016b], the covariance matrix of the communication signal and the subsampling matrix of the MIMO-MC radar are jointly optimized to minimize the interference at the radar receiver, subject to power and capacity constraints. MIMO-MC radars are shown particularly well suited for spectrum sharing [Li and Petropulu, 2015b, Li et al., 2016b]. The reason lies in the time-varying sparse sampling scheme applied at each radar receive antenna, which effectively modulates the communication–radar interference channel and increases its null space. This gives the opportunity to the communications system to transmit along that null space and thus avoid interfering with the radar. Later works [Li and Petropulu, 2016, 2017a,b] consider more realistic coexistence scenarios, extending the joint design to also include a radar precoder with consideration of signal-dependent clutter.

Inspired by the joint design–based approach, several works have been proposed for spectrum sharing of coexisting radar–communications systems. To address the fact that the interference generated by the radar toward the communication receiver is intermittent [Li and Petropulu, 2017b], the work in Zheng et al. [2018b] considers the coexistence between a communications system and a pulsed radar and formulates the communication rate as a compound rate by combining rates with and without radar interference. Then, the compound rate is optimized under power and radar signal-to-interference-plus-noise ratio (SINR) constraints. The work in Liu et al. [2017] extends the spectrum sharing ideas [Li et al., 2016b] to the coexistence of a MIMO radar and a multiantenna BS communicating with multiple MIMO users (MU-MIMO). Assuming imperfect knowledge of the interference channel, Liu et al. [2017] proposes a robust beamforming design for the BS to maximize the radar detection probability, while meeting the BS power budget and the SINR of the downlink (DL) users. In Liu et al. [2018a], the authors further exploit the known DL multiuser interference constructively to enhance the useful signal power. They propose a constructive interference-based beamforming design that improves the DL SINR significantly compared to Liu et al. [2017] under the same transmit power budget. The coexistence of multiple communications users and multiple radars is considered in [Cui et al., 2018], where a spatial precoder–decoder design based on interference alignment is proposed to achieve the desired multiplexing gain or diversity order.

In this chapter, we study cooperative spectrum sharing between MIMO-MC radars and point-to-point MIMO communications systems. We present the system model, the mathematical formulation of the joint design, and the solution of the optimization problem. The chapter consolidates the ideas presented in Li and Petropulu [2015a,b, 2016, 2017a,b] and Li et al. [2016a,b]. The remainder of this chapter is organized as follows. Section 8.2 provides some background on MIMO-MC radars. Section 8.3 presents the coexistence model between a MIMO-MC radar and a MIMIO communications system and outlines all assumptions made. Section 8.4 formulates the problem of spectrum sharing and provides efficient algorithms for solving it. Finally, Section 8.5 presents simulation results demonstrating the performance gains of cooperation, and Section 8.6 offers some concluding remarks.

8.2 MIMO Radars Using Sparse Sensing

In MIMO radars, each transmit (TX) antenna is connected to a different RF chain, which is fed by a different waveform. The receive (RX) antennas forward their measurements to a fusion center, where a "data matrix" is populated with the information received by each RX antenna. This matrix is then used in array processing methods to estimate target information. Sparse sensing techniques have gained huge popularity in application for MIMO radars in recent years, among which compressive sensing (CS) and MC-based MIMO radars [Candès and Plan, 2010] are widely studied. Compressive sensing-based MIMO (MIMO-CS) radars [Chen and Vaidyanathan, 2008b, Yu et al., 2010, Amin, 2014] achieve high localization performance with significantly fewer measurements by exploiting the target sparsity in range-Doppler-angle space. However, the basis mismatch problem arises in MIMO-CS radars due to the need of fitting a grid on the target space. On the other hand, MIMO-MC radars avoid such basis mismatch problems while achieving significantly reduced sample amount and hardware complexity [Sun et al., 2015]. It is known that, under certain conditions, a low-rank matrix can be reconstructed with proven accuracy based on a small set of its entries via MC techniques [Candès and Plan, 2010]. For a relatively small number of targets compared to the number of TX and RX antennas, MIMO-MC radars capitalize on the low-rank property of the data matrix to reduce the number of measurements that need to be sampled and recover the full data matrix later without signal-to-noise ratio (SNR) loss [Sun et al., 2015, Kalogerias and Petropulu, 2014, Sun and Petropulu, 2015].

Let us consider a MIMO radar system using uniform linear arrays (ULA) for both transmission and reception, with both arrays co-located. Let $M_{t,R}$ and $M_{r,R}$ be the number of TX and RX antennas, respectively, and d_t and d_r the inter-element spacing at the transmit and receive arrays, respectively. The radar transmits pulses, with pulse repetition interval (PRI) T_{PRI} and carrier wavelength λ_c. The radar operates in two phases; in the first phase, the TX antennas transmit waveforms and the RX antennas forward their measurements along with their sampling scheme to a fusion center. Suppose that in a given range bin there exist K far-field targets at distinct angles $\{\theta_k\}$, with reflection coefficients $\{\beta_k\}$ and Doppler shifts $\{\nu_k\}$. In the absence of clutter, the noisy data matrix at the fusion/control center can be formulated as [Sun et al., 2015, Kalogerias and Petropulu, 2014, Sun and Petropulu, 2015]

$$\mathbf{Y}_R = \mathbf{V}_r \boldsymbol{\Sigma} \mathbf{V}_t^T \mathbf{P}\mathbf{S} + \mathbf{W}_R, \tag{8.1}$$

where the mth row of $\mathbf{Y}_R \in \mathbb{C}^{M_{r,R} \times L}$ contains the L fast-time raw samples forwarded by the mth antenna [Richards, 2005]; $\mathbf{S} = [\mathbf{s}(1), \dots, \mathbf{s}(L)]$ is the waveform matrix, with $\mathbf{s}(l) = [s_1(l), \dots, s_{M_{t,R}}(l)]^T$ being the lth snapshot across the transmit antennas; the transmit waveforms are assumed to be orthogonal, i.e. it holds that $\mathbf{S}\mathbf{S}^H = \mathbf{I}_{M_{t,R}}$ [Sun et al., 2015]; $\mathbf{W}_R$ denotes additive noise; and $\mathbf{P} \in \mathbb{C}^{M_{t,R} \times M_{t,R}}$ denotes the transmit precoding matrix. $\mathbf{V}_t \triangleq [\mathbf{v}_t(\theta_1), \dots, \mathbf{v}_t(\theta_K)]$ and $\mathbf{V}_r \triangleq [\mathbf{v}_r(\theta_1), \dots, \mathbf{v}_r(\theta_K)]$, respectively, denote the transmit and receive steering matrix, with $\mathbf{v}_r(\theta) \in \mathbb{C}^{M_{r,R}}$, $\mathbf{v}_t(\theta) \in \mathbb{C}^{M_{t,R}}$ denoting, respectively, the transmit and receive steering vectors, defined as

$$\mathbf{v}_r(\theta) \triangleq \left[e^{-j2\pi 0 \vartheta^r}, \dots, e^{-j2\pi (M_{r,R}-1)\vartheta^r}\right]^T, \tag{8.2}$$

$$\mathbf{v}_t(\theta) \triangleq \left[e^{-j2\pi 0 \vartheta^t}, \dots, e^{-j2\pi (M_{t,R}-1)\vartheta^t}\right]^T, \tag{8.3}$$

where $\vartheta^r = d_r \sin(\theta)/\lambda_c$ and $\vartheta^t = d_t \sin(\theta)/\lambda_c$ denote the spatial frequencies in the receive and transmit arrays. Matrix $\boldsymbol{\Sigma}$ is defined as $\text{diag}([\beta_1 e^{j2\pi\nu_1}, \dots, \beta_K e^{j2\pi\nu_K}])$, where β_k and ν_k are the magnitude and phase of the reflection coefficient for kth target. $\mathbf{D} \triangleq \mathbf{V}_r \boldsymbol{\Sigma} \mathbf{V}_t^T$ is the target response matrix. At the fusion center, $\mathbf{Y}_R$ passes through the matched filters, after which target estimation is performed via standard array processing methods [Krim and Viberg, 1996].

When K is smaller than $M_{r,R}$ and L, the data matrix, $\mathbf{M} \triangleq \mathbf{DPS}$, is low rank and under certain conditions can be provably recovered based on a subset of its entries. This observation gave rise to MIMO-MC radars [Sun et al., 2015, Kalogerias and Petropulu, 2014, Sun and Petropulu, 2015], where each RX antenna subsamples the target returns (termed as sampling scheme #1 in Sun et al. [2015]) or performs matched filtering with a subset of transmit waveforms (termed as sampling scheme #2 in Sun et al. [2015]), and forwards the samples to the fusion center. Sampling scheme

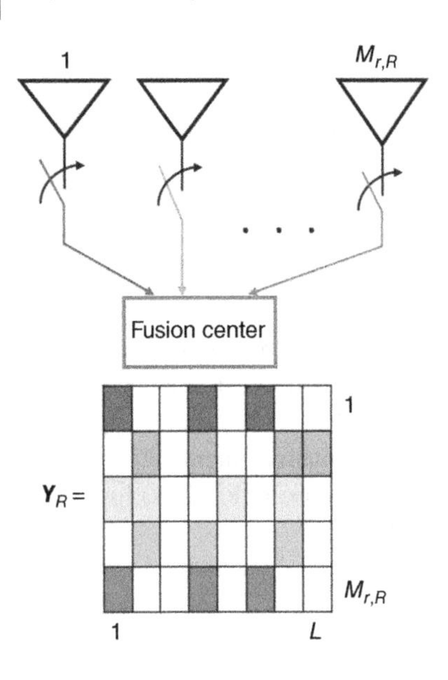

Figure 8.1 A collocated MIMO-MC radar system using random sampling at the receive antennas Source: Based on S. Sun et al., [2015].

#1 is illustrated in Figure 8.1, where a different random subsampling pattern is applied at each receive antenna. In the following, we assume that sampling scheme #1 occurs.

The partially filled data matrix at the fusion center can be mathematically expressed as follows:

$$\mathbf{\Omega} \odot \mathbf{Y}_R = \mathbf{\Omega} \odot (\mathbf{M} + \mathbf{W}_R), \tag{8.4}$$

where $\odot$ denotes the Hadamard product; $\mathbf{\Omega}$ is the subsampling matrix containing 0s and 1s. The subsampling rate p equals $\|\mathbf{\Omega}\|_0/(LM_{r,R})$, where $\|\mathbf{\Omega}\|_0$ denotes the number of 1s in $\mathbf{\Omega}$. When $p = 1$, the $\mathbf{\Omega}$ matrix is filled with 1s, and the MIMO-MC radar is identical to the traditional MIMO radar. At the fusion center, the completion of $\mathbf{M}$ can be achieved by the following nuclear norm minimization problem [Candès and Plan, 2010]:

$$\min_{\mathbf{M}} \|\mathbf{M}\|_* \quad \text{s.t. } \|\mathbf{\Omega} \odot \mathbf{M} - \mathbf{\Omega} \odot \mathbf{Y}_R\|_F \leq \delta, \tag{8.5}$$

where $\|\cdot\|_*$ denotes the matrix nuclear norm; $\delta > 0$ is a parameter determined by the sampled entries of the noise matrix, i.e. $\mathbf{\Omega} \odot \mathbf{W}_R$. It is important to note that, to preserve all the received target echo power, the data matrix $\mathbf{M}$ needs to be stably reconstructed with high accuracy. It is shown in Candès and Plan [2010] that the matrix recovery error is proportional to the noise level δ, given that the following conditions hold:

- $\mathbf{M}$ obeys incoherent property with parameters (μ_0, μ_1).
- $\mathbf{\Omega}$ corresponds to subsampling matrix entries uniformly at random with $m \triangleq M_{r,R}Lp \geq CKn \log n$, where $n \triangleq \max\{M_{r,R}, L\}$.

The *incoherence parameters* (μ_0, μ_1) of $\mathbf{M}$ are defined by

$$\mu_0 \geq \max(\mu(U), \mu(V)), \tag{8.6}$$

$$\mu_1 \sqrt{\frac{K}{M_{r,R}L}} \geq \left\| \sum_{k=1}^{K} \mathbf{U}_{\cdot k} \mathbf{V}_{\cdot k}^H \right\|_\infty, \tag{8.7}$$

where $\mathbf{U} \in \mathbb{C}^{M_{r,R} \times K}$ and $\mathbf{V} \in \mathbb{C}^{L \times K}$ contain the left and right singular vectors of $\mathbf{M}$; $\mathbf{U}_{\cdot k}$ denotes the kth column vector of $\mathbf{U}$; $\|\cdot\|_\infty$ denotes the maximum entry of the matrix. The coherence of subspace V spanned by basis matrix $\mathbf{V}$ is defined as

$$\mu(V) \triangleq \frac{L}{K} \max_{1 \le l \le L} \|\mathbf{V}_{l\cdot}\|^2 \in \left[1, \frac{L}{K}\right], \tag{8.8}$$

where $\mathbf{V}_{l\cdot}$ denotes the lth row vector of $\mathbf{V}$;

Let $n = \max\{M_{r,R}, L\}$. According to Candès and Recht [2009], assuming that matrix $\mathbf{M}$ is sampled uniformly at random, there exist constants C and c, such that if

$$m \ge C \max\left\{\mu_1^2, \mu_0^{1/2}\mu_1, \mu_0 n^{1/4}\right\} nK\gamma \log n \tag{8.9}$$

for some $\gamma > 2$, the minimizer of the nuclear norm problem is unique and equal to $\mathbf{M}$ with probability at least $1 - cn^{-\gamma}$.

The condition in (8.9) implies that the smaller the coherence parameters, the smaller the number of samples is needed for recovering the full matrix $\mathbf{M}$. Upper bounds of the coherence parameters can give us an idea of how small those parameters can be.

The upper bounds on the incoherence parameters of $\mathbf{M}$ for the case $\mathbf{P} = \mathbf{I}_{M_{t,R}}$ are given in the following theorem [Sun and Petropulu, 2015].

Theorem 8.1 ***(Incoherence property of* $\mathbf{M}$ *when* $\mathbf{P} = \mathbf{I}_{M_{t,R}}$*)*** *Let the minimum spatial frequency separation of the K targets be ξ_t and ξ_r w.r.t. the transmit and receive arrays. Denoting the Fejér kernel by $F_n(x) = \frac{1}{n}\frac{\sin^2(\pi n x)}{\sin^2(\pi x)}$, and for $d_t = d_r = \lambda_c/2$ and*

$$K \le \min\left\{\sqrt{\frac{M_{r,R}}{F_{M_{r,R}}(\xi_r)}}, \sqrt{\frac{M_{t,R}}{F_{M_{t,R}}(\xi_t)}}\right\}, \tag{8.10}$$

it holds that

$$\mu(U) \le \frac{\sqrt{M_{r,R}}}{\sqrt{M_{r,R}} - (K-1)\sqrt{F_{M_{r,R}}(\xi_r)}} \triangleq \mu_0^r. \tag{8.11}$$

Further, if every snapshot of the waveforms $\mathbf{S}_{\cdot l} \equiv \mathbf{s}(l)$ satisfies that

$$|\mathbf{S}_{\cdot l}^T \mathbf{v}_t(\theta)|^2 = \frac{M_{t,R}}{L}, \ \forall l \in \mathbb{N}_L^+, \theta \in \left[-\frac{\pi}{2}, \frac{\pi}{2}\right], \tag{8.12}$$

then $\mu(V)$ is upper bounded by

$$\mu(V) \le \frac{\sqrt{M_{t,R}}}{\sqrt{M_{t,R}} - (K-1)\sqrt{F_{M_{t,R}}(\xi_t)}} \triangleq \mu_0^t. \tag{8.13}$$

Consequently, the matrix $\mathbf{M}$ is incoherent with parameters $\mu_0 \triangleq \max\{\mu_0^r, \mu_0^t\}$ and $\mu_1 \triangleq \sqrt{K}\mu_0$ with probability 1.

The Fejér kernel achieves its maximum value at zero. As the minimum spatial separation of targets decreases, the value of the Fejér kernel increases. Then, the number of targets for which the theorem holds is smaller according to (8.10). The coherence bounds also increase according to (8.11) and (8.13). Therefore, the only way to increase K, i.e. accommodating more targets, and to reduce coherence bounds is to increase the number of transmit and receive antennas. It is shown in Sun and Petropulu [2015] that the coherence of matrix $\mathbf{M}$ is asymptotically optimal under the orthogonality condition of the radar waveforms in (8.12), i.e. $\mu(V) \to 1$ and $\mu(U) \to 1$ as $M_{t,R}$ and $M_{r,R}$ tend to infinity.

The orthogonality property of the radar waveforms in (8.12) was used to design waveforms with good incoherence properties. The work of Sun and Petropulu [2015] involves numerical optimization on the complex Stiefel manifold, which has high computational complexity. However, radar waveforms need to be updated frequently as security against adversaries, which makes the computational cost very high. It was also observed in Sun and Petropulu [2015] via simulations that using a random unitary matrix [Zyczkowski and Kus, 1994] as **S** resulted in performance very close to that of the optimum waveform.

The MC performance degrades severely when the SINR drops below 10 dB [Sun et al., 2015], which suggests that along with "good" radar waveforms, a precoder design for interference mitigation is very important. In the following, we consider a MIMO-MC radar that uses a random unitary matrix as the waveform matrix **S**, and a nontrivial precoder matrix **P** [Li and Petropulu, 2017a,b].

8.2.1 MIMO-MC Radar Using Random Unitary Matrix

A waveform matrix **S** that is random unitary [Zyczkowski and Kus, 1994] can be easily obtained through Gram–Schmidt orthogonalization of a random matrix, whose entries are independent and identically distributed (i.i.d.). The following theorem provides an upper bound on the incoherence parameter $\mu(V)$ of the noise free data matrix, **M**, when a random unitary waveform matrix **S** and a nontrivial radar precoder **P** are used.

Theorem 8.2 *Consider the MIMO-MC radar presented in Section 8.2, using a random unitary waveform matrix* **S**, *and with* **M** *as defined in (8.4). For any transmit precoder* **P** *such that for* $K_0 = Rank(\mathbf{M})$, *it holds that* $K_0 \leq K$, *and for an arbitrary transmit array geometry and target angles, the coherence of the right singular vector subspace of* **M** *is bounded as*

$$\mu(V) \leq \frac{K_0 + 2\sqrt{3K_0 \ln L} + 6 \ln L}{K_0} \triangleq \tilde{\mu}_0^t \tag{8.14}$$

with probability $1 - L^{-2}$, *and the coherence of subspace U obeys* $\mu(U) \leq \frac{K}{K_0}\mu_0^r$, *where* μ_0^r *is defined in Theorem 8.1.*

Proof: The following lemma is used in the proof.

Lemma 8.1 *Laurent and Massart [2000] Let* S_N *be a* χ^2 *random variable with N degrees of freedom. Then, for each* $t > 0$

$$\mathbb{P}\left[S_N - N \geq t\sqrt{2N} + t^2\right] \leq e^{-t^2/2}. \tag{8.15}$$

It is clear that K_0 is not larger than K. Recall that **M** has a compact singular value decomposition (SVD) given as

$$\mathbf{M} = \mathbf{U}\boldsymbol{\Gamma}\mathbf{V}^H, \tag{8.16}$$

where $\mathbf{U} \in \mathbb{C}^{M_{r,R} \times K_0}$ and $\mathbf{V} \in \mathbb{C}^{L \times K_0}$ contain the left and right singular vectors of **M**; $\boldsymbol{\Gamma} \in \mathbb{R}^{K_0 \times K_0}$ is diagonal containing the singular values. Consider the QR decomposition of $\mathbf{V}_r$ and $\mathbf{S}^T\mathbf{P}^T\mathbf{V}_t$, i.e.

$$\begin{aligned} \mathbf{V}_r &= \mathbf{Q}_r\mathbf{R}_r, \\ \mathbf{S}^T\mathbf{P}^T\mathbf{V}_t &= \mathbf{Q}_t\mathbf{R}_t, \end{aligned} \tag{8.17}$$

where $\mathbf{Q}_r \in \mathbb{C}^{M_{r,R} \times K}$ and $\mathbf{Q}_t \in \mathbb{C}^{L \times K_0}$ are with orthonormal columns, $\mathbf{R}_r \in \mathbb{C}^{K \times K}$ is upper triangular, and $\mathbf{R}_t \in \mathbb{C}^{K_0 \times K}$ has an upper staircase form. The matrix $\mathbf{R}_r\boldsymbol{\Sigma}\mathbf{R}_t^T \in \mathbb{C}^{K \times K_0}$ is full column rank with a compact SVD given by $\mathbf{U}_1\mathbf{\Gamma}_1\mathbf{V}_1^H$, where $\mathbf{U}_1 \in \mathbb{C}^{K\times K_0}$, $\mathbf{V}_1 \in \mathbb{C}^{K_0\times K_0}$, $\mathbf{U}_1^H\mathbf{U}_1 = \mathbf{V}_1^H\mathbf{V}_1 = \mathbf{I}_{K_0}$, and $\mathbf{\Gamma}_1$ is diagonal, containing the singular values of $\mathbf{R}_r\mathbf{\Sigma}\mathbf{R}_t^T$. Therefore, we have

$$\mathbf{M} = \mathbf{Q}_r\mathbf{U}_1\mathbf{\Gamma}_1\mathbf{V}_1^H\mathbf{Q}_t^T = \mathbf{Q}_r\mathbf{U}_1\mathbf{\Gamma}_1(\mathbf{Q}_t^*\mathbf{V}_1)^H, \tag{8.18}$$

which is a valid SVD of $\mathbf{M}$. The uniqueness of the singular values of a matrix indicates that $\mathbf{\Gamma} \equiv \mathbf{\Gamma}_1$. Therefore, we can choose $\mathbf{U} = \mathbf{Q}_r\mathbf{U}_1$ and $\mathbf{V} = \mathbf{Q}_t^*\mathbf{V}_1$. We have that

$$\begin{aligned}\mu(U) &= \frac{M_{r,R}}{K_0}\sup_{m\in\mathbb{N}^+_{M_{r,R}}} \|(\mathbf{Q}_r)_{m\cdot}\mathbf{U}_1\|_2^2 \\ &\le \frac{M_{r,R}}{K_0}\sup_{m\in\mathbb{N}^+_{M_{r,R}}} \|(\mathbf{Q}_r)_{m\cdot}\|_2^2 = \frac{K}{K_0}\mu_0^r,\end{aligned} \tag{8.19}$$

where μ_0^r is the upper bound on $\mu(U)$ defined in Theorem 8.1. We also have

$$\mu(V) = \frac{L}{K_0}\sup_{l\in\mathbb{N}_L^+} \|(\mathbf{Q}_t^*)_{l\cdot}\mathbf{V}_1\|_2^2 = \frac{L}{K_0}\sup_{l\in\mathbb{N}_L^+} \|(\mathbf{Q}_t)_{l\cdot}\|_2^2. \tag{8.20}$$

If K_0 is strictly smaller than K, we cannot represent $\mathbf{Q}_t$ in terms of $\mathbf{S}^T\mathbf{P}^T\mathbf{V}_t$ and $\mathbf{R}_t$ because of the singularity of $\mathbf{R}_t$. To get around this problem, we apply column permutations $\mathbf{F}$ on $\mathbf{R}_t$, so that the nonzero elements in $\mathbf{R}_t\mathbf{F} = (\mathbf{R}_1\ \mathbf{R}_2)$ appear in the beginning of each row. The resulting $\mathbf{R}_1 \in \mathbb{C}^{K_0\times K_0}$ is square, upper triangular, and invertible. The QR decomposition $\mathbf{S}^T\mathbf{P}^T\mathbf{V}_t$ can be rewritten as

$$\mathbf{S}^T\mathbf{P}^T\mathbf{V}_t\mathbf{F} = \mathbf{Q}_t\left(\mathbf{R}_1\ \mathbf{R}_2\right). \tag{8.21}$$

We can represent $\mathbf{Q}_t$ as

$$\mathbf{Q}_t = \mathbf{S}^T\mathbf{P}^T\mathbf{V}_t\mathbf{F}_{K_0}\mathbf{R}_1^{-1}, \tag{8.22}$$

where $\mathbf{F}_{K_0}$ denotes the first K_0 columns of $\mathbf{F}$. Substituting $\mathbf{Q}_t$ into $\mu(V)$, we obtain

$$\begin{aligned}\mu(V) &= \frac{L}{K_0}\sup_{l\in\mathbb{N}_L^+} \left\|(\mathbf{S}^T)_{l\cdot}\mathbf{P}^T\mathbf{V}_t\mathbf{F}_{K_0}\mathbf{R}_1^{-1}\right\|_2^2 \\ &= \frac{L}{K_0}\sup_{l\in\mathbb{N}_L^+} (\mathbf{S}^T)_{l\cdot}\mathbf{P}^T\mathbf{V}_t\mathbf{F}_{K_0}\mathbf{R}_1^{-1}\left(\mathbf{R}_1^{-1}\right)^H\mathbf{F}_{K_0}^H\mathbf{V}_t^H\mathbf{P}^*(\mathbf{S}^*)_{\cdot l}.\end{aligned} \tag{8.23}$$

It holds that

$$\begin{aligned}\mathbf{R}_1^{-1}\left(\mathbf{R}_1^{-1}\right)^H &= \left(\mathbf{R}_1^H\mathbf{R}_1\right)^{-1} \\ &= \left(\mathbf{R}_1^H\mathbf{Q}_t^H\mathbf{Q}_t\mathbf{R}_1\right)^{-1} \\ &= \left(\mathbf{F}_{K_0}^H\mathbf{V}_t^H\mathbf{P}^*\mathbf{S}^*\mathbf{S}^T\mathbf{P}^T\mathbf{V}_t\mathbf{F}_{K_0}\right)^{-1} \\ &= \left(\mathbf{F}_{K_0}^H\mathbf{V}_t^H\mathbf{P}^*\mathbf{P}^T\mathbf{V}_t\mathbf{F}_{K_0}\right)^{-1},\end{aligned} \tag{8.24}$$

where the last equality holds because $\mathbf{S}\mathbf{S}^H = \mathbf{I}_{M_{t,R}}$. Consider the QR decomposition of $\mathbf{P}^T\mathbf{V}_t\mathbf{F}_{K_0}$ given by

$$\mathbf{P}^T\mathbf{V}_t\mathbf{F}_{K_0} = \mathbf{Q}_a\mathbf{R}_a, \tag{8.25}$$

where $\mathbf{Q}_a \in \mathbb{C}^{M_{t,R}\times K_0}$ contains orthonormal columns, and $\mathbf{R}_a \in \mathbb{C}^{K_0\times K_0}$ is upper triangular and full rank. Substituting (8.24) and (8.25) into (8.23), we have

$$\begin{aligned}\mu(V) &= \frac{L}{K_0}\sup_{l\in\mathbb{N}_L^+} \mathbf{s}_l^T\mathbf{R}_a\left(\mathbf{R}_a^H\mathbf{R}_a\right)^{-1}\mathbf{R}_a^H\mathbf{s}_l^* \\ &= \frac{L}{K_0}\sup_{l\in\mathbb{N}_L^+} \mathbf{s}_l^T\mathbf{s}_l^* = \frac{L}{K_0}\sup_{l\in\mathbb{N}_L^+} \|\mathbf{s}_l\|_2^2,\end{aligned} \tag{8.26}$$

where $\mathbf{s}_l \triangleq \mathbf{Q}_a^T \mathbf{S}_{\cdot l}$, and the second equality holds because $\mathbf{R}_a$ is invertible. Based on Jiang [2006, Theorem 3], if $M_{t,R} = \mathcal{O}(L/\ln L)$, the entries of $\mathbf{S}$ can be approximated by i.i.d. Gaussian random variables with distribution $\mathcal{CN}(0, 1/L)$. Since $\mathbf{Q}_a$ has orthonormal columns, $\mathbf{s}_l \in \mathbb{C}^{K_0}, \forall l \in \mathbb{N}_L^+$ also contains i.i.d. Gaussian random variable with distribution $\mathcal{CN}(0, 1/L)$, and $L\|\mathbf{s}_l\|_2^2$ is distributed according to $\chi^2_{K_0}$. Based on Lemma 8.1 and setting $t = \sqrt{6 \ln L}$, it holds that

$$\mathbb{P}\left[L\|\mathbf{s}_l\|_2^2 \geq K_0 + 2\sqrt{3K_0 \ln L} + 6 \ln L\right] \leq L^{-3}. \tag{8.27}$$

Applying the union bound, we have

$$\mathbb{P}\left[\sup_{l \in \mathbb{N}_L^+} \|\mathbf{s}_l\|_2^2 \geq \frac{K_0 + 2\sqrt{3K_0 \ln L} + 6 \ln L}{L}\right] \leq L^{-2}. \tag{8.28}$$

Combining (8.26) and (8.28) gives

$$\mathbb{P}\left[\mu(V) \geq \frac{K_0 + 2\sqrt{3K_0 \ln L} + 6 \ln L}{K_0}\right] \leq L^{-2}. \tag{8.29}$$

From the derivation, the bound on $\mu(V)$ holds for any target angles, array geometry, and precoding matrix $\mathbf{P}$ as long as $\mathbf{P}^T \mathbf{V}_t \mathbf{F}_{K_0}$ is with full column rank K_0. This completes the proof of Theorem 8.2. ■

Based on Theorem 8.2, we have the following theorem for the incoherence parameters of $\mathbf{M}$.

Theorem 8.3 ***(Coherence of M with random unitary waveform matrix)*** *Consider the MIMO-MC radar presented in Section 8.2 with* $\mathbf{S}$ *being a random unitary matrix. For* $d_r = \lambda_c/2$, *arbitrary transmit array geometry, and for*

$$K \leq \sqrt{\frac{M_{r,R}}{F_{M_{r,R}}(\xi_r)}}, \tag{8.30}$$

the matrix $\mathbf{M}$ *is incoherent with parameters*

$$\mu_0 \triangleq \max\left\{\frac{K}{K_0}\mu_0^r, \tilde{\mu}_0^t\right\} \tag{8.31}$$

$$\mu_1 \triangleq \sqrt{K_0}\mu_0 \tag{8.32}$$

with probability $1 - L^{-2}$, *where* $\tilde{\mu}_0^t$ *is defined in Theorem 8.2.* μ_0^r *is the upper bound on* $\mu(U)$ *given in Theorem 8.1.*

The incoherence property of $\mathbf{M}$ *holds for any precoding matrix* $\mathbf{P}$ *such that the rank of* $\mathbf{M}$ *is* K_0.

Proof: The theorem can be proven by substituting the bounds on $\mu(U)$ and $\mu(V)$ in Theorem 8.1 with the bounds derived in Theorem 8.2. ■

Remarks on Theorems 8.2 *and* 8.3

1. If K_0 is $\mathcal{O}(\ln L)$, the upper bound $\tilde{\mu}_0^t > 1$ is a small constant $\mathcal{O}(1)$; this can be seen from the definition of $\tilde{\mu}_0^t$ in (8.14) by choosing $K_0 > \ln L$. Therefore, based on (8.14), the coherence parameter of $\mathbf{M}$ is close to 1, which means that $\mathbf{M}$ has a good incoherent property. A similar bound was provided on the coherence of the subspaces spanned by a random orthogonal basis in Candès and Recht [2009].
2. Unlike the results in Sun and Petropulu [2015, Theorem 2], the probabilistic bound on $\mu(V)$ is independent of the target angles and array geometry.

3. The above results hold for any random unitary matrix $\mathbf{S}$. The radar waveform can be changed periodically, which would be good for security reason, without affecting the MC performance.
4. The probabilistic bound on $\mu(V)$ in Theorem 8.2 is independent of $\mathbf{P}$ as long as the rank of $\mathbf{M}$ is K_0. This means that we can design $\mathbf{P}$, without affecting the incoherence property of $\mathbf{M}$, for the purpose of transmit beamforming and interference suppression. This key observation validates the feasibility of radar precoding-based spectrum sharing approaches for MIMO-MC radar and communications systems in the sequel. Note that this chapter focuses on the design of radar precoder in the spatial domain, not the waveform. The radar precoder will not affect the waveform ambiguity property in the time and Doppler domains.

8.3 Coexistence System Model

We consider a coexistence scenario as shown in Figure 8.2, where a MIMO-MC radar system and a MIMO communications system operate using the same carrier frequency. Note that the coexistence model is general, because when full sampling is adopted the MIMO-MC radar becomes equivalent to a traditional MIMO radar. In the coexistence system, $\mathbf{H} \in \mathbb{C}^{M_{r,C} \times M_{t,C}}$ denotes the communication channel, where $M_{r,C}$ and $M_{t,C}$ denote, respectively, the number of RX and TX antennas of the communications system. $\mathbf{G}_1 \in \mathbb{C}^{M_{r,C} \times M_{t,R}}$ and $\mathbf{G}_2 \in \mathbb{C}^{M_{r,R} \times M_{t,C}}$ denote the interference channels between the communications and radar systems. The radar operates in pulsed mode; in each pulse, it first transmits a short-pulse waveform, and then listens for target echoes for a much longer period. The duration of these two phases comprises the PRI. Figure 8.3 shows the radar–communication coexistence signal model during two periods of one radar PRI. At the communication receiver, radar interference is present only during the radar transmit period. On the other hand, the communication interference at the radar receiver is present during the entire radar PRI. In this chapter, we

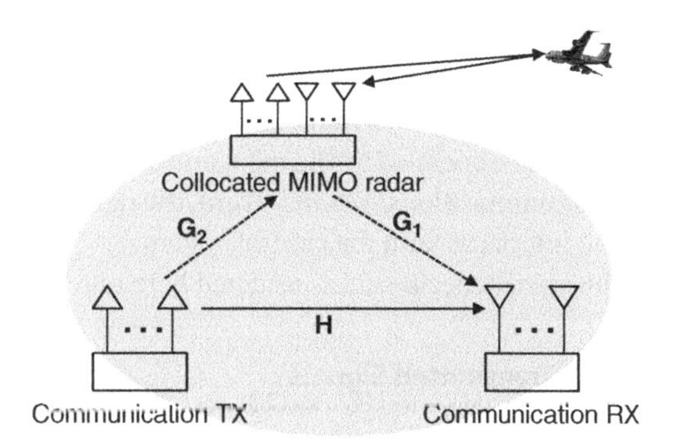

Figure 8.2 A MIMO communications system sharing spectrum with a MIMO radar system with collocated transmit and receive antennas. Source: Li et al. [2016b]/IEEE.

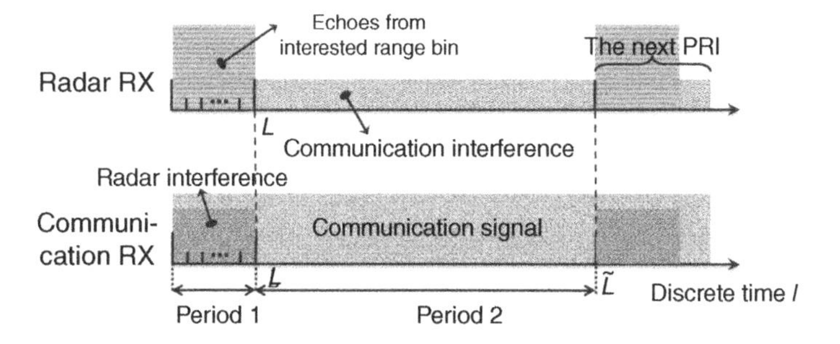

Figure 8.3 Radar–communication coexistence signal model during one radar PRI.

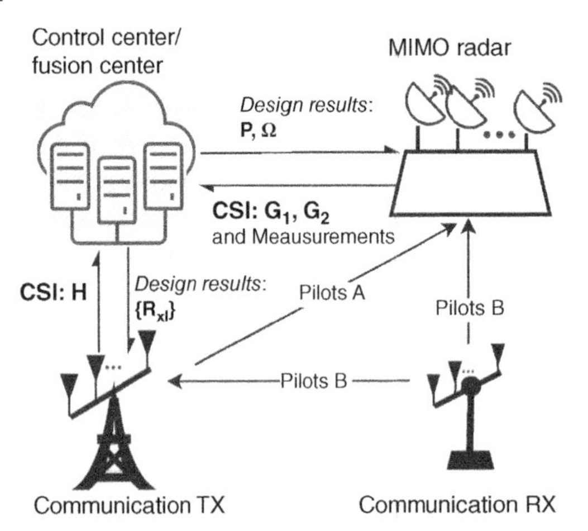

Figure 8.4 The spectrum sharing architecture. The cooperation is coordinated by the control center, a node with high computing power that also serves as the radar fusion center. The control center collects information from radar and communications systems, computes jointly optimal signaling schemes for both systems, and sends each scheme back to the corresponding system.

focus on joint radar and communication waveform design during Period 1. Readers can refer to Li et al. [2016a] for communication waveform design schemes that can reduce the interference to the radar receiver during the entire radar PRI.

Cooperative spectrum sharing can be implemented via the system architecture of Figure 8.4. The coordination of the cooperation is conducted by a control center, which collects information from the two systems, formulates, and solves an optimization problem, and passes to each system its optimal parameters. The control center can be thought of as an enhanced spectrum access system (SAS) used in the FCC release [FCC, 2015] and is connected to the radar–communications system via either a wireless link or a backhaul channel.

The control center can also integrate the functionality of the radar fusion center, i.e. target detection, estimation and tracking, and, specifically for the MIMO-MC radar, also reconstruction of the data matrix. There are several advantages of having a control center that encompasses the radar fusion center. First, a powerful all-in-one center greatly simplifies the complexity of the overall network. Second, radar operators, especially in military applications, are not willing to share information directly with civilian cellular systems out of security concerns. In such cases, the control center can be operated by the radar and enable cooperation while maintaining isolation of the radar and communications systems. Third, the radar and communications systems only need communication interfaces with the control system.

The coexistence model considered here relies on the following assumptions:

8.3.1 Transmitted Signals

It is assumed that the two systems transmit narrowband waveforms with the same symbol period. To evaluate the feasibility of radar and communications systems having the same symbol period, let us consider an S-band search and acquisition radar with range resolution equal to 300 m (a typical range resolution is between 100 and 600 m [Kopp, 2009, Web, 2015]). The corresponding radar subpulse duration is 2 μs. Communication symbol duration of 2 μs is quite typical in model cellular systems [Rappaport, 2001]. The transmitted signal is narrowband if the channel coherence bandwidth is larger than the signal bandwidth [Rappaport, 2001, Tse and Viswanath, 2005, Goldsmith, 2005]. In a macrocell, typical values for the channel coherence bandwidth are of the order of 1 MHz [Jover, 2015, LTE, 2010], which is larger than the signal bandwidth of 0.5 MHz (or symbol interval 2 μs). Thus, the narrowband assumption is typically valid.

If a higher signal bandwidth is needed, OFDM signaling can be used for both radar [Surender et al., 2010, Turlapaty and Jin, 2014] and communications systems [Jover, 2015, LTE, 2010].

Our coexistence model can still be valid on each OFDM carrier, over which the signal can be considered narrowband.

8.3.2 Fading

We assume that $\mathbf{H}$, $\mathbf{G}_1$, and $\mathbf{G}_2$ are flat fading, which is valid when the transmitted signals are narrowband. The flat-fading assumption is common practice in the radar–communications system coexistence literature [Sodagari et al., 2012, Babaei et al., 2013, Amuru et al., 2013, Khawar et al., 2014, Shahriar et al., 2015]. In addition, all channels are assumed to be block-fading over the radar PRI. For a radar with medium-pulse repetition frequency, the PRI is usually between 30 μs and 0.3 ms. The typical channel coherence time for 2.5 and 5.8 GHz carrier frequency ranges from 2 ms up to 200 ms [Andrews et al., 2007]. The channel coherence time is much larger than the radar PRI. As for the moving targets, the resulting Doppler shifts are usually assumed to be constant during one PRI [Richards, 2005, Li and Petropulu, 2015c]. Therefore, channel block fading is a reasonable assumption.

8.3.3 Channel State Information (CSI)

The channel $\mathbf{H}$ is assumed to be perfectly known at the communication transmitter. The channels $\mathbf{G}_1$ and $\mathbf{G}_2$ are also assumed to be perfectly known at the radar. The CSI estimation can be achieved using pilot channels [Sodagari et al., 2012, Filo et al., 2009] scheduled by the control center in time-division multiplexing (TDM) fashion. As a simple example, based on Figure 8.4, the communication transmitter, i.e. the BS, transmits a reference signal in pilot burst A, and this is used by the radar to estimate $\mathbf{G}_2$. The communication receiver, i.e. a user entity, transmits a reference signal in pilot burst B, and this is used by the BS and the radar to estimate $\mathbf{H}$ and $\mathbf{G}_1$, respectively, based on channel reciprocity [Rogalin et al., 2014]. All estimated CSI is sent to the control center by the radar and the BS, where it is used to jointly optimize the spatial multiplexing. Note that CSI estimation and feedback can be scheduled based on the channel coherence time, which is much larger than the radar PRI. Figure 8.5 shows a simplified schematic diagram for CSI estimation/feedback and receiving design results from the control center based on TDM. Existing techniques in cognitive radios and multiuser MIMO (MU-MIMO) [Zhang and Liang, 2008, Zhang et al., 2010, Kim and Giannakis, 2011, Lu et al., 2012, Phan et al., 2009, Du and Ratnarajah, 2013, Hou and Yang, 2011] can also be applied to reduce the overhead for CSI feedback.

8.3.4 The Radar Mode of Operation

We consider the target tracking scenario, in which the radar searches in particular directions of interest, given by set $\{\theta_k\}$, and a range bin of interest for targets with unknown radar cross-section

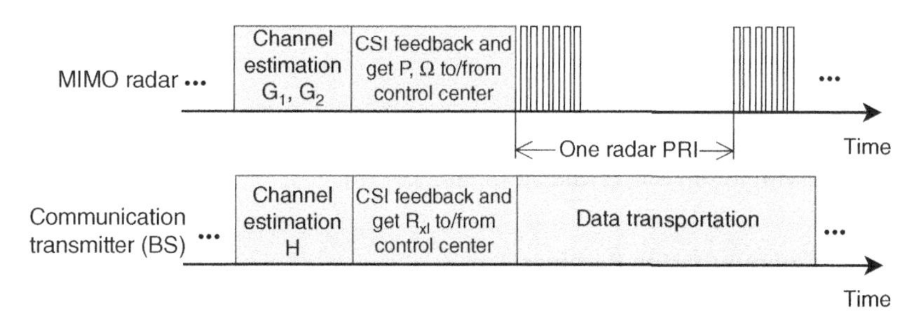

Figure 8.5 TDM-based CSI estimation and feedback and reception of design results from the control center.

(RCS) variances [Stoica et al., 2007, Cui et al., 2014]. In such scenarios, the target parameters have typically been obtained from previous tracking cycles and are used to optimize the transmission for better SINR performance [Richards, 2005].

Under the above assumptions, let us consider a target scene in a particular range bin as in Section 8.2 but with clutter. The baseband signal received by the radar receivers during L symbol durations in one radar PRI can be expressed as

$$\mathbf{\Omega} \odot \mathbf{Y}_R = \mathbf{\Omega} \odot \left(\underbrace{\mathbf{DPS}}_{\text{signal}} + \underbrace{\mathbf{CPS} + \mathbf{G}_2\mathbf{X}\mathbf{\Lambda}_2}_{\text{interference}} + \underbrace{\mathbf{W}_R}_{\text{noise}} \right), \tag{8.33}$$

and at the communication receiver as

$$\mathbf{Y}_C = \underbrace{\mathbf{HX}}_{\text{signal}} + \underbrace{\mathbf{G}_1\mathbf{PS}\mathbf{\Lambda}_1}_{\text{interference}} + \underbrace{\mathbf{W}_C}_{\text{noise}}. \tag{8.34}$$

In the above equations, $\mathbf{Y}_R$, $\mathbf{D}$, $\mathbf{P}$, $\mathbf{S}$, $\mathbf{W}_R$, and $\mathbf{\Omega}$ are defined in Section 8.2. Note that delay in the radar signal model is assumed known in current tracking cycle and properly compensated. The waveform-dependent interference **CPS** contains interferences from point scatterers (clutters or interfering objects). Suppose that there are K_c point clutters with angles $\{\theta_k^c\}$, reflection coefficients $\{\beta_k^c\}$ in the same range bin as the targets. $\mathbf{C} \triangleq \sum_{k=1}^{K_c} \beta_k^c \mathbf{v}_r(\theta_k^c)\mathbf{v}_t^T(\theta_k^c)$ is the clutter response matrix. $\mathbf{Y}_C$ and $\mathbf{W}_C$ denote the received signal and additive noise at the communication RX antennas, respectively. The columns of $\mathbf{X} \triangleq [\mathbf{x}(1), \dots, \mathbf{x}(L)]$ are codewords from the codebook of the communications system. We assume that $\mathbf{W}_{R/C}$ contains i.i.d. random entries distributed as $\mathcal{CN}(0, \sigma_{R/C}^2)$. The diagonal matrix $\mathbf{\Lambda}_i, i \in \{1,2\}$ contains the random phase offset $e^{j\alpha_{il}}$ between the MIMO-MC radar and the communications system at the lth symbol. The time-varying phase offsets are results of the random-phase jitters of the oscillators between the radar transmitter and the communication receiver and vice versa [Mudumbai et al., 2007, Li et al., 2016b]. Note that the Doppler shift will not be an issue for the design in this chapter. The radar signal model in (8.33) is for the fast-time samples received in one radar pulse. In the radar literature, the Doppler shift is usually assumed to be constant during one radar pulse [Richards, 2005, Chen and Vaidyanathan, 2008a, Li and Petropulu, 2015c, Yu et al., 2010]. Therefore, as shown in our signal model, the Doppler shift can be absorbed into the target RCS and does not affect our design.

The control center aims to protect the radar system and maximize the spectrum efficiency. In the following, we present a joint design of communication and radar transmissions, which will be implemented at the control center, so that we minimize the interference at the radar RX antennas for successful MC, while satisfying certain communications system requirements [Li and Petropulu, 2017b].

8.4 Cooperative Spectrum Sharing

In this section, we formulate the MIMO-MC radar and MIMO communication spectrum sharing problem and present an algorithm to solve it [Li and Petropulu, 2017b].

8.4.1 Interference at the Communication Receiver

The covariance of interference plus noise at the communication receiver equals

$$\mathbf{R}_{\text{Cin}} = \mathbf{G}_1\mathbf{\Phi}\mathbf{G}_1^H + \sigma_C^2\mathbf{I}, \tag{8.35}$$

where $\mathbf{\Phi} \triangleq \mathbf{P}\mathbf{P}^H/L$ is positive semidefinite. For $l \in \mathbb{N}_L^+$, the *instantaneous* information rate is unknown because the interference plus noise is not necessarily Gaussian due to the random-phase offset $e^{j\alpha_{il}}$. However, a lower bound of the rate equals [Diggavi and Cover, 2001]

$$\underline{C}(\mathbf{R}_{xl}, \mathbf{\Phi}) \triangleq \log_2 \left|\mathbf{I} + \mathbf{R}_{\text{Cin}}^{-1}\mathbf{H}\mathbf{R}_{xl}\mathbf{H}^H\right|. \tag{8.36}$$

The above bound is achieved when the codeword $\mathbf{x}(l)$, $l \in \mathbf{N}_L^+$ is distributed as $\mathcal{CN}(0, \mathbf{R}_{xl})$. The average communication rate over L symbols is as follows:

$$C_{\text{avg}}(\{\mathbf{R}_{xl}\}, \mathbf{\Phi}) \triangleq \frac{1}{L}\sum_{l=1}^{L}\underline{C}(\mathbf{R}_{xl}, \mathbf{\Phi}), \tag{8.37}$$

where $\{\mathbf{R}_{xl}\}$ denotes the set of all $\mathbf{R}_{xl}$'s.

8.4.2 Interference at the MIMO-MC Radar

Since the MIMO-MC radar subsamples $\mathbf{Y}_R$, only the sampled interference affects the MC performance. Let us define the *effective signal power* (ESP) and *effective interference power* (EIP) at the radar RX node, taking expectation over target–clutter reflection coefficients and radar–communication waveforms, i.e.

$$\begin{aligned}
\text{ESP} &\triangleq \mathbb{E}\left\{\text{tr}\left(\mathbf{\Omega} \odot (\mathbf{DPS})\left(\mathbf{\Omega} \odot (\mathbf{DPS})\right)^H\right)\right\} \\
&= \mathbb{E}\left\{\text{tr}\left\{\left[\sum_k \beta_k \mathbf{\Omega} \odot (\mathbf{D}_k\mathbf{PS})\right)\right]\left[\sum_k \beta_k \mathbf{\Omega} \odot (\mathbf{D}_k\mathbf{PS})^H\right]\right\}\right\} \\
&= \mathbb{E}\left\{\text{tr}\left\{\sum_k\sum_j \beta_k\beta_j \mathbf{\Omega} \odot (\mathbf{D}_k\mathbf{PS})\left[\mathbf{\Omega} \odot (\mathbf{D}_j\mathbf{PS})^H\right]\right\}\right\} \\
&= \text{tr}\left\{\sum_k\sum_j \mathbb{E}\{\beta_k\beta_j\}\left[\sum_l \mathbf{\Delta}_l\mathbf{D}_k\mathbf{P}\mathbb{E}\{\mathbf{s}_l\mathbf{s}_l^H\}\mathbf{P}^H\mathbf{D}_j^H\mathbf{\Delta}_l\right]\right\} \\
&\overset{(a)}{=} \text{tr}\left\{\sum_k \sigma_{\beta_k}^2\left[\sum_l \mathbf{\Delta}_l\mathbf{D}_k\mathbf{\Phi}\mathbf{D}_k^H\mathbf{\Delta}_l\right]\right\} \\
&\overset{(b)}{=} \text{tr}\left(\sum_k \sigma_{\beta_k}^2 \mathbf{\Delta}\mathbf{D}_k\mathbf{\Phi}\mathbf{D}_k^H\right) = \text{tr}\left(\mathbf{\Phi}\sum_k \sigma_{\beta_k}^2 \mathbf{D}_k^H\mathbf{\Delta}\mathbf{D}_k\right) \\
&= \text{tr}\left(\mathbf{\Phi}\sum_k \sigma_{\beta_k}^2 \mathbf{v}_t^*(\theta_k)\mathbf{v}_r^H(\theta_k)\mathbf{\Delta}\mathbf{v}_r(\theta_k)\mathbf{v}_t^T(\theta_k)\right) \\
&\overset{(c)}{=} pLM_{r,R}\,\text{tr}\left(\mathbf{\Phi}\sum_k \sigma_{\beta_k}^2 \mathbf{v}_t^*(\theta_k)\mathbf{v}_t^T(\theta_k)\right) \\
&= pLM_{r,R}\,\text{tr}\left(\mathbf{\Phi}\mathbf{D}_t\right)
\end{aligned} \tag{8.38}$$

$$\begin{aligned}
\text{EIP} &\triangleq \mathbb{E}\left\{\text{tr}\left(\mathbf{\Omega} \odot (\mathbf{CPS})(\mathbf{\Omega} \odot (\mathbf{CPS}))^H\right)\right\} \\
&\quad + \mathbb{E}\left\{\text{tr}\left(\mathbf{\Omega} \odot (\mathbf{G}_2\mathbf{X}\mathbf{\Lambda}_2)(\mathbf{\Omega} \odot (\mathbf{G}_2\mathbf{X}\mathbf{\Lambda}_2))^H\right)\right\} \\
&= pLM_{r,R}\,\text{tr}\left(\mathbf{\Phi}\mathbf{C}_t\right) + \sum_{l=1}^{L}\text{tr}\left(\mathbf{G}_{2l}\mathbf{R}_{xl}\mathbf{G}_{2l}^H\right),
\end{aligned} \tag{8.39}$$

where $\mathbf{D}_k \triangleq \mathbf{v}_r(\theta_k)\mathbf{v}_t^T(\theta_k)$ for $k \in \mathbf{N}_K^+$, $\mathbf{s}_l \triangleq \mathbf{s}(l)$, and $\mathbf{\Delta}_l = \text{diag}(\mathbf{\Omega}_{.l})$. (a) follows from the fact that $\mathbb{E}\{\beta_k\beta_j\} = \delta_{jk}\sigma_{\beta_k}^2$, where δ_{jk} denotes the Kronecker delta; (b) follows from the fact that $\mathbf{\Delta}_l = \mathbf{\Delta}_l\mathbf{\Delta}_l$ and

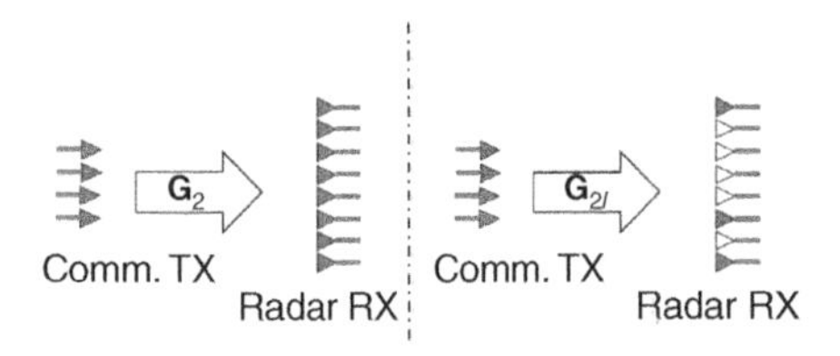

Figure 8.6 The subsampling at the radar receiver modulates the interference channel from the communication transmitter to the radar receiver $\mathbf{G}_2$. As shown in the left figure, the null space of $\mathbf{G}_2$ is typically empty; thus, the communications system transmission would always introduce interference to the radar. Due to the random subsampling, the null space of the modulated interference channel $\mathbf{G}_{2l}$ shown in the right figure becomes non-empty; thus, it becomes possible that the communication transmission introduces zero EIP to the radar receiver if the transmission is projected to the null space of $\mathbf{G}_{2l}$.

$\mathbf{\Delta} = \sum_{l=1}^{L} \mathbf{\Delta}_l$; (c) follows from the fact that $\mathbf{v}_r^H(\theta_k)\mathbf{\Delta}\mathbf{v}_r(\theta_k) = ||\mathbf{\Delta}||_1 = pLM_{r,R}$. Additionally, we have the following definitions: $\mathbf{D}_t = \sum_{k=1}^{K} \sigma_{\beta_k}^2 \mathbf{v}_t^*(\theta_k)\mathbf{v}_t^T(\theta_k)$, $\mathbf{C}_t = \sum_{k=1}^{K_c} \sigma_{\beta_k^c}^2 \mathbf{v}_t^*(\theta_k^c)\mathbf{v}_t^T(\theta_k^c)$, σ_{β_k} and $\sigma_{\beta_k^c}$ denote the standard deviation of β_k and β_k^c, respectively; $\mathbf{G}_{2l} \triangleq \mathbf{\Delta}_l\mathbf{G}_2$. The derivation for EIP is similar to that for ESP and is omitted for brevity. The above derivation assumes that each of the target and clutter reflection coefficient is an independent complex Gaussian variable with zero mean, which is widely considered in the literature [Cui et al., 2014, Chen et al., 2014, Chen and Vaidyanathan, 2008c].

The subsampling at the radar receiver modulates the interference channel $\mathbf{G}_2$ (see Figure 8.6). At sampling instance l, only the interference at radar RX antennas corresponding to 1s in $\mathbf{\Omega}_{\cdot l}$ is sampled. Thus, the effective interference channel during the lth symbol duration is $\mathbf{G}_{2l}$. To match the interference channel variation, the communications system should use adaptive transmission with symbol-dependent covariance matrix $\mathbf{R}_{xl}$ [Li et al., 2016b]. This would be the optimal approach; however, it involves high computational cost. A suboptimal alternative would be constant rate communication transmission, i.e. $\mathbf{R}_{xl} \equiv \mathbf{R}_x, \forall l \in \mathbb{N}_L^+$, outlined in Section 8.4.6.

Incorporating the expressions for effective target signal, interference, and additive noise, the *effective* radar SINR becomes

$$\text{ESINR} = \frac{\text{tr}\left(\mathbf{\Phi}\mathbf{D}_t\right)}{\text{tr}\left(\mathbf{\Phi}\mathbf{C}_t\right) + \sum_{l=1}^{L}\text{tr}\left(\mathbf{G}_{2l}\mathbf{R}_{xl}\mathbf{G}_{2l}^H\right)/(pLM_{r,R}) + \sigma_R^2}.$$

One can see that the ESINR depends on target and clutter information, interference channel $\mathbf{G}_2$, noise, communication TX covariance matrices $\{\mathbf{R}_{xl}\}$, the radar precoder $\mathbf{P}$ (embedded in $\mathbf{\Phi}$), and the radar subsampling scheme $\mathbf{\Omega}$. The joint design of $\{\mathbf{R}_{xl}\}$, $\mathbf{P}$, and $\mathbf{\Omega}$ is necessary to maximize the ESINR. In Theorem 8.3, we prove that the radar precoder $\mathbf{P}$ can be designed without affecting the incoherence property of $\mathbf{M}$.

8.4.3 Formulating the Design Problem at the Control Center

At the control center, the spectrum sharing problem can be formulated as follows:

$$(\mathbf{P}_1) \quad \max_{\{\mathbf{R}_{xl}\}\succeq 0, \mathbf{\Phi}\succeq 0, \mathbf{\Omega}} \quad \text{ESINR}\left(\{\mathbf{R}_{xl}\}, \mathbf{\Omega}, \mathbf{\Phi}\right),$$

$$\text{s.t.} \quad C_{\text{avg}}(\{\mathbf{R}_{xl}\}, \mathbf{\Phi}) \geq C, \tag{8.40a}$$

$$\sum_{l=1}^{L}\text{tr}\left(\mathbf{R}_{xl}\right) \leq P_C, L\text{tr}\left(\mathbf{\Phi}\right) \leq P_R, \tag{8.40b}$$

$$\operatorname{tr}\left(\mathbf{\Phi V}_k\right) \geq \xi \operatorname{tr}(\mathbf{\Phi}), \forall k \in \mathbb{N}_K^+, \tag{8.40c}$$

$$\mathbf{\Omega} \text{ is proper,} \tag{8.40d}$$

where $\mathbf{V}_k \triangleq \mathbf{v}_t^*(\theta_k)\mathbf{v}_t^T(\theta_k)$. The constraint of (8.40a) restricts the communication rate to be at least C, in order to support reliable communication and avoid service outage. The constraints of (8.40b) restrict the total communication and radar transmit power to be no larger than P_C and P_R, respectively. The constraints of (8.40b) restrict the power of the radar probing signal at directions of interest to be no smaller than the power achieved by the uniform precoding matrix $\frac{\operatorname{tr}(\mathbf{\Phi})}{M_{t,R}}\mathbf{I}$, i.e. $\mathbf{v}_t^T(\theta_k)\mathbf{\Phi v}_t^*(\theta_k) \geq \xi \mathbf{v}_t^T(\theta_k)\frac{\operatorname{tr}(\mathbf{\Phi})}{M_{t,R}}\mathbf{I v}_t^*(\theta_k) = \xi \operatorname{tr}(\mathbf{\Phi})$. Parameter $\xi \geq 1$ is used to control the beampattern at the target angles of interest. The purpose of this constraint is to ensure fairness across the multiple targets. The constraint in (8.40d) imposes the restrictions on the radar subsampling matrix $\mathbf{\Omega}$ such that it corresponds to a fixed subsampling rate p and has a large spectral gap. In the MC literature, $\mathbf{\Omega}$ is either a uniformly random subsampling matrix [Candès and Plan, 2010] or an adjacency matrix of a regular bipartite graph with a large spectral gap [Bhojanapalli and Jain, 2014]. The spectral gap of a matrix is defined as the difference between the largest singular value and the second largest singular value [Bhojanapalli and Jain, 2014].

In order for the control center to formulate and solve the problem of (8.40), it needs the following information:

- The communication and radar system CSI; estimation and feedback of CSI is discussed in Section 8.3.
- Target angles and clutter parameters $\{\sigma^2_{\beta_k^c}\}$ and $\{\theta_k^c\}$. Since the control center integrates the radar fusion center functionality, the target angles obtained from the previous tracking cycle will be available. In practice, the clutter parameters could be estimated when the targets are absent [Chen et al., 2014]. If $\{\sigma^2_{\beta_k}\}$ is not known, we can instead use a single-value σ_0^2 for all the targets. This choice effectively means that the objective treats all target directions equally. One possible choice for σ_0^2 is the smallest target RCS variance that could be detected by the radar. Note that the solution of $(\mathbf{P}_1)$ is independent on the specific value of σ_0^2.
- All parameters in the constraints. Parameters such as power budget and required communication rate can also be provided by the radar and communications systems.

Problem $(\mathbf{P}_1)$ is nonconvex w.r.t. optimization variable triple $\left(\{\mathbf{R}_x\}, \mathbf{\Omega}, \mathbf{\Phi}\right)$. In the following, in Section 8.4.4 we present an algorithm to find a local solution via alternating optimization, while in Section 8.4.5, we provide some insights into the feasibility and solution properties for $(\mathbf{P}_1)$ [Li and Petropulu, 2017b].

8.4.4 Solution to the Spectrum Sharing Problem Using Alternating Optimization

The alternating iterations w.r.t. $\{\mathbf{R}_{xl}\}$, $\mathbf{\Omega}$, and $\mathbf{\Phi}$ are discussed in the following three subsections.

8.4.4.1 The Alternating Iteration w.r.t. $\{\mathbf{R}_{xl}\}$

We first solve for $\{\mathbf{R}_{xl}\}$ while setting $\mathbf{\Omega}$ and $\mathbf{\Phi}$ to be equal to the solution from the previous iteration, i.e. we formulate the following problem:

$$\begin{aligned}
&(\mathbf{P}_\mathbf{R}) \min_{\{\mathbf{R}_{xl}\}\succeq 0} \sum_{l=1}^{L} \operatorname{tr}\left(\mathbf{G}_{2l}\mathbf{R}_{xl}\mathbf{G}_{2l}^H\right), \\
&\quad \text{s.t. } C_{\text{avg}}(\{\mathbf{R}_{xl}\}, \mathbf{\Phi}) \geq C, \sum_{l=1}^{L} \operatorname{tr}\left(\mathbf{R}_{xl}\right) \leq P_C.
\end{aligned} \tag{8.41}$$

Problem ($\mathbf{P_R}$) is convex and involves multiple matrix variables, the joint optimization with respect to which requires high computational complexity. The semidefinite matrix variables $\{\mathbf{R}_{xl}\}$ have $LM_{t,C}^2$ real scalar variables, which will result in complexity of $\mathcal{O}((LM_{t,C}^2)^{3.5})$ if an interior-point method [Boyd and Vandenberghe, 2004] is used. An efficient algorithm for solving the above problem can be implemented based on the Lagrangian dual decomposition [Boyd and Vandenberghe, 2004]. The Lagrangian of ($\mathbf{P_R}$) can be written as

$$\mathcal{L}(\{\mathbf{R}_{xl}\},\lambda_1,\lambda_2) = \sum_{l=1}^{L} \text{tr}\left(\mathbf{G}_{2l}\mathbf{R}_{xl}\mathbf{G}_{2l}^H\right) + \lambda_1 \left(\sum_{l=1}^{L} \text{tr}\left(\mathbf{R}_{xl}\right) - P_C \right) + \lambda_2 \left(C - C_{\text{avg}}(\{\mathbf{R}_{xl}\})\right),$$

where $\lambda_1 \geq 0$ and $\lambda_2 \geq 0$ are the dual variables associated with the transmit power and the communication rate constraints, respectively. The dual problem of ($\mathbf{P_R}$) is

$$(\mathbf{P_R}\text{-}D) \ \max_{\lambda_1,\lambda_2 \geq 0} g(\lambda_1, \lambda_2),$$

where $g(\lambda_1, \lambda_2)$ is the dual function defined as

$$g(\lambda_1, \lambda_2) = \inf_{\{\mathbf{R}_{xl}\}\succeq 0} \mathcal{L}(\{\mathbf{R}_{xl}\}, \lambda_1, \lambda_2).$$

The domain of the dual function, i.e. dom g, is $\lambda_1, \lambda_2 \geq 0$ such that $g(\lambda_1, \lambda_2) > -\infty$. It is also called dual feasible if $(\lambda_1, \lambda_2) \in$ dom g. The dual function $g(\lambda_1, \lambda_2)$ can be obtained by solving L independent subproblems for given λ_1 and λ_2, each of which can be written as follows:

$$(\mathbf{P_R}\text{-sub}) \min_{\mathbf{R}_{xl}\succeq 0} \ \text{tr}\left(\left(\mathbf{G}_2^H\boldsymbol{\Delta}_l\mathbf{G}_2 + \lambda_1\mathbf{I}\right)\mathbf{R}_{xl}\right) - \lambda_2 \log_2 \left|\mathbf{I} + \mathbf{R}_{wl}^{-1}\mathbf{H}\mathbf{R}_{xl}\mathbf{H}^H\right|. \tag{8.42}$$

Before giving the solution of ($\mathbf{P_R}$-sub), let us first state some observations.

Observation 1: The average capacity constraint should be active at the optimal point. This means that the achieved capacity is always C and $\lambda_2 > 0$. To show this, let us assume that the optimal point $\{\mathbf{R}_{xl}^*\}$ achieves $C_{\text{avg}}(\{\mathbf{R}_{xl}^*\}) > C$. Then, we can always shrink $\{\mathbf{R}_{xl}^*\}$ until the average capacity reduces to C, which would also reduce the objective. Thus, we end up with a contradiction. By complementary slackness, the corresponding dual variable is positive, i.e. $\lambda_2 > 0$.

Observation 2: $\left(\mathbf{G}_2^H\boldsymbol{\Delta}_l\mathbf{G}_2 + \lambda_1\mathbf{I}\right)$ is positive definite for all $l \in \mathbb{N}_L^+$. This can be shown by contradiction. Suppose that there exists l such that $\mathbf{G}_2^H\boldsymbol{\Delta}_l\mathbf{G}_2 + \lambda_1\mathbf{I}$ is singular. Then it must hold that $\mathbf{G}_2^H\boldsymbol{\Delta}_l\mathbf{G}_2$ is singular and $\lambda_1 = 0$. Therefore, we can always find a nonzero vector $\mathbf{v}$ lying in the null space of $\mathbf{G}_2^H\boldsymbol{\Delta}_l\mathbf{G}_2$. At the same time, it holds that $\mathbf{R}_{wl}^{-1/2}\mathbf{H}\mathbf{v} \neq 0$ with very high probability because $\mathbf{H}$ is a realization of the random channel. If we choose $\mathbf{R}_{xl} = \alpha\mathbf{v}\mathbf{v}^H$ and $\alpha \to \infty$, the Lagrangian $\mathcal{L}(\{\mathbf{R}_{xl}\}, 0, \lambda_2)$ will be unbounded from below, which indicates that $\lambda_1 = 0$ is not dual feasible. This means that λ_1 is strictly larger than 0 if $\mathbf{G}_2^H\boldsymbol{\Delta}_l\mathbf{G}_2$ is singular for any l. Thus, the claim is proven.

Based on the above observations, we have the following lemma:

Lemma 8.2 ***[Zhang et al., 2010] and [Kim and Giannakis, 2011]*** *For given feasible dual variables $\lambda_1, \lambda_2 \geq 0$, the optimal solution of ($\mathbf{P_R}$-sub) is given by*

$$\mathbf{R}_{xl}^*(\lambda_1, \lambda_2) = \boldsymbol{\Phi}_l^{-1/2}\mathbf{U}_l\boldsymbol{\Sigma}_l\mathbf{U}_l^H\boldsymbol{\Phi}_l^{-1/2}, \tag{8.43}$$

where $\boldsymbol{\Phi}_l \triangleq \mathbf{G}_2^H \boldsymbol{\Delta}_l \mathbf{G}_2 + \lambda_1 \mathbf{I}$; $\mathbf{U}_l$ *is the right singular matrix of* $\tilde{\mathbf{H}}_l \triangleq \mathbf{R}_{wl}^{-1/2} \mathbf{H} \boldsymbol{\Phi}_l^{-1/2}$; $\boldsymbol{\Sigma}_l = diag(\beta_{l1}, \ldots, \beta_{lr})$ *with* $\beta_{li} = (\lambda_2 - 1/\sigma_{li}^2)^+$, *r and* $\sigma_{li}, i = 1, \ldots, r$, *respectively, being the rank and the positive singular vales of* $\tilde{\mathbf{H}}_l$. *It also holds that*

$$\log_2 \left| \mathbf{I} + \mathbf{R}_{wl}^{-1} \mathbf{H} \mathbf{R}_{xl}^* \mathbf{H}^H \right| = \sum_{i=1}^{r} \left(\log(\lambda_2 \sigma_{li}^2) \right)^+. \tag{8.44}$$

Based on Lemma 8.2, the solution of $(\mathbf{P_R})$ can be obtained by finding the optimal dual variables λ_1^*, λ_2^*. The cooperative spectrum sharing problem $(\mathbf{P_R})$ can be solved via the procedure outlined in Algorithm 1. The convergence of Algorithm 1 is guaranteed by the convergence of the ellipsoid method [Bland et al., 1981]. The complexity of the dual decomposition-based algorithm is only linearly dependent on L.

Algorithm 1 Algorithm for the alternating iteration $(\mathbf{P_R})$

1: **Input:** $\boldsymbol{H}, \boldsymbol{G}_1, \boldsymbol{G}_2, \boldsymbol{\Omega}, P_t, C, \sigma_C^2$
2: **Initialization:** $\lambda_1 \geq 0, \lambda_2 \geq 0$
3: **repeat**
4: Calculate $\boldsymbol{R}_{xl}^*(\lambda_1, \lambda_2)$ according to (8.43) with the given λ_1 and λ_2.
5: Compute the subgradient of $g(\lambda_1, \lambda_2)$ as $\sum_{l=1}^{L} \text{tr}\left(\boldsymbol{R}_{xl}^*(\lambda_1, \lambda_2)\right) - P_t$ and $C - C_{\text{avg}}(\{\boldsymbol{R}_{xl}^*(\lambda_1, \lambda_2)\})$ respectively for λ_1 and λ_2.
6: Update λ_1 and λ_2 accordingly based on the ellipsoid method [Bland et al., 1981].
7: **until** λ_1 and λ_2 converge to a prescribed accuracy.
8: **Output:** $\boldsymbol{R}_{xl}^* = \boldsymbol{R}_{xl}^*(\lambda_1, \lambda_2)$

8.4.4.2 The Alternating Iteration w.r.t. Ω

Via simple algebraic manipulations, the EIP from the communication transmission can be reformulated as

$$\sum_{l=1}^{L} \text{tr}\left(\mathbf{G}_{2l} \mathbf{R}_{xl} \mathbf{G}_{2l}^H\right) \equiv \text{tr}\,(\boldsymbol{\Omega}^T \mathbf{Q}),$$

where the lth column of $\mathbf{Q}$ contains the diagonal entries of $\mathbf{G}_2 \mathbf{R}_{xl} \mathbf{G}_2^H$. With fixed $\{\mathbf{R}_{xl}\}$ and $\boldsymbol{\Phi}$, we can solve $\boldsymbol{\Omega}$ via

$$\min_{\boldsymbol{\Omega}} \ \text{tr}\,(\boldsymbol{\Omega}^T \mathbf{Q}), \quad \text{s.t. } \boldsymbol{\Omega} \text{ is proper.} \tag{8.45}$$

Recall that the sampling matrix $\boldsymbol{\Omega}$ is required to have a large spectral gap. However, it is difficult to incorporate such conditions in the above optimization problem. Based on the fact that row and column permutations of the sampling matrix would not affect its singular values and, thus the spectral gap, a suboptimal approach is to search for the best sampling scheme by permuting rows and columns of an initial sampling matrix $\boldsymbol{\Omega}^0$, i.e.

$$\min_{\boldsymbol{\Omega}} \ \text{tr}\,(\boldsymbol{\Omega}^T \mathbf{Q}), \quad \text{s.t. } \boldsymbol{\Omega} \in \wp(\boldsymbol{\Omega}^0), \tag{8.46}$$

where $\wp(\boldsymbol{\Omega}^0)$ denotes the set of matrices obtained by arbitrary row and/or column permutations. $\boldsymbol{\Omega}^0$ is generated with binary entries and $\lfloor pLM_{r,R} \rfloor$ ones, where $\lfloor x \rfloor$ denotes the largest integer smaller than or equal to x. Therefore, the constraint on the number of 1s in $\boldsymbol{\Omega}$ can also be satisfied. One good candidate for $\boldsymbol{\Omega}^0$ would be a uniformly random sampling matrix, as such matrix exhibits a large spectral gap with high probability [Bhojanapalli and Jain, 2014]. Multiple trials with different $\mathbf{\Omega}^0$s can be used to further improve the choice of $\mathbf{\Omega}$. However, the search space is very large since the cardinality of $\wp(\mathbf{\Omega}^0)$ is $M_{r,R}!L!$, where operator ! denotes the factorial. One can reduce the search space as follows Li and Petropulu [2017b]:

$$\min_{\mathbf{\Omega}} \text{ tr }(\mathbf{\Omega}^T\mathbf{Q}) \equiv \text{tr }(\mathbf{\Omega}\mathbf{Q}^T), \quad \text{s.t. } \mathbf{\Omega} \in \wp_r(\mathbf{\Omega}^0), \tag{8.47}$$

where $\wp_r(\mathbf{\Omega}^0)$ denotes the set of matrices obtained by arbitrary row permutations. The search space in (8.47) equals $M_{r,R}!$, i.e. the cardinality of $\wp_r(\mathbf{\Omega}^0)$, which is greatly reduced compared to that in (8.46). Furthermore, the following proposition shows that such reduction of search space comes without performance loss.

Proposition 8.4 For any $\mathbf{\Omega}^0$, searching for an $\mathbf{\Omega}$ in $\wp_r(\mathbf{\Omega}^0)$ can achieve the same EIP as searching in $\wp(\mathbf{\Omega}^0)$.

Proof: We can prove the proposition by showing that the EIP achieved by any $\mathbf{\Omega}_1 \in \wp(\mathbf{\Omega}^0)$ can also be achieved by a certain $\mathbf{\Omega}_2 \in \wp_r(\mathbf{\Omega}^0)$. For the pair $(\mathbf{\Omega}_1, \{\mathbf{R}_{xl}\})$, the same EIP can be achieved by the pair $(\mathbf{\Omega}_2, \{\tilde{\mathbf{R}}_{xl}\})$, where

- $\mathbf{\Omega}_2$ is constructed by performing on $\mathbf{\Omega}^0$ the row permutations performed from $\mathbf{\Omega}^0$ to $\mathbf{\Omega}_1$; and
- $\{\tilde{\mathbf{R}}_{xl}\}$ is a permutation of $\{\mathbf{R}_{xl}\}$ according to the column permutations performed from $\mathbf{\Omega}^0$ to $\mathbf{\Omega}_1$.

In other words, the column permutation on $\mathbf{\Omega}$ is unnecessary because $\{\mathbf{R}_{xl}\}$ will be automatically optimized to match the column pattern of $\mathbf{\Omega}$. The claim is proven. ∎

The problem in (8.47) aims to find the best one-to-one match between the rows of $\mathbf{\Omega}^0$ and the rows of $\mathbf{Q}$. Let us construct a cost matrix $\mathbf{C}^r \in \mathbb{R}^{M_{r,R}\times M_{r,R}}$ with $[\mathbf{C}^r]_{ml} \triangleq \mathbf{\Omega}^0_{m\cdot}(\mathbf{Q}_{l\cdot})^T$. The problem turns out to be a linear assignment problem with cost matrix $\mathbf{C}^c$, which can be solved efficiently in polynomial time $\mathcal{O}(M^3_{r,R})$ using the Hungarian algorithm [Kuhn, 1955].

8.4.4.3 The Alternating Iteration w.r.t. $\mathbf{\Phi}$

For the optimization of $\mathbf{\Phi}$ with fixed $\{\mathbf{R}_{xl}\}$ and $\mathbf{\Omega}$, the constraint in (8.40a) is nonconvex w.r.t. $\mathbf{\Phi}$. The first-order Taylor expansion of $\underline{C}(\mathbf{R}_{xl}, \mathbf{\Phi})$ at $\bar{\mathbf{\Phi}}$ is given by

$$\underline{C}(\mathbf{R}_{xl}, \mathbf{\Phi}) \approx \underline{C}(\mathbf{R}_{xl}, \bar{\mathbf{\Phi}}) - \text{tr}\left(\mathbf{A}_l(\mathbf{\Phi} - \bar{\mathbf{\Phi}})\right), \tag{8.48}$$

where $\mathbf{A}_l$ is defined as

$$\begin{aligned}\mathbf{A}_l &\triangleq -\left(\frac{\partial \underline{C}(\mathbf{R}_{xl}, \mathbf{\Phi})}{\partial \text{Re}(\mathbf{\Phi})}\right)^T_{\mathbf{\Phi}=\bar{\mathbf{\Phi}}} \\ &= \mathbf{G}_1^H[(\mathbf{G}_1\mathbf{\Phi}\mathbf{G}_1^H + \sigma_C^2\mathbf{I})^{-1} - (\mathbf{G}_1\mathbf{\Phi}\mathbf{G}_1^H + \sigma_C^2\mathbf{I} + \mathbf{H}\mathbf{R}_{xl}\mathbf{H}^H)^{-1}]\mathbf{G}_1 \;|_{\mathbf{\Phi}=\bar{\mathbf{\Phi}}}.\end{aligned} \tag{8.49}$$

The sequential convex programming technique is applied to solve $\mathbf{\Phi}$ by repeatedly solving the following approximate optimization problem:

$$\begin{aligned}(\mathbf{P}_{\mathbf{\Phi}}) \max_{\mathbf{\Phi}\succeq 0} \; & \frac{\text{tr }(\mathbf{\Phi}\mathbf{D}_t)}{\text{tr }(\mathbf{\Phi}\mathbf{C}_t) + \rho}, \\ \text{s.t. } & \text{tr}\,(\mathbf{\Phi}) \le P_R/L, \text{tr}\,(\mathbf{\Phi}\mathbf{A}) \le \tilde{C}, \\ & \text{tr}\left(\mathbf{\Phi}\mathbf{V}_k\right) \ge \xi \text{tr}\,(\mathbf{\Phi}), \forall k \in \mathbb{N}_K^+,\end{aligned} \tag{8.50}$$

where

$$\tilde{C} = \sum_{l=1}^{L} \left(\underline{C}(\mathbf{R}_{xl}, \bar{\mathbf{\Phi}}) + \text{tr}\,(\bar{\mathbf{\Phi}}\mathbf{A}_l) - C \right),$$
$$\mathbf{A} = \sum_{l=1}^{L} \mathbf{A}_l, \tag{8.51}$$
$$\rho = \frac{1}{pLM_{r,R}} \sum_{l=1}^{L} \text{tr}\left(\mathbf{R}_{xl}\mathbf{G}_2^H \mathbf{\Delta}_l \mathbf{G}_2\right) + \sigma_R^2.$$

$\tilde{C}$ and ρ are real positive constants w.r.t. $\mathbf{\Phi}$, and $\bar{\mathbf{\Phi}}$ is updated as the solution of the previous repeated problem. Problem (8.50) could be equivalently formulated as a semidefinite programming problem (SDP) via the Charnes–Cooper Transformation [Chen et al., 2014, Li and Ma, 2011].

$$\begin{aligned} \max_{\tilde{\mathbf{\Phi}} \succeq 0, \phi > 0} \quad & \text{tr}\,(\tilde{\mathbf{\Phi}}\mathbf{D}_t), \\ \text{s.t.} \quad & \text{tr}\,(\tilde{\mathbf{\Phi}}\mathbf{C}_t) = 1 - \phi\rho, \\ & \text{tr}\left(\tilde{\mathbf{\Phi}}\right) \leq \phi P_R / L, \text{tr}\left(\tilde{\mathbf{\Phi}}\mathbf{A}\right) \leq \phi\tilde{C}, \\ & \text{tr}\left(\tilde{\mathbf{\Phi}}(\mathbf{V}_k - \xi\mathbf{I})\right) \geq 0, \forall k \in \mathbb{N}_K^+. \end{aligned} \tag{8.52}$$

The optimal solution of (8.52), denoted by $(\tilde{\mathbf{\Phi}}^*, \phi^*)$, can be obtained by using any standard interior-point method-based SDP solver with a complexity of $\mathcal{O}((M_{t,R}^2)^{3.5})$. The solution of (8.50) is given by $\tilde{\mathbf{\Phi}}^*/\phi^*$. In each alternating iteration w.r.t. $\mathbf{\Phi}$, it is required to solve several iterations of SDP due to the sequential convex programming.

It is easy to show that the objective function, i.e. ESINR, is nondecreasing during the alternating iterations of $\{\mathbf{R}_{xl}\}$, $\mathbf{\Omega}$, and $\mathbf{\Phi}$ and is upper bounded. According to the monotone convergence theorem [Yeh, 2006], the alternating optimization is guaranteed to converge.

The cooperative spectrum sharing algorithm maximizing the effective radar SINR is summarized in Algorithm 2.

Algorithm 2 The overall spectrum sharing algorithm

1: **Input:** $\boldsymbol{D}_t, \boldsymbol{C}_t, \boldsymbol{H}, \boldsymbol{G}_1, \boldsymbol{G}_2, P_{C/R}, C, \sigma_{C/R}^2, \delta_1$
2: **Initialization:** $\mathbf{\Phi} = \frac{P_R}{LM_{t,R}}\boldsymbol{I}, \mathbf{\Omega} = \mathbf{\Omega}^0$.
3: **repeat**
4: Update $\{\boldsymbol{R}_{xl}\}$ by solving ($\mathbf{P}_\mathrm{R}$) using Algorithm 1 with fixed $\mathbf{\Omega}$ and $\mathbf{\Phi}$.
5: Update $\mathbf{\Omega}$ by solving (8.47) with fixed $\{\boldsymbol{R}_{xl}\}$ and $\mathbf{\Phi}$.
6: Update $\mathbf{\Phi}$ by solving a sequence of approximated SDP problem (8.50) with fixed $\{\boldsymbol{R}_{xl}\}$ and $\mathbf{\Omega}$.
7: **until** ESINR increases by amount smaller than δ_1
8: **Output:** $\{\boldsymbol{R}_{xl}\}, \mathbf{\Omega}, \boldsymbol{P} = \sqrt{L}\mathbf{\Phi}^{1/2}$.

8.4.5 Insight into the Feasibility and Solutions of the Spectrum Sharing Problem

In this subsection, we provide some key insights into the feasibility of ($\mathbf{P}_1$) and the rank of the solutions $\mathbf{\Phi}$ obtained by Algorithm 2.

8.4.5.1 Feasibility

A necessary condition on C for the feasibility of $(\mathbf{P}_1)$ w.r.t. $\{\mathbf{R}_{xl}\}$ is $C \leq C_{\max}(P_C)$ where

$$\begin{aligned} C_{\max}(P_C) \triangleq \max_{\{\mathbf{R}_{xl}\}\succeq 0} \frac{1}{L}\sum_{l=1}^{L}\log_2\left|\mathbf{I}+\sigma_C^{-2}\mathbf{H}\mathbf{R}_{xl}\mathbf{H}^H\right|, \\ \text{s.t.} \sum_{l=1}^{L}\text{tr}\left(\mathbf{R}_{xl}\right) \leq P_C. \end{aligned} \tag{8.53}$$

The above optimization problem is convex and has a closed-form solution based on water-filling [Tse and Viswanath, 2005]. It can be shown that $C_{\max}(P_C)$ is essentially the largest achievable communication rate when there is no interference from radar transmitters to the communication receivers. Note that $C = C_{\max}(P_C)$ will generate a non-empty feasible set for $\{\mathbf{R}_{xl}\}$ only if $\mathbf{G}_1\mathbf{\Phi}\mathbf{G}_1^H = \mathbf{0}$ (omitting the trivial case $\mathbf{\Phi} = \mathbf{0}$), i.e. the radar transmits in the null space of the interference channel $\mathbf{G}_1$ to the communication receivers.

A necessary condition on ξ for the feasibility of $(\mathbf{P}_1)$ w.r.t. $\mathbf{\Phi}$ is $\xi \leq \xi_{\max}$ where

$$\begin{aligned} \xi_{\max} \triangleq \max_{\mathbf{\Phi}\succeq 0,\xi\geq 0} \xi, \\ \text{s.t. tr}\left(\mathbf{\Phi}\mathbf{V}_k\right) \geq \xi\text{tr}\left(\mathbf{\Phi}\right), \forall k \in \mathbb{N}_K^+. \end{aligned} \tag{8.54}$$

Note that the above optimization problem is independent of tr $(\mathbf{\Phi})$. Without loss of generality, we assume that tr $(\mathbf{\Phi}) = 1$, based on which we have the following equivalent SDP formulation:

$$\begin{aligned} \xi_{\max} \triangleq \max_{\mathbf{\Phi}\succeq 0,\xi\geq 0} \xi,\ \text{s.t. tr}\left(\mathbf{\Phi}\right) = 1, \\ \text{tr}\left(\mathbf{\Phi}\mathbf{V}_k\right) \geq \xi, \forall k \in \mathbb{N}_K^+. \end{aligned} \tag{8.55}$$

It is easy to check that $\xi_{\max} \geq 1$, which can be achieved by set $(\mathbf{\Phi}, \xi)$ to be $(\mathbf{I}/M_{t,R}, 1)$.

The following proposition provides a sufficient condition for the feasibility of $(\mathbf{P}_1)$.

Proposition 8.5 If $C, \xi, P_C > 0, P_R > 0$ are chosen such that $C < C_{\max}(P_C)$ and $\xi \leq \xi_{\max}$, then $(\mathbf{P}_1)$ is feasible.

Proof: If $C < C_{\max}(P_C)$, the feasible set for $\{\mathbf{R}_{xl}\}$ determined by constraints in (8.40a) and (8.40b) $\mathcal{F}_{\{\mathbf{R}_{xl}\}}$ is non-empty as long as tr $(\mathbf{\Phi})$ is sufficiently small. If $\xi \leq \xi_{\max}$, the feasible set for $\mathbf{\Phi}$ determined by constraints in (8.40b) $\mathcal{F}_{\mathbf{\Phi}1}$ is non-empty and has no restriction on tr $(\mathbf{\Phi})$. If $\mathbf{\Phi} \in \mathcal{F}_{\mathbf{\Phi}1}$, then $\alpha\mathbf{\Phi} \in \mathcal{F}_{\mathbf{\Phi}1}, \forall\alpha > 0$. The overall feasible set for $\mathbf{\Phi}$, $\mathcal{F}_{\mathbf{\Phi}}$, is the intersection of feasible sets determined by (8.40a), (8.40b), and (8.40b). $\mathcal{F}_{\mathbf{\Phi}}$ is non-empty as long as $\mathcal{F}_{\mathbf{\Phi}1}$ and $\mathcal{F}_{\{\mathbf{R}_{xl}\}}$ are non-empty because we can choose any $\mathbf{\Phi} \in \mathcal{F}_{\mathbf{\Phi}1}$ and scale it down to make $(\mathbf{P}_1)$ feasible. The claim is proven. ■

8.4.5.2 The Rank of the Solutions $\mathbf{\Phi}$

We are also particularly interested in the rank of $\mathbf{\Phi}$, which is obtained using Algorithm 2. Since the sequential convex programming technique is used for solving $\mathbf{\Phi}$, it suffices to focus on the rank of the solution of $(\mathbf{P}_{\mathbf{\Phi}})$. To achieve this goal, we first introduce the following SDP problem:

$$\begin{aligned} \min_{\mathbf{\Phi}\succeq 0} \text{tr}\left(\mathbf{\Phi}\right),\ \text{s.t.}\ \text{tr}\left(\mathbf{\Phi}\mathbf{A}\right) \leq \tilde{C},\ \frac{\text{tr}\left(\mathbf{\Phi}\mathbf{D}_t\right)}{\text{tr}\left(\mathbf{\Phi}\mathbf{C}_t\right)+\rho} \geq \gamma, \\ \text{tr}\left(\mathbf{\Phi}\mathbf{V}_k\right) \geq 0, \forall k \in \mathbb{N}_K^+, \end{aligned} \tag{8.56}$$

where γ is a real positive constant. The following proposition relates the optimal solutions of problems (8.50) and (8.56).

Proposition 8.6 If γ in (8.56) is chosen to be the maximum achievable SINR of (8.50), denoted as SINR_{max}, the optimal $\mathbf{\Phi}$ of (8.56) is also optimal for (8.50).

Proof: Denote $\mathbf{\Phi}_1^*$ and $\mathbf{\Phi}_2^*$ the optimal solutions of (8.50) and (8.56), respectively. It is clear that $\mathbf{\Phi}_1^*$ is a feasible point of (8.56). This means that tr $(\mathbf{\Phi}_2^*) \leq$ tr $(\mathbf{\Phi}_1^*) \leq P_R$. Therefore, $\mathbf{\Phi}_2^*$ is a feasible point of (8.50). It holds that

$$\text{SINR}_{\text{max}} \equiv \frac{\text{tr}\,(\mathbf{\Phi}_1^*\mathbf{D}_t)}{\text{tr}\,(\mathbf{\Phi}_1^*\mathbf{C}_t) + \rho} \geq \frac{\text{tr}\,(\mathbf{\Phi}_2^*\mathbf{D}_t)}{\text{tr}\,(\mathbf{\Phi}_2^*\mathbf{C}_t) + \rho} \geq \text{SINR}_{\text{max}}. \tag{8.57}$$

It is only possible when all the equalities hold. In other words, $\mathbf{\Phi}_2^*$ is optimal for (8.50). This completes the proof. ■

To characterize the optimal solution of (8.56), we need the following key lemma:

Lemma 8.3 *Matrix* $\mathbf{A}_l$ *defined in* (8.49) *is positive semidefinite. In addition,* $\mathbf{A} = \sum_{l=1}^{L} \mathbf{A}_l$ *is also positive semidefinite.*

Proof: For simplicity of notation, we denote that $\mathbf{X} \triangleq \mathbf{G}_1\mathbf{\Phi}\mathbf{G}_1^H + \sigma_C^2\mathbf{I} \succ 0$ and $\mathbf{Y} \triangleq \mathbf{H}\mathbf{R}_{xl}\mathbf{H}^H \succeq 0$. Let us rewrite $\mathbf{A}_l$ as $\mathbf{A}_l = \mathbf{G}_1^H\left[\mathbf{X}^{-1} - (\mathbf{X}+\mathbf{Y})^{-1}\right]\mathbf{G}_1$. It is clear to see that $\mathbf{A}_l$ is Hermitian because both $\mathbf{X}^{-1}$ and $(\mathbf{X}+\mathbf{Y})^{-1}$ are Hermitian. It is sufficient to show that $\mathbf{Z} \triangleq \mathbf{X}^{-1} - (\mathbf{X}+\mathbf{Y})^{-1}$ is positive semidefinite. We have that

$$\mathbf{X}^{-1} - (\mathbf{X}+\mathbf{Y})^{-1} = \mathbf{X}^{-1}\mathbf{Y}(\mathbf{X}+\mathbf{Y})^{-1}, \tag{8.58}$$

which could be shown by right-multiplying $(\mathbf{X}+\mathbf{Y})$ on both sides of the equality. Since $\mathbf{X}$, $\mathbf{Y}$, and $\mathbf{Z}$ are Hermitian, we have

$$\mathbf{Z} = \mathbf{X}^{-1}\mathbf{Y}(\mathbf{X}+\mathbf{Y})^{-1} = (\mathbf{X}+\mathbf{Y})^{-1}\mathbf{Y}\mathbf{X}^{-1}. \tag{8.59}$$

Since $(\mathbf{X}+\mathbf{Y})^{-1}$ is invertible, there exists a unique positive-definite matrix $\mathbf{V}$, such that $(\mathbf{X}+\mathbf{Y})^{-1} = \mathbf{V}^2$. Simple algebra manipulation shows that

$$\begin{aligned}\mathbf{V}^{-1}\mathbf{Z}\mathbf{V}^{-1} &= (\mathbf{V}^{-1}\mathbf{X}^{-1}\mathbf{V}^{-1})(\mathbf{V}\mathbf{Y}\mathbf{V}) \\ &= (\mathbf{V}\mathbf{Y}\mathbf{V})(\mathbf{V}^{-1}\mathbf{X}^{-1}\mathbf{V}^{-1}),\end{aligned} \tag{8.60}$$

i.e. $\mathbf{V}^{-1}\mathbf{Z}\mathbf{V}^{-1}$ is a product of two commutable positive-semidefinite matrices $\mathbf{V}^{-1}\mathbf{X}^{-1}\mathbf{V}^{-1}$ and $\mathbf{V}\mathbf{Y}\mathbf{V}$. Therefore, $\mathbf{V}^{-1}\mathbf{Z}\mathbf{V}^{-1}$ and thus $\mathbf{Z}$ is positive semidefinite. We prove that $\mathbf{A}_l$ is semidefinite. Further, $\mathbf{A}$ is also semidefinite because it is the sum of L semidefinite matrices. ■

Based on Lemma 8.3, we prove the following result by following the approach in Li and Ma [2011]:

Proposition 8.7 Suppose that (8.56) is feasible when γ is set to SINR_{max}. Then, the following claims hold:

(1) Any optimal solution of (8.56) has rank at most K.
(2) All rank-K solutions $\mathbf{\Phi}_K^*$ of (8.56) have the same range space.
(3) Any solution $\mathbf{\Phi}_{K-}^*$ with rank less than K has range space such that Range $(\mathbf{\Phi}_{K-}^*) \subset$ Range $(\mathbf{\Phi}_K^*)$.
(4) Equations (8.50) and (8.52) always have solutions with rank at most K and with the same range space properties as that for (8.56).

Proof: Problem (8.56) is an SDP, whose Karush–Kuhn-Tucker (KKT) conditions [Boyd and Vandenberghe, 2004] are given as

$$\boldsymbol{\Psi} + \lambda_2 \mathbf{D}_t + \sum_{k=1}^{K} \nu_k \mathbf{V}_k = \mathbf{I} + \lambda_1 \mathbf{A} + \lambda_2 \gamma \mathbf{C}_t + \sum_{k=1}^{K} \nu_k \xi \mathbf{I}, \tag{8.61a}$$

$$\boldsymbol{\Psi}\boldsymbol{\Phi} = \mathbf{0}, \tag{8.61b}$$

$$\boldsymbol{\Psi} \succeq 0, \boldsymbol{\Phi} \succeq 0, \lambda_1 \geq 0, \lambda_2 \geq 0, \{\nu_k\} \geq 0, \tag{8.61c}$$

$$\mathrm{tr}\,(\boldsymbol{\Phi}\mathbf{D}_t) \geq \gamma \mathrm{tr}\,(\boldsymbol{\Phi}\mathbf{C}_t) + \gamma \rho, \tag{8.61d}$$

$$\mathrm{tr}\,(\boldsymbol{\Phi}\mathbf{V}_k) \geq 0, \forall k \in \mathbb{N}_K^+, \tag{8.61e}$$

where $\boldsymbol{\Psi} \succeq 0$, $\lambda_1 \geq 0$, $\lambda_2 \geq 0$, and $\{\nu_k\} \geq 0$ are dual variables. We can rewrite (8.61a) as follows:

$$\begin{aligned} &\mathrm{Rank}\,(\boldsymbol{\Psi}) + \mathrm{Rank}\left(\lambda_2 \mathbf{D}_t + \sum_{k=1}^{K} \nu_k \mathbf{v}_t^*(\theta_k)\mathbf{v}_t^T(\theta_k)\right) \\ &\geq \mathrm{Rank}\left(\mathbf{I} + \lambda_1 \mathbf{A} + \lambda_2 \gamma \mathbf{C}_t + \sum_{k=1}^{K} \nu_k \xi \mathbf{I}\right). \end{aligned} \tag{8.62}$$

Recall that $\mathbf{D}_t = \sum_k \sigma_{\beta_k}^2 \mathbf{v}_t^*(\theta_k)\mathbf{v}_t^T(\theta_k)$. It holds that $\lambda_2 \mathbf{D}_t + \sum_k \nu_k \mathbf{v}_t^*(\theta_k)\mathbf{v}_t^T(\theta_k)$ has rank at most K. Since $\mathbf{A}$ and $\mathbf{C}_t$ are positive semidefinite, the matrix on the right-hand side of (8.62) has full rank. Therefore, Rank $(\boldsymbol{\Psi})$ is not smaller than $M_{t,R} - K$. From (8.61b) and (8.61d), we conclude that any optimal solution $\boldsymbol{\Phi}$ must have rank at most K.

The second claim asserts that if there are multiple solutions with rank K, they all have the same range space. This can be proved using contradiction. Suppose that $\boldsymbol{\Phi}_1^*$ and $\boldsymbol{\Phi}_2^*$ are rank-K solutions of (8.56) and Range $(\boldsymbol{\Phi}_1^*) \neq$ Range $(\boldsymbol{\Phi}_2^*)$. Based on convex theory, any convex combination of $\boldsymbol{\Phi}_1^*$ and $\boldsymbol{\Phi}_2^*$, saying $\boldsymbol{\Phi}_3^* \triangleq \alpha\boldsymbol{\Phi}_1^* + (1-\alpha)\boldsymbol{\Phi}_2^*, \forall \alpha \in (0,1)$, is also a solution of (8.56). However, $\boldsymbol{\Phi}_3^*$ is with rank at least $K+1$, which contradicts the fact that any solution must have rank at most K. The third claim could also be proved using contradiction. Suppose that $\boldsymbol{\Phi}_1^*$ and $\boldsymbol{\Phi}_2^*$ are, respectively, rank-K solution and solution with rank smaller than K, and Range $(\boldsymbol{\Phi}_2^*)\backslash$Range $(\boldsymbol{\Phi}_1^*)$ is non-empty. Then, any convex combination of $\boldsymbol{\Phi}_1^*$ and $\boldsymbol{\Phi}_2^*$, saying $\boldsymbol{\Phi}_3^* \triangleq \alpha\boldsymbol{\Phi}_1^* + (1-\alpha)\boldsymbol{\Phi}_2^*, \forall \alpha \in (0,1)$, is also a solution of (8.56). However, $\boldsymbol{\Phi}_3^*$ is again with rank at least $K+1$, which contradicts the fact that any solution must have rank at most K.

The last claim on the solutions of (8.50) and (8.52) follows from Proposition 8.6. ∎

Proposition 8.7 indicates that the rank of the optimal precoding matrix will not be larger than the number of the targets.

8.4.6 Constant-rate Communication Transmission for Spectrum Sharing

Adaptive communication transmission for spectrum sharing methods involves high complexity. A suboptimal transmission approach of constant rate, i.e. $\mathbf{R}_{xl} \equiv \mathbf{R}_x, \forall l \in \mathbb{N}_L^+$, has a lower implementation complexity. In such cases, the spectrum sharing problem can be reformulated as

$$\begin{aligned} (\mathrm{P}_1') \quad &\max_{\mathbf{R}_x \succeq 0, \boldsymbol{\Phi} \succeq 0} \ \mathrm{ESINR}'(\mathbf{R}_x, \boldsymbol{\Omega}, \boldsymbol{\Phi}), \\ &\text{s.t.} \ \underline{C}(\mathbf{R}_x, \boldsymbol{\Phi}) \geq C, \\ &\quad L\,\mathrm{tr}\left(\mathbf{R}_x\right) \leq P_C, L\mathrm{tr}\,(\boldsymbol{\Phi}) \leq P_R, \\ &\quad \mathrm{tr}\left(\boldsymbol{\Phi}\mathbf{V}_k\right) \geq 0, \forall k \in \mathbb{N}_K^+, \end{aligned} \tag{8.63}$$

where

$$\mathrm{ESINR}' = \frac{\mathrm{tr}\left(\mathbf{\Phi}\mathbf{D}_t\right)}{\mathrm{tr}\left(\mathbf{\Phi}\mathbf{C}_t\right) + \mathrm{tr}\left(\mathbf{\Delta}\mathbf{G}_2\mathbf{R}_x\mathbf{G}_2^H\right)/(pLM_{r,R}) + \sigma_R^2} \tag{8.64}$$

and $\mathbf{\Delta} = \sum_{l=1}^{L} \mathbf{\Delta}_l$ is diagonal and with each entry equal to the number of 1s in the corresponding row of $\mathbf{\Omega}$. Similar techniques in Algorithm 2 can be used to solve $(\mathbf{P}_1')$.

We can see that $(\mathbf{P}_1')$ has much lower complexity because there is only one matrix variable for the communication transmission. However, the drawback of the constant-rate communication is that $\mathbf{R}_x$ cannot adapt to the variation of the effective interference channel $\mathbf{G}_{2l}$. On the other hand, the adaptive communication transmission considered in $(\mathbf{P}_1)$ can fully exploit the channel diversity introduced by the radar subsampling procedure. It will be seen in the simulations of Section 8.5.3 that the constant-rate transmission from the solution of (8.63) is inferior to the adaptive transmission from the solution of (8.40).

Another consequence is that the ESINR′ depends on $\mathbf{\Omega}$ only through $\mathbf{\Delta}$. Since $\mathbf{\Omega}$ is searched among the row permutations of a uniformly random sampling matrix, the number of 1s in each row of $\mathbf{\Omega}$ is close to pL or, equivalently, $\mathbf{\Delta}$ will be very close to the scaled identity matrix $pL\mathbf{I}$. To further reduce the complexity, the optimization w.r.t. $\mathbf{\Omega}$ in $(\mathbf{P}_1')$ is omitted because all row permutations of $\mathbf{\Omega}$ will result in a very similar ESINR′. From a different perspective, if the radar subsampling matrix $\mathbf{\Omega}$ is not available for the radar and communication cooperation, we can safely replace $\mathbf{\Delta}$ with $pL\mathbf{I}$ in the ESINR′. The above discussion asserts that for constant-rate communication transmission almost no performance degradation occurs due to the absence of the knowledge of $\mathbf{\Omega}$.

8.4.7 Traditional MIMO Radars for Spectrum Sharing

The traditional MIMO radars without subsampling can be considered a special case with $p = 1$, and, thus, there is no need for the MC. In such cases, the constant-rate communication transmission becomes the optimal scheme because the interference channel $\mathbf{G}_2$ stays as a constant for the period of L symbol time due to the block-fading assumption. The spectrum sharing problem has the same form as $(\mathbf{P}_1')$ with the objective function being

$$\mathrm{SINR} = \frac{\mathrm{tr}\left(\mathbf{\Phi}\mathbf{D}_t\right)}{\mathrm{tr}\left(\mathbf{\Phi}\mathbf{C}_t\right) + \mathrm{tr}\left(\mathbf{G}_2\mathbf{R}_x\mathbf{G}_2^H\right)/M_{r,R} + \sigma_R^2}. \tag{8.65}$$

Note that SINR $\approx$ ESINR′ because $\mathbf{\Delta} \approx pL\mathbf{I}$. Therefore, traditional MIMO radars can achieve approximately the same spectrum sharing performance as MIMO-MC radars when the communications system transmits at a constant rate. However, for MIMO-MC radars, the adaptive communication transmission and the radar subsampling matrix can be designed to achieve a significant radar SINR increase over the traditional MIMO radars. This advantageous flexibility is introduced by the sparse sensing (i.e. subsampling) in MIMO-MC radars. Performance results comparing MIMO-MC radars with different p values against the traditional MIMO radars are provided in Section 8.5.4.

8.5 Numerical Results

In this section, we provide simulation examples to quantify the performance of the above-described jointly designed spectrum sharing method for the coexistence of the MIMO-MC radars and communications systems.

Unless otherwise stated, we use the following default values for the system parameters. The MIMO radar system consists of collocated $M_{t,R} = 16$ TX and $M_{r,R} = 16$ RX antennas, respectively, forming transmit and receive half-wavelength ULAs. The radar waveforms are chosen from the rows of a random orthonormal matrix [Li and Petropulu, 2015b]. We set the length of the radar waveforms to $L = 16$. The wireless communications system consists of collocated $M_{t,C} = 4$ TX and $M_{r,C} = 4$ RX antennas, respectively, forming transmit and receive half-wavelength ULAs. For the communication capacity and power constraints, we take $C = 16$ bits/symbol and $P_C = 6400$ (the power is normalized by the additive noise power). The radar transmit power budget is $P_R = 1000 \times P_C$, which is typical for radar systems; high power is needed to combat path loss associated with far-field targets [Richards, 2005]. The additive white Gaussian noise variances are $\sigma_C^2 = \sigma_R^2 = 1$. There are three stationary targets with RCS variance $\sigma_{\beta 0}^2 = 0.5$, located in the far-field with a path loss of 30 dB. Clutter is generated by four point scatterers, all having the same RCS variance, σ_β^2, which is chosen according to the clutter-to-noise ratio (CNR) $10 \log \sigma_\beta^2/\sigma_R^2$. Some remarks are provided to justify the parameter selection. Based on these numbers, the possible range of SNR at the communication receiver is between 12 and 26 dB, which is supported by LTE systems [3GPP, 2009, Kawser et al., 2012]. Based on a simple link budget analysis, the chosen radar power budget corresponds to a per receive antenna SNR of about 23 dB when only additive noise is considered. For a typical radar system with a single antenna, operating with a probability of detection of 0.9 and a probability of false alarm of 10^{-6}, the required SNR is about 13.2 dB [Richards, 2005]. However, the actual SNR may be much smaller because spatial degrees of freedom are used to mitigate clutter and interference from the communications systems.

The channel $\mathbf{H}$ is modeled as Rayleigh fading, i.e. contains independent entries, distributed as $\mathcal{CN}(0,1)$. The interference channels $\mathbf{G}_1$ and $\mathbf{G}_2$ are modeled as Rician fading to simulate that there is a direct path between the radar and communications systems. This is a reasonable assumption as radars are often mounted high for their long-range sensing purpose. The power in the direct path is 0.1, and the variance of Gaussian components contributed by the scattered paths is 10^{-3}.

The performance metrics considered include the following:

- The radar *effective* SINR, i.e. the objective of the spectrum sharing problem
- The MC relative recovery error, defined as $\|\mathbf{M} - \widehat{\boldsymbol{M}}\|_F/\|\mathbf{M}\|_F$, where $\widehat{\boldsymbol{M}}$ is the completed data matrix at the radar fusion center
- The radar transmit beampattern, i.e. the transmit power for different azimuth angles $\mathbf{v}_t^T(\theta)\mathbf{P}\mathbf{v}_t^*(\theta)$
- The MUSIC pseudo-spectrum and the relative target RCS estimation root mean squared error (RMSE) obtained using the least squares estimation on the completed data matrix $\widehat{\boldsymbol{M}}$.

Monte Carlo simulations with 100 independent trials are carried out to get an average performance.

8.5.1 The Radar Transmit Beampattern and the MUSIC Spectrum

In this subsection, we present an example demonstrating the advantages of the above-described jointly designed radar precoding scheme compared to uniform precoding, i.e. $\mathbf{P} = \sqrt{LP_R/M_{t,R}}\mathbf{I}$, and NSP precoding, i.e. $\mathbf{P} = \sqrt{LP_R/M_{t,R}}\mathbf{V}\mathbf{V}^H$, where $\mathbf{V}$ contains the basis of the null space of $\mathbf{G}_1$ [Khawar et al., 2014]. For the joint-design-based scheme in (8.40), we choose $\xi = \lfloor \xi_{\max} \rfloor$. The target angles w.r.t. the array are, respectively, $-10°$, $15°$, and $30°$; the four point scatterers are at angles $-45°$, $-30°$, $10°$, and $45°$. The CNR is 30 dB. In this simulation, the direct path in $\mathbf{G}_1$ is generated as $\sqrt{0.1}\mathbf{v}_t(\phi)\mathbf{v}_t^H(\phi)$, where $\phi = 15°$, with $\mathbf{v}_t(\phi)$ being defined in (8.2). In other words, the communication receiver is taken at the same azimuth angle as the second target.

Recall that the NSP technique projects the radar waveform onto the null space of the interference channel $\mathbf{G}_2$ in order to avoid creating interference to the communication receiver. Because the null space and row space of a matrix are orthogonal to each other, there will be no radar power radiated along the null space of $\mathbf{G}_2$; thus, targets in those locations will be missed. The precoding approach presented here does not suffer from such problems because the precoding is computed via a joint design method instead of projecting to the null space of $\mathbf{G}_2$. The radar transmit beampattern and the spatial pseudo-spectrum obtained using the MUSIC algorithm are shown in Figure 8.7. The correspondingly achieved ESINR, MC relative recovery error, and relative target RCS estimation RMSE are listed in Table 8.1. We observe that the jointly designed precoding scheme achieves significant improvement in ESINR, MC relative recovery error, and target RCS estimation accuracy. As expected, the uniform precoding scheme just spreads the transmit power uniformly in all directions. The NSP precoding scheme achieves a similar beampattern as the uniform precoding scheme, with the exception of the deep null that the NSP places in the direction of the communication receiver. The null means that the transmit power toward the second target is severely attenuated and thus the probability of missing the second target is increased. We should note that neither the uniform nor the NSP precoding schemes are capable of clutter mitigation. From Figure 8.7, we observe that the jointly designed precoding scheme successfully focuses the transmit power toward the three targets and nullifies the power toward the point scatterers. The three targets can be accurately estimated from the pseudo-spectrum obtained by the joint design. Meanwhile, the communications system can still achieve the required rate by aligning its transmission along a channel subspace that does not interfere with the radar emissions. This significant advantage is enabled by the joint design of radar and communication transmissions.

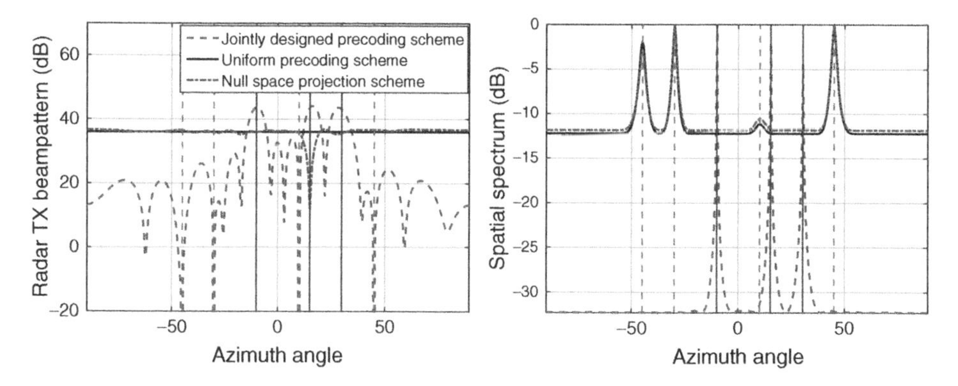

Figure 8.7 The radar transmit beampattern and the MUSIC spatial pseudo-spectrum for MIMO-MC radar and communication spectrum sharing. The true positions of the targets and clutters are labeled using solid and dashed vertical lines, respectively. Source: Li and Petropulu LiBo17TAES/IEEE.

Table 8.1 The radar ESINR, MC relative recovery errors, and the relative target RCS estimation RMSE for MIMO-MC radar and communication spectrum sharing

Precoding schemes	ESINR (dB)	MC relative recovery errors	Relative RCS est. RMSE
Joint design precoding	31.3	0.038	0.028
Uniform precoding	−44.3	1.00	1.000
NSP-based precoding	−46.3	1.00	0.995

8.5.2 Comparison of Different Levels of Cooperation

In this subsection, we compare several algorithms with different levels of radar and communication cooperation. The compared algorithms include the following:

- Uniform radar precoding and selfish communication: the radar transmit antennas use the trivial precoding, i.e. $\mathbf{P} = \sqrt{LP_R/M_{t,R}}\mathbf{I}$; and the communications system minimizes the transmit power to achieve certain average capacity without any concern about the interference it exerts on the radar system. This algorithm involves no radar and communication cooperation.
- NSP-based radar precoding and selfish communication: the radar transmit antennas use the fixed precoding, i.e. $\mathbf{P} = \sqrt{LP_R/M_{t,R}}\mathbf{V}\mathbf{V}^H$, while the selfish communication scheme is the same with the previous case.
- Uniform radar precoding and designing $\mathbf{R}_{xl}$ and $\boldsymbol{\Omega}$: only $\mathbf{R}_{xl}$ and $\boldsymbol{\Omega}$ are jointly designed to minimize the effective interference to the radar receiver.
- Design $\mathbf{P}$ and selfish communication: only the radar precoding matrix $\mathbf{P}$ is designed to maximize the radar ESINR.
- Jointly designed $\mathbf{P}$, $\mathbf{R}_{xl}$, and $\boldsymbol{\Omega}$ in (8.40).

We use the same values for all parameters as in the previous simulation except that the radar transmit power budget P_R changes from 51,200 to 2.56×10^6. Figure 8.8 shows the achieved ESINR, the MC relative recovery error, and the relative target RCS estimation RMSE. The algorithms that use trivial uniform and NSP-based radar precoding perform bad because the point scatterers are not properly mitigated. The scheme designing $\mathbf{P}$ could mitigate only the scatterers but the interference

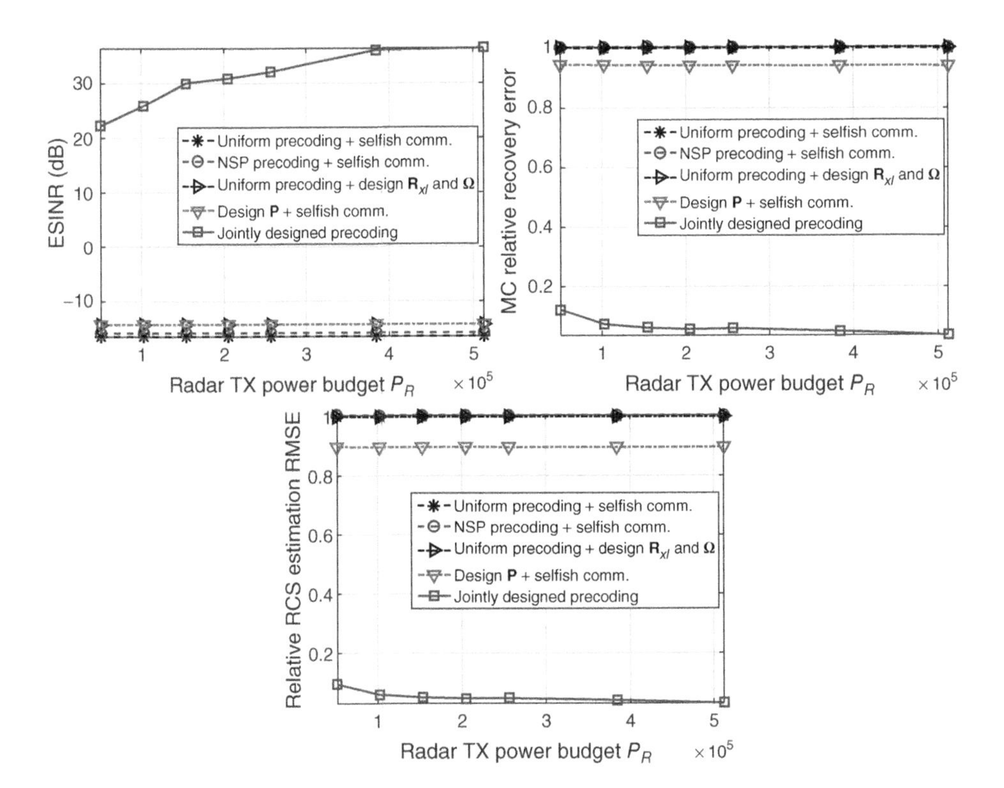

Figure 8.8 Comparison of spectrum sharing with different levels of cooperation between the MIMO-MC radar and the communications system under different P_R. Source: Li and Petropulu LiBo17TAES/IEEE.

from the communication transmission is not controlled. The joint design of $\mathbf{P}, \mathbf{R}_{xl}$, and $\mathbf{\Omega}$ simultaneously addresses the clutter and the mutual interference between the radar and the communications systems and thus achieves the best performance amongst all the algorithms. Notably, the proposed design achieves at least 20 dB ESINR, while other algorithms can only achieve ESINR lower than −10 dB. The performance gains come from high-level cooperation between the two systems.

8.5.3 Comparison Between Adaptive and Constant-rate Communication Transmissions

In this subsection, we evaluate the performance of two communication transmission schemes, namely, adaptive transmission with different $\mathbf{R}_{xl}$'s for all $l \in \mathbb{N}_L^+$, and constant-rate transmission with only one identical $\mathbf{R}_x$. We use the following parameter setting: $M_{t,R} = 16, M_{r,R} = M_{t,C} = 8, M_{r,C} = 2$, $C = 10$ bits/symbol, $P_C = 64$, and $P_R = 1000 \times P_C$. For the $\mathbf{G}_1$ and $\mathbf{G}_2$, Rayleigh fading is used with fixed $\sigma^2_{G_1}$ and varying $\sigma^2_{G_2}$. The results of ESINR, MC relative recovery error, and the relative target RCS estimation RMSE for different values of $\sigma^2_{G_2}$ are shown in Figure 8.9. The value of $\sigma^2_{G_2}$ varies from 0.05 to 0.5, which effectively simulates different distances between the communication transmitter and the radar receiver. It is clear that the adaptive communication transmission outperforms the constant-rate counterpart under various values of interference channel strength. As discussed in Section 8.4.6, the adaptive communication transmission can fully exploit the channel diversity of $\mathbf{G}_{2l}$ introduced by the radar subsampling procedure. The price for the performance advantages is high complexity. The average running times on a laptop

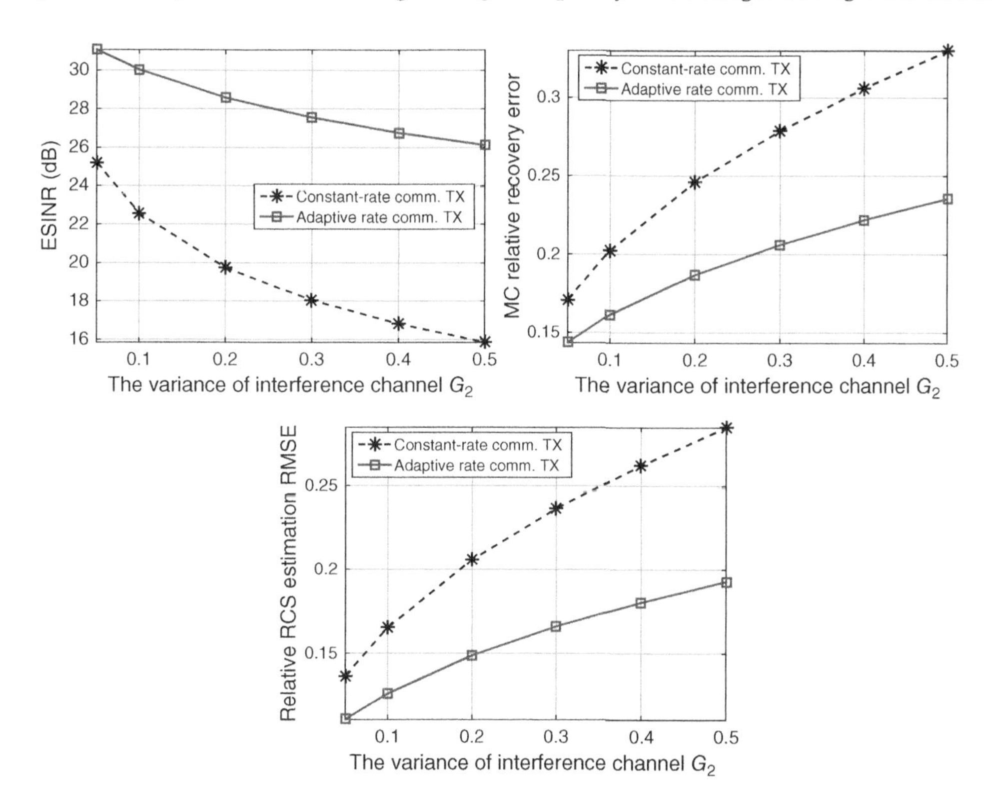

Figure 8.9 Comparison of spectrum sharing with adaptive and constant-rate communication transmissions under different levels of variance of the interference channel from the communication transmitter to the radar receiver. Source: Li and Petropulu LiBo17TAES/IEEE.

with Intel Core i5 Dual-Core 2.4 GHz CPU for the adaptive and constant-rate communication transmissions are, respectively, 15.6 and 4.8 seconds. The choice between these two transmission schemes can be made depending on the available computing resources.

8.5.4 Comparison Between MIMO-MC Radars and Traditional MIMO Radars

In this subsection, we present a simulation to show the advantages of MIMO-MC radars compared to the traditional full-sampled MIMO radars. The parameters are the same as those in previous simulation of Section 8.5.3 but with fixed $\sigma_{G_1}^2 = 0.3$ and $\sigma_{G_2}^2 = 1$, which indicates strong mutual interference, especially the interference from the communication transmitter to the radar receiver. The radar transmit power budget P_R is taken to be equal to $10 \times P_C$. We consider two targets; one is randomly located and the other is taken to be 25° away. We also consider four randomly located point scatterers. Figure 8.10 shows the results under different MIMO-MC subsampling rates p. Note that full sampling is used for the traditional MIMO radar. The MC relative recover error for the traditional radar is actually the output distortion-to-signal ratio. A smaller distortion-to-signal ratio corresponds to a larger output SNR. For ease of comparison, a black dashed line is used for the traditional MIMO radar. We observe that the MIMO-MC radar achieves better performance in ESINR than the traditional radar. This is because the communications system can effectively prevent its transmission from interfering the radar system when the number of actively sampled radar RX antennas is small, i.e. subsampling is small. In addition, the larger ESINR of the MIMO-MC radar results in a larger output SINR than that of the traditional radar. Furthermore, the MIMO-MC radar achieves better target RCS estimation accuracy than the traditional radar if its subsampling

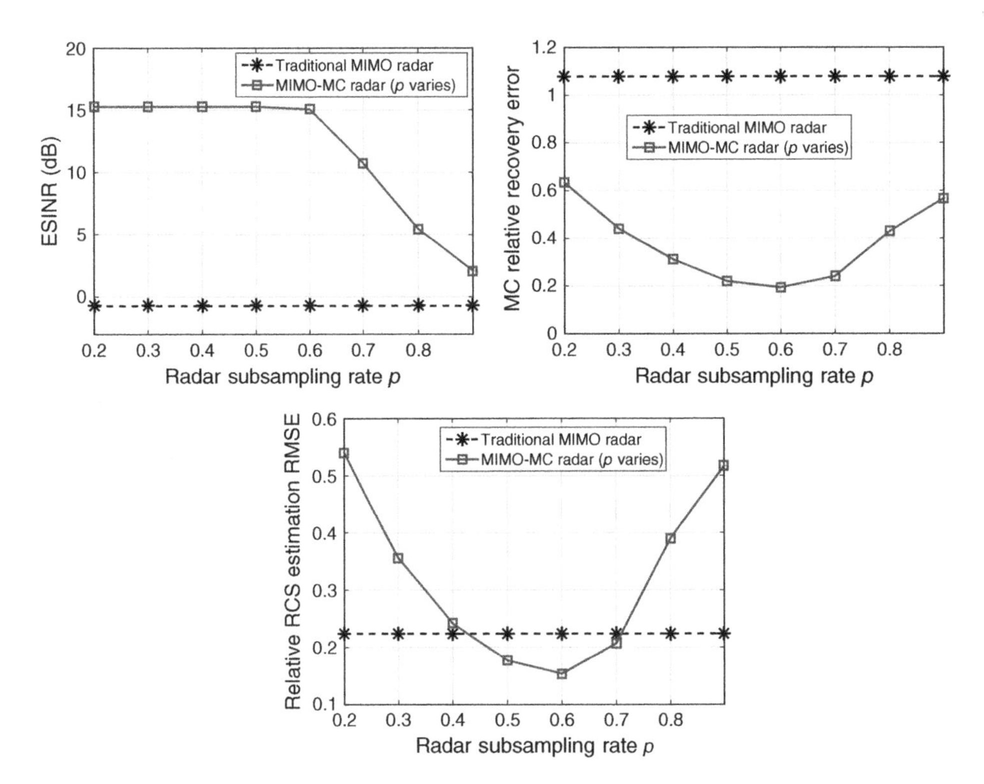

Figure 8.10 Comparison of spectrum sharing with traditional MIMO radars and MIMO-MC radars with different subsampling rates p. Source: Li and Petropulu LiBo17TAES/IEEE.

rate is between 0.4 and 0.7. For p larger than 0.7, the target RCS estimation accuracy achieved by the MIMO-MC radar is worse than that achieved by the traditional radar because small ESINRs for $p \geq 0.7$ introduce high distortion in the completed data matrix. The results in Figure 8.10 could be used to help the selection of radar subsampling rate p. For the best target RCS estimation accuracy, $p = 0.6$ is the best choice, while for the biggest savings in terms of samples and similar performance to traditional radars, $p = 0.4$ is the best choice. Note that the actual value for the best choice of p is scenario dependent. Since there is no closed-form solution for the joint design problem, it is difficult to provide a theoretical justification.

Based on these results, we conclude that MIMO-MC radars can coexist with communications systems and achieve better target RCS estimation than traditional radars while saving up to 60% data samples. Such significant advantages are introduced by the sparse sensing (i.e. subsampling) in MIMO-MC radars as discussed in Section 8.4.7.

8.6 Conclusions

In this chapter, we have considered the coexistence of a MIMO-MC radar and a wireless MIMO communications system by sharing a common carrier frequency. The radar transmits random unitary waveforms, and both radar and communications systems use precoders. The precoders and the radar subsampling scheme have been jointly designed by the control center to maximize the radar SINR while meeting certain rate and power constraints for the communications system. Random unitary waveforms can be easily generated and updated for waveform security. We should note that the presented joint-design-based spectrum sharing method can also be applied to traditional MIMO radars, which is a special case of MIMO-MC radars for $p = 1$.

We have shown via simulations that the cooperative design enables significant performance gains compared to a noncooperative design. The jointly designed spectrum sharing scheme successfully focuses the transmit power toward the targets and nullifies the power toward the clutter. It achieves significant improvement in ESINR, MC relative recovery error, and target RCS estimation accuracy. We have also compared the performance and complexity of the adaptive and the constant-rate communication transmission schemes for radar–communication spectrum sharing. Finally, we have provided simulation-based comparison of MIMO-MC radars and traditional MIMO radars coexisting with communications systems. We have observed that the MIMO-MC radar achieves better performance in terms of ESINR and output SNR. Our simulations suggest that MIMO-MC radars can coexist with communications systems and achieve better target RCS estimation than traditional radars, while saving up to 60% in data samples. Of course, these advantages come at increased computation complexity for MC.

We should note that the constraint requiring that the number of targets is smaller than the number of radar antennas results in an inefficient usage of the MIMO radar degrees of freedom. However, the high resolution of traditional MIMO radar is retained by MIMO-MC radars with a great reduction of sample and hardware complexity. The considered signal model is for narrowband waveforms. Broadband MIMO systems typically use OFDM waveforms [Gogineni et al., 2013]. In such cases, the joint design still applies to individual component carriers. This would substantially expand the application scenarios of the results presented in this chapter.

References

3GPP TS LTE evolved universal terrestrial radio access (E-UTRA) physical layer procedures. Technical Specification TS 36.213 V8.0, 3rd Generation Partnership Project Std, 2009.

Amendment of the commission's rules with regard to commercial operations in the 3550–3650 MHz band. Technical report, Federal Communications Commission (FCC), April 2015. https://apps.fcc.gov/edocs_public/attachmatch/FCC-15-47A1.pdf.

M. G. Amin. *Compressive sensing for urban radar*. CRC Press, 2014.

S. Amuru, R. M. Buehrer, R. Tandon, and S. Sodagari. MIMO radar waveform design to support spectrum sharing. In *IEEE Military Communication Conference*, pages 1535–1540, November 2013.

J. G. Andrews, A. Ghosh, and R. Muhamed. *fundamentals of WiMAX: understanding broadband wireless networking*. Prentice Hall PTR, Upper Saddle River, NJ, USA, 2007. ISBN 0132225522.

A. Aubry, A. De Maio, M. Piezzo, and A. Farina. Radar waveform design in a spectrally crowded environment via nonconvex quadratic optimization. *IEEE Transactions on Aerospace and Electronic Systems*, 50(2):1138–1152, 2014.

A. Aubry, A. De Maio, Y. Huang, M. Piezzo, and A. Farina. A new radar waveform design algorithm with improved feasibility for spectral coexistence. *IEEE Transactions on Aerospace and Electronic Systems*, 51(2):1029–1038, 2015.

Auction of 24 GHz upper microwave flexible use service licenses closes. Notice, Federal Communications Commission, 2019. https://www.fcc.gov/document/auction-102-closing-public-notice (Accessed: January 2021).

A. Babaei, W. H. Tranter, and T. Bose. A practical precoding approach for radar/communications spectrum sharing. In *8th International Conference on Cognitive Radio Oriented Wireless Networks*, pages 13–18, July 2013.

M. R. Bell, N. Devroye, D. Erricolo, T. Koduri, S. Rao, and D. Tuninetti. Results on spectrum sharing between a radar and a communications system. In *2014 International Conference on Electromagnetics in Advanced Applications (ICEAA)*, pages 826–829, 2014.

S. S. Bhat, R. M. Narayanan, and M. Rangaswamy. Bandwidth sharing and scheduling for multimodal radar with communications and tracking. In *IEEE Sensor Array and Multichannel Signal Processing Workshop*, pages 233–236, June 2012.

S. Bhojanapalli and P. Jain. Universal matrix completion. In *Proceedings of The 31st International Conference on Machine Learning*, pages 1881–1889, 2014.

M. Bica, K. W. Huang, V. Koivunen, and U. Mitra. Mutual information based radar waveform design for joint radar and cellular communication systems. In *2016 IEEE International Conference on Acoustics, Speech and Signal Processing (ICASSP)*, pages 3671–3675, March 2016.

R. G. Bland, D. Goldfarb, and M. J. Todd. The ellipsoid method: a survey. *Operations Research*, 29(6):1039–1091, 1981.

S. Boyd and L. Vandenberghe. *Convex optimization*. Cambridge University Press, 2004.

E. J. Candès and Y. Plan. Matrix completion with noise. *Proceedings of the IEEE*, 98(6):925–936, 2010.

E. J. Candès and B. Recht. Exact matrix completion via convex optimization. *Foundations of Computational Mathematics*, 9(6):717–772, 2009.

E. Cardillo and A. Caddemi. A review on biomedical MIMO radars for vital sign detection and human localization. *Electronics*, 9(9):1497, 2020.

C. Chen and P. P. Vaidyanathan. MIMO radar ambiguity properties and optimization using frequency-hopping waveforms. *IEEE Transactions on Signal Processing*, 56(12):5926–5936, 2008a.

C. Chen and P. P. Vaidyanathan. Compressed sensing in MIMO radar. In *Asilomar Conference on Signals, Systems and Computers*, pages 41–44, 2008b.

C. Chen and P. P. Vaidyanathan. *MIMO Radar spacetime adaptive processing and signal design*, pages 235–281. John Wiley & Sons, Inc., 2008c.

Z. Chen, H. Li, G. Cui, and M. Rangaswamy. Adaptive transmit and receive beamforming for interference mitigation. *IEEE Signal Processing Letters*, 21(2):235–239, 2014.

G. Cui, H. Li, and M. Rangaswamy. MIMO radar waveform design with constant modulus and similarity constraints. *IEEE Transactions on Signal Processing*, 62(2):343–353, 2014.

Y. Cui, V. Koivunen, and X. Jing. Interference alignment based spectrum sharing for MIMO radar and communication systems. In *2018 IEEE 19th International Workshop on Signal Processing Advances in Wireless Communications (SPAWC)*, pages 1–5, 2018. https://doi.org/10.1109/SPAWC.2018.8445973.

H. Deng and B. Himed. Interference mitigation processing for spectrum-sharing between radar and wireless communications systems. *IEEE Transactions on Aerospace and Electronic Systems*, 49(3):1911–1919, 2013.

S. N. Diggavi and T. M. Cover. The worst additive noise under a covariance constraint. *IEEE Transactions on Information Theory*, 47(7):3072–3081, 2001.

E. Drocella, J. Richards, R. Sole, F. Najmy, A. Lundy, and P. McKenna. 3.5 GHz exclusion zone analyses and methodology. Technical Report TR-15-517, US Dept. of Commerce, the National Telecommunications and Information Administration, 2015.

A. D. Droitcour, O. Boric-Lubecke, V. M. Lubecke, J. Lin, and G. T. A. Kovacs. Range correlation and I/Q performance benefits in single-chip silicon doppler radars for noncontact cardiopulmonary monitoring. *IEEE Transactions on Microwave Theory and Techniques*, 52(3):838–848, 2004. https://doi.org/10.1109/TMTT.2004.823552.

H. Du and T. Ratnarajah. Robust utility maximization and admission control for a MIMO cognitive radio network. *IEEE Transactions on Vehicular Technology*, 62(4):1707–1718, 2013.

FCC proposes innovative small cell use in 3.5 GHz band. News release, Federal Communications Commission (FCC), December 2012. https://apps.fcc.gov/edocs_public/attachmatch/DOC-317911A1.pdf.

M. Filo, A. Hossain, A. R. Biswas, and R. Piesiewicz. Cognitive pilot channel: enabler for radio systems coexistence. In *2nd International Workshop on Cognitive Radio and Advanced Spectrum Management*, pages 17–23, May 2009.

S. Gogineni, M. Rangaswamy, and A. Nehorai. Multi-modal OFDM waveform design. In *IEEE Radar Conference*, pages 1–5, April 2013.

A. Goldsmith. *Wireless communications*. Cambridge University Press, 2005.

A. Hassanien, M. G. Amin, Y. D. Zhang, and F. Ahmad. Signaling strategies for dual-function radar-communications: an overview. *IEEE Aerospace and Electronic Systems Magazine*, 31(10):36–45 2016.

E. Hossain, D. Niyato, and Z. Han. *Dynamic spectrum access and management in cognitive radio networks*. Cambridge University Press, 2009.

X. Hou and C. Yang. How much feedback overhead is required for base station cooperative transmission to outperform non-cooperative transmission? In *2011 IEEE International Conference on Acoustics, Speech and Signal Processing (ICASSP)*, pages 3416–3419. IEEE, 2011.

K. Huang, M. Bica, U. Mitra, and V. Koivunen. Radar waveform design in spectrum sharing environment: coexistence and cognition. In *Radar Conference (RadarCon), 2015 IEEE*, pages 1698–1703. IEEE, 2015.

I. Y. Immoreev, S. Samkov, and T.-H. Tao. Short-distance ultra wideband radars. *IEEE Aerospace and Electronic Systems Magazine*, 20(6):9–14, 2005. https://doi.org/10.1109/MAES.2005.1453804.

T. Jiang. How many entries of a typical orthogonal matrix can be approximated by independent normals? *The Annals of Probability*, 34(4):1497–1529, 2006. ISSN 00911798.

R. P. Jover. LTE PHY fundamentals. Technical report, 2015. https://www.slideshare.net/PrashantSengar/lte-phy-fundamentals-50510450(Accessed:July2015).

D. S. Kalogerias and A. P. Petropulu. Matrix completion in colocated MIMO radar: recoverability, bounds and theoretical guarantees. *IEEE Transactions on Signal Processing*, 62(2):309–321, 2014.

M. T. Kawser, B. Hamid, N. Hasan, M. S. Alam, and M. M. Rahman. Downlink SNR to CQI mapping for different multiple antenna techniques in LTE. *International Journal of Information and Electronics Engineering*, 2(5):757, 2012.

A. Khawar, A. Abdel-Hadi, and T. C. Clancy. Spectrum sharing between S-band radar and LTE cellular system: a spatial approach. In *IEEE International Symposium on Dynamic Spectrum Access Networks*, pages 7–14, April 2014.

A. Khawar, A. Abdelhadi, and T. C. Clancy. *MIMO radar waveform design for spectrum sharing with cellular systems: a MATLAB based approach*. Springer, 2016.

S. J. Kim and G. B. Giannakis. Optimal resource allocation for MIMO Ad Hoc cognitive radio networks. *IEEE Transactions on Information Theory*, 57(5):3117–3131, May 2011.

C. Kopp. Search and acquisition radars (S-band, X-band). Technical Report APA-TR-2009-0101, 2009. http://www.ausairpower.net/APA-Acquisition-GCI.html (Accessed: July 2015).

H. Krim and M. Viberg. Two decades of array signal processing research: the parametric approach. *IEEE Signal Processing Magazine*, 13(4):67–94, 1996.

H. W. Kuhn. The Hungarian method for the assignment problem. *Naval Research Logistics Quarterly*, 2(1–2):83–97, 1955.

P. Kumari, J. Choi, N. González-Prelcic, and R. W. Heath. IEEE 802.11ad-based radar: an approach to joint vehicular communication-radar system. *IEEE Transactions on Vehicular Technology*, 67(4):3012–3027, 2018. https://doi.org/10.1109/TVT.2017.2774762.

A. Lackpour, M. Luddy, and J. Winters. Overview of interference mitigation techniques between WiMAX networks and ground based radar. In *20th Annual Wireless and Optical Communications Conference*, pages 1–5, April 2011.

B. Laurent and P. Massart. Adaptive estimation of a quadratic functional by model selection. *Annals of Statistics*, 28(5):1302–1338, 2000.

Q. Li and W.-K. Ma. Optimal and robust transmit designs for MISO channel secrecy by semidefinite programming. *IEEE Transactions on Signal Processing*, 59(8):3799–3812, 2011.

B. Li and A. P. Petropulu. Radar precoding for spectrum sharing between matrix completion based MIMO radars and a MIMO communication system. In *IEEE Global Conference on Signal and Information Processing*, pages 737–741, December 2015a. https://doi.org/10.1109/GlobalSIP.2015.7418294.

B. Li and A. P. Petropulu. Spectrum sharing between matrix completion based MIMO radars and a MIMO communication system. In *IEEE International Conference on Acoustics, Speech and Signal Processing*, pages 2444–2448, April 2015b.

B. Li and A. P. Petropulu. Distributed MIMO radar based on sparse sensing: analysis and efficient implementation. *IEEE Transactions on Aerospace and Electronic Systems*, 51(4):3055–3070, October 2015c.

B. Li and A. P. Petropulu. MIMO radar and communication spectrum sharing with clutter mitigation. In *IEEE Radar Conference*, pages 1–6, May 2016.

B. Li and A. P. Petropulu. Matrix completion based MIMO radars with clutter and interference mitigation via transmit precoding. In *2017 IEEE International Conference on Acoustics, Speech and Signal Processing (ICASSP)*, pages 3216–3220, March 2017a.

B. Li and A. P. Petropulu. Joint transmit designs for coexistence of MIMO wireless communications and sparse sensing radars in clutter. *IEEE Transactions on Aerospace and Electronic Systems*, 53(6):2846–2864, 2017b.

B. Li, H. Kumar, and A. P. Petropulu. A joint design approach for spectrum sharing between radar and communication systems. In *IEEE International Conference on Acoustics, Speech and Signal Processing*, pages 3306–3310, March 2016a.

B. Li, A. P. Petropulu, and W. Trappe. Optimum co-design for spectrum sharing between matrix completion based MIMO radars and a MIMO communication system. *IEEE Transactions on Signal Processing*, 64(17):4562–4575, 2016b.

C. Li, Z. Peng, T. Huang, T. Fan, F. Wang, T. Horng, J. Muñoz-Ferreras, R. Gómez-García, L. Ran, and J. Lin. A review on recent progress of portable short-range noncontact microwave radar systems. *IEEE Transactions on Microwave Theory and Techniques*, 65(5):1692–1706, 2017. https://doi.org/10.1109/TMTT.2017.2650911.

X. Liang, H. Zhang, S. Ye, G. Fang, and T. A. Gulliver. Improved denoising method for through-wall vital sign detection using UWB impulse radar. *Digital Signal Processing*, 74:72–93, 2018.

F. Liu, C. Masouros, A. Li, and T. Ratnarajah. Robust MIMO beamforming for cellular and radar coexistence. *IEEE Wireless Communications Letters*, 6(3):374–377, 2017. https://doi.org/10.1109/LWC.2017.2693985.

F. Liu, C. Masouros, A. Li, T. Ratnarajah, and J. Zhou. Mimo radar and cellular coexistence: a power-efficient approach enabled by interference exploitation. *IEEE Transactions on Signal Processing*, 66(14):3681–3695, 2018a.

Q. Liu, H. Guo, J. Xu, H. Wang, A. Kageza, S. AlQarni, and S. Wu. Non-contact non-invasive heart and respiration rates monitoring with MIMO radar sensing. In *2018 IEEE Global Communications Conference (GLOBECOM)*, pages 1–6, 2018b.

F. Liu, C. Masouros, A. P. Petropulu, H. Griffiths, and L. Hanzo. Joint radar and communication design: applications, state-of-the-art, and the road ahead. *IEEE Transactions on Communications*, 68(6):3834–3862, 2020. https://doi.org/10.1109/TCOMM.2020.2973976.

G. Locke and L. E. Strickling. An assessment of the near-term viability of accommodating wireless broadband systems in the 1675-1710 MHz, 1755-1780 MHz, 3500-3650 MHz, and 4200-4220 MHz, 4380-4400 MHz bands. Technical Report TR-13-490, US Dept. of Commerce, the National Telecommunications and Information Administration, 2012.

LTE in a nutshell: The physical layer. White paper, Telesystem Innovations Inc., 2010.

L. Lu, X. Zhou, U. Onunkwo, and G. Y. Li. Ten years of research in spectrum sensing and sharing in cognitive radio. *EURASIP Journal on Wireless Communications and Networking*, 2012: 28, 2012.

M. Mercuri, P. J. Soh, G. Pandey, P. Karsmakers, G. A. E. Vandenbosch, P. Leroux, and D. Schreurs. Analysis of an indoor biomedical radar-based system for health monitoring. *IEEE Transactions on Microwave Theory and Techniques*, 61(5):2061–2068, 2013. https://doi.org/10.1109/TMTT.2013.2247619.

K. V. Mishra, M. R. B. Shankar, V. Koivunen, B. Ottersten, and S. A. Vorobyov. Toward millimeter-wave joint radar communications: a signal processing perspective. *IEEE Signal Processing Magazine*, 36(5):100–114, 2019. https://doi.org/10.1109/MSP.2019.2913173.

R. Mudumbai, G. Barriac, and U. Madhow. On the feasibility of distributed beamforming in wireless networks. *IEEE Transactions on Wireless Communications*, 6(5):1754–1763, 2007.

K. T. Phan, S. A. Vorobyov, N. D. Sidiropoulos, and C. Tellambura. Spectrum sharing in wireless networks via QoS-aware secondary multicast beamforming. *IEEE Transactions on Signal Processing*, 57(6):2323–2335, 2009.

F. Quaiyum, N. Tran, T. Phan, P. Theilmann, A. E. Fathy, and O. Kilic. Electromagnetic modeling of vital sign detection and human motion sensing validated by noncontact radar measurements. *IEEE Journal of Electromagnetics, RF and Microwaves in Medicine and Biology*, 2(1):40–47, 2018. https://doi.org/10.1109/JERM.2018.2807978.

Radtec Engineering Inc., Radar performance. Technical report, Radtec Engineering Inc., 2015. http://www.radar-sales.com/PDFs/Performance_RDR%26TDR.pdf (Accessed: July 2015).

T. Rappaport. *Wireless communications principles and practice*. Prentice Hall, 2001.

Realizing the full potential of government-held spectrum to spur economic growth. Technical report, The Presidents Council of Advisors on Science and Technology (PCAST), July 2012. http://www.dtic.mil/dtic/tr/fulltext/u2/a565091.pdf.

M. A. Richards. *Fundamentals of radar signal processing*. McGraw-Hill, New York, 2005.

R. Rogalin, O. Y. Bursalioglu, and H. Papadopoulos. Scalable synchronization and reciprocity calibration for distributed multiuser MIMO. *IEEE Transactions on Wireless Communications*, 13(4):1815–1831, 2014.

M. J. Rossini. The spectrum scarcity doctrine: a constitutional anachronism. *Southwestern Law Journal*, 39:827, 1985.

F. H. Sanders, R. L. Sole, J. E. Carroll, G. S. Secrest, and T. L. Allmon. Analysis and resolution of RF interference to radars operating in the band 2700–2900 MHz from broadband communication transmitters. Technical Report TR-13-490, US Dept. of Commerce, the National Telecommunications and Information Administration, 2012.

R. Saruthirathanaworakun, J. M. Peha, and L. M. Correia. Opportunistic sharing between rotating radar and cellular. *IEEE Journal on Selected Areas in Communications*, 30(10):1900–1910, 2012.

C. Shahriar, A. Abdelhadi, and T. C. Clancy. Overlapped-MIMO radar waveform design for coexistence with communication systems. In *IEEE Wireless Communications and Networking Conference*, pages 223–228, 2015.

X. Shang, J. Liu, and J. Li. Multiple object localization and vital sign monitoring using IR-UWB MIMO radar. *IEEE Transactions on Aerospace and Electronic Systems*, 56(6):4437–4450, 2020. https://doi.org/10.1109/TAES.2020.2990817.

K. Sjoberg, P. Andres, T. Buburuzan, and A. Brakemeier. Cooperative intelligent transport systems in europe: current deployment status and outlook. *IEEE Vehicular Technology Magazine*, 12(2):89–97, 2017. https://doi.org/10.1109/MVT.2017.2670018.

S. Sodagari, A. Khawar, T. C. Clancy, and R. McGwier. A projection based approach for radar and telecommunication systems coexistence. In *IEEE Global Telecommunication Conference*, pages 5010–5014, December 2012. https://doi.org/10.1109/GLOCOM.2012.6503914.

P. Stoica, J. Li, and Y. Xie. On probing signal design for MIMO radar. *IEEE Transactions on Signal Processing*, 55(8):4151–4161, 2007.

S. Sun and A. P. Petropulu. Waveform design for MIMO radars with matrix completion. *IEEE Journal on Selected Topics in Signal Processing*, 9(8):1400–1414, 2015.

S. Sun, W. Bajwa, and A. P. Petropulu. MIMO-MC radar: a MIMO radar approach based on matrix completion. *IEEE Transactions on Aerospace and Electronic Systems*, 51(3):1839–1852, 2015.

S. Sun, A. P. Petropulu, and H. V. Poor. MIMO radar for advanced driver-assistance systems and autonomous driving: advantages and challenges. *IEEE Signal Processing Magazine*, 37(4):98–117, 2020. https://doi.org/10.1109/MSP.2020.2978507.

S. C. Surender, R. M. Narayanan, and C. R. Das. Performance analysis of communications & radar coexistence in a covert UWB OSA system. In *IEEE Global Telecommunications Conference*, pages 1–5, 2010.

S. Sweeney, G. Benitez, and A. Maile, 2022. Verizon, AT&T delay 5G rollout around some airports after stark warnings from US airlines. ABC News. https://abcnews.go.com/Politics/verizon-att-delay-5g-rollout-airports-stark-warnings/story?id=82327471 (Accessed: March 2022).

Timeline for deployment of lte-v2x. 5gaa press information, 2018. http://5gaa.org/news/timeline-for-deployment-of-lte-v2x/ (Accessed: January 2021).

D. Tse and P. Viswanath. *Fundamentals of wireless communication*. Cambridge University Press, 2005.

A. Turlapaty and Y. Jin. A joint design of transmit waveforms for radar and communications systems in coexistence. In *IEEE Radar Conference*, pages 0315–0319, 2014.

E. Uhlemann. Time for autonomous vehicles to connect [connected vehicles]. *IEEE Vehicular Technology Magazine*, 13(3):10–13, 2018.

L. S. Wang, J. P. McGeehan, C. Williams, and A. Doufexi. Application of cooperative sensing in radar-communications coexistence. *IET Communications*, 2(6):856–868, 2008.

C-band nears 70b rockets above prior us spectrum auctions. News report, Fierce Wireless, 2020. https://www.fiercewireless.com/regulatory/c-band-clock-phase-auction-tops-charts-80-9b (Accessed: February 2021).

J. Yeh. Real analysis: theory of measure and integration second edition. In *Theory of Measure and Integration*. World Scientific Publishing Company, 2006.

Y. Yu, A. P. Petropulu, and H. V. Poor. MIMO radar using compressive sampling. *IEEE Journal on Selected Topics in Signal Processing*, 4(1):146–163, 2010.

R. Zhang and Y. Liang. Exploiting multi-antennas for opportunistic spectrum sharing in cognitive radio networks. *IEEE Journal on Selected Topics in Signal Processing*, 2(1):88–102, 2008.

R. Zhang, Y. Liang, and S. Cui. Dynamic resource allocation in cognitive radio networks. *IEEE Signal Processing Magazine*, 27(3):102–114, 2010.

Q. Zhao and B. M. Sadler. A survey of dynamic spectrum access. *IEEE Signal Processing Magazine*, 24(3):79–89, 2007.

L. Zheng, M. Lops, and X. Wang. Adaptive interference removal for uncoordinated radar/communication coexistence. *IEEE Journal on Selected Topics in Signal Processing*, 12(1):45–60, 2018a. https://doi.org/10.1109/JSTSP.2017.2785783.

L. Zheng, M. Lops, X. Wang, and E. Grossi. Joint design of overlaid communication systems and pulsed radars. *IEEE Transactions on Signal Processing*, 66(1):139–154, 2018b. https://doi.org/10.1109/TSP.2017.2755603.

L. Zheng, M. Lops, Y. C. Eldar, and X. Wang. Radar and communication coexistence: an overview: a review of recent methods. *IEEE Signal Processing Magazine*, 36(5):85–99, 2019. https://doi.org/10.1109/MSP.2019.2907329.

K. Zyczkowski and M. Kus. Random unitary matrices. *Journal of Physics A: Mathematical and General*, 27(12):4235, 1994.

9

Performance and Design for Cooperative MIMO Radar and MIMO Communications

Qian He[1,2], Zhen Wang[1,2], Junze Zhu[1,2], and Rick S. Blum[3]

[1] *Yangtze Delta Region Institute Quzhou, University of Electronic Science and Technology of China, Zhejiang, Quzhou, China*
[2] *Electronic Engineering Department, University of Electronic Science and Technology of China, Sichuan, Chengdu, China*
[3] *Electrical and Computer Engineering Department, Lehigh University, Bethlehem, PA, USA*

9.1 Introduction and Literature Review

With increasing demand for communications and radar systems along with increasingly limited spectrum, it has become an inevitable trend to develop integrated radar and communications systems. Well-thought-out procedures to ensure holistic operation of radar and communications systems can solve many of the problems existing in the simultaneous operation of radar equipment and communications equipment to greatly enhance the performance of both systems [Sturm and Wiesbeck, 2011]. Studies on integrated radar and communications systems date back to the 1970s; see for example Cager et al. [1978]. Many systems need to employ both radar and communications, e.g. aviation systems. The integration of radar and communications helps to reduce the weight, power consumption, economic cost, and electromagnetic interference of the overall system. Recent examples of such integration are found in automotive systems [Dokhanchi et al., 2019; Kumari et al., 2017, 2018, 2020; Duggal et al., 2020; Sit et al., 2018; Winkler and Detlefsen, 2007]. Integrated automotive radar and communications systems can improve transportation safety by enhancing the performance through optimized operation and information sharing. There is also great interest in integrated radar and communications for mobile/cellular communications systems, and this next-generation technology is called a perceptive mobile network (PMN) [Zhang et al., 2021; Liu et al., 2020c; Zheng et al., 2019; Gameiro et al., 2018; Chiriyath et al., 2017]. On the other hand, the rapid increase in the number of communications and radar devices expected in the growth of internet of things (IoT) technology will lead to even more competition for spectrum resources in the future. The study of radar and communications system integration can lessen the impact of this competition and reduce inefficient use of spectrum resources. Therefore, such studies are of extremely high value.

From the existing research, integrated radar and communications systems can fall under three types: (i) dual-functional radar-communications (DFRC) systems, (ii) coexisting radar and communications (CERC) systems, and (iii) cooperative CERC systems. The DFRC systems exhibit characteristics of high integration, where the radar and communications systems share the same

Signal Processing for Joint Radar Communications, First Edition.
Edited by Kumar Vijay Mishra, M. R. Bhavani Shankar, Björn Ottersten, and A. Lee Swindlehurst.

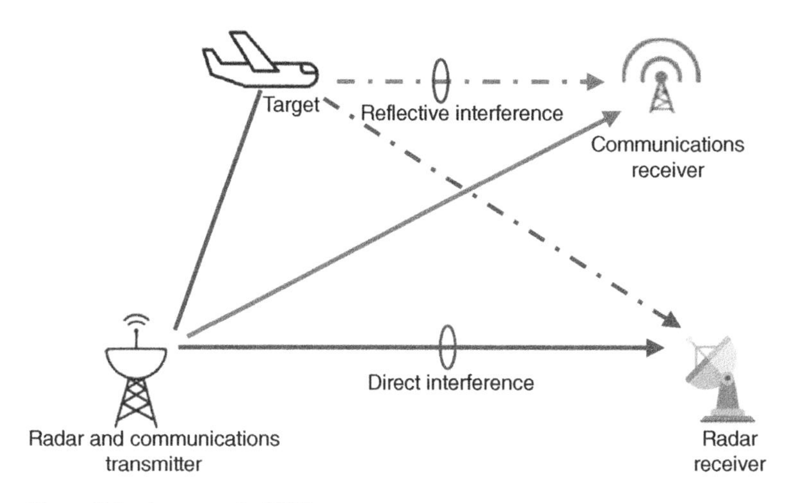

Figure 9.1 An example DFRC system.

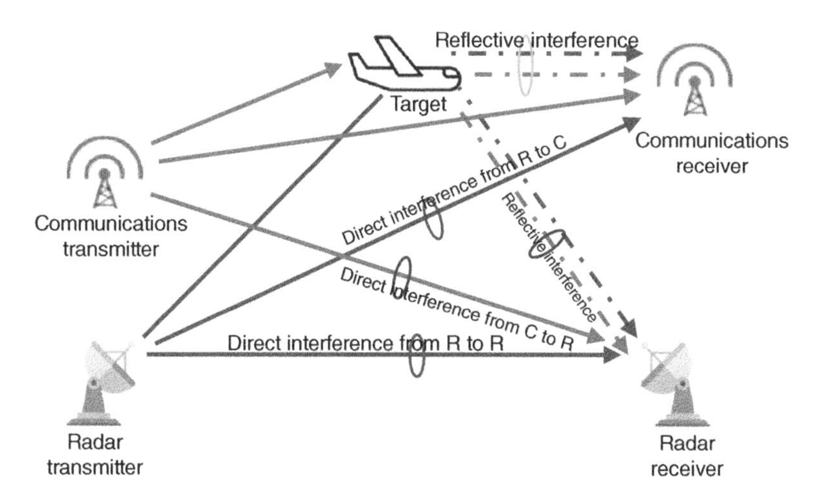

Figure 9.2 An example CERC system.

hardware platform at receiver or transmitter, and integrated signals can be designed to achieve both radar and communications functions. Figure 9.1 illustrates an example DFRC system where radar and communications system share a common transmitter. In CERC systems, the radar and communications systems are designed separately and share limited spectrum and/or other resources. A traditional noncooperative CERC system is illustrated in Figure 9.2. Unlike the traditional ones, recently proposed cooperative CERC systems use the so-called interference path to help improve the system performance. As discussed later in this chapter, the cases in Figures 9.1 and 9.2 can be extended to the multi-input multi-output (MIMO) case [Liu et al., 2020d].

Popular integrated radar and communications systems research topics are summarized in Figure 9.3. Many existing works on DFRC systems focus on beamforming and integrated waveform design, where both multiplexing waveforms and common waveforms have been investigated. Studies of CERC systems mainly include resource scheduling, transmitter design, receiver design, and transmitter–receiver co-design. Cooperative CERC systems have also attracted research attention.

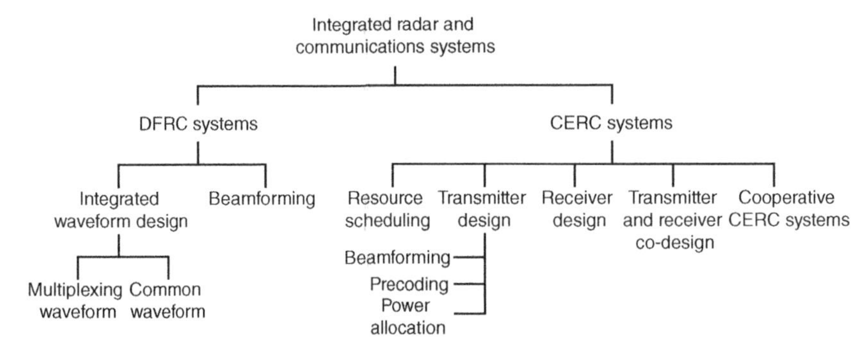

Figure 9.3 Popular research topics on integrated radar and communications systems.

9.1.1 Previous DFRC Approaches

The published work describing DFRC approaches have mainly focused on integrated signal processing for the radar and communications parts. This includes integrated waveform design and beamforming design.

9.1.1.1 Integrated Waveform Design

Two kinds of waveforms have been considered in integrated DFRC systems: multiplexing waveforms and common waveforms.

Multiplexing technology allows radar and communications to use the same hardware for the transmitter and share resources properly such that these signals are separable at the receivers. As the radar and communications signals can be separated, each function has little influence over the other. The resources shared can include space, frequency, and time. On the basis of realizing time–frequency division, code-division multiplexing [Lindenmeier et al., 2003] further allows separation via unique codes. In McCormick et al. [2017], the same radar array generates multifunction radar and communications emissions in space. The work in Roberton and Brown [2003] employs chirp signals modulated using different center frequencies for radar and spread spectrum communications. In Ravenscroft et al. [2017], Bao et al. [2019], orthogonal frequency-division multiplexing (OFDM) is employed such that some tones are allocated to radar, while others are allocated to communications. The authors of Takahara et al. [2012] use time-division multiplexing ultra-wide band (UWB) signals to complete the radar and communications tasks at different times. Pseudo-random codes [Xu et al., 2006], Oppermann sequences [Jamil et al., 2008], and complete complementary codes [Takase and Shinriki, 2010] have been considered to realize the orthogonality of radar and communications signals in the code domain.

When utilizing a common waveform, the integrated system employs a single waveform for both radar and communications. Two popular ways for designing the common waveform are the radar-centric and communications-centric designs, chosen to be more suitable for either radar or communications. The premise of radar-centric design is to first ensure the radar performance and then modulate a communications signal on the radar signal. For the radar-centric case, pulse position, pulse repetition rate, and other parameters of the radar waveform can be designed to encode communications information [Mealey, 2007]. Phase-shift keying [Nowak et al., 2016] and minimum frequency-shift keying [Chen et al., 2011, 2018] have been adopted to encode communications information in the transmitted radar waveform. Communications-centric designs are based on similar ideas. OFDM signals are widely used as the communications-centric common waveforms [Sen et al., 2011; Xia et al., 2015; Lellouch et al., 2016] since they are convenient for radar Doppler processing. An OFDM-chirp waveform is studied in Li et al. [2017], where phase modulation is employed to encode the communications information and the advantages of chirp signals in high-resolution radar are effectively exploited.

Recently, MIMO radar has been considered in DFRC systems with common waveforms, where proper signal processing is required for handling signals at multiple antennas. The authors in Hassanien et al. [2017] presented a method for embedding communications information into MIMO radar signals using frequency-hopping waveforms. An integrated system combining MIMO radar with OFDM communications has been designed in Liu et al. [2016]. The work in Gaudio et al. [2020] considers orthogonal time–frequency–space modulation for a DFRC system.

9.1.1.2 Beamforming

To improve the signal-to-interference-plus-noise ratio (SINR), beamforming has been studied for the DFRC system by focusing signal power while reducing interference in certain directions. Transmit beamforming to allow joint system operation that allows freedom in the choice of the modulation scheme with no loss in the communications data rate has been documented in Liu et al. [2020a,c]. In Liu et al. [2020a], a MIMO transmit beamformer is designed using semidefinite relaxation and zero-forcing under the SINR constraints. In addition to transmit beamformer design, receive beamforming has also been studied for DFRC systems. In Hassanien et al. [2018], the authors design adaptive beamforming at the receiver to separate the radar and communications signals. Compared with unilateral transmit or receive beamforming, transmit and receive beamformer co-design brings more flexibility. Hybrid analog–digital beamforming was studied in Cheng et al. [2020] to co-design the transmit and receive beamformers for completing both target detection and downlink communications. Further, in the context of beamforming, the problem of antenna selection for DFRC systems was addressed in Ahmed et al. [2020] by employing a re-weighted L1-norm minimization.

9.1.2 Previous CERC Approaches

While in DFRC systems the radar and communications share the same hardware at the transmit and/or receive end, in CERC systems the radar and communications employ independent hardware at both transmit and receive ends. Existing research on CERC includes resource scheduling, transmitter design, receiver design, and transmitter and receiver co-design.

9.1.2.1 Resource Scheduling

Through scheduling time, frequency, space, and other resources for coexisting but separate radar and communications systems, CERC systems can achieve both the radar and communications functions. The work in Bhat et al. [2012] explores bandwidth sharing between the radar and communications parts, where multi-objective optimization is used to arrive at the best available bandwidth sharing solutions. The work in Ghorbanzadeh et al. [2014] studies the feasibility of sharing the spectrum between sectorized cellular systems and stationary radars interfering with certain sectors of the communications. In Chen and Gu [2020], the authors study the time allocation problem for an integrated bi-static radar and communications system and analyze the radar detection accuracy. In Keskin et al. [2020], a time–frequency resource allocation scheme is proposed that assigns the OFDM time–frequency blocks to the radar and communications parts. In Linlong et al. [2021], in the context of heterogeneous radars and multitier communications, the power, dwell time, and shared bandwidth are allocated simultaneously to guarantee the radar performance and communications throughput.

9.1.2.2 Transmitter Design

Most work on designing a CERC transmitter focuses on reducing the interference between the radar and communications, among which beamforming, precoding, and power allocation are three of the commonly implemented methods.

In Liu et al. [2018], a power and interference minimization–based beamforming approach is proposed to maximize the MIMO radar detection probability while maintaining the required downlink MU-MIMO communications SINRs. The authors of Bao et al. [2019] propose a new two-step optimization algorithm to design MIMO OFDM radar transmit beamforming to maximize the radar input signal-to-noise ratio (SNR) while keeping the communications rate constant. In Li and Petropulu [2016], Li and Petropulu [2017], transmit precoding of both radar and communications is designed through a control center to maximize the radar SINR under certain communications rate and power constraints. Power allocation is also widely investigated for CERC systems. In Shi et al. [2018], the power minimization-based methods for radar waveform design are proposed to minimize the worst-case radar transmit power. The work in Wang and Li [2019] discusses the optimum power allocation solutions for both radar-centric and communications-centric designs, based on which the allocation-based design is proposed [Wang et al., 2019] to maximize the communications throughput under a minimum radar SINR constraint.

9.1.2.3 Receiver Design

In CERC systems, the radar and communications receivers each receive signals from both the radar and communications transmissions; thus, there have been papers on eliminating the influence of the other signal while completing the respective tasks of the two systems. In Geng et al. [2015], the coherent orthogonal phase-coding radar waveforms are received and extracted at the radar receiver for beamforming and interference mitigation processing in a coherent MIMO CERC system. The work in Kumari et al. [2018] develops single-end multi-frame radar receiver algorithms for target detection as well as range and velocity estimation for both single and multi-target scenarios. In Wang et al. [2020], an interference cancellation algorithm based on minimizing the atomic norm is proposed to address the communications-radar spectrum sharing and inter-system interference issues.

9.1.2.4 Transmitter and Receiver Co-design

Besides the transmitter design and receiver design, there is also literature on jointly designing the transmitter and receiver for better performance. For example, the radar transmit beamforming and communications transmit and receive beamforming are designed together to maximize the detection probability while maintaining communications quality in Singh et al. [2018]. In Biswas et al. [2020], the authors design the precoding MIMO radar waveforms and MIMO communications transmitter and receiver to minimize the sum mean-squared-error of the communications signals. In Cui et al. [2018], a spatial precoder–decoder design is presented based on interference alignment (IA) for managing the interference to optimize the performance of the radar generalized likelihood ratio test (GLRT) detector. The authors in Rihan and Huang [2018] design a two-tier alternating optimization spectrum sharing framework, which uses joint transmit and receive beamformers for both MIMO radar and MIMO communications based on an IA approach to mitigate the interference.

9.1.3 Cooperative CERC Systems

In the CERC systems discussed earlier, most of them regard the radar or communications signals as interference to each other. In fact, proper cooperation can bring enhanced performance for CERC systems.

In the cooperative CERC systems, communications signals can be exploited by the radar system to extract target information. In Bica et al. [2015], communications signals reflected by the target are exploited to improve the radar detection performance in a single-input and single-output (SISO) radar system and the target detector was derived. The authors in Bica and Koivunen [2017] show that the communications signals can help the radar improve delay estimation performance in the SISO case. The work in Bica and Koivunen [2019] utilizes target echoes from both radar and communications signals to complete target parameter estimation, and the radar waveform is optimized based on the Cramer-Rao bound (CRB) for a SISO system. In Zhang et al. [2019], an OFDM radar system exploits the reflected communications signals to help obtain higher radar resolution and reduce radar transmitted power in the SISO case. The GLRT detector is employed in Liu et al. [2020b] for a SISO system, where it is shown that the cooperation offers significant gains when the SNR associated with the direct path of the received communication is large enough.

Similarly, in cooperative CERC systems, if the radar findings can be utilized by the communications side, the communications performance can also be improved.

9.1.4 Overview of Remainder of the Chapter

As per the previous discussion, while DFRC systems have higher levels of integration, CERC systems have fewer constraints and thus more design freedom to enhance performance. Cooperative CERC systems have even greater potential to achieve better performance. Starting in Section 9.2, we present our own research on CERC systems. We investigate how the radar and communications parts of CERC systems can help each other, and we extend the analysis to MIMO radar and MIMO communications. On the radar side, the radar can employ target returns contributed from both the radar transmitters and communications transmitters to complete the radar task, which is equivalent to an *hybrid active–passive MIMO radar*. Target detection and localization are considered for the radar task, and the corresponding performances are analyzed. On the communications side, with the help of radar, not only are the communications signals received directly from communications transmitters utilized but also those reflected from the radar target are exploited to extract useful information, which is called *radar-aided MIMO communications*. We derive the corresponding MI to quantify the communications capability. The considered cooperative CERC system makes full use of the observations from both the communications and radar sides through cooperation and fully exploits the spatial dimension through MIMO technology. With proper signal processing at the receiving end, not only is the communications ability better than that of the traditional communications systems but also the performance of target detection and parameter estimation can surpass the limitations of the traditional radar.

In the majority of existing works on cooperative CERC systems mentioned in Section 9.1.3, communications signals and radar signals are first separated at the receive stations, and the separated signals are then used to complete the radar or communications task. However, there are scenarios where it may be difficult to separate the communications signals and radar signals for each of the propagation paths, for example in MIMO systems with widely separated antennas with correlated waveforms. Further, separating signals at the first stage might be suboptimal and lead to extra complexity. In this chapter, we consider received signals with the radar and communications signals naturally mixed together, based on which a cooperative CERC system is investigated. The remainder of this chapter is organized as follows. In Section 9.2, we develop a received signal model where the direct path and reflected path radar and communications returns are included in the observations. In Section 9.3, we study hybrid active–passive MIMO radar for cooperative CERC systems, including target detection and localization performance. Radar-aided MIMO communications for cooperative CERC systems is investigated in Section

9.4, where the communications MI is derived and compared with that of the traditional system. In Section 9.5, co-design of the radar and communications systems parts of a cooperative CERC system is described. Finally, conclusions are drawn in Section 9.6.

9.2 Cooperative CERC System Model

Consider a cooperative MIMO radar and communications system, where the radar part has M_R transmit antennas and N_R receive antennas, and the communications part has M_C transmit antennas and N_C receive antennas, all widely spaced, as illustrated in Figure 9.4. The complex bandpass signals emitted at the mth ($m = 1, \dots, M_R$) radar transmit antenna and the m'th ($m' = 1, \dots, M_C$) communications transmit antenna are $\sqrt{E_{R,m}}s_{R,m}(kT_s)$ and $\sqrt{E_{C,m'}}s_{C,m'}(kT_s)$, respectively, where $E_{R,m}$ and $E_{C,m'}$ denote the transmit power, T_s the sampling period, and $k(k = 1, \dots, K)$ an index numbering the different time samples.

9.2.1 Received Radar Signals

The complex bandpass signal received at the nth ($n = 1, \dots, N_R$) radar receive antenna at time instant kT_s is modeled as

$$\begin{aligned} r_{R,n}[k] = &\sum_{m=1}^{M_R} \sqrt{E_{R,m}}\zeta_{Rt,nm}s_{R,m}(kT_s - \tau_{Rt,nm}) \\ &+ \sum_{m=1}^{M_R} \sqrt{E_{R,m}}\zeta_{R,nm}s_{R,m}(kT_s - \tau_{R,nm}) \\ &+ \sum_{m'=1}^{M_C} \sqrt{E_{C,m'}}\zeta_{Ct,nm'}s_{C,m'}(kT_s - \tau_{Ct,nm'}) \\ &+ \sum_{m'=1}^{M_C} \sqrt{E_{C,m'}}\zeta_{C,nm'}s_{C,m'}(kT_s - \tau_{C,nm'}) + w_{R,n}[k] \end{aligned} \tag{9.1}$$

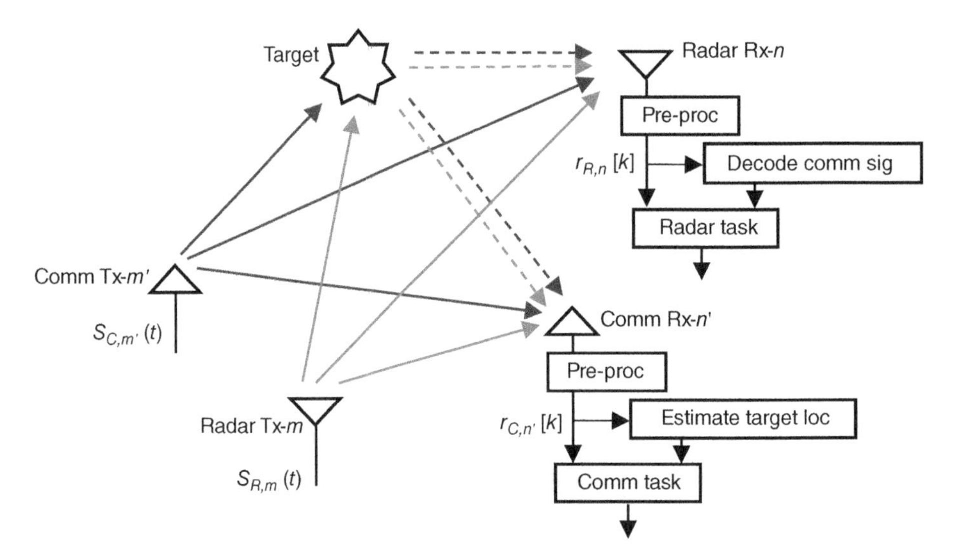

Figure 9.4 A cooperative MIMO radar and MIMO communications system. Source: [He et al., 2019]/IEEE.

in which the first four terms correspond to the target reflected path and direct path signals contributed by the radar transmissions and the target reflected path and direct path signals contributed by the communications transmissions, respectively. In (9.1), $\tau_{Rt,nm}$, $\tau_{R,nm}$, $\tau_{Ct,nm'}$, and $\tau_{C,nm'}$ denote the corresponding time delays, $\zeta_{Rt,nm}$, $\zeta_{R,nm}$, $\zeta_{Ct,nm'}$, and $\zeta_{C,nm'}$ denote the corresponding target reflection and direct path propagation coefficients (assumed known possibly via preprocessing), and $w_{R,n}[k]$ represents the clutter-plus-noise. Define the observation vector at the nth radar receiver as $\mathbf{r}_{R,n} = [r_{R,n}[1], \dots, r_{R,n}[K]]^\dagger$, where "$\dagger$" denotes transpose. The overall radar received signal vector can be written as

$$\mathbf{r}_R = [\mathbf{r}_{R,1}{}^\dagger, \dots, \mathbf{r}_{R,N_R}{}^\dagger]^\dagger = \mathbf{U}_{Rt}\mathbf{s}_{Rt} + \mathbf{U}_R\mathbf{s}_R + \mathbf{U}_{Ct}\mathbf{s}_{Ct} + \mathbf{U}_C\mathbf{s}_C + \mathbf{w}_R, \tag{9.2}$$

where $\mathbf{U}_\epsilon = \text{Diag}\{\mathbf{u}^\dagger_{\epsilon,1}[1], \mathbf{u}^\dagger_{\epsilon,1}[2], \dots, \mathbf{u}^\dagger_{\epsilon,N_R}[K]\}$, the subscript ϵ stands for Rt, R, Ct, and C in the first four terms in (9.2), respectively, and the operator Diag$\{\cdot\}$ denotes a block diagonal matrix

$$\mathbf{u}_{\epsilon,n}[k] = \left(u_{\epsilon,n1}[k], \dots, u_{\epsilon,nM_\eta}[k]\right)^\dagger, \tag{9.3}$$

$$u_{\epsilon,nm}[k] = \zeta_{\epsilon,nm}\sqrt{E_{\eta,m}}, \tag{9.4}$$

and $\mathbf{s}_\epsilon = (\mathbf{s}_{\epsilon,1}[1]^\dagger, \mathbf{s}_{\epsilon,1}[2]^\dagger, \dots, \mathbf{s}_{\epsilon,N_R}[K]^\dagger)^\dagger$ with

$$\mathbf{s}_{\epsilon,n}[k] = \left[s_{\eta,1}(kT_s - \tau_{\epsilon,n1}), \dots, s_{\eta,M_\eta}(kT_s - \tau_{\epsilon,nM_\eta})\right]^\dagger. \tag{9.5}$$

In (9.3)–(9.5), the subscript η stands for R for the first two terms in (9.2) and C for the third and fourth terms in (9.2). The clutter-plus-noise vector $\mathbf{w}_R = [\mathbf{w}_{R,1}{}^\dagger, \dots, \mathbf{w}_{R,N_R}{}^\dagger]^\dagger$ and $\mathbf{w}_{R,n} = \left(\mathbf{w}_{R,n}[1], \dots, \mathbf{w}_{R,n}[K]\right)^\dagger$, where $\mathbf{w}_R$ is Gaussian distributed with zero mean and covariance matrix $\mathbf{Q}_R$.

9.2.2 Received Communications Signals

Similarly, the complex bandpass signal received at the n'th communications receive antenna at time instant kT_s is given by

$$\begin{aligned}
r_{C,n'}[k] = &\sum_{m'=1}^{M_C} \sqrt{E_{C,m'}}\tilde{\zeta}_{Ct,n'm'}s_{C,m'}(kT_s - \tilde{\tau}_{Ct,n'm'}) \\
&+ \sum_{m'=1}^{M_C} \sqrt{E_{C,m'}}\tilde{\zeta}_{C,n'm'}s_{C,m'}(kT_s - \tilde{\tau}_{C,n'm'}) \\
&+ \sum_{m=1}^{M_R} \sqrt{E_{R,m}}\tilde{\zeta}_{Rt,n'm}s_{R,m}(kT_s - \tilde{\tau}_{Rt,n'm}) \\
&+ \sum_{m=1}^{M_R} \sqrt{E_{R,m}}\tilde{\zeta}_{R,n'm}s_{R,m}(kT_s - \tilde{\tau}_{R,n'm}) + w_{C,n'}[k],
\end{aligned} \tag{9.6}$$

where the first four terms correspond to the target reflected path and direct path signals contributed by the communications transmissions and the target reflected path and direct path signals contributed by the radar transmissions, respectively. In (9.6), $\tilde{\tau}_{Ct,n'm'}$, $\tilde{\tau}_{C,n'm'}$, $\tilde{\tau}_{Rt,n'm}$, and $\tilde{\tau}_{R,n'm}$ denote the corresponding time delays, $\tilde{\zeta}_{Ct,n'm'}$, $\tilde{\zeta}_{C,n'm'}$, $\tilde{\zeta}_{Rt,n'm}$, and $\tilde{\zeta}_{R,n'm}$ denote the corresponding target reflection and direct path propagation coefficients (assumed known possibly via preprocessing), and $w_{C,n'}[k]$ represents the clutter-plus-noise. Let $\mathbf{r}_{C,n'} = \left[r_{C,n'}(1), \dots, r_{C,n'}(K)\right]^\dagger$, then the overall received complex bandpass signal vector for the communications system can be written as

$$\mathbf{r}_C = [\mathbf{r}_{C,1}{}^\dagger, \dots, \mathbf{r}_{C,N_C}{}^\dagger]^\dagger = \tilde{\mathbf{U}}_{Ct}\tilde{\mathbf{s}}_{Ct} + \tilde{\mathbf{U}}_C\tilde{\mathbf{s}}_C + \tilde{\mathbf{U}}_{Rt}\tilde{\mathbf{s}}_{Rt} + \tilde{\mathbf{U}}_R\tilde{\mathbf{s}}_R + \mathbf{w}_C, \tag{9.7}$$

where $\tilde{\mathbf{U}}_{\epsilon} = \mathrm{Diag}\{\tilde{\mathbf{u}}_{\epsilon,1}^{\dagger}[1], \tilde{\mathbf{u}}_{\epsilon,1}^{\dagger}[2], \dots, \tilde{\mathbf{u}}_{\epsilon,N_C}^{\dagger}[K]\}$, $\tilde{\mathbf{u}}_{\epsilon,n'}[k] = \left(\tilde{u}_{\epsilon,n'1}[k], \dots, \tilde{u}_{\epsilon,n'M_{\eta}}[k]\right)^{\dagger}$, $\tilde{u}_{\epsilon,n'm}[k] = \zeta_{\epsilon,n'm}\sqrt{E_{\eta,m}}$, $\tilde{\mathbf{s}}_{\epsilon} = \left(\tilde{\mathbf{s}}_{\epsilon,1}[1]^{\dagger}, \tilde{\mathbf{s}}_{\epsilon,1}[2]^{\dagger}, \dots, \quad \tilde{\mathbf{s}}_{\epsilon,N_C}[K]^{\dagger}\right)^{\dagger}$, and $\tilde{\mathbf{s}}_{\epsilon,n'}[k] = [s_{\eta,1}(kT_s - \tilde{\tau}_{\epsilon,n'1}), \dots, s_{\eta,M_{\eta}}(kT_s - \tilde{\tau}_{\epsilon,n'M_{\eta}})]^{\dagger}$. The subscript ϵ represents Ct, C, Rt, and R in the first four terms of (9.7), respectively, and the subscript η represents C for the first two terms and R for the third and fourth terms in (9.7). The clutter-plus-noise vector at the communications receivers $\mathbf{w}_C = [\mathbf{w}_{C,1}{}^{\dagger}, \dots, \mathbf{w}_{C,N_C}{}^{\dagger}]^{\dagger}$ is assumed to have a zero-mean Gaussian distribution with covariance matrix $\mathbf{Q}_C$, where $\mathbf{w}_{C,n'} = (\mathbf{w}_{C,n'}[1], \dots, \mathbf{w}_{C,n'}[K])^{\dagger}$.

9.3 Hybrid Active–Passive Cooperative CERC MIMO Radar System

Under cooperation, the hybrid active–passive cooperative CERC MIMO radar system no longer regards the communications signals as interference. The cooperative system can not only employ the target returns of the traditional radar transmissions but also can employ the target return contributions from the communications transmissions to complete the radar task. Thus, the communications signals are incorporated in a similar way that a passive radar would incorporate a signal of opportunity. Thus, it is as if they are another radar transmitter.

9.3.1 Target Detection

This subsection considers the case where target detection is the radar task for cooperative CERC system. Using the information shared by the communications system, it is assumed that the radar receiver can decode and reconstruct the communications signal accurately [Richmond et al., 2016]. Assuming that the radar task for the hybrid active–passive MIMO radar system is to detect whether the target exists in the cell-under-test, according to the received radar signal model (9.2), the detection problem (H_0 : target absent versus H_1 : target present) can be described as Wang and He [2020]

$$\begin{aligned} H_0 &: \mathbf{r}_R = \mathbf{U}_R\mathbf{s}_R + \mathbf{U}_C\mathbf{s}_C + \mathbf{w}_R, \\ H_1 &: \mathbf{r}_R = \mathbf{U}_{Rt}\mathbf{s}_{Rt} + \mathbf{U}_R\mathbf{s}_R + \mathbf{U}_{Ct}\mathbf{s}_{Ct} + \mathbf{U}_C\mathbf{s}_C + \mathbf{w}_R. \end{aligned} \tag{9.8}$$

To simplify the analysis, consider that the communications signals have been decoded and reconstructed [Richmond et al., 2016], and the time delays $\tilde{\tau}_{Ct,n'm'}$, $\tilde{\tau}_{C,n'm'}$, $\tilde{\tau}_{Rt,n'm}$, and $\tilde{\tau}_{R,n'm}$ are known from preprocessing [He et al., 2019; Godrich et al., 2012]. Therefore, the received signal vector $\mathbf{r}_R$ obeys a Gaussian distribution under both hypotheses, and the probability density functions are, respectively,

$$p\left(\mathbf{r}_R|H_0\right) = \frac{1}{\pi^{KN_R}\det\left(\mathbf{Q_R}\right)}\exp\left\{-(\mathbf{r}_R - \mathbf{U}_R\mathbf{s}_R - \mathbf{U}_C\mathbf{s}_C)^H \mathbf{Q}_R^{-1}\left(\mathbf{r}_R - \mathbf{U}_R\mathbf{s}_R - \mathbf{U}_C\mathbf{s}_C\right)\right\} \tag{9.9}$$

and

$$\begin{aligned} p\left(\mathbf{r}_R|H_1\right) = \tfrac{1}{\pi^{KN_R}\det\left(\mathbf{Q_R}\right)}\exp\big\{&-(\mathbf{r}_R - \mathbf{U}_R\mathbf{s}_R - \mathbf{U}_{Rt}\mathbf{s}_{Rt} - \mathbf{U}_C\mathbf{s}_C - \mathbf{U}_{Ct}\mathbf{s}_{Ct})^H \\ &\mathbf{Q}_R^{-1}\left(\mathbf{r}_R - \mathbf{U}_R\mathbf{s}_R - \mathbf{U}_{Rt}\mathbf{s}_{Rt} - \mathbf{U}_C\mathbf{s}_C - \mathbf{U}_{Ct}\mathbf{s}_{Ct}\right)\big\}, \end{aligned} \tag{9.10}$$

where $\det(\cdot)$ represents the determinant operator. Then, the log-likelihood ratio is

$$\ln\frac{p\left(\mathbf{r}_R|H_1\right)}{p\left(\mathbf{r}_R|H_0\right)} = \mathbf{r}_R^H\mathbf{Q}_R^{-1}\left(\mathbf{U}_{Rt}\mathbf{s}_{Rt} + \mathbf{U}_{Ct}\mathbf{s}_{Ct}\right) + (\mathbf{U}_{Rt}\mathbf{s}_{Rt} + \mathbf{U}_{Ct}\mathbf{s}_{Ct})^H\mathbf{Q}_R^{-1}\mathbf{r}_R$$

$$- (\mathbf{U}_R\mathbf{s}_R + \mathbf{U}_C\mathbf{s}_C)^H \mathbf{Q}_R^{-1} \left(\mathbf{U}_{Rt}\mathbf{s}_{Rt} + \mathbf{U}_{Ct}\mathbf{s}_{Ct}\right)$$
$$- (\mathbf{U}_{Rt}\mathbf{s}_{Rt} + \mathbf{U}_{Ct}\mathbf{s}_{Ct})^H \mathbf{Q}_R^{-1} \left(\mathbf{U}_R\mathbf{s}_R + \mathbf{U}_C\mathbf{s}_C\right)$$
$$- (\mathbf{U}_{Rt}\mathbf{s}_{Rt} + \mathbf{U}_{Ct}\mathbf{s}_{Ct})^H \mathbf{Q}_R^{-1} \left(\mathbf{U}_{Rt}\mathbf{s}_{Rt} + \mathbf{U}_{Ct}\mathbf{s}_{Ct}\right), \tag{9.11}$$

in which all but the first two items are independent of $\mathbf{r}_R$. Including these terms in the test threshold, the test statistic can be obtained as

$$T_R = \mathbf{r}_R^H \mathbf{Q}_R^{-1} \left(\mathbf{U}_{Rt}\mathbf{s}_{Rt} + \mathbf{U}_{Ct}\mathbf{s}_{Ct}\right) + (\mathbf{U}_{Rt}\mathbf{s}_{Rt} + \mathbf{U}_{Ct}\mathbf{s}_{Ct})^H \mathbf{Q}_R^{-1} \mathbf{r}_R. \tag{9.12}$$

From (9.9)–(9.12), the distribution of T_R under H_0 and H_1 follows ($\mathcal{N}\left(\mu, \sigma^2\right)$ denotes a Gaussian probability density function with mean μ and variance σ^2):

$$T_R | H_0 \sim \mathcal{N}\left(\mu_0, \sigma^2\right), \quad T_R | H_1 \sim \mathcal{N}\left(\mu_1, \sigma^2\right), \tag{9.13}$$

where

$$\mu_0 = 2\Re\left\{(\mathbf{U}_R\mathbf{s}_R + \mathbf{U}_C\mathbf{s}_C)^H \mathbf{Q}_R^{-1} \left(\mathbf{U}_{Rt}\mathbf{s}_{Rt} + \mathbf{U}_{Ct}\mathbf{s}_{Ct}\right)\right\}, \tag{9.14}$$

$$\mu_1 = 2\Re\left\{(\mathbf{U}_R\mathbf{s}_R + \mathbf{U}_C\mathbf{s}_C + \mathbf{U}_{Rt}\mathbf{s}_{Rt} + \mathbf{U}_{Ct}\mathbf{s}_{Ct})^H \mathbf{Q}_R^{-1} \left(\mathbf{U}_{Rt}\mathbf{s}_{Rt} + \mathbf{U}_{Ct}\mathbf{s}_{Ct}\right)\right\}, \tag{9.15}$$

and

$$\sigma^2 = 2(\mathbf{U}_{Rt}\mathbf{s}_{Rt} + \mathbf{U}_{Ct}\mathbf{s}_{Ct})^H \mathbf{Q}_R^{-1} \left(\mathbf{U}_{Rt}\mathbf{s}_{Rt} + \mathbf{U}_{Ct}\mathbf{s}_{Ct}\right), \tag{9.16}$$

in which $\Re\{\cdot\}$ produces the real part of a complex number.

The false alarm probability is

$$P_{\text{FA}} = Pr\left(T_R > \beta | H_0\right) = Q\left(\frac{\beta - \mu_0}{\sigma}\right), \tag{9.17}$$

and the detection threshold β is

$$\beta = \sigma Q^{-1}\left(P_{\text{FA}}\right) + \mu_0, \tag{9.18}$$

where $Q(\cdot)$ represents the complementary distribution function of the standard Gaussian distribution, which is expressed as

$$Q(x) = \int_x^{\infty} \frac{1}{\sqrt{2\pi}} e^{-\frac{t^2}{2}} dt\,. \tag{9.19}$$

Thus, the radar target probability of detection (PD) for cooperative CERC system can be obtained as

$$P_D = Pr\left(T_R \geq \beta | H_1\right) = Q\left(\frac{\beta - \mu_1}{\sigma}\right) = Q\left(Q^{-1}\left(P_{\text{FA}}\right) + \frac{\mu_0 - \mu_1}{\sigma}\right). \tag{9.20}$$

9.3.2 Target Localization

Radar target detection and parameter estimation are two important functions for radar. The performance of target detection has been analyzed in the Section 9.3.1. Next, we consider the case where the task for the hybrid active–passive MIMO radar system is to locate a target in a two-dimensional space. Define an unknown parameter vector $\boldsymbol{\theta} = [x, y]^\dagger$ denoting the target location. In cooperative radar and communications systems, target returns received at the radar receiver due to the contributions from both radar and communications transmissions can be exploited for target localization, forming a hybrid active–passive MIMO radar network. Based on (9.10), the log likelihood function satisfies

$$\ln p\left(\mathbf{r}_R | \boldsymbol{\theta}\right) \propto -\left(\mathbf{r}_R - \mathbf{U}_R\mathbf{s}_R - \mathbf{U}_{Rt}\mathbf{s}_{Rt} - \mathbf{U}_C\mathbf{s}_C - \mathbf{U}_{Ct}\mathbf{s}_{Ct}\right)^H \mathbf{Q}_R^{-1}$$
$$\times \left(\mathbf{r}_R - \mathbf{U}_R\mathbf{s}_R - \mathbf{U}_{Rt}\mathbf{s}_{Rt} - \mathbf{U}_C\mathbf{s}_C - \mathbf{U}_{Ct}\mathbf{s}_{Ct}\right), \tag{9.21}$$

and the maximum likelihood (ML) estimate of $\boldsymbol{\theta}$ can be obtained using

$$\hat{\boldsymbol{\theta}}_{R,\mathrm{ML}} = \arg\max_{\boldsymbol{\theta}} \ln p\left(\mathbf{r}_R|\boldsymbol{\theta}\right). \tag{9.22}$$

Next, we compute the CRB to characterize the best achievable estimation performance. Define an intermediate parameter vector

$$\boldsymbol{\vartheta} = \left[\tau_{Rt,11}, \ldots, \tau_{Rt,N_RM_R}, \tau_{Ct,11}, \ldots, \tau_{Ct,N_RM_C}\right]^{\dagger}. \tag{9.23}$$

Under the standard regularity conditions [Sengijpta, 1995], the Fisher information matrix can be calculated as

$$[\mathbf{J}(\boldsymbol{\theta})]_{ij} = -\mathbb{E}\left[\frac{\partial^2 \ln p\left(\mathbf{r}_R|\boldsymbol{\theta}\right)}{\partial\theta_i\partial\theta_j}\right], \tag{9.24}$$

for $i,j = 1,2$. According to the chain rule, it can be obtained

$$\mathbf{J}(\boldsymbol{\theta}) = (\nabla_{\boldsymbol{\theta}}\boldsymbol{\vartheta}^{\dagger})\mathbf{J}(\boldsymbol{\vartheta})(\nabla_{\boldsymbol{\theta}}\boldsymbol{\vartheta}^{\dagger})^{\dagger}, \tag{9.25}$$

where $\nabla_{\boldsymbol{\theta}}$ is the gradient operator, $\nabla_{\boldsymbol{\theta}}\boldsymbol{\vartheta}^{\dagger} = [\mathbf{F} \quad \mathbf{G}]$,

$$\mathbf{F} = \begin{bmatrix} \frac{\partial\tau_{Rt,11}}{\partial x} & \frac{\partial\tau_{Rt,12}}{\partial x} & \cdots & \frac{\partial\tau_{Rt,N_RM_R}}{\partial x} \\ \frac{\partial\tau_{Rt,11}}{\partial y} & \frac{\partial\tau_{Rt,12}}{\partial y} & \cdots & \frac{\partial\tau_{Rt,N_RM_R}}{\partial y} \end{bmatrix}, \tag{9.26}$$

$$\mathbf{G} = \begin{bmatrix} \frac{\partial\tau_{Ct,11}}{\partial x} & \frac{\partial\tau_{Ct,12}}{\partial x} & \cdots & \frac{\partial\tau_{Ct,N_RM_C}}{\partial x} \\ \frac{\partial\tau_{Ct,11}}{\partial y} & \frac{\partial\tau_{Ct,12}}{\partial y} & \cdots & \frac{\partial\tau_{Ct,N_RM_C}}{\partial y} \end{bmatrix}, \tag{9.27}$$

and Sengijpta [1995]

$$\begin{aligned}\left[\mathbf{J}(\boldsymbol{\vartheta})\right]_{ij} &= -\mathbb{E}\left[\frac{\partial^2 \ln p\left(\mathbf{r}_R|\boldsymbol{\theta}\right)}{\partial\vartheta_i\partial\vartheta_j}\right] \\ &= 2\Re\left\{\frac{\partial(\mathbf{U}_{Rt}\mathbf{s}_{Rt} + \mathbf{U}_{Ct}\mathbf{s}_{Ct})^H}{\partial\vartheta_i}\mathbf{Q}_R^{-1}\frac{\partial(\mathbf{U}_{Rt}\mathbf{s}_{Rt} + \mathbf{U}_{Ct}\mathbf{s}_{Ct})}{\partial\vartheta_j}\right\}\end{aligned} \tag{9.28}$$

for $i,j = 1, \ldots, N_R(M_R + M_C)$, and $\mathbb{E}\{\cdot\}$ represents mathematical expectation. After manipulation, we obtain

$$\begin{aligned}\mathbf{J}(\boldsymbol{\theta}) = 2\Re\Big\{&\mathbf{F}\mathbf{J}_{\tau_{Rt}}{}^H\mathbf{Q}_R^{-1}\mathbf{J}_{\tau_{Rt}}\mathbf{F}^{\dagger} + \mathbf{G}\mathbf{J}_{\tau_{Ct}}{}^H\mathbf{Q}_R^{-1}\mathbf{J}_{\tau_{Rt}}\mathbf{F}^{\dagger} \\ &+ \mathbf{F}\mathbf{J}_{\tau_{Rt}}{}^H\mathbf{Q}_R^{-1}\mathbf{J}_{\tau_{Ct}}\mathbf{G}^{\dagger} + \mathbf{G}\mathbf{J}_{\tau_{Ct}}{}^H\mathbf{Q}_R^{-1}\mathbf{J}_{\tau_{Ct}}\mathbf{G}^{\dagger}\Big\},\end{aligned} \tag{9.29}$$

where $\mathbf{J}_{\tau_{Rt}} = \partial(\mathbf{U}_{Rt}\mathbf{s}_{Rt})/\partial\tau_{Rt}^{\dagger}$ and $\mathbf{J}_{\tau_{Ct}} = \partial(\mathbf{U}_{Ct}\mathbf{s}_{Ct})/\partial\tau_{Ct}^{\dagger}$. Accordingly, the CRB for the estimate of $\boldsymbol{\theta}$ is

$$\mathrm{CRB} = \mathbf{J}^{-1}(\boldsymbol{\theta}). \tag{9.30}$$

Assuming the estimation accuracy for the target position on each of the two axes has equal importance, define an overall estimation performance metric as the averaged root CRB (RCRB), as given by

$$\mathrm{RCRB} = \frac{1}{2}\left(\sqrt{\mathrm{CRB}_{1,1}} + \sqrt{\mathrm{CRB}_{2,2}}\right). \tag{9.31}$$

9.3.3 Comparison with Noncooperative Case

To illustrate the performance advantages for the cooperative system, we compare the performance of the cooperative case with the noncooperative case in this section.

9.3.3.1 Gain for Radar Target Detection

We first analyze the target detection performance of the noncooperative case (traditional active radar) where the detection problem can be described as

$$\begin{aligned} H_0 &: \mathbf{r}_R = \mathbf{U}_R\mathbf{s}_R + \mathbf{w}_R, \\ H_1 &: \mathbf{r}_R = \mathbf{U}_{Rt}\mathbf{s}_{Rt} + \mathbf{U}_R\mathbf{s}_R + \mathbf{w}_R. \end{aligned} \tag{9.32}$$

Using analysis similar to that in Section 9.3.1, the detection probability for the noncooperative case is

$$P_{D,\text{non}} = Q\left(Q^{-1}\left(P_{\text{FA}}\right) + \frac{\mu_{0,\text{non}} - \mu_{1,\text{non}}}{\sigma_{\text{non}}}\right), \tag{9.33}$$

where

$$\mu_{0,\text{non}} = 2\Re\left\{\mathbf{s}_R^H\mathbf{U}_R^H\mathbf{Q}_R^{-1}\mathbf{U}_{Rt}\mathbf{s}_{Rt}\right\}, \tag{9.34}$$

$$\mu_{1,\text{non}} = 2\Re\left\{(\mathbf{U}_R\mathbf{s}_R + \mathbf{U}_{Rt}\mathbf{s}_{Rt})^H\mathbf{Q}_R^{-1}\mathbf{U}_{Rt}\mathbf{s}_{Rt}\right\}, \tag{9.35}$$

and

$$\sigma_{\text{non}}^2 = 2\mathbf{s}_{Rt}^H\mathbf{U}_{Rt}^H\mathbf{Q}_R^{-1}\mathbf{U}_{Rt}\mathbf{s}_{Rt}. \tag{9.36}$$

Next, we use examples to illustrate the difference in radar performance between the cooperative and noncooperative cases. To define a general test setup that is easy to describe, we assume each of the radar and communications single antenna transmit and receive stations is located 70 km away from the origin of the coordinate system. Assume OFDM signals are adopted for communications transmission, so $s_{Cm'}(t) = \sum_{i=-\infty}^{\infty} s_{m'i}(t - iT')$ where

$$s_{m'i}(t) = \sum_{n=-N_f/2}^{N_f/2-1} a_{m'i}[n]e^{j2\pi n\Delta f t}p_{T'}(t), \tag{9.37}$$

in which $a_{m'i}[n]$ are data symbols, $p_{T'}(t)$ is a rectangular pulse with unit amplitude and width T', Δf is the frequency spacing between two adjacent subcarriers, N_f the number of subcarriers, and T' the pulse width. Let $\Delta f = 125$ Hz, $N_f = 6$, and $T' = 0.01$. Each of the communications transmit signals has identical power such that $E_{C,1} = \cdots = E_{C,M_C} = E_C$. The clutter-plus-noise is assumed white, so the covariance matrix is given by $\mathbf{Q}_C = \sigma_w^2\mathbf{I}$. The transmitted waveforms for the MIMO radar system are frequency spread single Gaussian pulse signals:

$$s_m(t) = (2/T^2)^{(1/4)}e^{-\pi t^2/T^2}e^{j2\pi m f_\Delta t}, \tag{9.38}$$

where f_Δ is the frequency offset between adjacent radar transmit signals and T the pulsewidth. Set $f_\Delta = 125$ Hz and $T = 0.01$. The power used for each radar transmit signal is identical so that $E_{R,1} = \cdots = E_{R,M_R} = E_R$. Assume the additive Gaussian clutter-plus-noise is white with covariance matrix $\mathbf{Q}_R = \sigma_w^2\mathbf{I}$. Suppose that a target is present at (50,30) m.

Denote the total transmit power of the CERC systems by E and the percentage of power assigned to radar by α_E, thus $M_R E_R = E\alpha_E$ and $M_C E_C = E(1 - \alpha_E)$. Define the signal to clutter-plus-noise ratio as $\text{SCNR} = E/\sigma_w^2$. Two antenna configurations are considered. In the first configuration, the MIMO radar system has $M_R = 2$ transmit antennas and $N_R = 3$ receive antennas, and the MIMO communications system has $M_C = 4$ transmit antennas and $N_C = 5$ receive antennas, as illustrated in Figure 9.5. In the second configuration, the MIMO radar system has $M_R = 4$ transmit antennas and $N_R = 6$ receive antennas, and the MIMO communications system has $M_C = 6$ transmit antennas and $N_C = 8$ receive antennas, as illustrated in Figure 9.6.

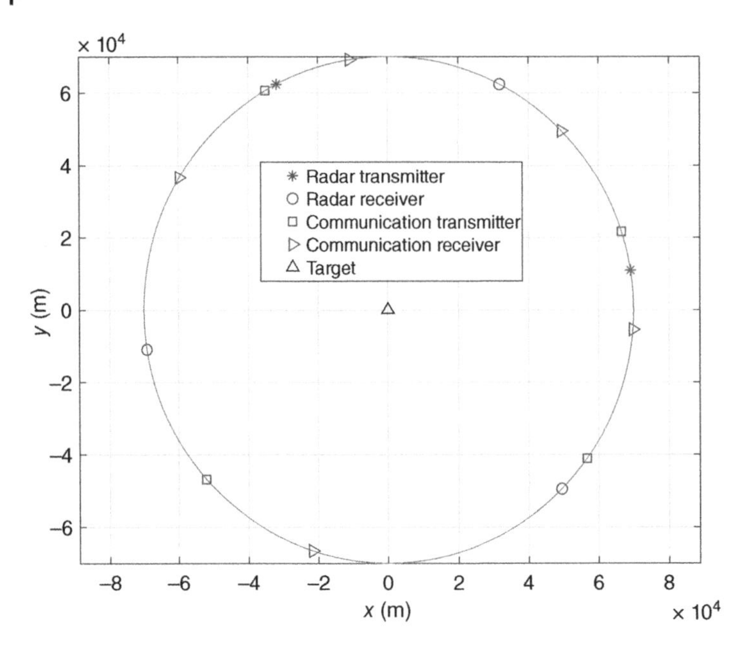

Figure 9.5 Parameter setup for cooperative CERC system with $M_R = 2$, $N_R = 3$, $M_C = 4$, and $N_C = 5$.

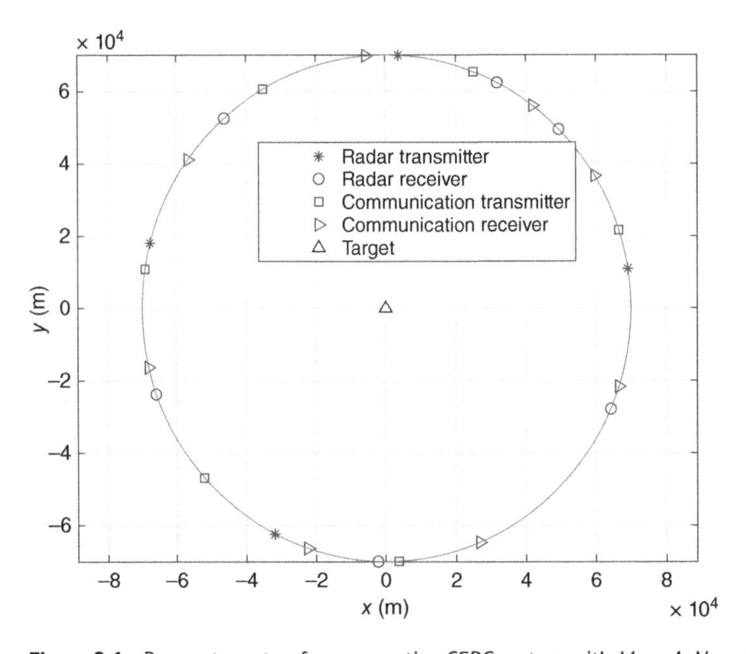

Figure 9.6 Parameter setup for cooperative CERC system with $M_R = 4$, $N_R = 6$, $M_C = 6$, and $N_C = 8$.

When the probability of false alarm is fixed at $P_{\text{FA}} = 10^{-3}$ and $\alpha_E = 0.2$, the radar target detection probability P_D (from (9.20) and (9.33)) is plotted versus SCNR for cooperative and noncooperative cases in Figure 9.7 for the two antenna configurations in Figures 9.5 and 9.6. It can be seen from the figure that for these two antenna configurations the target detection probability P_D for the cooperative cases is always larger than that for the noncooperative cases, which shows that the cooperation can significantly improve the radar target detection performance. Figure 9.8 shows the results for

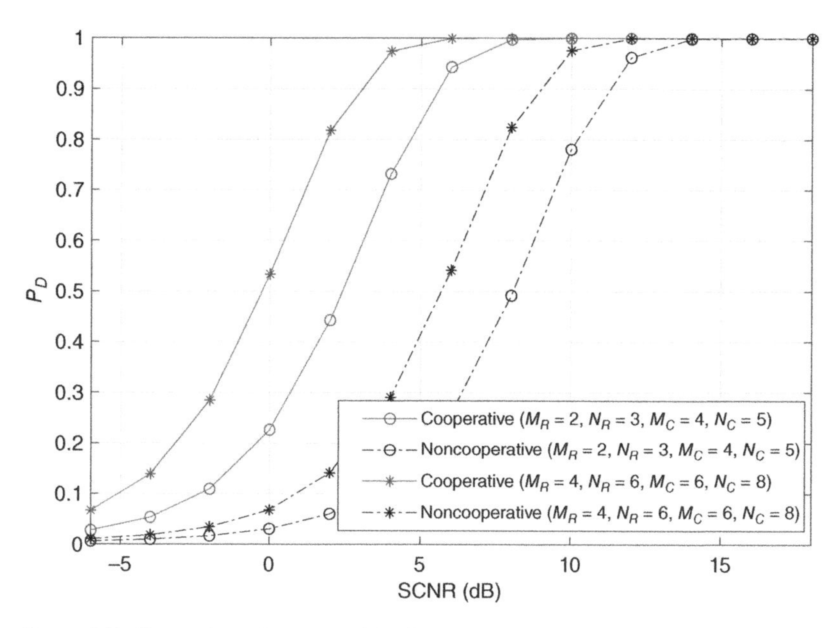

Figure 9.7 Target detection probability P_D ((9.20) and (9.33)) versus SCNR for the cooperative CERC system when $\alpha_E = 0.2$ and $P_{FA} = 10^{-3}$.

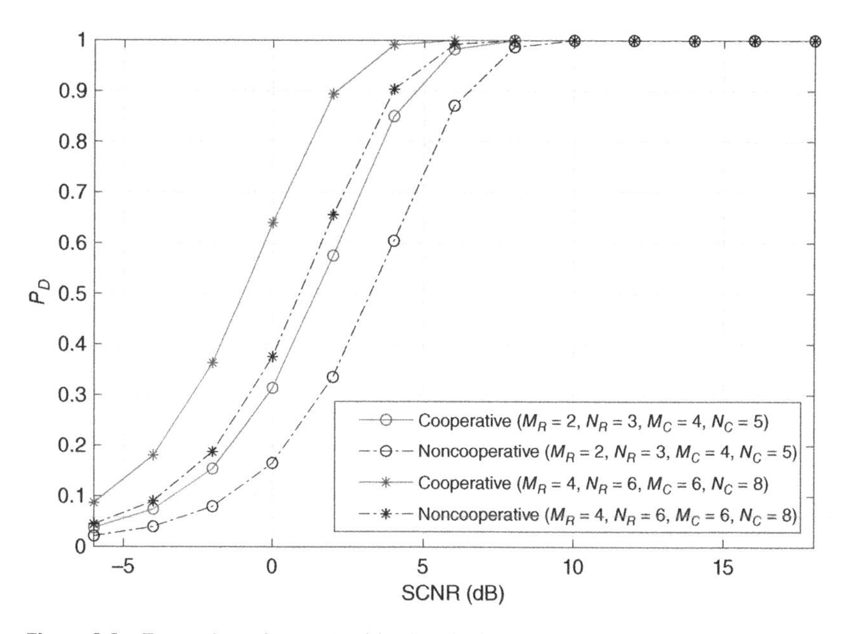

Figure 9.8 Target detection probability P_D ((9.20) and (9.33)) versus SCNR for the cooperative CERC system when $\alpha_E = 0.6$ and $P_{FA} = 10^{-3}$.

$\alpha_E = 0.6$ and a similar conclusion can be drawn. Taken together, Figures 9.7 and 9.8 show that as α_E increases, the detection performance for both cooperative and noncooperative cases increases, but the performance gain from cooperation decreases (the difference between the cooperative and noncooperative performance curves). This is because when $\alpha_E = 0$, only the communications signals are used as the transmitted signals for the cooperative hybrid active–passive radar system, and the radar transmission power of the noncooperative system is 0. In this case, the performance

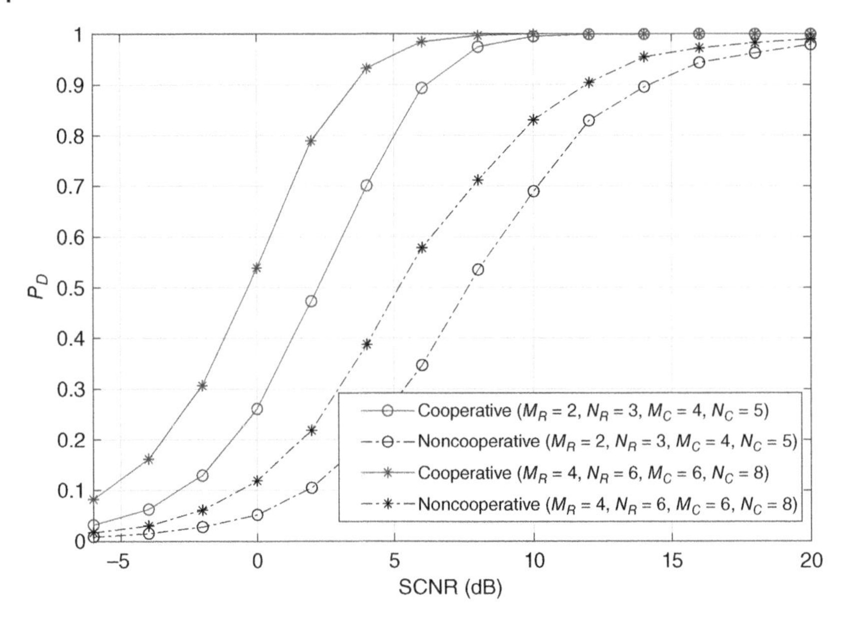

Figure 9.9 Target detection probability P_D ((9.20) and (9.33)) versus SCNR for the cooperative CERC system when $\alpha_E = 0.2$, $P_{FA} = 10^{-3}$, and random Gaussian reflection coefficients.

gain brought by cooperation is the largest. As α_E increases, the power of the active radar gradually increases. When the $\alpha_E = 1$, the cooperative system is equivalent to a pure active radar. So the radar detection performance is the best and the detection probabilities for the cooperative and noncooperative cases are the same.

In the previous theoretical derivations and simulations, we assumed pre-estimated reflection coefficients for simplicity and to provide a performance bound. Here, we provide numerical results for complex Gaussian reflection coefficients. Assume that all random reflection coefficients are independent of each other, the mean is the same as the previously assumed values (Figure 9.7) for the pre-estimated reflection coefficients, and the variance of all random reflection coefficients is 0.1. Under the same parameter settings except the reflection coefficients, we reproduce Figure 9.7 as Figure 9.9. It can be seen from Figure 9.9 that the results obtained under complex Gaussian reflection coefficients are similar to the results of Figure 9.7, which shows that the cooperation can significantly improve the radar target detection performance.

9.3.3.2 Gain for Radar Parameter Estimation

The target estimation performance of noncooperative case is discussed first. Based on the radar received signal model for the noncooperative case in (9.32), the CRB for the estimate of θ is

$$\mathrm{CRB}_{\mathrm{non}} = \left\{ 2\Re\left\{ \mathbf{F}\mathbf{J}_{\tau_{Rt}}{}^{H}\mathbf{Q}_R^{-1}\mathbf{J}_{\tau_{Rt}}\mathbf{F}^{\dagger} \right\} \right\}^{-1}, \tag{9.39}$$

where $\mathbf{F}$ and $\mathbf{J}_{\tau_{Rt}}$ are defined in 9.29.

Next, we use the same example as in Section 9.3.3.1 to illustrate the radar estimation performance difference between the cooperative and noncooperative cases. In Figure 9.10, the target position RCRBs from (9.29) and (9.39) are plotted versus SCNR when $\alpha_E = 0.2$ for cooperative and noncooperative cases for the two antenna configurations. We see that the resulting RCRBs for the cooperative case are always smaller those for the noncooperative case, which means that the cooperation is helpful in improving target localization performance for the radar task. Figure 9.11

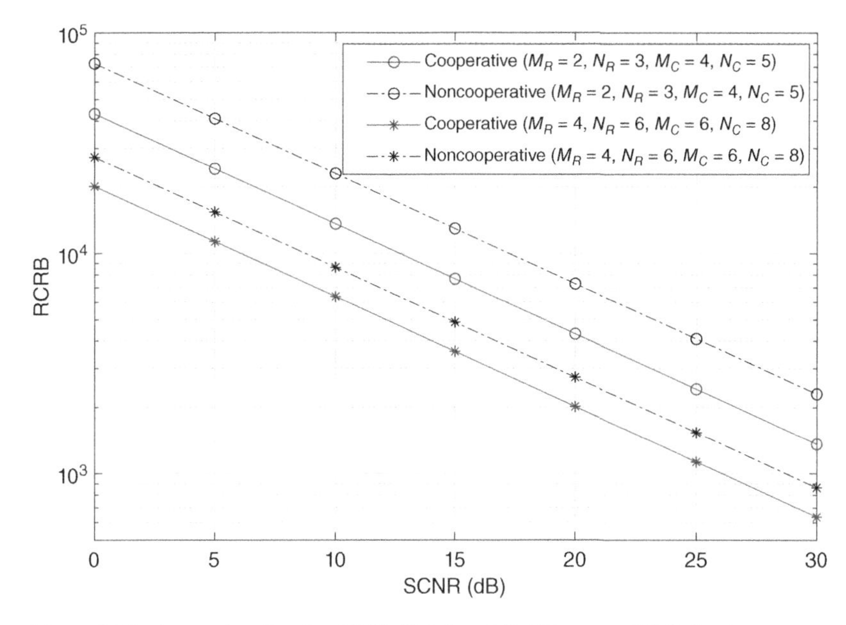

Figure 9.10 Target localization RCRB ((9.29) and (9.39)) versus SCNR for the cooperative CERC system when $\alpha_E = 0.2$.

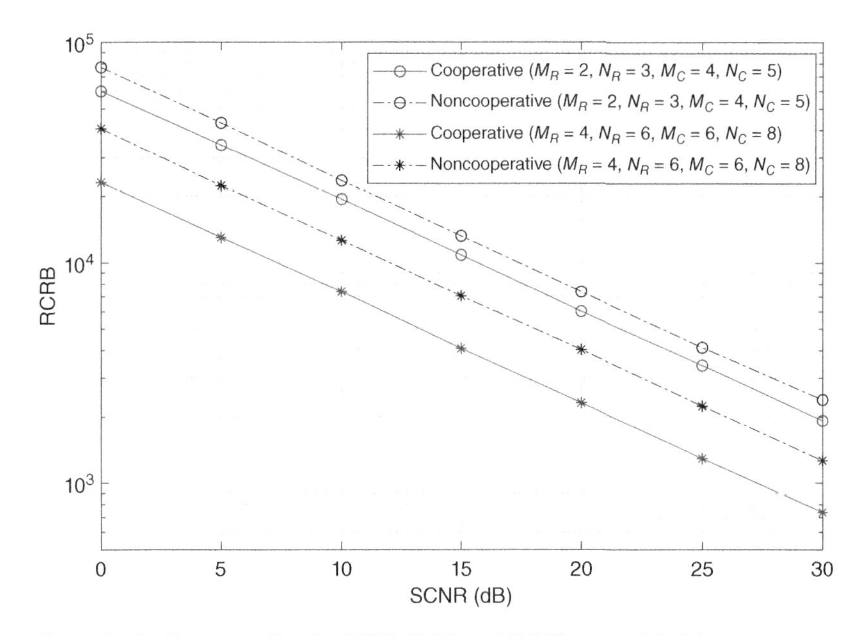

Figure 9.11 Target localization RCRB ((9.29) and (9.39)) versus SCNR for the cooperative CERC system when $\alpha_E = 0.2$ and random Gaussian reflection coefficients.

describes the RCRBs versus SCNR for random complex Gaussian reflection coefficients, and similar conclusions can be obtained.

Figure 9.12 shows the results for $\alpha_E = 0.6$ and a conclusion similar to Figure 9.10 can be drawn. Taken together, Figures 9.10 and 9.12 show that the RCRBs decrease as α_E increases, while the difference between the two RCRB curves decreases as α_E increases. Therefore, as α_E increases, the

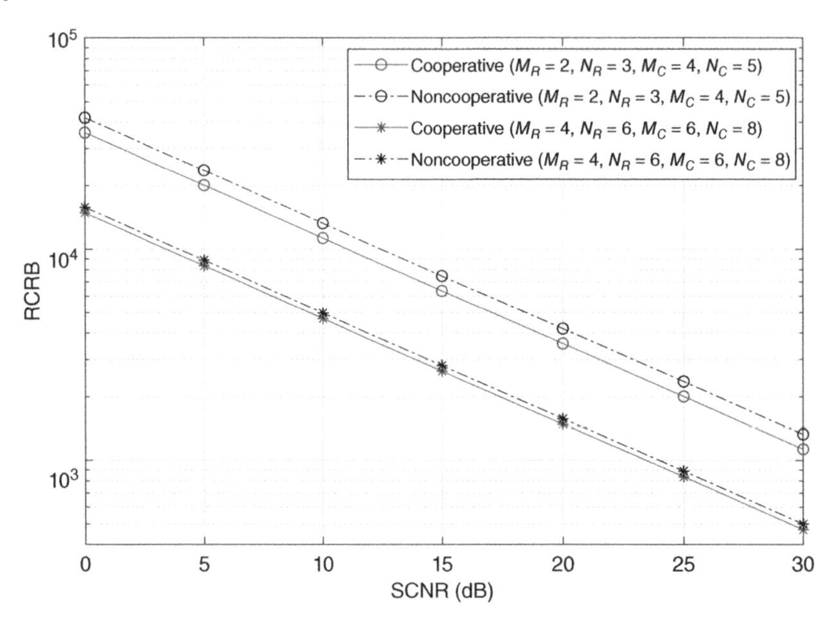

Figure 9.12 Target localization RCRB ((9.29) and (9.39)) versus SCNR for the cooperative CERC system when $\alpha_E = 0.6$.

estimation performance gets better, but the gain from cooperation gets smaller, which is similar to Figures 9.7 and 9.8.

It can be seen from Figure 9.7 through Figure 9.12 that the estimation performance and detection performance for the cooperative hybrid active–passive radar are better than that for the noncooperative cases. As α_E increases, the active radar gradually dominates, and the performance gains brought by cooperation gradually become smaller.

9.4 Radar-aided MIMO Communications in Cooperative CERC System

In a cooperative CERC system, the cooperative radar system can share its transmitted signals and antenna positions with the communications system. This information can be used by the communications receiver to estimate the target position $\boldsymbol{\theta}$. Utilizing the estimated $\boldsymbol{\theta}$, the target reflection due to radar transmission can be approximately eliminated from the received communications signals. Thus, the communications system can enhance its performance by cooperating with the radar system, which we call "radar-aided MIMO communications."

9.4.1 Mutual Information

In this section, we analyze the mutual information (MI) between the transmitted and received communications signals for the cooperative system to quantify communications capabilities. Using the signal model in (9.7) and assuming that the communications signals follow a Gaussian codebook (thus, they are Gaussian distributed) [Richmond et al., 2016], the ML estimate of the target position can be computed as

$$\hat{\boldsymbol{\theta}}_{C,\mathrm{ML}} = (\hat{x}_C, \hat{y}_C) = \arg\max_{\boldsymbol{\theta}} \ln p\left(\mathbf{r}_C|\boldsymbol{\theta}\right), \tag{9.40}$$

where

$$p\left(\mathbf{r}_C|\theta\right) = \frac{1}{\pi^{KN_C}\det\left(\mathbf{A}\right)}\exp\left\{-\left(\mathbf{r}_C - \tilde{\mathbf{U}}_R\tilde{\mathbf{s}}_R - \tilde{\mathbf{U}}_{Rt}\tilde{\mathbf{s}}_{Rt}\right)^H\mathbf{A}^{-1}\right.$$
$$\left.\left(\mathbf{r}_C - \tilde{\mathbf{U}}_R\tilde{\mathbf{s}}_R - \tilde{\mathbf{U}}_{Rt}\tilde{\mathbf{s}}_{Rt}\right)\right\}, \tag{9.41}$$

the covariance matrix of $\mathbf{r}_C$ is

$$\mathbf{A} = \mathbb{E}\left\{\left(\mathbf{r}_C - \tilde{\mathbf{U}}_R\tilde{\mathbf{s}}_R - \tilde{\mathbf{U}}_{Rt}\tilde{\mathbf{s}}_{Rt}\right)\left(\mathbf{r}_C - \tilde{\mathbf{U}}_R\tilde{\mathbf{s}}_R - \tilde{\mathbf{U}}_{Rt}\tilde{\mathbf{s}}_{Rt}\right)^H\right\} \tag{9.42}$$
$$= \tilde{\mathbf{U}}_C\tilde{\mathbf{S}}_c\tilde{\mathbf{U}}_C{}^H + \tilde{\mathbf{U}}_C\tilde{\mathbf{S}}_{ct}\tilde{\mathbf{U}}_{Ct}{}^H + \tilde{\mathbf{U}}_{Ct}\tilde{\mathbf{S}}_{ct}{}^H\tilde{\mathbf{U}}_C + \tilde{\mathbf{U}}_{Ct}\tilde{\mathbf{S}}_{ctt}\tilde{\mathbf{U}}_{Ct}{}^H + \mathbf{Q}_C,$$

$\tilde{\mathbf{S}}_C = \mathbb{E}\{\tilde{\mathbf{s}}_C\tilde{\mathbf{s}}_C{}^H\}$, $\tilde{\mathbf{S}}_{ct} = \mathbb{E}\{\tilde{\mathbf{s}}_C\tilde{\mathbf{s}}_{Ct}{}^H\}$, and $\tilde{\mathbf{S}}_{ctt} = \mathbb{E}\{\tilde{\mathbf{s}}_{Ct}\tilde{\mathbf{s}}_{Ct}{}^H\}$.

In (9.7), the third and fourth terms are from radar transmissions which are unrelated to the useful communications signals. The cooperation with the radar system just discussed can be employed to remove these signals. Considering that the radar signals and antenna positions are known to the communications system, the direct path received signals due to radar transmission can be eliminated directly. Note that the third term in the communications received signal (9.6) contains the time delay due to target reflection associated with the radar transmission, which can be approximated by using the ML estimate of the target position as

$$\hat{\tau}_{Rt,n'm} = \frac{1}{c}\left\{\left[\left(x_{R,m}^{Tx} - \hat{x}_C\right)^2 + \left(y_{R,m}^{Tx} - \hat{y}_C\right)^2\right]^{\frac{1}{2}}\right.$$
$$\left.+\left[\left(x_{C,n'}^{Rx} - \hat{x}_C\right)^2 + \left(y_{C,n'}^{Rx} - \hat{y}_C\right)^2\right]^{\frac{1}{2}}\right\}, \tag{9.43}$$

where $(x_{R,m}^{Tx}, y_{R,m}^{Tx})$ and $(x_{C,n'}^{Rx}, y_{C,n'}^{Rx})$ denote the positions of the mth radar transmitter and the n'th communications receiver, and c is the speed of light. Thus, we can write

$$\hat{\tau}_{Rt,n'm} = \tilde{\tau}_{Rt,n'm} + n_{Rt,n'm}, \tag{9.44}$$

where $n_{Rt,n'm}$ is the estimation error assumed to be Gaussian distributed [Chiriyath et al., 2016] and $\tilde{\tau}_{Rt,n'm}$ is the true delay in (9.6). Replacing $\tilde{\tau}_{Rt,n'm}$ in (9.6) with $\hat{\tau}_{Rt,n'm}$ can approximately eliminate the target return due to radar transmission. To model the error in this process, we adopt the method in Chiriyath et al. [2016] to obtain the following approximation:

$$s_{R,m}(kT_s - \tilde{\tau}_{Rt,n'm}) - s_{R,m}(kT_s - \tilde{\tau}_{Rt,n'm} - n_{Rt,n'm}) \tag{9.45}$$
$$\approx s_{R,m}^{(1)}(kT_s - \tilde{\tau}_{Rt,n'm})n_{Rt,n'm},$$

where the difference between the actual and estimated radar return waveforms can be replaced with a derivative for small fluctuations in delay. In (9.45), $s_{R,m}^{(1)}(t) = \partial s_{R,m}(t)/\partial t$ denotes the derivative of $s_{R,m}(t)$ with respect to t and $n_{Rt,n'm}$ the estimation error defined in (9.44). Therefore, using (9.45), after approximately eliminating the terms contributed by the radar from (9.7), the received signal vector at the communications receiver becomes

$$\mathbf{r}'_C = \mathbf{r}_C - \tilde{\mathbf{U}}_{Rt}\hat{\mathbf{s}}_{Rt} - \tilde{\mathbf{U}}_R\tilde{\mathbf{s}}_R = \tilde{\mathbf{U}}_{Ct}\tilde{\mathbf{s}}_{Ct} + \tilde{\mathbf{U}}_C\tilde{\mathbf{s}}_C + \tilde{\mathbf{V}}_{Rt}\tilde{\mathbf{n}}_{Rt} + \mathbf{w}_C, \tag{9.46}$$

where $\tilde{\mathbf{V}}_{Rt} = \text{Diag}\{\tilde{\mathbf{V}}_{Rt,1}, \dots, \tilde{\mathbf{V}}_{Rt,N_C}\}$ in which $\tilde{\mathbf{V}}_{Rt,n'} = \left(\tilde{\mathbf{v}}_{Rt,n'}[1], \dots, \tilde{\mathbf{v}}_{Rt,n'M_R}[K]\right)^{\dagger}$, $\tilde{\mathbf{v}}_{Rt,n'}[k] = (\tilde{v}_{Rt,n'1}[k], \dots, \tilde{v}_{Rt,n'M_R}[k])^{\dagger}$,

$$\tilde{v}_{Rt,n'm}[k] = \sqrt{E_{Rm}}\zeta_{Rt,n'm}s_{R,m}^{(1)}(kT_s - \tilde{\tau}_{Rt,n'm}), \tag{9.47}$$

and $\tilde{\mathbf{n}}_{Rt} = [\tilde{\mathbf{n}}_{Rt,1}{}^{\dagger}, \dots, \tilde{\mathbf{n}}_{Rt,N_C}{}^{\dagger}]^{\dagger}$ with $\tilde{\mathbf{n}}_{Rt,n'} = [\tilde{n}_{Rt,n'1}, \dots, \tilde{n}_{Rt,n'M_R}]^{\dagger}$.

Note that the first term in (9.46) contains the time delay caused by the target position, which can be calculated using the estimate of the target position as

$$\hat{\tau}_{Ct,n'm'} = \frac{1}{c}\left\{\left[\left(x^{Tx}_{C,m'} - \hat{x}_C\right)^2 + \left(y^{Tx}_{C,m'} - \hat{y}_C\right)^2\right]^{\frac{1}{2}} + \left[\left(x^{Rx}_{C,n'} - \hat{x}_C\right)^2 + \left(y^{Rx}_{C,n'} - \hat{y}_C\right)^2\right]^{\frac{1}{2}}\right\}, \tag{9.48}$$

where $(x^{Tx}_{C,m'}, y^{Tx}_{C,m'})$ denotes the positions of the m'th communications transmitter, and we can write

$$\hat{\tau}_{Ct,n'm'} = \tilde{\tau}_{Ct,n'm'} + n_{Ct,n'm'}, \tag{9.49}$$

where $n_{Ct,n'm'}$ is the estimation error assumed to be Gaussian distributed. Therefore, the unknown $\tilde{\tau}_{Ct,n'm'}$ that enters (9.46) through $\tilde{\mathbf{s}}_{Ct}$ can be replaced with $\hat{\tau}_{Ct,n'm'}$ and the resulting error could be modeled. To simplify analysis, this error is ignored, assuming that $n_{Ct,n'm'}$ is small enough.

In the resulting measurement vector $\mathbf{r}'_C$ in (9.46), noting that both the $\tilde{\mathbf{s}}_C$ and $\tilde{\mathbf{s}}_{Ct}$ come from communications signals that contain useful information to be communicated, the MI between $\mathbf{r}'_C$ and $(\tilde{\mathbf{s}}_C, \tilde{\mathbf{s}}_{Ct})$ can be computed as Yang and Blum [2007]

$$\begin{aligned} I\left(\mathbf{r}'_C, \tilde{\mathbf{s}}_C, \tilde{\mathbf{s}}_{Ct}\right) &= H\left(\mathbf{r}'_C\right) - H\left(\mathbf{r}'_C | \tilde{\mathbf{s}}_C, \tilde{\mathbf{s}}_{Ct}\right) \\ &= \log\det\left\{\mathbf{I} + \left(\tilde{\mathbf{V}}_{Rt}\mathbf{Q}_{Rt}\tilde{\mathbf{V}}_{Rt}{}^H + \mathbf{Q}_C\right)^{-1}\left(\tilde{\mathbf{U}}_C\tilde{\mathbf{S}}_c\tilde{\mathbf{U}}_C{}^H \right.\right. \\ &\quad \left.\left. + \tilde{\mathbf{U}}_C\tilde{\mathbf{S}}_{ct}\tilde{\mathbf{U}}_{Ct}{}^H + \tilde{\mathbf{U}}_{Ct}\tilde{\mathbf{S}}_{ct}{}^H\tilde{\mathbf{U}}_C + \tilde{\mathbf{U}}_{Ct}\tilde{\mathbf{S}}_{ctt}\tilde{\mathbf{U}}_{Ct}{}^H\right)\right\}, \end{aligned} \tag{9.50}$$

which provides a performance metric for communications, where $H(\cdot)$ denotes differential entropy and $\mathbf{Q}_{Rt} = \mathbb{E}\{\tilde{\mathbf{n}}_{Rt}\tilde{\mathbf{n}}^H_{Rt}\}$.

9.4.2 Comparison with Noncooperative Cases

In this section, we compare the communications MI in (9.50) for the cooperative and noncooperative cases to demonstrate the advantages of the cooperative systems. The noncooperative case considers a standard communications system that does not employ the reflected signals, such that the received signals can be written as

$$\mathbf{r}_C = \tilde{\mathbf{U}}_C\tilde{\mathbf{s}}_C + \mathbf{w}_C. \tag{9.51}$$

Thus, the MI for the noncooperative case can be computed as

$$I\left(\mathbf{r}_C, \tilde{\mathbf{s}}_C\right) = \log\det\left\{\mathbf{I} + \mathbf{Q}_C^{-1}\tilde{\mathbf{U}}_C\tilde{\mathbf{S}}_C\tilde{\mathbf{U}}_C^H\right\}. \tag{9.52}$$

Next, examples are used to illustrate the communications performance for the cooperative MIMO radar and MIMO communications systems. Consider the same cooperative system as in Section 9.3.3, under the same parameter settings except that the communications signals are assumed spatially white random Gaussian with autocorrelation $\mathbb{E}\{s_{C,m'}(kT_s)s_{C,m'}(k'T_s)\} = 0.9^{|(k-k')T_s|}$ [Zhang et al., 2016]. Figure 9.13 plots the MI (from (9.50) and (9.52)) versus SCNR for the cooperative and noncooperative cases and two antenna configurations when $\alpha_E = 0.2$. It is seen that the MI curves for the cooperative case are always above than those for the noncooperative case, showing a significant gain in terms of the MI for the communications task. Figure 9.14 plots the results for random complex Gaussian reflection coefficients, Figure 9.15 shows the results for $\alpha_E = 0.6$, and a conclusion similar to Figure 9.13 can be obtained. Taken together, Figures 9.13 and 9.15 show that both the MI and performance gain become smaller when α_E becomes larger.

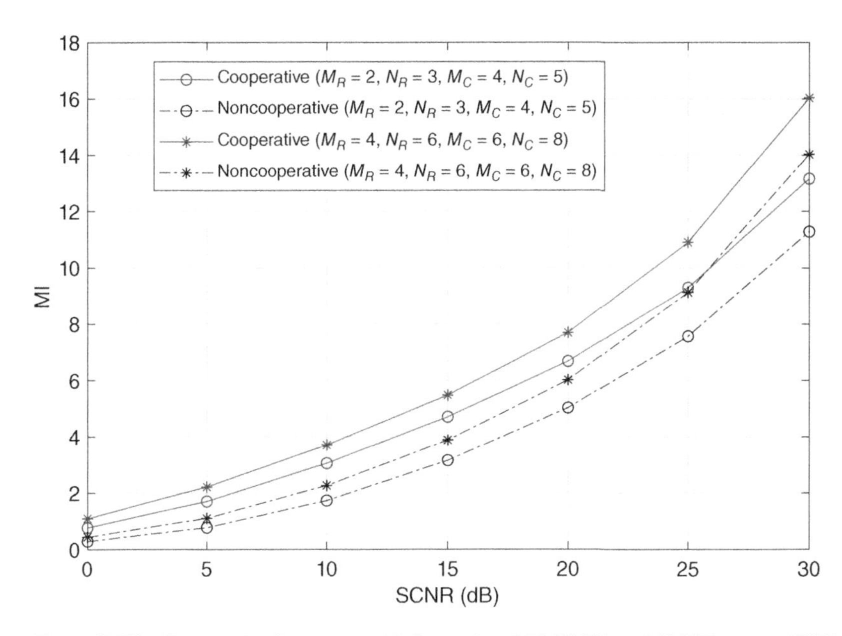

Figure 9.13 Communications mutual information (MI) ((9.50) and (9.52)) versus SCNR for a cooperative CERC system when $\alpha_E = 0.2$.

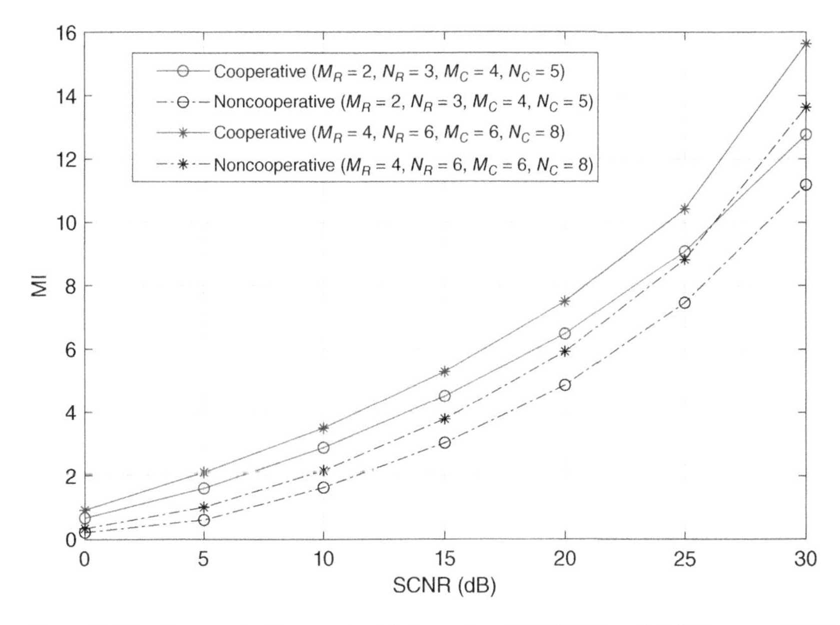

Figure 9.14 Communications mutual information (MI) ((9.50) and (9.52)) versus SCNR for a cooperative CERC system when $\alpha_E = 0.2$ and random Gaussian reflection coefficients.

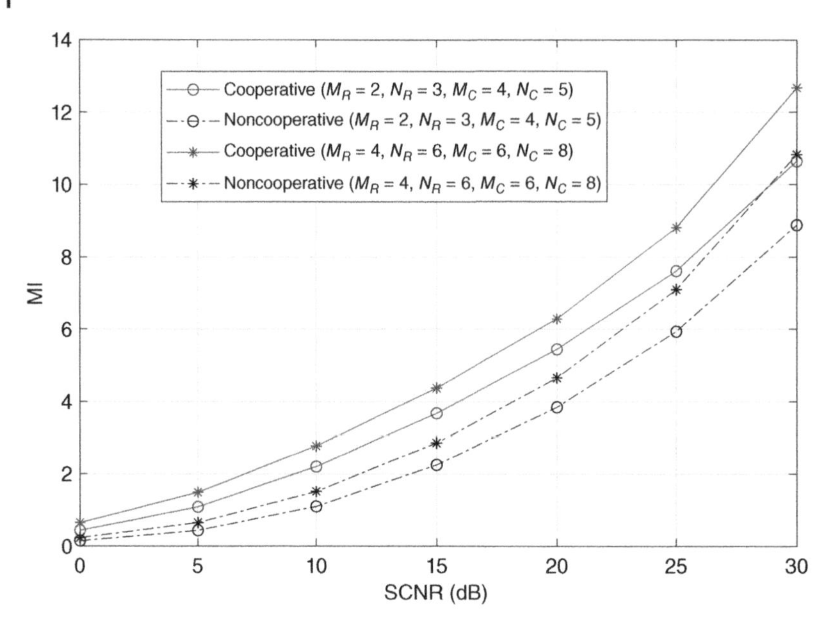

Figure 9.15 Communications mutual information (MI) ((9.50) and (9.52)) versus SCNR for a cooperative CERC system when $\alpha_E = 0.6$.

Considering the simulation results of Sections 9.3.3 and 9.4.2, we find that there is always a performance gain obtained when the radar and communications systems cooperate. For the cooperative systems, there is a trade-off between radar performance and communications efficiency.

9.5 Cooperative Radar and Communications System Co-design

From the previously presented results in (9.20), (9.30), and (9.50), we know that many system parameters affect the radar and communications performance. These system parameters include transmission waveforms, number and position of transmit and receive antennas, and signal power. Next, we discuss the co-design of cooperative MIMO radar and MIMO communications systems based on the performance metrics derived in Sections 9.3 and 9.4.

Let the parameters that need to be optimized be γ, which may be the radar and communications receive and transmit antenna positions $\gamma = \left[x^{Tx}_{R,m}, y^{Tx}_{R,m}, x^{Tx}_{C,m'}, y^{Tx}_{C,m'}, x^{Rx}_{R,n}, y^{Rx}_{R,n}, x^{Rx}_{C,n'}, y^{Rx}_{C,n'}\right]^{\dagger}$ or the radar and communications signal power $\gamma = [E_R, E_C]^{\dagger}$. To properly describe the joint radar and communications performance of a cooperative CERC system, a simple criterion is to sum a weighed combination of the radar performance and the communications performance, for some fixed weights β_R and β_C, to obtain

$$\mathbb{P}_{\text{co}}[\gamma] = \beta_R \mathbb{P}_R[\gamma] + \beta_C \mathbb{P}_C[\gamma], \tag{9.53}$$

which we call our comprehensive criterion. In (9.53), $\mathbb{P}_R$ represents the radar performance metric [detection probability or CRB], $\mathbb{P}_C$ denotes the communications performance metric [MI], and β_R and β_C represent the importance of radar and communications tasks, which can be adjusted to specific application scenarios respectively. Thus, the optimization problem can be described as

$$\begin{aligned} &\underset{\gamma}{\text{maximize}} \;\; \beta_R \mathbb{P}_R[\gamma] + \beta_C \mathbb{P}_C[\gamma], \\ &\text{subject to} \;\; \gamma \in C, \end{aligned} \tag{9.54}$$

where C is a set of constraints. When $\gamma = \left[x^{Tx}_{R,m}, y^{Tx}_{R,m}, x^{Tx}_{C,m'}, y^{Tx}_{C,m'}, x^{Rx}_{R,n}, y^{Rx}_{R,n}, x^{Rx}_{C,n'}, y^{Rx}_{C,n'}\right]^{\dagger}$, the above optimization problem can be used to optimize antenna positions, if movable antenna platforms are available, and when $\gamma = [E_R, E_C]^{\dagger}$, we can use (9.54) to solve the resource allocation of the cooperative system. Many other possible meaningful problems can be formulated.

9.5.1 Power Allocation Based on PD and MI

From the previous simulation results, it can be seen that there is a trade-off between the radar performance and the communications performance under different power allocation. Next, we take the power allocation as an example to analyze the co-design for the cooperative system in detail. First, the radar detection probability and communications MI are the employed performance metrics for the radar and communications task, respectively. The optimization problem can be described as

$$\begin{aligned} &\underset{\alpha_E}{\text{maximize}} \quad \beta_R P_D[\alpha_E] + \beta_C \text{MI}[\alpha_E], \\ &\text{subject to} \quad 0 \le \alpha_E \le 1. \end{aligned} \tag{9.55}$$

Next, we use the same example used in Section 9.3 to illustrate the power allocation for the cooperative system. Let $\beta_R = 1/P_{D,\max}$ and $\beta_C = \beta_{\text{co}}/\text{MI}_{\max}$, where $P_{D,\max} = \text{maximize}_{\,0\le\alpha_E\le1} P_D[\alpha_E]$, $\text{MI}_{\max} = \text{maximize}_{\,0\le\alpha_E\le1} \text{MI}[\alpha_E]$, and β_{co} is used to adjust the importance of the communications task relative to radar task. Thus, the term $\beta_R P_D[\alpha_E]$ represents the normalized detection probability and $\beta_C \text{MI}[\alpha_E]$ is β_{co} times the normalized MI. Figures 9.16–9.18 illustrate the comprehensive performance P_{co} versus α_E for two antenna configurations with different β_{co}. As can be seen from the figures, for all β_{co} and antenna configurations, the comprehensive performance of the cooperative system increases and then decreases with the increase of α_E. As β_{co} increases, the communications task becomes more and more important, and so the optimal α_E gradually decreases as expected. Thus, for cooperative CERC co-design, the coefficient β_{co} can be adjusted according to the specific application scenarios to properly describe the relative importance of

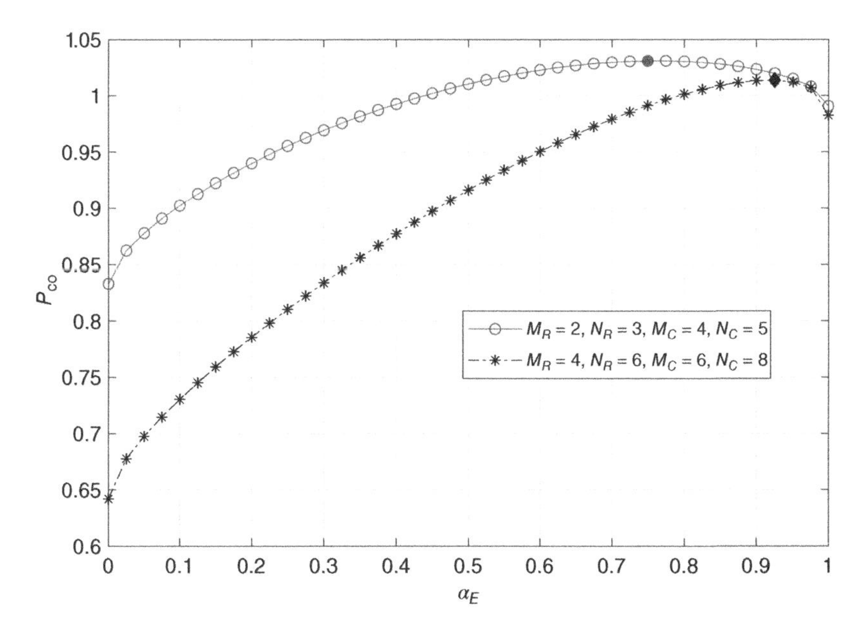

Figure 9.16 Comprehensive performance (9.55) versus α_E for a cooperative CERC system with $\beta_{\text{co}} = 0.2$.

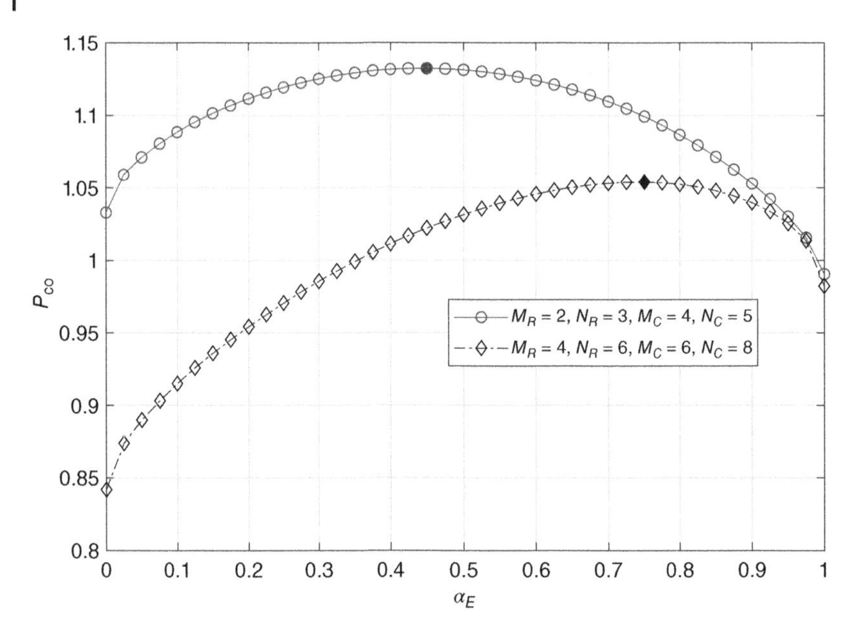

Figure 9.17 Comprehensive performance (9.55) versus α_E for a cooperative CERC system with $\beta_{co} = 0.4$.

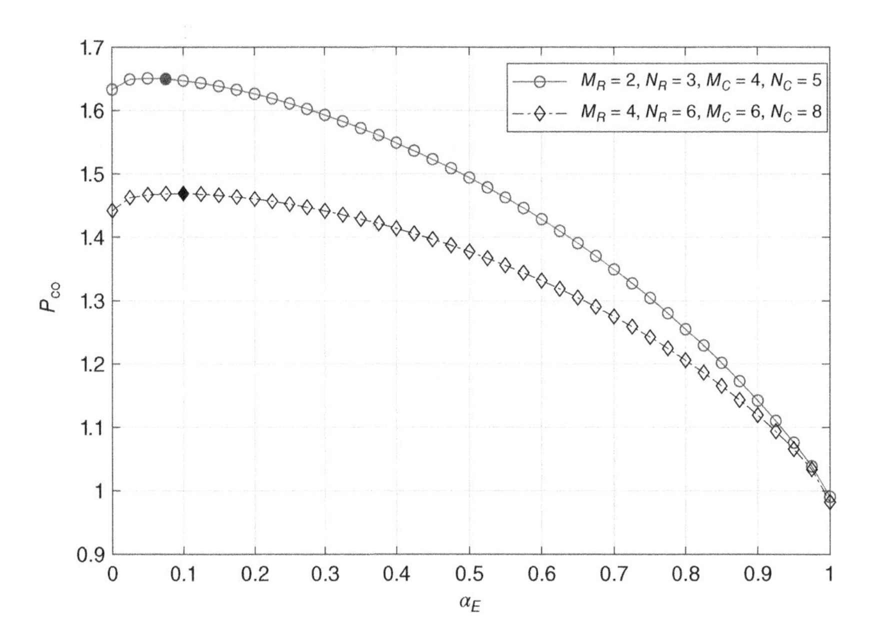

Figure 9.18 Comprehensive performance (9.55) versus α_E for a cooperative CERC system with $\beta_{co} = 1$.

radar and communications, and then the optimal power allocation can be obtained by solving the optimization problem in (9.55).

9.5.2 Power Allocation Based on CRB and MI

In Section 9.5.1, power allocation was determined using radar detection probability as the radar performance metric. This section considers power allocation using CRB as the metric for

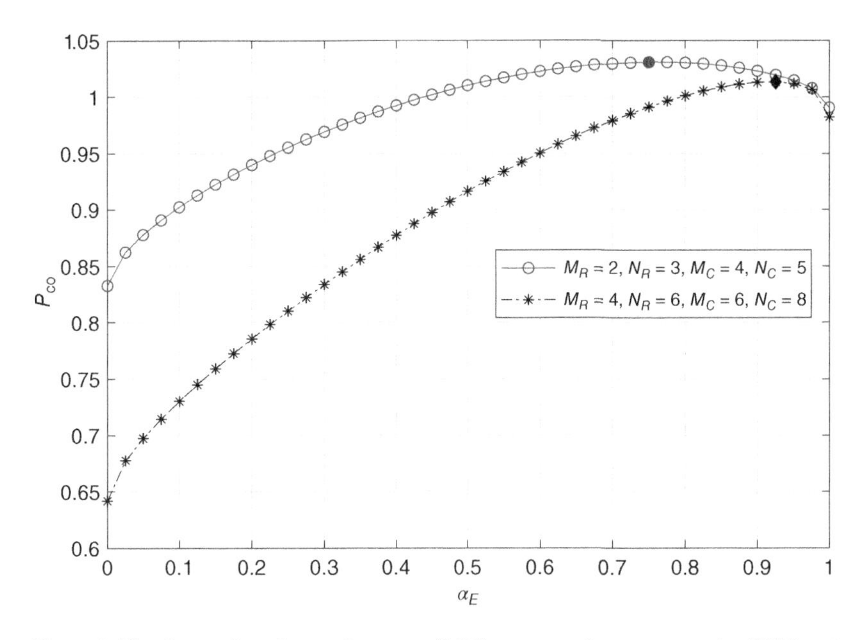

Figure 9.19 Comprehensive performance (9.56) versus α_E for a cooperative CERC system with $\beta_{co} = 0.2$.

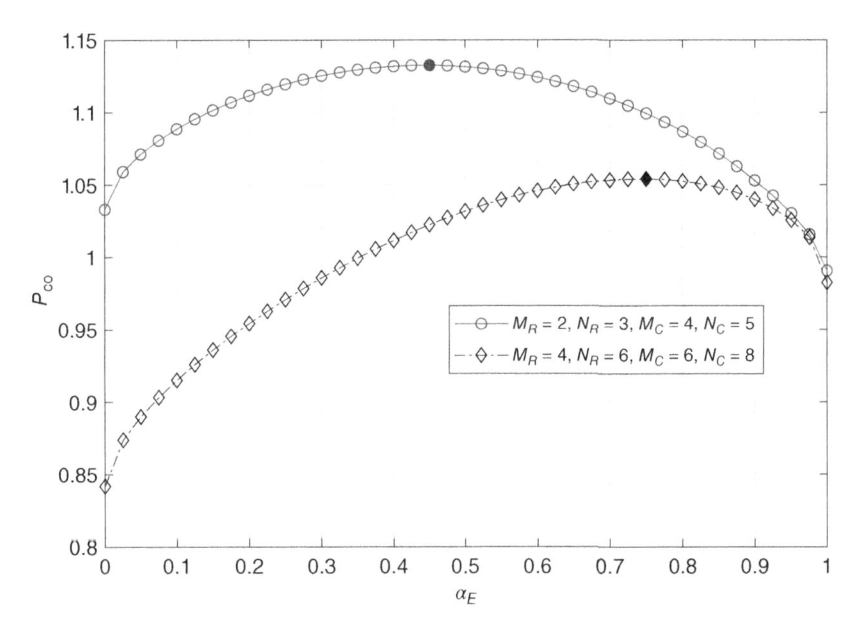

Figure 9.20 Comprehensive performance (9.56) versus α_E for a cooperative CERC system with $\beta_{co} = 0.4$.

radar performance. Since smaller CRB is better, the comprehensive performance metric can be expressed as

$$P_{co} = \beta_R \mathrm{RCRB}^{-1}[\alpha_E] + \beta_C \mathrm{MI}[\alpha_E]. \tag{9.56}$$

Let $\beta_R = 1/\mathrm{minimize}_{\,0 \leq \alpha_E \leq 1} \mathrm{RCRB}[\alpha_E]$. Figures 9.16–9.18 show the comprehensive performance P_{co} in (9.56) versus α_E for the two antenna configurations with $\beta_{co} = 0.2$, $\beta_{co} = 0.4$, and $\beta_{co} = 1$, respectively. Again, for all β_{co} and antenna configurations, the comprehensive performance of the

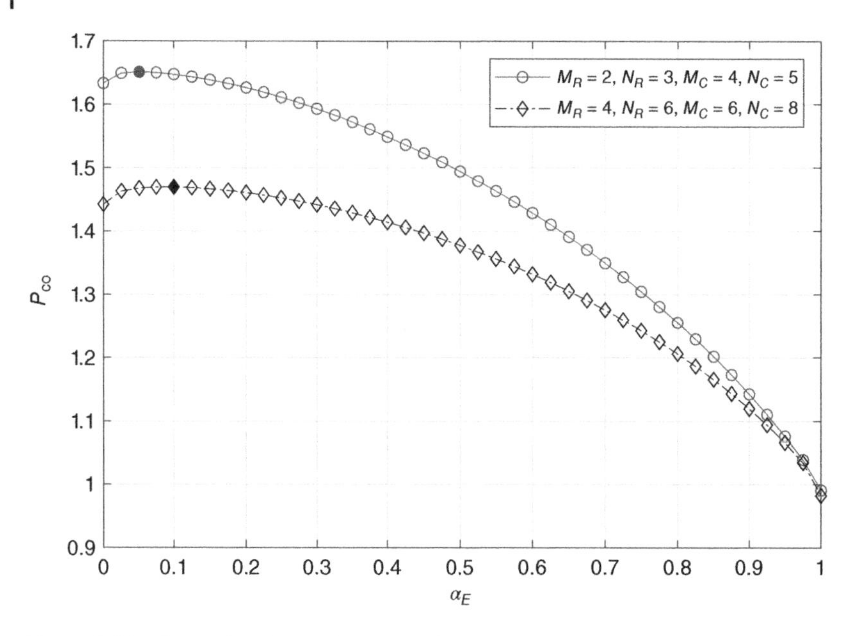

Figure 9.21 Comprehensive performance (9.56) versus α_E for a cooperative CERC system with $\beta_{co} = 1$.

cooperative system increases and then decreases with the increase of α_E. As the coefficient β_{co} increases, the optimal α_E gradually decreases as expected. Interestingly, by comparing Figure 9.16 with Figure 9.19, Figure 9.17 with Figure 9.20, and Figure 9.18 with Figure 9.21, respectively, under the same weighting coefficient β_{co}, the optimal power allocation obtained by the comprehensive performance P_{co} from (9.55) and (9.56) is almost the same. That is, no matter whether the detection performance or the estimation performance is used to measure the radar performance, the optimal power allocation obtained is almost the same.

9.6 Conclusions

In Section 9.1, this chapter provided a review of integrated radar and communications systems. Existing research on integrated radar and communications systems can be divided into three types: DFRC systems, CERC system, and cooperative CERC systems. The cooperative CERC systems consider information sharing between radar and communications, so both systems could benefit from each other instead of treating each others signals as interference. The rest of this chapter presented our own recent research, which focuses on cooperative CERC systems. A received signal model was developed for cooperative MIMO radar and MIMO communications systems in Section 9.2, which accounts for the target returns and direct path signals contributed from both the radar and communications transmissions.

On the radar side, the target returns contributed from both the radar transmitters and communications transmitters can be used to complete the radar task, forming a hybrid active–passive MIMO radar network. In Section 9.3, target detection and localization for the hybrid active–passive MIMO radar network were discussed. We obtained the detection probability and the CRB for target position for the cooperative system and compared them with those of the noncooperative case, which showed that cooperation helps to improve the radar performance.

In the cooperative CERC system, the communications system can extract information not merely from the directly received communications signals but also those reflected from the target, which is called “radar-aided MIMO communications. In Section 9.4, we analyzed the MI of the communications part for the cooperative system to quantify communications capability. By comparing the performance with the noncooperative case, it can be found that there is always a performance gain obtained when the radar and communications systems cooperate.

From the simulation results, it can be seen that there is a trade-off between the radar performance and the communications performance under different system parameters. Thus, the co-design for the cooperative radar and communications system was investigated in Section 9.5. We proposed a joint optimization formulation for cooperative systems and analyzed the power allocation problem based on the radar detection probability or CRB and communications MI and obtained the optimal power allocation.

References

A. Ahmed, S. Zhang, and Y. D. Zhang. Antenna selection strategy for transmit beamforming-based joint radar-communication system. *Digital Signal Processing*, 105:102768 2020.

D. Bao, G. Qin, J. Cai, and G. Liu. A precoding OFDM MIMO radar coexisting with a communication system. *IEEE Transactions on Aerospace and Electronic Systems*, 55(4):1864–1877, 2019. https://doi.org/10.1109/TAES.2018.2876736.

S. S. Bhat, R. M. Narayanan, and M. Rangaswamy. Bandwidth sharing and scheduling for multimodal radar with communications and tracking. In *2012 IEEE 7th Sensor Array and Multichannel Signal Processing Workshop (SAM)*, pages 233–236, 2012. https://doi.org/10.1109/SAM.2012.6250476.

M. Bica and V. Koivunen. Delay estimation method for coexisting radar and wireless communication systems. In *2017 IEEE Radar Conference (RadarConf)*, pages 1557–1561, 2017. https://doi.org/10.1109/RADAR.2017.7944455.

M. Bica and V. Koivunen. Radar waveform optimization for target parameter estimation in cooperative radar-communications systems. *IEEE Transactions on Aerospace and Electronic Systems*, 55(5):2314–2326, 2019. https://doi.org/10.1109/TAES.2018.2884806.

M. Bica, K. Huang, U. Mitra, and V. Koivunen. Opportunistic radar waveform design in joint radar and cellular communication systems. In *2015 IEEE Global Communications Conference (GLOBECOM)*, pages 1–7, 2015. https://doi.org/10.1109/GLOCOM.2015.7417624.

S. Biswas, K. Singh, O. Taghizadeh, and T. Ratnarajah. Design and analysis of FD MIMO cellular systems in coexistence with MIMO radar. *IEEE Transactions on Wireless Communications*, 19(7):4727–4743, 2020. https://doi.org/10.1109/TWC.2020.2986734.

R. Cager, D. LaFlame, and L. Parode. Orbiter Ku-Band integrated radar and communications subsystem. *IEEE Transactions on Communications*, 26(11):1604–1619, 1978. https://doi.org/10.1109/TCOM.1978.1094004.

Y. Chen and X. Gu. Time allocation for integrated Bi-static radar and communication systems. *IEEE Communications Letters*: 1033–1036, 2020. https://doi.org/10.1109/LCOMM.2020.3039054.

X. Chen, X. Wang, S. Xu, and J. Zhang. A novel radar waveform compatible with communication. In *2011 International Conference on Computational Problem-Solving (ICCP)*, pages 177–181, 2011. https://doi.org/10.1109/ICCPS.2011.6092272.

X. Chen, Z. Liu, Y. Liu, and Z. Wang. Energy leakage analysis of the radar and communication integrated waveform. *IET Signal Processing*, 12(3):375–382, 2018.

Z. Cheng, B. Liao, and Z. He. Hybrid transceiver design for dual-functional radar-communication system. In *2020 IEEE 11th Sensor Array and Multichannel Signal Processing Workshop (SAM)*, pages 1–5, 2020. https://doi.org/10.1109/SAM48682.2020.9104387.

A. R. Chiriyath, B. Paul, G. M. Jacyna, and D. W. Bliss. Inner bounds on performance of radar and communications co-existence. *IEEE Transactions on Signal Processing*, 64(2):464–474, 2016. https://doi.org/10.1109/TSP.2015.2483485.

A. R. Chiriyath, B. Paul, and D. W. Bliss. Radar-communications convergence: coexistence, cooperation, and co-design. *IEEE Transactions on Cognitive Communications and Networking*, 3(1):1–12, 2017. https://doi.org/10.1109/TCCN.2017.2666266.

Y. Cui, V. Koivunen, and X. Jing. Interference alignment based spectrum sharing for MIMO radar and communication systems. In *2018 IEEE 19th International Workshop on Signal Processing Advances in Wireless Communications (SPAWC)*, pages 1–5, 2018. https://doi.org/10.1109/SPAWC.2018.8445973.

S. H. Dokhanchi, B. S. Mysore, K. V. Mishra, and B. Ottersten. A mmwave automotive joint radar-communications system. *IEEE Transactions on Aerospace and Electronic Systems*, 55(3):1241–1260, 2019. https://doi.org/10.1109/TAES.2019.2899797.

G. Duggal, S. Vishwakarma, K. V. Mishra, and S. S. Ram. Doppler-resilient 802.11ad-based ultrashort range automotive joint radar-communications system. *IEEE Transactions on Aerospace and Electronic Systems*, 56(5):4035–4048, 2020. https://doi.org/10.1109/TAES.2020.2990393.

A. Gameiro, D. Castanheira, J. Sanson, and P. Monteiro. Research challenges, trends and applications for future joint radar communications systems. *Wireless Personal Communications*, 100, 05 2018. https://doi.org/10.1007/s11277-018-5614-8.

L. Gaudio, M. Kobayashi, G. Caire, and G. Colavolpe. On the effectiveness of OTFS for joint radar parameter estimation and communication. *IEEE Transactions on Wireless Communications*, 19(9):5951–5965, 2020. https://doi.org/10.1109/TWC.2020.2998583.

Z. Geng, H. Deng, and B. Himed. Adaptive radar beamforming for interference mitigation in radar-wireless spectrum sharing. *IEEE Signal Processing Letters*, 22(4):484–488, 2015. https://doi.org/10.1109/LSP.2014.2363585.

M. Ghorbanzadeh, A. Abdelhadi, and C. Clancy. A utility proportional fairness resource allocation in spectrally radar-coexistent cellular networks. In *2014 IEEE Military Communications Conference*, pages 1498–1503, 2014. https://doi.org/10.1109/MILCOM.2014.247.

H. Godrich, A. P. Petropulu, and H. V. Poor. Sensor selection in distributed multiple-radar architectures for localization: a knapsack problem formulation. *IEEE Transactions on Signal Processing*, 60(1):247–260, 2012. https://doi.org/10.1109/TSP.2011.2170170.

A. Hassanien, B. Himed, and B. D. Rigling. A dual-function MIMO radar-communications system using frequency-hopping waveforms. In *2017 IEEE Radar Conference (RadarConf)*, pages 1721–1725, 2017. https://doi.org/10.1109/RADAR.2017.7944485.

A. Hassanien, C. Sahin, J. Metcalf, and B. Himed. Uplink signaling and receive beamforming for dual-function radar communications. In *2018 IEEE 19th International Workshop on Signal Processing Advances in Wireless Communications (SPAWC)*, pages 1–5, 2018. https://doi.org/10.1109/SPAWC.2018.8445858.

Q. He, Z. Wang, J. Hu, and R. S. Blum. Performance gains from cooperative MIMO radar and MIMO communication systems. *IEEE Signal Processing Letters*, 26(1):194–198, 2019. https://doi.org/10.1109/LSP.2018.2880836.

M. Jamil, H. Zepernick, and M. I. Pettersson. On integrated radar and communication systems using Oppermann sequences. In *MILCOM 2008 - 2008 IEEE Military Communications Conference*, pages 1–6, 2008. https://doi.org/10.1109/MILCOM.2008.4753277.

M. F. Keskin, C. Aydogdu, and H. Wymeersch. Stepped-carrier OFDM V2V resource allocation for sensing and communication convergence. In *2020 14th European Conference on Antennas and Propagation (EuCAP)*, pages 1–5, 2020. https://doi.org/10.23919/EuCAP48036.2020.9135203.

P. Kumari, D. H. N. Nguyen, and R. W. Heath. Performance trade-off in an adaptive IEEE 802.11AD waveform design for a joint automotive radar and communication system. In *2017 IEEE*

International Conference on Acoustics, Speech and Signal Processing (ICASSP), pages 4281–4285, 2017. https://doi.org/10.1109/ICASSP.2017.7952964.

P. Kumari, J. Choi, N. Gonzlez-Prelcic, and R. W. Heath. IEEE 802.11ad-based radar: an approach to joint vehicular communication-radar system. *IEEE Transactions on Vehicular Technology*, 67(4):3012–3027, 2018. https://doi.org/10.1109/TVT.2017.2774762.

P. Kumari, S. A. Vorobyov, and R. W. Heath. Adaptive virtual waveform design for millimeter-wave joint communicationradar. *IEEE Transactions on Signal Processing*, 68:715–730, 2020. https://doi.org/10.1109/TSP.2019.2956689.

G. Lellouch, A. K. Mishra, and M. Inggs. Design of OFDM radar pulses using genetic algorithm based techniques. *IEEE Transactions on Aerospace and Electronic Systems*, 52(4):1953–1966, 2016. https://doi.org/10.1109/TAES.2016.140671.

B. Li and A. Petropulu. Spectrum sharing between MIMO-MC radars and communication systems. In *2016 50th Asilomar Conference on Signals, Systems and Computers*, pages 52–57, 2016. https://doi.org/10.1109/ACSSC.2016.7868992.

B. Li and A. P. Petropulu. Joint transmit designs for coexistence of MIMO wireless communications and sparse sensing radars in clutter. *IEEE Transactions on Aerospace and Electronic Systems*, 53(6):2846–2864, 2017. https://doi.org/10.1109/TAES.2017.2717518.

M. Li, W. Wang, and Z Zhi. Communication-embedded OFDM chirp waveform for delay-doppler radar. *IET Radar, Sonar and Navigation*, 12(3):353–360, 2017.

S. Lindenmeier, K. Boehm, and J. F. Luy. A wireless data link for mobile applications. *IEEE Microwave and Wireless Components Letters*, 13(8):326–328, 2003. https://doi.org/10.1109/LMWC.2003.815706.

W. Linlong, K. V. Mishra, S. M. R. Bhavani, and B. Ottersten. Resource allocation in heterogeneously-distributed joint radar-communications under asynchronous Bayesian tracking framework, 2021.

Y. Liu, G. Liao, and Z. Yang. Range and angle estimation for MIMO-OFDM integrated radar and communication systems. In *2016 CIE International Conference on Radar (RADAR)*, pages 1–4, 2016. https://doi.org/10.1109/RADAR.2016.8059539.

F. Liu, C. Masouros, A. Li, T. Ratnarajah, and J. Zhou. MIMO radar and cellular coexistence: a power-efficient approach enabled by interference exploitation. *IEEE Transactions on Signal Processing*, 66(14):3681–3695, 2018. https://doi.org/10.1109/TSP.2018.2833813.

X. Liu, T. Huang, N. Shlezinger, Y. Liu, J. Zhou, and Y. C. Eldar. Joint transmit beamforming for multiuser MIMO communications and MIMO radar. *IEEE Transactions on Signal Processing*, 68: 3929–3944, 2020a. https://doi.org/10.1109/TSP.2020.3004739.

Y. Liu, G. Liao, S. Zhu, Z. Yang, Y. Chen, and X. Zhang. Performance improvement in a coexistent radar and communications system. In *2020 IEEE 11th Sensor Array and Multichannel Signal Processing Workshop (SAM)*, pages 1–5, 2020b. https://doi.org/10.1109/SAM48682.2020.9104303.

F. Liu, C. Masouros, A. P. Petropulu, H. Griffiths, and L. Hanzo. Joint radar and communication design: applications, state-of-the-art, and the road ahead. *IEEE Transactions on Communications*, 68(6):3834–3862, 2020c. https://doi.org/10.1109/TCOMM.2020.2973976.

J. Liu, K. V. Mishra, and M. Saquib. Co-designing statistical MIMO radar and in-band full-duplex multi-user MIMO communications, 2020d.

P. M. McCormick, B. Ravenscroft, S. D. Blunt, A. J. Duly, and J. G. Metcalf. Simultaneous radar and communication emissions from a common aperture, Part II: Experimentation. In *2017 IEEE Radar Conference (RadarConf)*, pages 1697–1702, 2017. https://doi.org/10.1109/RADAR.2017.7944480.

R. M. Mealey. A method for calculating error probabilities in a radar communication system. *IEEE Transactions on Space Electronics and Telemetry*, 9(2):37–42, 2007. https://doi.org/10.1109/TSET.1963.4337601.

M. Nowak, M. Wicks, Z. Zhang, and Z. Wu. Co-designed radar-communication using linear frequency modulation waveform. *IEEE Aerospace and Electronic Systems Magazine*, 31(10):28–35, 2016. https://doi.org/10.1109/MAES.2016.150236.

B. Ravenscroft, P. M. McCormick, S. D. Blunt, J. Jakabosky, and J. G. Metcalf. Tandem-hopped OFDM communications in spectral gaps of FM noise radar. In *2017 IEEE Radar Conference (RadarConf)*, pages 1262–1267, 2017. https://doi.org/10.1109/RADAR.2017.7944398.

C. D. Richmond, P. Basu, R. E. Learned, J. Vian, A. Worthen, and M. Lockard. Performance bounds on cooperative radar and communication systems operation. In *2016 IEEE Radar Conference (RadarConf)*, pages 1–6, 2016. https://doi.org/10.1109/RADAR.2016.7485101.

M. Rihan and L. Huang. Optimum co-design of spectrum sharing between MIMO radar and MIMO communication systems: an interference alignment approach. *IEEE Transactions on Vehicular Technology*, 67(12):11667–11680, 2018. https://doi.org/10.1109/TVT.2018.2872917.

M. Roberton and E. R. Brown. Integrated radar and communications based on chirped spread-spectrum techniques. In *IEEE MTT-S International Microwave Symposium Digest, 2003*, Volume 1, pages 611–614, 2003. https://doi.org/10.1109/MWSYM.2003.1211013.

S. Sen, G. Tang, and A. Nehorai. Multiobjective optimization of OFDM radar waveform for target detection. *IEEE Transactions on Signal Processing*, 59(2):639–652, 2011. https://doi.org/10.1109/TSP.2010.2089628.

S. K. Sengijpta. Fundamentals of statistical signal processing: estimation theory. *Technometrics*, 37:465–466 1995.

C. Shi, F. Wang, M. Sellathurai, J. Zhou, and S. Salous. Power minimization-based robust OFDM radar waveform design for radar and communication systems in coexistence. *IEEE Transactions on Signal Processing*, 66(5):1316–1330, 2018. https://doi.org/10.1109/TSP.2017.2770086.

K. Singh, S. Biswas, T. Ratnarajah, and F. A. Khan. Transceiver design and power allocation for full-duplex MIMO communication systems with spectrum sharing radar. *IEEE Transactions on Cognitive Communications and Networking*, 4(3):556–566, 2018. https://doi.org/10.1109/TCCN.2018.2830758.

Y. L. Sit, B. Nuss, and T. Zwick. On mutual interference cancellation in a MIMO OFDM multiuser radar-communication network. *IEEE Transactions on Vehicular Technology*, 67(4):3339–3348, 2018. https://doi.org/10.1109/TVT.2017.2781149.

C. Sturm and W. Wiesbeck. Waveform design and signal processing aspects for fusion of wireless communications and radar sensing. *Proceedings of the IEEE*, 99(7):1236–1259, 2011. https://doi.org/10.1109/JPROC.2011.2131110.

H. Takahara, K. Ohno, and M. Itami. A study on UWB radar assisted by inter-vehicle communication for safety applications. In *2012 IEEE International Conference on Vehicular Electronics and Safety (ICVES 2012)*, pages 99–104, 2012. https://doi.org/10.1109/ICVES.2012.6294272.

H. Takase and M. Shinriki. A dual-use system for radar and communication with complete complementary codes. In *11th International Radar Symposium*, pages 1–4, 2010.

Z. Wang and Q. He. Target detection and mutual information improvement for cooperative MIMO radar-communication systems. *Journal of Signal Processing*, 36(10):1654–1661, 2020. https://doi.org/10.16798/j.issn.1003-0530.2020.10.004. (in Chinese).

F. Wang and H. Li. Joint power allocation for radar and communication co-existence. *IEEE Signal Processing Letters*, 26(11):1608–1612, 2019. https://doi.org/10.1109/LSP.2019.2941087.

F. Wang, H. Li, and M. A. Govoni. Power allocation and co-design of multicarrier communication and radar systems for spectral coexistence. *IEEE Transactions on Signal Processing*, 67(14):3818–3831, 2019. https://doi.org/10.1109/TSP.2019.2920598.

C. Wang, J. Tong, G. Cui, X. Zhao, and W. Wang. Robust interference cancellation for vehicular communication and radar coexistence. *IEEE Communications Letters*, 24(10):2367–2370, 2020. https://doi.org/10.1109/LCOMM.2020.3006111.

V. Winkler and J. Detlefsen. Automotive 24 GHz pulse radar extended by a DQPSK communication channel. In *2007 European Radar Conference*, pages 138–141, 2007. https://doi.org/10.1109/EURAD.2007.4404956.

X. Xia, T. Zhang, and L. Kong. MIMO OFDM radar IRCI free range reconstruction with sufficient cyclic prefix. *IEEE Transactions on Aerospace and Electronic Systems*, 51(3):2276–2293, 2015. https://doi.org/10.1109/TAES.2015.140477.

S. J. Xu, Y. Chen, and P. Zhang. Integrated radar and communication based on DS-UWB. In *2006 3rd International Conference on Ultrawideband and Ultrashort Impulse Signals*, pages 142–144, 2006. https://doi.org/10.1109/UWBUS.2006.307182.

Y. Yang and R. S. Blum. MIMO radar waveform design based on mutual information and minimum mean-square error estimation. *IEEE Transactions on Aerospace and Electronic Systems*, 43(1):330–343, 2007. https://doi.org/doi: 10.1109/TAES.2007.357137.

X. Zhang, H. Li, J. Liu, and B. Himed. Joint delay and Doppler estimation for passive sensing with direct-path interference. *IEEE Transactions on Signal Processing*, 64(3):630–640, 2016. https://doi.org/10.1109/TSP.2015.2488584.

Z. Zhang, Z. Du, and W. Yu. Information theoretic waveform design for OFDM radar-communication coexistence in Gaussian mixture interference. *IET Radar, Sonar and Navigation*, 13(11):2063–2070, 2019.

A. Zhang, M. L. Rahman, X. Huang, Y. J. Guo, S. Chen, and R. W. Heath. Perceptive mobile networks: cellular networks with radio vision via joint communication and radar sensing. *IEEE Vehicular Technology Magazine*, 16(2):20–30, 2021. https://doi.org/10.1109/MVT.2020.3037430.

L. Zheng, M. Lops, Y. C. Eldar, and X. Wang. Radar and communication coexistence: an overview: a review of recent methods. *IEEE Signal Processing Magazine*, 36(5):85–99, 2019. https://doi.org/10.1109/MSP.2019.2907329.

Part III

Networking and Hardware Implementations

10

Frequency-Hopping MIMO Radar-based Data Communications

Kai Wu, Jian A. Zhang, Xiaojing Huang, and Yingjie J. Guo

University of Technology Sydney, Global Big Data Technologies Centre, and School of Electrical and Data Engineering, Faculty of Engineering and IT, Sydney, New South Wales, Australia

10.1 Introduction

Enabled by the advancement in radio frequency (RF) technologies and signal processing, the convergence of multifunctional RF systems becomes increasingly promising and is envisioned as a key feature of future sixth-generation (6G) networks [You et al., 2021]. Among numerous RF systems, including wireless communications, radio sensing, mobile computing, localization, etc., the first two have achieved significant progress recently in their integration. This is evidenced by several timely overview, survey, and tutorial papers [Zheng et al., 2019; Rahman et al., 2020; Liu et al., 2020a; Hassanien et al., 2019; Akan and Arik, 2020; Mishra et al., 2019; Luong et al., 2021].

Driven by the spectrum scarcity and cost saving, coexistence between radar and communications has been extensively investigated in the past few years, with focus on mitigating the interference between the two RF functions [Zheng et al., 2019]. Thanks to the shared commonalities in terms of signal processing algorithms, hardware and, to some extent, system architecture, joint communications and radar/radio sensing (JCAS), also referred to as dual-function radar–communications (DFRC), is emerging as an effective solution to integrating wireless communications and radio sensing [Rahman et al., 2020]. Substantially different from the coexistence of the two RF functions, JCAS/DFRC aims to use a single transmitted signal for both communications and sensing, enabling a majority of the transmitter modules and receiver hardware to be shared.

The design of JCAS/DFRC can be communication-centric or radar-centric [Liu et al., 2020a]. The former performs radar sensing using ubiquitous communication signals, e.g. IEEE 802.11p Surender and Narayanan [2011] and IEEE 802.11ad Kumari et al. [2019], whereas the latter embeds information bits into existing radar waveforms or specifically optimized dual-function waveforms [Liu et al., 2020a]. Regarding the communication-centric JCAS, it is worth mentioning a recently proposed perceptive mobile network (PMN) [Rahman et al., 2020]. Integrating sensing function into mobile networks, the PMN is envisaged to revolutionize the future fifth-generation (5G) and 6G networks by offering ubiquitous sensing capabilities for numerous smart applications, such as smart cities/factories. Compared with their communication-centric counterparts, radar-centric designs, which are also typically referred to as DFRC, can have superior sensing performance, given the sensing-dedicated waveforms.

Many recent DFRC designs lean toward using multiple-input multiple-output (MIMO) radar [Hassanien et al., 2016b, 2019; Liu et al., 2020a; Barneto et al., 2021], potentially due to the

Signal Processing for Joint Radar Communications, First Edition.
Edited by Kumar Vijay Mishra, M. R. Bhavani Shankar, Björn Ottersten, and A. Lee Swindlehurst.

following reasons. *First*, MIMO radar has gained increasing popularity in the radar community given its advantages of better spatial resolution, improved interference rejection capability, improved parameter identifiability, and enhanced flexibility in beam pattern design, compared with the conventional (phased) array radar [Chen and Vaidyanathan, 2008]. *Second*, in addition to the spatial degree-of-freedom (DoF), MIMO radar also provides extra DoFs in the waveform, when embedding communication information. The spatial DoF can also be provided by a conventional array radar, while the waveform DoF cannot. The waveform DoF provided by MIMO radar enables more information to be embedded for DFRC and hence higher communication data rate to be achieved [Hassanien et al., 2019; Liu et al., 2020a]. Besides information embedding, MIMO waveform optimization for dual functions is also a hot topic; see, e.g., Liu et al. [2020b].

In the context of DFRC, some researchers optimize the beam pattern of a MIMO radar to perform conventional modulations, e.g. phase shift keying (PSK) and amplitude shift keying, by designing the sidelobes in the MIMO radiation patterns [Hassanien et al., 2016a; Wang et al., 2019]. Other researchers optimize radar waveform to perform nontraditional modulations, such as waveform shuffling and code shift keying [Hassanien et al., 2018]. An overview of different signaling strategies/schemes based on MIMO radars can be found in [Hassanien et al. 2019]. As popularly used in the literature, the term signaling strategy/scheme is referred to as the way communication information is embedded in the radar waveform. Most previous DFRC designs embed one communication symbol within one or multiple radar pulses. The symbol rate is hence limited by radar pulse repetition frequency (PRF).

Since the first work on the frequency-hopping[1] MIMO (FH-MIMO) radar-based DFRC [Hassanien et al., 2017], such DFRC designs have attracted extensive attention recently [Eedara et al., 2018, 2019; Eedara and Amin, 2019; Eedara et al., 2020; Baxter et al., 2018; Eedara and Amin, 2020], as using FH-MIMO radars can break the abovementioned limit and substantially increase the communication symbol rate to multiples of (e.g. 15 times) PRF. Pioneering in conceiving the novel DFRC architecture based on FH-MIMO radar (FH-MIMO DFRC), the research reported in Hassanien et al. [2017], Eedara et al. [2018], Eedara et al. [2019], Eedara and Amin [2019], Eedara et al. [2020], Baxter et al. [2018], and Eedara and Amin [2020] focuses on analyzing the impact of information modulation on the radar ranging performance. In these works, little attention is paid to effective implementation of data communications. Some latest research reported in Wu et al. [2021a], Wu et al. [2020c], and Wu et al. [2020d] develops methods to address the practical issues, e.g. channel estimation and synchronization, in FH-MIMO DFRC.

In this chapter, we introduce the novel DFRC architecture. We first illustrate the essential aspects, e.g. signal model and information embedding/demodulation, based on some ideal assumptions. Then, we provide some latest designs addressing practical issues, e.g. timing offset and channel estimation, in FH-MIMO DFRC. Moreover, we highlight some challenging issues that need to be solved for FH-MIMO DFRC. Note that this chapter is focused on the particular DFRC using FH-MIMO radars. For the basic principles of DFRC, readers are referred to the chapter "Principles of Dual-Function Radar Communication Systems" by Prof. A. Hassanien and Prof. Moeness Amin. Moreover, interested readers may also refer to the chapter "Radar-Aided Communication" by Prof. Nuria Gonzalez-Prelcic on how radar sensing can be used to improve communication performance.

---

1 Frequency hopping, also known as frequency agility, is a technique conventionally used by radar to improve the survivability in the presence of jammers, as rapid frequency changes can make potential jammers hard or slow to follow and react. In modern applications, frequency hopping can also be employed to allow multiple radar platforms to operate simultaneously without interfering with each other too much. This feature is similar to frequency hopping in communications systems. Moreover, frequency hopping is preferred by some radar systems, e.g. weather radars, for frequency diversity.

10.2 System Diagram and Signal Model

Consider the FH-MIMO DFRC illustrated in Figure 10.1a. The FH-MIMO radar has colocated transmitter (Tx) and receiver (Rx). This configuration makes the angle-of-departure (AoD) of a target with respect to (w.r.t.) the Tx identical to the angle-of-arrival (AoA) w.r.t. the Rx. In addition to target illumination, the radar performs downlink communication with a user end (UE). As shown in Figure 10.1a, there is a line-of-sight (LoS) path between the radar and UE with the AoD denoted by ϕ and path gain β. A non-line-of-sight (NLoS) is also illustrated in the figure. For illustration convenience, we assume for now the path gain of any NLoS path is negligibly small compared with that of LoS.

10.2.1 Signal Model of FH-MIMO Radar

The FH-MIMO radar of interest is pulse-based, as illustrated in Figure 10.1b. In each pulse repetition interval (PRI), the radar Tx is only turned on for a short time and then turned off in the rest of the PRI. When its Tx is off, the radar has the Rx turned on for echo reception. The ratio between the on-time of radar Tx and PRI is generally referred to as duty cycle.

Remark 10.1 Using separate transmitter and receiver arrays is common in MIMO radars, particularly when an extended array aperture is expected for improved spatial resolution. Consider that the transmitter is an N-antenna uniform linear array (ULA) with the antenna spacing of $\lambda/2$. Here, λ denotes wavelength. If N orthogonal waveforms are transmitted and the receiver has M antennas with the antenna spacing of $N\lambda/2$, then a virtual array with MN antennas ($\lambda/2$ antenna spacing) can be obtained. Namely, we only use $M+N$ antennas to obtain an MN-antenna virtual array. As mentioned, such aperture extension is subject to waveform orthogonality. We refer interested readers to [Li and Stoica 2008] for more details on orthogonal MIMO radars.

As shown in Figure 10.1b, each pulse of an FH-MIMO radar is divided into H sub-pulses, also called *hops* Baxter et al. [2018]. In each sub-pulse, the radar transmits a single-tone signal.[2] The center frequency of the radar-transmitted signal, as indicated by the third digit in each table unit in Figure 10.1b, changes randomly across hops and antennas.

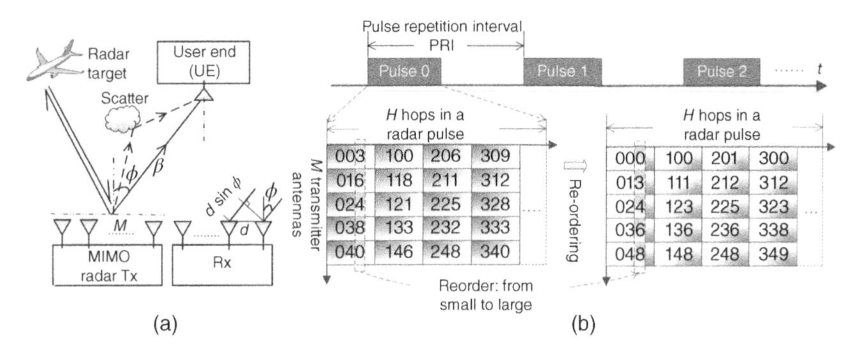

Figure 10.1 (a) Illustration on the system diagram of an FH-MIMO DFRC. (b) The signal frame structure for the downlink communication in Figure 10.1a.

2 The chirp signal can also be used; refer to [Nusenu et al. 2018]. In this chapter, we employ the relatively simple single-tone signal so as to focus on introducing the essential aspects of FH-MIMO DFRC.

By dividing the frequency band evenly into K sub-bands, the center frequency of the kth sub-band is given by $\frac{kB}{K}$ $(k = 0, 1, \dots, K-1)$. Let $M(< K)$ denote the number of antennas in the radar Tx, and then M out of K sub-bands are selected to be the hopping frequencies at a hop, one for each antenna. Denote the index of the sub-band selected at hop h and antenna m as k_{hm} that satisfies $k_{hm} \in \{0, 1, \dots, K-1\}$. To ensure the waveform orthogonality of the FH-MIMO radar, the following conditions are generally helpful:[3] [Eedara et al., 2018; Chen and Vaidyanathan, 2008]

$$k_{hm} \neq k_{hm'} \ (\forall m \neq m', \ \forall h); \quad \Delta = BT/K \in \mathbb{I}_+, \tag{10.1}$$

where T is hop duration, $\mathbb{I}_+$ denote the set of positive integers, and Δ is an auxiliary variable (introduced to simplify sequential expressions). At hop h, the mth antenna of the radar Tx transmits the following signal after frequency up-conversion and digital-to-analog conversion:

$$s_{hm}(n) = e^{\mathrm{j}2\pi(k_{hm}B/K)\times(nT/N)} = e^{\mathrm{j}2\pi k_{hm}\Delta n/N} \ (n = 0, 1, \dots, N-1), \tag{10.2}$$

where the last result is obtained by plugging (10.1) in the middle result.

Remark 10.2 Under the constraints given in (10.2), any two antennas transmit integer multiples of cycles of single-tone signals with different center frequencies at any hop. Mathematically, we have $\sum_{n=0}^{N-1} s_{hm}^*(n)s_{hm'}(n) = 0$ for $\forall h, m, m'$, where m and m' are indexes of two different antennas, and $(\cdot)^*$ takes complex conjugation. As mentioned in Remark 10.1, such orthogonality is necessary to achieve a virtual array aperture of up to MM_r numbers of antennas, where M_r denotes the number of antennas at the radar Rx. However, we notice that the orthogonality is only available at nonzero time delays. From a MIMO radar perspective, we can design the FH code to achieve low auto- and cross-correlations. For radar-enabled data communication, the orthogonality is important in the sense that it facilitates the separation of the signals transmitted by different Tx antennas of the radar, as to be elaborated on next.

10.2.2 Information Embedding at Radar

10.2.2.1 PSK Embedding in Radar Waveform

To convey information, a modulation term, denoted by F_{hm}, can be multiplied onto $s_{hm}(t)$, as given in (10.2). Conventional modulation schemes, e.g. quadrature amplitude modulation (QAM) and PSK, can be employed in FH-MIMO DFRC. However, if QAM or other amplitude-dependent modulations are used, the radar Tx can back off from its full transmission capacity. This is not desirable in long-distance target detection. To this end, PSK modulation can be most preferred for FH-MIMO DFRC. Given a J-bit ($J \geq 1$) PSK constellation $\Omega_J = \left\{0, \frac{2\pi}{2^J}, \dots, \frac{2\pi(2^J-1)}{2^J}\right\}$, we can take $F_{hm} = e^{\mathrm{j}\varpi_{hm}}$ $(\varpi_{hm} \in \Omega_J)$ for hop h and antenna m. Note that F_{hm} is constant in a hop duration. This is necessary to preserve the waveform orthogonality discussed in Remark 10.2. Under PSK embedding, the radar transmitted signal can be modified based on (10.2), specifically,

$$\tilde{s}_{hm} = e^{\mathrm{j}2\pi k_{hm}\Delta n/N} e^{\mathrm{j}\varpi_{hm}} \ (\varpi_{hm} \in \Omega_J; \ n = 0, 1, \dots, N-1). \tag{10.3}$$

10.2.2.2 Frequency-hopping Code Selection (FHCS)

Different from PSK, frequency-hopping code selection (FHCS) exploits the combinations of hopping frequencies in each radar hop to convey communication information [Baxter et al., 2018]. Since $M(< K)$ out of K frequencies are selected per hop, there are C_K^M different combinations

3 If communication is not integrated into the FH-MIMO radar system, it is possible to synthesize a set of FH codes for strong waveform orthogonality without enforcing the condition in (10.1).

of hopping frequencies. These combinations can be used as constellation symbols. Since only combination is of interest in FHCS, the specific orders of selected hopping frequencies per radar hop do not affect information modulation. For example, $\{k_{h0} = 1, k_{h1} = 2, k_{h2} = 3\}$ and $\{k_{h0} = 2, k_{h1} = 3, k_{h2} = 1\}$ represent the same information symbols in FHCS.

10.2.2.3 Frequency-hopping permutation selection (FHPS)

In addition to HFCS that selects different sets of frequencies, we may further apply antenna permutation that allocates the selected frequencies to different antennas. However, the demodulation of frequency-hopping permutation selection (FHPS) is complicated, and its performance is sensitive to the channel difference between antennas, which may be insufficient to achieve good demodulation performance. Hence, we only consider PSK and FHCS next.

10.2.3 Information Demodulation

Consider a single-antenna UE that receives the radar signal through an LoS path. Below, we first illustrate the signal model at UE and then the information (de)modulation for an ideal DFRC. By "ideal," we mean the necessary information required for information demodulation at the UE is available. In this chapter, we focus on the downlink communication, i.e. from a radar to a UE. The reverse uplink communication is generally more challenging in DFRC [Hassanien et al., 2019] and has not been well addressed yet.

Let ϕ denote the AoD from the radar Tx to the UE antenna. The communication channel can be depicted as $\beta\mathbf{a}(\phi)$, where β is the LoS path gain and $\mathbf{a}(\phi)$ is the steering vector in the direction of ϕ. For a pseudo-static channel, it is reasonable to assume that the channel parameters, i.e. β and ϕ, are assumed to be independent of time in one radar pulse. Assuming perfect timing, the baseband signal received by the communication Rx at hop h is

$$y_h(n) = \beta \sum_{m=0}^{M-1} e^{-j\pi m \sin\phi} e^{j\varpi_{hm}} e^{j2\pi k_{hm}\Delta n/N} + \xi(n), \tag{10.4}$$

where $\xi(n)$ denotes an additive white Gaussian noise (AWGN). Note that a ULA is assumed in (10.4) for the radar transmitter. We see from (10.4) that the radar-transmitted signals are superimposed in the time domain. Nevertheless, the waveform orthogonality enforced by (10.1) can be used to separate those signals in the frequency domain. Taking the N-point discrete Fourier transform (DFT) of $y_h(n)$ yields[4]

$$Y_h(l) = \sum_{n=0}^{N-1} y_h(n) e^{-j2\pi ln/N} = \beta e^{-jm\pi \sin\phi} e^{j\varpi_{hm}} \delta(l - l_{hm}), \quad l_{hm} = k_{hm}\Delta, \tag{10.5}$$

where $\delta(l)$ is the Dirac delta function. Specifically, we have $\delta(0) = 1$ and $\delta(l) = 0$ for $\forall l \neq 0$. In (10.5), δ-function is obtained due to the integer values of both k_{hm} and Δ. Since $k_{hm} \neq k_{hm'}$ ($\forall m \neq m'$), there are M peaks in $Y_h(l)$. Therefore, by identifying the peaks, the following M values can be obtained:

$$Y_{hp} = Y_h(l_{hm}) = \beta e^{-jm\pi \sin\phi} e^{j\varpi_{hm}}, \quad p = 0, 1, \ldots, M-1. \tag{10.6}$$

The pth peak is not necessarily contributed by the pth antenna, since the hopping frequencies are randomly distributed over antennas in each radar hop.

---

4 Here, we assume that the Doppler frequency caused by the relative motion between the DFRC transmit/receive platform and the communication receiver is negligible.

Table 10.1 Critical information for PSK demodulation in FH-MIMO DFRC.

(I1)	k_{hm}	Index of the sub-band used by radar Tx antenna m at each hop h
(I2)	β and ϕ	Communication channel between radar and UE
(I3)	—	Accurate timing

10.2.3.1 PSK Demodulation in FH-MIMO DFRC

In order to demodulate ϖ_{hm}, two information can be necessary according to (10.6): (I1) the index of the sub-band selected by antenna m at hop h, i.e. k_{hm}; and (I2) the communication channel, i.e. β and ϕ. With (I1) available at a UE, the l_{hm} used in (10.5) becomes known to the UE; as a result, the UE knows which peak in $|Y_h(l)|$ is contributed by which radar Tx antenna. Then, using (I2), the channel disturbance, $\beta e^{-\mathrm{j}m\pi \sin\phi}$, can be readily suppressed by the UE to estimate ϖ_{hm}. From the above elaboration, we summarize the necessary information required for PSK demodulation in Table 10.1. The item (I3) is an implicit yet crucial information applied in the above ideal demodulation. Without (I3), there would be inter-antenna and inter-hop interference, where the former is caused due to the violation of the second constraint given in (10.1) and the later is because the sampled hop can span over two actual hops. In fact, none of the information listed in Table 10.1 is trivial to obtain. We provide methods to address their acquisitions in Section 10.3.

10.2.3.2 FHCS Demodulation

Under FHCS modulation, the signal received by a UE can still be expressed as $y_h(n)$ given in (10.4), except that the value of ϖ_{hm} is of no interest in FHCS. Provided that (I3) in Table 10.1 is satisfied, the DFT of $y_h(n)$ is $Y_h(l)$ given in (10.5). To demodulate an FHCS symbol, the UE can simply estimate the set of k_{hm} from $Y_h(l)$ through identifying the indexes of the peaks in $|Y_h(l)|$.

Note that the mapping between the peak index and k_{hm} does not affect FHCS demodulation since only the combinations of hopping frequencies are used as information symbols. The benefit of using the combinations for information modulation is that (I1) and (I2), as required for PSK demodulation, are not necessary for FHCS demodulation. This substantially simplifies the UE receiving. However, this is not without disadvantage. In particular, the number of information bits can be conveyed by FHCS relies on the values of M (the number of antennas in radar Tx) and K (the number of radar sub-bands available). The maximum value of C_K^M is achieved at $M = \frac{K}{2}$ (provided this is an integer). This restricts the maximum communication data rate of FHCS.

10.3 Practical FH-MIMO DFRC

In this section, we introduce some recent results facilitating more practical FH-MIMO DFRC in LoS scenarios. Unlike the ideal DFRC introduced in Section 10.2.2, we do not assume perfect timing offset and the availability of communication channel in this section. To this end, the information demodulation at the UE requires the estimation of both timing offset and channel.

There can be a nonzero timing offset in practice, particularly when a synchronization link between radar and UE is unavailable or not accurate sufficiently. In a packet-based communication, the UE can attain a coarse timing by performing conventional methods like energy or correlation-based detection [Liu et al., 2019]. Provided a nonzero timing offset, a sampled hop spans over two actual hops, as illustrated in Figure 10.2. Let η denote the timing offset. Given the sampling interval T_s, η leads to a delay of $L_\eta (= \lfloor \frac{\eta}{T_s} \rceil)$ samples, where $\lfloor x \rceil$ rounds x to the nearest

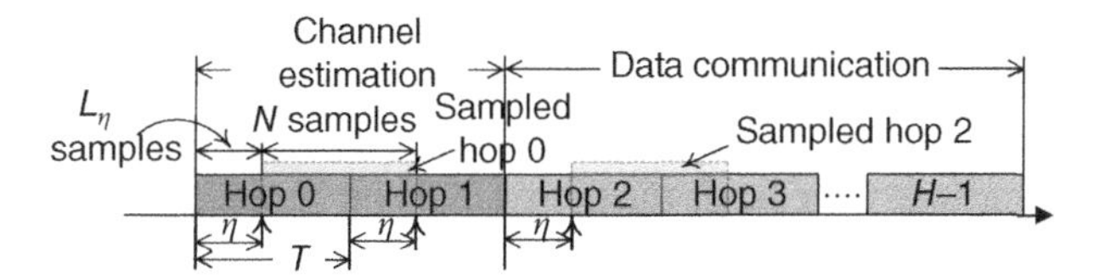

Figure 10.2 Illustration of timing offset at UE, where the UE-sampled hops are delayed versions of radar-transmitted hops Wu et al. [2021a, Fig. 1(b)]. Source: ©2021 IEEE.

integers. According to Figure 10.2, we can write the L_η-delayed hop h based on $y_h(n)$ given in (10.4), specifically,

$$y_h(n) = \xi(n) + \beta \times \begin{cases} \sum_{m=0}^{M-1} e^{j\varpi_{hm}} e^{-j\pi m \sin\phi} e^{j2\pi k_{hm}\Delta(n+L_\eta)/N}, & n \in \mathcal{N}_1, \\ \sum_{m=0}^{M-1} e^{j\varpi_{(h+1)m}} e^{-j\pi m \sin\phi} e^{j2\pi k_{hm}\Delta(n-N+L_\eta)/N}, & n \in \mathcal{N}_2, \end{cases} \tag{10.7}$$

where $\mathcal{N}_1 = \{0, 1, \dots, N - L_\eta - 1\}$ and $\mathcal{N}_2 = \{N - L_\eta, \dots, N - 1\}$.

To estimate the timing offset L_η, several waveform constraints can be helpful [Wu et al., 2021a], as summarized in Table 10.2. Provided (C1) is satisfied, taking the N-point DFT of $y_0(n)$ leads to

$$\begin{aligned} Y_0(l) = \sum_{n=0}^{N-1} y_0(n) e^{-j\frac{2\pi l n}{N}} &= N\beta \sum_{m=0}^{M-1} e^{-j\pi m \sin\phi} e^{j2\pi \frac{k_{hm}B\eta}{K}} \\ &\times e^{-j\frac{\pi(N-1)l}{N}} \delta\left(l - k_{hm}\Delta\right) e^{j\pi(N-1)k_{hm}\Delta/N} + \Xi(l), \end{aligned} \tag{10.8}$$

where $\Xi(l)$ denotes the DFT of the AWGN, and (C3) has been applied. Thanks to (C1), the waveform orthogonality is preserved for the first sampled hop, as evidenced by the delta function in the above result. Let l_{0m} $(m = 0, 1, \dots, M - 1)$ denote the index of the M peaks in $|Y_0(l)|$. Applying (C2), the estimates of k_{0m} can be obtained as

$$\hat{k}_{0m} = l_{0m}/\Delta, \quad m = 0, 1, \dots, M - 1. \tag{10.9}$$

Extracting the complex values of the M peaks in $Y_0(l)$, we obtain

$$Y_m = Y(l_{0m}) = N\beta e^{-j\frac{2\pi m u}{M}} \omega_\eta^{k_{0m}} + \Xi(k_{0m}\Delta), \tag{10.10}$$

where the intermediate variables u and ω_η are defined as

$$u \triangleq M \sin\phi/2, \quad \omega_\eta \triangleq e^{j2\pi\frac{B\eta}{K}}. \tag{10.11}$$

The way u is defined is to enable accurate estimation of ϕ. The estimations of ω_η and u will be illustrated in the subsequent subsections. Before that, some discussions are provided about the constraint (C2) given in Table 10.1.

Discussion on (C2): Based on (C2), the hopping frequencies used by the radar in hop 0 are estimated by the UE, as illustrated in (10.9). This is a desirable feature in FH-MIMO DFRC since the radar does not have to alert UEs often about the changes in hopping frequencies. (From the radar side, this change can frequently happen for better safety.) In the spirit of DFRC, we can always have

Table 10.2 Waveform constraints for LoS FHX-MIMO DFRC.

(C1)	$k_{0m} = k_{1m}$ $(\forall m)$	The same hopping frequency sequence is used by the first two hops
(C2)	$k_{hm} < k_{hm'}$ $(\forall h, m, m')$	Index of the sub-band used by radar Tx antenna m is always smaller than that by antenna $m'(> m)$ for any hop h
(C3)	$\varpi_{0m} = 0$ $(\forall m)$	No information is modulated in the first two hops, so as to use them for parameter estimations

(C2) enforced at the radar, so that the UE can readily estimate the hopping frequencies used by the radar, as done in (10.9). Imposing (C2) at the radar is equivalent to reordering the waveform per hop, as illustrated in Figure 10.1. We see that the third digits in any column are in a deterministic order after reordering. Notably, this reordering does not change the range ambiguity function of the FH-MIMO radar, as proved in Wu et al. [2020a]. However, it can change the cross-correlation levels between the individual waveforms. This variation in cross-correlation will result in the so-called range sidelobe modulation; please refer to Eedara et al. [2020] for more details on this issue.

10.3.1 Two Estimators for Timing Offset

Depending on the hopping frequency sequence, two estimators, referred to as the coherent accumulation estimation (CAE) and the Chinese remainder estimator (CRE), are developed in [Wu et al. 2021a]. Here, we summarize their steps in Tables 10.3 and 10.4. While the steps are straightforward, the development behind them is somewhat nontrivial. Interested readers may refer to [Wu et al. 2021a] for more details. Some insights into the suitability of the two estimators for different scenarios are provided next.

10.3.1.1 Comparison Between CAE and CRE

The two estimators have different asymptotic performance as a function of the signal-to-noise ratio (SNR). Let γ denote the SNR, namely, the ratio between the received signal power and the

Table 10.3 Coherent accumulation estimator (CAE) for estimating ω_η.

Input: Y_m given in (10.10)

Step (1) Take the ratio of adjacent Y_m, leading to $\check{Y}_m = Y_m / Y_{m+1} = e^{j\frac{2\pi u}{M}} \omega_\eta^{k_{0m} - k_{0(m+1)}}$, $m = 0, 1, \ldots, M-2$
Step (2) Taking the ratio of adjacent $\check{Y}_m$ yields $\bar{Y}_m = \check{Y}_m / \check{Y}_{m+1} = \omega_\eta^{\kappa_m} = e^{j\kappa_m \angle\omega_\eta}$, $m = 0, \ldots, M-3$, where $\kappa_m \triangleq k_{0m} - 2k_{0(m+1)} + k_{0(m+2)}$
Step (3) Based on (10.9) and the above definition of κ_m, identify the set of indexes of radar Tx antennas satisfying $\kappa_{\tilde{m}} \neq 0$, i.e. $\mathcal{M} = \{\forall \tilde{m}\}$, s.t. $\kappa_{\tilde{m}} \neq 0$
Step (4) Further identify the indexes in $\mathcal{M}$ satisfying that $|\kappa_{\bar{m}}| = 1$. The set of such indexes are collected by $\bar{\mathcal{M}} = \{\forall \bar{m}\} \subseteq \mathcal{M}$, s.t. $|\kappa_{\bar{m}}| = 1 \; \forall \bar{m} \in \mathcal{M}$
Step (5) Return the estimate: $\measuredangle\bar{\omega}_\eta = \angle\left(\frac{1}{\bar{M}} \sum_{\bar{m} \in \bar{\mathcal{M}}} \left(\Re\{\bar{Y}_{\bar{m}}\} + j\kappa_{\bar{m}} \Im\{\bar{Y}_{\bar{m}}\}\right)\right)$,[a] where $\angle()$ takes angle, while $\measuredangle()$ denotes the estimate of an angle; and $\Re\{\}$ and $\Im\{\}$ take real and imaginary parts, respectively

a) $\bar{M}$ denotes the cardinality of $\bar{\mathcal{M}}$.

Table 10.4 Chinese remainder theorem estimator (CRE) for estimating ω_η.

Step (1) Run Steps (1)–(3) in Table 10.3.
Step (2) Based on $\mathcal{M}$ obtained in Step (3) of Table 10.3, construct the following set $\check{\mathcal{M}} = \{\forall \check{m}\}$ satisfying that $\mathcal{G}\{|\kappa_{\check{m}}|\} = 1$, $|\kappa_{\check{m}}| \neq 1$ ($\forall \check{m} \in \mathcal{M}$) and $\check{M} \geq 2$,[a] where $\mathcal{G}$ yields the greatest common divisor
Step (3) Calculate the possible estimates corresponding to each $\check{m}$ in $\check{\mathcal{M}}$, yielding $\measuredangle\omega_\eta(d_{\check{m}}) = \frac{\angle\bar{Y}_{\check{m}} + 2d_{\check{m}}\pi}{\kappa_{\check{m}}}$, where $d_{\check{m}} (= 0, \pm 1, \ldots)$ is the ambiguity degree
Step (4) Return the estimate: $\measuredangle\check{\omega}_\eta = \frac{1}{\check{M}} \sum_{\check{m} \in \check{\mathcal{M}}} \measuredangle\omega_\eta(d^*_{\check{m}})$, where the ambiguity degree $d^*_{\check{m}}$ $\forall \check{m}$ is identified such that the estimates $\measuredangle\omega_\eta(d^*_{\check{m}})$ $\forall \check{m} \in \check{\mathcal{M}}$ are approximately identical

a) $\check{M}$ denotes the cardinality of $\check{\mathcal{M}}$.

communication Rx noise power. Based on (10.7), we have $\gamma = |\beta|^2/\sigma_n^2$, where $\sigma_n^2 = \mathbb{E}\{|\xi(n)|^2\}$ is the noise variance of $\xi(n)$. As derived in [Wu et al. 2021a, Proposition 1], the mean squared error lower bounds (MSELBs)[5] of CAE and CRE are given in (10.12) and (10.13), respectively,

$$\bar{\sigma}_\eta^2 = 3/(\bar{M}L\gamma), \tag{10.12}$$

$$\breve{\sigma}_\eta^2 = \frac{1}{\breve{M}^2}\sum_{\breve{m}\in\breve{\mathcal{M}}} 3/(\kappa_{\breve{m}}^2 L\gamma). \tag{10.13}$$

We see from (10.12) and (10.13) that the accuracy of both estimators are dependent on hopping frequencies. Specifically, $\bar{\sigma}_\eta^2$ decreases when the number of ones in $\bar{\mathcal{M}}$ (i.e. $\bar{M}$) increases. Thus, the MSELB for CAE has a lower limit, i.e. $\underline{\bar{\sigma}_\eta^2} = \frac{3}{(M-2)N\gamma} \leq \bar{\sigma}_\eta^2$, where $(M-2)$ is the maximum value that $\bar{M}$ can take. In contrast, the accuracy of CRE depends on the number and the values of the co-prime elements in $\breve{\mathcal{M}}$, as obtained in Step (2) of Table 10.4. It is easy to see that $\breve{\mathcal{M}} = \{2,3\}$ with $\breve{M} = 2$ is the smallest set with the minimum co-prime numbers. Substituting $\breve{\mathcal{M}} = \{2,3\}$ into (10.13), we obtain the upper limit of $\breve{\sigma}_\eta^2$, as given by $\overline{\breve{\sigma}_\eta^2} = \frac{2}{L\gamma} \times \frac{\frac{1}{4}+\frac{1}{9}}{4} = \frac{3}{L\gamma} \times \frac{13}{144} \geq \breve{\sigma}_\eta^2$. Note that if $\underline{\bar{\sigma}_\eta^2} > \overline{\breve{\sigma}_\eta^2}$ then $\bar{\sigma}_\eta^2 > \breve{\sigma}_\eta^2$ is assured. Moreover, $\underline{\bar{\sigma}_\eta^2} > \overline{\breve{\sigma}_\eta^2}$ leads to $\frac{1}{M-2} > \frac{13}{144}$, and further $M \leq 13$. The above illustration is formally summarized in the following theorem.

Theorem 10.1 *For a uniform linear array with the antenna spacing of half a wavelength, provided the number of antennas at the radar Tx satisfies $M \leq 13$, CRE always has a better asymptotic performance than CAE, i.e. $\bar{\sigma}_\eta^2 > \breve{\sigma}_\eta^2$.*

The above comparison is established for the asymptotic performance of the two estimators in high SNR regions, where the ambiguity degree $d_{\breve{m}}^*$ required for CRE can be reliably identified; see Step (4) of Table 10.4. In low SNR regions, however, the correct identification of $d_{\breve{m}}^*$ cannot be ensured, which degrades the estimation accuracy of CRE. This issue does not exist for CAE, which does not suffer from estimation ambiguity. In addition, when $\bar{M}$ is large, the coherent accumulation in Step (5) of Table 10.3 can help improve the estimation SNR of CAE. In this sense, CAE is more suited to low SNR regions, compared with CRE. As a matter of fact, there is an SNR threshold of γ, denoted by γ_{T}, satisfying: if $\gamma > \gamma_{\mathrm{T}}$, CRE is more accurate than CAE; otherwise, CAE is better. This threshold can be determined for specific systems based on simulations.

10.3.2 Estimating Communication Channel

Dividing both sides of (10.10) by $\hat{\omega}_\eta^{\hat{k}_{0m}}$ leads to

$$Z_m = Y_m / \hat{\omega}_\eta^{\hat{k}_{0m}} = N\beta e^{-j\frac{2\pi m u}{M}}, \tag{10.14}$$

where the ϕ-related variable, u, is defined in (10.11) and the noise term is dropped to focus on algorithm illustration. Note in (10.14) that $\omega_\eta^{\hat{k}_{0m}}$ is assumed to be fully suppressed so that we can focus on formulating the estimation method for u.

We see from (10.14) that u can be regarded as a discrete frequency, and hence u estimation is turned into the frequency estimation of a sinusoidal signal Z_m $(m = 0, 1, \dots, M-1)$. The frequency estimator that we developed recently in Wu et al. [2020b] can be applied here for u estimation.

5 Note that CRLB is the lower limit of the MSELB provided here; while, according to Reggiannini [1997], CRLB is not applicable to estimators, like the proposed CAE and CRE, which estimate a random phase with a finite support $[-\pi, \pi)$.

In overall, the estimator first searches for the DFT peak of the sinusoidal signal Z_m to obtain a coarse estimation of u and then interpolates the DFT coefficients around the peak to refine the estimation. Taking the DFT of Z_m w.r.t. m leads to $z_{m'} = \sum_{m=0}^{M-1} Z_m e^{-j\frac{2\pi m m'}{M}}$. By identifying the peak of $|z_{m'}|$, a coarse estimation of u can be obtained as $\frac{\tilde{m}}{M}$, where $\tilde{m}$ is the index of the peak. The true value of u can be written as $\frac{\tilde{m}+\zeta}{M}$ with $\zeta(\in[-0.5, 0.5])$ being a fractional frequency residual. By estimating ζ, the coarse u estimate can be refined.

We can estimate ζ recursively from the interpolated DFT coefficients around $\tilde{m}$. Initially, we set $\zeta = 0$ and calculate the interpolated DFT at the discrete frequency $\tilde{m} \pm \epsilon + \zeta$, where $\epsilon = \min\{M^{-\frac{1}{3}}, 0.32\}$ Wu et al. [2020b, Eq. (23)] is an auxiliary variable of the u estimation algorithm. It has been proved in Serbes [2019] that the above value of ϵ leads to an efficient estimator in the sense of approaching CRLB. The interpolated DFT coefficients, denoted by $z_{\pm}$, can be calculated as $z_{\pm} = \sum_{m=0}^{M-1} Z_m e^{-j\frac{2\pi m(\tilde{m}+\zeta\pm\epsilon)}{M}}$. An update of ζ can be obtained using $z_{\pm}$, i.e.

$$\zeta = \frac{\epsilon \cos^2(\pi\epsilon)}{1 - \pi\epsilon \cot(\pi\epsilon)} \times \Re\{\rho\} + \zeta, \tag{10.15}$$

where ζ on the right-hand side (RHS) is the old value and $\rho = \frac{z_+ - z_-}{z_+ + z_-}$. Use the new value of ζ to update $z_{\pm}$, which is then used, as above, for δ update. By updating ζ three times in overall, the algorithm can generally converge [Serbes, 2019]. The final estimate of u is obtained as $\hat{u} = (\tilde{m} + \zeta)/M$. Substituting $\hat{u}$ into (10.11) and (10.14), the ϕ and β estimations are

$$\hat{\phi} = \arcsin \hat{u}\lambda/(Md), \quad \hat{\beta} = \frac{1}{MN} \sum_{m=0}^{M-1} Z_m e^{j\frac{2\pi m\hat{u}}{M}}. \tag{10.16}$$

10.3.3 Numerical Evaluation

In this subsection, numerical examples are provided to demonstrate a narrowband FH-MIMO DFRC in LoS scenarios. Unless otherwise specified, the FH-MIMO radar is configured according to Table 10.5. Based on the above parameters, the following hopping frequency sequences are simulated and compared:

$$\text{(optimal for) CAE: } \bar{\mathbf{k}} = [0, 1, 3, 4, 6, 7, 9, 10, 12, 13]^T,$$
$$\text{(optimal for) CRE: } \check{\mathbf{k}} = [0, 1, 2, 3, 4, 5, 6, 7, 17, 19]^T,$$
$$\text{suboptimal: } \mathbf{k}^* = [0, 1, 3, 4, 6, 7, 9, 10, 17, 19]^T.$$

These sequences are used in the first two hops of each radar pulse so that the UE can estimate the timing offset and channel, as illustrated in Sections 10.3.1 and 10.3.2. Take $\mathbf{k}^*$ for an example. We have $k_{h0} = 0$, $k_{h1} = 1$, $k_{h2} = 3$, $k_{h3} = 4$, and so on, where $h = 0$ and 1. The sequence $\bar{\mathbf{k}}$ is optimal for CAE in the sense that the MSELB of CAE, as given in (10.12), is minimized at $\forall\gamma$. Similarly, $\check{\mathbf{k}}$ is optimal in the sense that the mean squared error (MSE) of CRE, as given in (10.13), is minimized. The sequence $\mathbf{k}^*$ is suboptimal in the sense that the MSELB of CAE is minimized subject to that two elements in a hopping frequency sequence are reserved for facilitating CRE. In addition, based on Step (2) in Table 10.4 and Step (4) in Table 10.3, we can validate that only CAE can be performed when $\bar{\mathbf{k}}$ is used, only CRE can be performed when $\check{\mathbf{k}}$ is used, while either estimator can be performed when $\mathbf{k}^*$ is used. To this end, the applicability of the estimators is achieved at the cost of optimality.

Figure 10.3 plots the MSE of $\angle\omega_\eta$ estimation against the received SNR γ at the communication receiver. We see that CRE has a much better high SNR performance than CAE. To obtain the MSE of 10^{-5}, an SNR improvement of 15 dB can be achieved by CRE, compared with CAE. We also see that CAE outperforms CRE in low SNR regions. At $\gamma = 5$ dB, CAE reduces the MSE by

Table 10.5 Simulation parameters.

Par.	Value	Description	Par.	Value	Description
M	10	Number of radar Tx antennas	K	20	Number of radar sub-bands
B	100 MHz	Radar bandwidth	f_s	$2B$	Sampling frequency
T	0.8 μs	Time duration of a hop	β	$e^{j\angle\beta}$	LoS path gain, $\angle\beta \sim \mathcal{U}^{a)}_{[-\pi,\pi]}$
ϕ	20°	LoS AoD from radar Tx to UE	γ	—	SNR measured based on (10.4)

a) $\mathcal{U}_{[a,b]}$ denotes the uniform distribution between a and b.

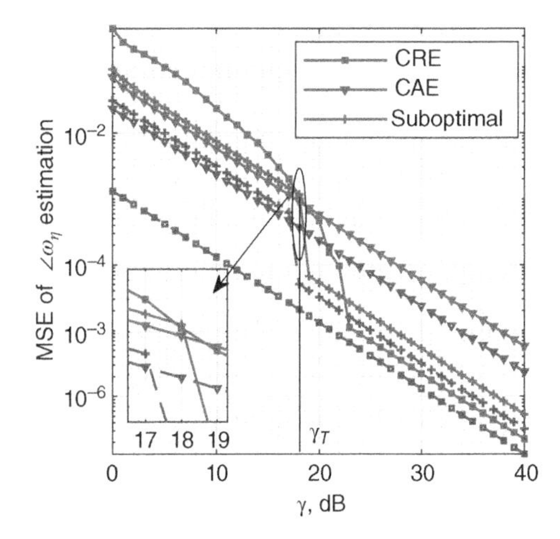

Figure 10.3 MSE of $\angle\omega_\eta$ estimation against γ, where dash curves are MSELBs corresponding to the solid curves with the same markers. Source: The figure is modified from Wu et al. [2021a, Fig. 4(a)]. ©2021 IEEE.

81.3%($= \frac{0.1-0.01867}{0.1} \times 100\%$) compared with CRE. We further see that CRE and CAE are able to asymptotically approach their MSELBs given in (10.12) and (10.13), respectively.

By comparing CRE and CAE, the SNR threshold $\gamma_T = 18$ dB can be obtained from the zoomed-in turning point. To this end, we can perform CAE and CRE below and above γ_T, respectively, meanwhile using the suboptimal $\mathbf{f}^*$. We see from Figure 10.3 that the suboptimal $\mathbf{f}^*$ provides the suboptimal estimation accuracy in the whole SNR regions. Nevertheless, we see that the suboptimal $\mathbf{f}^*$ improves the estimation accuracy obviously over CRE in low SNR regions ($\gamma \leq \gamma_T$), and substantially outperforms CAE in high SNR regions ($\gamma > \gamma_T$). It is noteworthy that the improvement achieved by the suboptimal $\mathbf{f}^*$ across the whole SNR region is based on a single hop. In contrast, neither CRE nor CAE can work for the whole SNR region.

Figure 10.4 observes the u estimation accuracy against γ, where the $\angle\omega_\eta$ estimations obtained from Figure 10.3 are used for calculating Z_m in (10.14). We see that the u estimation accuracy is closely dependent on the $\angle\omega_\eta$ estimations. In particularly, the suboptimal $\mathbf{f}^*$ enables the MSE of u estimation to outperform those achieved by CRE and CAE in low and high SNR regions, respectively. In the low SNR region of $\gamma \leq \gamma_T$, the suboptimal $\mathbf{f}^*$ produces almost the same u estimation performance as CAE; moreover, the deviation of the squared error achieved by the suboptimal $\mathbf{f}^*$ is much lower than that achieved by CRE. On one hand, this validates again the superiority of the proposed suboptimal $\mathbf{f}^*$ over $\bar{\mathbf{f}}$ and $\breve{\mathbf{f}}$ in terms of achieving a high-accuracy ω_η across the whole SNR region; and on the other hand, this validates the robustness of the proposed u estimation to the estimation error of CAE and CRE, even in low SNR regions.

Figure 10.5 plots the MSE of $\hat{\beta} \times \beta^*$ against γ, where the u estimations obtained in Figure 10.4 are used in (10.16). We see that, owing to the high accuracy of $\angle\omega_\eta$ and u, the estimate $\hat{\beta}$ is very close to the true value. From the zoomed-in subfigure, we see that at $\gamma = 20$ dB the MSE of $\hat{\beta} \times \beta^*$

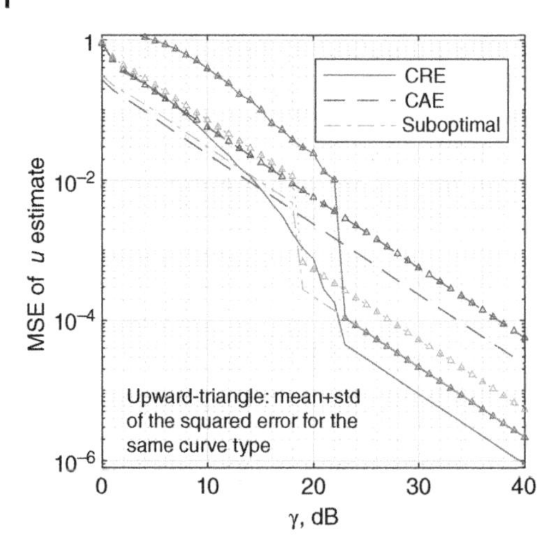

Figure 10.4 MSE of u estimation against γ, where the $\angle\omega_\eta$ estimations from Figure 10.3 are used to calculate Z_m in (10.14) (as required for u estimation). Source: The figure is from Wu et al. [2021a, Fig. 5]. ©2021 IEEE.

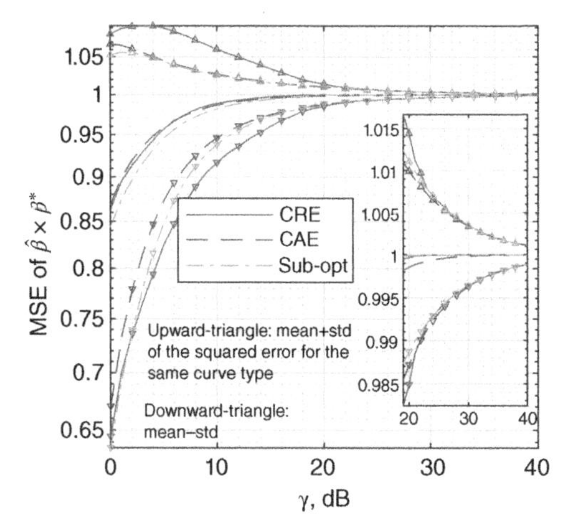

Figure 10.5 MSE of $\hat{\beta} \times \beta^*$ against γ, where the u estimations in Figure 10.4 are used to estimate $\hat{\beta}$ as given in (10.16) and β^* is the conjugate of β. Source: The figure is from Wu et al. [2021a, Fig. 6]. ©2021 IEEE.

approaches 1 with an error of less than 0.001. We also see that, due to the different performance index from that used in Figures 10.3 and 10.4, the MSEs achieved by the three estimators are not as differentiable as they are in the previous two figures. Nevertheless, from the deviation of their squared errors, we see that CAE and the suboptimal $\mathbf{f}^*$ can produce better β estimations compared with CRE (particularly in low SNR regions), as consistent with Figures 10.3 and 10.4.

Figure 10.6 compares the SERs achieved based on the ideal timing offset and channel parameters and the estimated ones, where the estimations obtained at $\gamma = 15$ dB from Figures 10.3–10.5 are used. In the x-axis of Figure 10.6, J denotes the number of information bits per symbol (hop). The values of J for the three modulation schemes are analyzed below:

- BPSK is short binary PSK. Given the radar parameters in Table 10.5, each hop conveys $M = 10$ BPSK symbols, and hence the overall number of bits conveyed per radar hop is $J = 10$ bits.
- FHCS uses the combinations of hopping frequencies as information symbols, as illustrated in Section 10.2.2. Given $M = 10$ antenna and $K = 20$ sub-bands, there are $C_{20}^{10} = 184756$ combinations of hopping frequencies. Thus, we have $J = \lfloor \log_2^{184756} \rfloor = 17$ bits for FHCS.
- PFHCS simultaneously uses FHCS and BPSK per radar hop, leading to $J = 17 + 10 = 27$ bits.

From Figure 10.6, we see that due to the larger number of symbol bits of the new constellation PFHCS, its SER performance is improved substantially, compared with that achieved

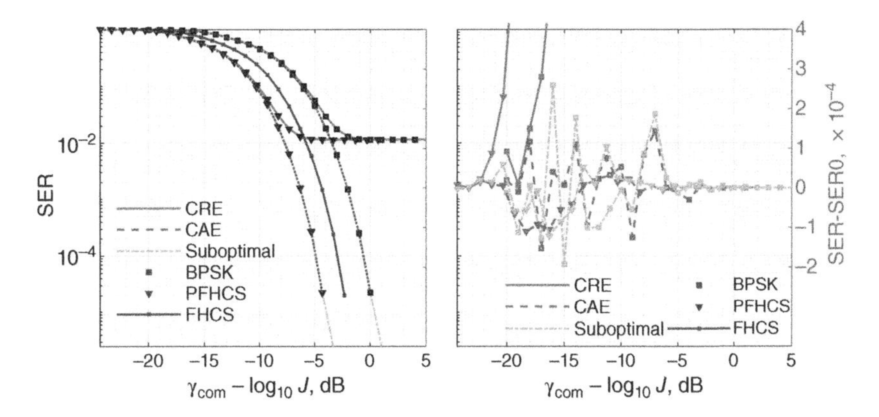

Figure 10.6 SER performance, where SER0 denotes the SER evaluated based on the ideal channel, γ_{com} denotes the communication SNR in the region of [−10, 15] dB. In the legend, PFHCS represents the case where FHCS and BPSK are jointly used. Note that the x-axis has a similar physical meaning to E_b/N_0.

by BPSK and FHCS. When ($\gamma_{com} - 10\log_{10} J$) takes about −5 dB, PFHCS reduces the SER by 99.41 %($= \frac{0.045\ 25 - 0.000\ 265}{0.045\ 25}$) and 95.48 %($= \frac{0.005\ 868 - 0.000\ 265}{0.005\ 868}$), compared with BPSK and FHCS, respectively. We also see the large channel estimation error incurred by CRE makes the SERs of BPSK and PFHCS converge to 10^{-2}, whereas the small estimation error of the suboptimal estimator and CAE produces the continuously decreasing SERs against ($\gamma_{com} - 10\log_{10} J$). From the right y-axis, we further see that CRE incurs a much larger SER difference compared with the suboptimal estimator and CAE, as consistent with Figures 10.3–10.5. It is noteworthy that the improvement on the communication performance demonstrated in Figure 10.6 is achieved by both the new FH-MIMO radar waveform and the proposed estimation methods for timing offset and channel.

10.4 Discussion

As a newly conceived DFRC architecture based on FH-MIMO radar, there are still many challenging issues to be addressed effectively. In this section, we discuss the major challenges and suggest potential solutions for future research.

10.4.1 New Signaling Strategies Enabling Higher Data Rates

As shown in Figure 10.6, the overall data rate achieved by FH-MIMO DFRC is 27 Mbps, under the condition that BPSK and FHCS is combined, the radar has 10 antennas and the bandwidth of 100 MHz, and the UE has a single antenna. To provide a communication link with higher speed using FH-MIMO DFRC, new signaling strategies would be required.

We remark that the information conveyed in the hopping frequencies has not been fully explored yet. In light of FHCS, the permutations of hopping frequencies can be employed to convey information bits [Wu et al., 2020c]. Using permutations is expected to increase the data rate dramatically, since the number of permutations can be huge. The number of permutations that permutes $M(=10)$ out of $K(=20)$ sub-bands can be larger than 670 billion, which, in theory, is able to convey up to 39 bits per radar hop. This increases the number of bits per hop by $12(= 39 - 27)$ bits compared with using PFHCS; refer to Figure 10.6. This improvement of data rate, however, requires a proper demodulation method to be developed at the UE. One challenge is to deal with the reduced Euclidean distances between the permutations. Another challenge is to develop

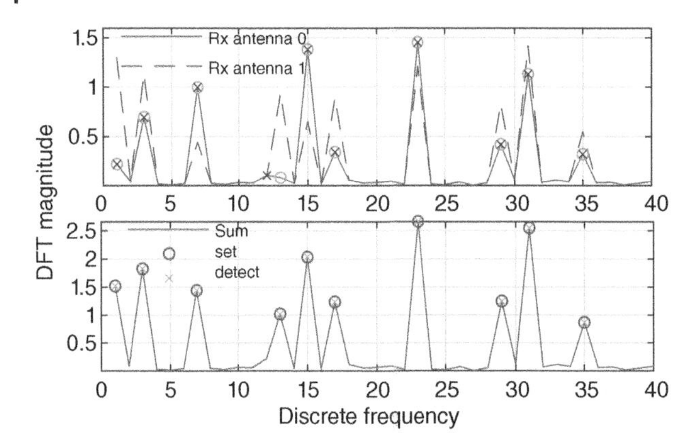

Figure 10.7 Illustration of the severe signal attenuation caused by multipath fading, where the Rician channel is considered with the Rician factor of 5 dB and the LoS path takes the unit power. Two Rx antennas with the spacing of half a wavelength are considered at the UE, each with an independent AWGN of −10 dB noise power. Source: The figure is from Wu et al. [2021b, Fig. 2]. ©2021 IEEE.

effective receiving schemes for UE to reliably demodulate the permutation-based information symbol, particularly when the issues discussed sequentially are present.

10.4.2 Multipath Channel

Thus far, only the LoS-dominated AWGN channel is considered for FH-MIMO DFRC. In practice, multipath fading can happen, when an airborne UE, e.g. unmanned aerial vehicle (UAV), flies at a low altitude. Under the multipath fading, the signal can be severely attenuated, when arriving at the UE-Rx.

Figure 10.7 illustrates the signal attenuation caused by multipath fading by plotting $|Y_0(l)|$, namely, the magnitude of the DFT of the UE-received signal within a radar hop, where $Y_0(l)$ is given in (10.8) and an accurate timing is assumed here. Based on the waveform constraints given in Table 10.2, the mth DFT peak corresponds to the signal from Tx antenna m ($m = 0, 1, \dots, M-1$) at the radar. We see from the figure that severe attenuation of the signal transmitted by the third Tx antenna is observed on the first Rx antenna at the UE. We also see from the upper subfigure that the hopping frequency of the fourth antenna is incorrectly detected. This incorrect detection will lead to incorrect demodulation of not only hopping frequency-based modulations, e.g. FHCS, but also phase-based modulations, e.g. PSK. For the latter modulation, the signal used for demodulation is extracted based on the identified indexes of the DFT peaks (that are used for FHCS demodulation).

A viable solution to combating the multipath fading in FH-MIMO DFRC is equipping the UE with multiple Rx antennas and exploit the antenna diversity. This is demonstrated in Figure 10.7. The upper subfigure also plots the magnitude of the DFT of the signal received by a different Rx antenna. Clearly, the attenuation caused by multipath fading to the signal transmitted by the fourth antenna is greatly relieved. By combining the signals of the two antennas incoherently, the sum signal can be used to detect the hopping frequency more reliably; see the right subfigure in Figure 10.7. It is worth mentioning that improving the detecting performance of the hopping frequencies also substantially benefits the SER performance of demodulating phase-based modulations. With the benefits of using multi-antenna receivers at the UE recognized, we should also bear in mind that effective signal reception also relies on the accurate estimation of timing offset and channel parameters. These practical issues need to be properly addressed for multi-antenna configurations.

10.4.3 Security Issues

MIMO radar-based DFRC can have a security issue, as underlined in some recent studies [Su et al., 2020]. The security issue can affect the FH-MIMO DFRC as well, since the underlying FH-MIMO radar has an omnidirectional radiation pattern given omnidirectional antenna elements [Wu et al., 2020c]. To this end, provided that the signal framework used between the radar and the legitimate UE is known by an eavesdropper, the radar-transmitted signal can be demodulated by an eavesdropper using the method provided in Section 10.3. Here, we consider the situation where the conventional secret key–based encryption either fails or is not performed. It is not impossible for the secret key–based encryption to be "cracked" with the fast developments and advances in computing power devices [Hamamreh et al., 2019]. Moreover, secret key–based encryption may not always be preferable, particularly for applications that are delay-sensitive, power-limited, and processing-restricted [Hamamreh et al., 2019].

As an effective supplement to upper-layer security techniques, physical layer security has attracted increasing attention recently in 5G and beyond communications [Wu et al., 2019], and can be a promising solution to achieving a secure FH-MIMO DFRC. Specifically, the secrecy enhancement method developed in Eltayeb et al. [2017] may be tailored for FH-MIMO DFRC. In short, the method randomly selects a subset of (at least two) antennas in a transmitter array and half of the antennas in the subset multiply their beamforming weights by -1. Note that the larger the number of antennas in the subset, the greater secrecy enhancement can be achieved, which, however, leads to smaller communication data rates. Through the above processing, AWGNs are injected to the whole spatial region except for the direction of a targeted receiver. The method may be tailored for FH-MIMO DFRC, however, at the cost of reducing the data rate. This is a typical trade-off between the secrecy rate and data rate of a communications system. Last, we point out that secrecy analysis and optimization is also an interesting work that has not been carried out for FH-MIMO DFRC yet. Existing work on MIMO communication-radar, e.g. Deligiannis et al. [2018], may provide good references for this line of researches.

10.4.4 FH-MIMO DFRC Experiment

Although some real-life experiments/demonstrations on DFRC have been performed; see, e.g. Wong et al. [2021], measurements and demonstrations of *FH-MIMO DFRC* have not been reported yet in the open literature. This calls for interested researchers in both radar and communication fields to carry out such experiments. On the other hand, it is noticeable from the literature, e.g. Sit et al. [2016] and Frankford et al. [2010], that the software-defined radio (SDR) platforms are often used for validating MIMO radar waveforms (not FH-MIMO though). As a cost- and time-efficient method, it can be convenient to use SDR platforms to validate FH-MIMO DFRC as well. In particular, there are two popular SDR platforms. The open-source GNURadio, e.g. ETTUS X310, can support 4.05 GHz carrier frequency and 100 MHz bandwidth Sit et al., [2016]. A more powerful SDR platform is developed by the Ohio State University (OSU), which supports up to 500 MHz instantaneous bandwidth and a center frequency tunable in 2 to 18 GHz Frankford et al., [2010]. Either X310 or OSU-SDR only supports two transmitter and two receiver chains per platform. Thus, we may need to jointly use several SDRs to allow more transceivers in practice.

10.4.5 Conclusions

In this chapter, FH-MIMO DFRC was introduced comprehensively. The signal model and information embedding/demodulation methods were first illustrated. Then, the working mechanism of

FH-MIMO DFRC was systematically elaborated on in an LoS scenario yet with numerous practical (imperfect) factors taken into account. Remaining issues in this relatively young DFRC system were highlighted, which, hopefully, can inspire future studies.

References

O. B. Akan and M. Arik. Internet of radars: sensing versus sending with joint radar-communications. *IEEE Communications Magazine*, 58(9):13–19, 2020. https://doi.org/10.1109/MCOM.001.1900550.

C. B. Barneto, S. D. Liyanaarachchi, M. Heino, T. Riihonen, and M. Valkama. Full duplex radio/radar technology: the enabler for advanced joint communication and sensing. *IEEE Wireless Communications*, 28(1):82–88, 2021. https://doi.org/10.1109/MWC.001.2000220.

W. Baxter, E. Aboutanios, and A. Hassanien. Dual-function MIMO radar-communications via frequency-hopping code selection. In *2018 52nd Asilomar Conference on Signals, Systems, and Computers*, pages 1126–1130, October 2018. https://doi.org/10.1109/ACSSC.2018.8645212.

C. Chen and P. P. Vaidyanathan. MIMO radar ambiguity properties and optimization using frequency-hopping waveforms. *IEEE Transactions on Signal Processing*, 56(12):5926–5936, 2008. ISSN 1941-0476. https://doi.org/10.1109/TSP.2008.929658.

A. Deligiannis, A. Daniyan, S. Lambotharan, and J. A. Chambers. Secrecy rate optimizations for MIMO communication radar. *IEEE Transactions on Aerospace and Electronic Systems*, 54(5):2481–2492, 2018. https://doi.org/10.1109/TAES.2018.2820370.

I. P. Eedara and M. G. Amin. Dual function FH MIMO radar system with DPSK signal embedding. In *2019 27th European Signal Processing Conference (EUSIPCO)*, pages 1–5, September 2019. https://doi.org/10.23919/EUSIPCO.2019.8902743.

I. P. Eedara and M. G. Amin. Performance comparison of dual-function systems embedding phase-modulated signals in FH radar (conference presentation). In *Signal Processing, Sensor/Information Fusion, and Target Recognition XXIX*, Volume 11423, page 114230U. International Society for Optics and Photonics, 2020.

I. P. Eedara, A. Hassanien, M. G. Amin, and B. D. Rigling. Ambiguity function analysis for dual-function radar communications using PSK signaling. In *52nd Asilomar Conference on Signals, Systems, and Computers*, pages 900–904, October 2018. https://doi.org/10.1109/ACSSC.2018.8645328.

I. P. Eedara, M. G. Amin, and A. Hassanien. Analysis of communication symbol embedding in FH MIMO radar platforms. In *2019 IEEE Radar Conference (RadarConf)*, pages 1–6, April 2019. https://doi.org/10.1109/RADAR.2019.8835532.

I. P. Eedara, M. G. Amin, and A. Hassanien. Controlling clutter modulation in frequency hopping MIMO dual-function radar communication systems. In *IEEE International Radar Conference (RADAR)*, pages 466–471, 2020.

M. E. Eltayeb, J. Choi, T. Y. Al-Naffouri, and R. W. Heath. Enhancing secrecy with multiantenna transmission in millimeter wave vehicular communication systems. *IEEE Transactions on Vehicular Technology*, 66(9):8139–8151, 2017.

M. T. Frankford, N. Majurec, and J. T. Johnson. Software-defined radar for MIMO and adaptive waveform applications. In *2010 IEEE Radar Conference*, pages 724–728, 2010. https://doi.org/10.1109/RADAR.2010.5494525.

J. M. Hamamreh, H. M. Furqan, and H. Arslan. Classifications and applications of physical layer security techniques for confidentiality: a comprehensive survey. *IEEE Communication Surveys and Tutorials*, 21(2):1773–1828, 2019. https://doi.org/10.1109/COMST.2018.2878035.

A. Hassanien, M. G. Amin, Y. D. Zhang, and F. Ahmad. Dual-function radar-communications: information embedding using sidelobe control and waveform diversity. *IEEE Transactions on Signal Processing*, 64(8):2168–2181, 2016a. ISSN 1941-0476. https://doi.org/10.1109/TSP.2015.2505667.

A. Hassanien, M. G. Amin, Y. D. Zhang, and F. Ahmad. Signaling strategies for dual-function radar communications: an overview. *IEEE Aerospace and Electronic Systems Magazine*, 31(10):36–45, 2016b. ISSN 1557-959X. https://doi.org/10.1109/MAES.2016.150225.

A. Hassanien, B. Himed, and B. D. Rigling. A dual-function MIMO radar-communications system using frequency-hopping waveforms. In *2017 IEEE Radar Conference (RadarConf)*, pages 1721–1725. IEEE, 2017.

A. Hassanien, E. Aboutanios, M. G. Amin, and G. A. Fabrizio. A dual-function MIMO radar-communication system via waveform permutation. *Digital Signal Processing*, 83:118–128, 2018.

A. Hassanien, M. G. Amin, E. Aboutanios, and B. Himed. Dual-function radar communication systems: a solution to the spectrum congestion problem. *IEEE Signal Processing Magazine*, 36(5):115–126, 2019. ISSN 1558-0792. https://doi.org/10.1109/MSP.2019.2900571.

P. Kumari, S. A. Vorobyov, and R. W. Heath. Adaptive virtual waveform design for millimeter-wave joint communication-radar. *IEEE Transactions on Signal Processing*, 1, 2019:715–730. ISSN 1941-0476. https://doi.org/10.1109/TSP.2019.2956689.

J. Li and P. Stoica. *MIMO radar signal processing*. John Wiley & Sons, 2008.

F. Liu, A. Garcia-Rodriguez, C. Masouros, and G. Geraci. Interfering channel estimation in radar-cellular coexistence: how much information do we need? *IEEE Transactions on Wireless Communications*, 18(9):4238–4253, 2019.

F. Liu, C. Masouros, A. P. Petropulu, H. Griffiths, L. Hanzo et al. Joint radar and communication design: applications, state-of-the-art, and the road ahead. *IEEE Transactions on Communications*, 68(6):3834–3862, 2020a.

J. Liu, K. V. Mishra, and M. Saquib. Co-designing statistical MIMO radar and in-band full-duplex multi-user MIMO communications. *arXiv preprint arXiv:2006.14774*, 2020b.

N. C. Luong, X. Lu, D. T. Hoang, D. Niyato, and D. I. Kim. Radio resource management in joint radar and communication: a comprehensive survey. *IEEE Communication Surveys and Tutorials*, 2021, 23:780–814.

K. V. Mishra, M. R. B. Shankar, V. Koivunen, B. Ottersten, and S. A. Vorobyov. Toward millimeter-wave joint radar communications: a signal processing perspective. *IEEE Signal Processing Magazine*, 36(5):100–114, 2019.

S. Y. Nusenu, W.-Q. Wang, and H. Chen. Dual-function MIMO radar-communications employing frequency-hopping chirp waveforms. *Progress In Electromagnetics Research*, 64:135–146, 2018.

Md. L. Rahman, J. A. Zhang, K. Wu, X. Huang, Y. J. Guo, S. Chen, and J. Yuan. Enabling joint communication and radio sensing in mobile networks–a survey. *arXiv preprint arXiv:2006.07559*, 2020.

R. Reggiannini. A fundamental lower bound to the performance of phase estimators over Rician-fading channels. *IEEE Transactions on Communications*, 45(7):775–778, 1997. ISSN 1558-0857. https://doi.org/10.1109/26.602582.

A. Serbes. Fast and efficient sinusoidal frequency estimation by using the DFT coefficients. *IEEE Transactions on Communications*, 67(3):2333–2342, 2019. ISSN 0090-6778. https://doi.org/10.1109/TCOMM.2018.2886355.

Y. L. Sit, B. Nuss, S. Basak, M. Orzol, W. Wiesbeck, and T. Zwick. Real-time 2D+ velocity localization measurement of a simultaneous-transmit OFDM MIMO radar using software defined radios. In *2016 European Radar Conference (EuRAD)*, pages 21–24, 2016.

N. Su, F. Liu, and C. Masouros. Secure radar-communication systems with malicious targets: integrating radar, communications and jamming functionalities. *IEEE Transactions on Wireless Communications*, 83–95, 2020. https://doi.org/10.1109/TWC.2020.3023164.

S. C. Surender and R. M. Narayanan. UWB noise-OFDM netted radar: physical layer design and analysis. *IEEE Transactions on Aerospace and Electronic Systems*, 47(2):1380–1400, 2011.

X. Wang, A. Hassanien, and M. G. Amin. Dual-function MIMO radar communications system design via sparse array optimization. *IEEE Transactions on Aerospace and Electronic Systems*, 55(3):1213–1226, 2019. ISSN 2371-9877. https://doi.org/10.1109/TAES.2018.2866038.

D. M. Wong, B. K. Chalise, J. Metcalf, and M. Amin. Information decoding and SDR implementation of DFRC systems without training signals. In *ICASSP 2021 - 2021 IEEE International Conference on Acoustics, Speech and Signal Processing (ICASSP)*, pages 8218–8222, 2021. https://doi.org/10.1109/ICASSP39728.2021.9413379.

K. Wu, W. Ni, T. Su, R. P. Liu, and Y. J. Guo. Efficient Angle-of-Arrival estimation of lens antenna arrays for wireless information and power transfer. *IEEE Journal on Selected Areas in Communications*, 37(1):116–130, 2019. ISSN 0733-8716. https://doi.org/10.1109/JSAC.2018.2872363.

K. Wu, Y. J. Guo, X. Huang, and R. W. Heath. Accurate channel estimation for frequency-hopping dual-function radar communications. In *2020 IEEE International Conference on Communications Workshops (ICC Workshops)*, pages 1–6, 2020a.

K. Wu, W. Ni, J. A. Zhang, R. P. Liu, and Y. J. Guo. Refinement of optimal interpolation factor for DFT interpolated frequency estimator. *IEEE Communications Letters*, 782–786, 2020b. ISSN 2373-7891. https://doi.org/10.1109/LCOMM.2019.2963871.

K. Wu, J. A. Zhang, X. Huang, and Y. J. Guo. Integrating secure and high-speed communications into frequency hopping MIMO radar. *arXiv preprint arXiv:2009.13750*, 2020c.

K. Wu, J. A. Zhang, X. Huang, and Y. J. Guo. Frequency-hopping mimo radar-based communications: an overview. *arXiv preprint arXiv:2010.09257 (accepted for publication)*, 2020d.

K. Wu, J. A. Zhang, X. Huang, Y. J. Guo, and R. W. Heath. Waveform design and accurate channel estimation for frequency-hopping mimo radar-based communications. *IEEE Transactions on Communications*, 69(2):1244–1258, 2021a. https://doi.org/10.1109/TCOMM.2020.3034357.

K. Wu, J. A. Zhang, X. Huang, Y. J. Guo, and J. Yuan. Reliable frequency-hopping mimo radar-based communications with multi-antenna receiver. *IEEE Transactions on Communications*, 69(8):5502–5513, 2021b. https://doi.org/10.1109/TCOMM.2021.3079270.

X. You et al. Towards 6G wireless communication networks: vision, enabling technologies, and new paradigm shifts. *Science China Information Sciences*, 64(1):1–74, 2021.

L. Zheng, M. Lops, Y. C. Eldar, and X. Wang. Radar and communication coexistence: an overview: a review of recent methods. *IEEE Signal Processing Magazine*, 36(5):85–99, 2019.

11

Optimized Resource Allocation for Joint Radar-Communications

Ammar Ahmed[1] *and Yimin D. Zhang*[2]

[1] *Advanced Safety & User Experience, Aptiv, Agoura Hills, CA, USA*
[2] *Department of Electrical and Computer Engineering, Temple University, Philadelphia, PA, USA*

11.1 Introduction

Spectrum sharing between radar and communication systems has recently gained significant attention due to the ever-increasing demand of spectral resources [Luong et al., 2020; Zheng et al., 2019; Mishra et al., 2019; Labib et al., 2017; Hassanien et al., 2016b; Cohen et al., 2018]. In this context, some radar-communication spectrum sharing studies consider spectrum efficiency enhancement by embedding communication information in conventional radar waveforms, such as continuous phase modulation [Roberton and Brown, 2003; Dokhanchi et al., 2019], differential quadrature phase shift keying (QPSK) modulation [Sahin et al., 2017], and frequency modulation [Saddik et al., 2007] schemes. Suitable radar waveform designs enabling communication operations have also attracted significant research attention [Chiriyath et al., 2019; Liu et al., 2018; Liu et al., 2020; Amuru et al., 2013].

Although different terminologies have been employed in the literature for a variety of concurrent radar and communication operations, these systems can be broadly classified into three different types. The first type of such systems employs passive radar that exploits the sources of opportunity that are transmitted from communication or broadcast systems. Although passive radar does not have control on the waveform due to the non-cooperative nature of the illuminating sources, it enjoys the inherent secrecy attributed to the receive-only nature of the radar and the richness of multiple sources [Griffiths and Baker, 2017; Zhang et al., 2017]. The second type of spectrum sharing system employs coexisting radar and communication operations where both radar and communication subsystems cooperate with each other to mitigate their mutual interference [Saruthirathanaworakun et al., 2012; Bliss, 2014; Paul et al., 2017; Mahal et al., 2017; Bică and Koivunen, 2019a; Wang and Li, 2019]. On the other hand, spectrum sharing objectives can be significantly simplified in the case where both radar and communication functions are performed at the same physical platform. Joint radar-communication (JRC) systems are a common example of such third type of spectrum sharing systems that exploit the same hardware and waveform resources to satisfy the objectives of both radar and communication subsystems [Euzière et al., 2015; Hassanien et al., 2016a; Liu et al., 2018; Kumari et al., 2018; Ahmed et al., 2018]. Although resource allocation

Signal Processing for Joint Radar Communications, First Edition.
Edited by Kumar Vijay Mishra, M. R. Bhavani Shankar, Björn Ottersten, and A. Lee Swindlehurst.

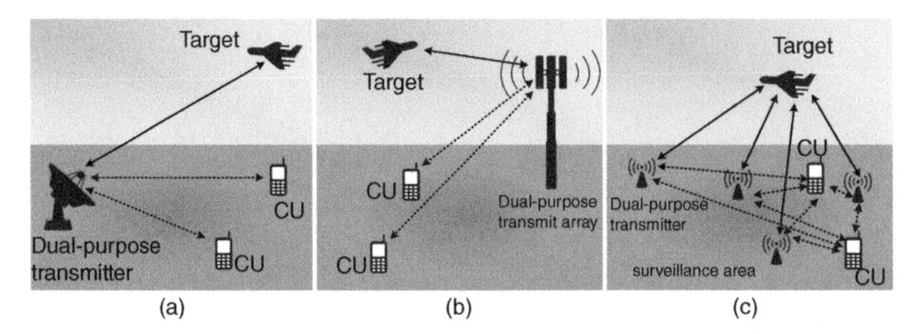

Figure 11.1 Illustration of three different types of JRC systems: (a) single transmit antenna-based JRC, (b) transmit array-based JRC, and (c) distributed JRC.

strategies for all the three types of radar-communication spectrum sharing systems have been discussed in the literature, we limit our focus to the JRC systems in this chapter.

We consider optimized resource allocation for the three most commonly used types of JRC schemes, as shown in Figure 11.1. As an example, we show one radar target and $U = 2$ communication units (CUs) located within the vicinity of the JRC system. In these situations, the first type of JRC system consists of one transmit antenna that broadcasts dual-purpose waveforms for the radar and communication operations. The resource allocation objective for such a system is to optimize the transmit energy of the dual-purpose waveform based on the propagation channels of radar and CUs [Ahmed et al., 2019a; Bică and Koivunen, 2019b; Wang et al., 2019a]. The second type of JRC system employs a transmit antenna array that directs a high-energy beam toward the surveillance region of the radar, whereas the CUs are assumed to be located in the sidelobe region. We consider a multi-antenna beamforming-based system in which the number of hardware up-conversion chains are less than the number of available antennas. Considering the availability of a large number of antennas compared to the number of available hardware chains, optimal antenna selection is considered an important resource allocation objective along with the overall power optimization of such JRC systems [Wang et al., 2019b; Ahmed et al., 2020a; Ahmed et al., 2020b]. The third type of JRC system consists of a distributed JRC network [Ahmed et al., 2019b; Shi et al., 2020; Liu et al., 2021]. For such a system, our objective is to optimize the transmit power for different JRC transmitters in order to improve the overall JRC system performance by minimizing the target localization error and maximizing the communication capacity limit [Ahmed et al., 2019b].

Notations: Lowercase and uppercase bold characters are used to, respectively, denote vectors and matrices. $(\cdot)^{\mathrm{T}}$ and $(\cdot)^{\mathrm{H}}$, respectively, represent the transpose and the Hermitian transpose operators. Moreover, $|\cdot|$, $\|\cdot\|_0$, $\|\cdot\|_1$, and $\|\cdot\|_2$ denote the absolute value, ℓ_0-, ℓ_1-, and ℓ_2-norms, respectively. The notation $\mathbf{1}_{L\times 1}$ represents the L-length column vector of all ones, whereas $\mathbf{I}_L$, $\mathbf{0}_L$, and $\mathbf{1}_L$ respectively represent the identity matrix, matrix of all zeros, and matrix of all ones, each having a dimension of $L \times L$. diag(x) denotes a diagonal matrix whose diagonal elements consist of the vector $\mathbf{x}$. Furthermore, tr$(\cdot)$ denotes the matrix trace, det$(\cdot)$ is the matrix determinant, and $\mathbb{E}[\cdot]$ represents the expectation operator. The notation $\odot$ shows the Hadamard product, $*$ is the convolution operator, and $\mathrm{j} = \sqrt{-1}$ represents the imaginary number, whereas $\min_u(\Psi(u))$ and $\max_u(\Psi(u))$, respectively, represent the minimum and maximum of function $\Psi(u)$ with respect to u. Finally, $\mathtt{h}(x)$ denotes the differential entropy of the random variable x, $\mathtt{I}(x; y)$ denotes the mutual information (MI) between random variables x and y, whereas log$(\cdot)$ denotes base-2 logarithm throughout this chapter.

11.2 Single Transmitter-Based JRC System

Consider a JRC system consisting of a single dual-purpose transmit antenna responsible for transmitting orthogonal frequency-division multiplexing (OFDM) waveforms shared for both radar and communication purposes [Ahmed et al., 2019a; Bică and Koivunen, 2019b; Wang et al., 2019a; Sturm et al., 2009]. For simplicity, we consider a single target and two communication users in the vicinity of JRC system. The communication channel gains and the radar target response are assumed to vary with frequency. In practice, frequency-selective fading in communication channels arises from dispersive channels, whereas radar cross section (RCS) typically varies with the frequency, especially in the Rayleigh and Mie regions. Depending on the desired mode of operation, the OFDM subcarriers can either be shared by radar and communications for both functions [Ahmed et al., 2019a] or be divided between radar and communications so that each function exclusively uses its separate subcarriers [Bică and Koivunen, 2019b]. In this chapter, we discuss shared subcarrier utilization, as shown in Figure 11.2, such that all subcarriers are primarily used by the radar, whereas these subcarriers are also simultaneously allocated to different CUs, enabling a secondary communication operation.

11.2.1 Signal Model

Let the transmit dual-purpose waveform consist of OFDM signal $\mathbf{x}$ having length L such that $K \leq L$ subcarriers are exploited. The transmit waveform is represented as:

$$\mathbf{x} = \mathbf{F}_{\text{IDFT}}\mathbf{s}, \tag{11.1}$$

where $\mathbf{F}_{\text{IDFT}}$ is the $L \times K$ inverse discrete Fourier transform (IDFT) matrix such that its kth column is given by $\mathbf{F}_{\text{IDFT}}^{(k)} = \frac{1}{\sqrt{L}}\left[1, \exp(j2\pi f_k/L), \ \dots, \ \exp(j2\pi(L-1)f_k/L)\right]^{\text{T}}$ with f_k denoting the frequency of kth OFDM subcarrier. Moreover, $\mathbf{s} = [s_1, \dots, s_K]^{\text{T}}$ is a $K \times 1$ vector whose elements correspond to the amplitudes and phases of the respective OFDM waveforms. Each subcarrier can exploit a digital modulation scheme for information transfer. In this chapter, we consider the QPSK. The information is carried in the phase of each subcarrier, such that the phase of s_k represents the desired phase for the kth subcarrier and $\xi_k = \mathbb{E}\left[|s_k|^2\right]$ represents the subcarrier power.

The total transmit power by the JRC transmitter can be expressed as:

$$P_{\text{total}} = \mathbb{E}\left[\mathbf{x}^{\text{H}}\mathbf{x}\right] = \mathbb{E}\left[\mathbf{s}^{\text{H}}\mathbf{F}_{\text{IDFT}}^{\text{H}}\mathbf{F}_{\text{IDFT}}\mathbf{s}\right] = \mathbb{E}\left[\mathbf{s}^{\text{H}}\mathbf{s}\right] = \sum_{k=1}^{K}\xi_k = \text{tr}\{\boldsymbol{\Xi}\}, \tag{11.2}$$

where $\mathbf{F}_{\text{IDFT}}^{\text{H}}\mathbf{F}_{\text{IDFT}} = \mathbf{I}_K$ is evident from the orthonormality of the OFDM subcarriers, and $\boldsymbol{\Xi} = \text{diag}\{\boldsymbol{\xi}\}$ with $\boldsymbol{\xi} = [\xi_1, \dots, \xi_K]^{\text{T}}$.

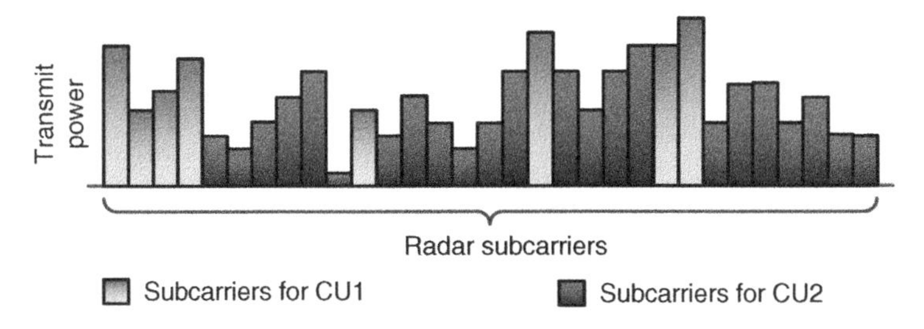

Figure 11.2 Illustrative example of subcarrier allocation and power distribution strategy for JRC system.

Our objective is to optimize the JRC system based on the MI between the transmitted and the received signals at radar and communication subsystems. For the radar subsystem, such optimization ensures that the target signal-to-noise ratio (SNR) is maximized under the same transmit energy and bandwidth constraint, resulting in the best possible target detection, classification, and estimation performance [Bell, 1993]. The communication subsystem also enjoys improved communication capacity by exploiting MI maximization [Cover and Thomas, 2006; Wen et al., 2016; Dang et al., 2021].

11.2.1.1 Radar Subsystem

The transmitted OFDM signal is reflected by the target that has a frequency-sensitive response and reaches the radar receiver. Let h_k denote the radar channel coefficient, which accounts for both the RCS and the path loss, corresponding to the kth subcarrier. The target response to all the K subcarriers can be expressed in a vector form as $\mathbf{h} = [h_1, \dots, h_K]^{\mathrm{T}}$. The corresponding radar channel impulse response is given by $\tilde{\mathbf{h}} = \mathbf{F}_{\mathrm{IDFT}}\mathbf{h}$. Therefore, the target-reflected signal received at the radar can be represented as follows:

$$\tilde{\mathbf{y}}_{\mathrm{rad}} = \tilde{\mathbf{h}} * \mathbf{x} + \tilde{\mathbf{n}}, \tag{11.3}$$

where $\tilde{\mathbf{n}}$ is the zero-mean complex additive white Gaussian noise vector. All K subcarriers of the received signal can be recovered by discrete Fourier transform (DFT) of Eq. (11.3). The received radar signal takes the following form [Ahmed et al., 2019a; Bică and Koivunen, 2016]:

$$\mathbf{y}_{\mathrm{rad}} = \mathbf{H}\mathbf{s} + \mathbf{n}, \tag{11.4}$$

where $\mathbf{H} = \mathrm{diag}(\mathbf{h})$, and $\mathbf{n} \backsim \mathcal{CN}(\mathbf{0}_K, \mathbf{\Sigma}_{\mathrm{n}})$ is the $K \times 1$ vector containing the DFT of $\tilde{\mathbf{n}}$ where $\mathbf{\Sigma}_{\mathrm{n}} = \mathrm{diag}\{\sigma^2_{\mathrm{n},1}, \dots, \sigma^2_{\mathrm{n},K}\}$ and $\sigma^2_{\mathrm{n},k}$ is the noise power of the kth subcarrier. The radar channel is assumed to follow zero-mean complex Gaussian distribution, whereas the transmit signal $\mathbf{x}$ is deterministic. This implies that the $\mathbf{y}_{\mathrm{rad}}$ can be modeled as a zero-mean Gaussian signal whose covariance matrix is expressed as:

$$\mathbb{E}\left[\mathbf{y}_{\mathrm{rad}}\mathbf{y}^{\mathrm{H}}_{\mathrm{rad}}\right] = \mathbb{E}\left[\mathbf{H}\mathbf{s}\mathbf{s}^{\mathrm{H}}\mathbf{H}^{\mathrm{H}} + \mathbf{n}\mathbf{n}^{\mathrm{H}}\right]. \tag{11.5}$$

Assuming mutual independence of radar channels for all the K subcarriers, we get $\mathbf{\Sigma}_{\mathrm{h}} = \mathbb{E}[\mathbf{H}\ \mathbf{H}^{\mathrm{H}}] = \mathrm{diag}(\sigma^2_{\mathrm{h}_1}, \dots, \sigma^2_{\mathrm{h}_K})$, where $\sigma^2_{\mathrm{h}_k}$ is the radar power gain for the kth subcarrier. In this case, Eq. (11.5) can be rewritten as:

$$\mathbb{E}\left[\mathbf{y}_{\mathrm{rad}}\mathbf{y}^{\mathrm{H}}_{\mathrm{rad}}\right] = \mathbf{\Xi}\mathbf{\Sigma}_{\mathrm{h}} + \mathbf{\Sigma}_{\mathrm{n}}. \tag{11.6}$$

Therefore, the MI between the dual-purpose transmit waveform and the target response $\mathbf{h}$ is given by [Cover and Thomas, 2006; Ahmed et al., 2019a]:

$$\begin{aligned}
\mathrm{I}\left(\mathbf{y}_{\mathrm{rad}};\mathbf{h}|\mathbf{s}\right) &= \mathrm{h}(\mathbf{y}_{\mathrm{rad}}|\mathbf{s}) - \mathrm{h}(\mathbf{y}_{\mathrm{rad}}|\mathbf{h},\mathbf{s}) = \mathrm{h}(\mathbf{y}_{\mathrm{rad}}|\mathbf{s}) - \mathrm{h}(\mathbf{n}) \\
&= \log\left[(\pi e)^K \det\left(\mathbf{\Xi}\mathbf{\Sigma}_{\mathrm{h}} + \mathbf{\Sigma}_{\mathrm{n}}\right)\right] - \log\left[(\pi e)^K \det\left(\mathbf{\Sigma}_{\mathrm{n}}\right)\right] \\
&= \log\left(\det\left(\mathbf{\Xi}\mathbf{\Sigma}_{\mathrm{h}} + \mathbf{\Sigma}_{\mathrm{n}}\right)\right) - \log\det\left(\mathbf{\Sigma}_{\mathrm{n}}\right).
\end{aligned} \tag{11.7}$$

Since $\mathbf{\Xi}\mathbf{\Sigma}_{\mathrm{h}}$ is a diagonal matrix, we can express its determinant as the product of its diagonal entries, thus yielding:

$$\mathrm{I}(\mathbf{y}_{\mathrm{rad}};\mathbf{h}|\mathbf{s}) = \log\left(\prod_{k=1}^{K}\frac{\xi_k\sigma^2_{\mathrm{h}_k} + \sigma^2_{\mathrm{n},k}}{\sigma^2_{\mathrm{n},k}}\right) = \sum_{k=1}^{K}\log\left(1 + \frac{\xi_k\sigma^2_{\mathrm{h}_k}}{\sigma^2_{\mathrm{n},k}}\right). \tag{11.8}$$

11.2.1.2 Communication Subsystem

Similar to Eq. (11.4) for the radar subsystem, the OFDM signal reaching the uth CU is expressed as:

$$\mathbf{y}_{\text{com},u} = \mathbf{G}_u\mathbf{s} + \mathbf{m}_u, \quad u = 1, \dots, U, \tag{11.9}$$

where $\mathbf{G}_u = \text{diag}(\mathbf{g}_u)$ with $\mathbf{g}_u = [g_{u,1}, \dots, g_{u,K}]^{\text{T}}$ denoting the channel coefficients of the K subcarriers associated with the uth CU. In addition, the noise term $\mathbf{m}_u \backsim \mathcal{CN}(\mathbf{0}_K, \mathbf{\Sigma}_{\text{m}_u})$ is independent of $\mathbf{G}_u$, where $\mathbf{\Sigma}_{\text{m}_u} = \text{diag}\{\sigma^2_{\text{m}_{u,1}}, \dots, \sigma^2_{\text{m}_{u,K}}\}$. Furthermore, the statistical properties of the communication channels are known to be $\mathbf{g}_u \sim \mathcal{CN}(\mathbf{0}_K, \mathbf{\Sigma}_{\text{g}_u})$ such that $\mathbf{\Sigma}_{\text{g}_u} = \text{diag}\{\sigma^2_{\text{g}_{u,1}}, \dots, \sigma^2_{\text{g}_{u,K}}\}$. Finally, the communication MI, which is analogous to the Shannon capacity, can be expressed as Cover and Thomas [2006], Ahmed et al. [2019a]:

$$\begin{aligned}\text{I}\left(\mathbf{y}_{\text{com},u}; \mathbf{g}_u|\mathbf{s}\right) &= \text{h}(\mathbf{y}_{\text{com},u}|\mathbf{s}) - \text{h}(\mathbf{y}_{\text{com},u}|\mathbf{g}_u, \mathbf{s}) \\ &= \text{h}(\mathbf{y}_{\text{com},u}|\mathbf{s}) - \text{h}(\mathbf{m}_u).\end{aligned} \tag{11.10}$$

Note that the MI maximization is analogous to communication capacity maximization. Therefore, MI in Eq. (11.10) serves as a quality metric for the communication subsystem [Cover and Thomas, 2006; Wen et al., 2016; Dang et al., 2021]. Using the statistical representation of $\mathbf{g}_u$ and $\mathbf{m}_u$, and the fact that $\Xi\mathbf{\Sigma}_{\text{g}_u}$ is diagonal, Eq. (11.10) takes the following form:

$$\begin{aligned}\text{I}(\mathbf{y}_{\text{com},u}; \mathbf{g}_u|\mathbf{s}) &= \log\left(\det\left(\Xi\mathbf{\Sigma}_{\text{g}_u} + \mathbf{\Sigma}_{\text{m}_u}\right)\right) - \log\left(\det\left(\mathbf{\Sigma}_{\text{m}_u}\right)\right) \\ &= \log\left[\prod_{k=1}^{K}\frac{\xi_k\sigma^2_{\text{g}_{u,k}} + \sigma^2_{\text{m}_{u,k}}}{\sigma^2_{\text{m}_{u,k}}}\right] = \sum_{k=1}^{K}\log\left(1 + \frac{\xi_k\sigma^2_{\text{g}_{u,k}}}{\sigma^2_{\text{m}_{u,k}}}\right).\end{aligned} \tag{11.11}$$

11.2.2 Optimal Power Distribution and Subcarrier Allocation

Now we optimize the power distribution for dual-purpose OFDM subcarriers by considering MI as the performance metric. The MI maximization for radar problems is important in order to detect the presence of a weak target such that a prior knowledge of its frequency-dependent scattering characteristics is available. The SNR improvement by employing MI maximization can significantly improve target detection performance in such cases [Bell, 1993]. Similarly, the communication subsystems can also enjoy improved performance by MI maximization as it is analogous to maximizing the Shannon capacity [Cover and Thomas, 2006]. The target and channel information can be either their instantaneous values or their statistical values [Ahmed et al., 2019a; Bică and Koivunen, 2019b].

We consider that all subcarriers are used by the radar and simultaneously shared with the CUs as shown in Figure 11.1a. The subcarrier allotment to the CUs is done in a mutually exclusive manner to enable multiple access communication. Two optimization strategies are considered. The former is a radar-centric approach in which the supreme precedence is given to the radar objective. In this case, we optimize the signal power allocated to each subcarrier depending on the radar channels, and the allotment of these subcarriers to the CUs is considered subsequently. On the other hand, the latter approach is a cooperative scheme that incorporates the information of both radar target response and communication channels. When optimizing the signal power allocated to each subcarrier in this strategy, the radar shows some flexibility on the maximum possible MI it can achieve in order to improve the communication system performance.

11.2.2.1 Radar-Centric Design

In the scenario of radar-centric design, the supreme precedence is given to the radar objective by maximizing the MI for radar given by Eq. (11.8); however, this approach does not guarantee that the communication objectives are satisfied. Since the MI in Eq. (11.8) is a concave function of ξ, the resulting convex optimization takes the following form:

$$\begin{aligned}
&\underset{\xi}{\text{maximize}} \quad \sum_{k=1}^{K} \log\left(1 + \frac{\xi_k \sigma_{\mathrm{h}_k}^2}{\sigma_{\mathrm{n},k}^2}\right), \\
&\text{subject to} \quad \mathbf{1}_{K\times 1}^{\mathrm{T}} \xi \leq P_{\mathrm{total}_{\max}}, \\
&\qquad\qquad\quad \mathbf{0}_{K\times 1} \leq \xi \leq \xi_{\max},
\end{aligned} \tag{11.12}$$

where $\xi_{\max} = [\xi_{1\max}, \dots, \xi_{K\max}]^{\mathrm{T}}$ and $\xi_{k\max}$ is the maximum allowable transmit power for the kth subcarrier. The constraints in problem (11.12) emphasize that the total transmit power of all subcarriers is bounded by the total available power, whereas the transmit power of each subcarrier is bounded by the maximum possible power for the corresponding individual subcarrier.

In the radar-centric approach, the power allocation of subcarriers is only dependent on target's radar channel conditions, whereas the communication channels play no role in determining the transmit power of individual subcarriers. The ODFM subcarriers, whose individual powers for the optimal radar operation are determined by exploiting (11.12), are further allocated to different CUs in a mutually exclusive manner without modifying their respective powers. For this purpose, we exploit a subcarrier selection approach by employing mixed-integer linear programming (MILP) method. In order to effectively utilize all the subcarriers and avoid mutual interference between the CUs, each subcarrier is used by exactly one CU. Several different criteria can be considered for such power allocation. In this chapter, we consider two criteria for subcarrier allocation that respectively maximize the sum and the worst-case communication MI of the CUs. The first criterion allocates the subcarriers to individual CUs such that the sum communication MI is maximized, as follows:

$$\begin{aligned}
&\underset{\mathbf{w}_k, \forall k}{\text{maximize}} \quad \sum_{u=1}^{U} \sum_{k=1}^{K} w_{u,k} \log\left(1 + \frac{\xi_k \sigma_{\mathrm{g}_{u,k}}^2}{\sigma_{\mathrm{m}_{u,k}}^2}\right), \\
&\text{subject to} \quad \mathbf{1}_{U\times 1}^{\mathrm{T}} \mathbf{w}_k = 1, \quad w_{u,k} \in \{0,1\}, \quad \forall u, \forall k,
\end{aligned} \tag{11.13}$$

where $w_{u,k}$ is a binary selection variable, which takes a value of 1 when the kth subcarrier is allocated to the uth CU and 0 otherwise, and $\mathbf{w}_k = [w_{1,k}, \dots, w_{U,k}]^{\mathrm{T}}$. The constraint $\mathbf{1}_{U\times 1}^{\mathrm{T}} \mathbf{w}_k = 1, \forall k$ ensures that each individual subcarrier is dedicated to only one CU. Note that the power of each subcarrier is already determined in (11.12) and the above optimization only allocates the subcarriers to the individual CUs to maximize the sum communication MI for all the U CUs.

When maximizing the sum MI in problem (11.13), it is possible that some CUs having poor channel conditions are completely ignored. In this case, most of the subcarriers are dedicated to the CUs with good channel conditions, whereas the CUs with poor channel conditions are widely ignored or are not allocated any subcarriers. This can be an undesirable situation if the CUs of critical importance are completely deprived of the available resources. In such cases, one approach is to consider individual capacity constraints in the problem (11.13) ensuring that the minimum required capacity for each CU be satisfied. Such a strategy is to exploit the worst-case capacity optimization by maximizing the worst-case MI for each CU. This ensures that each CU is served with fair resources, irrespective of their channel conditions.

As a result, the second criterion that maximizes the worst-case communication MI using the optimized radar power ξ in (11.12) can be given as the following maximin problem:

$$\begin{aligned}
&\underset{\mathbf{w}_k, \forall k}{\text{maximize}} \quad \min_{u} \left[\sum_{k=1}^{K} w_{u,k} \log\left(1 + \frac{\xi_k \sigma_{\mathrm{g}_{u,k}}^2}{\sigma_{\mathrm{m}_{u,k}}^2}\right)\right], \\
&\text{subject to} \quad \mathbf{1}_{U\times 1}^{\mathrm{T}} \mathbf{w}_k = 1, \quad w_{u,k} \in \{0,1\}, \quad \forall u, \forall k.
\end{aligned} \tag{11.14}$$

The optimization problem (11.14) can also be equivalently written as follows:

$$\begin{aligned} &\underset{\mathbf{w}_k, \forall k}{\text{maximize}} \quad \gamma_{\text{worst}}, \\ &\text{subject to} \quad \sum_{k=1}^{K} w_{u,k} \log\left(1 + \frac{\xi_k \sigma^2_{\mathrm{g}_{u,k}}}{\sigma^2_{\mathrm{m}_{u,k}}}\right) \geq \gamma_{\text{worst}}, \quad \forall u, \\ &\mathbf{1}^{\mathrm{T}}_{U\times 1}\mathbf{w}_k = 1, \quad w_{u,k} \in \{0,1\}, \quad \forall u, \forall k. \end{aligned} \tag{11.15}$$

Note that the MILP optimization problems (11.13)–(11.15) can be solved using the popular solvers like Mosek [MOSEK ApS, 2019] and Gurobi [Gurobi Optimization, 2020].

Worst-case optimization approach is important for the CUs that cannot tolerate being ignored in case they have poor channel conditions. However, it is noted that this optimization tends to enable equal communication MI for all the CUs irrespective of their channel conditions and might drain significant power in the poor communication channels, rendering the overall communication performance to be very low. Therefore, the design engineers must be cautious when selecting the performance metrics for the JRC systems.

11.2.2.2 Cooperative Design

Unlike the radar-centric design described in Section 11.2.2.1 where the power of each subcarrier is optimized solely based on the radar objectives, the cooperative design considered in this subsection allows the radar to show some flexibility on the maximum possible MI it can achieve. As a result, the communication channels can be considered in the optimization, resulting in a substantial improvement in the communication performance.

The cooperative design strategy takes two steps. First, the optimization problem (11.12) is exploited to determine the maximum MI α_{opt} that can be achieved by the radar. The radar subsystem then decides its flexibility parameter γ_{flex} whose value varies between 0 and 1, where a higher γ_{flex} favors the radar objectives. In this way, the radar function allows the dual-purpose transmitter to vary the power allocation such that the radar MI does not fall below $\gamma_{\text{flex}}\alpha_{\text{opt}}$.

The initial values of the subcarrier allocation coefficients $w_{u,k}$ can be either chosen randomly or optimized by (11.13) or (11.15). The following optimization then achieves the acceptable radar objective while maximizing the sum communication MI:

$$\begin{aligned} &\underset{\xi}{\text{maximize}} \quad \sum_{u=1}^{U}\sum_{k=1}^{K} w_{u,k} \log\left(1 + \frac{\xi_k \sigma^2_{\mathrm{g}_{u,k}}}{\sigma^2_{\mathrm{m}_{u,k}}}\right), \\ &\text{subject to} \quad \sum_{k=1}^{K} \log\left(1 + \frac{\xi_k \sigma^2_{\mathrm{h}_k}}{\sigma^2_{\mathrm{n},k}}\right) \geq \gamma_{\text{flex}}\alpha_{\text{opt}}, \\ &\mathbf{1}^{\mathrm{T}}_{K\times 1}\xi \leq P_{\text{total}_{\max}}, \\ &\mathbf{0}_{K\times 1} \leq \xi \leq \xi_{\max}. \end{aligned} \tag{11.16}$$

In the above optimization, the radar sacrifices some of the MI by a factor of γ_{flex}. The optimal value of ξ obtained from (11.16) is fed back to (11.13) or (11.15), depending upon the desired type of communication optimization strategies. The optimization problem (11.16) for power distribution and problem (11.13) or (11.15) for subcarrier allocation are repeated iteratively until convergence so that there is no significant change in the achieved subcarrier allocation and power distribution. This optimization strategy allows the JRC system to consider both radar and communication channels while distributing the power among different subcarriers.

11.2.3 Simulation Results

We consider one radar target and two CUs located within the vicinity of the JRC system. The JRC transmitter exploits a dual-purpose OFDM waveform consisting of 64 subcarriers. Our objective is

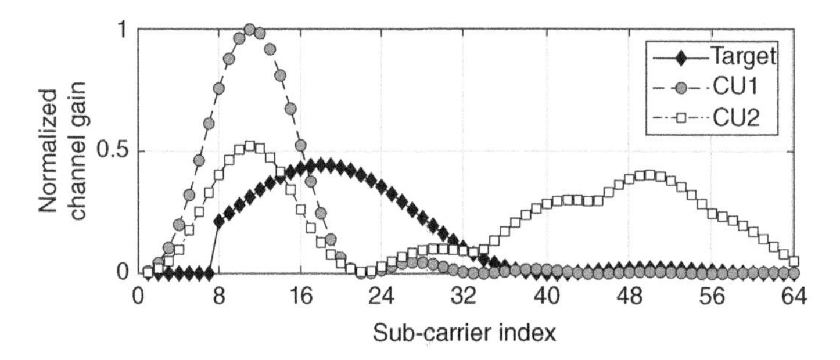

Figure 11.3 Normalized channel gains for target and CUs.

to optimize the power allocation for each subcarrier by considering the radar and communication channels. Figure 11.3 shows the normalized gains for target and communication channels, respectively expressed as $\sigma_{\mathrm{h}_k}/\sigma_{\mathrm{n}_k}$ and $\sigma_{\mathrm{g}_{u,k}}/\sigma_{\mathrm{m}_{u,k}}$, for $k = 1,2$. The maximum subcarrier power and the allowed total maximum power are normalized to 20 units and 200 units, respectively. We use the Mosek solver [MOSEK ApS, 2019] within the CVX toolbox [Grant and Boyd, 2014] to solve all the optimization problems involving integer variables.

First, we discuss the radar-centric design. Figure 11.4 shows the power allocation for radar using the radar-centric in problem (11.12) that maximizes the MI for radar. It can be observed that most of the power is allocated to the subcarriers that have a high target RCS. Figure 11.4a shows the results for maximum sum communication MI optimization using problem (11.13). It can be observed that the first CU achieves higher communication MI compared to the second CU as most of the power is allocated in the subcarriers with better radar channel conditions. On the other hand, the second CU is allocated the subcarriers where communication channel gains are very low, resulting in low communication MI for the second CU. Figure 11.4b depicts the optimized results using the worst-case optimization (11.15). We can see that more subcarriers are now allocated to the second

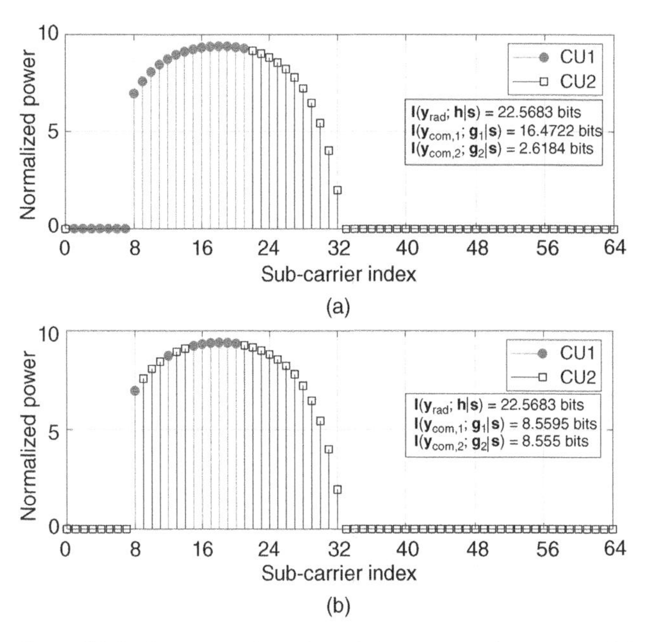

Figure 11.4 Radar-centric design of single transmitter-based JRC system. (a) Maximum sum communication MI and (b) maximum worst-case communication MI.

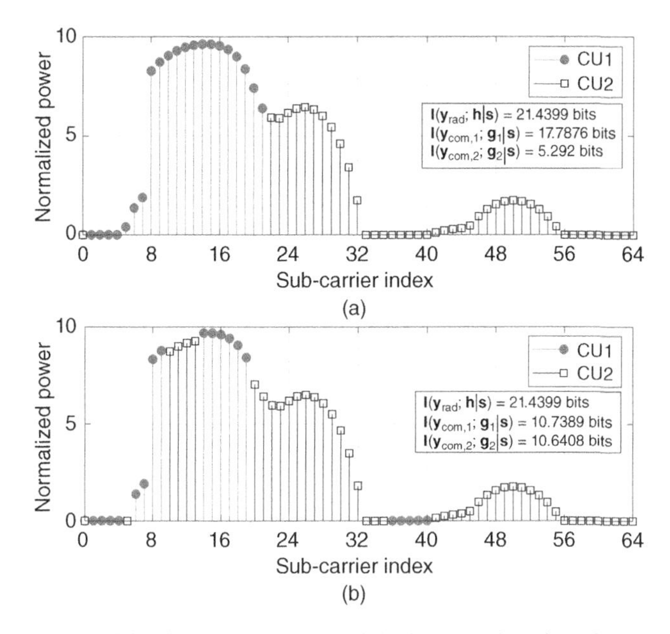

Figure 11.5 Cooperative design of single transmitter-based JRC system. (a) Maximum sum communication MI and (b) maximum worst-case communication MI.

CU as it has poor channel conditions and requires more subcarriers to achieve the same communication MI as the first CU. This results in better communication performance for the second CU; however, the overall communication MI is lower than that in Figure 11.4a.

Next, we consider the cooperative radar-communication design. For this purpose, the radar's objective is to achieve $\gamma_{\text{flex}} = 95\%$ of the maximum possible radar MI. The sacrifice of 5% MI by the radar enables the JRC system to allocate more power solely in the communication-favored subcarriers. However, since the radar's objective is the primary task of the JRC system, precedence is still provided to the radar by keeping $\gamma_{\text{flex}} = 95\%$, which is closer to unity. Figure 11.5a shows the power allocation and subcarrier distribution for the case of maximum communication MI. Note that although the radar MI is reduced compared to Figure 11.4, the overall communication MI is improved. Similarly, Figure 11.5b illustrates the worst-case optimization for the cooperative scenario. Again, we can observe that the cooperative strategy maximizes the worst-case communication MI for both communication receivers at the expense of reduced overall communication MI. Therefore, it can be observed that the cooperative design of the JRC system is suitable for both radar and communication subsystems. This design not only maintains 95% of the maximum possible radar MI but also results in more than 20% improvement in the sum communication MI for the CUs.

11.3 Transmit Array-Based JRC System

In this section, we consider the transmit array-based JRC system that transmits a high-energy beam toward the radar's desired area of surveillance. The CUs are assumed to be located in the sidelobe region of the radar. The system exploits different beamforming weight vectors over time that produce a similar mainbeam profile but project distinct sidelobe amplitude levels and phases toward the CUs. The transmitted sidelobe gain and phase, along with the waveform diversity, serve as the basis of communication information transfer in this system [Hassanien et al., 2016a,b; Ahmed et al., 2018].

11.3.1 Signal Model

Consider a JRC system that consists of an M-element transmit linear array of antennas. The array response vector in the direction θ can be given by $\mathbf{a}(\theta) = [e^{j2\pi d_1 \sin(\theta)/\lambda}, \ldots, e^{j2\pi d_M \sin(\theta)/\lambda}]^{\mathrm{T}}$, where λ is the signal wavelength and d_m represents the location of the mth antenna for $m = 1, \ldots, M$. The radar surveillance (mainlobe) region, sidelobe region, and the transition region are, respectively, denoted by Θ_{rad}, Θ_{sl}, and Θ_{trans}. The angular direction of the uth CU with respect to the JRC transmitter is given by θ_u, for $u = 1, \ldots, U$ and $\theta_u \in \Theta_{\mathrm{com}} \subset \Theta_{\mathrm{sl}}$.

The JRC system exploits N beamforming weight vectors where the vector $\mathbf{u}_n$, $1 \leq n \leq N$, satisfying the radar and communication objectives can be obtained from the following minimax optimization [Ahmed et al., 2018; Hassanien et al., 2016a; Hassanien and Amin 2022, for details on Chapter 2]:

$$\begin{aligned}
&\underset{\mathbf{u}_n}{\text{minimize}}\ \max_{\theta_r} \left(\left|G_{\mathrm{rad}} e^{j\varphi(\theta_r)} - \mathbf{u}_n^{\mathrm{H}} \mathbf{a}(\theta_r)\right|\right), && \theta_r \in \Theta_{\mathrm{rad}}, \\
&\text{subject to } \left|\mathbf{u}_n^{\mathrm{H}} \mathbf{a}(\theta_\varepsilon)\right| \leq \varepsilon_{\mathrm{sl}}, && \theta_\varepsilon \in \Theta_{\mathrm{sl}}, \\
&\qquad \mathbf{u}_n^{\mathrm{H}} \mathbf{a}(\theta_u) = e^{j\phi_{n,u}} \Delta_{n,u}, && 1 \leq u \leq U,
\end{aligned} \tag{11.17}$$

where G_{rad} represents the desired mainlobe gain, $\varphi(\cdot)$ denotes a feasible mainlobe phase profile, and $\varepsilon_{\mathrm{sl}}$ is the maximum allowable sidelobe level of the radar. Each beamforming weight vector transmits quadrature amplitude modulation (QAM) symbols with distinct gains and phases toward the uth CU, respectively, denoted as $\Delta_{n,u}$ and $e^{j\phi_{n,u}}$, which represent the transmitted communication information. In order to use Q amplitudes and P phases toward each CU, the JRC system will be required to extract $N = (PQ)^U$ unique beamforming weight vectors using (11.17) [Ahmed et al., 2018]. The number of communication directions must be smaller than the number of available sensors M because otherwise problem (11.17) becomes infeasible. Moreover, parameter $\varepsilon_{\mathrm{sl}}$ should be carefully chosen to warrant a feasible solution. In addition to ascertaining the abovementioned feasibility considerations, problem (11.17) is feasible provided that the communication directions are distinct and well separated.

The JRC system exploits K dual-purpose orthogonal waveforms which serve the objectives of both radar and communication subsystems. These orthogonal waveforms are represented by $\psi_1(t)$, $\psi_2(t), \ldots, \psi_K(t)$ and satisfy:

$$\frac{1}{T}\int_0^T \psi_{k_1}(t)\psi_{k_2}^*(t-\tau)\,\mathrm{d}t = \delta\left(k_1 - k_2\right)\delta(t-\tau), \quad k_1, k_2 = 1, \ldots, K, \tag{11.18}$$

where T represents the pulse duration of radar, t is the fast time, $\psi_{k_2}(t-\tau)$ represents the time-delayed version of $\psi_{k_2}(t)$ delayed by time τ ($< T$), and $\delta(\cdot)$ is the Kronecker delta function. The use of different orthogonal waveforms during different radar pulses enables the JRC system to employ waveform diversity for the transmission of communication information.

When the nth beamforming weight vector and the kth waveform are used, the transmit signal of the JRC can be expressed as follows:

$$\mathbf{x}(t) = \mathbf{u}_n \psi_k(t). \tag{11.19}$$

As such, $\mathbf{x}(t)$ not only satisfies the radar mainlobe objective but also projects a QAM symbol with amplitude $\Delta_{n,u}$ and phase $e^{j\phi_{n,u}}$ toward the uth CU using the waveform $\psi_k(t)$. Thus, the JRC system transmits the desired communication information at a given time by varying the transmit gain, phase, and waveform toward the CUs [Ahmed et al., 2018]. Such a variation of communication information is triggered by the selection of desired beamforming weight vectors extracted from the optimization problem (11.17).

11.3.2 Power Optimization and Sensor Allocation

Many modern systems are equipped with a large number of antennas than the available number of up-conversion hardware chains [Mehanna et al., 2013; Ahmed et al., 2020a; Wang et al., 2019b]. Each hardware chain can consist of the components like low-noise amplifiers, digital-to-analog converter, and signal processors. Therefore, the selection of most suitable antennas for the up-conversion chains and optimized distribution of the transmit power are two important resource allocation objectives in beamforming-based JRC systems. An optimized transmit array-based JRC system exploits the beamforming weight vectors such that the total number of selected antennas and the transmit power are minimized. Since these two objectives may be conflicting, the precedence is given to the antenna selection objective. The main concept of resource allocation in transmit array-based JRC systems is illustrated in Figure 11.6.

11.3.2.1 Resource Allocation for Individual Beamformers

A naive approach for resource allocation in a transmit array-based JRC system is to optimize all the beamforming weight vectors individually. Although this strategy works well in theory, it results in frequent antenna switching when different beamforming weight vectors require different antenna selection patterns. Note that the switching of beamforming weight vectors is crucial to transmitting different communication information in the form of distinct gains and phases toward the CUs. This leads to increased implementation complexity as different antennas might be required by different beamforming weight vectors. Moreover, this scheme collectively requires more antennas to be activated because the union of the selected antennas for different beamformers results in a large number of functional antennas.

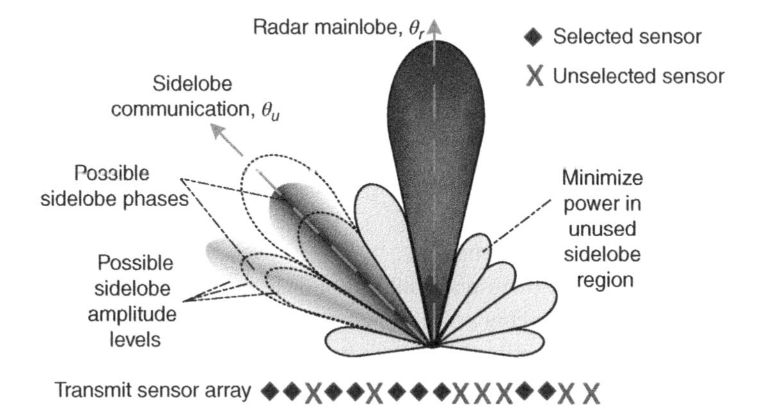

Figure 11.6 Sensor selection for JRC system ("X" denotes unselected antenna positions).

We modify the optimization problem (11.17) to obtain the following multi-objective resource optimization problem:

$$\begin{aligned}
\underset{\mathbf{u}_n}{\text{minimize}} \quad & \|\mathbf{u}_n\|_2^2 + \gamma_{\text{tune}} \|\mathbf{u}_n\|_0, \\
\text{subject to} \quad & |G_{\text{rad}} e^{j\varphi(\theta_r)} - \mathbf{u}_n^{\text{H}} \mathbf{a}(\theta_r)| \leq \gamma_{\text{tol}}, \quad \theta_r \in \Theta_{\text{rad}}, \\
& |\mathbf{u}_n^{\text{H}} \mathbf{a}(\theta_\varepsilon)| \leq \varepsilon_{\text{sl}}, \quad \theta_{\text{sl}} \in \Theta_{\text{sl}}, \\
& \mathbf{u}_n^{\text{H}} \mathbf{a}(\theta_u) = \Delta_{n,u} e^{j\phi_{n,u}}, \quad u = 1, \ldots, U.
\end{aligned} \tag{11.20}$$

The ℓ_0-norm-based non-convex objective tends to minimize the number of selected antennas, whereas the ℓ_2-norm-based objective tries to minimize the total transmit power. Here, γ_{tune} acts as the tuning parameter, which trades off between the two resource optimization objectives, and γ_{tol} is the maximum bearable tolerance for the desired radar mainlobe profile. The ℓ_0-norm in (11.20) can be relaxed by exploiting the ℓ_1-norm, albeit a weaker and indirect measure of sparsity [Candès et al., 2008]. In doing so, the antennas with a higher beamforming gain are more penalized compared to those with a lower beamforming gain due to the nature of ℓ_1-norm, thus yielding a suboptimal sparsity solution. This issue can be mitigated by introducing a re-weighting function [Candès et al., 2008] along with the ℓ_1-norm-based objective function, resulting in the following optimization [Ahmed et al., 2020a]:

$$\begin{aligned}
\underset{\mathbf{u}_n}{\text{minimize}} \quad & \|\mathbf{u}_n\|_2^2 + \gamma_{\text{tune}} \left\|\mathbf{v}_n^{(i)} \odot \mathbf{u}_n\right\|_1, \\
\text{subject to} \quad & |G_{\text{rad}} e^{j\varphi(\theta_r)} - \mathbf{u}_n^{\text{H}} \mathbf{a}(\theta_r)| \leq \gamma_{\text{tol}}, \quad \theta_r \in \Theta_{\text{rad}}, \\
& |\mathbf{u}_n^{\text{H}} \mathbf{a}(\theta_\varepsilon)| \leq \varepsilon_{\text{sl}}, \quad \theta_{\text{sl}} \in \Theta_{\text{sl}}, \\
& \mathbf{u}_n^{\text{H}} \mathbf{a}(\theta_u) = \Delta_{n,u} e^{j\phi_{n,u}}, \quad u = 1, \ldots, U,
\end{aligned} \tag{11.21}$$

where the superscript (i) denotes the ith iteration. For the first iteration, $\mathbf{v}_n^{(1)} = \mathbf{1}_{M\times 1}$. The optimization problem (11.21) is executed in an iterative manner such that the re-weighting coefficient vector $\mathbf{v}_n^{(i+1)}$ is updated using the beamforming weight vector $\mathbf{u}_n$ after the ith iteration. The weight update for the mth element of the weighting function $\mathbf{v}_n$, denoted by $v_{m,n}$, can be expressed as Candès et al. [2008] and Ahmed et al. [2020a]:

$$v_{n,m} = \begin{cases} \dfrac{1}{|u_{n,m}|}, & \text{if } |u_{n,m}| > 0, \\ 1/\zeta, & \text{if } |u_{n,m}| = 0, \end{cases} \tag{11.22}$$

where $u_{n,m}$ is the mth element of $\mathbf{u}_n$, and ζ should be set slightly smaller than the least expected nonzero magnitudes of $\mathbf{u}_n$. The solution of such iterative ℓ_1-norm-based optimization approaches to its ℓ_0-norm-based counterpart within few (e.g. 2–15) iterations [Candès et al., 2008].

11.3.2.2 Joint Resource Allocation for Multiple Beamformers

An intelligent strategy for transmit array-based JRC resource allocation is to jointly optimize the resources for all the beamforming weight vectors concurrently. As such, this approach leads to the power optimization and antenna selection profile that shares the same set of antennas for all the beamformers. In this strategy, antenna switching is not required, and only beamforming weights are changed while switching between one beamformer and the other. All the beamforming weight vectors share the same set of antennas, satisfy the mainlobe requirements, and transmit different QAM symbols to the CUs. By achieving this joint antenna selection profile, the JRC system avoids antenna switching that would have been mandatory when performing resource allocation based on individual beamformers, which generally use a different set of antennas.

The joint resource allocation can be performed by employing the mixed $\ell_{1,2}$-norm-based group-sparsity concept [Yuan and Lin, 2006; Ahmed et al., 2020a; Mehanna et al., 2013]. The mixed $\ell_{1,2}$-norm is defined as:

$$\|\mathbf{U}\|_{1,2} = \sum_{m=1}^{M}\left(\sum_{n=1}^{N}|u_{n,m}|^2\right)^{1/2}, \tag{11.23}$$

where $\mathbf{U} = [\mathbf{u}_1, \dots, \mathbf{u}_N]$ and $u_{n,m}$ represents the element on the mth row and the nth column of $\mathbf{U}$. The minimization of the above mixed $\ell_{1,2}$-norm encourages group sparsity. The ℓ_2-norm part treats all weights equally in the dimension of the beamformer index to yield the grouped weighting for each antenna. On the other hand, the ℓ_1-norm part encourages sparsity in sensor dimension for all the beamforming vectors using the accumulated ℓ_2-norm across the beamformer index.

The multi-objective optimization for joint power minimization and antenna selection takes the following form [Ahmed et al., 2020a]:

$$\begin{aligned}
\underset{\mathbf{u}_n}{\text{minimize}} \quad & \sum_{n=1}^{N}\|\mathbf{u}_n\|_2^2 + \gamma_{\text{tune}}\|\mathbf{U}\|_{1,2}, & \\
\text{subject to} \quad & |G_{\text{rad}}e^{j\varphi(\theta_r)} - \mathbf{u}_n^{\text{H}}\mathbf{a}(\theta_r)| \le \gamma_{\text{tol}}, & \theta_r \in \Theta_{\text{rad}}, \\
& |\mathbf{u}_n^{\text{H}}\mathbf{a}(\theta_\varepsilon)| \le \varepsilon_{\text{sl}}, & \theta_{\text{sl}} \in \Theta_{\text{sl}}, \\
& \mathbf{u}_n^{\text{H}}\mathbf{a}(\theta_u) = \Delta_{n,u}e^{j\phi_{n,u}}, & u = 1, \dots, U.
\end{aligned} \tag{11.24}$$

Note that the above optimization problem extracts all the N beamforming weight vectors simultaneously compared to the optimization problem (11.21), which needs to be solved independently for each beamforming weight vector. The mixed $\ell_{1,2}$-norm enforces the group sparsity [Yuan and Lin, 2006], resulting in the selection of exactly the same antennas for the beamformers; however, the weight vector for each beamformer is distinct as each vector serves different communication requirements. Similar to the optimization problem (11.21), the group sparsity in problem (11.24) can also be enhanced democratically by exploiting a similar re-weighting function for the mth antenna as follows [Ahmed et al., 2020a]:

$$v_m = \begin{cases} \left(\sum_{n=1}^{N}|u_{n,m}|^2\right)^{-1/2}, & \text{if } \sum_{n=1}^{N}|u_{n,m}|^2 > 0, \\ 1/\zeta, & \text{if } \sum_{n=1}^{N}|u_{n,m}|^2 = 0. \end{cases} \tag{11.25}$$

The resulting group-sparsity-based optimization that jointly produces all N beamforming weight vectors is given as:

$$\begin{aligned}
\underset{\mathbf{u}_n}{\text{minimize}} \quad & \sum_{n=1}^{N}\|\mathbf{u}_n\|_2^2 + \gamma_{\text{tune}}\left\|\mathbf{v}^{(i)} \odot \mathbf{U}\right\|_{1,2}, & \\
\text{subject to} \quad & |G_{\text{rad}}e^{j\varphi(\theta_r)} - \mathbf{u}_n^{\text{H}}\mathbf{a}(\theta_r)| \le \gamma_{\text{tol}}, & \theta_r \in \Theta_{\text{rad}}, \\
& |\mathbf{u}_n^{\text{H}}\mathbf{a}(\theta_\varepsilon)| \le \varepsilon_{\text{sl}}, & \theta_{\text{sl}} \in \Theta_{\text{sl}}, \\
& \mathbf{u}_n^{\text{H}}\mathbf{a}(\theta_u) = \Delta_{n,u}e^{j\phi_{n,u}}, & u = 1, \dots, U,
\end{aligned} \tag{11.26}$$

where $\mathbf{v}^{(i)}$ denotes the re-weighting function obtained in the ith iteration and

$$\|\mathbf{v} \odot \mathbf{U}\|_{1,2} = \sum_{m=1}^{M}\left(\sum_{n=1}^{N}|v_m u_{n,m}|^2\right)^{1/2} = \sum_{m=1}^{M} v_m\left(\sum_{n=1}^{N}|u_{n,m}|^2\right)^{1/2}. \tag{11.27}$$

The optimization problem (11.26) is also solved in an iterative manner until the convergence is achieved.

11.3.3 Simulation Results

We consider a transmit beamforming-based JRC system consisting of $M = 25$ element uniform linear antenna array with inter-sensor spacing of half wavelength. The radar objective is to transmit the mainbeam at a constant gain of 0 dB between -6° and 6°. The JRC system serves two CUs located in the sidelobe region at angles 30° and 50°, respectively. The maximum allowable sidelobe level ϵ_{sl} is set to −20 dB. For the convenience of illustrating the transmit beampatterns, we consider binary phase shift keying (BPSK) signaling where the JRC transmit array has an objective to transmit two different amplitude levels of −20 and −25 dB toward both CUs, resulting in four distinct beamforming weight vectors. The multi-objective coefficient γ_{tune} is set to 0.9, which gives more precedence to antenna selection task. We use CVX toolbox [Grant and Boyd, 2014] to solve all the optimization problems.

We plot the beamforming gains for the first scheme in Figure 11.7a where four individual beamforming weight vectors are synthesized separately by exploiting optimization problem (11.21). Figure 11.7b shows the respective antenna selection profile of the four beamformers resulting after 20 iterations. It can be observed that all these beamforming weight vectors use only 15 antennas; however, the antenna selection profile of each beamformer is different. Therefore, antenna switching will be required when the beamformers are switched between the two users. The combined antenna selection profile for these beamformers, by selecting all the antennas required by the four beamformers, is illustrated in Figure 11.7c. It is evident that overall 18 antennas will be required for radar and communication operations using this design strategy.

For the second scheme, we extract all the beamformers collectively by exploiting group-sparsity -based approach in optimization problem (11.26). The beamforming gains of these beamformers can be observed in Figure 11.8a. The antenna selection profile with respect to the iteration count is illustrated in Figure 11.8b, which shows that only 15 antennas are used by this approach, which is less than the total number of 18 antennas used in Figure 11.7c. Moreover, it is observed that the re-weighted iterative algorithm converges quickly as we do not see any change in antenna selection profile after the second iteration for this case.

We observe that the individual as well as the joint optimization strategies, respectively, represented by the optimization problems (11.21) and (11.26), result in almost the same power utilization; however, the joint extraction of beamformers is more efficient because it requires overall a less number of hardware chains and avoids frequent antenna switching.

11.4 Distributed JRC System

In this section, we consider a distributed JRC network where each dual-purpose JRC transmitter emits orthogonal dual-purpose waveforms. The radar subsystem works as a distributed MIMO radar that localizes and tracks the target within the vicinity of the JRC system. For simplicity, the JRC network is assumed to be synchronized and fully connected, such that the processed radar data is available to the fusion center in the form of estimated target parameters. Moreover, the JRC transmitters exploit waveform diversity for transmitting communication information to the CUs [Ahmed et al., 2019b; Shi et al., 2020]. The objective of the fusion center is to enable optimized power allocation for different JRC transmitters depending on the estimated target parameters and communication channel gains.

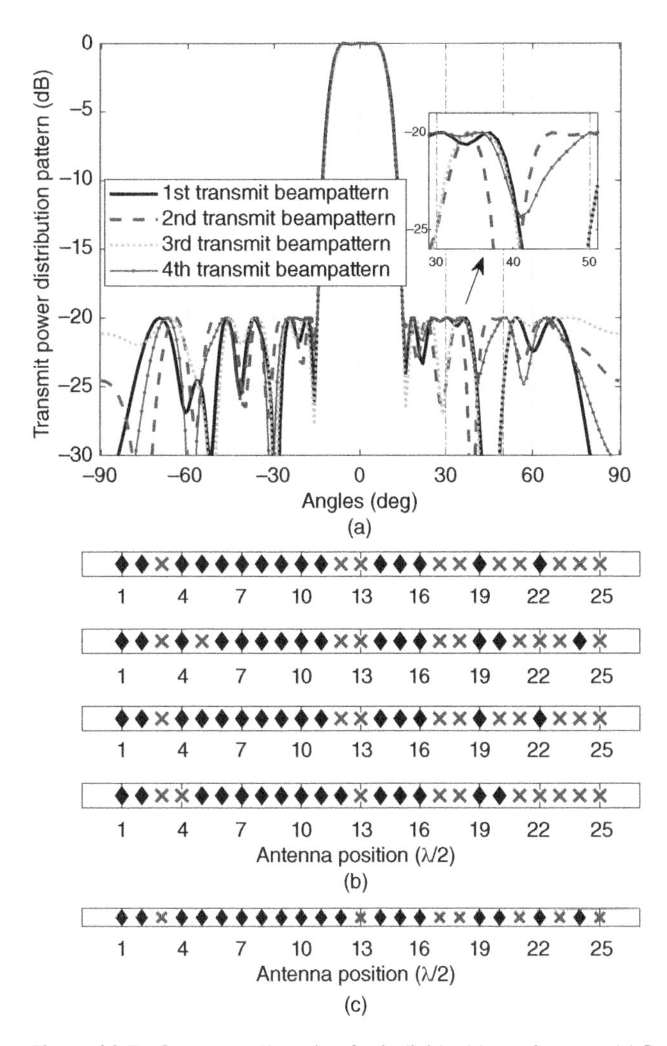

Figure 11.7 Resource allocation for individual beamformers. (a) Beamforming gains, (b) antenna selection profile for all beamformers, and (c) combined antenna selection profile.

11.4.1 Signal Model

Consider a narrowband distributed MIMO radar system consisting of M dual-purpose omnidirectional transmitters and N radar-only receivers. The dual-purpose transmitters, radar-only receivers, and CUs are assumed to be located arbitrarily in a two-dimensional (2-D) coordinate system. The primary objective of the distributed JRC is to track the location (x, y) of a point target in the 2-D coordinate system. A coarse estimate of the target's RCS and position is assumed available from the previous cycles of radar data. Each JRC transmitter can use K_m mutually exclusive unit power orthogonal waveforms, such that $\sum_{m=1}^{M} K_m = K$. During a radar pulse, each of the M JRC transmitters radiates an orthogonal waveform selected from the set of its available waveforms. The time-varying selection of different waveforms at different times enables the JRC system to transmit communication information to CUs. Moreover, we assume that the waveform transmitted

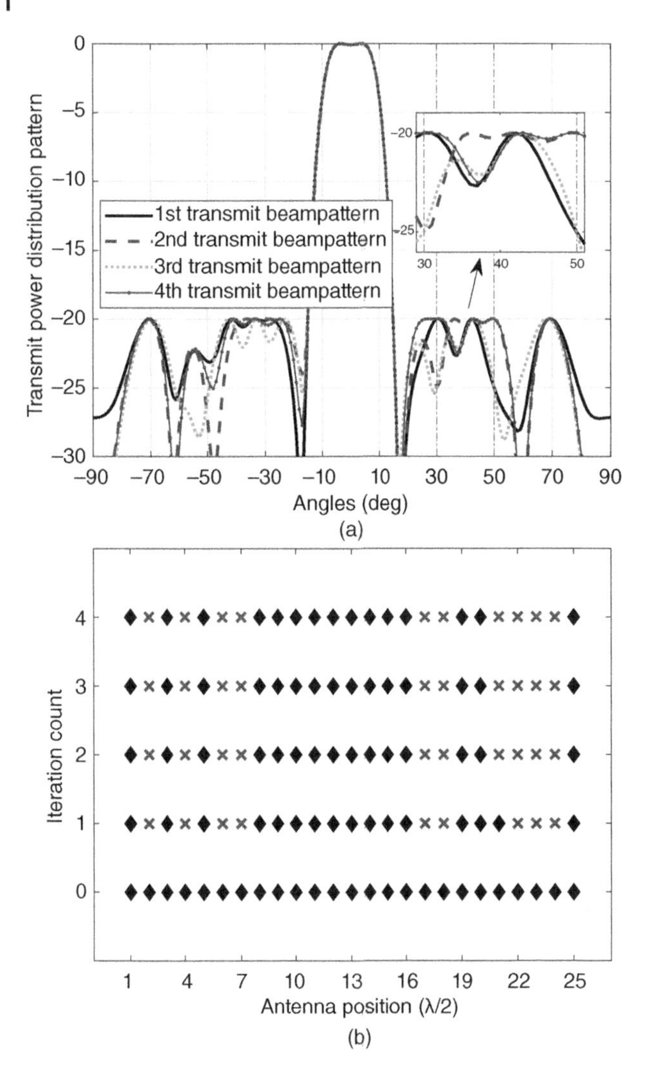

Figure 11.8 Resource allocation for multiple beamformers. (a) Beamforming gains and (b) antenna selection profile with respect to iteration count.

from each transmitter is broadcast to all CUs located in the vicinity of the JRC transmitters. Let $\mathbf{p}_{\text{tx}} = [p_{1_{\text{tx}}}, p_{2_{\text{tx}}}, \dots, p_{M_{\text{tx}}}]^{\text{T}}$ be the $M \times 1$ vector containing the transmit powers of the JRC transmitters. In addition, $\mathbf{p}_{\text{tx,max}} = [p_{1_{\text{tx,max}}}, p_{2_{\text{tx,max}}}, \dots, p_{M_{\text{tx,max}}}]^{\text{T}}$ and $\mathbf{p}_{\text{tx,min}} = [p_{1_{\text{tx,min}}}, p_{2_{\text{tx,min}}}, \dots, p_{M_{\text{tx,min}}}]^{\text{T}}$ are $M \times 1$ vectors, respectively, representing the corresponding maximum and the minimum possible transmit powers. Furthermore, we denote $P_{\text{total}_{\text{max}}} \geq \sum_{m=1}^{M} p_{m_{\text{tx,max}}}$ as the maximum allowable power to be transmitted from the JRC transmitters collectively. Figure 11.1c illustrates the distributed JRC network.

11.4.1.1 Radar Subsystem

The radar signal corresponding to the mth JRC transmitter and received at the nth radar-only receiver is given by:

$$s_{m,n}(t) = \sqrt{\alpha_{m,n} p_{m_{\text{tx}}}} h_{m,n} s_m(t - \tau_{m,n}) + w_{m,n}(t), \tag{11.28}$$

where $s_m(t)$ represents one of the K available orthogonal waveforms, $\alpha_{m,n}$ is the signal variation due to path loss effects, $h_{m,n}$ denotes the target RCS for the propagation path from the mth JRC transmitter to the nth radar-only receiver, and $w_{m,n}(t) \sim \mathcal{CN}(0, \sigma_w^2)$ represents the circularly symmetric zero-mean complex white Gaussian noise. The propagation delay due to the path from the mth transmitter to the nth receiver is denoted as $\tau_{m,n} = (D_{m\mathrm{tx}} + D_{n\mathrm{rx}})/c$, where $D_{m\mathrm{tx}}$ and $D_{n\mathrm{rx}}$ are the range to target from the mth transmitter and that from the nth receiver, respectively, and c is the propagation velocity of the electromagnetic waves. The path loss factor takes the form of $\alpha_{m,n} \propto D_{m_{\mathrm{tx}}}^{-2} D_{n_{\mathrm{rx}}}^{-2}$. Moreover, let $\mathbf{h} = [h_{1,1}, h_{1,2}, \ldots, h_{1,N}, h_{2,1}, \ldots, h_{M,N}]^{\mathrm{T}}$ be the $MN \times 1$ vector of all bistatic RCS of the targets.

The radar performance can be evaluated in terms of the Cramer-Rao bound (CRB), which represents the lower bound of the mean squared error of the target's location estimates, expressed as Godrich et al. [2010, 2011]:

$$\eta = \sigma_{x,y}(\mathbf{p}_{\mathrm{tx}}) = \frac{\mathbf{q}^{\mathrm{T}}\mathbf{p}_{\mathrm{tx}}}{\mathbf{p}_{\mathrm{tx}}^{\mathrm{T}}\mathbf{A}\mathbf{p}_{\mathrm{tx}}}, \tag{11.29}$$

where $\mathbf{q} = \mathbf{q}_a + \mathbf{q}_b$, $\mathbf{A} = \mathbf{q}_a\mathbf{q}_b^{\mathrm{T}} - \mathbf{q}_c\mathbf{q}_c^{\mathrm{T}}$, $\mathbf{q}_a = [q_{a_1}, q_{a_2}, \ldots, q_{a_M}]^{\mathrm{T}}$, $\mathbf{q}_b = [q_{b_1}, q_{b_2}, \ldots, q_{b_M}]^{\mathrm{T}}$ and $\mathbf{q}_c = [q_{c_1}, q_{c_2}, \ldots, q_{c_M}]^{\mathrm{T}}$. Here,

$$\begin{aligned}
q_{a_m} &= \varrho_m \sum_{n=1}^{N} \alpha_{m,n} \left|h_{m,n}\right|^2 \left(\frac{x_m - x}{D_{m_{\mathrm{tx}}}} + \frac{x_n - x}{D_{n_{\mathrm{rx}}}}\right)^2, \\
q_{b_m} &= \varrho_m \sum_{n=1}^{N} \alpha_{m,n} \left|h_{m,n}\right|^2 \left(\frac{y_m - x}{D_{m_{\mathrm{tx}}}} + \frac{y_n - x}{D_{n_{\mathrm{rx}}}}\right)^2, \\
q_{c_m} &= \varrho_m \sum_{n=1}^{N} \alpha_{m,n} \left|h_{m,n}\right|^2 \left(\frac{x_m - x}{D_{m_{\mathrm{tx}}}} + \frac{x_n - x}{D_{n_{\mathrm{rx}}}}\right)\left(\frac{y_m - x}{D_{m_{\mathrm{tx}}}} + \frac{y_n - x}{D_{n_{\mathrm{rx}}}}\right), \\
\varrho_m &= 8\pi^2 \mathcal{B}_m^2/(\sigma_w^2 c^2),
\end{aligned}$$

and $\mathcal{B}_m$ is the effective bandwidth of the signal transmitted from the mth transmitter.

11.4.1.2 Communication Subsystem

The signal transmitted by the mth JRC transmitter and received directly at the uth CU is expressed as:

$$s_{m,u}(t) = \sqrt{\beta_{m,u} p_{m_{\mathrm{tx}}}} g_{m,u} s_m(t - \overline{\tau}_{m,u}) + \overline{w}_u(t), \tag{11.30}$$

where $g_{m,u}$ denotes the complex channel gain, $\overline{\tau}_{m,u}$ is the propagation delay, $\beta_{m,u} \propto \overline{D}_{m,u}^{-2}$ incorporates the path loss effects, and $\overline{D}_{m,u}$ is the distance between the mth transmitter and the uth CU. We assume that $\overline{w}_u(t) \sim \mathcal{CN}(0, \overline{\sigma}_u^2)$ is circularly complex white Gaussian noise at the uth receiver whose statistics are known at the transmitter. Moreover, the signals reflected by the target and received at the CUs are assumed to be negligible due to the two-way path loss effects and limited RCS. The channel state information, denoted as $g_{m,u}, \forall m, u$, is assumed to be known at the radar fusion center.

The communication performance is evaluated in terms of the achieved Shannon's capacity. The data rate from the mth transmitter to the uth receiver is expressed in terms of the Shannon's capacity given as:

$$\mathfrak{R}_{m,u} = \log\left(1 + \frac{\left|g_{m,u}\right|^2 p_{m_{\mathrm{tx}}}}{\Gamma_{m,u}\overline{\sigma}_u^2}\right) = \log\left(1 + \frac{p_{m,tx}}{\kappa_{m,u}}\right), \tag{11.31}$$

where $\Gamma_{m,u} \geq 1$ represents the SNR gap, which translates the loss in the data rate into the loss in the SNR and is determined by the coding scheme [Goldsmith, 2005]. Without the term $\Gamma_{m,u}$, Eq. (11.31) represents the theoretical maximum achievable capacity limit. As there are no ideal coding schemes available to achieve this rate in general, $\Gamma_{m,u}$ accounts for an equivalent loss of SNR corresponding to the achieved communication capacity by a specific coding scheme for a given SNR. Furthermore, $\kappa_{m,u} = \Gamma_{m,u}\overline{\sigma}_u^2/\left|g_{m,u}\right|^2$ and the sum data rate can be calculated as $\mathfrak{R} = \sum_{m=1}^{M}\sum_{u=1}^{U}\mathfrak{R}_{m,u}$ in terms of bits/s/Hz per radar pulse.

Information embedding can be accomplished by exploiting waveform diversity. If $\tilde{K}$ is the number of mutually exclusive waveforms assigned to each transmitter, i.e. $K_m = \tilde{K}, \forall m$, the total number of bits transmitted from the distributed JRC MIMO system during one radar pulse is $M \log \tilde{K}$ provided that the dictionaries are nonoverlapping and all transmitters are active with a sufficiently high SNR.

The overall signal received at the uth CU from all transmitters can be expressed as:

$$s_u(t) = \sum_{m=1}^{M} \sqrt{\beta_{m,u} p_{m_{tx}} g_{m,u}} s_m(t - \overline{\tau}_{m,u}) + \overline{w}_u(t). \tag{11.32}$$

Performing matched filtering at the CUs to synthesize the embedded information yields:

$$\begin{aligned} y_u(k) &= \frac{1}{T}\int_0^T \psi_k^*(t + i\Delta t) s_{m,u}(t) dt \\ &= \begin{cases} \sqrt{\beta_{m,u} p_{m_{tx}} g_{m,u}} + w_{u,k}(t), & \text{if } \psi_k \text{ transmitted,} \\ w_{u,k}(t), & \text{otherwise,} \end{cases} \end{aligned} \tag{11.33}$$

where Δt is the time delay defining the time resolution of delay matched filtering, i is a nonnegative integer with $0 \leq i \leq T/\Delta t$, and $w_{u,k}(t)$ is the noise output. Since the sensor association of $\psi_k(t)$ is known, waveform diversity can be employed to transmit the communication information from different JRC transmitters to the CUs.

11.4.2 Optimal Allocation for Power and Sensors

11.4.2.1 Radar-centric Operation

The optimal power allocation for radar-centric operation can be obtained by minimizing the total transmit power for the distributed JRC system such that desirable localization accuracy, described in terms of the desired CRB η_{desired}, is achieved as follows [Godrich et al., 2011; Ahmed et al., 2019b]:

$$\begin{aligned} \underset{p_{tx}}{\text{minimize}} \quad & \mathbf{1}_{1\times M}\mathbf{p}_{tx}, \\ \text{subject to} \quad & \mathbf{p}_{tx,\min} \leq \mathbf{p}_{tx} \leq \mathbf{p}_{tx,\max}, \\ & \sigma_{x,y}(\mathbf{p}_{tx}) \leq \eta_{\text{desired}}. \end{aligned} \tag{11.34}$$

The last inequality constraint in the optimization problem (11.34) can be relaxed by using Eq. (11.29) to the following convex form [Godrich et al., 2011]:

$$\begin{aligned} \underset{p_{tx}}{\text{minimize}} \quad & \mathbf{1}_{1\times M}\mathbf{p}_{tx}, \\ \text{subject to} \quad & \mathbf{p}_{tx,\min} \leq \mathbf{p}_{tx} \leq \mathbf{p}_{tx,\max}, \\ & \mathbf{q} - \eta_{\text{desired}}\mathbf{A}\mathbf{p}_{tx} \leq \mathbf{0}. \end{aligned} \tag{11.35}$$

The solution of the convex optimization problem (11.35) yields the optimized transmit power vector $\mathbf{p}_{tx,opt}$. However, by minimizing the total transmit power of the JRC system needed to achieve the desired localization performance, this strategy results in unacceptable communication capacity.

On the other hand, best localization performance can be achieved by employing the following optimization problem [Godrich et al., 2011]:

$$\begin{aligned} \underset{\mathbf{p}_{\text{tx}}}{\text{maximize}} \quad & \mathbf{q}^{\text{T}}\mathbf{p}_{tx}, \\ \text{subject to} \quad & \mathbf{p}_{\text{tx,min}} \leq \mathbf{p}_{\text{tx}} \leq \mathbf{p}_{\text{tx,max}}, \\ & \mathbf{1}_{1\times M}\mathbf{p}_{\text{tx}} \leq P_{\text{total}_{\max}}. \end{aligned} \tag{11.36}$$

From $\mathbf{p}_{\text{tx}}$ obtained from the above optimization problem, we can calculate the best-case localization CRB η_{opt} using Eq. (11.29). Contrary to the optimization problem (11.35), the problem (11.36) uses all the power available to the JRC and is more suitable for communication operation due to the higher potential for communication MI.

11.4.2.2 Communication-centric Operation

We optimize the power allocation for communication-centric operation by exploiting the conventional water-filling approach [Goldsmith, 2005]. The optimal power allocation for the maximum allowable transmit power is achieved by solving the following equation simultaneously for all CUs:

$$\mathbf{U}\begin{bmatrix} \mathbf{p}_{\text{tx}} \\ X_u \end{bmatrix} = \begin{bmatrix} P_{\text{total}_{\max}} \\ \boldsymbol{\kappa}_u \end{bmatrix}, \tag{11.37}$$

where $\mathbf{U} = \begin{bmatrix} \mathbf{1}_{M\times 1}^{\text{T}} & 0 \\ \mathbf{I}_M & -\mathbf{1}_{M\times 1} \end{bmatrix}$, $\boldsymbol{\kappa}_u = -\left[\kappa_{1,u}, \kappa_{2,u}, \ldots, \kappa_{M,u}\right]^{\text{T}}$, and X_u represents the water-filling power level. Equation (11.37) may provide different optimal power distributions for different CUs depending on their channel side information. To avoid negative power solutions in Eq. (11.37), we rewrite Eq. (11.37) for all the CUs as the following constrained least-square optimization problem [Ahmed et al., 2019b]:

$$\begin{aligned} \underset{\mathbf{p}_{\text{tx}}, X_u, \forall u}{\text{minimize}} \quad & \sum_{r=u}^{U} \left| \mathbf{V}\begin{bmatrix} \mathbf{p}_{\text{tx}} \\ X_u \end{bmatrix} - \boldsymbol{\kappa}_u \right|_2^2, \\ \text{subject to} \quad & \mathbf{p}_{\text{tx,min}} \leq \mathbf{p}_{\text{tx}} \leq \mathbf{p}_{\text{tx,max}}, \\ & \mathbf{1}^{\text{T}}\mathbf{p}_{\text{tx}} \leq P_{\text{total}_{\max}}, \\ & X_u \geq 0, \quad u = 1,2,\ldots,U, \end{aligned} \tag{11.38}$$

where $\mathbf{V} = [\mathbf{I}_M, -\mathbf{1}_{M\times 1}]$. The optimization problem (11.38) is convex. Unlike the optimization problems (11.34) and (11.35) where the minimum power required for satisfactory radar operation is extracted, the problem (11.38) utilizes the maximum allowable power and distributes it with respect to the channel quality for all the CUs. For a given maximum power budget $P_{\text{total}_{\max}}$, the optimization problem (11.38) tends to maximize the water-filling level X_r, thus resulting in higher power allocation for better channel conditions.

11.4.2.3 Cooperative Operation

The optimal power allocation extracted from the optimization problems (11.36) and (11.38), respectively, designed for radar-only and communication-only operations, are not favorable for the optimal joint operation of both subsystems. As such, the minimum power required for radar operation extracted from problem (11.35) may not establish an acceptable communication performance. Moreover, the optimization problem (11.36) only considers radar objectives and ignores the communication conditions. On the other hand, the problem (11.38) is independent of radar channels and, therefore, cannot provide optimal radar performance.

We can add the radar performance constraint in the optimization problem (11.38) to obtain the following modified convex optimization problem [Ahmed et al., 2019b] that considers the objectives of both radar and communication subsystems:

$$\begin{aligned}
\underset{\mathbf{p}_{\text{tx}}, X_u, \forall u}{\text{minimize}} \quad & \sum_{u=1}^{U} \left\| \mathbf{V} \begin{bmatrix} \mathbf{p}_{\text{tx}} \\ X_u \end{bmatrix} - \boldsymbol{\kappa}_u \right\|_2^2, \\
\text{subject to} \quad & \mathbf{p}_{\text{tx,min}} \leq \mathbf{p}_{\text{tx}} \leq \mathbf{p}_{\text{tx,max}}, \\
& \mathbf{q} - \eta_{\text{desired}} \mathbf{A} \mathbf{p}_{\text{tx}} \leq \mathbf{0}, \\
& \mathbf{1}^{\text{T}} \mathbf{p}_{\text{tx}} \leq P_{\text{total}_{\max}}, \\
& X_u \geq 0, \quad u = 1, 2, \dots, U,
\end{aligned} \tag{11.39}$$

where η_{desired} is the desired localization error, which should have a higher value compared to η_{opt} determined from optimization problem (11.36) and Eq. (11.29). We define the flexibility ratio of the radar as $\gamma_{\text{flex}} = \eta_{\text{desired}}/\eta_{\text{opt}}$ whose value varies between 0 and 1, with higher value favoring the radar objective. The optimization problem (11.39) not only achieves the desired radar localization accuracy but also provides favorable power output for the communication operation by allocating more power for better communication channels.

11.4.3 Simulation Results

We consider a distributed JRC system consisting of $M = 5$ isotropic transmitters located in the 2-D space at coordinates (150, 1950), (200, 750), (1100, 1150), (1750, 350), and (1950, 1200) m. The system is also equipped with $N = 5$ radar-only receivers located at (150, 1050), (400, 350), (1050, 1900), (1450, 100), and (1850, 900) m. A point target is assumed to be located at (1000, 1000) m, whereas two CUs are located at (200, 250) and (1100, 350) m, respectively. Each transmitter is allowed a maximum transmit power of 60 W. The overall maximum allowable power $P_{\text{total}_{\max}}$ for the JRC system is 100 W. The vector of distributed radar coefficients is assumed to be $\mathbf{h}$ uniformly distributed from 0.9 to 1, we get $|\mathbf{h}| = [0.961, 0.914, 0.964, 0.972, 0.900, 0.912, 0.945, 0.957, 0.985, 0.951, 0.941, 0.934, 0.977, 0.977, 0.985, 0.963, 0.945, 0.913, 0.956, 0.909, 0.905, 0.971, 0.981, 0.968, 0.909]^{\text{T}}$. The parameter ϱ_m is considered to be 8.773×10^5 for all $1 \leq m \leq M$. Moreover, we consider $\boldsymbol{\kappa}_1 = -[1/0.75, 1/95, 1/0.02, 1/0.95, 1/0.55]^{\text{T}}$ and $\boldsymbol{\kappa}_2 = -[1/0.2, 1/0.95, 1/0.02, 1/0.75, 1/0.75]^{\text{T}}$, which indicates that both CUs have experienced deep fading with the third JRC transmitter. We can calculate the path loss coefficients from the distances between the transmitters and the receiver. It can be observed that the third transmitter of the JRC system is the most suitable for radar task; however, it is the least important for communication operation due to the deep fade with both CUs. The simulation scenario is shown in Figure 11.9. The CVX toolbox [Grant and Boyd, 2014] is used to solve all optimization problems.

Table 11.1 illustrates the optimized power allocation for all the three JRC strategies under consideration. The communication performance for the two CUs is illustrated in terms of their sum Shannon capacity. Using the radar-centric design in problem (11.36), we observe that most of the transmit power is allocated to the third transmitter because it provides the best target localization accuracy due to its lowest path loss effects. On the other hand, the third transmitter has poorest communication channel conditions with all the CUs, rendering it unsuitable for communication operation. This is also evident from the results of communication-centric design, which does not allocate any power to this transmitter. The results from the radar-centric and communication-centric designs show that both of these strategies work optimally only for one of the two subsystems. The radar-centric scheme provides a poor sum Shannon capacity of 12.51 bits/s/Hz, whereas the communication-centric scheme provides a high localization

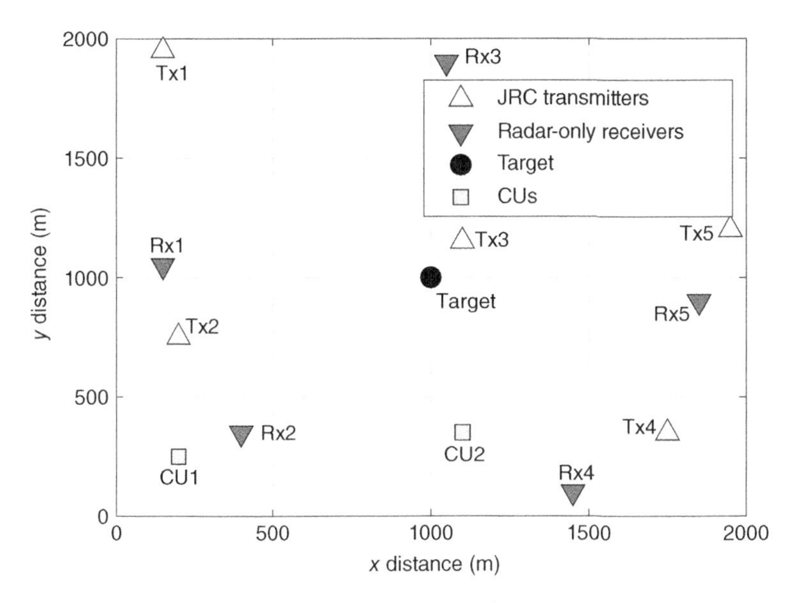

Figure 11.9 Simulation layout for distributed JRC system.

Table 11.1 Power allocation for distributed JRC system for $M = N = 5$ and $U = 2$, $P_{\text{total}_{\max}} = 100$ W, $\eta_{\text{desired}} = 7.0\,\text{m}^2$, $\eta_{\text{opt}} = 6.57\,\text{m}^2$.

	Radar-centric (11.36)	**Communication-centric (11.38)**	**Cooperative (11.39)**
$\mathbf{p}_{\text{tx}}$ (W)	$\begin{bmatrix} 0.0 \\ 0.0 \\ 60.0 \\ 40.0 \\ 0.0 \end{bmatrix}$	$\begin{bmatrix} 23.3 \\ 26.0 \\ 0.0 \\ 25.4 \\ 25.3 \end{bmatrix}$	$\begin{bmatrix} 10.5 \\ 11.0 \\ 57.6 \\ 11.1 \\ 9.8 \end{bmatrix}$
P_{total} (W)	100	100	100
η (m^2)	6.57	205.84	6.89
$\mathfrak{R}$ (bits/s/Hz)	12.51	40.45	30.58

error of $205.84\,\text{m}^2$. These issues have been mitigated by exploiting the cooperative optimization problem (11.39), which provides a desired trade-off between the performance of both radar and communication operations. We see that the Shannon capacity achieved by the cooperative design is 30.58 bits/s/Hz, whereas the localization error is $6.89\,\text{m}^2$. Therefore, it can be observed that the cooperative JRC design approach provides suitable performance for both radar and communication subsystems.

11.5 Conclusions

We presented resource allocation strategies for popular JRC systems. Three different scenarios are considered. First, we discussed single antenna-based JRC systems that exploit OFDM subcarriers to concurrently perform radar and communication objectives. The resource allocation problem for this case was considered in terms of the MI between the transmitted JRC signals and the received signals at the radar receiver and CUs. We noted that the joint cooperative subcarrier selection and power allocation enabled concurrent optimization of both radar and communication operations, in contrast to the designs that consider either radar or communication objective only. Second, we analyzed the resource allocation problem for antenna array-based JRC systems in terms of total transmit power and antenna selection. It was noted that the antenna selection profiles that avoid frequent sensor switching can result in the lowest overall sensor utilization for such systems, yielding simpler and efficient designs. Finally, power optimization for distributed JRC network was considered. We used the localization error bound in terms of CRB and Shannon capacity as the radar and communication performance metrics, respectively. It was observed that a cooperative design that considers both radar and communication objectives yields favorable performance for both radar and communication subsystems. While a number of contributions have been made toward optimized resource allocation for JRC systems, this topic remains to be further explored for a broader range of constraints, scenarios, and emerging JRC strategies.

References

A. Ahmed, Y. D. Zhang, and Y. Gu. Dual-function radar-communications using QAM-based sidelobe modulation. *Digital Signal Processing*, 82:166–174, 2018.

A. Ahmed, Y. D. Zhang, A. Hassanien, and B. Himed. OFDM-based joint radar-communication system: optimal sub-carrier allocation and power distribution by exploiting mutual information. In *Asilomar Conference on Signals, Systems, and Computers*, pages 559–563, Pacific Grove, CA, November 2019a.

A. Ahmed, Y. D. Zhang, and B. Himed. Distributed dual-function radar-communication MIMO system with optimized resource allocation. In *IEEE Radar Conference*, pages 1–5, Boston, MA, April 2019b.

A. Ahmed, S. Zhang, and Y. D. Zhang. Antenna selection strategy for transmit beamforming-based joint radar-communication system. *Digital Signal Processing*, 105:102768, 2020a.

A. Ahmed, S. Zhang, and Y. D. Zhang. Optimized sensor selection for joint radar-communication systems. In *IEEE International Conference on Acoustics, Speech and Signal Processing*, pages 4682–4686, May 2020b.

S. Amuru, M. R. Buehrer, R. Tandon, and S. Sodagari. MIMO radar waveform design to support spectrum sharing. In *IEEE Military Communications Conference*, pages 1535–1540, San Diego, CA, November 2013.

M. R. Bell. Information theory and radar waveform design. *IEEE Transactions on Information Theory*, 39(5):1578–1597, 1993.

M. Bică and V. Koivunen. Generalized multicarrier radar: models and performance. *IEEE Transactions on Signal Processing*, 64(17):4389–4402, 2016.

M. Bică and V. Koivunen. Radar waveform optimization for target parameter estimation in cooperative radar-communications systems. *IEEE Transactions on Aerospace and Electronic Systems*, 55(5):2314–2326, 2019a.

M. Bică and V. Koivunen. Multicarrier radar-communications waveform design for RF convergence and coexistence. In *IEEE International Conference on Acoustics, Speech and Signal Processing*, pages 7780–7784, Brighton, UK, May 2019b.

D. W. Bliss. Cooperative radar and communications signaling: the estimation and information theory odd couple. In *IEEE Radar Conference*, pages 50–55, May 2014.

E. J. Candès, M. B. Wakin, and S. P. Boyd. Enhancing sparsity by reweighted ℓ_1 minimization. *Journal of Fourier Analysis and Applications*, 14(5):877–905, 2008.

A. Chiriyath, S. Ragi, H. Mittelmann, and D. Bliss. Radar waveform optimization for joint radar communications performance. *Electronics*, 8:1498, 2019.

D. Cohen, K. V. Mishra, and Y. C. Eldar. Spectrum sharing radar: coexistence via Xampling. *IEEE Transactions on Aerospace and Electronic Systems*, 54(3):1279–1296, 2018.

T. M. Cover and J. A. Thomas. *Elements of information theory*. Wiley-Interscience, 2006.

S. Dang, S. Guo, B. Shihada, and M.-S. Alouini. Information-theoretic analysis of OFDM with subcarrier number modulation. *IEEE Transactions on Information Theory*, 67(11):7338–7354, 2021.

S. H. Dokhanchi, M. R. B. Shankar, K. V. Mishra, and B. Ottersten. A mmWave automotive joint radar-communications system. *IEEE Transactions on Aerospace and Electronic Systems*, 55(3):1241–1260, 2019.

J. Euzière, R. Guinvar'h, I. Hinostroza, B. Uguen, and R. Gillard. Time modulated array for dual function radar and communication. In *IEEE International Symposium on Antennas and Propagation USNC/URSI National Radio Science Meeting*, pages 806–807, October 2015.

H. Godrich, A. M. Haimovich, and R. S. Blum. Target localization accuracy gain in MIMO radar-based systems. *IEEE Transactions on Information Theory*, 56(6):2783–2803, 2010.

H. Godrich, A. P. Petropulu, and H. V. Poor. Power allocation strategies for target localization in distributed multiple-radar architectures. *IEEE Transactions on Signal Processing*, 59(7):3226–3240, 2011.

A. Goldsmith. *Wireless communications*. Cambridge University Press, 2005.

M. Grant and S. Boyd. CVX: Matlab software for disciplined convex programming, version 2.1. http://cvxr.com/cvx, March 2014.

H. D. Griffiths and C. J. Baker. *An introduction to passive radar*. Artech House. 2017.

A. Hassanien and M. G. Amin. Principles of dual-function radar-communication systems. In *Signal Processing for Joint Radar Communications*, edited by Kumar Vijay Mishra, M. R. Bhavani Shankar, Björn Ottersten and A. Lee Swindlehurst, 25–49, 2022.

Gurobi Optimization. *Gurobi Optimizer Reference Manual*, 2020. http://www.gurobi.com.

A. Hassanien, M. G. Amin, Y. D. Zhang, and F. Ahmad. Dual-function radar-communications: information embedding using sidelobe control and waveform diversity. *IEEE Transactions on Signal Processing*, 64(8):2168–2181, 2016a.

A. Hassanien, M. G. Amin, Y. D. Zhang, and F. Ahmad. Signaling strategies for dual-function radar communications: an overview. *IEEE Aerospace and Electronic Systems Magazine*, 31(10):36–45, 2016b.

P. Kumari, J. Choi, N. González-Prelcic, and R. W. Heath. IEEE 802.11ad-based radar: an approach to joint vehicular communication-radar system. *IEEE Transactions on Vehicular Technology*, 67(4):3012–3027, 2018.

M. Labib, V. Marojevic, A. F. Martone, J. H. Reed, and A. I. Zaghloui. Coexistence between communications and radar systems: a survey. *URSI Radio Science Bulletin*, 2017 (362):74–82, 2017.

F. Liu, L. Zhou, C. Masouros, A. Li, W. Luo, and A. Petropulu. Toward dual-functional radar-communication systems: optimal waveform design. *IEEE Transactions on Signal Processing*, 66(16):4264–4279, 2018.

F. Liu, C. Masouros, A. P. Petropulu, H. Griffiths, and L. Hanzo. Joint radar and communication design: applications, state-of-the-art, and the road ahead. *IEEE Transactions on Communications*, 68(6):3834–3862, 2020.

J. Liu, K. V. Mishra, and M. Saquib. Co-designing statistical MIMO radar and in-band full-duplex multi-user MIMO communications. *ArXiv*, https://arxiv.org/abs/2006.14774, October 2021.

N. C. Luong, X. Lu, D. T. Hoang, D. Niyato, and D. I. Kim. Radio resource management in joint radar and communication: a comprehensive survey. *ArXiv*, https://arxiv.org/abs/2007.13146, July 2020.

J. A. Mahal, A. Khawar, A. Abdelhadi, and T. C. Clancy. Spectral coexistence of MIMO radar and MIMO cellular system. *IEEE Transactions on Aerospace and Electronic Systems*, 53(2):655–668, 2017.

O. Mehanna, N. D. Sidiropoulos, and G. B. Giannakis. Joint multicast beamforming and antenna selection. *IEEE Transactions on Signal Processing*, 61(10):2660–2674, 2013.

K. V. Mishra, M. R. B. Shankar, V. Koivunen, B. Ottersten, and S. A. Vorobyov. Toward millimeter-wave joint radar communications: a signal processing perspective. *IEEE Signal Processing Magazine*, 36(5):100–114, 2019.

MOSEK ApS. *The MOSEK Optimization Toolbox for MATLAB Manual. Version 9.0.*, 2019. http://docs.mosek.com/9.0/toolbox/index.html.

B. Paul, A. R. Chiriyath, and D. W. Bliss. Survey of RF communications and sensing convergence research. *IEEE Access*, 5:252–270, 2017.

M. Roberton and E. R. Brown. Integrated radar and communications based on chirped spread-spectrum techniques. In *IEEE MTT-S International Microwave Symposium Digest*, Volume 1, pages 611–614, July 2003.

G. N. Saddik, R. S. Singh, and E. R. Brown. Ultra-wideband multifunctional communications/radar system. *IEEE Transactions on Microwave Theory and Techniques*, 55(7):1431–1437, 2007.

C. Sahin, J. Jakabosky, P. M. McCormick, J. G. Metcalf, and S. D. Blunt. A novel approach for embedding communication symbols into physical radar waveforms. In *IEEE Radar Conference*, pages 1498–1503, Seattle, WA, May 2017.

R. Saruthirathanaworakun, J. M. Peha, and L. M. Correia. Opportunistic sharing between rotating radar and cellular. *IEEE Journal on Selected Areas in Communications*, 30(10):1900–1910, 2012.

C. Shi, Y. Wang, F. Wang, S. Salous, and J. Zhou. Power resource allocation scheme for distributed MIMO dual-function radar-communication system based on low probability of intercept. *Digital Signal Processing*, 106:102850, 2020.

C. Sturm, T. Zwick, and W. Wiesbeck. An OFDM system concept for joint radar and communications operations. In *IEEE Vehicular Technology Conference*, pages 1–5, Barcelona, Spain, April 2009.

F. Wang and H. Li. Joint power allocation for radar and communication co-existence. *IEEE Signal Processing Letters*, 26(11):1608–1612, 2019.

F. Wang, H. Li, and M. A. Govoni. Power allocation and co-design of multicarrier communication and radar systems for spectral coexistence. *IEEE Transactions on Signal Processing*, 67(14):3818–3831, 2019a.

X. Wang, A. Hassanien, and M. G. Amin. Dual-function MIMO radar communications system design via sparse array optimization. *IEEE Transactions on Aerospace and Electronic Systems*, 55(3):1213–1226, 2019b.

M. Wen, X. Cheng, M. Ma, B. Jiao, and H. V. Poor. On the achievable rate of OFDM with index modulation. *IEEE Transactions on Signal Processing*, 64(8):1919–1932, 2016.

M. Yuan and Y. Lin. Model selection and estimation in regression with grouped variables. *Journal of the Royal Statistical Society Series B (Statistical Methodology)*, 68:49–67, 2006.

Y. D. Zhang, M. G. Amin, and B. Himed. Structure-aware sparse reconstruction and applications to passive multi-static radar. *IEEE Aerospace and Electronic Systems Magazine*, 32(2):68–78, 2017.

L. Zheng, M. Lops, Y. Eldar, and X. Wang. Radar and communication coexistence: an overview: a review of recent methods. *IEEE Signal Processing Magazine*, 36(5):85–99, 2019.

12

Emerging Prototyping Activities in Joint Radar-Communications

M. R. Bhavani Shankar[1], *Kumar Vijay Mishra*[2], *and Mohammad Alaee-Kerahroodi*[1]

[1] *Interdisciplinary Centre for Security, Reliability and Trust (SnT), University of Luxembourg, Luxembourg*
[2] *United States CCDC Army Research Laboratory, Adelphi, MD, USA*

12.1 Motivation

In recent years, radars that share their spectrum with wireless communications have gained significant research interest [Dokhanchi et al., 2019, Duggal et al., 2020, Elbir et al., 2021]. Such a joint radar-communications (JRC) system is being considered for its potential advantages including efficient spectrum and hardware utilization and enhanced situational awareness in applications not limited to automotive. The JRC paradigm is also being considered in the sixth-generation (6G) deployment under *integrated sensing and communications.* While lower frequencies offer fertile ground for the implementation of JRC systems because of spectrum congestion and hardware (HW) availability [Paul et al., 2016], the wide bandwidth millimeter-wave (abbmmWave) bands are interesting from the perspective of the emerging high resolution and high data rate applications [Mishra et al., 2019a].

From a system design objective, the JRC techniques need to devise ways toward improved integration of both sensing and communications systems, wherein finite radio resources must meet the demands of coexistence [Wu et al., 2022] or co-design [Liu et al., 2022] regimes. This multifunction approach seeks to develop architectures and algorithms including signaling strategies, receiver processing, and side information to support their simultaneous operation while meeting appropriately chosen design metrics. Many of these aspects have been addressed in the literature and have been concisely captured in the earlier chapters. These seminal works, as in many of the classical signal processing and communications (SPCOM) methodologies, have largely taken a model-based approach to the optimization and evaluation of the proposed methodologies for waveforms and receivers. This approach allows for tractability of analysis and eases design. Typical assumptions include the Gaussian modeling of signals, noise and channels, linearity in responses, device uncertainties, isotropic scattering models, absence of clutter, instantaneous availability of information, among others. In many a circumstance, these assumptions do not hold, thereby requiring a re-evaluation of the efficiency of these techniques. As a case in point, power amplifiers are not essentially linear, certain types of device noise like shot/impulse noise are not Gaussian and so is signal-correlated quantization noise in low-bit systems [Kumari et al., 2020, Pace, 2000, Tsui, 2004]. Software simulations incorporating these perturbations do offer certain guarantees, but these simulations are based on models for perturbations, which, again, may suffer from modeling inaccuracies. At mmWave, special considerations are required for designing wideband receivers [Mishra, 2012, Tsui, 2010, Daniels et al., 2017].

Signal Processing for Joint Radar Communications, First Edition.
Edited by Kumar Vijay Mishra, M. R. Bhavani Shankar, Björn Ottersten, and A. Lee Swindlehurst.

Toward stepping into the next stage of technological maturity, it becomes essential to validate the JRC concepts, even in controlled, but representative scenarios. This warrants *prototyping* an essential aspect of disseminating SPCOM research of late [Mishra et al., 2019b]. It has been well understood that prototyping enables development teams to explore concepts, understand technical challenges, specify product requirements, and reduce uncertainty. In this context, it is desired that the HW prototyping of JRC systems achieves the following objectives [Jensen, 2018, Kunicina et al., 2020]:

- Ascertain the feasibility of JRC algorithmic implementation on HW platform.
- Identify and mitigate errors and incorporate missing functionality at an early stage; typical errors would be in synchronization, to gain control, and in alignment, and functionalities beyond the physical layer may be required (e.g. channel feedback protocol).
- Incorporate the device and scenario modeling errors implicitly.
- Gain confidence in the achieved JRC performance and pitch it for consideration in large-scale demonstrations.

12.2 Prototyping: General Principles and Categorization

The following principles are generally considered toward the development of a HW prototype [Kunicina et al., 2020]:

1. **Identification of the prototyping objectives**: The first step is a clear enumeration of the demonstration objectives from the prototyping. Clearly, considering all the aspects of the invention is desirable; however, several factors including the resources and budget impact this choice. Another aspect is the potential use of this prototype in valorization. Such an objective listing provides an overview of the device and platform requirements for demonstration.
2. **Prototype high-level design**: Based on the objectives, the components and the experimental setup are identified and a high-level design is carried out to ascertain feasibility. This identifies and eliminates system-level uncertainties and consolidates the interfaces among components and software.
3. **Prototype realization and evaluation**: Subsequently, the prototype is developed and standard unit and integration tests performed. An elaborate testing is undertaken in the chosen scenario, collecting necessary metrics and performing statistical analysis. These are then compared to system-level software simulators for conformance.
4. **Feedback and refinement**: The evaluation provides valuable feedback and data for refining the prototype; significant deviations from the software simulator are of particular importance. This stage also provides the researchers the confidence to pursue further toward enhancing and creating a minimum viable product (MVP) and/or enhance the Technology Readiness Level (TRL) of their invention.

Several categorizations of prototypes exist when considering the product development cycles; one such categorization is as follows[1]:

1. **Proof of concept prototypes**. This is typically the first attempt at prototyping with the purpose of proving the technical feasibility of the research idea.
2. **Looks-like prototypes**. This mock-up depicts the *visual aspects* of the final product but does not contain working parts. It enables an understanding of the physical appearance of the product.

---

1 Based on *Four Types of Prototypes and Which is Right for You?*: Voler Systems, https://www.volersystems.com/blog/design-tips/4-types-prototypes-right, Last Accessed 02 November 2021.

Table 12.1 Design aspects for prototyping.

Prototyping aspect	Influencing parameters
Regulatory	Spectrum allocation, transmission power
Hardware	Requirements including reconfigurability, cost, interfaces
Form factor	Portability, availability of ICs
Power	Battery operated (for portable), socket powered
Software	Operating system, embedded firmware, APIs
Durability	Handled at different temperatures, humidity, personnel

3. **Functional or works-like prototypes**. These prototypes are used for testing the key functionalities without the looks of the final product. It complements the look-like prototypes.
4. **Engineering prototypes**. This would be the final development prior to full production; it is undertaken to ensure quality and manufacturability.

Table 12.1 summarizes the different design aspects for prototyping.[2] The prototyping relevant to this chapter is the **Proof of Concept.** The prototypes described in the sequel use existing materials, parts, and components to prove that the idea of JRC is feasible under the controlled environment of a lab or a small-scale deployment. Within the proof of concept, several possibilities exist:

- **Full-fledged HW testbed with over-the-air (OTA) testing**. In this mode, the data acquisition, storage, and processing is undertaken in HW, developed on an appropriate platform and embedded software. It uses over-the-air (OTA) transmissions to achieve the goal. Such a development enables real-time demonstration using stand-alone HW, thereby indicating an advanced prototyping. The HW is typically enabled by widely available software defined radio (SDR) platforms offering varying degrees of programmability on the included field programmable gate array (FPGA).
- **Full-fledged HW testbed without over-the-air (OTA) testing**. In many situations, the OTA transmissions are replaced by wired connections with the impact of channel emulated on the FPGA. These situations arise when (i) license to transmit in a particular band is restricted, unavailable, or pending (e.g. in Luxembourg, 2400–2483.5 MHz Band is reserved only for Short Range Devices with a maximum effective isotropic radiated power (EIRP) of 10 mW, which precludes from outdoor applications); (ii) there is a need to avoid radio-frequency (RF) interference to other setups; (iii) due to the lack of device and/or experimental setup capability to achieve desired goals when performing OTA transmissions. A case in point regarding the last situation is the use of low-cost sub-6GHz SDRs with 100–150 MHz bandwidth for radar applications. These require tens of meters between the antennas and the target to demonstrate any meaningful result on range resolution. This requirement is clearly not met in limited areas with OTA transmissions.
- **Hardware in the loop**. In situations where the HW lacks the needed computation power, in early stages of prototyping or in the use of commercial-off-the-shelf (COTS) components, it would be necessary to include processing on a powerful host, typically a personal computer. While the host computation in software enables easier programming in widely used MATLAB®,[3]

2 Adapted from *Types of Hardware Prototypes: Requirements and Validation*, Cadence PCB Design and Analysis Blog, https://resources.pcb.cadence.com/blog/2019-types-of-hardware-prototypes-requirements-and-validation, last accessed November 2, 2021.
3 https://www.mathworks.com

LabVIEW®,[4] C/C++, etc., the interface between the HW and SW may offer bottlenecks based on the rate of acquisition. In sub 6 GHz-based SDR platforms, the bandwidths are the order of 100–200 MHz, thereby allowing seamless transfer of data from the HW platform for host processing resulting in a real-time application. On the other hand, in COTS solutions like TI AWR, Infenion RXS*, the bandwidths are in the range of a 1–4 GHz; using these COTS for developing applications on the module itself other than those provided by the manufacturer tends to be rather difficult. Further, porting such high speed data onto a computer is also difficult thereby rendering real-time applications infeasible.

12.3 JRC Prototypes: Typical Features and Functionalities

Communication prototypes have been considered for long and have got a fillip from the standardization activity involving academia and research alike. Similarly, radar prototyping in academia has gathered pace, of late, with the development of low-cost COTS sensors. These prototypes demonstrate a number of features and functionalities that are characterized below.

12.3.1 Operational Layer

The communication protocol stack is defined by the seven Open Systems Interconnection (OSI) layers and HW-oriented prototypes focus on the lowest Physical (PHY) layer and the one above, the Medium Access Control (MAC) layers. At the PHY layer, the prototypes implement a chosen air-interface (e.g. LTE/A, 5G-NR, etc.) including the associated frame format, baseband processing, and associated RF operations. Some prototypes focus only on waveform (e.g. OFDM, Single Carrier), while keeping the framing to a minimum; prototypes dealing with end-to-end PHY performance include complete framing. On the other hand, prototypes with integrated MAC layer implement the associated functionalities such as packet scheduling, ordering, transmissions/re-transmissions, handshaking, backoff. Finally, the higher layers are typically software-oriented and integrated using known interfaces. As a case in point, open-source (GNU GPLv2) discrete-event network simulators are available in C++ for Network and higher layers, e.g. NS3 Simulator, https://www.nsnam.org.

12.3.2 Operational Frequencies

Despite the maturity of resource reuse, the limited bandwidth in lower frequencies, particularly below 6 GHz, precludes the complete exploitation of the 5G potential as well as meeting the 5G requirements. This also reduces the range resolution in radars, thereby precluding emerging applications requiring cm-length accuracy. Further, the equipment at lower frequencies are bulky as well. These call for the use of mmWave spectrum due to the large available bandwidths and high degree of miniaturization. However, a significant communication prototyping activity has focused on the lower frequency bands due to availability of low-cost hardware platforms. On the other hand, academic radar prototypes in lower frequencies are hard to find for reasons mentioned above. Further, the communication prototypes are in debt at mmWave frequencies. Infact, the implementation in such frequencies is significantly complicated compared to sub-6 GHz systems due to the following [Bang et al., 2021]:

---

4 https://www.ni.com/en-us/shop/labview.htm

- Use of highly integrated manufacturing technologies and large-scale phased arrays that contain many RF analog components.
- Expensive mmWave HW components.
- mmWave analog components suffer greatly from different HW imperfections and constraints that severely affect system performance, including phase noise, power amplifier nonlinearities, and inphase and quadrature (IQ) imbalance.

Nonetheless, significant research and investment from academia and industry is being made to create flexible and scalable HW solutions [Gul et al., 2016]. This coupled with advances in SDR platforms, novel HW/SW interfaces, and HW abstraction as well as easier programming of HW devices have motivated the design of 5G HW prototypes in mmWave bands. Several issues arise when implementing *flexible and simple* HW prototypes in mmWave [Gul et al., 2016]; these impact RF front ends, data conversion, storage, processing and distribution, processing algorithms, real-time operation, among others. On a similar note, mmWave COTS modules from **TI, Infenion** offering starter-kits with the flexibility of on-chip processing or raw data acquisition have enhanced the uptake of radar prototyping, mainly for indoor scenarios.

12.3.3 MIMO Single-User Architectures

Since two decades, the use of multiple antennas in wireless communications has been the norm. The simple prototypes demonstrate multiple input multiple output (MIMO) technology for the case of point-to-point link wherein signals received on the multiple antennas can be jointly processed. In this context, the classical VBLAST from Bell labs offered the first prototype operating at a carrier frequency of 1.9 GHz, and a symbol rate of 24.3 ksymbols/seconds, in a bandwidth of 30 kHz. The prototype exploits an antenna array, with results reported for 8 transmit and 12 receive antennas [Wolniansky et al., 1998]. Subsequently, significant prototyping efforts focused on single user or point-point PHY links emulating communication between a transmitter and a particular receiver. These designs assumed existence of orthogonal allocation of resources (e.g. time, bandwidth) across different users (receivers). This led to the sufficiency of prototypes emulating a single-user link on Additive White Gaussian Noise (AWGN) channels to be representative of the system performance.

12.3.4 MIMO Multiuser (MU-MIMO) Architectures

On the other hand, with an increasing demand for data inducing a scarcity of resources, the reuse of resources to serve multiple users offers an attractive alternative, albeit, at the cost of interference among them. In order to manage the interference, it is essential for the transmitter to know the channel conditions to the served users, necessitating a feedback from different terminals. In addition to requiring multiple receiver HW prototypes demodulating and decoding data corresponding to different users, the multiuser prototype should also support channels for information flow from each of the user to the transmitter. This return link implementation requires protocol support to manage and synchronize feedback. It should be noted that such an implementation for single-user channels is simpler due to a single return link.

MU-MIMO has become an essential feature of communication prototypes emulating functionalities of 3G standard and beyond. The functionalities implemented in these prototypes include multiple precoding techniques, power allocation, and user scheduling at the transmitter, receive beamforming/equalization, synchronization, channel, and transmission quality feedback.

12.3.5 Massive MIMO

A recent trend is to employ a large number of antennas at the base-station (BS) to exploit spatial diversity combat small-scale fading. This massive MIMO technology has attracted significant interest due to the potential gains offered by theory. However, practical implementation is in debt, partly due to many critical practical issues that need to be addressed. An increased number of interconnections, much larger processing loads, and enhanced system complexity and cost have been listed as some of the challenges that need be addressed [Edfors et al., 2016].

Massive MIMO prototyping differs from the conventional approach both in analog as well as digital processing. With an increase in the number of antennas, the analog components feeding those antennas also increase; to ensure a cost-effective solution, low-cost components would be used, which tend to be nonideal. These need to be considered during prototyping to ensure adequate reflection of actual performance. In particular, the antenna and analog front-end imperfections, lack of nonreciprocity and calibration for time division duplexing (TDD) schemes, and limited computational power needs to be included. From a digital processing, a typical choice would include SDRs and additional computational resources to ensure flexibility and real-time operations. To exploit the gains offered by a large number of antennas, sample-level time synchronization and phase alignment among the different streams become crucial. Another important aspect is large amounts of data that need to be acquired, stored, and moved around different processing units; the underlying HW platform should be able to handle this efficiently. Finally, the baseband processing should implement transmission and reception techniques that could be rather involved.

12.3.6 Phased Array and MIMO Radar

Phased-array radar systems have been considered since long originating from the military applications toward detecting and tracking small objects at significantly large distances by concentrating power in a sharp beam that is electronically steered at high rates [O'Connor, 2019]. With the proliferation of radar in civilian applications including automotive and indoor sensing, there has been the inclusion of additional constraints including (i) wide-angle operation to have an increased coverage, (ii) wide-band processing leveraging on the high available bandwidth in contrast to traditional narrow-band systems, and (iii) restrictions on size, weight, cost, and power. Several avenues have been considered toward addressing one or many of these problems. To reduce the cost of the RF components for beamforming and signal distribution, RF Systems on a Chip (RFSoC) have been considered to minimize cost and offer implementation of SP algorithms in real-time [Fagan et al., 2018]. A prototype of RFSoC with 16 analog and digital channels has a software-defined receiver and waveform generator along with real-time adaptive beamforming in an S-band phased array radar. The focus of the research has been to demonstrate the potential improvement for size weight and power [SWaP], real-time signal processing capacity, and advanced design processes for rapid algorithm implementation. The preproduction RFSoC prototype demonstrates the potential value of RFSoC and rapid algorithm development for next generation radar systems.

In this context, the current developments have focused on the design and development of large-scale phased array antennas with tunable phase shifter.

12.3.7 Summary

With the increased interest, support, and availability of HW, prototyping, particularly in research, has seen an increased uptake in the recent years. In the case of wireless communications, Table 12.2 presents some representative examples of different prototypes to highlight the changes in HW and

Table 12.2 Examples of wireless communication prototypes.

Prototype	WARP	SAMURAI	KU Leuven
Reference	WARP Project [2021]	Cattoni et al. [2012]	KU Leuven [2021]
Protocol	802.11, OFDM	LTE Rel 8 TM6.	LTE-like (TDD)
Architecture	SISO-SU	MU-MIMO	Massive MIMO
Transmit antennas	1	4 (Express MIMO)	68
Frequency	2.4/5 GHz	250 MHz?-3.8 GHz[a)]	400 MHz–4.4 GHz
Layers	PHY, MAC	PHY, MAC	PHY
HW	Virtex-6	Virtex 5 LX330	Kintex 7
	LX240T FPGA	Virtex 5 LX110T	
SW	C	C	LabVIEW

a) https://cordis.europa.eu/docs/projects/cnect/6/257626/080/deliverables/001-ACROPOLISD52v10.pdf

Table 12.3 Typical components of radar prototyping.

HW platform	Enabling software	Potential functionality	Possible waveforms
USRP NI-29xx	LabVIEW NXG	MIMO Radar	FMCW
sub 6 GHz	MATLAB	Cognitive radar	PMCW
National	Qt C++	Coexistence	CW
Instruments	Python	JRC	Custom
COTS Modules	MATLAB	Distributed sensing	FMCW
(AWR, IWR, ...)	Python	Imaging	
24, 60, 77–81 GHz	Code composer studio	Tracking	
Texas Inst.	C++	Interference analysis	
FPGA boards	VHDL	Cognitive radar	FMCW
mmWave	ISE/Vivado	Customized MIMO radar	PMCW
Xilinx	Python	DoA finding	CW
	HLS/C++	Customized SP units	Custom

SW. The current trend has been on the 5G prototyping with several initiatives worldwide on developing testbeds and is numerous to list here. An useful resource could be the IEEE Future Networks Initiative [FNI] which curates a directory of available 5G and beyond networking testbeds for use by both academic and industry research groups.[5]

On a similar trend, there have been a number of radar testbeds in research institutions implementing various radar waveforms (FMCW/PMCW/Pulsed/CW) and undertaking different radar tasks. Unlike the wireless communications, lack of a standardized approach results in different prototypes using different configurations. In this context, rather than an elaborate listing, Table 12.3 collects the key components from modern radar prototyping.

5 https://futurenetworks.ieee.org/testbeds

12.4 JRC Prototyping

The previous sections provided an overview and some examples of prototyping in wireless communications and radar. This section, builds on the previous section and brings out additional aspects needed for JRC prototyping. The architecture of the JRC prototype depends on the topology considered. As discussed in the previous chapters, the two predominant topologies are – Coexistence and Codesign that exist with regards to the transmission. On the other hand, with regards to the radar reception, the receiver can be colocated with the transmitter (monostatic) or separated (bistatic). In this chapter, we focus on the prevalant monostatic radar implementations. The bistatic operations require additional synchronization and is currently under investigation for the JRC. The implications of the chosen architecture on the prototyping are detailed below.

12.4.1 Coexistence

In this topology, the radar and communication portions are implemented as separate functionalities as shown in Figure 12.1. To foster coexistence, the following cognitive paradigms can be considered.

1. **Bilateral operation**. This operation is depicted in Figure 12.1a. Here both the systems adapt their transmission to minimize mutual interference. To enable this, an interface link between the two systems enabling high-level information exchange (uni/bi directional) is needed and an adaptation module that acts on the information to enable coexistence.
2. **Unilateral operation**. This operation is depicted in Figure 12.1b. In this setting, a cognitive module operates in either or both of the systems that sense the onset of interference and adapt the transmission and reception unilaterally, without additional information exchange. This operation arises in the opportunistic use of spectrum.

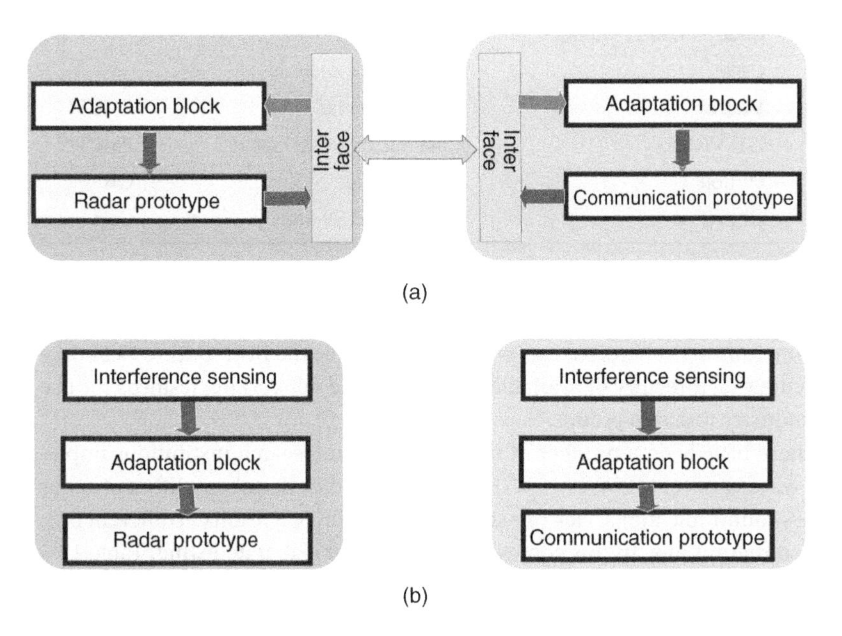

Figure 12.1 Typical coexistence architectures depicting the mode of enforcing coexistence. (a) Bilateral mode of operation. (b) Unilateral mode of operation.

In this context, the design critically depends on the nature of operation considered above. In particular, the following aspects need to be carefully considered for *Bilateral Operation*.

1. **Information to be exchanged**. The amount of information, its rate of exchange, the sensitivity of the information, and the latency play a fundamental part in achieving. Typical information to be exchanged involves spatial, temporal, and spectral resource allocation elements such as the power, frequency bands, timeslots, and direction of transmission. Unless the resources are large in number (e.g. massive MIMO), exchange of these quantities requires limited bits. Further, the rate of update of these quantities depends on the changes in scenario and can be rather infrequent for nearly static or prescanned (known trajectory, user behavior, etc.) setups.
2. **Interface link design**. Based on the information to be exchanged, a communication link and an underlying protocol need to be implemented. This would involve appropriate air-interface based on link-budget and appropriate handshaking protocol. Both wireless and wired designs can be considered; while the former is elegant and is representative of the actual setup, it needs to address additionally the spectral allocation for the interference link. A wired setup, on the other hand, enables quick testing of the system idea.
   From a HW device perspective, the interface link design entails the need to accommodate additional transmission and reception ports at either end along with appropriate RF front-ends. Dedicated digital processing toward link maintenance (e.g. setup, handshaking) at either end is also needed.
3. **Adaptation block**. This digital processing block needs to be implemented at radar and/or communication prototypes to adapt the transmission and reception. The operation of this block needs to be synchronized with that of the underlying prototypes, e.g. adaptation needs to be carried out at a predefined intervals which need to be synchronized either with the transmission frame of the communication or the coherent processing interval of the radar.

With regards to the *Unilateral Operation*, the host system (radar or communication) needs to incorporate a *sensing block* that determines the reuse of resources by the incumbent system and an *adaptation block* to act on this information. This block could incorporate an energy detector determining the incident energy in spatial, temporal, or spectral domains. Such a block need not be aware of the specifics of the transmission from the incumbent system, thereby simplifying information exchange overhead. From a HW perspective, incorporating such a sensing block would require antenna ports and relevant RF blocks to sense incident waves and a digital processing unit to incorporate the sensing algorithm. The output of the sensing algorithm would be used in the adaptation block mentioned earlier.

12.4.2 Codesign

The codesign prototype involves a single transmitter supporting both functionalities; the bulk of the design lies in the digital processing to derive appropriate transmission strategies with the HW (e.g. power, bandwidth) being dimensioned to ensure that the quality of service is met. Differently from the coexistence, the *codesign* aspect needs to consider the impact of self-interference. Particularly, when the codesign JRC transmitter is emitting a waveform, it could cause significant self-interference saturating the front-end impacting not only the radar but also the received communication signal in case of bidirectional transmission. Research on exploiting in-band self-interference cancelation using principles from full-duplex operations has been pursued, and some prototypes have been considered [Barneto et al., 2021]. An alternative, but less elegant way, is to isolate the transmit and receive functionalities.

Having presented the general setup of the JRC prototypes, the chapter now delves deeper and discusses an example prototype on JRC coexistence in detail based on unilateral operation. Other prototypes would be discussed in the sequel.

12.5 Coexistence JRC Prototype

In this section, a JRC prototype enabling spectrum sharing between the radar and communication is presented. The communication system is assumed to be the primary user of the spectrum and the MIMO radar uses these frequencies in an opportunistic fashion. In this context, the radar performs unilateral operations to sense the spectrum use and exploit it with minimal interference to the communication link. To enable this coexistence, following the blocks in Figure 12.1b, the radar module designs:

1. **Sensing block** involves a spectrum-sensing application where the granularity of the spectrum block is determined by the resource allocation unit of the communication system.
2. **Adaptation block** exploits the output of the spectrum sensing and adapts the MIMO waveform *on-the-fly*. The sensing and adaptation lead to a *Cognitive Radar* setup where the transmission is adapted based on the environment.

The focus of this section will be on the prototyping aspects. Details on the waveform design aspects can be obtained from Alaee-Kerahroodi et al. [2022].

12.5.1 Architecture

The prototype consists of three components – application frameworks – as depicted in Figure 12.2 along with a photograph of the setup in the lab Figure 12.3 [Alaee-Kerahroodi et al., 2022]. The three components use Universal Software Radio Peripheral (USRP)s for the transmission and reception of the wireless RF signals, the characteristics of which are presented in Table 12.4. The transmission is validated using Rohde and Schwarz spectrum analyzer. The components and their associated hardware (HW) are presented as follows:

1. Long Term Evolution (LTE) Application implemented by National Instruments on USRP 2974 and used for LTE communications
2. Spectrum sensing implemented on USRP B210
3. Cognitive Multiple Input Multiple Output (MIMO) radar implemented on USRP 2944R for cognitive MIMO

Table 12.4 Hardware characteristics of the proposed prototype.

Parameters	B210	2974/2944R
Frequency range	70 MHz–6 GHz	10 MHz–6 GHz
Max. input power	−15 dBm	+10 dBm
Max. output power	10 dBm	20 dBm
Noise figure	8 dB	5–7 dB
Bandwidth	56 MHz	160 MHz
DACs	61.44 MS/s, 12 bits	200 MS/s, 16 bits
ADCs	61.44 MS/s, 12 bits	200 MS/s, 14 bits

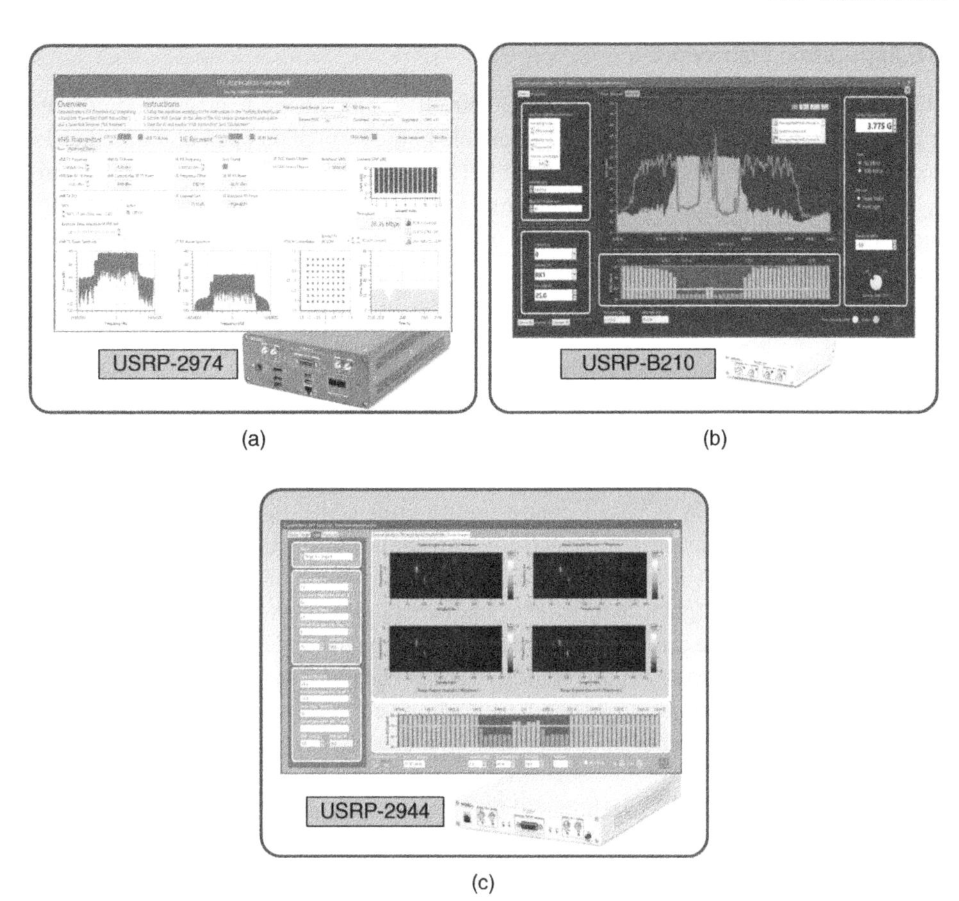

Figure 12.2 Components or application frameworks of the prototype: LTE application developed by NI, spectrum sensing and cognitive MIMO radar applications presented in this chapter Alaee-Kerahroodi et al. [2022].

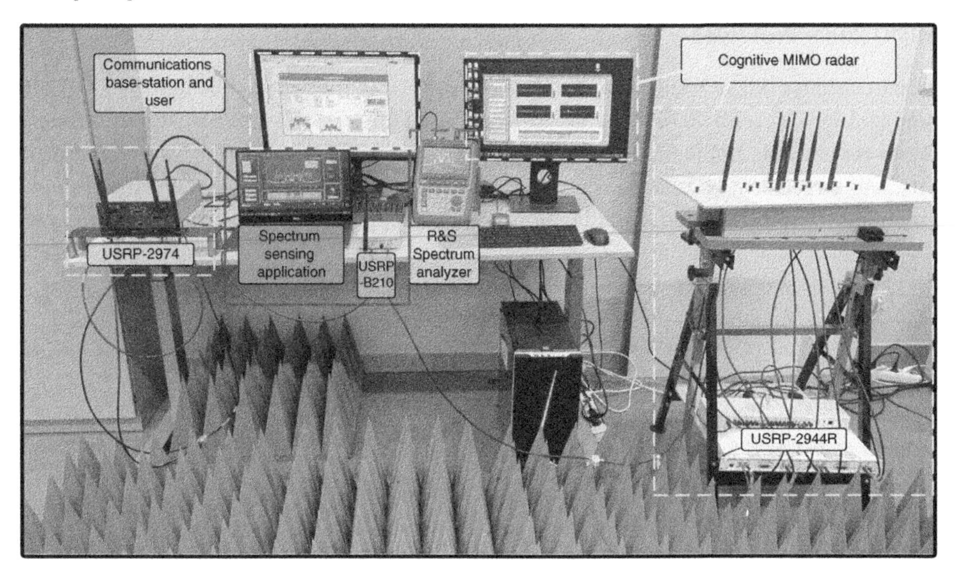

Figure 12.3 A photograph of the proposed coexistence prototype. The photo shows communication Base Station (BS) and user, spectrum sensing, and cognitive MIMO radar systems.

12.5.2 LTE Application Framework

The LabVIEW-based LTE Application Framework (Figure 12.2a) provides a real-time LTE physical layer implementation; this add-on software is available in the form of an open and modifiable source-code [National Instruments Corp., 2021]. Table 12.5 lists particular 3GPP LTE features to which the generated physical layer frame is compliant. A basic implementation of MAC enabling packet-based data transmission along with a framework for rate adaptation is also implemented.

The National Instruments (NI)-USRP 2974 has two independent Radio-Frequency (RF) chains and the application framework supports single antenna links. In this context, the considered prototype emulates a Single Input Single Output (SISO) communication link between a BS and communications user on the two different RF chains of the same USRP.

12.5.3 Spectrum-Sensing Application

An application based on LabVIEW NXG 3.1 connecting to Ettus USRP B210 (Figure 12.2b) has been developed for continuous sensing of the spectrum. This application exhibits flexibility and can update different parameters *on-the-fly*, e.g. averaging modes, window type, energy detection threshold, and the USRP configurations (gain, channel, start frequency). Herein, the center frequency can be set to any arbitrary value in the range 70 MHz to 6 GHz, and the bandwidth spanned can be set to values in the interval [50–100] MHz. In this context, it should be noted that USRP B210 provides 56 MHz of real-time bandwidth by using AD9361 RFIC direct-conversion transceiver. However, efficient implementation enables the developed application to analyze larger bandwidths by sweeping the spectrum.

The spectrum-sensing application determines the spectrum occupancy with a resolution of 1 MHz by computing the energy in the band and a subsequent thresholding [Alaee-Kerahroodi et al., 2022]. It then obtains a frequency occupancy chart and transfers this to the cognitive MIMO radar application through a network connection (LAN/Wi-Fi). The spectrum sensing is developed on a separate USRP and hence, in this context, it can be used as a stand-alone application.

12.5.4 MIMO Radar Prototype

The developed cognitive MIMO radar application framework is illustrated in Figure 12.4. The licensed band at 3.78 GHz with 40 MHz bandwidth was used for transmission [Alaee-Kerahroodi et al., 2022]. In this flexible implementation, the parameters related to the radar waveform, processing units, and targets can be updated even during the operational phase to adapt the system to the environment. The MIMO radar application was developed using LabVIEW NXG 3.1 and

Table 12.5 LTE features of the considered application framework.

Features	Features
Closed-loop Over-The-Air (OTA) operation	20 MHz bandwidth
channel state and ACK/NACK feedback	$\leq$ 75 Mbps data-rate
5-frame structure	QPSK, 16-QAM, and 64-QAM
Physical Downlink Shared Channel (PDSCH)	FDD
Physical Downlink Control Channel (PDCCH)	TDD
Channel estimation	Zero forcing equalization

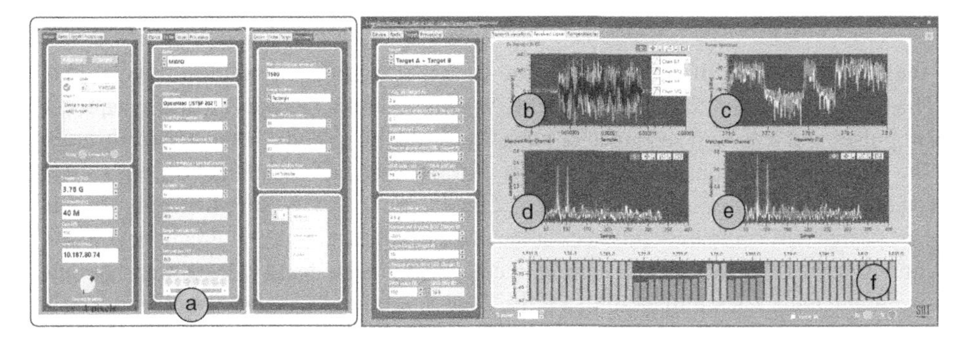

Figure 12.4 Developed cognitive MIMO radar application: (a) user interface for setting the parameters of the HW device, radar, and the processing aspects; (b) baseband I/Q signals from the two receive channels; (c) spectrum of the received signal from the two receive channels; (d) output of filters matched to the two transmit waveforms on the first receive channel; (e) output of the filters matched to two transmit waveforms on the second receive channel; (f) side information from the energy detector of the spectrum-sensing application.

Table 12.6 Features of the developed cognitive MIMO radar.

Parameters	MIMO radar
Center frequency	70 MHz–6 GHz
Bandwidth	1–80 MHz
Processing units	Matched filtering, range-Doppler processing
Window type	Rectangle, Hamming, Blackman, etc.
Averaging mode	Coherent integration (FFT)
Transmit waveforms	Random-polyphase, Frank, Golomb, Random-Binary, Barker, m-Sequence, Gold, Kasami, Up-LFM, Down-LFM, and the optimized sequences

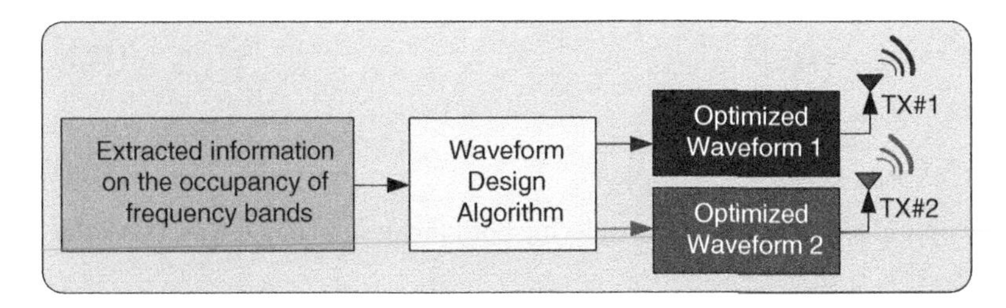

Figure 12.5 Block diagram of the developed cognitive MIMO radar transmitter. Information about occupied frequency bands provided by the spectrum-sensing application is used to design 2 waveforms for JRC coexistence [Alaee-Kerahroodi et al., 2022].

connected to the NI-USRP 2944R comprising 2×2 MIMO RF transceiver with a programmable Kintex-7 FPGA. Table 12.6 details the features and flexibility of the developed application.

The transmit operations of the developed cognitive MIMO radar is depicted in Figure 12.5. Based on a perusal of the spectrum occupancy list transmitted by the spectrum-sensing application over a wireless network, the radar optimizes the transmit waveform, details of which are provided in the sequel.

12.5.5 Waveform Design

Central to the unilateral mode of coexistence is the design of the radar waveform, herein, chosen as the Phase-Modulated Continuous Wave (PMCW) type. Herein, the waveform is a sequence of modulated pulses and a certain sequence of pulses form the pulse code or the waveform to be designed. The duration of the pulse and the length of the pulse code are obtained from system design. Through the flexibility offered in the design of modulation, PMCW waveform offers higher design degrees of freedom compared to other classical approaches such as the Frequency Modulated Continuous Wave (FMCW). In this setting, a colocated narrow-band MIMO radar system, with M transmit antennas, each transmitting a pulse code or a waveform sequence of length N in the fast-time domain is considered. Denote by matrix $\mathbf{X} \in \mathbb{C}^{M \times N} \triangleq [\mathbf{x}_1^T, \dots, \mathbf{x}_M^T]^T$, the transmitted set of sequences in baseband, where the vector $\mathbf{x}_m \triangleq [x_{m,1}, \dots, x_{m,N}]^T \in \mathbb{C}^N$ indicates the N samples of the mth transmitter ($m \in \{1, \dots, M\}$). The aim of the waveform design is twofold:

1. To design a set of transmit sequences having small cross-correlation among each other, to enable their separation at the receiver, and to enhance the angular resolution property using the concept of MIMO radars. To focus on the HW aspects, the detailed modeling of the cross-correlation is omitted with details available in Alaee-Kerahroodi et al. [2022]. It suffices to know that there exists a cost function $g_c(\mathbf{X})$ whose minimization ensures the required property on cross-correlation.
2. In addition, the waveforms need to be designed to have certain spectral properties to ensure utilization of unused frequencies while limiting the power transmitted in frequencies that are occupied by the communication signals. This way, the radar can use the full bandwidth, keep the interference low, and work without additional knowledge of the location(s) of the communication terminals. As above, the details are available in Alaee-Kerahroodi et al. [2022], and it suffices to mention the existence of a cost function $g_s(\mathbf{X})$ that ensures adherence to the desired spectral properties.

Additionally, the PMCW code is drawn from constant modulus constellation to enable efficient power amplification. Define $\Omega_\infty = [0, 2\pi)$, and $\Omega_L = \left\{0, \frac{2\pi}{L}, \dots, \frac{2\pi(L-1)}{L}\right\}$, to a continuous and discrete set of phase values, respectively. To ensure constant modulus, one of the constraints defined below is imposed on the design,

$$C_1 \triangleq \{\mathbf{X} \mid x_{m,n} = e^{j\phi_{m,n}}, \phi_{m,n} \in \Omega_\infty\}, \tag{12.1}$$

or

$$C_2 \triangleq \{\mathbf{X} \mid x_{m,n} = e^{j\phi_{m,n}}, \phi_{m,n} \in \Omega_L\}. \tag{12.2}$$

To this end, a bi-objective optimization problem, considering both the desired objectives, takes the form,

$$\begin{cases} \min\limits_{\mathbf{X}} & g_s(\mathbf{X}), g_c(\mathbf{X}) \\ \text{s.t.} & C_1 \text{ or } C_2. \end{cases} \tag{12.3}$$

Such problems are time-consuming to solve; instead, scalarization is a well-known technique that converts the bi-objective optimization problem to a single-objective problem through a weighted sum of the objective functions. While scalarization spans the solutions of bi-objective function only in certain settings, such a formulation is nonetheless pursued for the ease of its implementation. The scalarization subsequently leads to the following Pareto-optimization problem,

$$\mathcal{P} \begin{cases} \min\limits_{\mathbf{X}} & g(\mathbf{X}) \triangleq \theta g_s(\mathbf{X}) + (1-\theta) g_c(\mathbf{X}) \\ \text{s.t.} & C_1 \text{ or } C_2, \end{cases} \tag{12.4}$$

The coefficient $\theta \in [0,1]$ is a weight factor effecting trade-off between radar performance and coexistence. The problem in (12.4) is rather difficult to solve and suboptimal, yet fast and effective, iterative solutions are presented in Alaee-Kerahroodi et al. [2022]. These iterative methods naturally support the *on-the-fly* waveform adaptation for co-existence whence system parameters can be changed in between iterations without interrupting the flow.

12.5.6 Adaptive Receive Processing

The developed cognitive MIMO radar receiver is depicted in Figure 12.6. The sampling at the receiver starts at the onset of a trigger initiated by a transmitter to indicate the onset of transmission. A classical MIMO radar operation is undertaken where, in each receive channel, two filters matched to each of the two transmitted waveforms is implemented using the FFT-based technique. Consequently, a total of four-matched filter outputs are obtained at the receiver and four range-Doppler plots corresponding to the receive channels and transmitting waveforms are obtained by implementing FFT in the slow-time dimension.

The receiver processing unit adapts the matched filter as and when the transmit waveform is changed.

12.5.7 Experimental Setup

In the sequel, a selection of experiments conducted using the developed prototype are presented and the HW results are analyzed. Given the frequency of operation and the bandwidths used, adequate separation is needed for target resolution. In the absence of a large experimental facility,

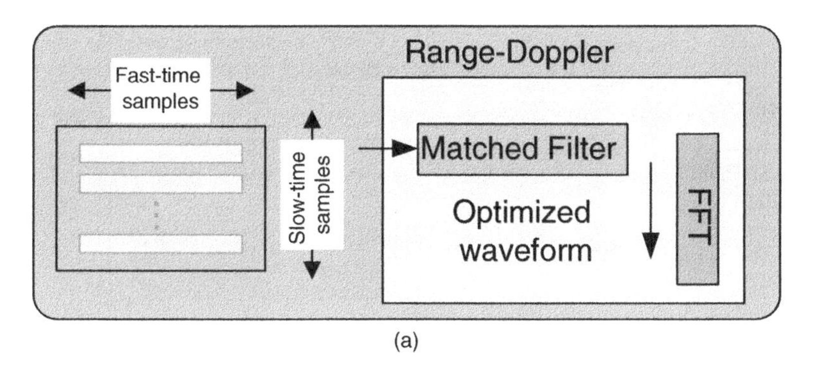

(a)

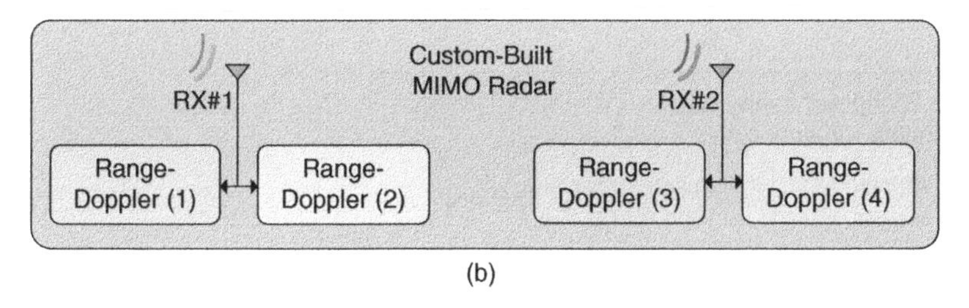

(b)

Figure 12.6 In (a), the matched filter coefficients are updated for appropriate filtering in the fast-time domain. Consequently, the range-Doppler plots are calculated after performing Fast Fourier Transform (FFT) in the slow-time dimension. The part related to (a) will be repeated in (b) for every transmit waveform. (a) Processing unit for range-Doppler creation related to every optimized waveform. (b) Block diagram of the receiver of the developed cognitive MIMO radar application.

for the practical applicability of the developed prototype and for the verification, a complete OTA evaluation is discarded. Instead, the connections shown in Figure 12.7 using RF cables and splitters/combiners are setup, and the performance measured in a controlled environment.

12.5.7.1 Target Generation

The transmit waveforms are attenuated using 30 dB attenuators highlighted in Figure 12.7. This attenuated signal is further shifted in time, frequency, and spatial dimension in software to create a reflection from a target as depicted in Figure 12.8. While a number of reflections, one corresponding to each target, can be generated, two targets are considered in the JRC prototype for ease of implementation. The target reflections are further perturbed with the communications interference.

12.5.7.2 Transmit Waveforms

Two choices are made available for the transmit waveforms: they could either be selected based on the options in Table 12.6 or obtained from the adaptation in Section 12.5.5. During its execution, input parameters to optimize the waveforms provided by the Graphical User Interface (GUI) to MATLAB, and the optimized set of sequences are returned to the application through the GUI. LabVIEW G dataflow application is used to develop the remaining processing blocks of the radar system including matched filtering, Doppler processing, and scene generation. Tables 12.7 and 12.8 summarize the parameters used for radar and targets in this experiment.

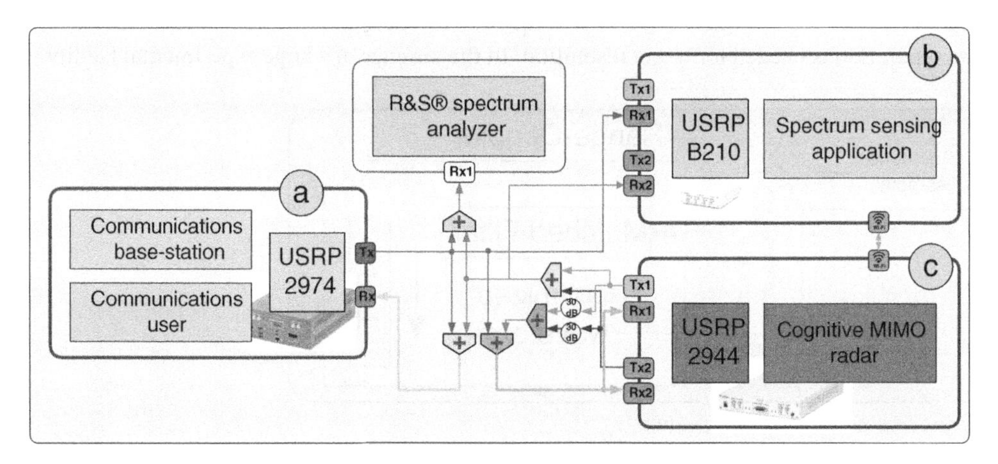

Figure 12.7 Diagram depicting the connection of the JRC coexistence prototype [Alaee-Kerahroodi et al., 2022].

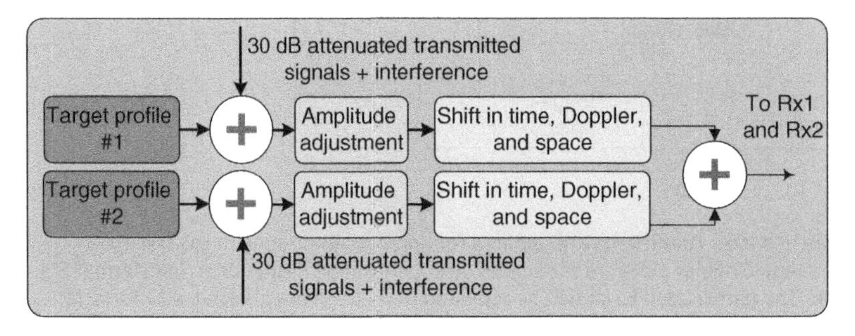

Figure 12.8 Schematic depicting target generation [Alaee-Kerahroodi et al., 2022].

Table 12.7 Radar experiment parameter setup.

Parameters	Value
Center frequency	2 GHz
Real-time bandwidth	40 MHz
Transmit and receive channels	2×2
Transmit power	10 dBm
Transmit code length	400
Duty cycle	50%
Range resolution	3.7 m
Pulse repetition interval	20 μs

Table 12.8 Radar target parameter setup.

Parameters	Target 1	Target 2
Range delay (μs)	2	2.6
Normalized Doppler (Hz)	0.2	−0.25
Angle (°)	25	15
Attenuation (dB)	30	35

Table 12.9 Communication link parameter setup.

Parameters	Value
Communication Modulation and Coding Schemes (MCS)	MCS0 (QPSK 0.12) MCS10 (16QAM 0.33) MCS17 (64QAM 0.43)
Center frequency (Tx and Rx)	2 GHz
Bandwidth	20 MHz

12.5.7.3 Communication Setup

For the LTE communications, the downlink between a SISO BS and a single user with one antenna is established. It should however be noted that the experiments can be also be performed with uplink LTE. LabVIEW LTE framework offers the possibility to vary the modulation and coding schemes (MCS) of physical downlink shared channel (PDSCH), used for the transport of data between the BS and the user, from 0 to 28 with the constellation size increasing from quadrature phase shift keying (QPSK) to 64QAM National Instruments Corp. [2022]. Table 12.9 indicates the experimental parameters for the communications.

12.5.8 Performance Evaluation

12.5.8.1 Representative Results

From Section 12.5.5 and following [Alaee-Kerahroodi et al., 2022], it is clear that when $\theta = 0$, the optimized waveforms are unable to exhibit notches at the undesired frequencies. The notches appear gradually with an increase in θ, with the deepest notch appearing when $\theta = 1$. However,

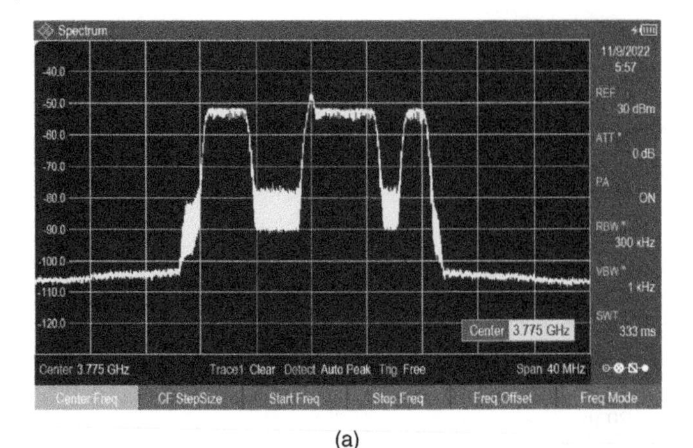

(a)

(b)

Figure 12.9 (a) Validation of the developed spectrum sensor using a commercial product for LTE downlink spectrum. ((a) LTE spectrum at R&H spectrum analyzer. (b) Spectrum at the developed application.) (b) Resulting spectrum of communications (yellow) and radar (dotted-red) at the developed two-channel spectrum sensing application [Alaee-Kerahroodi et al., 2022].

when $\theta = 1$, the cross-correlation is highest (phased-array), thereby diminishing the target resolution properties. The cross-correlation decreases with θ, with the minimum being at $\theta = 0$. Thus, an appropriate selection of θ leads to an optimized trade-off between spectral shaping and orthogonality.

Toward demonstrating the effect of an optimized radar waveform in a coexistence scenario, $\theta = 0.75$ is selected [Alaee-Kerahroodi et al., 2022]. Tables 12.8 and 12.9, respectively, report the values are used for radar and communications. When transmitting a set of $M = 2$ waveforms, each of length $N = 400$, the radar occupies a bandwidth of 40 MHz with nulls to be obtained adaptively based on the feedback from the spectrum-sensing application.

However, the LTE communications framework has 20 MHz bandwidth at its disposal for transmission. For full bandwidth transmission, the radar will not be able to perform satisfactorily due to the imposition of limited transmit power as a result of interference constraints. To simulate a meaningful spectrum coexistence scenario, some portions of this bandwidth are not used by the communications to enable their use by the radar. Toward this, the flexibility of the LTE application is exploited and a resource block allocation of 1111111111110000000111111 (4 physical resource blocks/bit) is considered. Here, an entry “1” at a position indicates the use of the corresponding time-bandwidth resources in the LTE application framework.

The spectrum of this LTE downlink is measured with the developed spectrum-sensing application as depicted in Figure 12.9. This figure serves two purposes:

1. Validates the developed spectrum analyzer application with a commercial product
2. Indicates the achievement of the desired spectrum shaping objective

When unaware of the spectrum use by the communications, the radar maximizes its target resolution capabilities by transmitting optimized sequences with $\theta = 0$ and consuming the entire bandwidth. This leads to significant mutual interference, disrupting the operations of both radar and communications, as shown in part (a) of Figures 12.10 and 12.11, thereby hampering their coexistence. In this situation, using the optimized waveforms obtained by $\theta = 0.75$ improves the performance of both systems, as illustrated in part (b) of Figures 12.10 and 12.11.

12.5.8.2 Signal to Interference Plus Noise Ratio (SINR) Performance Analysis

To further analyze the performance of the proposed prototype, the Signal to Interference plus Noise Ratio (SINR) of the two targets is calculated for radar, while the PDSCH throughput calculated by the LTE application framework is reported. The experimentation methodology is highlighted as follows:

Step-1: **No radar transmission**. In the step, the LTE PDSCH throughput for different MCS, i.e. MCS0, MCS10, and MCS 17, are collected. For each MCS, 5,10, 15, and 20 dBm of LTE transmit power are used in accordance with link-budget.

Step-2: **No LTE transmission**. In the stage, the radar utilizes the entire bandwidth and generates optimized waveform by setting $\theta = 0$. The resulting received signal to noise ratio (SNR) for the two targets is obtained numerically as the ratio of the peak power of the detected targets

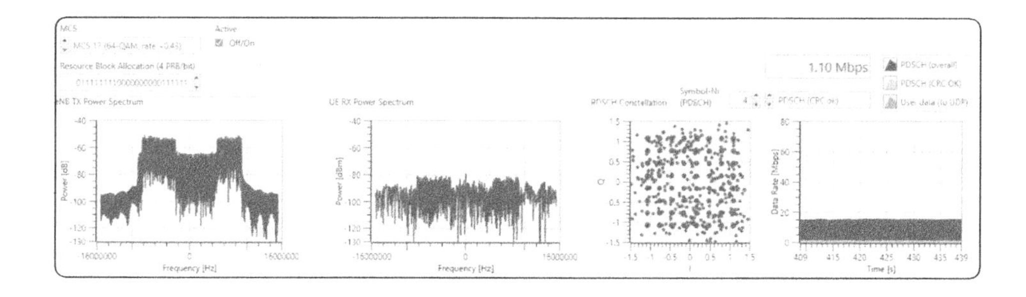

(a)

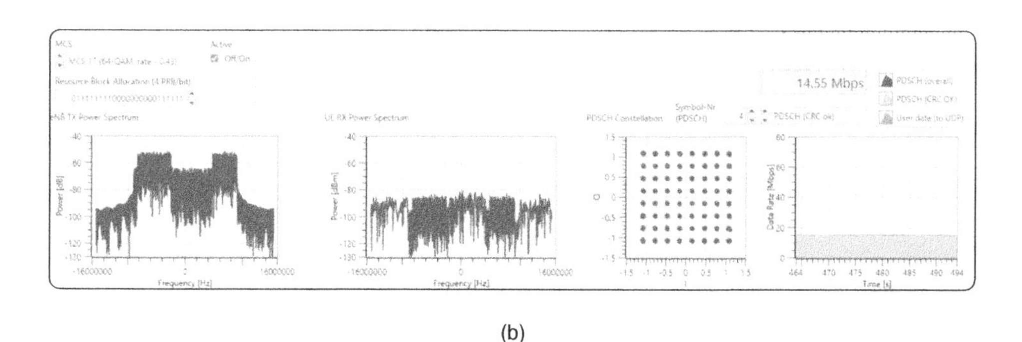

(b)

Figure 12.10 (a) Radar is transmitting random-phase sequences on the communications frequency band leading to a reduction in throughput of the latter. (b) Transmission of optimized MIMO radar waveforms enhances the performance of communications. (a) LTE in the presence of radar interference that occupies entire band by utilizing optimized sequences ($\theta = 0$). (b) LTE in the presence of optimized sequences for the coexistence scenario ($\theta = 0.75$).

to the average power of the cells in the vicinity of the target location (on the range-Doppler map).

Step-3: **Concurrent transmission**. A set of optimized radar waveforms for $\theta = 0$ are transmitted along with the LTE waveform, thereby causing significant mutual interference. The PDSCH throughput as well as the SINR of Target-1 and Target-2 are noted for MCS0, MCS10, and MCS17 and at 5,10, 15, and 20 dBm power levels keeping the radar transmit power fixed. To avoid bias in the results, the experiment is repeated five times for each setting and the performance indicators are averaged.

Step-4: **Coexistence**. Step-3 is now repeated, but with the waveforms optimized for $\theta = 0.75$.

Figure 12.12 reports the obtained PDSCH throughput, which is representative of the number of successfully decoded bits [Alaee-Kerahroodi et al., 2022]. These results show that the link throughput degrades in the presence of radar interference; the degradation being higher for higher MCS, since MCS are susceptible to variations in SINR. Subsequently, the LTE throughput

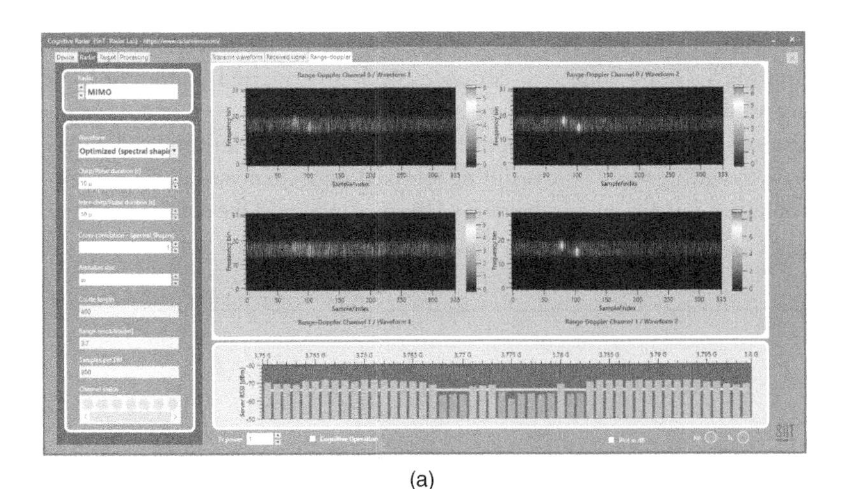

(a)

(b)

Figure 12.11 (a) MIMO radar utilizing optimized waveforms and not taking cognizance of communications transmissions ($\theta = 0$) results in interference from communications which degrades the radar detection performance. (b) MIMO radar utilizing optimized sequences for enabling coexistence ($\theta = 0.75$) resulting in the enhanced performance of radar.

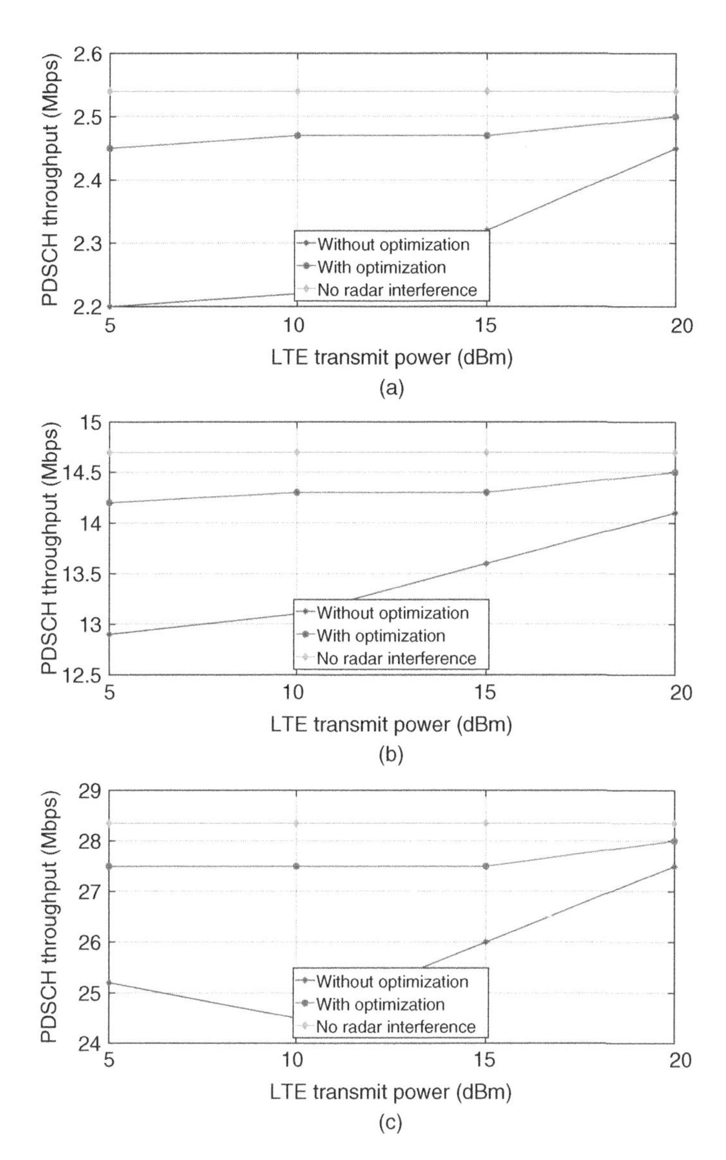

Figure 12.12 PDSCH throughput of LTE under radar interference for different MCS: (a) MCS 0 (QPSK 0.12), (b) MCS 10 (16QAM 0.33), (c) MCS 17 (64QAM 0.43). Radar interference reduces PDSCH throughput, but cognitive spectrum sensing followed by spectral shaping of the radar waveform improves PDSCH throughput for all LTE MCS. Source: Alaee-Kerahroodi et al. [2022].

improves when the radar optimizes its waveform with $\theta = 0.75$ and the improvement is prominent in the higher MCS.

Figure 12.13 illustrates the SINR performance of the two targets in the presence of LTE interference. SINR improves for both targets when the radar optimizes the transmit waveforms by setting $\theta = 0.75$. Interestingly, for high LTE transmission power (15 and 20 dBm), the improvement is higher; this results from the avoidance of the occupied LTE bands. As a case in point, when the communications system is transmitting with a power of 20 dBm, the use of optimized waveforms enhances the target SINR in excess of 7 dB over all the MCS values. Further, the achieved SINR of Target-1 and Target-2 in the absence of the LTE interference serves as the benchmark. These values

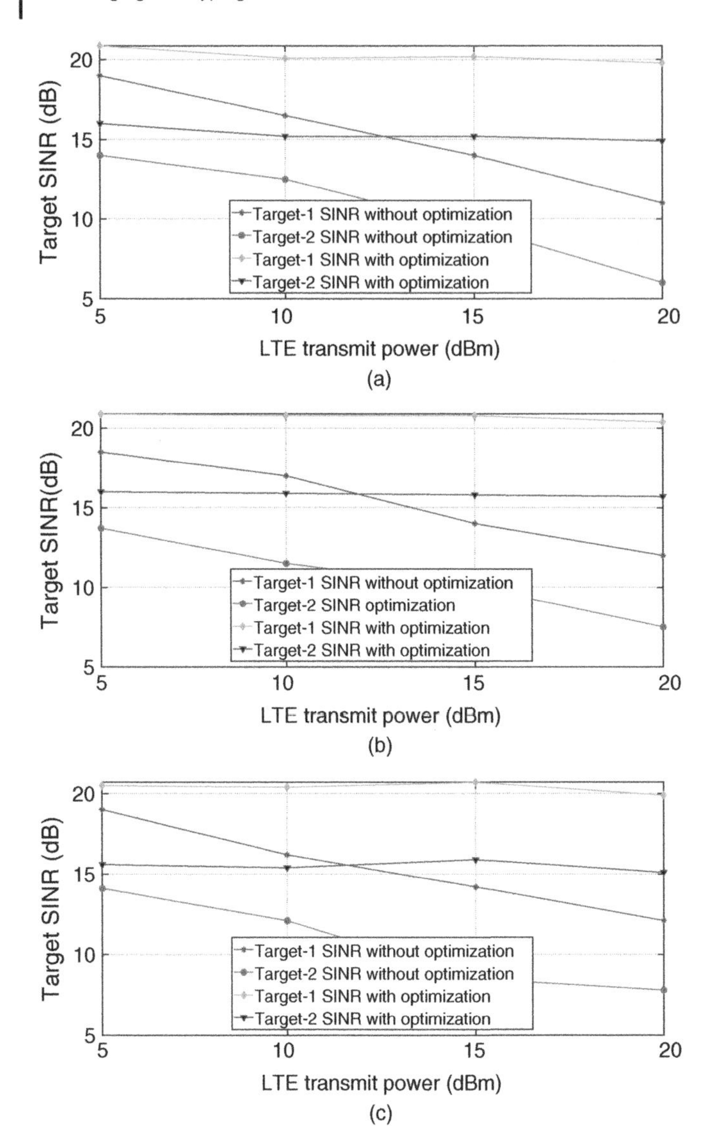

Figure 12.13 Target SINR under interference from downlink LTE link for different MCS: (a) MCS 0 (QPSK 0.12), (b) MCS 10 (16QAM 0.33), (c) MCS 17 (64QAM 0.43). forms are used. An upper bound on Signal to Noise Ratio (SNR) for the first and second targets was 22 and 17 dB, respectively, in the absence of communications interference.

are is 22 and 17 dB, respectively. Finally, as a consequence of considering different attenuation paths for the two targets (kindly refer Table 12.7), their measured SINRs are different.

12.6 Other JRC Prototypes

12.6.1 Low-rate Coexistence Prototype

In Cohen et al. [2017], a spectral coexistence prototype based on Xampling (SpeCX) was mentioned. It employed a low-rate ADC which filters the received signal to predetermined frequencies before taking pointwise samples. These compressed samples, or "Xamples," contain the information needed to recover the desired signal parameters. The SpeCX prototype consists of separate units of a cognitive radio receiver and an emulated cognitive radar transceiver. At the heart of the cognitive radio system lies a modulated wideband converter (MWC) that achieves the lower sampling rate bound and implements the sub-Nyquist analog front-end receiver. The card first splits the wideband signal into 4 hardware channels, with an expansion factor of 5, yielding 20 virtual channels after digital expansion. In each channel, the signal is then mixed with a periodic sequence with 20 MHz bandwidth that is generated on a dedicated FPGA. The sequences are chosen as truncated versions of Gold Codes [Gold, 1967], commonly used in telecommunication and satellite navigation. These were heuristically found to give good detection results in the MWC system, primarily due to small bounded cross-correlations within a set. This is useful when multiple devices are broadcasting in the same frequency range.

Next, the modulated signal passes through an analog anti-aliasing low-pass filter (LPF). Specifically, a Chebyshev LPF of seventh order with a cut-off frequency (−3 dB) of 50 MHz was chosen for the implementation. Finally, the low rate analog signal is sampled by a National Instruments© ADC operating at sampling rate of 120 MHz (with intended oversampling), leading to a total sampling rate of 480 MHz. The digital receiver is implemented on a National Instruments PXIe-1065 computer with DC coupled ADC. Since the digital processing is performed at the low rate of 120 MHz, very low computational load is required in order to achieve real-time recovery. MATLAB and LabVIEW platforms are used for the various digital recovery operations. The sampling matrix is computed only once off-line using a calibration process.

The prototype is fed with RF signals composed of up to 5 real communication transmissions, namely 10 spectral bands with total bandwidth occupancy of up to 200 MHz and varying support, with Nyquist rate of 6 GHz. Specifically, to test the system's support recovery capabilities, an RF input is generated using vector signal generators (VSGs), each producing a modulated data channel with individual bandwidth of up to 20 MHz, and carrier frequencies ranging from 250 MHz up to 3.1 GHz. The input transmissions then go through an RF combiner, resulting in a dynamic multiband input signal that enables fast carrier switching for each of the bands. This input is specially designed to allow testing the system's ability to rapidly sense the input spectrum and adapt to changes, as required by modern cognitive radio and shared spectrum standards, e.g. in the SSPARC program. The system's effective sampling rate, equal to 480 MHz, is only 8% of the Nyquist rate and 2.4 times the Landau rate. This rate constitutes a relatively small oversampling factor of 20% with respect to the theoretical lower sampling bound. The main advantage of the Xampling framework, demonstrated here, is that sensing is performed in real-time from sub-Nyquist samples for the entire spectral range, which results in substantial savings in both computational and memory complexity.

Support recovery is digitally performed on the low rate samples. The prototype successfully recovers the support of the communications transmitted bands. Once the support is recovered, the signal itself can be reconstructed from the sub-Nyquist samples in real time. The reconstruction does not require interpolation to the Nyquist rate, and the active transmissions are recovered at the low rate of 20 MHz, corresponding to the bandwidth of the aliased slices. The prototype's digital recovery stage is further expanded to support decoding of common communication modulations, including BPSK, QPSK, QAM, and OFDM. There are no restrictions regarding the modulation type, bandwidth or other signal parameters, since the baseband information is exactly reconstructed regardless of its respective content.

By combining both spectrum sensing and signal reconstruction, the MWC prototype serves as two separate communications devices. The first is a state-of-the-art cognitive radio that performs real-time spectrum sensing at sub-Nyquist rates, and the second is a unique receiver able to decode

Table 12.10 SpaCor prototype functionalities.

Baseband waveform generation	Radar echo generation
OTA signaling	Radar echo reception
Frequency band waveform transmission	Communication signal reception

multiple data transmissions simultaneously, regardless of their carrier frequencies, while adapting to spectral changes in real time.

The cognitive radar system included a custom made sub-Nyquist radar receiver board composed of four parallel channels which sample four distinct bands of the radar signal spectral content. In each channel, the transmitted band with bandwidth 80 kHz is filtered, demodulated to baseband, and sampled at 250 kHz (with intentional oversampling). This way four sets of consecutive Fourier coefficients are acquired. After sampling, the spectrum of each channel output is computed via FFT and the 320 Fourier coefficients are used for digital recovery of the delay-Doppler map. The prototype simulates transmission of 50 pulses toward 9 targets. The cognitive radar transmits over 4 bands, selected according to an optimization procedure and occupying 3.2% of the traditional wideband radar bandwidth.

12.6.2 Index Modulation Based Codesign Prototype

A particular prototype involving the codesign of the radar and communication functionality from a system level is presented in Ma et al. [2021]. The system considers different bands for the radar and communication systems as well as separate waveforms; thus, there is no interference among the two systems and the waveform selection is flexible. However, the use of multiband signaling leads to enhanced system complexity. Herein, the authors being in the concept of codesign and devise and demonstrate a spatial modulation-based communication-radar (SpaCoR) system. Central to the work is the use of Index Modulation through the generalized spatial modulation (GSM) wherein the index of antenna element chosen is determined by the information bits to be transmitted; such a design, while reducing the number of active RF chains, also induces spatial agility. In this context, the authors in Ma et al. [2021] restrict an antenna element to transmit either the radar or communication waveform. To enhance the performance of such a system, an enhanced allocation of antennas to waveforms is presented in Ma et al. [2021] – leading to a codesign.

To demonstrate the feasibility of their system, the authors have implemented the *SpaCoR prototype* using a dedicated hardware. This versatile prototype features aspects enable its reuse in different dual function systems:

The overall structure of the prototype combines the following:

1. Host processing on a PC server offering a GUI for setting the relevant parameter, generating the waveforms and processing the received signals. Once the parameters are set, the JRC waveform is generated by the PC application at the onset of the experiment. Subsequently, the JRC waveform is transferred to the transmit board discussed next.
2. A transmitter board comprising a FPGA board to realize a high speed data interface between the digital PC generated waveforms and the digital-to-analog convertor (DAC), 4DSP FMC216 DAC cards, and an up-conversion card for RF transmission.
3. A 2D digital antenna array with 16 elements with the ability to control each element independently. The antenna works with carrier frequency 5.1 GHz has a bandwidth of 80 MHz and

the size of each patch is 1.8 cm ×1.3 cm. The horizontal and the vertical distance between two horizontally and the two vertically adjacent elements are both 2.7 cm. The array is used in the configuration with eight transmit and eight receive antennas [Ma et al., 2021].
4. A receiver board converting the passband analog echoes to baseband digital streams for further processing on the host. The receiver board consists of a VC707 FPGA board, two FMC168 analog-to-digital convertors (ADCs) cards, and a RF down-convertor board.
5. A radar echo generator (REG) uses the transmitted waveform and generates the reflected echoes corresponding to moving radar targets in an OTA setup. This unit consists of a Rhode & Schwarz FSW signal and spectrum analyzer, which captures the received waveform, and a Rhode & Schwarz SWM200 A VSG, to add the setup delays and Doppler shifts to the observed waveform for re-transmission. The transmit and receive antennas are of dimensions 5 cm × 5 cm.

Experiments on the prototype are carried out with a radar pulse of 30 μs communication baud rate of 0.4 MSps (Mega symbols per second), radar frequency band of 5.06–5.11 GHz, and the communication band of 5.11–5.14 GHz. The 16 bit DAC and ADCs are used with DAC update rate of 312.5 MSps and ADC sampling rate of 250 MSps. The results from Ma et al. [2021] indicate the spatial agility induced by the GSM transmission over fixed antenna allocations. In particular, it improves the angular resolution and reduces the sidelobe level in the transmit beam pattern as well the Bit-error rate (BER) performance of the communication system.

12.7 Conclusion

This chapter presents the need for JRC prototyping and discusses the requirements for combining existing communication or radar prototypes to enable joint functionality. A coexistence prototype is detailed, and two other are summarized. These prototypes and their demonstration bring the research in this emerging field already closer to the practitioners to enable early cross-fertilization of ideas and incorporation in upcoming initiatives like 6G. In addition, there have been a number of other important developments in the field of prototyping which is not captured in this book chapter. An useful initiative toward collating the requisite information in a single place is the Demonstration and Datasets Working Group [WG5] of the Integrated Sensing and Communications Emerging Technology Initiative from IEEE Communication Society [https://isac.committees.comsoc.org/demonstrations-datasets].

References

M. Alaee-Kerahroodi, E. Raei, S. Kumar, and M.R. Bhavani Shankar. Cognitive radar waveform design and prototype for coexistence with communications. *IEEE Sensors Journal*, 22(10):9787–9802, 2022. doi: 10.1109/JSEN.2022.3163548.

J. Bang, H. Chung, J. Hong, H. Seo, J. Choi, and S. Kim. Millimeter-wave communications: recent developments and challenges of hardware and beam management algorithms. *IEEE Communications Magazine*, 59(8):86–92, 2021. doi: 10.1109/MCOM.001.2001010.

C.B. Barneto, S.D. Liyanaarachchi, M. Heino, T. Riihonen, and M. Valkama. Full duplex radio/radar technology: the enabler for advanced joint communication and sensing. *IEEE Wireless Communications*, 28(1):82–88, 2021. doi: 10.1109/MWC.001.2000220.

A.F. Cattoni, H.T. Nguyen, J. Duplicy, D. Tandur, B. Badic, R. Balraj, F. Kaltenberger, I. Latif, A. Bhamri, G. Vivier, I.Z. Kovácsk, and P. Horváth. Multi-user mimo and carrier aggregation in 4G systems: the samurai approach. In *2012 IEEE Wireless Communications and Networking Conference Workshops (WCNCW)*, pages 288–293, 2012. doi: 10.1109/WCNCW.2012.6215508.

D. Cohen, K.V. Mishra, and Y.C. Eldar. Spectrum sharing radar: coexistence via Xampling. *IEEE Transactions on Aerospace and Electronic Systems*, 54(3):1279–1296, 2017.

R.C. Daniels, E.R. Yeh, and R.W. Heath. Forward collision vehicular radar with IEEE 802.11: feasibility demonstration through measurements. *IEEE Transactions on Vehicular Technology*, 67(2):1404–1416, 2017.

S.H. Dokhanchi, M.R.R. Bhavani Shankar, K.V. Mishra, and B. Ottersten. A mmWave automotive joint radar-communications system. *IEEE Transactions on Aerospace and Electronic Systems*, 55(3):1241–1260, 2019.

G. Duggal, S. Vishwakarma, K.V. Mishra, and S.S. Ram. Doppler-resilient 802.11ad-based ultrashort range automotive joint radar-communications system. *IEEE Transactions on Aerospace and Electronic Systems*, 56(5):4035–4048, 2020.

O. Edfors, L. Liu, F. Tufvesson, N. Kundargi, and K. Nieman. Massive mimo for 5G: theory, implementation and prototyping. In F.-L. Luo and C.J. Zhang, editors, *Signal processing for 5G: algorithms and implementations*, Chapter 9, pages 191–230. John Wiley & Sons, 2016.

A.M. Elbir, K.V. Mishra, and S. Chatzinotas. Terahertz-band joint ultra-massive MIMO radar-communications: model-based and model-free hybrid beamforming. *IEEE Journal of Special Topics in Signal Processing*, 15(6):1468–1483, 2021.

R. Fagan, F.C. Robey, and L. Miller. Phased array radar cost reduction through the use of commercial RF systems on a chip. In *2018 IEEE Radar Conference (RadarConf18)*, pages 0935–0939, 2018. doi: 10.1109/RADAR.2018.8378686.

R. Gold. Optimal binary sequences for spread spectrum multiplexing (Corresp.). *IEEE Transactions on Information Theory*, 13(4):619–621, 1967.

M. Gul, E. Ohlmer, A. Aziz, W. McCoy, and Y. Rao. Millimeter waves for 5G: from theory to practice. In F.-L. Luo and C.J. Zhang, editors, *Signal processing for 5G: algorithms and implementations*, Chapter 14, pages 321–353. John Wiley & Sons, 2016.

L.S. Jensen. *Design by prototyping in hardware start-ups*. Department of Mechanical Engineering, Technical University of Denmark, Lyngby, Denmark, 2018. https://backend.orbit.dtu.dk/ws/portalfiles/portal/189015354/S256_Lasse_Skovgaard_Jensen_PhD_Thesis.pdf.

P. Kumari, A. Mezghani, and R.W. Heath. A low-resolution ADC proof-of-concept development for a fully-digital millimeter-wave joint communication-radar. In *IEEE International Conference on Acoustics, Speech and Signal Processing*, pages 8619–8623, 2020.

N. Kunicina, A. Zabasta, A. Patlins, I. Bilic, and J. Peksa. Prototyping process in education and science. In *2020 IEEE 61th International Scientific Conference on Power and Electrical Engineering of Riga Technical University (RTUCON)*, pages 1–6, 2020. doi: 10.1109/RTUCON51174.2020.9316550.

KU Leuven. KU Leuven massive MIMO 5G Testbed, 2021. https://www.esat.kuleuven.be/telemic/research/NetworkedSystems/infrastructure/massive-mimo-5g.

J. Liu, K.V. Mishra, and M. Saquib. Co-designing statistical MIMO radar and in-band full-duplex multi-user MIMO communications. *IEEE Transactions on Aerospace and Electronic Systems*, 2022. doi: 10.48550/arXiv.2006.14774.

D. Ma, N. Shlezinger, T. Huang, Y. Shavit, M. Namer, Y. Liu, and Y.C. Eldar. Spatial modulation for joint radar-communications systems: design, analysis, and hardware prototype. *IEEE Transactions on Vehicular Technology*, 70(3):2283–2298, 2021. doi: 10.1109/TVT.2021.3056408.

K.V. Mishra. Frequency diversity wideband digital receiver and signal processor for solid-state dual-polarimetric weather radars. Master's thesis, Colorado State University, 2012.

K.V. Mishra, M.R.R. Bhavani Shankar, V. Koivunen, B. Ottersten, and S.A. Vorobyov. Toward millimeter wave joint radar-communications: a signal processing perspective. *IEEE Signal Processing Magazine*, 36(5):100–114, 2019a.

K.V. Mishra, Y.C. Eldar, E. Shoshan, M. Namer, and M. Meltsin. A cognitive sub-Nyquist MIMO radar prototype. *IEEE Transactions on Aerospace and Electronic Systems*, 56(2):937–955, 2019b.

National Instruments Corp. Overview of the LabVIEW communications application frameworks, 2021. https://www.ni.com/en-gb/innovations/white-papers/14/overview-of-the-labview-communications-application-frameworks.html.

National Instruments Corp. NI Labview LTE framework, 2022. https://www.ni.com/en-us/support/documentation/supplemental/16/labview-communications-lte-application-framework-2-0-and-2-0-1.html.

D. O'Connor. Early days of phased array radars for ballistic missile detection and tracking. In *2019 International Radar Conference (RADAR)*, pages 1–5, 2019. doi: 10.1109/RADAR41533.2019.171388.

P.E. Pace. *Advanced techniques for digital receivers*. Artech House, 2000.

B. Paul, A.R. Chiriyath, and D.W. Bliss. Survey of RF communications and sensing convergence research. *IEEE Access*, 5:252–270, 2016.

J.B. Tsui. *Digital techniques for wideband receivers*, volume 2. SciTech Publishing, 2004.

J.B.Y. Tsui. *Special design topics in digital wideband receivers*. Artech House, 2010.

WARP Project. The WARP project, 2021. http://warpproject.org.

P.W. Wolniansky, G.J. Foschini, G.D. Golden, and R.A. Valenzuela. V-blast: an architecture for realizing very high data rates over the rich-scattering wireless channel. In *1998 URSI International Symposium on Signals, Systems, and Electronics. Conference Proceedings (Cat. No.98EX167)*, pages 295–300, 1998. doi: 10.1109/ISSSE.1998.738086.

L. Wu, K.V. Mishra, M.R.R. Bhavani Shankar, and B. Ottersten. Resource allocation in heterogeneously-distributed joint radar-communications under asynchronous Bayesian tracking framework. *IEEE Journal on Selected Areas in Communications*, 40(7):2026–2042, 2022.

13

Secrecy Rate Maximization for Intelligent Reflective Surface-Assisted MIMO Communication Radar

Sisai Fang[1], *Gaojie Chen*[1], *Sangarapillai Lambotharan*[2], *Cunhua Pan*[3], *and Jonathon A. Chambers*[4]

[1] *University of Surrey, 5GIC & 6GIC, Institute for Communication Systems, Department of Electrical and Electronic Engineering (ICS), Guildford, UK*
[2] *Loughborough University, School of Mechanical, Manufacturing and Electrical Engineering, Leicestershire, Loughborough, UK*
[3] *Queen Mary University of London, School of Electronic Engineering and Computer Science, London, UK*
[4] *University of Leicester, School of Engineering, Leicestershire, Leicester, UK*

13.1 Introduction

In the last two decades, multiple-input multiple-output (MIMO) technology has been shown to be very powerful in wireless communications for enhancing capacity and coverage and incorporated in 4G and 5G standards. Motivated by this success, MIMO technology in radar has attracted considerable attention recently in academia and industries. Compared to phased-array radars, which transmit scaled versions of the same waveform [Fenn et al., Mar., 2000], a MIMO radar emits orthogonal waveforms via multiple antennas, which can enhance the virtual aperture and also provide a better diversity due to different radar cross sections seen from various aspect angles [Li and Stoica, 2009]. There are two main MIMO radar configurations: the first one is based on colocated antennas and called mono-static radar [Li and Stoica, 2007], and the second one employs widely separated antennas and termed distributed or bi-static/multi-static radar [Haimovich et al., 2008]. These advantages have led many researchers to apply this technology in a range of problems including beamforming, waveform design, target detection optimization, and radar imaging [Fuhrmann et al., 2010; Duly et al., 2013; Tajer et al., 2010].

Growing demand for information exchange over mobile networks is creating an explosion of digital traffic volumes in emerging wireless networks, which leads to an increasing demand on radio spectrum sharing technology. Within this backdrop, coexistence of radar and wireless communications has been receiving significant interest to reduce the demand for new spectrum [Griffiths et al., 2013; Bliss, 2014]. A pioneering joint radar-communications system was introduced by Blunt et al. [2010] to embed the information into the radar signal. An efficient exploitation of the assigned spectrum and cross-interference minimization was investigated in Aubry et al. [2015]. Spatial filtering and adaptive radar beamforming were proposed to maximize the signal-to-interference plus noise ratio (SINR) in Deng and Himed [2013] and Geng et al. [2015], respectively. Hassanien et al. [2016a] proposed a novel dual-function scheme, which employs the same transmit and receive antennas for simultaneous radar operation and information transfer to legitimate receivers (LRs). A set of unique radar waveforms with an embedded communication data signal is conceived while satisfying the

Signal Processing for Joint Radar Communications, First Edition.
Edited by Kumar Vijay Mishra, M. R. Bhavani Shankar, Björn Ottersten, and A. Lee Swindlehurst.

radar performance. Since multisensor radar systems can control variations in the sidelobe level (SLL) toward a specific spatial direction, dual-function radar–communications (DFRC) systems can not only maintain a radar function in the main lobe of the signal but also realize data communications in the sidelobes. For example, Euzière et al. [2014] utilized sparse time-modulated array (STMA) and phase-only synthesis time-modulated array (POSTMA) to control the SLL toward the desired direction. A DFRC system employing sidelobe control of the transmit beamforming and waveform diversity was developed by Hassanien et al. [2016b], where two transmit weight vectors were designed to carry multiple simultaneously transmitted orthogonal waveforms, embedding a sequence of information bits. Motivated by the benefits of MIMO technology, Qian et al. [2018] investigated the joint design of a MIMO radar with colocated antennas and a MIMO communications system, to maximize the SINR at the radar receiver while satisfying the achievable rate for the communications system. Liu et al. [2018] developed a series of optimization-based transmit beamforming approaches for a joint MIMO radar–communications system, which is considered as a dual-functional platform that can simultaneously transmit probing signals to the targets and serve multiple downlink users. More recently, Liu et al. [2020] proposed a joint transmit beamforming model for a dual-function MIMO radar and multiuser MIMO communication transmitter, where the joint transmitter utilizes jointly precoded individual communication and radar waveforms.

Information security is key for future wireless communications networks. Especially, physical layer security (PLS) is a complementary technology to cryptographic methods because of its low complexity and direct exploitation of the physical propagation medium for creating confidential communication links in the presence of malicious eavesdroppers. The authors of Krikidis et al. [2009], Dong et al. [2010], and Ding et al. [2012] utilized cooperative transmission to enhance secrecy communication. Buffer-aided relay selection was proposed to improve the PLS in cooperative networks by Chen et al. [2014] and Nie et al. [2020]. Friendly jammers have been introduced in Zheng et al. [2011] to confuse the eavesdroppers through transmission of interfering signals toward them, which will increase the PLS of a wiretap fading channel. Full-duplex technology was used to mitigate the SNR at the passive eavesdroppers [Chen et al., 2015]. Due to the additional degrees of freedom and the diversity gain offered by the MIMO technology, Khisti and Wornell [2010] designed secret MIMO communications systems and analyzed the capacity of the Gaussian wiretap channel model when there were multiple antennas at the transmitter, intended receiver, and eavesdropper. Alageli et al. [2020] investigated the maximization of secrecy rate against a multiple-antenna eavesdropper. Chu et al. [2015] used a multiple-antenna cooperative jammer to induce further interference to the eavesdropper and to maximize the secrecy rate at the legitimate receiver. Nguyen et al. [2018] proposed two artificial noise (AN)-aiding schemes to secure confidential information in a massive MIMO network. Except for the consideration of traditional ground communications, Zhu et al. [2018] applied a jamming strategy to enhance the secrecy performance of unmanned aerial vehicle (UAV) networks over the millimeter-wave band. Then, Tang et al. [2019] proposed an enhancement of security by using friendly jamming in the presence of UAV eavesdroppers. Motivated by the benefits of the PLS, secrecy enhancement has also been considered in DFRC systems for target detection in the presence of a potential eavesdropper. For example, Deligiannis et al. [2018] utilized MIMO radar to transmit the jamming and desired data signals simultaneously to confuse an eavesdropper while preserving radar target detection criteria. Several optimization problems were studied, including secrecy rate maximization, target return SNR maximization, and transmit power minimization. A joint passive radar and communications system was considered by Chalise and Amin [2018] to balance a trade-off between the SINR at the radar receiver and the information secrecy rate. To guarantee the secrecy of the legitimate user in the communications system, an optimization problem was designed to maximize the SNR at the passive radar receiver while keeping the secrecy rate above a certain threshold. Su et al. [2021] investigated the

PLS in a MIMO DFRC system, where the radar targets were considered as potential eavesdroppers that eavesdropped the communication signals intended for the legitimate users. To ensure transmission secrecy, AN was utilized at the transmitter and the optimization was aimed at minimizing the SNR at the eavesdropping target receiver, while guaranteeing a specific SINR at the legitimate users.

More recently, the intelligent reflecting surface (IRS) concept has been emerging as a potential enabling technology for enhancing physical layer performance of next-generation wireless communications networks [Liaskos et al., 2018]. Different from traditional transmission and reception techniques, IRSs are metasurfaces consisting of an array of discrete reflective elements, which can manipulate incident electromagnetic waves according to a desired functionality, such as reflection in an intended direction, polarization, or filtering. PLS was introduced for the first time in IRS-assisted systems by Cui et al. [2019] and Shen et al. [2019]. The authors proposed a secrecy rate maximization problem, by jointly designing the transmit beamforming and the IRS reflection coefficients. The importance of AN in IRS-assisted communications networks was investigated, and the performance was analyzed in Guan et al. [2020]. More recently, Hong et al. [2020] considered an IRS-assisted AN-aided secure MIMO wireless communications system, where the base station (BS), the legitimate information receiver, and the eavesdropper are equipped with multiple antennas. Fang et al. [2020] maximized the secrecy rate by jointly designing the trajectory and optimum power in a secure IRS-assisted UAV system. To maximize the secrecy rate of an IRS-assisted system, the transmit precoding matrix at the BS, covariance matrix of AN and phase shifts at the IRS were jointly optimized subject to various constraints such as the transmit power and the unit modulus of IRS phase shifts. Dong and Wang [2020] enhanced the secrecy rate without the channel state information of the eavesdropper.

In this chapter, we consider a DFRC system that consists of a tracking radar with multiple antennas, a legitimate communication receiver and a target which also acts as an eavesdropper and is equipped with multiple antennas. The aim of the proposed radar system is to achieve the desired target detection performance and to transmit information to a legitimate receiver securely. To hinder the ability of the eavesdropper to decode the information and to improve the target detection, we propose to transmit AN in addition to the information signal. Thus, we design the beam steering vectors for both the communication signal and the distortion signal by solving two optimizations, namely, secrecy rate maximization and transmit power minimization. In addition, motivated by the benefits of IRS, we propose a novel IRS-assisted radar system to further enhance the secrecy rate and the target detection performance. In this case, we jointly optimize the transmit beamforming vectors at the radar and the phase shifts of the IRS. However, both the secrecy rate maximization and the transmit power minimization problems are non-convex. In order to reformulate the optimization problems as convex problems, we introduce the block coordinate descent (BCD) method, followed by a Lagrangian multiplier method for optimizing the transmit power and the majorization–minimization (MM) algorithm for optimizing the phase shifts of the IRS. Taylor series approximations are used to convert the SINR equation into convex forms. Finally, simulation results have been generated to demonstrate the potential IRS for enhancing secrecy rate in a DFRC system.

13.1.1 Notation

We use bold lowercase letters and bold uppercase letters to denote column vectors and matrices, respectively. $\mathbf{a}^H$ gives the Hermitian of the vector $\mathbf{a}$, $\mathbf{a}^T$, and $\mathbf{a}^*$ denote its transpose and conjugate operators, respectively. $[\mathbf{A}]_{i,j}$ is the element located on the ith row and jth column of matrix $\mathbf{A}$. $\mathbf{C} \odot \mathbf{D}^T$ is the Hadamard product of $\mathbf{C}$ and $\mathbf{D}$. The trace of a matrix $\mathbf{A}$ is represented by $\mathrm{Tr}(\mathbf{A})$, $|\mathbf{A}|$

denotes the determinant of **A**. Re $\{\cdot\}$ and arg $\{\cdot\}$ denote the real part of a complex value and the extraction of phase information. $\mathbf{I}_M$ stands for the $M \times M$ identity matrix. diag $\{\cdot\}$ and $(\cdot)^*$ denote the operator for diagonalization and the optimal value, respectively. The notation $[x]^+$ stands for $\max\{x, 0\}$.

13.2 System Model

We consider an IRS-assisted DFRC system that consists of a radar with multiple antennas, an IRS with M reflective elements, a legitimate receiver with multiple antennas, and a malicious target, as shown in Figure 13.1. We assume that the target which also acts as an eavesdropper employs multiple antennas for intercepting the data transmission between the radar and its LR. The two major aims for the radar are to secure a certain detection criterion for the target and to synchronously transmit information to its legitimate receiver, while hindering the ability of the eavesdropping target to intercept the information signal. To achieve these objectives, in addition to the radar–communication signal, the radar emits AN as a jamming signal to confuse the target. However, both the radar signal and the AN will be used constructively at the radar receiver for the detection of the target.

The radar consists of N_R colocated transmit/receive antennas. The legitimate receiver and the eavesdropping target are equipped with N_C and N_E antennas, respectively. To accomplish target detection and data communication, the radar emits a modulated waveform $s_1(t)$, which consists of L_1 information bits, as follows:

$$s_1(t) = \sum_{i=0}^{L_1-1} \delta_{1i} m(t - iT), \tag{13.1}$$

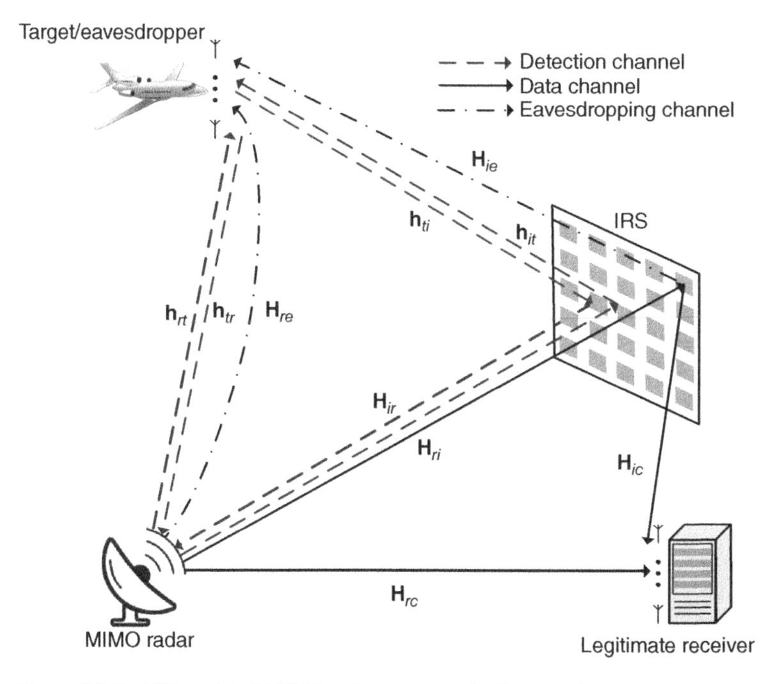

Figure 13.1 IRS-assisted MIMO radar–communications system.

where δ_{1i} denotes the ith information bit, $m(t) = \sum_{i=0}^{L_m-1} c_i \Pi(t - iT_c)$ defines the modulation waveform, c_i denotes the ith chip value of the modulation sequence, $\Pi(t)$ represents the rectangular pulse of duration T_c, $T = L_m T_c$ represents the information bit duration and L_m is the spreading gain as defined in Verdu [1998]. To interrupt the eavesdropper, the radar simultaneously transmits a pseudo-random distortional waveform $s_2(t)$, which is designed as:

$$s_2(t) = \sum_{i=0}^{L_2-1} \delta_{2i} \Pi(t - iT_c), \tag{13.2}$$

where δ_{2i} is the ith random bit of the distortional pseudo-random sequence and L_2 denotes the length of the sequence. Similar to Deligiannis et al. [2018], both $s_1(t)$ and $s_2(t)$ are assumed to have unit variance; however, the power of these signals will be controlled by the norm of the beamformer vectors. Besides, we assume L_1 and L_2 are large enough so that both the information signal and distortion signal have almost zero autocorrelation and correlation values, which will generate the desired range-Doppler ambiguity function for the radar.

13.2.1 MIMO Radar with IRS

We use the IRS to enhance the quality of signal transmission to the legitimate users, which indirectly allows additional degrees of freedom for the injection of AN to distort signals arriving at the eavesdropper. The channel coefficient matrices/vectors in the system are written as $\mathbf{H}_{xy}/\mathbf{h}_{xy}$, $x, y \in \{i, r, t, e, c\}$ (from node x to node y), where i, r, t, e, c denote the IRS, MIMO radar, target, eavesdropper, and the legitimate receiver, respectively. The channel coefficient matrices/vectors are shown as follows:

$$\mathbf{h}_{tr} = \mathbf{h}_{rt}^T = \mathbf{b}_1(\phi_1) \in \mathbb{C}^{N_R \times 1}, \tag{13.3a}$$

$$\mathbf{H}_{re} = \mathbf{b}_2(\phi_2)\, \mathbf{b}_1(\phi_1)^T \in \mathbb{C}^{N_E \times N_R}, \tag{13.3b}$$

where

$$\mathbf{b}_1(\phi_1) = \left[1, e^{j\frac{2\pi}{\lambda} d_R \sin\phi_1}, \ldots, e^{j\frac{2\pi}{\lambda}(N_R-1) d_R \sin\phi_1}\right] \in \mathbb{C}^{N_R \times 1}, \tag{13.4a}$$

$$\mathbf{b}_2(\phi_2) = \left[1, e^{j\frac{2\pi}{\lambda} d_E \sin\phi_2}, \ldots, e^{j\frac{2\pi}{\lambda}(N_E-1) d_E \sin\phi_2}\right] \in \mathbb{C}^{N_E \times 1}. \tag{13.4b}$$

The elements of $\mathbf{H}_{rc} \in \mathbb{C}^{N_C \times N_R}$, $\mathbf{H}_{ic} \in \mathbb{C}^{M \times N_C}$, $\mathbf{H}_{ri} \in \mathbb{C}^{M \times N_R}$, and $\mathbf{H}_{ir} \in \mathbb{C}^{N_R \times M}$ are generated using zero-mean circularly symmetric independent and identically distributed complex Gaussian random variables, and $\mathbf{h}_{ti} = \mathbf{h}_{it}^T \in \mathbb{C}^{M \times 1}$. In addition, d_R and d_E are the distances between the two adjacent elements/antennas of the MIMO radar and the eavesdropper, respectively, and λ denotes the wavelength of the carrier signal. ϕ_1 and ϕ_2 are the angle of the departure of the signal from the radar to the target and the angle of arrival of the signal from the radar to the array of the antennas of the eavesdropper, respectively. We have assumed both the radar and the eavesdropper's antennas are linearly spaced one-dimensional array of antennas and placed on the same plane. For other possible configurations and for $\mathbf{H}_{ie} \in \mathbb{C}^{M \times N_E}$, three-dimensional geometry with both azimuth and elevation angles should be considered.

Hence, the transmit signal at the MIMO radar in the tth time slot is given by

$$\mathbf{x}(t) = \mathbf{w}_1 s_1(t) + \mathbf{w}_2 s_2(t), \tag{13.5}$$

where $\mathbf{w}_1 \in \mathbb{C}^{N_R \times 1}$ and $\mathbf{w}_2 \in \mathbb{C}^{N_R \times 1}$ are the beam steering vectors for the legitimate information signal and the distortion signal, respectively. The signal received at the MIMO radar[1] can be

1 For the DFRC systems without an IRS, the first three entries in (1.6) and the first entry in (1.11) and (1.12) associated with the reflecting channels will not be needed.

expressed as

$$\begin{aligned}\mathbf{y}(t) &= \alpha_1 \mathbf{h}_{tr}\mathbf{h}_{it}\boldsymbol{\Theta}\mathbf{H}_{ri}\mathbf{x}\left(t-\tau_{ritr}\right) + \alpha_2 \mathbf{H}_{ir}\boldsymbol{\Theta}\mathbf{h}_{ti}\mathbf{h}_{rt}\mathbf{x}\left(t-\tau_{rtir}\right) \\ &+ \alpha_3 \mathbf{H}_{ir}\boldsymbol{\Theta}\mathbf{h}_{ti}\mathbf{h}_{it}\boldsymbol{\Theta}\mathbf{H}_{ri}\mathbf{x}\left(t-\tau_{ritir}\right) + \alpha \mathbf{h}_{tr}\mathbf{h}_{rt}\mathbf{x}\left(t-\tau_{rtr}\right) + \mathbf{n}_r(t),\end{aligned} \tag{13.6}$$

where $\boldsymbol{\Theta} = \text{diag}\left\{e^{j\theta_1}, \dots, e^{j\theta_M}\right\} \in \mathbb{C}^{M\times M}$ denotes the phase shifts matrix of the IRS; $\tau_{ritr} = \tau_{rtir}$, τ_{ritir}, and τ_{rtr} denote the signal propagation delays of the radar–IRS-target–radar, radar–target–IRS–radar, radar–IRS–target–IRS–radar and the radar–target–radar links, respectively, and $\mathbf{n}_r(t)$ denotes the additive white Gaussian noise (AWGN) at the radar, α_1, α_2, α_3, and α are the complex amplitudes corresponding to the radar cross section (RCS) including the free space propagation loss at various angles of incidence. Due to three distinct path delays $\tau_{ritr} = \tau_{rtir}$, τ_{ritir}, and τ_{rtr}, the range-Doppler map may capture two ghost targets in addition to the true target location. However, the distance between the target and the radar via the IRS is larger than the direct distance between the target and the radar. In addition, considerable amount of energy will also be lost when the signals reflect on the IRS. Therefore, it is possible to distinguish the true target from the ghost targets, and our aim is to maximize the SINR corresponding to the detection of only the true target. Please also note that as the IRS is at a fixed position and not moving, the false target due to signal path from the radar to the IRS and back to the radar receiver will appear close to the zero Doppler in the range-Doppler map (similar to clutters). Hence, this can be easily differentiated from the moving target, and it will not interfere with the estimation of the target of interest. Please note that a complete radar signal model in Eq. (1.6) should include Doppler shifts, which will be estimated using the fast Fourier transform (FFT) as part of obtaining the range-Doppler map. As our focus is only to demonstrate enhancement of performance through the IRS and as we consider only SINR for the target detection, we have omitted the Doppler shift in Eq. (1.6). Hence, the first three terms of (13.6) will not contribute toward the following SINR associated with the detection of the true target:

$$\text{SINR}_r = \frac{\text{Tr}\left(\mathbf{H}_r\mathbf{w}_1\mathbf{w}_1^H\mathbf{H}_r^H\right) + \text{Tr}\left(\mathbf{H}_r\mathbf{w}_2\mathbf{w}_2^H\mathbf{H}_r^H\right)}{\varepsilon_1 + \varepsilon_2 + \sigma_r^2}, \tag{13.7}$$

where

$$\mathbf{H}_r = \alpha \mathbf{h}_{tr}\mathbf{h}_{tr}^H, \tag{13.8}$$

and $\varepsilon_1 = \frac{\text{Tr}(\mathbf{H}_r\mathbf{w}_1\mathbf{w}_1^H\mathbf{H}_r^H)}{L}$, $\varepsilon_2 = \frac{\text{Tr}(\mathbf{H}_r\mathbf{w}_2\mathbf{w}_2^H\mathbf{H}_r^H)}{L}$ represent the residual interference when the received signal is match-filtered with $s_1(t)$ and $s_2(t)$, respectively, Deligiannis et al. [2018]. σ_r^2 is the noise power at the target, L denotes the radar matched filtering sequence length. Since we assume that L is arbitrarily large, ε_1 and ε_2 can be neglected for the rest of this chapter.

The data rates at the legitimate user and the eavesdropper can be expressed as follows:

$$R_c = \log\left|\mathbf{I} + \left(\mathbf{H}_c\mathbf{w}_1\mathbf{w}_1^H\mathbf{H}_c^H\right)\left(\frac{\mathbf{H}_c\mathbf{w}_2\mathbf{w}_2^H\mathbf{H}_c^H}{L_m} + \sigma_c^2\mathbf{I}\right)^{-1}\right|, \tag{13.9}$$

$$R_e = \log\left|\mathbf{I} + \left(\mathbf{H}_e\mathbf{w}_1\mathbf{w}_1^H\mathbf{H}_e^H\right)\left(\frac{\mathbf{H}_e\mathbf{w}_2\mathbf{w}_2^H\mathbf{H}_e^H}{L_m} + \sigma_e^2\mathbf{I}\right)^{-1}\right|, \tag{13.10}$$

where

$$\mathbf{H}_c = \beta_1\mathbf{H}_{ic}^H\boldsymbol{\Theta}\mathbf{H}_{ri} + \beta_2\mathbf{H}_{rc}, \tag{13.11}$$

$$\mathbf{H}_e = \gamma_1\mathbf{H}_{ie}^H\boldsymbol{\Theta}\mathbf{H}_{ri} + \gamma_2\mathbf{H}_{re}, \tag{13.12}$$

and σ_c^2, σ_e^2 are the power of noise at the legitimate user and the eavesdropper; $\beta_1, \beta_2, \gamma_1, \gamma_2$ denote propagation loss variables. Then, the secrecy rate can be expressed as follows:

$$\mathrm{SR} = \left[R_c - R_e\right]^+. \tag{13.13}$$

The secrecy rate depends on the channel $\mathbf{H}_e$. When the eavesdropper employs only a single antenna, this channel can be estimated using the knowledge of the angles of arrival of the signals. When the eavesdropper employs multiple antennas, we need to know inter element spacing and the orientation of the array of antennas at the eavesdropper. In this case, our secrecy rate can be viewed as an upper bound of the achievable secrecy rate. The optimization framework for the maximization of the secrecy rate and the minimization of the transmit power is proposed in the following section.

13.3 System Optimizations

We optimize the transmit beamforming vectors for the radar–communication signals and the AN jointly with the IRS coefficients. We consider two problem formulations, namely, the secrecy rate maximization and the transmit power minimization.

13.3.1 Secrecy Rate Maximization

In this subsection, we aim to maximize the secrecy rate by optimizing the phase shifts of the IRS $\boldsymbol{\Theta}$, the beam steering vectors for the legitimate information signal $\mathbf{w}_1$ and the distortion signal $\mathbf{w}_2$ iteratively. The optimization problem is formulated as follows:

$$\begin{aligned}
(\mathbf{P1}):\ & \underset{\boldsymbol{\Theta},\mathbf{w}_1,\mathbf{w}_2}{\text{maximize}}\ \mathrm{SR},\\
& \text{subject to}\ \ \mathrm{Tr}\left(\mathbf{H}_r\left(\mathbf{w}_1\mathbf{w}_1^H + \mathbf{w}_2\mathbf{w}_2^H\right)\mathbf{H}_r^H\right) \geqslant \sigma_r^2\gamma_r,\\
& \qquad\qquad\ \ \mathrm{Tr}\left(\mathbf{w}_1\mathbf{w}_1^H + \mathbf{w}_2\mathbf{w}_2^H\right) \leqslant P_{\max},\\
& \qquad\qquad\ \ \left|e^{j\theta_m}\right| = 1, m = 1,2,\ldots M,
\end{aligned}$$

where γ_r and $P_{\max}$ are the SINR threshold and the maximum transmit power of the radar, respectively. The secrecy rate expression in (**P1**) can be written as

$$\begin{aligned}
\mathrm{SR}' &= \log\left|\mathbf{I} + \mathbf{H}_c\mathbf{w}_1\mathbf{w}_1^H\mathbf{H}_c^H\left(L_m^{-1}\mathbf{H}_c\mathbf{w}_2\mathbf{w}_2^H\mathbf{H}_c^H + \sigma_c^2\mathbf{I}\right)^{-1}\right|\\
&\quad + \log\left|\sigma_e^2\mathbf{I} + L_m^{-1}\mathbf{H}_e\mathbf{w}_2\mathbf{w}_2^H\mathbf{H}_e^H\right|\\
&\quad - \log\left|\mathbf{H}_e\mathbf{w}_1\mathbf{w}_1^H\mathbf{H}_e^H + \sigma_e^2\mathbf{I} + L_m^{-1}\mathbf{H}_e\mathbf{w}_2\mathbf{w}_2^H\mathbf{H}_e^H\right|\\
&= \underbrace{\log\left|\mathbf{I} + \mathbf{H}_c\mathbf{w}_1\mathbf{w}_1^H\mathbf{H}_c^H\left(L_m^{-1}\mathbf{H}_c\mathbf{w}_2\mathbf{w}_2^H\mathbf{H}_c^H + \sigma_c^2\mathbf{I}\right)^{-1}\right|}_{f_1}\\
&\quad + \underbrace{\log\left|\mathbf{I} + \mathbf{H}_e\mathbf{w}_2\mathbf{w}_2^H\mathbf{H}_e^H\left(\sigma_e^2L_m\mathbf{I}\right)^{-1}\right|}_{f_2}\\
&\quad \underbrace{- \log\left|\mathbf{I} + \sigma_e^{-2}\mathbf{H}_e\left(\mathbf{w}_1\mathbf{w}_1^H + L_m^{-1}\mathbf{w}_2\mathbf{w}_2^H\right)\mathbf{H}_e^H\right|}_{f_3}.
\end{aligned} \tag{13.14}$$

The function f_1 in (13.14) can be reformulated by exploiting the relationship between the data rate and the mean-square error (MSE) for the optimal decoding vector. Specifically, with the linear decoding vector $\mathbf{u}_c \in \mathbb{C}^{N_c \times 1}$, the MSE of estimation is given by

$$E_c\left(\mathbf{u}_c, \mathbf{w}_1, \mathbf{w}_2\right) \triangleq \left(\mathbf{u}_c^H \mathbf{H}_c \mathbf{w}_1 - 1\right)\left(\mathbf{u}_c^H \mathbf{H}_c \mathbf{w}_1 - 1\right)^H + \mathbf{u}_c^H \left(L_m^{-1} \mathbf{H}_c \mathbf{w}_2 \mathbf{w}_2^H \mathbf{H}_c^H + \sigma_c^2 \mathbf{I}\right) \mathbf{u}_c. \tag{13.15}$$

By introducing an auxiliary variable $w_c > 0$, we have

$$\begin{aligned} f_1 &= \underset{w_c>0,\mathbf{u}_c}{\text{maximize}}\ h_1\left(\mathbf{u}_c, \mathbf{w}_1, \mathbf{w}_2, w_c\right) \\ &\triangleq \underset{w_c>0,\mathbf{u}_c}{\text{maximize}}\ \log w_c - w_c E_c\left(\mathbf{u}_c, \mathbf{w}_1, \mathbf{w}_2\right) + 1, \end{aligned} \tag{13.16}$$

the optimal $\mathbf{u}_c^*$ and w_c^* for the maximum value (similar to Shi et al. [2015]) of $h_1\left(\mathbf{u}_c, \mathbf{w}_1, \mathbf{w}_2, w_c\right)$ are given by

$$\begin{aligned} \mathbf{u}_c^* &= \arg\max_{\mathbf{u}_c}\ h_1\left(\mathbf{u}_c, \mathbf{w}_1, \mathbf{w}_2, w_c\right) \\ &= \left(L_m^{-1} \mathbf{H}_c \mathbf{w}_2 \mathbf{w}_2^H \mathbf{H}_c^H + \sigma_I^2 \mathbf{I} + \mathbf{H}_c \mathbf{w}_1 \mathbf{w}_1^H \mathbf{H}_c^H\right)^{-1} \mathbf{H}_c \mathbf{w}_1, \end{aligned} \tag{13.17}$$

$$\begin{aligned} w_c^* &= \arg\max_{w_c>0}\ h_1\left(\mathbf{u}_c, \mathbf{w}_1, \mathbf{w}_2, w_c\right) = \left[E_c\left(\mathbf{u}_c^*, \mathbf{w}_1, \mathbf{w}_2\right)\right]^{-1} \\ &= \Big[\left(\mathbf{u}_c^{*H} \mathbf{H}_c \mathbf{w}_1 - 1\right)\left(\mathbf{u}_c^{*H} \mathbf{H}_c \mathbf{w}_1 - 1\right)^H \\ &\quad + \mathbf{u}_c^{*H}\left(L_m^{-1} \mathbf{H}_c \mathbf{w}_2 \mathbf{w}_2^H \mathbf{H}_c^H + \sigma_c^2 \mathbf{I}\right) \mathbf{u}_c^*\Big]^{-1}. \end{aligned} \tag{13.18}$$

Similarly, by introducing $\mathbf{u}_e \in \mathbb{C}^{N_E \times 1}$ and auxiliary variables $w_e > 0, \mathbf{W}_z \succcurlyeq 0, \mathbf{W}_z \in \mathbb{C}^{N_E \times N_E}$, we have

$$\begin{aligned} f_2 &= \underset{w_e>0,\mathbf{u}_e}{\text{maximize}}\ h_2\left(\mathbf{u}_e, \mathbf{w}_2, w_e\right) \\ &\triangleq \underset{w_e>0,\mathbf{u}_e}{\text{maximize}}\ \log w_e - w_e E_e\left(\mathbf{u}_e, \mathbf{w}_2\right) + N_R, \end{aligned} \tag{13.19}$$

$$\begin{aligned} f_3 &= \underset{\mathbf{W}_z \succcurlyeq 0}{\text{maximize}}\ h_3\left(\mathbf{w}_1, \mathbf{w}_2, \mathbf{W}_z\right) \\ &\triangleq \underset{\mathbf{W}_z \succcurlyeq 0}{\text{maximize}}\ \log\left|\mathbf{W}_z\right| - \text{Tr}\left(\mathbf{W}_z \mathbf{E}_z\left(\mathbf{w}_1, \mathbf{w}_2\right)\right) + N_E, \end{aligned} \tag{13.20}$$

and the optimal values for the maximum $h_2\left(\mathbf{u}_e, \mathbf{w}_2, w_e\right)$ and $h_3\left(\mathbf{w}_1, \mathbf{w}_2, \mathbf{W}_z\right)$ are given by

$$\begin{aligned} \mathbf{u}_e^* &= \arg\max_{\mathbf{u}_e} h_2\left(\mathbf{u}_e, \mathbf{w}_2, w_e\right) \\ &= \left(L_m \sigma_e^2 \mathbf{I} + \mathbf{H}_e \mathbf{w}_2 \mathbf{w}_2^H \mathbf{H}_e^H\right)^{-1} \mathbf{H}_e \mathbf{w}_2, \end{aligned} \tag{13.21}$$

$$\begin{aligned} w_e^* &= \arg\max_{w_e>0} h_2\left(\mathbf{u}_e, \mathbf{w}_2, w_e\right) = \left[E_e\left(\mathbf{u}_e^*, \mathbf{w}_2\right)\right]^{-1} \\ &= \left[\left(\mathbf{u}_e^{*H} \mathbf{H}_e \mathbf{w}_2 - 1\right)\left(\mathbf{u}_e^{*H} \mathbf{H}_e \mathbf{w}_2 - 1\right)^H + \mathbf{u}_e^{*H} L_m \sigma_e^2 \mathbf{u}_e^*\right]^{-1}, \end{aligned} \tag{13.22}$$

$$\mathbf{W}_z^* = \arg\max_{\mathbf{W}_z \succcurlyeq 0} h_3\left(\mathbf{w}_1, \mathbf{w}_2, \mathbf{W}_z\right) = \left[\mathbf{E}_z\left(\mathbf{w}_1, \mathbf{w}_2\right)\right]^{-1}, \tag{13.23}$$

where

$$\mathbf{E}_z\left(\mathbf{w}_1, \mathbf{w}_2\right) \triangleq \mathbf{I} + \sigma_e^{-2} \mathbf{H}_e\left(\mathbf{w}_1 \mathbf{w}_1^H + L_m^{-1} \mathbf{w}_2 \mathbf{w}_2^H\right) \mathbf{H}_e^H. \tag{13.24}$$

Then (13.14) can be transformed to

$$\text{SR}' = \underset{\mathbf{W}_z \succcurlyeq 0, w_c, w_e > 0, \mathbf{u}_c, \mathbf{u}_e}{\text{maximize}}\ \text{SR}''(\mathbf{T}), \tag{13.25}$$

where

$$\mathrm{SR}''(\mathbf{T}) = h_1(\mathbf{u}_c, \mathbf{w}_1, \mathbf{w}_2, w_c) + h_2(\mathbf{u}_e, \mathbf{w}_2, w_e) + h_3(\mathbf{w}_1, \mathbf{w}_2, \mathbf{W}_z), \tag{13.26}$$

and $\mathbf{T} = [\mathbf{u}_c, w_c, \mathbf{u}_e, w_e, \mathbf{W}_z, \mathbf{w}_1, \mathbf{w}_2, \mathbf{\Theta}]$. Then, (**P1**) can be reformulated as follows:

$$\begin{aligned}
(\mathbf{P2}):\ & \underset{\mathbf{W}_z \succcurlyeq 0, w_c, w_e > 0, \mathbf{u}_c, \mathbf{u}_e, \mathbf{w}_1, \mathbf{w}_2, \mathbf{\Theta}}{\text{maximize}} \quad \mathrm{SR}''(\mathbf{T}), \\
& \text{subject to } \ \mathrm{Tr}\left(\mathbf{H}_r\left(\mathbf{w}_1\mathbf{w}_1^H + \mathbf{w}_2\mathbf{w}_2^H\right)\mathbf{H}_r^H\right) \geqslant \sigma_r^2\gamma_r, \\
& \qquad \mathrm{Tr}\left(\mathbf{w}_1\mathbf{w}_1^H + \mathbf{w}_2\mathbf{w}_2^H\right) \leqslant P_{\max}, \\
& \qquad \left|e^{j\theta_m}\right| = 1, \quad m = 1,2,\ldots M.
\end{aligned}$$

To solve (**P2**), we apply the BCD method. Firstly, given $\mathbf{w}_1$, $\mathbf{w}_2$, and $\mathbf{\Theta}$, we update $\mathbf{u}_c$, w_c, $\mathbf{u}_e$, w_e, $\mathbf{W}_z$ by using (13.17), (13.18), (13.21), (13.22), and (13.23), respectively. Secondly, for a given $\mathbf{\Theta}$, we update $\mathbf{w}_1$ and $\mathbf{w}_2$. Finally, we update $\mathbf{\Theta}$, given $\mathbf{u}_c$, w_c, $\mathbf{u}_e$, w_e, $\mathbf{W}_z$, $\mathbf{w}_1$, $\mathbf{w}_2$ as follows:

(1) Optimize phase shifts $\mathbf{\Theta}$ with given $\mathbf{w}_1$ and $\mathbf{w}_2$.

Given $\mathbf{w}_1$ and $\mathbf{w}_2$, the secrecy rate function can be transformed to

$$\begin{aligned}
\mathrm{SR} = & -\mathrm{Tr}(\mathbf{\Theta}\mathbf{A}) - \mathrm{Tr}\left(\mathbf{\Theta}^H\mathbf{A}^H\right) - \mathrm{Tr}\left(\mathbf{\Theta}^H\mathbf{C}_1\mathbf{\Theta}\mathbf{D}_1\right) \\
& - \mathrm{Tr}\left(\mathbf{\Theta}^H\mathbf{C}_2\mathbf{\Theta}\mathbf{D}_2\right) - \mathrm{Tr}\left(\mathbf{\Theta}^H\mathbf{C}_3\mathbf{\Theta}\mathbf{D}_3\right) - b = -g(\mathbf{\Theta}),
\end{aligned} \tag{13.27}$$

where $\mathbf{A}$, $\mathbf{C}_1$, $\mathbf{C}_2$, $\mathbf{C}_3$, $\mathbf{D}_1$, $\mathbf{D}_2$, $\mathbf{D}_3$, and b can be found in Appendix 13.A.

The secrecy rate maximization problem can be formulated as:

$$\begin{aligned}
(\mathbf{P3}):\ & \underset{\mathbf{\Theta}}{\text{minimize}} \ \ g(\mathbf{\Theta}), \\
& \text{subject to} \quad \left|e^{j\theta_m}\right| = 1, \quad m = 1,2,\ldots,M.
\end{aligned}$$

By exploiting the matrix properties of (1.10.6) in Zhang [2017], the trace operators can be removed, by defining $\boldsymbol{\phi} = \left[e^{j\theta_1},\ldots,e^{j\theta_M}\right]^T$ we have

$$\mathrm{Tr}(\mathbf{\Theta}\mathbf{A}) = \boldsymbol{\phi}^T\mathbf{a}, \tag{13.28a}$$

$$\mathrm{Tr}\left(\mathbf{\Theta}^H\mathbf{A}^H\right) = \mathbf{a}^H\left(\boldsymbol{\phi}^*\right), \tag{13.28b}$$

$$\mathrm{Tr}\left(\mathbf{\Theta}^H\mathbf{C}_1\mathbf{\Theta}\mathbf{D}_1\right) = \boldsymbol{\phi}^H\left(\mathbf{C}_1 \odot \mathbf{D}_1^T\right)\boldsymbol{\phi}, \tag{13.28c}$$

$$\mathrm{Tr}\left(\mathbf{\Theta}^H\mathbf{C}_2\mathbf{\Theta}\mathbf{D}_2\right) = \boldsymbol{\phi}^H\left(\mathbf{C}_2 \odot \mathbf{D}_2^T\right)\boldsymbol{\phi}, \tag{13.28d}$$

$$\mathrm{Tr}\left(\mathbf{\Theta}^H\mathbf{C}_3\mathbf{\Theta}\mathbf{D}_3\right) = \boldsymbol{\phi}^H\left(\mathbf{C}_3 \odot \mathbf{D}_3^T\right)\boldsymbol{\phi}, \tag{13.28e}$$

where $\mathbf{a} = \left[[\mathbf{A}]_{11},\ldots,[\mathbf{A}]_{M,M}\right]^T$ is a vector gathering the diagonal elements of matrix $\mathbf{A}$, then (**P3**) can be rewritten as follows:

$$\begin{aligned}
(\mathbf{P4}):\ & \underset{\boldsymbol{\phi}}{\text{minimize}} \ \ \boldsymbol{\phi}^H\mathbf{\Xi}\boldsymbol{\phi} + \boldsymbol{\phi}^T\mathbf{a} + \mathbf{a}^H\left(\boldsymbol{\phi}^*\right), \\
& \text{subject to} \quad \left|e^{j\theta_m}\right| = 1, \quad m = 1,2,\ldots,M,
\end{aligned}$$

where $\mathbf{\Xi} = \mathbf{C}_1 \odot \mathbf{D}_1^T + \mathbf{C}_2 \odot \mathbf{D}_2^T + \mathbf{C}_3 \odot \mathbf{D}_3^T$. Since $\mathbf{\Xi}$ is a semidefinite matrix, (**P4**) can be rewritten as follows:

$$\begin{aligned}
(\mathbf{P5}):\ & \underset{\boldsymbol{\phi}}{\text{minimize}} \ \ f(\boldsymbol{\phi}) \triangleq \boldsymbol{\phi}^H\mathbf{\Xi}\boldsymbol{\phi} + 2\,\mathrm{Re}\left\{\boldsymbol{\phi}^H\left(\mathbf{a}^*\right)\right\}, \\
& \text{subject to} \quad \left|e^{j\theta_m}\right| = 1, \quad m = 1,2,\ldots,M,
\end{aligned}$$

which can be solved by the semidefinite relaxation (SDR) method via transforming the unimodulus constraint into a rank-one constraint. However, the computational complexity is high for the SDR method. Thus, we propose to solve (**P5**) efficiently by the MM algorithm. For any given solution $\boldsymbol{\phi}^t$ at the tth iteration and for any feasible $\boldsymbol{\phi}$, we have

$$\boldsymbol{\phi}^H \boldsymbol{\Xi} \boldsymbol{\phi} \leqslant \boldsymbol{\phi}^H \mathbf{X} \boldsymbol{\phi} - 2\,\mathrm{Re}\left\{\boldsymbol{\phi}^H (\mathbf{X} - \boldsymbol{\Xi}) \boldsymbol{\phi}^t\right\} + \left(\boldsymbol{\phi}^t\right)^H (\mathbf{X} - \boldsymbol{\Xi}) \boldsymbol{\phi}^t \triangleq y\left(\boldsymbol{\phi} | \boldsymbol{\phi}^t\right), \tag{13.29}$$

where $\mathbf{X} = \lambda_{\max} \mathbf{I}_M$ and $\lambda_{\max}$ is the maximum eigenvalue of $\boldsymbol{\Xi}$. We reconstruct the objective function as follows:

$$g\left(\boldsymbol{\phi} | \boldsymbol{\phi}^t\right) = y\left(\boldsymbol{\phi} | \boldsymbol{\phi}^t\right) + 2\,\mathrm{Re}\left\{\boldsymbol{\phi}^H \mathbf{a}^*\right\}. \tag{13.30}$$

Since $\boldsymbol{\phi}^H \mathbf{X} \boldsymbol{\phi} = M \lambda_{\max}$ is a constant, we can rebuild the problem as follows:

$$\begin{aligned} (\mathbf{P6}) : \ & \underset{\boldsymbol{\phi}}{\text{minimize}} \ f(\boldsymbol{\phi}) \triangleq 2\,\mathrm{Re}\left\{\boldsymbol{\phi}^H \mathbf{q}^t\right\}, \\ & \text{subject to} \quad \left|e^{j\theta_m}\right| = 1, \quad m = 1,2,\ldots,M, \end{aligned}$$

where $\mathbf{q}^t = \left(\lambda_{\max} \mathbf{I}_M - \boldsymbol{\Xi}\right) \boldsymbol{\phi}^t - \mathbf{a}^*$. The optimal solution for (**P6**) is given by

$$\boldsymbol{\phi}^{t+1} = e^{j \arg(\mathbf{q}^t)}. \tag{13.31}$$

The details of the MM algorithm are provided in Algorithm 1.

Algorithm 1 MM algorithm

1 Initialize $m = 0$ and the maximum number of iterations m_{max}.
   Compute the objective function in (**P4**) as $f\left(\boldsymbol{\phi}^0\right)$.
2 **while** the required accuracy is not reached and $m \leqslant m_{\max}$ **do**:
3   Compute $\mathbf{q}^m = \left(\lambda_{\max} \mathbf{I}_M - \boldsymbol{\Xi}\right) \boldsymbol{\phi}^m - \mathbf{a}^*$, and update $\boldsymbol{\phi}^{m+1} = e^{j \arg(\mathbf{q}^m)}$.
4   Compute the objective function in (**P4**) as $f\left(\boldsymbol{\phi}^{m+1}\right)$.
5   Compute $\frac{|f(\boldsymbol{\phi}^{m+1}) - f(\boldsymbol{\phi}^m)|}{f(\boldsymbol{\phi}^{m+1})}$.
6 **end while**

---

(2) Optimize $\mathbf{w}_1$ and $\mathbf{w}_2$ with given phase shifts $\boldsymbol{\Theta}$.
For a given $\boldsymbol{\Theta}$, the secrecy rate function can be reformulated as:

$$\begin{aligned} g\left(\mathbf{w}_1, \mathbf{w}_2\right) = & \ \mathbf{w}_1^H \mathbf{H}_{w_1} \mathbf{w}_1 - w_c \mathbf{u}_c^H \mathbf{H}_c \mathbf{w}_1 - w_c \mathbf{w}_1^H \mathbf{H}_c^H \mathbf{u}_c \\ & + \mathbf{w}_2^H \mathbf{H}_{w_2} \mathbf{w}_2 - w_e \mathbf{u}_e^H \mathbf{H}_e \mathbf{w}_2 - w_e \mathbf{w}_2^H \mathbf{H}_e^H \mathbf{u}_e, \end{aligned} \tag{13.32}$$

where

$$\mathbf{H}_{w_1} - \mathbf{H}_c^H \mathbf{u}_c w_c \mathbf{u}_c^H \mathbf{H}_c + \sigma_e^{-2} \mathbf{H}_e^H \mathbf{W}_z \mathbf{H}_e, \tag{13.33}$$

$$\mathbf{H}_{w_2} = L_m^{-1} \mathbf{H}_c^H \mathbf{u}_c w_c \mathbf{u}_c^H \mathbf{H}_c + \mathbf{H}_e^H \mathbf{u}_e w_e \mathbf{u}_e^H \mathbf{H}_e + \left(L_m \sigma_e^2\right)^{-1} \mathbf{H}_e^H \mathbf{W}_z \mathbf{H}_e. \tag{13.34}$$

Then, the secrecy rate maximization problem can be formulated as:

$$\begin{aligned} (\mathbf{P7}) : \ & \underset{\mathbf{w}_1, \mathbf{w}_2}{\text{minimize}} \ g\left(\mathbf{w}_1, \mathbf{w}_2\right), \\ & \text{subject to} \quad \mathrm{Tr}\left(\mathbf{w}_1 \mathbf{w}_1^H + \mathbf{w}_2 \mathbf{w}_2^H\right) \leqslant P_{\max}, \\ & \qquad\qquad\quad \mathrm{Tr}\left(\mathbf{H}_r \left(\mathbf{w}_1 \mathbf{w}_1^H + \mathbf{w}_2 \mathbf{w}_2^H\right) \mathbf{H}_r^H\right) \geqslant \sigma_r^2 \gamma_r. \end{aligned}$$

However, the second constraint of (**P7**) is non-convex; thus, we transform it using the first-order Taylor series approximation as follows:

$$\mathrm{Tr}\left(\mathbf{H}_r \left(\mathbf{w}_1 - \tilde{\mathbf{w}}_1\right) \tilde{\mathbf{w}}_1^H \mathbf{H}_r^H\right) + \mathrm{Tr}\left(\mathbf{H}_r \tilde{\mathbf{w}}_1 \left(\mathbf{w}_1 - \tilde{\mathbf{w}}_1\right)^H \mathbf{H}_r^H\right)$$

$$
\begin{aligned}
&+\operatorname{Tr}\left(\mathbf{H}_r\tilde{\mathbf{w}}_1\tilde{\mathbf{w}}_1^H\mathbf{H}_r^H\right) + \operatorname{Tr}\left(\mathbf{H}_r\left(\mathbf{w}_2-\tilde{\mathbf{w}}_2\right)\tilde{\mathbf{w}}_2^H\mathbf{H}_r^H\right) \\
&+\operatorname{Tr}\left(\mathbf{H}_r\tilde{\mathbf{w}}_2\left(\mathbf{w}_2-\tilde{\mathbf{w}}_2\right)^H\mathbf{H}_r^H\right) + \operatorname{Tr}\left(\mathbf{H}_r\tilde{\mathbf{w}}_2\tilde{\mathbf{w}}_2^H\mathbf{H}_r^H\right) \geqslant \sigma_r^2\gamma_r,
\end{aligned}
\tag{13.35}
$$

where $\tilde{\mathbf{w}}_1$ and $\tilde{\mathbf{w}}_2$ are the solutions in the previous iteration. Then, **(P7)** is a quadratically constrained quadratic program (QCQP) problem and can be solved by CVX. However, the computational complexity is high for QCQP; therefore, we propose the Lagrangian multiplier method to reduce the complexity. Firstly, the Lagrangian function of **(P7)** is obtained as

$$
\begin{aligned}
\mathcal{L}\left(\mathbf{w}_1,\mathbf{w}_2,\lambda_{\text{sr}}\right) \triangleq\ & \mathbf{w}_1^H\mathbf{H}_{w_1}\mathbf{w}_1 - w_c\mathbf{u}_c^H\mathbf{H}_c\mathbf{w}_1 - w_c\mathbf{w}_1^H\mathbf{H}_c^H\mathbf{u}_c \\
& + \mathbf{w}_2^H\mathbf{H}_{w_2}\mathbf{w}_2 - w_e\mathbf{u}_e^H\mathbf{H}_e\mathbf{w}_2 - w_e\mathbf{w}_2^H\mathbf{H}_e^H\mathbf{u}_e \\
& + \lambda_{\text{sr}}\left(\operatorname{Tr}\left(\mathbf{w}_1\mathbf{w}_1^H + \mathbf{w}_2\mathbf{w}_2^H\right) - P_{\max}\right) \\
=\ & \mathbf{w}_1^H\left(\mathbf{H}_{w_1} + \lambda_{\text{sr}}\mathbf{I}\right)\mathbf{w}_1 + \mathbf{w}_2^H\left(\mathbf{H}_{w_2} + \lambda_{\text{sr}}\mathbf{I}\right)\mathbf{w}_2 \\
& - w_c\mathbf{w}_1^H\mathbf{H}_c^H\mathbf{u}_c - w_c\mathbf{u}_c^H\mathbf{H}_c\mathbf{w}_1 - w_e\mathbf{w}_2^H\mathbf{H}_e^H\mathbf{u}_e \\
& - w_e\mathbf{u}_e^H\mathbf{H}_e\mathbf{w}_2 - \lambda_{\text{sr}}P_{\max}.
\end{aligned}
\tag{13.36}
$$

Furthermore, the second constraint of **(P7)** can be transformed to (13.37) by defining $b\left(\tilde{\mathbf{w}}_1,\tilde{\mathbf{w}}_2\right) = \sigma_r^2\gamma_r + \tilde{\mathbf{w}}_1^H\mathbf{H}_r^H\mathbf{H}_r\tilde{\mathbf{w}}_1 + \tilde{\mathbf{w}}_2^H\mathbf{H}_r^H\mathbf{H}_r\tilde{\mathbf{w}}_2$.

$$
2\operatorname{Re}\left\{\mathbf{w}_1^H\mathbf{H}_r^H\mathbf{H}_r\tilde{\mathbf{w}}_1 + \mathbf{w}_2^H\mathbf{H}_r^H\mathbf{H}_r\tilde{\mathbf{w}}_2\right\} \geqslant b\left(\tilde{\mathbf{w}}_1,\tilde{\mathbf{w}}_2\right). \tag{13.37}
$$

Then, the dual problem of **(P7)** is given by

$$
\begin{aligned}
\textbf{(P8)}:\ & \underset{\lambda_{\text{sr}}}{\text{maximize}}\ \ h\left(\lambda_{\text{sr}}\right), \\
& \text{subject to}\ \ \lambda_{\text{sr}} \geqslant 0,
\end{aligned}
$$

where

$$
\begin{aligned}
h\left(\lambda_{\text{sr}}\right) &\triangleq \underset{\mathbf{w}_1,\mathbf{w}_2}{\text{minimize}}\ \mathcal{L}\left(\mathbf{w}_1,\mathbf{w}_2,\lambda_{\text{sr}}\right), \\
\text{subject to}\ \ & 2\operatorname{Re}\left\{\mathbf{w}_1^H\mathbf{H}_r^H\mathbf{H}_r\tilde{\mathbf{w}}_1 + \mathbf{w}_2^H\mathbf{H}_r^H\mathbf{H}_r\tilde{\mathbf{w}}_2\right\} \geqslant b\left(\tilde{\mathbf{w}}_1,\tilde{\mathbf{w}}_2\right).
\end{aligned}
\tag{13.38}
$$

Note that problem (13.38) is a linearly constrained convex quadratic optimization problem. This can be solved in a closed form by using the Lagrangian dual decomposition method as follows:

$$
\mathbf{w}_1^* = \left(\mathbf{H}_{w_1} + \lambda_{\text{sr}}\mathbf{I}\right)^{-1}\left(\mathbf{H}_c^H\mathbf{u}_c w_c + \mu_{\text{sr}}^*\mathbf{H}_r^H\mathbf{H}_r\tilde{\mathbf{w}}_1\right), \tag{13.39}
$$

$$
\mathbf{w}_2^* = \left(\mathbf{H}_{w_2} + \lambda_{\text{sr}}\mathbf{I}\right)^{-1}\left(\mathbf{H}_e^H\mathbf{u}_e w_e + \mu_{\text{sr}}^*\mathbf{H}_r^H\mathbf{H}_r\tilde{\mathbf{w}}_2\right), \tag{13.40}
$$

where

$$
\mu_{\text{sr}}^* = \frac{\max\left\{b\left(\tilde{\mathbf{w}}_1,\tilde{\mathbf{w}}_2\right) - 2\operatorname{Re}\left\{\begin{array}{l}\tilde{\mathbf{w}}_1^H\mathbf{H}_r^H\mathbf{H}_r\left(\mathbf{H}_{w_1} + \lambda_{\text{sr}}\mathbf{I}\right)^{-1}\mathbf{H}_c^H\mathbf{u}_c w_c \\ +\tilde{\mathbf{w}}_2^H\mathbf{H}_r^H\mathbf{H}_r\left(\mathbf{H}_{w_2} + \lambda_{\text{sr}}\mathbf{I}\right)^{-1}\mathbf{H}_e^H\mathbf{u}_e w_e\end{array}\right\}, 0\right\}}{2\operatorname{Re}\left\{\begin{array}{l}\tilde{\mathbf{w}}_1^H\mathbf{H}_r^H\mathbf{H}_r\left(\mathbf{H}_{w_1} + \lambda_{\text{sr}}\mathbf{I}\right)^{-1}\mathbf{H}_r^H\mathbf{H}_r\tilde{\mathbf{w}}_1 \\ +\tilde{\mathbf{w}}_2^H\mathbf{H}_r^H\mathbf{H}_r\left(\mathbf{H}_{w_2} + \lambda_{\text{sr}}\mathbf{I}\right)^{-1}\mathbf{H}_r^H\mathbf{H}_r\tilde{\mathbf{w}}_2\end{array}\right\}}. \tag{13.41}
$$

The optimal λ_{sr} can be found according to $\operatorname{Tr}\left(\mathbf{w}_1^*\mathbf{w}_1^{*H} + \mathbf{w}_2^*\mathbf{w}_2^{*H}\right) = P_{\max}$ via the bisection method as described in Algorithm 2 (note that we should check whether $\lambda_{\text{sr}} = 0$ satisfies the total power constraint before starting the bisection search.). The overall algorithm for the secrecy rate maximization is given in Algorithm 3.

Algorithm 2 Bisection method

1 Initialize $0 \leqslant \lambda^l \leqslant \lambda^u$ and the accuracy ε_b.
2 **while** $|\lambda^u - \lambda^l| > \varepsilon_b$ **do**:
3 Compute $\lambda = \frac{\lambda^l + \lambda^u}{2}$, μ^* of (13.41), $\mathbf{w}_1$, $\mathbf{w}_2$ of (13.39), (13.40), and evaluate $\text{Tr}\left(\mathbf{w}_1\mathbf{w}_1^H + \mathbf{w}_2\mathbf{w}_2^H\right)$.
4 **if** $\text{Tr}\left(\mathbf{w}_1\mathbf{w}_1^H + \mathbf{w}_2\mathbf{w}_2^H\right) \geqslant P_{\max}$
5 $\lambda^l = \lambda$.
6 **else**
7 $\lambda^u = \lambda$.
8 **end**
9 **end while**

Algorithm 3 Secrecy rate maximization

1 Initialize the variables $\mathbf{w}_1^0$, $\mathbf{w}_2^0$, $\boldsymbol{\Theta}^0$. Set $t = 0$ and the accuracy for iteration ε_{sr}.
2 **while** the required accuracy is not reached **do**:
3 Given $\mathbf{w}_1^t$, $\mathbf{w}_2^t$, $\boldsymbol{\Theta}^t$, compute the secrecy rate in (13.14) as $\text{SR}\left(\mathbf{w}_1^t, \mathbf{w}_2^t, \boldsymbol{\Theta}^t\right)$ and evaluate $\mathbf{u}_c^t$, w_c^t, $\mathbf{u}_e^t$, w_e^t, $\mathbf{W}_z^t$ according to (13.17), (13.18), (13.21)-(13.23), respectively.
4 Given $\mathbf{u}_c^t$, w_c^t, $\mathbf{u}_e^t$, w_e^t, $\mathbf{W}_z^t$, $\mathbf{w}_1^t$, $\mathbf{w}_2^t$, optimize $\boldsymbol{\Theta}^{t+1}$ by the MM algorithm.
5 Given $\mathbf{u}_c^t$, w_c^t, $\mathbf{u}_e^t$, w_e^t, $\mathbf{W}_z^t$, $\boldsymbol{\Theta}^{t+1}$, optimize $\mathbf{w}_1^{t+1}$, $\mathbf{w}_2^{t+1}$ by the bisection method.
6 Compute $\frac{\left|\text{SR}\left(\mathbf{w}_1^{t+1}, \mathbf{w}_2^{t+1}, \boldsymbol{\Theta}^{t+1}\right) - \text{SR}\left(\mathbf{w}_1^{t}, \mathbf{w}_2^{t}, \boldsymbol{\Theta}^{t}\right)\right|}{\text{SR}\left(\mathbf{w}_1^{t+1}, \mathbf{w}_2^{t+1}, \boldsymbol{\Theta}^{t+1}\right)}$.
7 $t = t + 1$.
8 **end while**

13.3.2 Transmit Power Minimization

In this subsection, we aim to minimize the transmit power at the radar while satisfying the target detection criterion and the secrecy rate, by jointly optimizing the phase shifts of the IRS $\boldsymbol{\Theta}$ and the transmit beamforming vectors $\mathbf{w}_1$ and $\mathbf{w}_2$.

Firstly, we formulate the problem as follows:

$$
\begin{aligned}
(\mathbf{P}9):\ &\underset{\mathbf{w}_1,\mathbf{w}_2,\boldsymbol{\Theta}}{\text{minimize}}\ \ \text{Tr}\left(\mathbf{w}_1\mathbf{w}_1^H + \mathbf{w}_2\mathbf{w}_2^H\right),\\
&\text{subject to}\ \ \text{Tr}\left(\mathbf{H}_r\left(\mathbf{w}_1\mathbf{w}_1^H + \mathbf{w}_2\mathbf{w}_2^H\right)\mathbf{H}_r^H\right) \geqslant \sigma_r^2\gamma_r,\\
&\qquad\qquad\ \ \text{SR}\left(\mathbf{w}_1, \mathbf{w}_2, \boldsymbol{\Theta}\right) \geqslant \kappa_r,\\
&\qquad\qquad\ \ \left|e^{j\theta_m}\right| = 1, \quad m = 1,2,\ldots,M,
\end{aligned}
$$

where κ_r is the secrecy rate threshold. We apply the Lagrangian multiplier method to obtain the Lagrangian function of (**P**9) as follows:

$$
\begin{aligned}
f\left(\mathbf{w}_1, \mathbf{w}_2, \boldsymbol{\Theta}, \lambda_p\right) &= \text{Tr}\left(\mathbf{w}_1\mathbf{w}_1^H + \mathbf{w}_2\mathbf{w}_2^H\right) + \lambda_p\left(\kappa_r - \text{SR}\left(\mathbf{w}_1, \mathbf{w}_2, \boldsymbol{\Theta}\right)\right)\\
&= \text{Tr}\left(\mathbf{w}_1\mathbf{w}_1^H + \mathbf{w}_2\mathbf{w}_2^H\right) + \lambda_p\kappa_r + \lambda_p\left(\mathbf{w}_1^H\mathbf{H}_{w_1}\mathbf{w}_1 + \mathbf{w}_2^H\mathbf{H}_{w_2}\mathbf{w}_2\right)\\
&\quad - \lambda_p w_c\left(\mathbf{w}_1^H\mathbf{H}_c^H\mathbf{u}_c + \mathbf{u}_c^H\mathbf{H}_c\mathbf{w}_1\right) - \lambda_p w_e\left(\mathbf{w}_2^H\mathbf{H}_e^H\mathbf{u}_e + \mathbf{u}_e^H\mathbf{H}_e\mathbf{w}_2\right) \qquad (13.42)\\
&= \mathbf{w}_1^H\left(\lambda_p\mathbf{H}_{w_1} + \mathbf{I}\right)\mathbf{w}_1 + \mathbf{w}_2^H\left(\lambda_p\mathbf{H}_{w_2} + \mathbf{I}\right)\mathbf{w}_2 + \lambda_p\kappa_r\\
&\quad - \lambda_p w_c\left(\mathbf{w}_1^H\mathbf{H}_c^H\mathbf{u}_c + \mathbf{u}_c^H\mathbf{H}_c\mathbf{w}_1\right) - \lambda_p w_e\left(\mathbf{w}_2^H\mathbf{H}_e^H\mathbf{u}_e + \mathbf{u}_e^H\mathbf{H}_e\mathbf{w}_2\right).
\end{aligned}
$$

Then the dual problem of (**P9**) is given by

$$\begin{aligned}(\mathbf{P10}): \ &\underset{\lambda_p}{\text{maximize}} \ \ h\left(\lambda_p\right),\\ &\text{subject to} \ \ \lambda_p \geqslant 0,\end{aligned}$$

where

$$\begin{aligned} h\left(\lambda_p\right) &\triangleq \underset{\mathbf{w}_1,\mathbf{w}_2,\boldsymbol{\Theta}}{\text{minimize}} \, f\left(\mathbf{w}_1,\mathbf{w}_2,\boldsymbol{\Theta},\lambda_p\right),\\ \text{subject to} \quad & 2\,\text{Re}\left\{\mathbf{w}_1^H\mathbf{H}_r^H\mathbf{H}_r\tilde{\mathbf{w}}_1 + \mathbf{w}_2^H\mathbf{H}_r^H\mathbf{H}_r\tilde{\mathbf{w}}_2\right\} \geqslant b\left(\tilde{\mathbf{w}}_1,\tilde{\mathbf{w}}_2\right).\end{aligned} \tag{13.43}$$

We apply the BCD-MM algorithm to solve problem (13.43); the beam steering vectors $\mathbf{w}_1$ and $\mathbf{w}_2$ can be solved in a closed form by using the Lagrangian dual decomposition method as follows:

$$\mathbf{w}_1^* = \left(\lambda_p\mathbf{H}_{w_1} + \mathbf{I}\right)^{-1}\left(\lambda_p\mathbf{H}_c^H\mathbf{u}_c w_c + \mu_p^*\mathbf{H}_r^H\mathbf{H}_r\tilde{\mathbf{w}}_1\right), \tag{13.44}$$

$$\mathbf{w}_2^* = \left(\lambda_p\mathbf{H}_{w_2} + \mathbf{I}\right)^{-1}\left(\lambda_p\mathbf{H}_e^H\mathbf{u}_e w_e + \mu_p^*\mathbf{H}_r^H\mathbf{H}_r\tilde{\mathbf{w}}_2\right), \tag{13.45}$$

where

$$\mu_p^* = \frac{\max\left\{b\left(\tilde{\mathbf{w}}_1,\tilde{\mathbf{w}}_2\right) - 2\,\text{Re}\left\{\begin{array}{c}\tilde{\mathbf{w}}_1^H\mathbf{H}_r^H\mathbf{H}_r\left(\lambda_p\mathbf{H}_{w_1}+\mathbf{I}\right)^{-1}\lambda_p\mathbf{H}_c^H\mathbf{u}_c w_c\\ +\tilde{\mathbf{w}}_2^H\mathbf{H}_r^H\mathbf{H}_r\left(\lambda_p\mathbf{H}_{w_2}+\mathbf{I}\right)^{-1}\lambda_p\mathbf{H}_e^H\mathbf{u}_e w_e\end{array}\right\},0\right\}}{2\,\text{Re}\left\{\begin{array}{c}\tilde{\mathbf{w}}_1^H\mathbf{H}_r^H\mathbf{H}_r\left(\lambda_p\mathbf{H}_{w_1}+\mathbf{I}\right)^{-1}\mathbf{H}_r^H\mathbf{H}_r\tilde{\mathbf{w}}_1\\ +\tilde{\mathbf{w}}_2^H\mathbf{H}_r^H\mathbf{H}_r\left(\lambda_p\mathbf{H}_{w_2}+\mathbf{I}\right)^{-1}\mathbf{H}_r^H\mathbf{H}_r\tilde{\mathbf{w}}_2\end{array}\right\}}. \tag{13.46}$$

The optimal phase shifts of the IRS $\boldsymbol{\Theta}^*$ can be obtained via the MM algorithm similar to the secrecy rate maximization problem. The algorithm for the transmit power minimization is given in Algorithm 4.

Algorithm 4 Transmit power minimization

1 Initialize the variables $\mathbf{w}_1^0$, $\mathbf{w}_2^0$, $\boldsymbol{\Theta}^0$. Set $t = 0$ and accuracy for iterations ε_p.
2 **while** the required accuracy is not reached **do**:
3 Given $\mathbf{w}_1^t$, $\mathbf{w}_2^t$, $\boldsymbol{\Theta}^t$, compute the transmit power as $\text{P}\left(\mathbf{w}_1^t,\mathbf{w}_2^t\right)$ and evaluate $\mathbf{u}_c^t$, w_c^t, $\mathbf{u}_e^t$, w_e^t, $\mathbf{W}_z^t$ according to (13.17), (13.18), (13.21)–(13.23), respectively.
4 **while** $|\lambda^u - \lambda^l| > \varepsilon_b$
5 $\lambda = \frac{\lambda^l+\lambda^u}{2}$, compute μ_p^* in (13.46), and evaluate $\mathbf{w}_1^{t+1}$, $\mathbf{w}_2^{t+1}$ in (13.44) and (13.45), respectively.
6 Optimize $\boldsymbol{\Theta}^{t+1}$ by the MM algorithm, and then compute $\text{SR}\left(\mathbf{w}_1^{t+1},\mathbf{w}_2^{t+1},\boldsymbol{\Theta}^{t+1}\right)$ in (13.14).
7 **if** $\text{SR}\left(\mathbf{w}_1^{t+1},\mathbf{w}_2^{t+1},\boldsymbol{\Theta}^{t+1}\right) \geqslant \kappa_r$
8 $\lambda^u = \lambda$
9 **else**
10 $\lambda^l = \lambda$
11 **end while**
12 Compute $\frac{|\text{P}(\mathbf{w}_1^{t+1},\mathbf{w}_2^{t+1}) - \text{P}(\mathbf{w}_1^t,\mathbf{w}_2^t)|}{\text{P}(\mathbf{w}_1^{t+1},\mathbf{w}_2^{t+1})}$.
13 $t = t + 1$.
14 **end while**

13.4 Simulation Results

We consider a system as in Figure 13.1, which consists of an IRS panel, a radar, a multiple antenna communication receiver, and an eavesdropping target. The radar consists of ten transmit/receive antennas ($N_R = 10$). The legitimate receiver has five receive antennas ($N_C = 5$) and the eavesdropping target consists of ten receive antennas ($N_E = 10$). The maximum number of the elements in the IRS panel is set to one hundred ($M = 100$). We assume that the tracking radar has information regarding the approximate location of the target. More specifically, the direction of the target seen from the radar is assumed to be known as $\phi_1 = 72°$. The direction of the radar from the eavesdropper is assumed to be $\phi_2 = -85°$. Hence, the channel from the radar to the target eavesdropper $\mathbf{H}_{re}$ is obtained using (13.3b). The legitimate receiver and IRS channel coefficient matrices $\mathbf{H}_{rc}$, $\mathbf{H}_{ic}$, and $\mathbf{H}_{ri}$ can be obtained using conventional pilot-based estimation. In our simulations, elements of $\mathbf{H}_{rc}$, $\mathbf{H}_{ic}$, and $\mathbf{H}_{ri}$ were generated using zero-mean circularly symmetric independent and identically distributed complex Gaussian random variables. Assuming that the geometry of the eavesdropper's antennas is known, we computed $\mathbf{H}_{ie}$, using the angle of arrival of the signal. However, the absence of this knowledge will not severely affect the secrecy rate performance as the corresponding channel gain would be very small because the IRS is placed in such a way to maximize the gain toward the legitimate receivers. The RCS coefficient and the propagation loss variables are fixed as $\alpha = 0.1$, $\beta_1 = 0.1$, $\beta_2 = 1$, $\gamma_1 = 0.01$, $\gamma_2 = 1$. The variance of the background white Gaussian noise at the radar receiver array, the legitimate receiver, and the eavesdropping target are set equal to unity ($\sigma_r^2 = \sigma_c^2 = \sigma_e^2 = 1$), and the spreading gain of the modulation waveform is fixed to 8 bits ($L_m = 8$).

13.4.1 Secrecy Rate Maximization

Figure 13.2 shows three examples of convergence behavior for $M = 10$, $M = 50$, and the case with No-IRS for $P_{\max} = 10\,\text{W}$ and SINR threshold $\gamma_r = 5\,\text{dB}$. As seen, the secrecy rates of all the three cases increase fast with the iterations, which demonstrates the efficiency of the proposed BCD-MM algorithm. It is evident from Figure 13.3 that the secrecy rate increases with the transmit power.

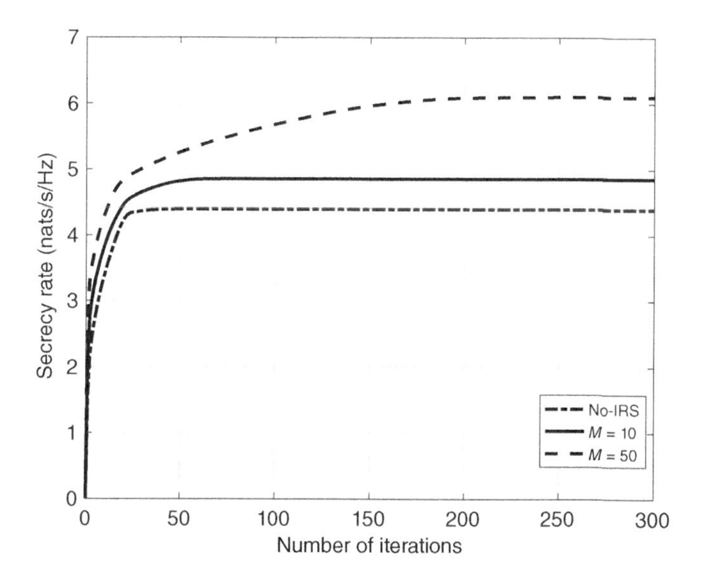

Figure 13.2 Convergence behavior of the secrecy rate maximization problem.

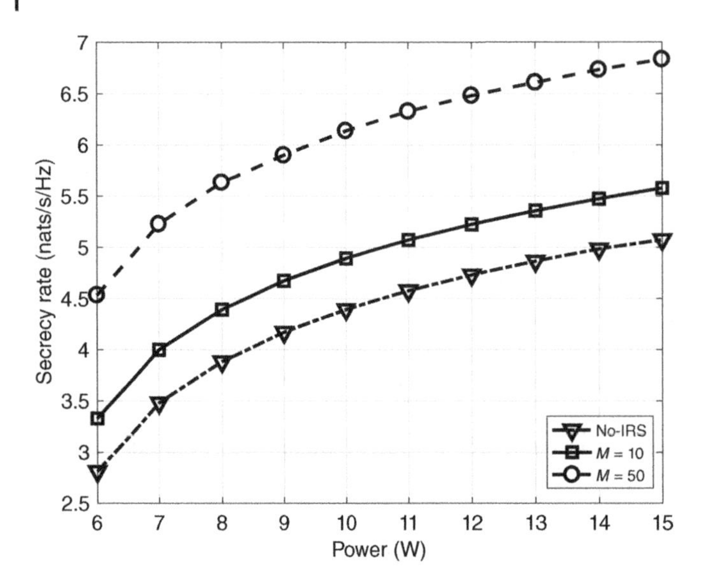

Figure 13.3 Achieved secrecy rate against different maximum power limit for the secrecy rate maximization problem.

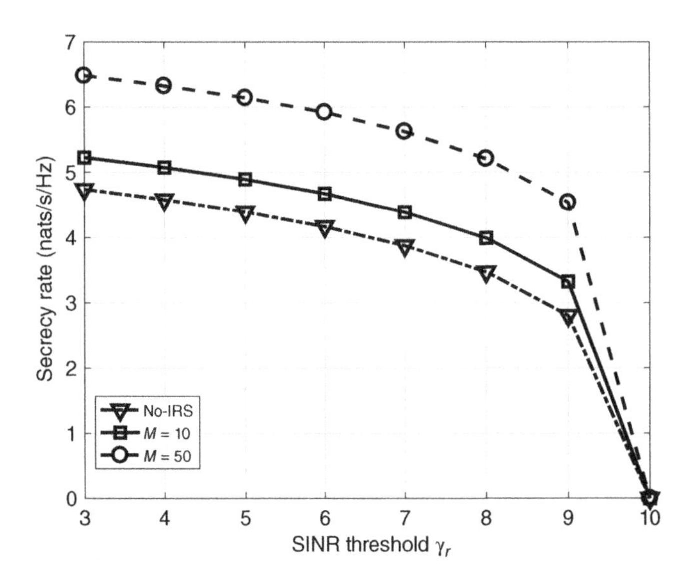

Figure 13.4 Achieved secrecy rate against different SINR thresholds for the secrecy rate maximization problem.

In addition, the secrecy rate increases as the number of IRS elements increases. It can be observed that a larger number of elements M guarantees a higher secrecy rate for any transmit power $P_{\max}$. Figure 13.4 illustrates the impact of SINR threshold γ_r on the secrecy rate. The secrecy rate is improved significantly with the assistance of the IRS; for example, secrecy rate improvement of 1.74 nats/s/Hz is seen for $\gamma_r = 3$ dB and $M = 50$, compared to the case of No-IRS. However, when $\gamma_r = 10$ dB, the secrecy rates of all the three cases reach 0. This is because the SINR constraint becomes dominant over the transmit power constraint as γ_r grows. The impact of the number of elements M on the maximum secrecy rate is shown in Figure 13.5, for $P_{\max} = 10$ W and $\gamma_r = 5$ dB. We

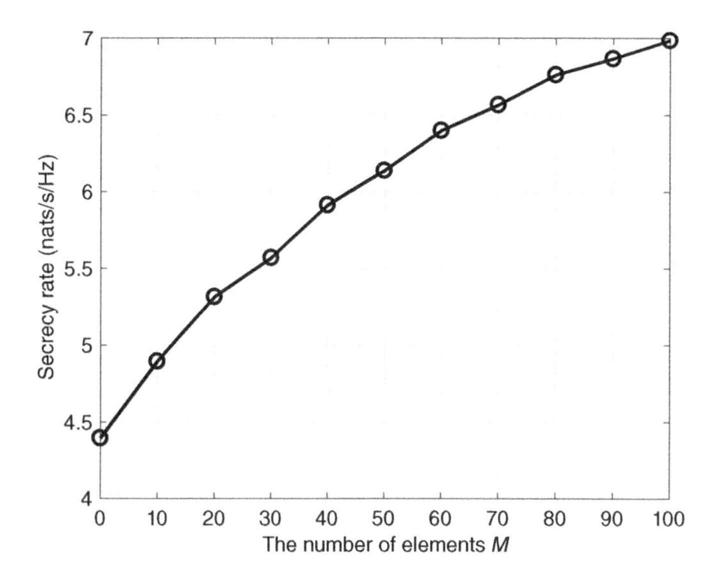

Figure 13.5 Achieved secrecy rate against different number of IRS elements M for the secrecy rate maximization problem.

observe that the secrecy rate increases with the number of elements M by the proposed BCD-MM algorithm; for example, 6.99 nats/s/Hz can be achieved when $M = 100$, which is a 59% improvement compared to the case without an IRS.

13.4.2 Transmit Power Minimization

Figure 13.6 shows convergence of the algorithm when the objective of the optimization is the minimization of transmit power, $M = 10$, $M = 50$ and the case with No-IRS. The secrecy rate threshold was set as $\kappa_r = 3$ nats/s/Hz and the SINR threshold was set as $\gamma_r = 5$ dB. As seen, the transmit powers of all the three cases converge faster with the iterations and reach a steady state

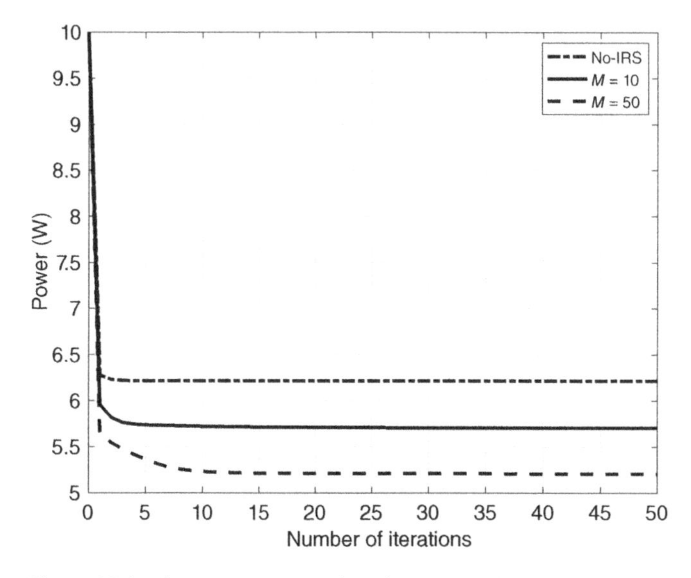

Figure 13.6 Convergence behavior of the transmit power minimization problem.

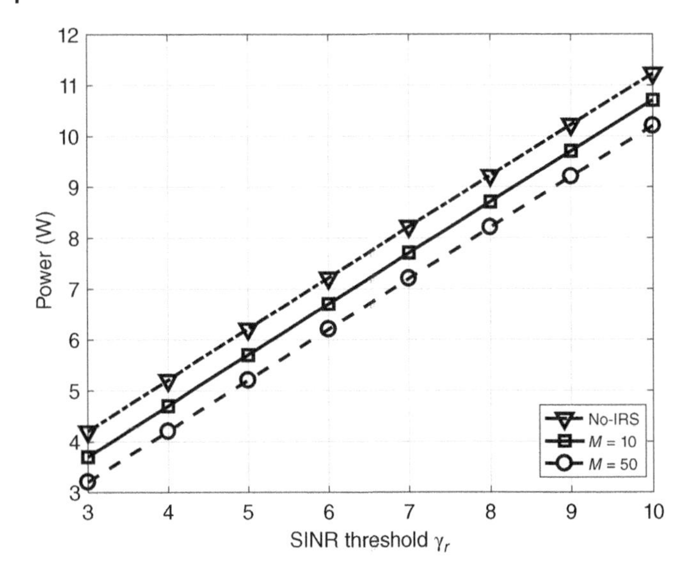

Figure 13.7 Achieved transmit power against the different SINR thresholds γ_r for the transmit power minimization problem.

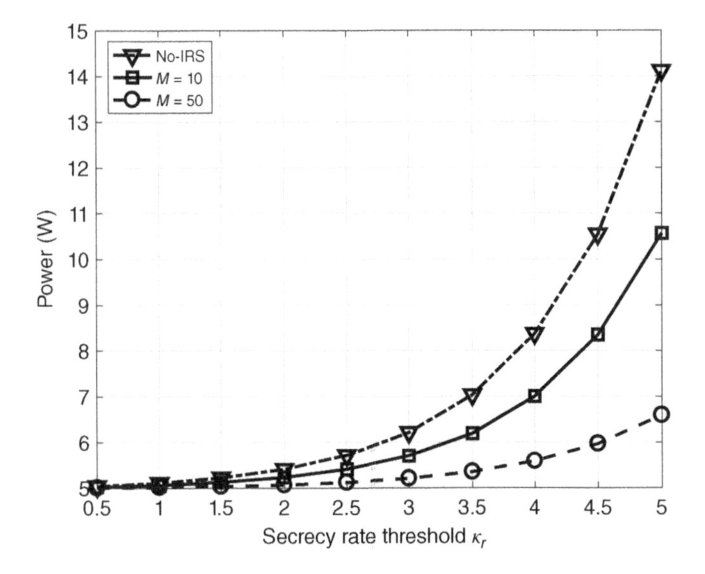

Figure 13.8 Achieved transmit power against different secrecy rate thresholds κ_r for the transmit power minimization problem.

after 30 iterations via the proposed BCD-MM algorithm. It is evident from Figure 13.7 that the transmit power increases linearly with the SINR threshold γ_r. Moreover, it is evident that the transmit power can be reduced with the help of the IRS, for example from 11.23 W (No-IRS) to 10.71 and 10.21 W ($M = 10$, $M = 50$) for $\gamma_r = 10\,\text{dB}$. Also, it is apparent from Figure 13.8 that the transmit power is reduced greatly when the IRS is used; for example, the transmit power is reduced from 14.12 to 6.59 W with $\gamma_r = 5\,\text{dB}$, $\kappa_r = 5\,\text{nats/s/Hz}$ when an IRS with $M = 50$ elements is employed.

13.5 Conclusion

We have investigated the benefits of employing an IRS in a DFRC system for enhancing secrecy rate of radar communications in the presence of an eavesdropping target. The model consists of a MIMO radar, a legitimate receiver, a target with eavesdropping capability, and an IRS. We have proposed two optimization techniques. The first one aimed at the joint design of transmit waveform beamforming vectors and IRS elements for enhancing the secrecy rate of radar communications. The second one focused on minimizing the transmit power subject to achieving certain secrecy rate for radar communications and SINR for the target detection. As the original problem turned out to be non-convex, we have performed a number of approximations and transformations including BCD method and the first-order Taylor series approximation. To reduce computational complexity, Lagrangian dual decomposition method and MM algorithms were employed. The simulation results confirmed that the use of IRS can significantly enhance secrecy rate and minimize the total transmit power compared to a scheme that does not employ an IRS. Specifically, the secrecy rate is increased by 59% from 4.39 to 6.99 nats/s/Hz; and the transmit power is decreased by 53% from 14.12 to 6.59 W.

13.A Appendix

13.A.1 Derivation of (13.27)

By substituting $\mathbf{H}_c$ in (13.11>) and $\mathbf{H}_e$ in (13.12) into (13.26), we can transform the secrecy rate function as follows:

$$\begin{aligned}
\mathrm{SR} = & -w_c\mathbf{u}_c^H\left(\beta_1\mathbf{H}_{ic}^H\mathbf{\Theta}\mathbf{H}_{ri}+\beta_2\mathbf{H}_{rc}\right)\mathbf{W}\left(\beta_1\mathbf{H}_{ri}^H\mathbf{\Theta}^H\mathbf{H}_{ic}+\beta_2\mathbf{H}_{rc}^H\right)\mathbf{u}_c - w_c\mathbf{u}_c^H\sigma_c^2\mathbf{u}_c \\
& - w_e\mathbf{u}_e^H\left(\gamma_1\mathbf{H}_{ie}^H\mathbf{\Theta}\mathbf{H}_{ri}+\gamma_2\mathbf{H}_{re}\right)\mathbf{W}_2\left(\gamma_1\mathbf{H}_{ri}^H\mathbf{\Theta}^H\mathbf{H}_{ie}+\gamma_2\mathbf{H}_{re}^H\right)\mathbf{u}_e - w_e\mathbf{u}_e^H\sigma_e^2\mathbf{u}_e \\
& + w_c\mathbf{u}_c^H\left(\beta_1\mathbf{H}_{ic}^H\mathbf{\Theta}\mathbf{H}_{ri}+\beta_2\mathbf{H}_{rc}\right)\mathbf{w}_1 + w_c\mathbf{w}_1^H\left(\beta_1\mathbf{H}_{ri}^H\mathbf{\Theta}^H\mathbf{H}_{ic}+\beta_2\mathbf{H}_{rc}^H\right)\mathbf{u}_c \\
& + w_e\mathbf{u}_e^H\left(\gamma_1\mathbf{H}_{ie}^H\mathbf{\Theta}\mathbf{H}_{ri}+\gamma_2\mathbf{H}_{re}\right)\mathbf{w}_2 + w_e\mathbf{w}_2^H\left(\gamma_1\mathbf{H}_{ri}^H\mathbf{\Theta}^H\mathbf{H}_{ie}+\gamma_2\mathbf{H}_{re}^H\right)\mathbf{u}_e \\
& - \mathrm{Tr}\left(\sigma_e^{-2}\mathbf{W}_z\left(\gamma_1\mathbf{H}_{ie}^H\mathbf{\Theta}\mathbf{H}_{ri}+\gamma_2\mathbf{H}_{re}\right)\mathbf{W}\left(\gamma_1\mathbf{H}_{ri}^H\mathbf{\Theta}^H\mathbf{H}_{ie}+\gamma_2\mathbf{H}_{re}^H\right)\right) \\
& + \log w_c + \log w_e + \log\left|\mathbf{W}_z\right| - w_c - w_e - \mathrm{Tr}\left(\mathbf{W}_z\right) + 1 + N_R + N_E \\
= & -\mathrm{Tr}\left(\mathbf{\Theta}\mathbf{A}\right) - \mathrm{Tr}\left(\mathbf{\Theta}^H\mathbf{A}^H\right) - \mathrm{Tr}\left(\mathbf{\Theta}^H\mathbf{C}_1\mathbf{\Theta}\mathbf{D}_1\right) - \mathrm{Tr}\left(\mathbf{\Theta}^H\mathbf{C}_2\mathbf{\Theta}\mathbf{D}_2\right) \\
& - \mathrm{Tr}\left(\mathbf{\Theta}^H\mathbf{C}_3\mathbf{\Theta}\mathbf{D}_3\right) - d \\
= & -g\left(\mathbf{\Theta}\right),
\end{aligned} \tag{13.A.1}$$

where

$$\begin{aligned}
\mathbf{A} = & \mathbf{H}_{ri}\mathbf{W}\mathbf{H}_{rc}^H\beta_2^H\mathbf{u}_c w_c\mathbf{u}_c^H\beta_1\mathbf{H}_{ic}^H - \mathbf{H}_{ri}\mathbf{w}_1 w_c\mathbf{u}_c^H\beta_1\mathbf{H}_{ic}^H \\
& + \mathbf{H}_{ri}\mathbf{W}_2^H\gamma_2^H\mathbf{H}_{re}^H\mathbf{u}_e w_e\mathbf{u}_e^H\gamma_1\mathbf{H}_{ie}^H - \mathbf{H}_{ri}\mathbf{w}_2 w_e\mathbf{u}_e^H\gamma_1\mathbf{H}_{ie}^H \\
& + \sigma_e^{-2}\mathbf{H}_{ri}\mathbf{W}\gamma_2^H\mathbf{H}_{re}^H\mathbf{W}_z\gamma_1\mathbf{H}_{ie}^H,
\end{aligned} \tag{13.A.2a}$$

$$\mathbf{C}_1 = \mathbf{H}_{ic}\mathbf{u}_c w_c\mathbf{u}_c^H\beta_1\mathbf{H}_{ic}^H, \tag{13.A.2b}$$

$$\mathbf{C}_2 = \gamma_1\mathbf{H}_{ie}\mathbf{u}_e w_e\mathbf{u}_e^H\mathbf{H}_{ie}^H, \tag{13.A.2c}$$

$$\mathbf{C}_3 = \sigma_e^{-2}\gamma_1\mathbf{H}_{ie}\mathbf{W}_z\mathbf{H}_{ie}^H, \tag{13.A.2d}$$

$$\mathbf{D}_1 = \beta_1^H\mathbf{H}_{ri}\mathbf{W}\mathbf{H}_{ri}^H, \tag{13.A.2e}$$

$$\mathbf{D}_2 = \gamma_1^H \mathbf{H}_{ri} \mathbf{W}_2 \mathbf{H}_{ri}^H, \tag{13.A.2f}$$
$$\mathbf{D}_3 = \gamma_1^H \mathbf{H}_{ri} \mathbf{W} \mathbf{H}_{ri}^H, \tag{13.A.2g}$$

$$\begin{aligned} b = {} & w_c \mathbf{u}_c^H \beta_2 \mathbf{H}_{rc} \mathbf{W} \boldsymbol{\beta}_2^H \mathbf{H}_{rc}^H \mathbf{u}_c + w_e \mathbf{u}_e^H \gamma_2 \mathbf{H}_{re} \mathbf{W}_2 \gamma_2^H \mathbf{H}_{re}^H \mathbf{u}_e \\ & - w_c \mathbf{u}_c^H \beta_2 \mathbf{H}_{rc} \mathbf{w}_1 - w_c \mathbf{w}_1^H \beta_2^H \mathbf{H}_{rc}^H \mathbf{u}_c + w_c \sigma_c^2 \mathbf{u}_c^H \mathbf{u}_c + w_c \\ & - w_e \mathbf{u}_e^H \gamma_2 \mathbf{H}_{re} \mathbf{w}_2 - w_e \mathbf{w}_2^H \gamma_2^H \mathbf{H}_{re}^H \mathbf{u}_e + w_e \mathbf{u}_e^H \sigma_e^2 L_m \mathbf{u}_e + w_e \\ & + w_c \mathbf{u}_c^H \beta_2 \mathbf{H}_{rc} \mathbf{W} \boldsymbol{\beta}_2^H \mathbf{H}_{rc}^H \mathbf{u}_c + w_e \mathbf{u}_e^H \gamma_2 \mathbf{H}_{re} \mathbf{W}_2 \gamma_2^H \mathbf{H}_{re}^H \mathbf{u}_e \\ & + \mathrm{Tr}\left(\mathbf{W}_z\right) + \mathrm{Tr}\left(\sigma_e^{-2} \mathbf{W}_z \gamma_2 \mathbf{H}_{re} \mathbf{W} \gamma_2^H \mathbf{H}_{re}^H\right) - 1 - N_R - N_E \\ & + \mathrm{Tr}\left(\sigma_e^{-2} \mathbf{W}_z \gamma_2 \mathbf{H}_{re} \mathbf{W} \gamma_2^H \mathbf{H}_{re}^H\right) - \log w_c - \log w_e - \log \left|\mathbf{W}_z\right|, \end{aligned} \tag{13.A.2h}$$

$$\mathbf{W}_2 = \mathbf{w}_2 \mathbf{w}_2^H, \tag{13.A.2i}$$
$$\mathbf{W} = \mathbf{w}_1 \mathbf{w}_1^H + L_m^{-1} \mathbf{w}_2 \mathbf{w}_2^H, \tag{13.A.2j}$$

which completes the derivation.

References

M. Alageli, A. Ikhlef, F. Alsifiany, M. A. M. Abdullah, G. Chen, and J. Chambers. Optimal downlink transmission for cell-free SWIPT massive MIMO systems with active eavesdropping. *IEEE Transactions on Information Forensics and Security*, 15:1983–1998, 2020.

A. Aubry, A. De Maio, Y. Huang, M. Piezzo, and A. Farina. A new radar waveform design algorithm with improved feasibility for spectral coexistence. *IEEE Transactions on Aerospace and Electronic Systems*, 51(2):1029–1038, 2015.

D. W. Bliss. Cooperative radar and communications signaling: the estimation and information theory odd couple. *IEEE Radar Conference, Cincinnati, OH, USA*, pages 50–55, 2014.

S. D. Blunt, M. R. Cook, and J. Stiles. Embedding information into radar emissions via waveform implementation. *2010 International Waveform Diversity and Design Conference, WDD*, pages 195–199, 2010.

B. K. Chalise and M. G. Amin. Performance tradeoff in a unified system of communications and passive radar: a secrecy capacity approach. *Digital Signal Processing*, 82:282–293, 2018.

G. Chen, Z. Tian, Y. Gong, Z. Chen, and J. A. Chambers. Max-ratio relay selection in secure buffer-aided cooperative wireless networks. *IEEE Transactions on Information Forensics and Security*, 9(4):719–729, 2014.

G. Chen, Y. Gong, P. Xiao, and J. A. Chambers. Physical layer network security in the full-duplex relay system. *IEEE Transactions on Information Forensics and Security*, 10(3):574–583, 2015.

Z. Chu, K. Cumanan, Z. Ding, M. Johnston, and S. Y. Le Goff. Secrecy rate optimizations for a MIMO secrecy channel with a cooperative jammer. *IEEE Transactions on Vehicular Technology*, 64(5):1833–1847, 2015.

M. Cui, G. Zhang, and R. Zhang. Secure wireless communication via intelligent reflecting surface. *IEEE Wireless Communications Letters*, 8(5):1410–1414, 2019.

A. Deligiannis, A. Daniyan, S. Lambotharan, and J. A. Chambers. Secrecy rate optimizations for MIMO communication radar. *IEEE Transactions on Aerospace and Electronic Systems*, 54(5):2481–2492, 2018.

H. Deng and B. Himed. Interference mitigation processing for spectrum-sharing between radar and wireless communications systems. *IEEE Transactions on Aerospace and Electronic Systems*, 49(3):1911–1919, 2013.

Z. Ding, K. K. Leung, D. L. Goeckel, and D. Towsley. On the application of cooperative transmission to secrecy communications. *IEEE Journal on Selected Areas in Communications*, 30(2):359–368, 2012.

L. Dong and H. M. Wang. Enhancing secure MIMO transmission via intelligent reflecting surface. *IEEE Transactions on Wireless Communications*, 19(11):7543–7556, 2020.

L. Dong, Z. Han, A. P. Petropulu, and H. V. Poor. Improving wireless physical layer security via cooperating relays. *IEEE Transactions on Signal Processing*, 58(3 PART 2):1875–1888, 2010.

A. J. Duly, D. J. Love, and J. V. Krogmeier. Time-division beamforming for MIMO radar waveform design. *IEEE Transactions on Aerospace and Electronic Systems*, 49(2):1210–1223, 2013.

J. Euzière, R. Guinvarc'h, M. Lesturgie, B. Uguen, and R. Gillard. Dual function radar communication Time-modulated array. *2014 International Radar Conference, Lille, France*, 2014.

S. Fang, G. Chen, and Y. Li. Joint optimization for secure intelligent reflecting surface assisted UAV networks. *IEEE Wireless Communications Letters*, 10(2): 276–280, 2020. https://doi.org/10.1109/LWC.2020.3027969.

A. J. Fenn, D. H. Temme, W. P. Delaney, and W. E. Courtney. The development of phased array radar technology. *Lincoln Laboratory Journal*, 12(2):321–340, 2000.

D. R. Fuhrmann, J. P. Browning, and M. Rangaswamy. Signaling strategies for the Hybrid MIMO phased-array radar. *IEEE Journal on Selected Topics in Signal Processing*, 4(1):66–78, 2010.

Z. Geng, H. Deng, and B. Himed. Adaptive radar beamforming for interference mitigation in radar-wireless spectrum sharing. *IEEE Signal Processing Letters*, 22(4):484–488, 2015.

H. Griffiths, S. Blunt, L. Cohen, and L. Savy. Challenge problems in spectrum engineering and waveform diversity. *IEEE Radar Conference, Ottawa, ON, Canada*, 2013.

X. Guan, Q. Wu, and R. Zhang. Intelligent reflecting surface assisted secrecy communication: is artificial noise helpful or not? *IEEE Wireless Communications Letters*, 9(6):778–782, 2020.

A. M. Haimovich, R. S. Blum, and L. J. Cimini. MIMO radar with widely separated antennas. *IEEE Signal Processing Magazine*, 25(1):116–129, 2008.

A. Hassanien, M. G. Amin, Y. D. Zhang, and F. Ahmad. Signaling strategies for dual-function radar-communications: an overview. *IEEE Aerospace and Electronic Systems Magazine*, 31(10):36–45, 2016a.

A. Hassanien, M. G. Amin, Y. D. Zhang, and F. Ahmad. Dual-function radar-communications: information embedding using sidelobe control and waveform diversity. *IEEE Transactions on Signal Processing*, 64(8):2168–2181, 2016b.

S. Hong, C. Pan, H. Ren, K. Wang, and A. Nallanathan. Artificial-noise-aided secure MIMO wireless communications via intelligent reflecting surface. *IEEE Transactions on Communications*, 68(12):7851–7866, 2020.

A. Khisti and G. W. Wornell. Secure transmission with multiple antennas-part II: the MIMOME wiretap channel. *IEEE Transactions on Information Theory*, 56(11):5515–5532, 2010.

I. Krikidis, J. S. Thompson, and S. Mclaughlin. Relay selection for secure cooperative networks with jamming. *IEEE Transactions on Wireless Communications*, 8(10):5003–5011, 2009.

J. Li and P. Stoica. MIMO radar with colocated antennas. *IEEE Signal Processing Magazine*, 24(5):106–114, 2007.

J. Li and P. Stoica. *MIMO radar signal processing*. Wiley, New Jersey, 2009.

C. Liaskos, S. Nie, A. Tsioliaridou, A. Pitsillides, S. Ioannidis, and I. Akyildiz. A new wireless communication paradigm through software-controlled metasurfaces. *IEEE Communications Magazine*, 56(9):162–169, 2018.

F. Liu, C. Masouros, A. Li, H. Sun, and L. Hanzo. MU-MIMO communications with MIMO radar: from co-existence to joint transmission. *IEEE Transactions on Wireless Communications*, 17(4):2755–2770, 2018.

X. Liu, T. Huang, N. Shlezinger, Y. Liu, J. Zhou, and Y. C. Eldar. Joint transmit beamforming for multiuser MIMO communications and MIMO radar. *IEEE Transactions on Signal Processing*, 68:3929–3944, 2020.

N. Nguyen, H. Q. Ngo, T. Q. Duong, H. D. Tuan, and K. Tourki. Secure massive MIMO with the artificial noise-aided downlink training. *IEEE Journal on Selected Areas in Communications*, 36(4):802–816, 2018.

Y. Nie, X. Lan, Y. Liu, Q. Chen, G. Chen, L. Fan, and D. Tang. Achievable rate region of energy harvesting based secure two-way buffer-aided relay networks. *IEEE Transactions on Information Forensics and Security, to appear*, 16:1610–1625 2020. https://doi.org/10.1109/TIFS.2020.3039047.

J. Qian, M. Lops, L. Zheng, X. Wang, and Z. He. Joint system design for coexistence of MIMO radar and MIMO communication. *IEEE Transactions on Signal Processing*, 66(13):3504–3519, 2018.

H. Shen, W. Xu, S. Gong, Z. He, and C. Zhao. Secrecy rate maximization for intelligent reflecting surface assisted multi-antenna communications. *IEEE Communications Letters*, 23(9):1488–1492, 2019.

Q. Shi, W. Xu, J. Wu, E. Song, and Y. Wang. Secure beamforming for MIMO broadcasting with wireless information and power transfer. *IEEE Transactions on Wireless Communications*, 14(5):2841–2853, 2015.

N. Su, F. Liu, and C. Masouros. Secure radar-communication systems with malicious targets: integrating radar, communications and jamming functionalities. *IEEE Transactions on Wireless Communications*, 20(1):83–95, 2021.

A. Tajer, G. H. Jajamovich, X. Wang, and G. V. Moustakides. Optimal joint target detection and parameter estimation by MIMO radar. *IEEE Journal on Selected Topics in Signal Processing*, 4(1):127–145, 2010.

J. Tang, G. Chen, and J. P. Coon. Secrecy performance analysis of wireless communications in the presence of UAV jammer and randomly located UAV eavesdroppers. *IEEE Transactions on Information Forensics and Security*, 14(11):3026–3041, 2019.

S. Verdu. *Multiuser detection*. Cambridge University Press, Cambridge, UK, 1998.

X. D. Zhang. *Matrix analysis and applications*. Cambridge University Press, 2017.

G. Zheng, L. Choo, and K. Wong. Optimal cooperative jamming to enhance physical layer security using relays. *IEEE Transactions on Signal Processing*, 59(3):1317–1322, 2011.

Y. Zhu, G. Zheng, and M. Fitch. Secrecy rate analysis of UAV-enabled mmWave networks using Matérn hardcore point processes. *IEEE Journal on Selected Areas in Communications*, 36(7):1397–1409, 2018.

14

Privacy in Spectrum Sharing Systems with Applications to Communications and Radar

Konstantinos Psounis[1] *and Matthew A. Clark*[2]

[1] *University of Southern California, Los Angeles, CA, USA*
[2] *The Aerospace Corporation, El Segundo, CA, USA*

14.1 Introduction

Radar and wireless communication systems rely on access to the radio frequency (RF) spectrum to function. Bandwidth-hungry wireless networks such as cellular and Wi-Fi are now placing unprecedented demand on RF spectrum bands. Unfortunately, desirable frequency ranges are limited, and there is no unencumbered spectrum for new services. Replacing legacy technologies is time consuming and expensive, meaning that rapid introduction of a new technology requires improvement in spectrum sharing. Joint operation of radar and communication systems will depend on successful spectrum sharing with other radar, wireless communication networks, and potentially other types of RF services.

How to best employ spectrum sharing technologies remains an open question. With the introduction of cognitive radio, smart devices that sense their spectrum environment and dynamically adjust their operations to avoid interference were thought to offer the solution to the apparent spectrum crunch.[1] Decentralized cognitive radio solutions face challenges such as the "hidden node problem," and difficulty with remediation of misbehaving devices [Carroll et al., 2012]. Centralized solutions have been introduced, e.g. to facilitate sharing in television white spaces (TVWS) [Deb et al., 2009]. In this setting, users interface directly with spectrum access systems (SASs), which maintain databases of spectrum policy and user information. Centralized sharing offers improved efficiency through resource optimization [Wang et al., 2016], simplified RF devices [Han et al., 2016] and are increasingly seen as a key tool for spectrum sharing [Holdren et al., 2012].

The U.S. Citizens Broadband Radio Service (CBRS) offers an example of centralized spectrum sharing involving radar and communication systems, enabled by dynamic SASs in the 3550–3700 MHz band [Federal Communications Commission, 2015, 2016]. SASs enable new entrants access to the band, sharing with priority incumbent systems, which largely consist of radar. The SAS interfaces with spectrum users and employs an infrastructure of spectrum sensors, called the environmental sensing capability (ESC). Based on user inputs and measurements, the SAS determines suitable protections to prevent harmful interference to priority/primary users (PUs). These protections are enforced by the SAS when granting spectrum access to secondary users (SUs), e.g. by limiting the total aggregate interference power caused by SUs assigned to frequencies that are co-channel with a PU.

1 http://www.cnn.com/2010/TECH/mobile/11/05/gahran.mobile.spectrum.crunch/index.html

Signal Processing for Joint Radar Communications, First Edition.
Edited by Kumar Vijay Mishra, M. R. Bhavani Shankar, Björn Ottersten, and A. Lee Swindlehurst.

Many of the PUs in 3550–3700 MHz are government entities, e.g. military radars. Considering the databases of information held by the proposed SASs, incumbent users have raised concerns about maintaining the privacy of their operations [Atkins, 2015]. The SAS may need information on user locations, frequencies, time of use, and susceptibility to interference. Any of these parameters may be considered sensitive by the operators of both communication and radar systems and should be protected from exposure to potential adversaries. An adversary may make inference attacks on user information by passively observing the sharing system. A more powerful adversary may be able to hack into the SAS and observe stored information directly. User privacy may be preserved for PUs by obfuscating the information reported to the SAS and by obfuscating the allocations made by the SAS to SUs. Typical cyber security approaches as well as security against eavesdroppers covered in Chapter 13 are not alone sufficient as the normal operation of the SAS may allow an adversary to deduce critical aspects of a PU operation. For example, an adversary may legitimately operate a number of cell phones within a secondary network and leverage assignments from the SAS to make inferences about the characteristics of a military radar. The accuracy of the data communicated between users and the SAS will impact both user privacy and spectrum use efficiency. Coarse precision will reveal less PU information, but require more conservative access to the spectrum by SUs in order to avoid harmful interference. As a result, there is a potential trade-off between the privacy of the PUs and the utility of the shared spectrum.

In this chapter, we explore the characteristics of this privacy–performance trade-off. We evaluate privacy in terms of inference attacks, i.e. adversaries that observe the system without disrupting it in an attempt to learn information about the users. To measure privacy, we explore metrics that operate on the joint probability distributions of adversary observations and underlying user states, providing approaches that are agnostic to adversary implementations and specific user privacy concerns. We also measure privacy by explicitly modeling adversary inference attacks and comparing them with the true user state, enabling intuitive metrics that are typically more meaningful to end users.

We evaluate privacy and performance in the context of a generalized spectrum sharing framework for dynamic SASs with real-time user-SAS communication, a spectrum sensing infrastructure, and robustness to uncertainty. This framework can be applied to evaluate heuristic privacy preservation strategies. It also enables us to formulate the user privacy optimization problem, allowing us to model a range of specific user privacy concerns and system utilities. This also allows us to derive the theoretically optimal privacy preservation strategy. Though its implementation may be intractable in practice, the conditions on the optimal solution can help to inform improvements to the joint design of radar, communication, and spectrum sharing systems.

New radar and communication technologies will depend on spectrum sharing systems. Government organizations, corporations, and even private citizens may require privacy protections while participating in such systems. Effective privacy tools will help to address user privacy concerns and promote broad adoption of spectrum sharing systems, enabling increased access and efficient use of the RF spectrum.

In Section 14.2, we introduce the spectrum sharing system model. Section 14.3 then discusses the adversary models and metrics applicable to the utility–privacy trade-off, which we formally analyze in Section 14.4. Recognizing tractability issues with formal solutions to the optimization of privacy and utility, we review practical privacy strategies in Section 14.5 to include common heuristic methods. Motivated by sharing in CBRS, we demonstrate the methodology through a simulation case study in Section 14.6, where several scenarios involving various user requirements and sharing approaches are examined. Finally, we offer a brief summary including a discussion of open questions and future research directions in Section 14.7.

Throughout this chapter, we use the mathematical notation applied in other chapters, e.g. bold and lowercase to denote vectors, bold and uppercase to denote matrices, and italic to denote variables. In addition, we use subscripts in two ways: to index the elements of vectors and matrices and to distinguish scalar variables that are similar in nature. Similarly, superscripts will be used to distinguish related vectors and matrices. Calligraphic font denotes sets, and a underbar will denote random variables.

14.2 Spectrum Sharing Systems

A spectrum sharing system should enable the compatible operation of PU and SU networks. To establish a general, rigorous framework for analyzing the privacy–utility trade-off in spectrum sharing settings, we consider a system of n networked users, e.g. PU and SU. The system operates m functions, which take user data as input and provide some utility to the users and to the overall system, but also potentially expose user privacy. Functions can be user specific, e.g. obfuscation functions for user data, or system-wide functions residing at a central entity, e.g., a SAS function computing power assignments for SUs in the context of CBRS spectrum sharing. The general problem is how to design these functions to practically achieve a near-optimal privacy–utility trade-off.

The general framework is illustrated in Figure 14.1a. Throughout this chapter, we will refer to a specific application of this framework in the CBRS spectrum sharing scenario to serve as a concrete example. This is illustrated in Figure 14.1b. In CBRS, a SAS operates via direct interface with users, an ESC, and a process to determine spectrum allocations for SUs. The following subsections offer a detailed model of each component. In addition to the notation described in Section 14.1, we use a tilde accent mark, " ~ ", to distinguish PU ground truth variables, i.e. variables that contain sensitive, non-obfuscated PU data. Further, we use a circumflex accent mark, " ^ ", to distinguish variables or distributions that are computed as part of the adversary inference attack. To simplify the example and notation, we will also limit analysis to PU privacy, such that distinguishing between

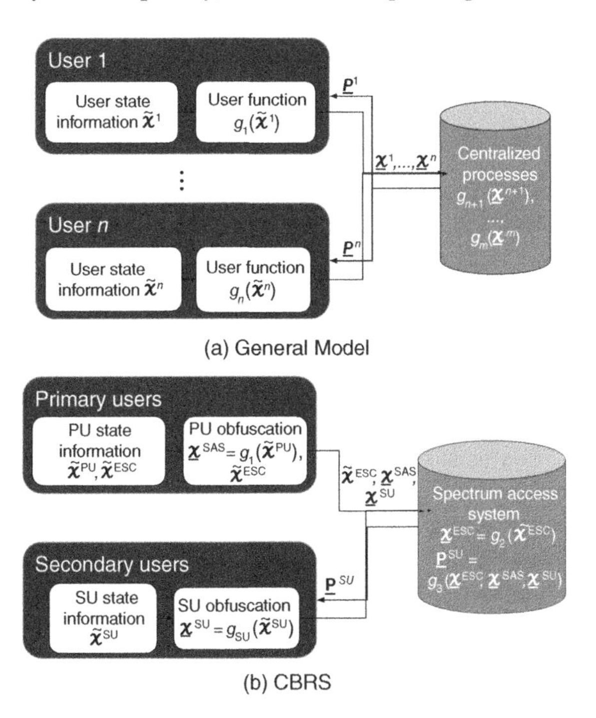

Figure 14.1 Spectrum sharing system models.

true, obfuscated, and adversary estimated parameters is only necessary for PU parameters, and not SU parameters.

14.2.1 SU–SAS Interface

An SU requests an allocation from the SAS by sending information on its present location and RF parameters via a connection that does not rely on spectrum from the SAS. SU devices, indexed from 1 to n_{SU} and referred to as SUs for brevity from this point on, will each provide a range of useful transmission powers, a set of frequency tuning ranges, and a range of useful bandwidths. To simplify the notation, we will assume these SU restrictions are captured in a set of feasible assignments $\mathcal{P}$. The SAS stores all SU locations in a set we denote $\mathcal{L}^{\text{SU}} \subseteq \mathcal{L}$, where $\mathcal{L}$ is the set of all discretized locations in the considered region. To allow the access system to take advantage of frequency-dependent scheduling, an SU may also send channel state information for the links in the SU network. Assuming n_c discrete frequency channels, the SAS may store all SU reported channel gains in an array $\mathbf{G}^{\text{SU}} = [\mathbf{G}^{\text{SU}}_{i\ell}]$, where $\ell, 1 \leq \ell \leq n_c$, is the frequency channel index and i, $1 \leq i \leq n_{\text{SU}}$, is the index for the ith SU. To allow the access system to take advantage of antenna systems, an SU may also send information describing their antenna characteristics, to include the number of transmit elements in the antenna array, $\mathbf{a}^{Stx} = [\mathbf{a}^{Stx}_i]$, and the number of elements at the intended receiver, $\mathbf{a}^{Srx} = [\mathbf{a}^{Srx}_i]$. In this case, the SU network may also send the matrix of complex channel state information $\mathbf{H}^{\text{SU}}_i \in \mathbb{C}^{\mathbf{a}^{Srx}_i \times \mathbf{a}^{Stx}_i}$. For brevity, we will refer to the set of all SU parameters as $\mathcal{X}^{\text{SU}} = \{\mathcal{L}^{\text{SU}}, \mathbf{G}^{\text{SU}}, \mathcal{P}, \mathbf{a}^{Stx}, \mathbf{a}^{Srx}, \mathbf{H}^{\text{SU}}\}$.

To ensure concepts are clear, we focus on a specific CBRS example where the PUs operate military radars and the SUs consist of cellular network operators. Suppose the SU transmissions are from the cellular user equipment (UE) to base station (BS) receivers. BS transmissions to UE are assumed to occur at another frequency or time and can be treated analogously. SU transmissions are restricted to specific frequency, power, and bandwidth ranges due to device hardware limitations as well as application-specific requirements. $\mathbf{G}^{\text{SU}}$ specifies the estimated channel gains on the UE-to-BS transmission path.

14.2.2 PU–SAS Interface

For each frequency channel, the locations of PUs operating in the region are described by a set denoted $\tilde{\mathcal{L}}^{\text{PU}} \subseteq \mathcal{L}$, with a potentially obfuscated set $\underline{\mathcal{L}}^{\text{SAS}}$ provided directly to the SAS database, and where there are $\tilde{n}_{\text{PU}}$ operational PUs. PUs will operate with transmission powers $\tilde{\mathbf{p}}^{\text{PU}} \in \mathbb{R}_+^{\tilde{n}_{\text{PU}}}$. A potentially obfuscated minimum $\underline{\mathbf{p}}^{\text{SAS}}$ should be provided to the SAS operators to enable reliable detection by the ESC. In our military radar example, all PUs are both transmitters and receivers. In general, PU transmitters and receivers may be physically separated. In this case, a policy should be identified to the SAS operator to infer potential receiver locations based on detection of the transmitter locations.

The PUs have interference protection criteria, $\tilde{\boldsymbol{\phi}} \in \mathbb{R}_+^{\tilde{n}_{\text{PU}}}$ and $\tilde{\boldsymbol{\psi}} \in \mathbb{R}_+^{\tilde{n}_{\text{PU}}}$, where $\tilde{\phi}_j$ is a harmful received interference power threshold for the jth PU, $1 \leq j \leq \tilde{n}_{\text{PU}}$, and $0 < \tilde{\psi}_j < 1$ is a reliability requirement for the PU, i.e. the maximum probability that the threshold given by $\tilde{\phi}_j$ can be exceeded. This reliability parameter accounts for inherent uncertainty due to the ESC detection process, any obfuscation strategies, and imperfect SAS prediction of aggregate interference from the SUs. In our example, $\tilde{\phi}_j$ could be the received interference power that would prevent a radar from detecting a target. Since a practical radar will have some baseline probability of missed detection even without SU interference, selecting $\tilde{\psi}_j$ to be significantly less than this baseline probability will ensure that sharing spectrum with SUs will have a negligible effect on the overall performance of the radar.

A PU may also provide information describing their antenna characteristics, to include the number of transmit elements in the array, $\tilde{\mathbf{a}}^{Ptx} = [\tilde{\mathbf{a}}_j^{Ptx}]$, and the number of elements at the intended receiver, $\tilde{\mathbf{a}}^{Prx} = [\tilde{\mathbf{a}}_j^{Prx}]$. Further, SUs may be required to operate pilot tones or else will orchestrate periodic training periods such that the PUs may estimate channel state information for the interference channel between each SU and PU pair, denoted $\tilde{\mathbf{H}}_{ji}^{\mathrm{PU}} \in \mathbb{C}^{\tilde{\mathbf{a}}_j^{Prx} \times \tilde{\mathbf{a}}_i^{Stx}}$.

We will refer to the set of PU system and operational ground truth parameters as $\tilde{\mathcal{X}}^{\mathrm{PU}} = \{\tilde{\mathcal{L}}^{\mathrm{PU}}, \tilde{\mathbf{p}}^{\mathrm{PU}}, \tilde{\boldsymbol{\phi}}, \tilde{\boldsymbol{\psi}}, \tilde{\mathbf{a}}^{Ptx}, \tilde{\mathbf{a}}^{Prx}, \tilde{\mathbf{H}}^{\mathrm{PU}}\}$. Some or all of these parameters may be used to derive information that the PUs will report to the SAS. We denote $\underline{\mathcal{X}}^{\mathrm{SAS}} = \{\underline{\mathcal{L}}^{\mathrm{SAS}}, \underline{\mathbf{p}}^{\mathrm{SAS}}, \underline{\boldsymbol{\phi}}^{\mathrm{SAS}}, \underline{\boldsymbol{\psi}}^{\mathrm{SAS}}, \underline{\mathbf{a}}^{Ptx}, \underline{\mathbf{a}}^{Prx}, \underline{\mathbf{H}}^{\mathrm{PU}}\}$ as the set of potentially obfuscated information received by the SAS from the PU interface. To aid in selecting a reporting strategy, we assume the PU may estimate the SU utility that will be achieved for any given PU protection criteria. This estimate may be based on feedback from the SAS, from the SU operators, or based on public knowledge of typical SU use cases. Denote u_{SU} as the function computing the estimated SU utility from the PU reported state $\underline{\mathcal{X}}^{\mathrm{SAS}}$. Similarly, we assume the PU may estimate the likelihood a selected reporting strategy will result in harmful interference at one or more PU receivers with the function u_{PU}. These estimates may be provided by the SAS, or else may be estimated by the PU directly based on typical SU operations.

14.2.3 ESC Measurements and Reporting

The ESC is used to detect operational PUs, enabling the SAS to enforce appropriate interference protections. The ESC may also serve other functions, such as identification of misbehaving SUs, but such functions do not affect the privacy-performance trade-off and are beyond the scope of this chapter. We denote the PU information potentially detectable by the ESC on each frequency channel and at any given time with the set $\tilde{\mathcal{X}}^{\mathrm{ESC}} = \{\tilde{\mathcal{L}}^{\mathrm{ESC}}, \tilde{\mathbf{p}}^{\mathrm{PU}}\}$. Here, $\tilde{\mathcal{L}}^{\mathrm{ESC}} \subseteq \tilde{\mathcal{L}}^{\mathrm{PU}}$ since some PUs may be receive-only and undetectable by the ESC.

ESC measurements are sent to a centralized node for estimation of the PU state, denoted $\underline{\mathcal{X}}^{\mathrm{ESC}} = \{\underline{\mathcal{L}}^{\mathrm{ESC}}, \underline{\mathbf{p}}^{\mathrm{ESC}}\}$. The performance of the ESC depends on the physical parameters of the network as well as the estimation algorithm used to interpret the sensor measurements [Clark and Psounis, 2017a, 2018a]. We will assume the region is partitioned into a grid of discrete sensing cells and any given ESC implementation can be characterized by the probability of missed detection, denoted p_{md}, i.e. the probability that the ESC fails to detect a PU in a particular sensing cell, as well as the probability of false alarm, denoted p_{fa}, i.e. the probability that the ESC detects a PU present in a sensing cell where no PU is actually operating.

14.2.4 SAS Assignments to SUs

The SAS manages n_c frequency channels and grants access to SUs over discrete time slots, where the duration of these slots is chosen as a trade between efficiency and complexity of the system. Assignments to SUs must protect the n_{PU} PU locations identified in $\underline{\mathcal{X}}^{\mathrm{SAS}}$ and in $\underline{\mathcal{X}}^{\mathrm{ESC}}$ from harmful interference.

The SAS may protect PUs by controlling SU transmit power levels to limit total interference power arriving at PU locations. In this case, the SAS will assume a propagation model with uncertainty when predicting channel gains between SUs and detected PU locations, i.e. $\underline{\mathbf{G}}^{\mathrm{PU}} = [\underline{\mathbf{G}}_{ij\ell}^{\mathrm{PU}}] \in \mathbb{R}^{n_{\mathrm{SU}} \times \underline{n}_{\mathrm{PU}} \times n_c}$ is the random array for the channel gains between each PU (e.g. radar) and SU device (e.g. UE) with indices ℓ, i, and j corresponding to the frequency channel, the SU, and the PU, respectively.

For each time slot, the SAS will allocate spectrum to SUs to maximize an SU utility function, U_{SU}, subject to robust constraints protecting the PUs from harmful interference. Given PU interference protection criteria, a SAS power assignment function should return maximum transmit power allocations for each SU–channel pair as an array $\underline{\mathbf{P}}^{\text{SU}} = [\underline{\mathbf{P}}_{i\ell}^{\text{SU}}] \in \mathbb{R}_{+}^{n_{\text{SU}} \times n_c}$ such that

$$\Pr\left(\sum_{\ell=1}^{n_c}\sum_{i=1}^{n_{\text{SU}}} \underline{\mathbf{P}}_{i\ell}^{\text{SU}}\underline{\mathbf{G}}_{ij\ell}^{\text{PU}} \geq \boldsymbol{\phi}_j^{\text{SAS}}\right) \leq \boldsymbol{\psi}_j^{\text{SAS}}, \quad j \in \{1, \dots, n_{\text{PU}}\}, \tag{14.1}$$

where Pr (.) denotes probability. A power assignment of zero excludes an SU from transmitting in the corresponding frequency channels during this time slot. In this way, this assignment function acts as an admission control, channel assignment, and power assignment function. The selection of this function as well as any obfuscation involved in the reporting of $\underline{\mathcal{X}}^{\text{SAS}}$ and $\underline{\mathcal{X}}^{\text{ESC}}$ will affect the utility of the spectrum for the SUs and the PUs' privacy. The inequality (14.1) can also be viewed as a constraint on required PU utility, U_{PU}, considering protection from interference is the utility provided to PUs by the SAS.

Alternatively to SU power assignments, if PUs and SUs employ antenna arrays, then the SAS may control harmful interference to PUs through spatial multiplexing. Specifically, through the assignment of precoders to SUs, the SAS may mitigate both SU-to-PU and PU-to-SU interference. For the latter, the SAS will need to estimate the PU-to-SU interference channel, $\mathbf{H}_{ij}^{\text{BS}} \in \mathbb{C}^{\mathbf{a}_i^{Srx} \times \mathbf{a}_j^{Ptx}}$ based on the PU and SU inputs. We explore the topic of precoder schemes in Section 14.6.1, considering their performance and privacy. In practice, due to the need to minimize latency, spatial multiplexing-based sharing may require that the PU operate the SAS as part of its data fusion center and may not allow for operation of the SAS by a separate third party. Table 14.1 summarizes the notation introduced in this section along with notation that will be introduced in Section 14.3.

14.3 User Privacy in Spectrum Sharing

Evaluating user privacy first requires definition of adversary threat models, metrics to quantify privacy, and potential obfuscation mechanisms. Here, we describe the role of these fundamental privacy components, which will allow us to formally analyze PU privacy in the following section. While we focus on the CBRS example, protecting the privacy of PU radars, e.g. in terms of operating locations and susceptibility to interference, the approaches are equally applicable to other conceivable spectrum sharing scenarios, such as PU communication systems sharing with SU radars.

14.3.1 Adversary Threat Models

Denote adversary observations of the spectrum sharing system with the set $\mathcal{Y} \in \mathbb{Y}$, where $\mathbb{Y}$ is assumed to be discrete and finite.[2] Modeling the PU state as a random set $\mathcal{X} \in \mathbb{X}$, an adversary makes an inference attack by estimating $\hat{P}_{\underline{\tilde{\mathcal{X}}}^{\text{PU}}|\underline{\mathcal{Y}}}(\mathcal{X}|\mathcal{Y})$, a distribution for the probability that the true PU state, $\underline{\tilde{\mathcal{X}}}^{\text{PU}} = \mathcal{X}$, given the observation $\mathcal{Y}$. $\hat{P}$ is used to highlight that the adversary estimate may differ from the true distribution if the adversary is suboptimal. Adversaries may have different levels of access to information in the sharing system, e.g. due to the potential for an adversary to hack components of the system. We will consider two adversary cases.

2 A discrete and finite $\mathbb{Y}$ is consistent with our assumption that the SAS will operate on discretized parameters to support digital processing. In scenarios where the adversary can observe an analog parameter directly, we can still apply this assumption as an effective approximation with quantization to an arbitrary level of precision.

Table 14.1 Notation.

Symbol	Parameter	Symbol	Parameter
$\mathbf{a}^{Prx}$	PU receive antenna elements	$\mathbf{a}^{Ptx}$	PU transmit antennas elements
$\mathbf{a}^{Srx}$	SU receive antennas elements	$\mathbf{a}^{Stx}$	SU transmit antennas elements
c_{PU}	SU utility threshold	c_{SU}	PU utility threshold
$\mathbf{G}^{\text{PU}}$	SU–PU channel gain	$\mathbf{G}^{\text{SU}}$	SU–SU channel gain
$\mathbf{H}^{\text{BS}}$	PU-to-SU complex channel state	$\mathbf{H}^{\text{PU}}$	SU-to-PU complex channel state
$\mathbf{H}^{\text{SU}}$	SU–SU complex channel state	$\mathcal{L}$	Set of locations in the region
$\tilde{\mathcal{L}}^{\text{PU}}$	Set of PU locations	$\mathcal{L}^{\text{SU}}$	Set of SU locations
$\lambda_{\mathcal{X}}$	Multiplier for simplex constraint	μ_{SU}	Multiplier for SU utility
$\mu_{\mathcal{X}}$	Multiplier for PU utility	$\mu_{\mathcal{X},\mathcal{Y}}$	Multiplier for simplex constraint
$\tilde{n}_{\text{PU}}$	Number of operating PUs	n_{PU}	SAS estimated number of PUs
n_{SU}	Number of SU transmitters	$\mathcal{P}$	Permissible SU allocations
$\mathbf{p}^{\text{ESC}}$	ESC detected PU power	$\tilde{\mathbf{p}}^{\text{PU}}$	PU transmit powers
$\mathbf{P}^{\text{SU}}$	SAS allocations to SUs	$\hat{P}_{\underline{\tilde{\mathcal{X}}^{\text{PU}}},\underline{\mathcal{Y}}}$	Adversary estimated PU state distribution
$\tilde{\boldsymbol{\phi}}$	PU interference power threshold	$\tilde{\psi}$	Max probability $\tilde{\boldsymbol{\phi}}$ is exceeded
U_{PU}	PU utility function	U_{SU}	SU utility function
u_{PU}	PU estimated harmful interference	u_{SU}	PU estimated SU utility
$\tilde{\mathcal{X}}^{\text{ESC}}$	ESC detectable PU information	$\mathcal{X}^{\text{ESC}}$	ESC estimated PU information
$\tilde{\mathcal{X}}^{\text{PU}}$	PU state information	$\mathcal{X}^{\text{SAS}}$	PU reported information
$\mathcal{X}^{\text{SU}}$	SU state information	$\mathcal{Y}$	Adversary observations

Adversary case #1: Observation of SU assignments: SUs may be exposed to private PU information through the spectrum allocations granted by the access system. An adversary that can hack into the SU network can make inference attacks on PUs based on samples of $\mathcal{X}^{\text{SU}}$ collected over time. Alternatively, an adversary may operate their own SU devices and receive legitimate assignments from the SAS, providing the basis for an inference attack.

Adversary case #2: Direct observation of the SAS: A more powerful adversary may be able to hack into the SAS directly, potentially eavesdropping on operational data the PUs send to the SAS, collecting samples of $\mathcal{X}^{\text{SAS}}$ as well as observing the measurements of the ESC. The adversary estimate in this case will depend on the design parameters of the ESC and any obfuscation strategies employed by PUs in their information reporting.

We might also consider an adversary that hacks into the PU systems via the SAS interface. For PUs that only store obfuscated parameters on their system, the adversary does not gain any information by hacking these PU systems. In that case, the privacy analysis is identical to that of case # 2. For PUs that rely on the ESC instead of reporting their information, there is no communication between them and the SAS system. Any attempt to hack into them is not specific to spectrum sharing and thus is beyond the scope of this chapter.

The adversary inference attack can be modeled as a standard Bayesian inference problem. However, with this model, the optimal implementation of an inference attack is intractable in most practical scenarios due to the very large number of possible observations and PU states [Clark and Psounis, 2018a,b]. In such scenarios, we can study heuristic adversary inference approaches. For example, in the case study in Section 14.6, we apply machine learning to estimate the PU state from observations available to the adversary. This approach allows us to model an adversary inference attack with fixed computing resources but cannot guarantee whether or not other practical adversary implementations may exist that are more effective.

14.3.2 Privacy Metrics

Privacy is not trivial to quantify and effective metrics for one setting are not necessarily well suited for another. Recent analysis of statistical databases and location-based services leverage differential privacy for the strong privacy guarantees it offers [Dwork, 2006; Dwork and Smith, 2009]. Specifically, let K be a randomized function applied to the input dataset, e.g. an obfuscation mechanism. K provides ϵ,δ-differential privacy if for all datasets D_1 and D_2 that differ in at most one element and $\forall S \in \text{range}(K)$,

$$\Pr[K(D_1) \in S] \leq e^{\epsilon} \Pr[K(D_2) \in S] + \delta, \tag{14.2}$$

where $\Pr[.]$ denotes probability. For D_1 and D_2 that differ by only one entry, ϵ and δ bound the probability that the output of K will differ for either input. Differentially private mechanisms that achieve small ϵ and δ make it harder to infer the input (D_1 versus D_2) by observing the output. This is a strong guarantee since it doesn't make any assumptions about the computation power and auxiliary information available to the adversary and it holds for the worst case of all possible inputs. However, known practical implementations have been shown to unacceptably degrade utility in some applications [Huang et al., 2017b]. While differential privacy may be applicable to some applications related to spectrum sharing scenarios, such as crowd-sourced spectrum sensing [Jin and Zhang, 2018], achieving significant levels of differential privacy for PUs operating with a SAS would require users to accept harmful interference [Clark and Psounis, 2016, 2018b]. Application of differential privacy to spectrum sharing scenarios typically requires some relaxation of the definition and weakening of the privacy guarantees. For example, Dong et al. [2018] apply differential privacy to spectrum sharing by using a preprocessing step to remove any SUs that would likely cause harmful interference. However, this preprocessing degrades the privacy guarantees that would normally accompany their differential privacy motivated strategy, which also is shown to significantly degrade utility for the SUs. Similarly, Liu et al. [2020] propose a time-windowed variation of differential privacy for the spectrum sharing scenario. The practical protections offered by these versions of differential privacy with weakened guarantees warrant further study.

Several recent works have sought to aggregate user data to preserve privacy via anonymization among a large number of users [Li et al., 2012; Gao et al., 2012; Wang and Zhang, 2015; Grissa et al., 2017; Xing et al., 2018]. k-anonymity is applied as the metric in this case, where a spectrum sharing system and associated obfuscation mechanisms achieve k-anonymity if any given user's data is indistinguishable from the data of $k-1$ other users' data [Niu et al., 2014]. While k-anonymity can be effective in scenarios with a large number of independent users, anonymization may not be achievable or meaningful in cases where users may be very sparse or are operated by a single organization. Both these conditions are true for military PUs in CBRS.

As an option more generally applicable to spectrum sharing scenarios, we can measure the potential effectiveness of the adversary estimated distribution $\hat{P}_{\underline{\tilde{\mathcal{X}}}^{\text{PU}}|\underline{\mathcal{Y}}}(\mathcal{X}|\mathcal{Y})$ by leveraging the concept of mutual information from information theory [Cover and Thomas, 2012], defined as

$$I(\underline{\tilde{\mathcal{X}}}^{\text{PU}}, \underline{\mathcal{Y}}) = \sum_{\mathcal{Y} \in \mathbb{Y}} \sum_{\mathcal{X} \in \mathbb{X}} P_{\underline{\tilde{\mathcal{X}}}^{\text{PU}}, \underline{\mathcal{Y}}}(\mathcal{X}, \mathcal{Y}) \log \left(\frac{P_{\underline{\tilde{\mathcal{X}}}^{\text{PU}}, \underline{\mathcal{Y}}}(\mathcal{X}, \mathcal{Y})}{P_{\underline{\tilde{\mathcal{X}}}^{\text{PU}}}(\mathcal{X}) P_{\underline{\mathcal{Y}}}(\mathcal{Y})} \right). \tag{14.3}$$

Mutual information measures the dependence of two random variables. Two independent random variables have zero mutual information, and nothing can be learned about one random variable by observing the other. Mutual information has proven to be an effective metric for the design of MIMO radar precoders, and even for co-design of MIMO radar and MIMO communication systems [Liu et al., 2021]. Here, to achieve privacy, we will seek to minimize mutual information between the adversary observations and the sensitive user data, noting that larger mutual information implies a greater potential for an adversary to form an effective estimate of the PU state $\tilde{\mathcal{X}}^{\text{PU}}$ from observations $\mathcal{Y}$. Mutual information is agnostic to any specific adversary implementation and is generally applicable to any potential PU privacy concern, e.g. location, time-of-use, frequency use, and system parameter privacy. It should correlate well with any other privacy metric of particular interest. However, mutual information may only be useful as a relative measure since it is not directly tied to an intuitive or physical measurement of privacy, e.g. distance for location privacy, and thus may be difficult to relate to real-world user privacy requirements.

Another approach is to explicitly model the adversary estimate and quantify the expected error as in Shokri et al. [2011]. For a set of observations $\mathcal{Y}$ and true PU state $\tilde{\mathcal{X}}^{\text{PU}}$, we can define the expected error as

$$\sum_{\mathcal{X}\in\mathbb{X}} \hat{P}_{\underline{\tilde{\mathcal{X}}^{\text{PU}}}|\underline{\mathcal{Y}}}(\mathcal{X}|\mathcal{Y})\|\mathcal{X} - \tilde{\mathcal{X}}^{\text{PU}}\|, \tag{14.4}$$

where $\|\mathcal{X} - \tilde{\mathcal{X}}^{\text{PU}}\|$ denotes some measure of distance between the candidate PU state and the true state. Considering that $\mathcal{X}$ will consist of multiple tuples of unlike variables, the most generally meaningful distance measure is not obvious. For example, with location privacy, the right metric should relate to physical distance, whereas if PUs are concerned about adversaries attempting to decode users' signals, the information theoretic metrics used in Chen et al. [2014] may be more appropriate.

We focus on the issue of location privacy for our case study in Section 14.6, though time and frequency use privacy could be treated analogously. We will apply a measure for the average distance error, that is, the average of the distances between each possible location in $\mathcal{X}$ and a real one in $\tilde{\mathcal{X}}^{\text{PU}}$ from Clark and Psounis [2018b]. We normalize the estimated probabilities for each location to sum up to one using a Voronoi diagram approach and partition the region into $\tilde{n}_{\text{PU}}$ subregions such that each partition consists of the locations from the original region that are closest to one location in $\tilde{\mathcal{X}}^{\text{PU}}$. Denote these subregions $\mathcal{L}^i$ for $i \in \{1, \dots, \tilde{n}_{\text{PU}}\}$ corresponding to the indices for the PU entries that generate each subregion. Let ℓ_j be the true location of the jth PU. Then, we define the average distance error as

$$\frac{1}{\tilde{n}_{\text{PU}}}\sum_{j=1}^{\tilde{n}_{\text{PU}}}\sum_{\ell\in\mathcal{L}^i} \|\ell - \ell_j\| \frac{\hat{P}(\ell \in \underline{\tilde{\mathcal{L}}}^{\text{PU}}|\mathcal{Y})}{\sum_{\ell'\in\mathcal{L}^i} \hat{P}(\ell' \in \underline{\tilde{\mathcal{L}}}^{\text{PU}}|\mathcal{Y})}, \tag{14.5}$$

where $\hat{P}(\ell \in \underline{\tilde{\mathcal{L}}}^{\text{PU}}|\mathcal{Y})$ is the adversary estimate of the probability that a location $\ell \in \mathcal{L}$ is contained in the true PU set $\tilde{\mathcal{L}}^{\text{PU}}$ given the observations $\mathcal{Y}$. The norm in this case is Euclidean distance.

14.3.3 Obfuscation Mechanisms

The effectiveness of adversary inference attacks can be limited by obfuscation mechanisms in the spectrum sharing system architecture. Obfuscation may be inherent to uncertainty in the output of individual components or may be explicitly added to data passed at the interface between components. In our specific CBRS PU privacy example, obfuscation can be achieved by the PU directly, by the ESC, by the SAS, or any combination thereof.

Obfuscation mechanism #1: ESC sensor measurements: Sensor measurements are noisy, and even an optimal estimator will experience missed detections and false alarms. These errors will introduce uncertainty for any adversary observing the SAS directly or observing derived SU allocations.

Obfuscation mechanism #2: PU added obfuscation: The PU may explicitly obfuscate its reporting, translating $\tilde{\mathcal{X}}^{\mathrm{PU}}$ to $\underline{\mathcal{X}}^{\mathrm{SAS}}$. This may include false PU entries, missed detections, false alarms, random noise, or any other random mapping from the true PU state to reported parameters.

Obfuscation mechanism #3: SAS allocations to SUs: SAS computation of SU allocations may have some inherent privacy since the process may be intractable to invert. Obfuscation may also be explicitly added, e.g. by randomly perturbing allocated transmit powers or beamforming precoders. This will affect PU privacy in the first adversary case, where the adversary observes SU allocations.

14.4 Optimal Privacy and Performance

To achieve the competing goals of user privacy and utility, we can tackle the design of spectrum sharing system functions with a multi-objective optimization problem. One may treat utility as the objective function and account for privacy as a constraint, do the opposite, or treat both utility and privacy in the objective function. These approaches all result in equivalent problems, and the form can be selected for convenience given the specific application. Since it is often more intuitive to define thresholds for user utility than for privacy when it comes to real-world applications, in what follows we treat user utility as a constraint on the optimization of user privacy. Thresholds on the utility may be parametrically varied to explore the privacy-utility trade-off. Let $h(\mathcal{Y}, \mathcal{X})$ be the function that returns some metric quantifying user privacy. We formally state this privacy optimization problem as

$$\underset{g_1,\dots,g_m \in \mathbb{G}}{\text{maximize}} \quad h\left(\underline{\mathcal{Y}}, \underline{\mathcal{X}}\right), \tag{14.6a}$$

$$\text{subject to} \quad U_1(\underline{\mathcal{Y}}, \underline{\mathcal{X}}) \geq u_1, \dots, U_k(\underline{\mathcal{Y}}, \underline{\mathcal{X}}) \geq u_k \tag{14.6b}$$

$$\underline{\mathcal{Y}}_1 = g_1(\underline{\mathcal{X}}_1), \dots, \underline{\mathcal{Y}}_n = g_n(\underline{\mathcal{X}}_n) \tag{14.6c}$$

$$\underline{\mathcal{X}} \in \mathbb{X}, \underline{\mathcal{Y}} \in \mathbb{Y}, \underline{\mathcal{X}}_i \subseteq \underline{\mathcal{X}} \cup \underline{\mathcal{Y}}, \underline{\mathcal{Y}}_i \subseteq \underline{\mathcal{Y}}, \tag{14.6d}$$

where $U_i, i \in \{1, \dots, k\}$ denote utility functions for the system. k will depend on the application setting, e.g. in cases where each user has their own utility, $k = n$ while in cases where there is a single-system utility, $k = 1$. As an example, consider spectrum sharing where PUs have a function to report their requirements for interference protection and share (obfuscated) PU locations, SUs have a function to request access to the spectrum, and the sharing system has a function to grant SU access, e.g. by assigning allowable transmission power levels. The utility of the PUs will depend on the likelihood of experiencing interference, while the utility of the SUs will depend on the amount of spectrum they are permitted to access. Problem (14.6) provides a general framework that can be applied to rigorous comparison of heuristics in specific problem settings without further limitations or assumptions.

In CBRS, the potential obfuscation mechanisms described in Section 14.3.3 closely relate to the principal functions of the spectrum sharing architecture, i.e. the design of the ESC, the user reporting strategy, and the SAS spectrum assignment strategy. For the purpose of optimization, we will apply mutual information as our privacy metric. Achieving low mutual information between the PU ground truth $\tilde{\mathcal{X}}^{\mathrm{PU}}$ and the adversary observations will be predictive of achieving good user privacy while also being agnostic to any specific adversary inference attack model. This is useful both for tractability of analysis, considering that a closed-form model for an adversary inference attack is often not available, and also for ensuring that we identify a design that will be effective for any inference attack that can be conceived of by an adversary.

Consider the CBRS use case with a SAS that makes power assignments to SUs. Let g_1, g_2, and g_3 be functions corresponding to the three obfuscation mechanisms described, abstracting the necessary processes of the SAS along with any explicitly added obfuscation. We can view the trade-off between SU utility and PU privacy in the context of a formal optimization using our prior notation (see Table 14.1), i.e.

$$\underset{g_1,g_2,g_3}{\text{minimize}} \quad I\left(\underline{\mathcal{Y}}; \underline{\tilde{\mathcal{X}}}^{\text{PU}}\right), \tag{14.7a}$$

$$\text{subject to} \quad \mathbb{E}\{U_{\text{SU}}(\underline{\mathbf{P}}^{\text{SU}}, \mathbf{G}^{\text{SU}}, \mathbf{H}^{\text{SU}}, \underline{\mathbf{H}}^{\text{BS}}, \mathcal{P})\} \geq c_{\text{SU}} \tag{14.7b}$$

$$\underline{\mathcal{X}}^{\text{SAS}} = g_1(\tilde{\mathcal{X}}^{\text{PU}}, u_{\text{SU}}, u_{\text{PU}}) \tag{14.7c}$$

$$\underline{\mathcal{X}}^{\text{ESC}} = g_2(\tilde{\mathcal{X}}^{\text{ESC}}) \tag{14.7d}$$

$$\underline{\mathbf{P}}^{\text{SU}} = g_3(\underline{\mathcal{X}}^{\text{SAS}}, \underline{\mathcal{X}}^{\text{ESC}}, \mathcal{X}^{\text{SU}}) \tag{14.7e}$$

$$U_{\text{PU}}(\underline{\mathbf{P}}^{\text{SU}}, \underline{\mathbf{G}}^{\text{PU}}, \tilde{\mathbf{H}}^{\text{PU}}, \tilde{\boldsymbol{\phi}}, \tilde{\boldsymbol{\psi}}) \geq c_{\text{PU}}. \tag{14.7f}$$

Problem (14.7) maximizes PU privacy by minimizing mutual information between the true PU state, $\underline{\tilde{\mathcal{X}}}^{\text{PU}}$, and adversary observations, $\underline{\mathcal{Y}}$. Optimization is over the three obfuscation mechanisms, subject to (14.7b), a constraint on the expected utility of the SU network meeting threshold c_{SU}, and to a constraint on the PU utility (14.7f), which ensures the interference to the PUs is held to a sufficiently low level. The obfuscation functions in (14.7d)–(14.7e) will affect the adversary observation, interference to PUs, and the SU assignments, which affect the PU privacy, PU utility, and SU utility, respectively.

The optimization in problem (14.7) is over functional spaces. With the assumption that the state spaces $\mathbb{X}$ and $\mathbb{Y}$ are discrete and finite (though they may be very large), we can treat the random mapping performed by the obfuscation functions as corresponding arrays representing conditional probability distributions. Specifically, the combination of $g_1(.)$ and $g_2(.)$ can be modeled together as $P_{\underline{\mathcal{X}}^{\text{ESC}},\underline{\mathcal{X}}^{\text{SAS}}|\underline{\tilde{\mathcal{X}}}^{\text{PU}}}(\mathcal{X}', \mathcal{X}''|\mathcal{X})$, where $\mathcal{X}, \mathcal{X}', \mathcal{X}'' \in \mathbb{X}$, i.e. the probability that the state $\mathcal{X}'$ is detected by the ESC and the state $\mathcal{X}''$ is reported to the SAS given that the true state is $\mathcal{X}$. Similarly, all three obfuscation functions can be modeled as $P_{\underline{\mathbf{P}}^{\text{SU}}|\underline{\tilde{\mathcal{X}}}^{\text{PU}}}(\mathbf{P}|\mathcal{X})$, where $\mathbf{P} \in \mathcal{P}$, corresponding to the probability that the SUs are granted allocation $\mathbf{P}$ given the true PU state $\mathcal{X}$. More generally, $P_{\underline{\mathcal{Y}}|\underline{\tilde{\mathcal{X}}}^{\text{PU}}}$ is the conditional distribution for the adversary observation given the true PU state in any spectrum sharing setting. In practice, writing down these probability distributions explicitly is still intractable due to the large number of possible states. Even so, we find this approach analytically useful for mapping (14.7) to an equivalent problem, i.e.

$$\underset{P_{\underline{\mathcal{Y}}|\underline{\tilde{\mathcal{X}}}}}{\text{minimize}} \quad I\left(\underline{\mathcal{Y}}; \underline{\tilde{\mathcal{X}}}^{\text{PU}}\right)$$

$$\text{subject to} \quad \sum_{\mathcal{Y} \in \mathbb{Y}} P_{\underline{\mathcal{Y}}}(\mathcal{Y}) U'_{\text{SU}}(\mathcal{Y}) \geq c'_{\text{SU}} \tag{14.8a}$$

$$\sum_{\mathcal{Y} \in \mathbb{Y}} P_{\underline{\mathcal{Y}}|\underline{\tilde{\mathcal{X}}}^{\text{PU}}}(\mathcal{Y}|\mathcal{X}) U'_{\text{PU}}(\mathcal{X}, \mathcal{Y}) \leq c'_{\text{PU}}, \forall \mathcal{X} \in \mathbb{X} \tag{14.8b}$$

$$P_{\underline{\mathcal{Y}}|\underline{\tilde{\mathcal{X}}}^{\text{PU}}}(\mathcal{Y}|\mathcal{X}) \geq 0, \forall \mathcal{Y} \in \mathbb{Y}, \mathcal{X} \in \mathbb{X} \tag{14.8c}$$

$$\sum_{\mathcal{Y} \in \mathbb{Y}} P_{\underline{\mathcal{Y}}|\tilde{\underline{\mathcal{X}}}^{\text{PU}}}(\mathcal{Y}|\mathcal{X}) = 1, \forall \mathcal{X} \in \mathbb{X}, \tag{14.8d}$$

where the SU utility, $U'_{\text{SU}}[\mathcal{Y}]$, and the probability of interference, $U'_{\text{PU}}[\mathcal{X}, \mathcal{Y}]$, are functions of the realizations of true states and observations, and can be computed before the optimization. Along with thresholds on the user utilities c'_{SU} and c'_{PU}, that can be algebraically derived from their counterparts in problem [14.7] [Clark and Psounis, 2020], they parameterize the optimization. $P_{\underline{\mathcal{Y}}|\tilde{\underline{\mathcal{X}}}^{\text{PU}}}$ is either $P_{\underline{\mathbf{P}}^{\text{SU}}|\tilde{\underline{\mathcal{X}}}^{\text{PU}}}$ or $P_{\underline{\mathcal{X}}^{\text{ESC}},\underline{\mathcal{X}}^{\text{SAS}}|\tilde{\underline{\mathcal{X}}}^{\text{PU}}}$, depending on the adversary case of concern. Equations (14.8a) and (14.8b) correspond to the utility and interference constraints, i.e. (14.7b) and (14.7f), respectively. Constraints (14.8c) and (14.8d) are simplex constraints ensuring that $P_{\underline{\mathcal{Y}}|\tilde{\underline{\mathcal{X}}}^{\text{PU}}}$ yields a valid probability distribution. Since mutual information is convex with respect to the conditional probability distribution [Salamatian et al., 2015], and all of the constraints are linear, problem (14.8) is convex. Therefore, for any strictly feasible instance of (14.8), we can compute the Karush-Kuhn-Tucker (KKT) conditions [Boyd and Vandenberghe, 2004] on the optimal solution. Per Clark and Psounis [2020], the KKT conditions can be applied to show that the optimal solution for problem (14.8) satisfies

$$P^*_{\underline{\mathcal{Y}}|\tilde{\underline{\mathcal{X}}}^{\text{PU}}}(\mathcal{Y}|\mathcal{X}) = \frac{P^*_{\underline{\mathcal{Y}}}(\mathcal{Y}) \exp(\mu^*_{\text{SU}} U'_{\text{SU}}(\mathcal{Y}) - U'_{\text{PU}}(\mathcal{X}, \mathcal{Y})\mu^*_{\mathcal{X}} / P_{\tilde{\underline{\mathcal{X}}}^{\text{PU}}}(\mathcal{X}))}{\sum_{\mathcal{Y}' \in \mathbb{Y}} P^*_{\underline{\mathcal{Y}}}(\mathcal{Y}') \exp(\mu^*_{\text{SU}} U'_{\text{SU}}(\mathcal{Y}') - U'_{\text{PU}}(\mathcal{Y}', \mathcal{X})\mu^*_{\mathcal{X}} / P_{\tilde{\underline{\mathcal{X}}}^{\text{PU}}}(\mathcal{X}))}, \tag{14.9}$$

where $*$ denotes optimality. μ_{SU} and $\mu_{\mathcal{X}}$ are Lagrange multipliers for the SU utility and PU interference constraints. Clark and Psounis [2020] show that a closed-form solution to (14.9) does not exist in general. Of course, in practical cases, we cannot expect to even enumerate all possible PU states, let alone solve a general polynomial where each state corresponds to a term. However, we can use this result to inform the development of more practical strategies.

To implement a solution to problem (14.8) in practice, we must map the resulting $P_{\underline{\mathcal{Y}}|\tilde{\underline{\mathcal{X}}}^{\text{PU}}}$ back to the obfuscation mechanisms g_1, g_2, and g_3 in problem (14.7). For a power assignment SAS under adversary case #1, where the adversary observes SU devices, $P_{\underline{\mathcal{Y}}|\tilde{\underline{\mathcal{X}}}^{\text{PU}}} = P_{\underline{\mathbf{P}}^{\text{SU}}|\tilde{\underline{\mathcal{X}}}^{\text{PU}}}$ can be implemented in g_3 at the SAS, with g_1 and g_2 as pass-through functions, i.e. $g_1(\mathcal{X}) = g_2(\mathcal{X}) = \mathcal{X}$, corresponding to truthful reporting by the PU. In adversary case #2, where the SAS is not trusted, the solution $P_{\underline{\mathcal{X}}^{\text{ESC}},\underline{\mathcal{X}}^{\text{SAS}}|\tilde{\underline{\mathcal{X}}}^{\text{PU}}}$ can be implemented in the PU reporting g_1, with g_2 set to always return 0, i.e. no ESC implementation, and the SAS assignment strategy, g_3, set to maximize SU utility. However, the optimal strategy requires knowledge of the SU state, $\mathcal{X}^{\text{SU}}$, but these parameters are not generally available to the PU. In an alternative architecture, the SAS could pass $\mathcal{X}^{\text{SU}}$ to the PU to enable near-optimally obfuscated reporting. In cases where this is not desirable, e.g. due to SU privacy concerns or other practical limitations, the PU must approximate SU utility, u_{SU}, and probability of interference, u_{PU}. The PU may be able to estimate these values in several ways [Wang et al., 2018], e.g. based on observation of typical SU operations in the region, and may improve its estimates over time with feedback from the SAS on achieved SU utility and its own observations of interference.

14.5 Practical Privacy Preservation

Given the intractability of the optimal solution to the optimization problem identified in Section 14.4 for typical spectrum sharing scenarios, many practitioners resort to heuristics tailored to the specific application. In this section, we review three of the prevailing strategies in the literature prior to considering specific case studies in the following section.

14.5.1 Data Perturbation

The most common obfuscation techniques in the literature add random noise, false entries or otherwise randomly perturb the user data. In spectrum sharing, data perturbation may include reporting the existence of false PU locations to the SAS or adding random noise to the PU parameters reported for real PUs. The SAS may also add random noise to the assignments granted to SUs. Care must be taken in crafting data perturbation strategies to ensure data is not perturbed in a way that unacceptably increases the risk of harmful interference to PUs. As a simple example, randomly removing a true PU location from the reported data would significantly increase the risk of harmful interference for that PU.

Creating more variability in the true user data can also serve to increase user privacy, i.e. we can increase privacy by increasing the uncertainty in the adversary's a priori information for the user data distribution. For example, a PU might employ frequency hopping such that an adversary cannot readily infer current channel use from observations of past use. Similarly, a faster moving PU may make it more difficult for an adversary attempting to estimate the PU's location.

Robertson et al. [2013] and Mosbah et al. [2017] effectively apply data perturbation strategies to preserve PU privacy with a SAS. These analyses are provided in the context of a SAS with binary decisions where either any given channel is available to an SU at a particular time slot or it is not. This assumption limits the applicability of the analysis and results to PU time-of-use privacy. Both this time-of-use privacy problem and the falsification/obfuscation strategies used can be considered special cases of the spectrum sharing privacy framework presented in this chapter.

Other recent works have studied PU privacy preserving obfuscation methods in a more generalized SAS setting [Clark and Psounis, 2016, 2018a,b; Bahrak et al., 2014; Gao et al., 2013; Vaka et al., 2016; Rajkarnikar et al., 2017]. Motivated by the intuition that randomization should increase privacy, and with varying levels of generality in their models, these works apply perturbation strategies such as adding false PU entries or randomly reducing SU transmit power allocations. Rajkarnikar et al. [2017] derive conditions for when adding false PU entries uniformly over a region offers better performance over increasing the size of an exclusion zone around true PU locations. Bhattarai et al. [2018] formulate an optimization problem for an SU power reduction strategy, which is feasible to solve numerically with a specific adversary model and where SAS query responses are limited to a small set. These works identify optimal solutions for spectrum sharing constrained problems, e.g. when obfuscation is limited to predetermined strategies with restrictions on the SAS operation for tractability of the analysis and numerical computation.

14.5.2 Allocation Reporting Codebook

The allocation reporting codebook (ARC) algorithm is another method to design privacy strategies based on characteristics of the user data set [Clark and Psounis, 2020]. ARC was inspired by the spectrum sharing setting and based on three observations regarding the numerator of (14.9) as follows:

1. The probability that $\mathcal{Y}$ is reported, $P_{\underline{\mathcal{Y}}|\underline{\tilde{\mathcal{X}}}^{\text{PU}}}(\mathcal{Y}|\mathcal{X})$, should exponentially increase with $u_{\text{SU}}(\mathcal{Y})$, the SU utility offered by $\mathcal{Y}$.
2. For a given true PU state $\mathcal{X}$, the probability that $\mathcal{Y}$ is reported should exponentially decrease with $u_{\text{PU}}(\mathcal{X}, \mathcal{Y})$, the probability that reporting this state will cause harmful interference to the PUs.
3. The probability that $\mathcal{Y}$ is reported for a given $\mathcal{X}$ should linearly increase with $P_{\underline{\mathcal{Y}}}(\mathcal{Y})$, the probability that $\mathcal{Y}$ is reported for all other states, i.e. we should reuse the same reported states to the extent practical.

For this last observation, it is certainly intractable to compute the probability of an observation for each possible PU state. As an approximation, ARC samples the space of adversary observations, $\mathbb{Y}$, and constructs a codebook of possible observations, $\mathbb{C} \subseteq \mathbb{Y}$. At each time slot, for the given true PU and SU topologies, ARC computes the SU utility and probability of interference to PUs that would result for any selected codeword in the codebook and weights the codewords exponentially. ARC then randomly selects a codeword according to the weights, which determines the obfuscated decision, e.g. the data reported by the PU or the allocations made by the SAS to the SUs.

Algorithm 1 provides pseudocode for ARC. For a true PU state, ARC assigns a probability of reporting each codeword in the codebook according to exponential weights on the SU utility and harmful interference. μ_{SU} corresponds to the SU utility constraint and can be viewed as a design parameter for trading between SU utility and PU privacy. Codebook size is also a design parameter that potentially offers improved performance at the cost of complexity. ARC also solves for the $\mu_{\mathcal{X}}$ that ensures the constraint on PU interference is satisfied and achieves a complexity of $O(s(u_{SU} + u_{PU}))$. When SU utility, u_{SU}, and interference, u_{PU}, can be computed in polynomial time, which is the case for sum-rate utility and the probabilistic channel uncertainty models we consider, ARC runs in polynomial time with respect to the size of the codebook, s, and the number of PUs and SUs, offering a generally efficient solution that can be applied to large problems. ARC has been shown to offer obfuscation that approximates the optimal solution [Clark and Psounis, 2020].

Algorithm 1 Allocation and reporting codebook (ARC).

1: given μ_{SU}, Codebook Size $= s$
2: **for** i=1:s-1 **do**
3: Randomly generate $\mathcal{Y}_i \in \mathbb{Y}$ and compute $u_{SU}(\mathcal{Y}_i)$.
4: **end for**
5: **for** True PU state $\mathcal{X}$ at each time slot **do**
6: $\mathcal{Y}_s \in \mathbb{Y}$ such that $u_{PU}(\mathcal{X}, \mathcal{Y}_s) \leq c_{PU}$
7: Compute $u_{PU}(\mathcal{X}, \mathcal{Y}_i)$ for $i \in \{1, \dots, s\}$.
8: Find $\mu_{\mathcal{X}}$ such that $\sum_{i=1}^{s} w_i U'_{PU}(\mathcal{X}, \mathcal{Y}_i) \leq c_{PU}$,
$w_i = \frac{\exp(\mu_{SU} U'_{SU}(\mathcal{Y}_i) - U'_{PU}(\mathcal{X}, \mathcal{Y}_i)\mu_{\mathcal{X}})}{\sum_{j=1}^{s} \exp(\mu_{SU} U'_{SU}(\mathcal{Y}_j) - U'_{PU}(\mathcal{X}, \mathcal{Y}_j)\mu_{\mathcal{X}})}$.
9: Randomly sample i_r from $i \in \{1, \dots, s\}$ according to weights w_i.
10: **return** $\mathcal{Y}_{i_r}$
11: **end for**

The methodology used to develop ARC may be applied to other spectrum sharing scenarios with minor adaptations. Utility constraints and private functions may be modified or added without changing the approach. In the spectrum sharing scenarios we consider, PU reporting and SAS assignments to SUs are the functions we wish to make private. SU sum-rate and PU interference are two utility constraints. In another setting, we may wish to account for tertiary users (TUs), e.g. the General Authorized Access users in CBRS. We may consider SU privacy optimization in this setting, with utility constraints accounting for PU harmful interference, SU utility, and TU utility. Further, SU and TU utilities need not be limited to the sum-rate expressions we applied, but may be selected to account for other factors, e.g. fairness.

14.5.3 Generative Adversarial Privacy

Recently, methodical schemes that learn from user data to generate obfuscation strategies have been applied to mobile user data settings. Leveraging recent advancements in generative adversarial networks (GANs) [Goodfellow et al., 2014; Camino et al., 2018; Mirza and Osindero, 2014], generative adversarial privacy (GAP) can be employed in a system to learn how to discern private features from the dataset and then how to cleverly obfuscate the data such that these private features are difficult to discern [Huang et al., 2017a]. A GAP system positions a privatizer and a simulated adversary against each other and iteratively trains each via deep learning networks until the privatizer and the adversary converge to an equilibrium. GAP has been applied to mobile user data settings [Clark et al., 2019] and is effective in spectrum sharing scenarios.

14.5.4 Homomorphic Encryption

As an alternative or in conjunction with methods to obfuscate the data shared within a spectrum sharing system, encryption methods may be employed in the system architecture to restrict the information that is revealed. Specifically, homomorphic encryption techniques, such as the Paillier cryptosystem, have been proposed to allow users to provide encrypted versions of their data to the SAS and even other users, allowing those users to conduct arithmetic operations on the encrypted data without any knowledge of the underlying sensitive data [Dou et al., 2017; Cheng et al., 2020; Grissa et al., 2021]. These techniques can help to avoid inadvertent leakage of information though they alone cannot address the issue of PU privacy loss due to inference attacks from the information that must be exposed to the SUs, such as SU assignments from the SAS.

14.6 Spectrum Sharing Case Studies with Radar Primary Users

To demonstrate the analysis tools and privacy preservation schemes explored in the prior sections, we now consider their application in real-world settings motivated by practical spectrum sharing scenarios such as CBRS.

14.6.1 Protecting Radar in Co-designed Radar-Communication Systems

For systems with antenna arrays, spectrum sharing through spatial multiplexing may be effective. Specifically, spectrum sharing between multiple-input-multiple-output (MIMO) radar and MIMO communication systems has been the subject of recent research [Mahal et al., 2017; Zhao and Swami, 2007; Saruthirathanaworakun et al., 2012]. Existing works mostly offer designs that optimize the objective of one system or the other, such as the null space projection precoding schemes of Sodagari et al. [2012], Babaei et al. [2013], Khawar et al. [2016], and Mahal et al. [2017].

Li et al. [2016a,b] and Li and Petropulu [2017] propose co-design of radar and communication systems in a setting where spectrum sharing is moderated by a controller, effectively a SAS, which collects information from the two systems and designs precoders for them to mitigate interference. The precoders contain information about the two systems, which may raise privacy concerns. For example, consider an SU smartphone coexisting with a PU military radar; the SU is assigned a precoding matrix to control the interference that it generates toward the PU. An adversary observing the SU precoder could potentially reverse engineer the precoder and obtain information about the PU, e.g. the radar's location. Of course, by using dedicated equipment, one could localize the radar based on its high power. Here, however, the possibility of localizing a radar with no additional equipment potentially poses a much more extensive threat to the privacy of radar operations. Suppose the SU and PU are stationary, and the PU radar antennas transmit orthogonal waveforms. The measurements of all receive antennas are forwarded to a fusion center for processing and target estimation. Although in theory the fusion center and SAS could operate independently, in practice, the SAS may need to be integrated with the MIMO radar fusion center to reduce latency and limit the risk of information transmission to an untrusted node.

We apply our privacy framework to evaluate whether the information contained in the exchanged precoders can pose a privacy risk for the PU. The adversary, disguised as a smartphone operator, performs an inference attack on the radar location based on the SAS provided precoder. We'll explicitly model adversary inference attacks using a machine learning approach, assuming an adversary that partitions the radar's operational region into cells, and proceeds by training a separate classifier for each cell, based on either historical or simulated precoder matrix training data. Once the classifiers of all cells are trained, for every new precoder observation, the adversary can attempt to determine the cell in which the radar is located. PU location privacy will depend on the amount of information the precoder reveals about the radar location. To gain more insight into potential privacy exposure, we also examine mutual information, as discussed in Section 14.3.2, between the precoders sent to smartphones and the radar location.

14.6.1.1 Mitigating Interference with Precoders

Figure 14.2 illustrates our scenario of an SU smartphone communicating with a local base station and potentially interfering with the PU radar. The base station downlink transmission is in another frequency band and thus does not create interference. The MIMO radar also may cause interference to the base station receiver through direct line-of-sight, reflected and/or refracted paths. The interference for the ith SU channel matrix is directly related to the radar location. Molisch [2012] provides a model for the Rician fading channel as

$$\mathbf{H}^{\mathrm{PU}} = \gamma\left(\sqrt{\frac{k}{1+k}}\mathbf{S}_{\mathrm{LOS}} + \sqrt{\frac{1}{1+k}}\mathbf{S}_{\mathrm{NLOS}}\right), \tag{14.10}$$

where $\gamma = \frac{\sqrt{e_x}\lambda}{4\pi d\sqrt{\mathbf{a}_i^{Stx}}}$ with e_x representing the transmitting energy, λ the carrier wavelength, d the distance between the PU and SU, and k the Rician factor.

$$\mathbf{S}_{\mathrm{LOS}} = \mathbf{e}_r(\omega_r)\mathbf{e}_t(\omega_t)^H \tag{14.11}$$

and $\mathbf{S}_{\mathrm{NLOS}}$ is a matrix of i.i.d. complex Gaussian random variables with zero mean and unit variance. $\omega_t = \sin(\phi_t)$, $\omega_r = \sin(\phi_r)$ are the angle of incidence of the line-of-sight path on the transmit and receive uniform linear arrays. The transmit and receiving steering vectors are given by

$$\begin{aligned}\mathbf{e}_t(\omega_t) &= \left[1, e^{-j\frac{2\pi\Delta_t}{\lambda}\omega_t}, \dots, e^{-j(\mathbf{a}_i^{Stx}-1)\frac{2\pi\Delta_t}{\lambda}\omega_t}\right]^T,\\ \mathbf{e}_r(\omega_r) &= \left[1, e^{-j\frac{2\pi\Delta_r}{\lambda}\omega_r}, \dots, e^{-j(\mathbf{a}_j^{Prx}-1)\frac{2\pi\Delta_r}{\lambda}\omega_r}\right]^T,\end{aligned} \tag{14.12}$$

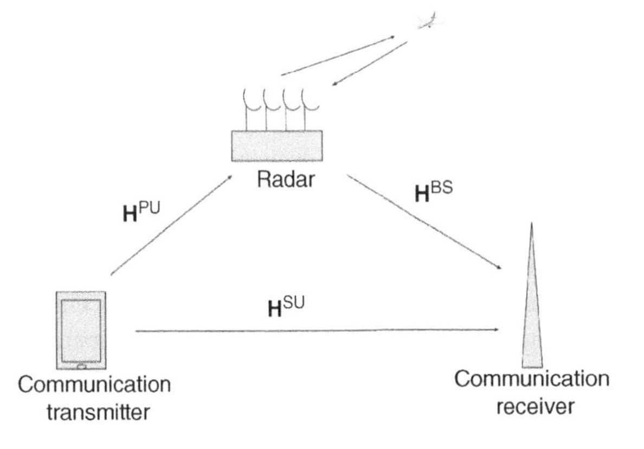

Figure 14.2 MIMO radar and MIMO communication system interference. Source: Hilli et al. [2019] © 2019, IEEE.

where Δ_t and Δ_r are the transmit and receive antenna spacing, respectively. $[.]^T$ denotes the matrix transpose. In order to mitigate the interference to the radar, we consider two separate precoder schemes for the SU.

Null space precoder: We consider a zero-forcing precoder [Sodagari et al., 2012; Babaei et al., 2013; Khawar et al., 2016] that mitigates interference by having the SU operate in the null space of the interference channel, $\mathbf{H}^{\text{PU}}$, i.e.

$$\mathbf{P}_n^{\text{SU}} = \text{null}(\mathbf{H}^{\text{PU}}). \tag{14.13}$$

To ensure that a null space exists, the number of radar receive antennas needs to be smaller than the number of communication transmit antennas.

Optimized precoder: Li et al. [2016a,b] and Li and Petropulu [2017] apply a precoder that aims to minimize the interference power to the radar, subject to meeting power and rate constraints. The precoder $\mathbf{P}_o^{\text{SU}} = \sqrt{\mathbf{R}_x}$ is the solution to the optimization problem

$$\underset{\mathbf{R}_x}{\text{minimize}} \quad \text{Tr}\{\mathbf{H}^{\text{PU}}\mathbf{R}_x(\mathbf{H}^{\text{PU}})^H\}, \tag{14.14}$$

$$\text{subject to} \quad \text{Tr}\{\mathbf{R}_x\} \leq p_t, \tag{14.15}$$

$$\log|\mathbf{I} + \mathbf{R}_{\text{in}}^{-1}\mathbf{H}^{\text{BS}}\mathbf{R}_x(\mathbf{H}^{\text{BS}})^H| \geq c_{\text{SU}}, \tag{14.16}$$

where $\mathbf{R}_x$ is the covariance matrix of the transmitted codewords and $(\mathbf{H}^{\text{PU}})^H$ denotes the conjugate transpose of $\mathbf{H}^{\text{PU}}$. p_t is the transmit power budget of the smartphone and c_{SU} is the minimum communication rate. $\mathbf{R}_{\text{in}}$ is the interference plus noise covariance at the communication receiver, which is assumed to be known. Tr(.) denotes the matrix trace, and $\mathbf{I}$ denotes the Identity matrix.

14.6.1.2 Inference Attacks Using the Precoders

Dimas et al. [2019] model an adversary that searches a 2 km × 2 km area for a radar and attempts to use the precoders to prioritize the search. Specifically, the adversary partitions the area into 500 m × 500 m cells and makes a binary decision as to whether the radar is located in each cell. The adversary interfaces with the SAS, claiming to operate $n_{\text{SU}} = 5$ smartphones. The smartphone coordinates are uniformly chosen from a subarea in the region consistent with the appearance that the smartphones are operating with a local base station. A single PU radar system operates a MIMO array with $a^{Prx} = a^{Ptx} = 6$ antennas and the ith MIMO smartphone operates with a base station using $\mathbf{a}_i^{Stx} = \mathbf{a}_i^{Srx} = 8$ antennas.

The adversary uses the precoding matrices sent to every colluding smartphone as features to a classification problem. A classifier is trained for every cell to determine whether a radar is present in the cell. Training is conducted on a balanced set of labeled data that the adversary could have created through simulation or through historical data. Dimas et al. [2019] consider both Naive Bayes and SVM classifiers. The adversary applies the trained classifiers to observed precoders, such that the collection of classifier outputs provides an estimate for radar operations in the area.

Figure 14.3 plots the adversary's receiver operating characteristic (ROC) for a single cell classifier. True positives represent the cases in which the classifier correctly determined that the radar was located in the cell, while false positives represent the cases in which the classifier decided the radar was present when in reality it was not. Figure 14.3 includes results for a baseline case for comparison where the adversary observes the interference channel, $\mathbf{H}^{\text{PU}}$ directly. Observation of the interference channel enables the adversary to accurately predict the presence of a radar in a cell. Observing the null space precoder creates a more challenging problem for the adversary, resulting in poorer estimation accuracy. Finally, the optimized precoder creates an even more

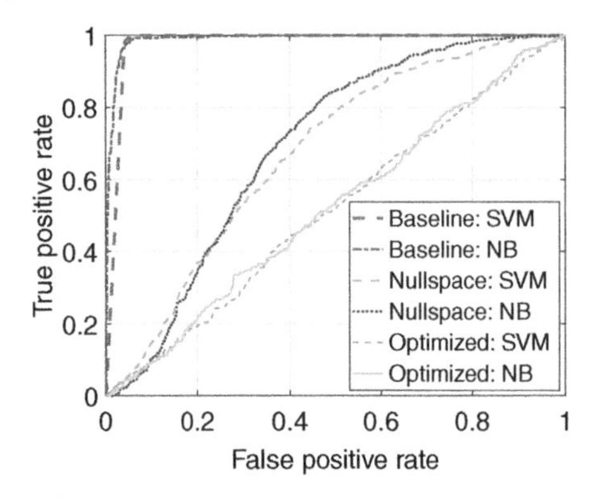

Figure 14.3 Adversary classifier ROC. Source: Dimas et al. [2019] © 2019, IEEE.

challenging problem for the adversary, producing an almost diagonal ROC curve, which essentially corresponds to random adversary guessing and indicates that the optimized precoder offers better privacy for the radar operator.

Mutual information between the true radar location and the adversary observation is estimated numerically for the two precoding approaches and the baseline case using the K-means clustering algorithm [Krishna and Murty, 1999] for a scenario with a single adversary smartphone. Figure 14.4 plots this estimate for a variable number of communication transmit and radar receive antennas. Consistent with the accuracy of the machine learning adversary's estimate, the estimated mutual information shows that the baseline observation of the interference channel, $\mathbf{H}^{\mathrm{PU}}$, reveals the most information about the radar location, while the optimized precoder, $\mathbf{P}_o^{\mathrm{SU}}$, reveals the least. That the baseline observation reveals the most information is also a consequence of the data processing inequality, i.e. in a Markov chain, processing cannot increase information [Cover and Thomas, 2012].

Increasing the number of transmit antennas at the communication system results in an increase in mutual information for the null space precoder and the baseline direct observation of the interference channel. This can be justified by the respective increase in the column space of the interference channel, which directly affects the size of the null space precoder as well. On the other hand, in Figure 14.4b we see a reduction in mutual information as the number of receive antennas at the radar increases. The optimized precoder appears to yield the least mutual information. As the solution of a constrained optimization problem, the optimized precoder depends on many

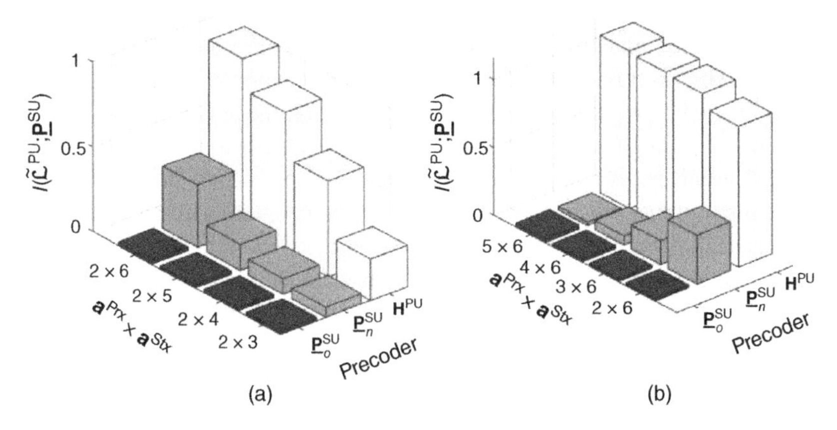

Figure 14.4 MI between precoders and radar positions for a varying number of (a) transmit and (b) receive antennas. Source: Dimas et al. [2019] © 2019, IEEE.

factors, which will make it more difficult for an adversary to discern the underlying radar location based only on an observation of the precoder.

14.6.1.3 Precoder Obfuscation

Despite the relatively strong privacy improvement offered by the optimized precoder relative to the null space precoder, this evaluation is specific to the adversary's ability to identify the radar's presence within a particular cell. If instead, the PU is concerned about the adversary's ability to estimate the angle toward the radar relative to the SU position, we should evaluate that inference attack separately, and not assume privacy against one kind of inference attack infers privacy against others.

The radar location can be defined by two parameters relative to the SU: the distance from the SU to the radar (d) and the angle between the radar and the communication unit (θ). Hilli et al. [2019] showed that the optimized precoder for a single SU does not reveal any information about the distance to the radar. Specifically, in (14.14), $\mathbf{H}^{\text{PU}}$ is the only matrix that has information about the radar location. Substituting (14.10) into the objective function of (14.14), we have

$$\mathbf{P}_o^{\text{SU}} = \min_{\mathbf{R}_x} \gamma^2 \text{Tr}\left\{\left(\sqrt{\frac{k}{1+k}}\mathbf{S}_{\text{LOS}} + \sqrt{\frac{1}{1+k}}\mathbf{S}_{\text{NLOS}}\right)\right.$$
$$\left.\mathbf{R}_x\left(\sqrt{\frac{k}{1+k}}\mathbf{S}_{\text{LOS}} + \sqrt{\frac{1}{1+k}}\mathbf{S}_{\text{NLOS}}\right)^H\right\}. \tag{14.17}$$

Equation (14.17) indicates that the optimal value of $\mathbf{R}_x$ does not change when the distance (i.e. γ) changes. Thus, the distance cannot be estimated from the optimal value of $\mathbf{R}_x$ based on the precoder assigned to a single SU.

On the other hand, $\mathbf{H}^{\text{PU}}$ depends on $\mathbf{S}_{\text{LOS}}$, which, based on (14.11), has a direct relationship to the angle between the radar and the cellphone. The objective function in (14.14) can be rewritten as

$$\text{Tr}\{\mathbf{H}^{\text{PU}}\mathbf{R}_x(\mathbf{H}^{\text{PU}})^H\} = \text{Tr}\{(\mathbf{H}^{\text{PU}})^H\mathbf{H}^{\text{PU}}\mathbf{R}_x\}. \tag{14.18}$$

Hilli et al. [2019] go on to show that for known antenna spacing, e.g. $\Delta t = \frac{\lambda}{2}$, the objective function can be written as

$$\text{Tr}\{(\mathbf{H}^{\text{PU}})^H\mathbf{H}^{\text{PU}}\mathbf{R}_x\} = \gamma^2 \left[\sum_{\ell=1}^{\mathbf{a}_i^{Stx}} \mathbf{a}_j^{Prx}(\mathbf{R}_x)_{\ell,\ell} + \frac{2k}{1+k}\mathbf{a}_j^{Prx} \underbrace{\sum_{\ell=1}^{\mathbf{a}_i^{Stx}-1}\sum_{q>\ell}^{\mathbf{a}_i^{Stx}} |[\mathbf{R}_x]_{\ell,q}| \cos\left[\pi[\ell-q]\omega_t + \angle[\mathbf{R}_x]_{\ell,q}\right]}_{L[\theta]}\right]. \tag{14.19}$$

From the PU and SAS perspective, (14.19) should be minimized with respect to $\mathbf{R}_x$ to limit interference. This corresponds to the aforementioned optimized precoder. From the adversary perspective, note that in (14.19), $L[\theta]$ is the only term that has information regarding the radar angle. If the PU takes the optimized precoder approach, the adversary can identify the angle that minimizes $L[\theta]$ as an estimate for the actual angle toward the PU radar. Due to the low dimensionality of the problem, and the limited range of possible values for θ, an adversary may effectively find the minimizing θ through brute force. Figure 14.5a illustrates such an inference attack against the optimized precoder for a case with $A_i^{Stx} = 10$. $L[\theta]$ is plotted and its minimum value can be seen to offer a close estimate to the real radar angle, indicated with the red vertical line. While the optimized precoder

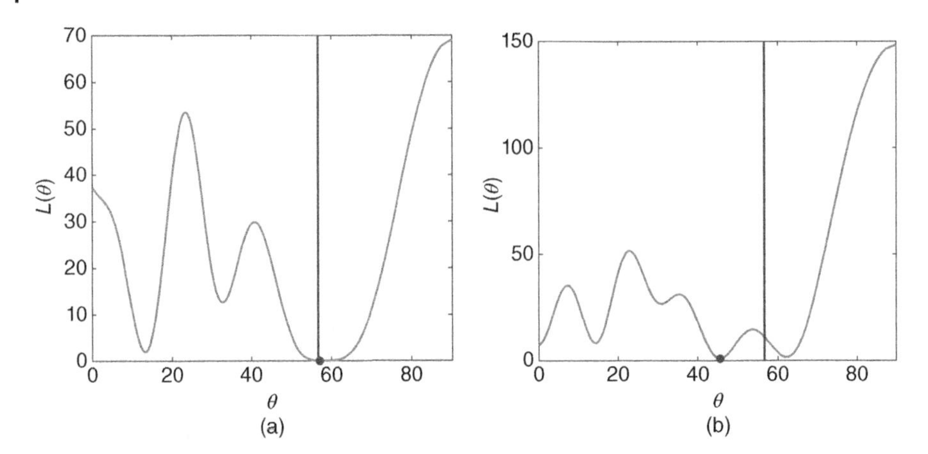

Figure 14.5 Adversary estimated radar angle. The vertical line represents the true angle of the radar, while the circular marker represents the estimated angle obtained by minimizing $L[\theta]$. [a] Interference optimized precoder and [b] obfuscated precoder. Source: Hilli et al. [2019] © 2019, IEEE.

appeared to offer the PU significant privacy protection in our prior example, this result highlights that other scenarios could pose a significant privacy breach for the PU. For example, if the adversary were to operate several SUs dispersed throughout the region of interest, rather than concentrated around a single base station, the estimated angles from each SU to the PU radar could allow the adversary to triangulate the radar location with high accuracy.

As a heuristic approach to obfuscate the optimized precoder and thwart this adversary inference attack, Hilli et al. [2019] propose forcing the gradient of $L[\theta]$ to be bigger than zero at the actual radar angle. This prevents the actual radar angle from being the minimum of $L[\theta]$, rendering this adversary inference attack less effective. This can be accomplished by adding the gradient of $L[\theta]$ at ω_t to the objective of the precoder optimization, i.e.

$$\begin{aligned} \underset{\mathbf{R}_x}{\text{minimize}} \quad & \text{Tr}\{\mathbf{H}^{\text{PU}}\mathbf{R}_x[\mathbf{H}^{\text{PU}}]^H\} \\ & -h \sum_{\ell=1}^{\mathbf{a}_i^{Stx}-1} \sum_{q>\ell}^{\mathbf{a}_i^{Stx}} [\ell - q]|[\mathbf{R}_x]_{\ell,q}| \sin[\pi[\ell - q]\omega_t + \angle[\mathbf{R}_x]_{\ell,q}], \\ \text{subject to} \quad & \text{Tr}\{\mathbf{R}_x\} \leq p_t, \\ & \log|\mathbf{I} + \mathbf{R}_{in}^{-1}\mathbf{H}^{\text{BS}}\mathbf{R}_x[\mathbf{H}^{\text{BS}}]^H| \geq c_{\text{SU}}, \end{aligned} \tag{14.20}$$

where $h = \bar{h}\frac{-2\pi k \mathbf{a}_j^{Prx}}{1+k}$, and $\bar{h}$ is a positive scalar. $\bar{h}$ acts as a means to trade between goals of minimizing interference to the PU and maximizing PU privacy. Hilli et al. [2019] show that the addition of the gradient of $L[\theta]$ preserves the convexity of (14.20) such that the optimization may be treated with standard techniques. Figure 14.5b shows the estimated angle using the precoder matrix designed according to (14.20) for the same scenario as in Figure 14.5a. The true angle to the radar no longer shows a clear correspondence to the minimum of $L[\theta]$, limiting the efficacy of the adversary inference attack.

As another example, consider a single PU MIMO radar with four receive and six transmit antennas. Consider a single SU MIMO transmitter with 10 transmit antennas, transmitting to a base station with seven receive antennas. Both systems operate at a carrier frequency of 3.55 GHz and suppose $\gamma = 1$, noting that this will not affect the final adversary estimate. For 5000 Monte Carlo trials, a uniform random radar angle is selected between 0° and 90°, and the decoder matrix is designed using both the optimal precoder and the obfuscated precoder. The estimated angles are

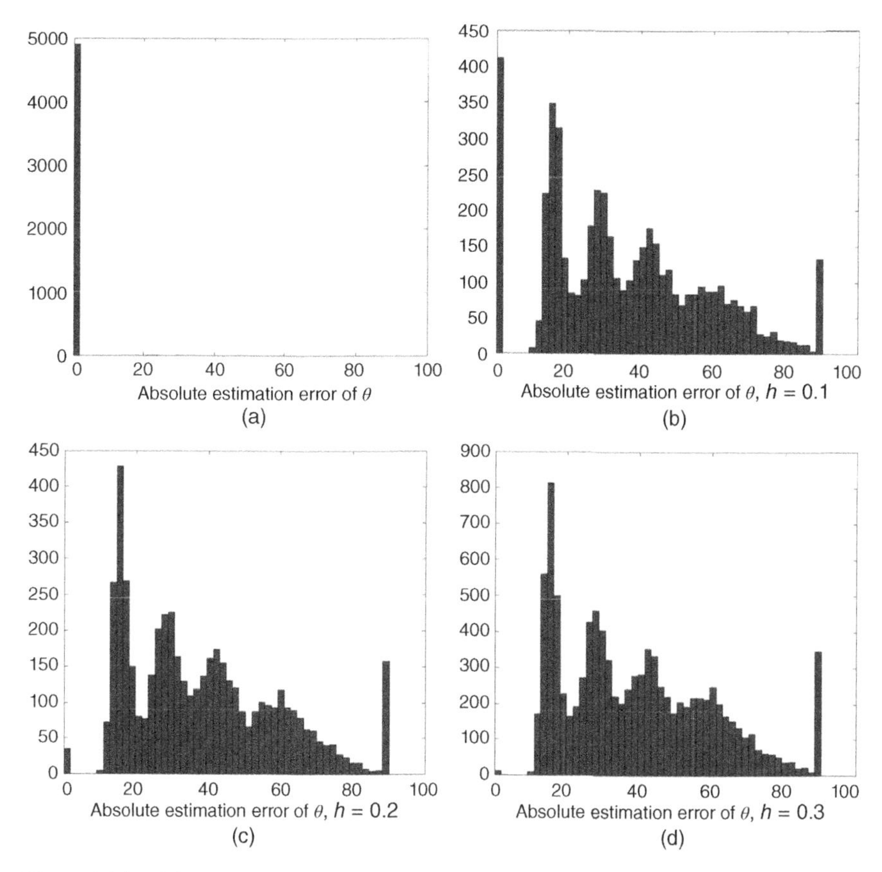

Figure 14.6 Adversary estimation error against [a] the interference optimal precoder, and the obfuscated precoder with [b] $h = 0.1$, [c] $h = 0.2$, and [d] $h = 0.3$. Source: Hilli et al. [2019] © 2019, IEEE.

obtained by brute force, and the absolute errors between the estimated angles and the real radar angles are calculated, along with the interference powers for both approaches.

Figure 14.6 shows the absolute error between the estimated angle and the real radar angle for the optimal precoder and the proposed approach using different values of $\bar{h}$. For the optimal precoder, the absolute estimation error is mostly small, corresponding to the adversary effectively determining the direction to the radar in nearly every trial. For the obfuscated precoder, fewer trials show such high adversary accuracy, and the accuracy decreases with the increase of parameter $\bar{h}$. Obfuscating the radar location comes at the cost of increasing the interference power. Figure 14.7 shows the interference power for the optimal precoder and the obfuscated precoder approaches. The optimal precoder effectively minimizes interference power, while the interference power increases with the increase of $\bar{h}$ for the obfuscated precoder. Thus, the obfuscated precoder offers the PU a means to protect its privacy at the expense of accepting increased interference, where the PU may control this trade-off through the adjustment of $\bar{h}$.

14.6.2 Protecting Radar Through Power Assignments to Secondary Users

The joint design of radar and communication system precoders may be impractical for some spectrum sharing settings due to the need for low latency reporting of complete channel state information and updating of precoder assignments. While such a system offers great potential

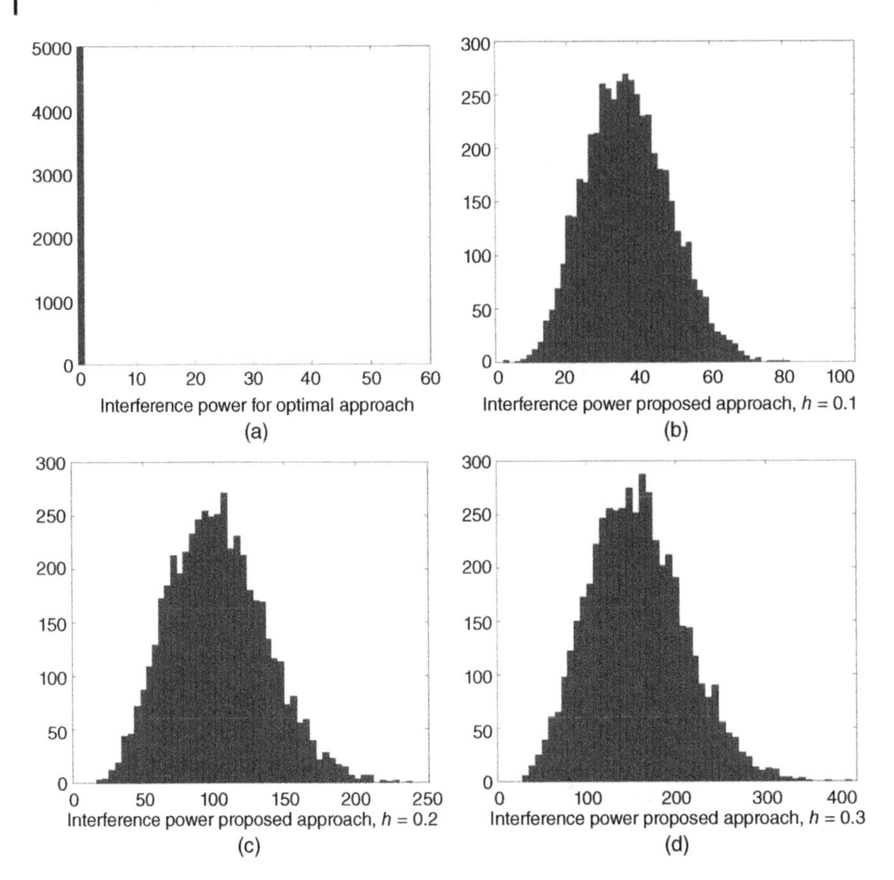

Figure 14.7 Interference power into the PU with [a] the interference optimal precoder, and the obfuscated precoder with [b] $h = 0.1$, [c] $h = 0.2$, and [d] $h = 0.3$. Source: Hilli et al. [2019] © 2019, IEEE.

to effectively exploit constrained RF spectrum bands through spatial multiplexing, it may place an unacceptably high burden on the design and operation of both user and SAS systems. As an alternative, interference may be mitigated by control of the assignment of frequencies to users and upper limits on transmit power. Consistent with the spectrum sharing model described in Section 14.2, this approach requires exchange of user information, which poses a potential privacy concern for users, prompting the need for privacy preservation.

14.6.2.1 Privacy Preservation with Stationary Users

For the case of a stationary PU, e.g. a fixed radar, first consider the case where the adversary makes inference attacks on the PU location based on observing the locations reported to the SAS, e.g. the adversary has hacked the SAS and may observe PU reported information directly. The PU may consider intuitive heuristics to protect their privacy. One such heuristic is for the PU to report multiple false locations in addition to their true location. Consider two deployments of PUs from Clark and Psounis [2018b]. In the first, the PUs are uniformly placed in the same 20 km by 20 km region as the SUs. In the second, the PUs are uniformly placed in a 100 km by 100 km region and the SUs are uniformly placed in a 20 km by 20 km corner of the region. The latter deployment is obviously more favorable for SU utility, and it is also more representative of some real-world sharing scenarios. For example, in CBRS, we would expect to find large concentrations of SUs in population centers, whereas many military operations would be more likely to take place in somewhat remote areas. Figure 14.8 plots the sample average SU sum-rate utility in a network of 1000 UE versus the

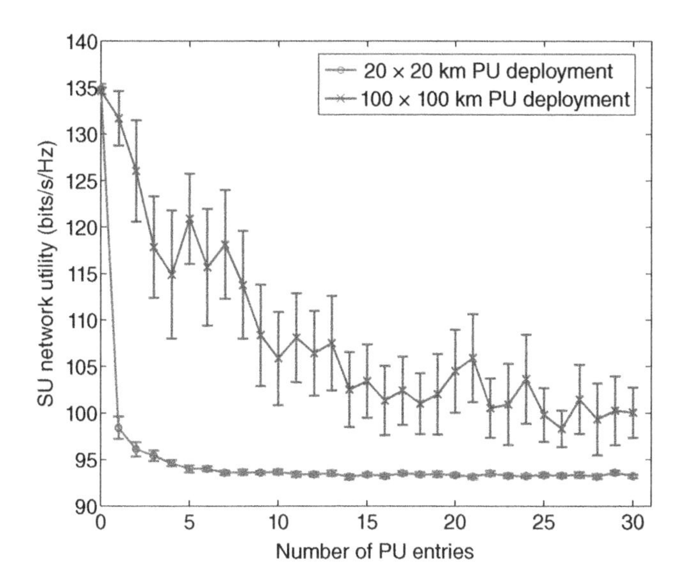

Figure 14.8 SU utility versus number of PU locations [error bars denote standard deviation]. Source: Clark and Psounis [2018b] © 2018, IEEE.

number of PU locations reported, averaged over 40 randomly generated topologies. Inserting false PU locations offers a direct way to trade SU utility for PU privacy, though the SU utility is, not surprisingly, much more limited when the PUs and SUs are concentrated in the same region.

Suppose the adversary does not have direct access to the PU reports directly, but does have access to SU assignments, e.g. by masquerading as a legitimate operator of SU devices. Let the adversary divide a 20 km by 20 km region into 4 km by 4 km cells. There are four PU locations [$n_p = 4$], and power assignments to SUs are made leveraging the EIPA algorithm from Clark and Psounis [2017b] to approximately maximize user utility subject to an interference constraint on the PUs. In an attempt to complicate adversary inference attacks and preserve PU privacy, the power assignments resulting from running the EIPA algorithm are then randomly and uniformly reduced by up to 20 dB. Figure 14.9 provides a visualization of the adversary estimate in this scenario for each cell with the given topology and based on a single observation of SAS power assignments to 40 adversary-controlled SUs. Figure 14.9b shows the optimal estimate, based on Bayesian inference, after just a single observed time slot. The adversary is able to immediately pinpoint all four locations, with some uncertainty about the possibility of a fifth location. Figure 14.9c shows the adversary estimate resulting from a particle filter implementation [Clark and Psounis, 2018b]. The areas around the actual PU locations are tracked by the particle filter, but exact locations are not precisely identified. While less accurate than the optimal estimate, the partical filter is more computationally efficient and will more readily scale to larger estimation problems.

Since the PUs are stationary, the adversary may collect multiple observations of SU assignments over time to improve on its estimate. Figure 14.10a, plots the average distance error of the adversary estimate [as defined in Section 14.3.2] over time for the optimal and the particle filter estimation methods. The optimal estimator is able to reduce the average distance error to about 200 m after a single observation and then reduces it to zero, identifying the PU locations exactly, after a few minutes of observations. With the particle filter, the adversary is able to steadily reduce the error during the first 10 minutes of observation, reaching an error of about 5 km, until it is able to pinpoint the PU locations exactly from an observation about 50 minutes into the simulation. The figure also

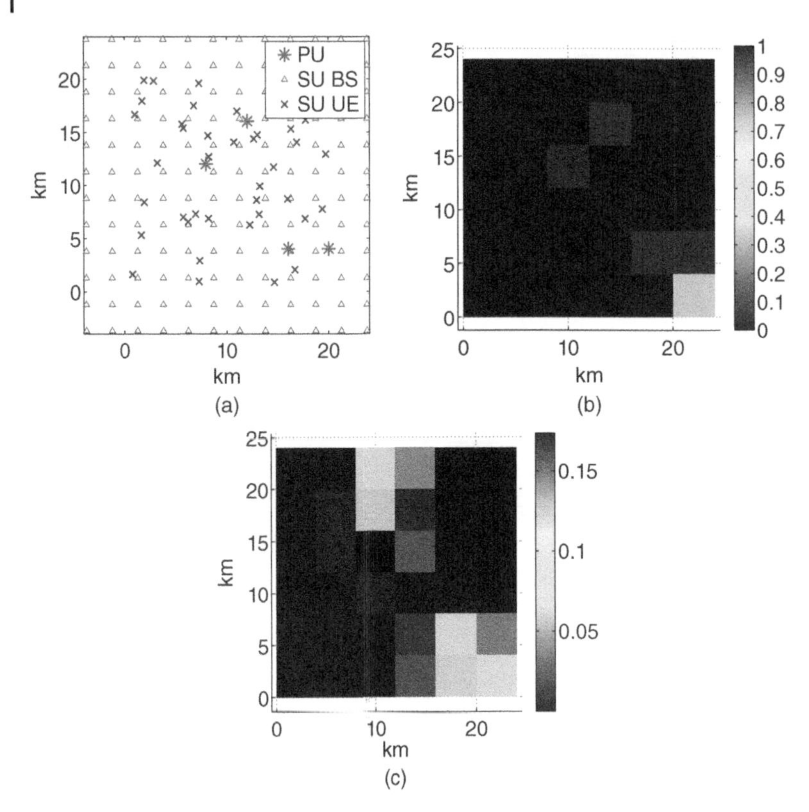

Figure 14.9 Topology and adversary estimates for fixed PUs. [a] Topology, [b] optimal estimate, and [c] PF estimate. Source: Clark and Psounis [2018b] © 2018, IEEE.

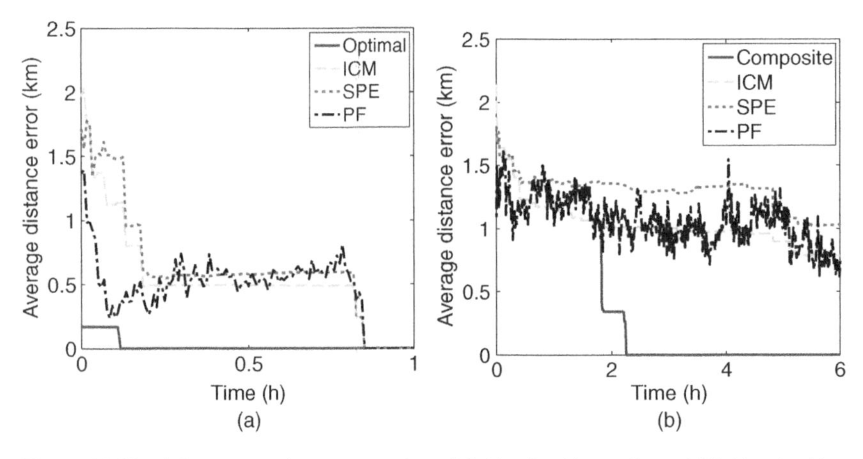

Figure 14.10 Adversary estimates over time. [a] 4 km by 4 km cells and [b] 1 km by 1 km cells. Source: Clark and Psounis [2018b] © 2018, IEEE.

includes two other heuristic adversary estimation techniques, referred to as ICM and SPE in Clark and Psounis [2018b], but these are generally outperformed by the particle filter.

In this case, an average distance error of zero indicates that the adversary is able to identify exactly which of the 4 km by 4 km cells contain a PU. It does not mean that the adversary is necessarily able to discern the location of any given PU within the cell. We consider a higher adversary resolution of 1 km by 1 km cells in Figure 14.10b. Here the particle filter implementation is able to quickly determine PU locations to an average distance error of about 1 km, slowly improving on the estimate over time. This estimator is unable to pinpoint the PU locations as it did in the low resolution case, despite a much longer duration of adversary observation [six hours], but with the higher resolution, this actually corresponds to a more accurate estimate in practice. Another consequence of the finer adversary resolution is that direct computation of the optimal estimator is impractical and we instead provide results for a composite approach of heuristically eliminating locations until few enough remain to apply the optimal estimator. Since the PUs are static, the adversary can take multiple observations to conduct the pruning phase and progressively reduce the candidate set down to a manageable size for computation of the optimal estimate. Once this pruning phase completes after a little less than two hours, we see that the optimal estimator is able to identify the PU locations exactly.

The plots illustrate the progression of the adversary estimate over time and the corresponding loss of PU privacy. A minimum level of privacy can be maintained by providing false PU entries to the SAS, where this privacy is maintained even against a powerful adversary that can hack into the SAS directly, or for an adversary that can observe SU assignments indefinitely. The assignment process of the SAS and the addition of randomization also offers some privacy, but we can see that this effect is only temporary, even against suboptimal estimators, and will not benefit a PU that remains static indefinitely.

14.6.2.2 Privacy Preservation with Mobile Users

PU movement introduces some additional potentially sensitive aspects of PU operations that could pose a privacy concern. Recording a history of the PU movement path may allow an adversary to infer identity based on locations visited. The trajectory of the movement may also allow an adversary to infer the intended destination of the PUs. The trajectory history, or track, introduces the opportunity for new metrics, e.g. how to measure the distance between an estimated track and a ground truth track. The subject of multiple target tracking has been well studied in the literature [Bell et al., 2013].

Motivated by mobile shipborne radar PUs in CBRS, consider two PUs moving relatively slowly at 10 kph in a 20 km by 20 km region as in Clark and Psounis [2018b]. The SAS will attempt to preserve PU privacy by randomizing SU power assignments by as much as 10 dB. Figure 14.11a plots the PU movement tracks along with the estimated tracks of an adversary using a particle filter to produce estimates. We can see qualitatively that the adversary is able to well approximate the PU tracks.

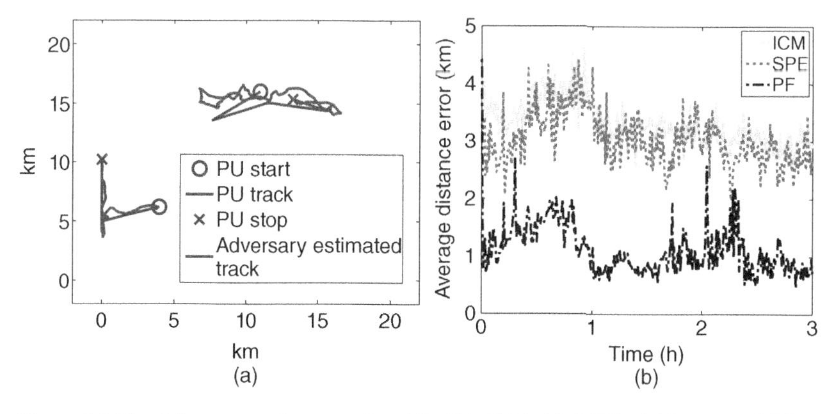

Figure 14.11 Adversary estimates of mobile PUs [10 kph]. [a] Track history and [b] average distance error. Source: Clark and Psounis [2018b] © 2018, IEEE.

In Figure 14.11b, we plot the average distance error over time and see that the particle filter is able to track the PU typically to about 1 km accuracy. While the mobility of the PUs prevents the adversary from steadily reducing privacy as in the fixed PU case, this does not prevent the adversary from achieving a good estimate.

Various heuristic methods can be applied to increase privacy, including greater randomization of adversary assignments and the addition of false PUs that would generate additional tracks to attempt to confuse the adversary. The key consideration for privacy with mobile users is that an adversary will not be able to indefinitely improve the accuracy of their estimate. This holds for other PU parameters as well. In general, if the private parameters of the PU change fast enough, the adversary will not improve the accuracy of their estimate with multiple observations; however, they may learn about the change in PU behavior over time, which could introduce new privacy concerns for the PU, e.g. exposure of their historical location track versus just their instantaneous position. The interested reader can find further exploration of PU mobility in Clark and Psounis [2018b].

14.6.2.3 Small-Scale Privacy Comparison

Let us now compare the effectiveness of obfuscation strategies more rigorously in the SAS power assignment setting, first with a small-scale scenario, where computation of an optimal strategy is feasible and can be applied for comparison. Consider a case of PU location privacy with four possible PU locations and two fixed SUs operating in a 10 km by 10 km region. A SAS determines power assignments for the SUs, with assignments ranging from −40 to +20 dBm in 4 dBm increments. The SUs are assumed to transmit to a receiver with a −100 dBm noise floor at a random distance of up to 2.5 km, approximately modeling typical UE transmissions to BSs in a cellular network. SU transmissions are orthogonal and do not interfere with each other, e.g. as a result of a scheduler in the SU network. The PU is protected with a harmful interference power threshold of −114 dBm, typical for a ground-based military radar in CBRS [Drocella et al., 2015]. We use a breakpoint model [Molisch, 2012] for mean channel gain with a free space model out to the breakpoint and a path loss exponent of 4 applied beyond the breakpoint. For SU utility, we will consider the sum-rate throughput, computed as the sum Shannon capacity of all SU assignments.

Suppose the adversary observes the allocations granted to the SUs, i.e. $P_{\underline{\mathcal{Y}}|\tilde{\underline{\mathcal{X}}}^{\text{PU}}} = P_{\underline{\mathbf{P}}^{\text{SU}}|\tilde{\underline{\mathcal{X}}}^{\text{PU}}}$. We implement the optimal privacy strategy at the SAS by applying a convex solver to [14.8], precomputing parameters $U'_{\text{SU}}[\mathcal{Y}]$ and $U'_{\text{PU}}[\mathcal{X}, \mathcal{Y}]$ for every possible PU state and SU allocation. We also implement the ARC algorithm in the SAS by first constructing a random codebook of SU allocations, then, for each possible PU state, compute the conditional probability of selecting a codeword in the codebook according to the exponential weights returned by ARC as described in Section 14.5.2. μ_{SU} can be selected to trade between PU privacy and SU utility. By computing mutual information and sum SU Shannon capacity for different privacy strategies, we quantitatively compare, in Figure 14.12, the performance of ARC, the optimal strategy, several heuristic perturbation strategies, and a Gaussian noise mechanism. Recall that zero mutual information corresponds to maximum privacy, such that operating in the top-left corner of the plot is ideal. The optimal strategies are plotted parametrically for different values of the required SU utility, c_{SU}. ARC is plotted with three different codebook sizes, 4, 8, and 16, and for each codebook size, is plotted parametrically for different values of μ_{SU}. For the perturbation strategies, we consider the approach of adding 0, 1, or 2 false PU entries to randomly select a reported state from the true state. Note an interpolation line is included in the figures for visualization, but only the operating points with markers are achievable for this strategy. We also plot strategies where false alarms and missed detections are applied to each possible PU location in the state, which can be viewed as modeling the performance of an ESC. We separately plot missed detection rates of 0%, 0.1% ,and 0.5%, and

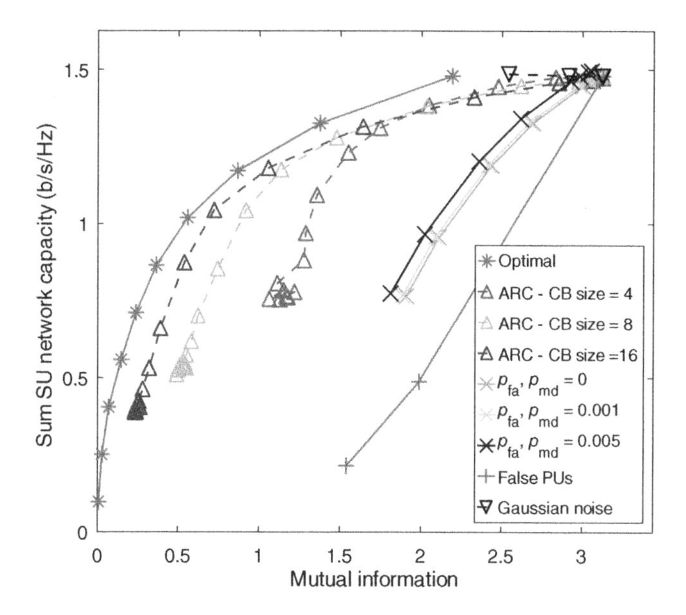

Figure 14.12 Mutual information after SAS obfuscation. Source: Clark and Psounis [2020] © 2020, IEEE.

for each missed detection rate, we parametrically plot false alarm rates from 0% to 30%. Finally, with the Gaussian mechanism, a common strategy for achieving differential privacy, Gaussian noise is added to the SU power assignments, parametrically plotting over noise variances from 0.3 to 4 dB.

As the codebook size increases, ARC offers an increasingly close approximation of the optimal privacy strategy, with μ_{SU} offering an effective way to trade between SU utility and PU privacy. The false entry strategy rapidly degrades SU utility as additional false entries are included. The Gaussian mechanism does not offer significant privacy protection for the variances plotted here. This is true both in terms of mutual information and in terms of differential privacy. The most private case only achieves $[\epsilon, \delta]$-differential privacy with $\epsilon = 150$ and $\delta = 10^{-5}$, where such a large ϵ indicates a negligible level of differential privacy [Dwork and Smith, 2009]. This mechanism increases harmful interference to the PUs by increasing some SU power assignments. The highest variance case plotted has an 18% chance of causing harmful interference to at least one PU. Increasing the noise variance to achieve greater privacy would further increase the already unacceptably high probability of interference. Introducing missed detections and false alarms offer a somewhat better trade-off than the false entries. Both are significantly outperformed by the optimal and ARC, even with the smallest codebook size. Note that the top right marker for the false PU strategy corresponds to no obfuscation, i.e. the deterministic, utility maximizing strategy. By comparison, ARC can offer greater than a 50% improvement in privacy with comparable SU utility.

Similar observations hold when ARC is applied to the case of an adversary that may have hacked the SAS such that it may observe the state reported by the PUs and/or the ESC directly. Further, to demonstrate that mutual information is a general predictor for other privacy metrics that may be more intuitively meaningful for specific use cases and adversary estimation schemes, let us consider the average distance error of an adversary estimate, see (14.5), to evaluate location privacy. Figure 14.13 plots the SU utility as a function of the average distance error between the adversary inferred locations and the true PU locations assuming an optimal Bayesian inference attack by the adversary. In this case, a larger average distance error corresponds to better PU privacy and operating in the upper-right portion of the plot is ideal. The same obfuscation strategies considered in Figure 14.12 are again plotted, depicting similar relative performance as before, although there

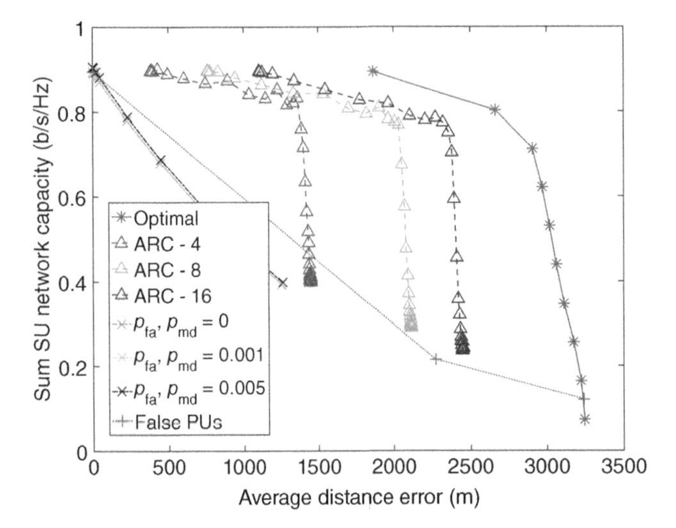

Figure 14.13 Adversary error after PU obfuscation. Source: Clark and Psounis [2020] © 2020, IEEE.

is a more significant performance gap between the optimal strategy and ARC with the largest codebook. This can be explained by considering that ARC in this case has still been implemented to approximately optimize privacy in terms of mutual information, not average distance error. Naturally, operating with no obfuscation loses all PU privacy against an adversary that can hack into the SAS directly. However, ARC can cause the adversary to misidentify the PU locations by an average of 1.2 km with negligible impact on SU utility, while other obfuscation approaches will reduce SU utility by more than 50% to achieve the same level of privacy.

14.6.2.4 Large-Scale Privacy Comparison

To assess if our findings hold in larger scenarios, consider a 20 km by 20 km region with 36 SU base station receivers supporting UE operating with transmit powers from −40 to +20 dBm in 1 dB increments. Otherwise, PUs and SUs will use the same radio parameters that were used in Section 14.6.2.3. We will focus on a single SU frequency channel where each base station will attempt to schedule one UE per time slot to minimize self-interference among the SU network. BSs are deployed on a fixed grid with inter-site spacing of 3.3 km and with UE deployed randomly around each BS. Zero, one, two, or three real PUs are also deployed randomly in each topology. A typical topology is illustrated in Figure 14.14.

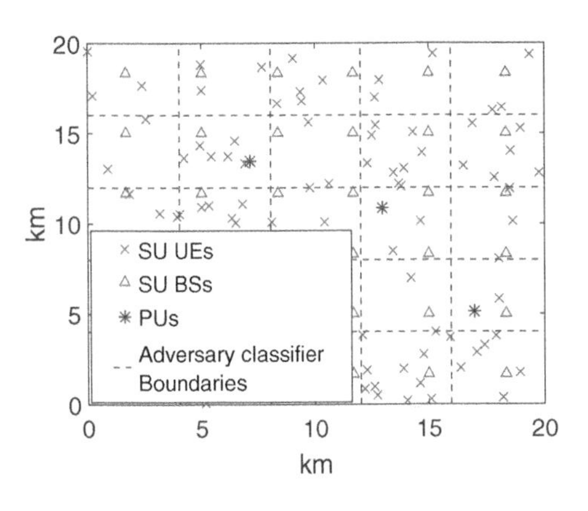

Figure 14.14 Example topology for CBRS case study. Source: Clark and Psounis [2020] © 2020, IEEE.

Due to the size of the scenario, we cannot enumerate all the possible PU and SU states. This precludes us from computing the optimal privacy strategy or the optimal adversary inference attack. Instead, we will apply ARC as an algorithm intended to approximate the optimal strategy. We will also apply and compare with the performance of strategies involving false PUs as well as false alarm/missed detection strategies. We model a suboptimal adversary with fixed resources to compute an inference attack, estimating the PU locations. Specifically, we again leverage a machine learning approach, training classifiers to produce the estimate. As before, the adversary partitions the region into 4 km by 4 km square estimation cells and separately trains a binary classifier on each cell to estimate the probability that a PU is present in the cell.

Suppose the adversary may observe SU state information [locations and channel gain measurements for the UE-BS channel] and power assignments from the SAS at a single time slot. In practice, the adversary could continue to reduce PU privacy with additional observations over multiple time slots as we have considered in Section 14.6.2.1. However, considering a single time slot keeps the computational burden of modeling the adversary estimate more manageable, while still enabling a relative comparison of different privacy strategies. We separately train adversaries for each privacy strategy considered, producing 16 000 random topologies to train the classifiers on the PU ground truth and resulting SU assignments. 4000 more topologies are then generated to evaluate the performance of the adversary, where the classifiers return estimates of the PU locations, serving as an explicit adversary inference attack.

Figure 14.15 plots the average Shannon capacity per SU versus the average distance error of the adversary's estimate. A larger average distance error corresponds to a poorer adversary estimate and thus more PU privacy. Because we are evaluating a 20 km by 20 km region, and the adversary is making estimates with a 4 km by 4 km resolution, the minimum and maximum average distance error is approximately 2 and 10 km, respectively. We plot ARC with a codebook size of 400 and various values of μ_{SU}. We plot the other strategies used in Section 14.6.2.3 for comparison, along with the strategy of randomly reducing SU power assignments according to a uniform random variable as we applied in Sections 14.6.2.1 and 14.6.2.2. The maximum of the uniform random reduction is set at 0, 3, 5, or 10 dB. Note this strategy does not appear to offer any noticeable improvement in privacy. The false entry strategy performs similarly to the small-scale case, offering good privacy, but adding even one false entry significantly reduces SU utility. The false alarm and missed detection strategy allows for some privacy at higher SU utilities and compares favorably with ARC in the low SU utility regime. In the high SU utility regime, ARC continues to offer a significant privacy advantage, increasing the average distance error of the adversary estimate by 400% with negligible impact on the SU utility. Compared with the false alarm and missed detection strategy, ARC offers up to a 40% increase in SU utility at a comparable level of privacy.

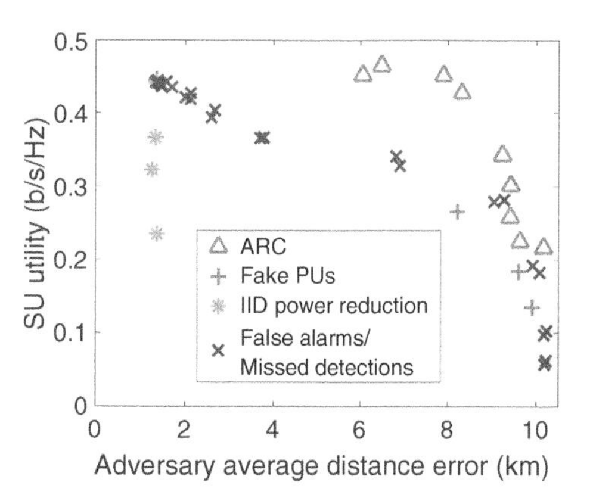

Figure 14.15 SU utility versus average distance error in CBRS. Source: Clark and Psounis [2020] © 2020, IEEE.

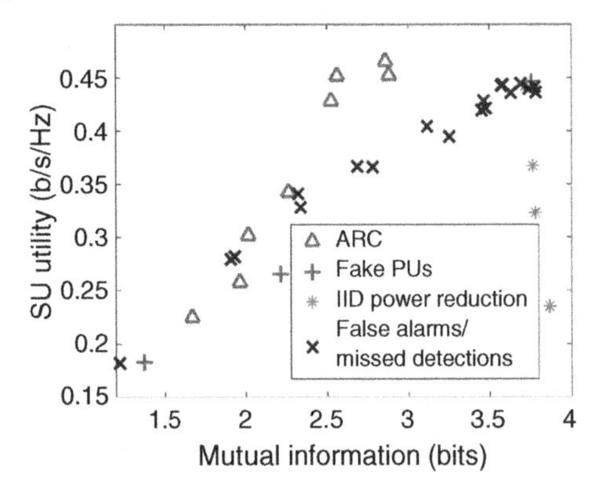

Figure 14.16 SU utility versus estimated mutual information in CBRS. Source: Clark and Psounis [2020] © 2020, IEEE.

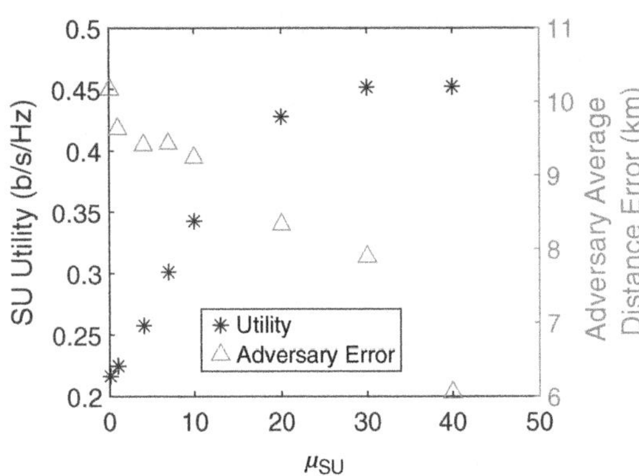

Figure 14.17 SU utility and PU privacy versus μ_{SU}. Source: Clark and Psounis [2020] © 2020, IEEE.

Since we cannot enumerate all possible states with a problem of this size, we cannot compute mutual information directly. As an approximation, we estimate mutual information between the PU locations and the SU assignments from the empirical distribution given by the 20 000 topologies used to train the machine learning adversary. Specifically, we apply k-means clustering to group each sample of these high-dimensional random variables into one of 120 bins and compute mutual information on the discrete distribution given by the normalized count of the number of samples in each bin. This estimated mutual information is plotted in Figure 14.16, where we again compare ARC with the perturbation strategies. Recall that less mutual information implies more privacy, and observe that the relative performance of the strategies is qualitatively similar to the comparison with the average distance error metric and the smaller scale experiments. Specifically, ARC can reduce the estimated mutual information by about 35% with negligible loss of SU utility.

We also consider what values for μ_{SU} should be selected to meet performance and privacy goals. In Figure 14.17, we plot both SU utility and average distance error for the heuristic in the simulated scenario as a function of the parameter μ_{SU}. We verify that in the large-scale scenario, μ_{SU} continues to provide a viable mechanism to trade SU utility for PU privacy. Note that for μ_{SU} between 0 and 20, there is nearly a linear trade-off. For $\mu_{SU} \geq 20$, a small amount of utility may be traded for substantial gains in privacy.

14.7 Summary and Future Directions

In this chapter, we have examined a formal model for privacy and performance analysis of robust centralized spectrum sharing systems, encompassing passive sensing and direct user reporting. Sharing can be accomplished through mitigating interference with the control of power assignments granted to users, and also through the design of precoders to exploit spatial multiplexing for multi-antenna systems. Such sharing systems will serve as necessary infrastructure for the expansion of new radar and communication systems. Failure to adequately address privacy could result in incumbent users impeding the expansion of spectrum sharing and opening of bands to new types of systems and users. Failure to adequately address privacy while also accounting for user utility will also impede new spectrum access if the new bands offer little benefit to prospective users.

Analyzing privacy requires careful consideration of what adversary observations are anticipated and what parameters are considered private by the users. With a properly constructed user privacy optimization problem, theoretically optimal privacy preserving strategies offer significant potential for improvement in the privacy-performance trade-off with respect to many common intuitive heuristics. Given challenges in implementing optimal strategies, schemes designed to approximate the optimal, such as ARC, can offer a practical method to preserve PU privacy. ARC closely approximates the optimal strategy in small-scale test cases. In larger scale cases, ARC was demonstrated as an effective method to maintain user privacy while offering greater utility to users compared to common heuristic perturbation algorithms often applied in the literature. Application and assessment of privacy preservation algorithms in spectrum sharing settings remains an open area of study. Other methods such as GAP might be applied and prove to offer new advantages in spectrum sharing scenarios. Further, the interplay of achievable privacy-utility trade-offs with different encryption architectures has not been fully explored in the literature.

References

P. Atkins. Re: Commercial operations in the 3550-3650 MHz band (GN Docket No. 12-354). *National Telecommunications and Information Administration Letter to the FCC*, March 2015.

A. Babaei, W. H. Tranter, and T. Bose. A nullspace-based precoder with subspace expansion for radar/communications coexistence. In *GLOBECOM, 2013 IEEE*, pages 3487–3492. IEEE, 2013.

B. Bahrak, S. Bhattarai, A. Ullah, J. M. J. Park, J. Reed, and D. Gurney. Protecting the primary users' operational privacy in spectrum sharing. In *IEEE DySPAN*, pages 236–247, April 2014. https://doi.org/10.1109/DySPAN.2014.6817800.

K. Bell, T. Corwin, L. Stone, and R. Streit. *Bayesian multiple target tracking, 2nd edition*. Artech House, 2013.

S. Bhattarai, P. R. Vaka, and J. Park. Thwarting location inference attacks in database-driven spectrum sharing. *IEEE Transactions on Cognitive Communications and Networking*, 4(2):314–327, 2018. ISSN 2332-7731.

S. Boyd and L. Vandenberghe. *Convex optimization*. Cambridge University Press, New York, NY, USA, 2004. ISBN 0521833787.

R. Camino, C. Hammerschmidt, and R. State. Generating multi-categorical samples with generative adversarial networks. *arXiv preprint arXiv:1807.01202*, 2018.

J. Carroll, G. Sanders, F. Sanders, and R. Sole. Case study: investigation of interference into 5 GHz weather radars from unlicensed national information infrastructure devices. *US Dept. of Commerce, NTIA Report TR-12-486*, June 2012.

Y. Chen, J. Zhang, X. Wang, X. Tian, W. Wu, F. Wu, and C. W. Tan. Secrecy capacity scaling of large-scale cognitive networks. MobiHoc, pages 125–134, New York, NY, USA, 2014. ACM.

Q. Cheng, D. N. Nguyen, E. Dutkiewicz, and M. Mueck. Preserving honest/dishonest users' operational privacy with blind interference calculation in spectrum sharing system. *IEEE Transactions on Mobile Computing*, 19(12):2874–2890, 2020. https://doi.org/10.1109/TMC.2019.2936377.

M. Clark and K. Psounis. Can the privacy of primary networks in shared spectrum be protected? In *IEEE INFOCOM*, pages 1–9, April 2016.

M. Clark and K. Psounis. Designing sensor networks to protect primary users in spectrum access systems. In *2017 13th Annual Conference on Wireless On-demand Network Systems and Services (WONS)*, pages 112–119, February 2017a.

M. A. Clark and K. Psounis. Equal interference power allocation for efficient shared spectrum resource scheduling. *IEEE Transactions on Wireless Communications*, 16(1):58–72, January 2017b. ISSN 1536-1276. https://doi.org/10.1109/TWC.2016.2618376.

M. Clark and K. Psounis. Achievable privacy-performance tradeoffs for spectrum sharing with a sensing infrastructure. In *2018 14th Annual Conference on Wireless On-demand Network Systems and Services (WONS)*, pages 103–110, February 2018a.

M. A. Clark and K. Psounis. Trading utility for privacy in shared spectrum access systems. *IEEE/ACM Transactions on Networking*, 26(1):259–273, February 2018b. ISSN 1063-6692. https://doi.org/10.1109/TNET.2017.2778260.

M. Clark and K. Psounis. Optimizing primary user privacy in spectrum sharing systems. *IEEE/ACM Transactions on Networking*, 28(2):533–546, 2020. https://doi.org/10.1109/TNET.2020.2967776.

L. Clark, M. Clark, K. Psounis, and P. Kairouz. Privacy utility trades in wireless data via optimization and learning. In *Information Theory and Applications Workshop (ITA)*, 2019.

T. M. Cover and J. A. Thomas. *Elements of information theory*. John Wiley & Sons, 2012.

S. Deb, V. Srinivasan, and R. Maheshwari. Dynamic spectrum access in DTV whitespaces: design rules, architecture and algorithms. In *MobiCom '09*, pages 1–12, New York, NY, USA, 2009. https://doi.org/10.1145/1614320.1614322.

A. Dimas, M. A. Clark, B. Li, K. Psounis, and A. P. Petropulu. On radar privacy in shared spectrum scenarios. In *ICASSP*, pages 7790–7794, May 2019. https://doi.org/10.1109/ICASSP.2019.8682745.

X. Dong, Y. Gong, J. Ma, and Y. Guo. Protecting operation-time privacy of primary users in downlink cognitive two-tier networks. *IEEE Transactions on Vehicular Technology*, 67(7):6561–6572, 2018. ISSN 0018-9545.

Y. Dou, K. Zeng, H. Li, Y. Yang, B. Gao, K. Ren, and S. Li. p^2 -sas: Privacy-preserving centralized dynamic spectrum access system. *IEEE Journal on Selected Areas in Communications*, 35(1):173–187, 2017. ISSN 0733-8716. https://doi.org/10.1109/JSAC.2016.2633059.

E. Drocella, J. Richards, R. Sole, F. Najmy, A. Lundy, and P. McKenna. 3.5 GHz exclusion zone analyses and methodology. *US Dept. of Commerce, NTIA Report 15-517*, June 2015.

C. Dwork. Differential privacy. In *Proceedings of the International Colloquium Automata, Languages and Programming*, pages 1–12, 2006.

C. Dwork and A. Smith. Differential privacy for statistics: what we know and what we want to learn. *Journal of Privacy and Confidentiality*, 1 (2):135–154, 2009. https://doi.org/10.29012/jpc.v1i2.570.

Federal Communications Commission. *Report and Order and Second Further Notice of Proposed Rulemaking*. 15-47, GN Docket No. 12-354, April 2015.

Federal Communications Commission. *Order on Reconsideration and Second Report and Order*. 16-55, GN Docket No. 12-354, May 2016.

Z. Gao, H. Zhu, S. Li, S. Du, and X. Li. Security and privacy of collaborative spectrum sensing in cognitive radio networks. *IEEE Wireless Communications*, 19(6):106–112, 2012. ISSN 1536-1284. https://doi.org/10.1109/MWC.2012.6393525.

Z. Gao, H. Zhu, Y. Liu, M. Li, and Z. Cao. Location privacy in database-driven cognitive radio networks: attacks and countermeasures. In *2013 Proceedings IEEE INFOCOM*, pages 2751–2759, April 2013. https://doi.org/10.1109/INFCOM.2013.6567084.

I. Goodfellow, J. Pouget-Abadie, M. Mirza, B. Xu, D. Warde-Farley, S. Ozair, A. Courville, and Y. Bengio. Generative adversarial nets. In *Advances in Neural Information Processing Systems*, pages 2672–2680, 2014.

M. Grissa, B. Hamdaoui, and A. A. Yavuza. Location privacy in cognitive radio networks: a survey. *IEEE Communication Surveys and Tutorials*, 19(3):1726–1760, 2017. ISSN 1553-877X. https://doi.org/10.1109/COMST.2017.2693965.

M. Grissa, A. A. Yavuz, B. Hamdaoui, and C. Tirupathi. Anonymous dynamic spectrum access and sharing mechanisms for the CBRS band. *IEEE Access*, 9:33860–33879, 2021. https://doi.org/10.1109/ACCESS.2021.3061706.

Y. Han, E. Ekici, H. Kremo, and O. Altintas. Spectrum sharing methods for the coexistence of multiple RF systems: a survey. *Ad Hoc Networks*, 53:53–78, 2016. ISSN 1570-8705.

A. A. Hilli, A. Petropulu, and K. Psounis. MIMO radar privacy protection through gradient enforcement in shared spectrum scenarios. In *DySPAN*, pages 1–5, 2019. https://doi.org/10.1109/DySPAN.2019.8935749.

J. Holdren, et al. Realizing the full potential of government held spectrum to spur economic growth. *President's Council of Advisors on Science and Technology Report to the President*, 15–30 2012. https://obamawhitehouse.archives.gov/sites/default/files/microsites/ostp/pcast_spectrum_report_final_july_20_2012.pdf.

C. Huang, P. Kairouz, X. Chen, L. Sankar, and R. Rajagopal. Context-aware generative adversarial privacy. *CoRR*, abs/1710.09549, 2017a. http://arxiv.org/abs/1710.09549.

C. Huang, P. Kairouz, X. Chen, L. Sankar, and R. Rajagopal. Context-aware generative adversarial privacy. *Entropy*, 19(12), 2017b. ISSN 1099-4300. https://doi.org/10.3390/e19120656. http://www.mdpi.com/1099-4300/19/12/656.

X. Jin and Y. Zhang. Privacy-preserving crowdsourced spectrum sensing. *IEEE/ACM Transactions on Networking*, 26(3):1236–1249, 2018. ISSN 1063-6692. https://doi.org/10.1109/TNET.2018.2823272.

A. Khawar, A. Abdelhadi, and T. C. Clancy. Coexistence analysis between radar and cellular system in LoS channel. *IEEE Antennas and Wireless Propagation Letters*, 15:972–975, 2016.

K. Krishna and M. N. Murty. Genetic K-means algorithm. *IEEE Transactions on Systems, Man, and Cybernetics Part B: Cybernetics*, 29(3):433–439, 1999.

B. Li and A. P. Petropulu. Joint transmit designs for coexistence of MIMO wireless communications and sparse sensing radars in clutter. *IEEE Transactions on Aerospace and Electronic Systems*, 53(6):2846–2864, 2017.

B. Li, H. Kumar, and A. P. Petropulu. A joint design approach for spectrum sharing between radar and communication systems. In *ICASSP*, pages 3306–3310. IEEE, 2016a.

B. Li, A. P. Petropulu, and W. Trappe. Optimum co-design for spectrum sharing between matrix completion based MIMO radars and a MIMO communication system. *IEEE Transactions on Signal Processing*, 64(17):4562–4575, 2016b.

S. Li, H. Zhu, Z. Gao, X. Guan, Kai Xing, and X. Shen. Location privacy preservation in collaborative spectrum sensing. In *2012 Proceedings IEEE INFOCOM*, pages 729–737, March 2012. https://doi.org/10.1109/INFCOM.2012.6195818.

J. Liu, C. Zhang, B. Lorenzo, and Y. Fang. DPavatar: A real-time location protection framework for incumbent users in cognitive radio networks. *IEEE Transactions on Mobile Computing*, 19(3):552–565, 2020. https://doi.org/10.1109/TMC.2019.2897099.

J. Liu, K. V. Mishra, and M. Saquib. Co-designing statistical MIMO radar and in-band full-duplex multi-user MIMO communications, 2021.

J. A. Mahal, A. Khawar, A. Abdelhadi, and T. C. Clancy. Spectral coexistence of MIMO radar and MIMO cellular system. *IEEE Transactions on Aerospace and Electronic Systems*, 53(2):655–668, 2017.

M. Mirza and S. Osindero. Conditional generative adversarial nets. *CoRR*, abs/1411.1784, 2014. http://arxiv.org/abs/1411.1784.

A. F. Molisch. *Wireless communications*, Volume 34. John Wiley & Sons, 2012.

A. B. Mosbah, T. A. Hall, M. Souryal, and H. Afifi. An analytical model for inference attacks on the incumbent's frequency in spectrum sharing. In *IEEE DySPAN*, pages 1–2, March 2017. https://doi.org/10.1109/DySPAN.2017.7920770.

B. Niu, Q. Li, X. Zhu, G. Cao, and H. Li. Achieving k-anonymity in privacy-aware location-based services. In *IEEE INFOCOM*, pages 754–762, April 2014. https://doi.org/10.1109/INFOCOM.2014.6848002.

N. Rajkarnikar, J. M. Peha, and A. Aguiar. Location privacy from dummy devices in database-coordinated spectrum sharing. In *IEEE DySPAN*, pages 1–10, March 2017. https://doi.org/10.1109/DySPAN.2017.7920796.

A. Robertson, J. Molnar, and J. Boksiner. Spectrum database poisoning for operational security in policy-based spectrum operations. In *IEEE MILCOM*, pages 382–387, November 2013. https://doi.org/10.1109/MILCOM.2013.72.

S. Salamatian, A. Zhang, F. du Pin Calmon, S. Bhamidipati, N. Fawaz, B. Kveton, P. Oliveira, and N. Taft. Managing your private and public data: bringing down inference attacks against your privacy. *IEEE Journal on Selected Topics in Signal Processing*, 9(7):1240–1255, 2015. https://doi.org/10.1109/JSTSP.2015.2442227.

R. Saruthirathanaworakun, J. M. Peha, and L. M. Correia. Opportunistic sharing between rotating radar and cellular. *IEEE Journal on Selected Areas in Communications*, 30(10):1900–1910, 2012.

R. Shokri, G. Theodorakopoulos, J. Y. Le Boudec, and J. P. Hubaux. Quantifying location privacy. In *2011 IEEE Symposium on Security and Privacy*, pages 247–262, May 2011. https://doi.org/10.1109/SP.2011.18.

S. Sodagari, A. Khawar, T. C. Clancy, and R. McGwier. A projection based approach for radar and telecommunication systems coexistence. In *GLOBECOM*, pages 5010–5014. IEEE, 2012.

P. R. Vaka, S. Bhattarai, and J. M. Park. Location privacy of non-stationary incumbent systems in spectrum sharing. In *GLOBECOM*, pages 1–6, December 2016. https://doi.org/10.1109/GLOCOM.2016.7841962.

W. Wang and Q. Zhang. Privacy-preserving collaborative spectrum sensing with multiple service providers. *IEEE Transactions on Wireless Communications*, 14(2):1011–1019, 2015.

W. Wang, Y. Chen, Q. Zhang, and T. Jiang. A software-defined wireless networking enabled spectrum management architecture. *IEEE Communications Magazine*, 54(1):33–39, 2016. ISSN 0163-6804. https://doi.org/10.1109/MCOM.2016.7378423.

J. Wang, S. M. Errapotu, Y. Gong, L. Qian, R. Jäntti, M. Pan, and Z. Han. Data-driven optimization based primary users' operational privacy preservation. *IEEE Transactions on Cognitive Communications and Networking*, 4(2):357–367, 2018. ISSN 2332-7731. https://doi.org/10.1109/TCCN.2018.2837876.

L. Xing, Q. Ma, J. Gao, and S. Chen. An optimized algorithm for protecting privacy based on coordinates mean value for cognitive radio networks. *IEEE Access*, 6:21971–21979, 2018. ISSN 2169-3536. https://doi.org/10.1109/ACCESS.2018.2822839.

Q. Zhao and A. Swami. A survey of dynamic spectrum access: signal processing and networking perspectives. In *ICASSP*, Volume 4, pages IV–1349. IEEE, 2007.

Epilogue

Joint radar-communications (JRC) systems are playing an increasingly important role in various aspects of future wireless engineering. Their remarkably high range of interdisciplinary applications is provoking a blurring of boundaries between different engineering communities: radar systems, remote sensing, signal processing, wireless communications, microwave theory, electromagnetics, antenna design, wave propagation, information theory, vehicular systems, meteorology, control theory, instrumentation, photonics, and circuit theory. This has also led to a resurgence of interest in the modern as well as the classical signal processing theory and hardware. While substantial progress has been made recently in many important aspects of JRC, the field remains ripe with opportunities and challenges.

The 14 chapters of this book stand testimony to: (a) the breadth of JRC literature allowing different interpretations based on applications, (b) the importance of the topic that has caught the imagination of different communities of internationally renowned researchers, and (c) the rich problems that still need active investigation. The book has aimed to collate a number of interesting topics, which by themselves present a sparsely painted canvas based on the current work. There continue to exist debates on whether to follow a unified framework to the design, optimization, and operationalization of such systems that offers a generic solution or to consider a use-case–based approach offering finer solutions with limited scope. The elements that go into the design of such systems, including the frequency bands of operation, waveform to be used, protocol to be followed, processing and overheads, further diversify the investigation. A multi-volume collection would be needed to collate even the important results here.

The integration of sensing and communications, while leading to significant advantages in key domains, also poses the challenge of solving issues from both systems. To minimize the number of challenges, there has been an effort to realize issues similar in both systems so as to offer similar technology solutions. Thus, in addition to the extension of scope, be it in terms of use-cases or modeling and solution methodologies, the evolution of several interesting technologies impacting both the systems needs to be considered. The key issues arising in both include the need for a larger contiguous spectrum, extension of coverage, and efficient reuse of available resources. We list here some major upcoming JRC technologies.

JRC at higher bands: As sensing becomes ubiquitous, its proliferation in daily lives has increased and, it is being largely aimed at indoor high-resolution applications. With the densification of wireless communications, including through heterogeneous deployment of systems, short-range high data rate applications seem to be the new frontier. These require high bandwidths, and a migration to higher bands needs to be considered. In this context, a terahertz (THz) band offers itself as an interesting proposition, provided technological maturity is achieved. While this book has

Signal Processing for Joint Radar Communications, First Edition.
Edited by Kumar Vijay Mishra, M. R. Bhavani Shankar, Björn Ottersten, and A. Lee Swindlehurst.

largely focused on mmWaves, the use of THz for joint sensing and communications opens avenues both for multi-channel signal processing as well as for high frequency analog/digital electronics to address the challenges therein including short channel coherence times, high propagation loss, large bandwidth induced perturbation, among others.

IRS-aided JRC: There has been a growing concern about electromagnetic transmissions, and in the context of green systems, there has been an effort to reduce the power and number of active transmissions in many areas. This poses issues with the link quality and precludes use of active repeaters. Further, in a dense urban or indoor environment, a direct path to the intended object seldom exists, thus impacting both radar and communication systems. The concept of intelligent reflective surfaces (IRS), which has been restricted to the antenna and propagation community, has been picked up by the signal processing community to address the aforementioned challenges. Since such surfaces are passive reflectors of the electromagnetic waves, they are agnostic to radar or communications, thus offering an ideal opportunity to use them in joint radar and communications. The research on modeling of joint radar and communications with such systems, their design, placement and optimization for a joint system design, as well as minimization of system overheads for their operationalization, is ongoing and would have a significant impact on the reach of joint radar and communications.

Distributed JRC: Another common issue relates to distributed systems. Wireless communications standardization supports multipoint communications to offer a higher data rate or stable connection or both. Radar applications need multiple sensors to offer a better field of view; they are essential in indoor applications where the scene is dense and targets often occlude. A distributed joint radar communication system adds another layer of complexity to system design through the side link communications, while it impacts the algorithm design due to spatially varying channels. It also raises queries on the limits of performance and scaling laws.

Full-duplex JRC: An operational full-duplex system, aside from the implementation and component complexity, offers a twofold increase in the available resources for communications. It also minimizes the need for the use of a separate antenna in continuous wave radar systems. Several efforts have been undertaken to build full-duplex communication systems and analyze their performance in the presence of residual self-interference. The impact of such a technology on the performance of relevant joint radar and communication architectures is worth an assessment given the high likelihood of the deployment of such systems.

Other modalities: Another aspect worth noting is the scope of joint radar communications being limited to radio frequency applications in this work. Both systems are well capable of meeting their objectives in other domains. For example, visible light communications may be combined with visible light positioning. Molecular communications and sensing is another bio-inspired interdisciplinary field having significant ramifications in many related domains. These seem to be exciting times for an exciting topic, and this book has revealed only the tip of the iceberg.

It is expected that the JRC state of the art will be frequently updated in the years to come. Therefore, it is more or less evident that any book on JRC is predetermined to be incomplete. It will also be evident that we have selected for inclusion in the book a set of topics based on our own preferences, reflected by our own experience, from among a wide spectrum of modern signal processing approaches. Nevertheless, our intention is to provide a solid package of theoretical techniques, making the book valuable for JRC practitioners.

Adelphi, Maryland, US — Kumar Vijay Mishra
Luxembourg — M. R. Bhavani Shankar

Index

Note: *Italicized* and **bold** page numbers refer to figures and tables, respectively.

Signal Processing for Joint Radar Communications, First Edition.
Edited by Kumar Vijay Mishra, M. R. Bhavani Shankar, Björn Ottersten, and A. Lee Swindlehurst.

d

e

f

g

q

Printed and bound by CPI Group (UK) Ltd, Croydon, CR0 4YY

16/04/2024

14484173-0004